GENOMES 5

Genomes 5 has been completely revised and updated. It is a thoroughly modern textbook about genomes and how they are investigated. As with previous *Genomes* editions, techniques come first, then genome anatomies, followed by genome function, and finally genome evolution. The genomes of all types of organism are covered: viruses, bacteria, fungi, plants, and animals, including humans and other hominids.

Genome sequencing and assembly methods have been thoroughly revised to include new developments in long-read DNA sequencing. Coverage of genome annotation emphasizes genome-wide RNA mapping, with CRISPR-Cas 9 and GWAS methods of determining gene function covered. The knowledge gained from these techniques forms the basis of the chapters that describe the three main types of genomes: eukaryotic, prokaryotic (including eukaryotic organelles), and viral (including mobile genetic elements). Coverage of genome expression and replication is truly genomic, concentrating on the genome-wide implications of DNA packaging, epigenome modifications, DNA-binding proteins, noncoding RNAs, regulatory genome sequences, and protein-protein interactions. Also included are examples of the applications of metabolomics and systems biology. The final chapter is on genome evolution, including the evolution of the epigenome, using genomics to study human evolution, and using population genomics to advance plant breeding. Established methods of molecular biology are included if they are still relevant today and there is always an explanation as to why the method is still important.

Genomes 5 is the ideal text for upper-level courses focused on genomes and genomics.

Key Features

- A highly accessible and well-structured book with chapters organized into four parts to aid navigation

- Superb artwork illustrates the key concepts and mechanisms

- Each chapter has a set of short answer questions and in-depth problems to test the reader's understanding of the material

- Thoroughly up to date with references to the latest research from the 2020s

- Upon registering, instructor access to a guidance sheet for the book's in-depth problems

GENOMES 5

T. A. Brown

CRC Press
Taylor & Francis Group
Boca Raton London New York

CRC Press is an imprint of the
Taylor & Francis Group, an **informa** business

Garland Science
Taylor & Francis Group

Cover designed by Matthew McClements

Fifth edition published 2023
by CRC Press
6000 Broken Sound Parkway NW, Suite 300, Boca Raton, FL 33487-2742

and by CRC Press
4 Park Square, Milton Park, Abingdon, Oxon, OX14 4RN

CRC Press is an imprint of Taylor & Francis Group, LLC

© 2023 Terry Brown

Library of Congress Cataloging-in-Publication Data

Names: Brown, T. A. (Terence A.) author.
Title: Genomes 5 / T.A. Brown.
Other titles: Genomes five
Description: Fifth edition. | Boca Raton, FL : CRC Press, 2023. | Includes
bibliographical references and index.
Identifiers: LCCN 2022058795 (print) | LCCN 2022058796 (ebook) | ISBN
9780367678661 (hbk) | ISBN 9780367674076 (pbk) | ISBN 9781003133162 (ebk)
Subjects: LCSH: Genomes.
Classification: LCC QH447 .B76 2023 (print) | LCC QH447 (ebook) | DDC
572.8/6--dc23/eng/20221209
LC record available at https://lccn.loc.gov/2022058795
LC ebook record available at https://lccn.loc.gov/2022058796

ISBN: 978-0-367-67866-1 (hbk)
ISBN: 978-0-367-67407-6 (pbk)
ISBN: 978-1-003-13316-2 (ebk)

MIX
Paper from
responsible sources
FSC
www.fsc.org   FSC® C013056

DOI: 10.1201/9781003133162

Typeset in Utopia
by Deanta Global Publishing Services, Chennai, India

Printed and bound in Great Britain by
TJ Books Limited, Padstow, Cornwall

Access the Instructor Resources by visiting www.routledge.com/9780367674076 and then clicking 'Support Materials', whereupon you can register as qualified instructor in order to gain access

CONTENTS

PREFACE

Our knowledge of genomes continues to advance at a remarkable pace due in large part to the development of more efficient and more powerful sequencing methods. The ease and low cost of short-read methods mean that it is now feasible to sequence multiple versions of the same genome, an example being the 1,000,000 human genomes project that is scheduled to reach completion in 2028. Improvement in long-read methods has led to the first truly complete sequence of the human genome and is underlying ambitious plans to obtain high-quality sequences for every eukaryotic species on the planet. Equally dramatic advances are being made in methodology for studying transcriptomes and proteomes, giving us new insights into the way in which the information contained in a DNA sequence is utilized during genome expression, and in particular emphasizing the importance of RNA splicing and chemical modification as a means of increasing the biological potential of a genome.

Genomes 5 retains the overall structure of the previous editions, with the chapters divided into four parts, on genome sequencing and annotation, genome anatomies, genome expression, and genome replication and evolution. The order of chapters remains unchanged, with the one exception that recombination is now described before mutation and repair, a change enabling the central role of recombination in DNA repair to be discussed in a more satisfactory manner. Importantly, as knowledge increases, it is becoming more possible to describe the processes of transcription and translation from a genome-wide perspective, rather than simply through an examination of the expression of individual genes. I hope that in *Genomes 5* I have managed to convey the importance of the genome as a single entity rather than simply a collection of genes.

I would like to thank Chuck Crumly and Jordan Wearing of Garland Science for their continued enthusiasm for *Genomes* and their help during the planning and writing stages, and Matthew McClements for his splendid artwork. As with the previous editions, *Genomes 5* would not have been finished without the support of my wife, Keri. The acknowledgment in the first edition that 'if you find this book useful then you should thank Keri, not me, because she is the one who ensured that it was written,' is equally true for the fifth edition.

A NOTE TO THE READER

I have tried to make the fifth edition of *Genomes* as user-friendly as possible. The book therefore includes a number of devices intended to help the reader and to make the book an effective teaching and learning aid.

Organization of the book

Genomes 5 is divided into four parts:

> **Part 1 – How genomes are studied** begins with an orientation chapter that introduces the reader to genomes, transcriptomes, and proteomes, and then, in Chapter 2, moves on to the methods, centered on PCR and cloning, that were used in the pre-genome era to examine individual genes.

The techniques that are used for constructing genetic and physical maps, which are still important in many genome projects, are then described in Chapter 3, followed in Chapter 4 by the methodology for obtaining DNA sequences and assembling reads into draft and finished genomes. Two chapters are then devoted to analysis of genome sequences, Chapter 5 on the annotation of a genome by identification of genes and other features, and Chapter 6 on functional analysis of the genes that are discovered.

Part 2 – **Genome anatomies** surveys the anatomies of the various type of genome that are found on our planet. Chapter 7 covers eukaryotic nuclear genomes with emphasis on the human genome, partly because of the importance of the human genome in so many areas of research, but also because our genome is the best-studied of all those for which sequences are available. Chapter 8 deals with the genomes of prokaryotes and of eukaryotic organelles, the latter included here because of their prokaryotic origins, and Chapter 9 describes virus genomes and mobile genetic elements, these being grouped together because some types of mobile element are related to virus genomes.

Part 3 – **How genomes are expressed** describes how the biological information contained in a genome is utilized by the cell within which that genome resides. Chapter 10 addresses the important issue of how the packaging of DNA into chromatin affects expression of different parts of the genome, and Chapter 11 then describes the central role that DNA-binding proteins play in expressing those parts of the genome that are active at a particular time. Chapter 12 moves on to the transcriptome, describing how transcriptomes are studied, their compositions, and how a cell's transcriptome is synthesized and maintained. Chapter 13 gives an equivalent description of proteomics and the proteome, and Chapter 14 concludes this part of the book by exploring how the genome acts within the context of the cell and organism, responding to extracellular signals and driving the biochemical changes that underlie differentiation and development.

Part 4 – **How genomes replicate and evolve** links DNA replication, recombination, and mutation with the gradual evolution of genomes over time. In Chapters 15–17, the molecular processes responsible for replication, recombination, mutation, and repair are described, and in Chapter 18 the ways in which these processes are thought to have shaped the structures and genetic contents of genomes over evolutionary time are considered. Chapter 18 then ends with a small number of case studies to illustrate how molecular phylogenomics and population genomics are being used in research and biotechnology.

Learning aids

Each chapter has a set of Short Answer Questions and In-Depth Problems as well as an annotated Further Reading list. At the end of the book, there is an extensive Glossary.

Short Answer Questions require 50- to 500-word answers. The questions cover the entire content of each chapter in a fairly straightforward manner, and most can be marked simply by checking each answer against the relevant part of the text. A student can use the short answer questions to work systematically through a chapter, or can select individual ones in order to evaluate their ability to answer questions on specific topics. The short answer questions could also be used in closed-book tests.

In-Depth Problems require a more detailed answer. They vary in nature and in difficulty, the simplest requiring little more than a literature survey, the intention of these particular problems being that the student advances their learning a few stages from where *Genomes 5* leaves off. Other problems require that the student evaluates a statement or a hypothesis, based on their understanding of the material in the book, possibly supplemented by reading around the subject. These problems will, hopefully, engender a certain amount of thought and critical awareness. A few problems are difficult, in some cases to the extent that there is no solid answer to the question posed. These are designed to stimulate debate and speculation, which stretches the knowledge of each student and forces them to think carefully about their statements. The in-depth problems can be tackled by students working individually, or alternatively can form the starting point for a group discussion.

Further reading lists at the end of each chapter include those research papers, reviews, and books that I look on as the most useful sources of additional material. My intention throughout *Genomes 5* has been that students should be able to use the reading lists to obtain further information when writing extended essays or dissertations on particular topics. Research papers are therefore included, but only if their content is likely to be understandable to the average reader of the book. Emphasis is also placed on accessible reviews, one strength of these general articles being the context and relevance that they provide to a piece of work. The reading lists are divided into sections reflecting the organization of information in the chapter, and in some cases I have appended a few words summarizing the particular value of each item, to help the reader decide which ones they wish to seek out. In some cases, Further reading also includes URLs for databases and other online resources relevant to the material covered in a chapter.

The **Glossary** defines every term that is highlighted in bold in the text, along with a number of additional terms that the reader might come across when referring to books or articles in the reading lists. The Glossary therefore provides a quick and convenient means by which the reader can remind themselves of the technical terms relevant to the study of genomes, and also acts as a revision aid to make sure those definitions are clearly understood during the minutes of uncertainty that many students experience immediately before an exam.

INSTRUCTOR RESOURCES

The images from this book are available to registered instructors in two convenient formats: PowerPoint and PDF.

Also available is an instructor's guide to the book's in-depth problems, which can be downloaded as a PDF.

To download these resources, instructors must register at the 'Instructor Resources Download Hub.' This can be found on the book's product page under the section 'Instructor Resources:' https://www.routledge.com/Genomes-5/Brown/p/book/9780367674076

ACKNOWLEDGEMENTS

The Author and Publisher of *Genomes 5* gratefully acknowledge the contribution of the following reviewers in the development of this edition.

David Baillie, Simon Fraser University; Linda Bonen, University of Ottawa; Hugh Cam, Boston College; Yuri Dubrova, University of Leicester; Bart Eggen, University of Groningen; Robert Fowler, San José State University; Sidney Fu, George Washington University; Adrian Hall, Sheffield Hallam University; Lee Hwei Huih, Universiti Tunku Abdul Rahman; Glyn Jenkins, Aberystwyth University; Julian M. Ketley, University of Leicester; Torsten Kristensen, University of Aarhus; Gerhard May, University of Dundee; Mike McPherson, University of Leeds; Isidoro Metón, Universitat de Barcelona; Gary Ogden, St. Mary's University; Paul Overvoorde, Macalester College; John Rafferty, University of Sheffield; Andrew Read, University of Manchester; Joaquin Cañizares Sales, Universitat Politècnica de València; Michael Schweizer, Heriot-Watt University; Eric Spana, Duke University; David Studholme, Exeter University; John Taylor, University of Newcastle; Gavin Thomas, University of York; Matthew Upton, Plymouth University; Guido van den Ackerveken, Utrecht University; Vassie Ware, Lehigh University; Wei Zhang, Illinois Institute of Technology.

PART 1

HOW GENOMES ARE STUDIED

GENOMES, TRANSCRIPTOMES, AND PROTEOMES

Life as we know it is specified by the **genomes** of the myriad organisms with which we share the planet. Every organism possesses a genome that contains the **biological information** needed to construct and maintain a living example of that organism. Most genomes, including the human genome and those of all other cellular life forms, are made of **DNA** (deoxyribonucleic acid), but a few viruses have **RNA** (ribonucleic acid) genomes. DNA and RNA are **polymeric** molecules made up of chains of monomeric subunits called **nucleotides**. Each molecule of DNA comprises two **polynucleotides** wound around one another to form the famous **double helix**, the two strands held together by chemical bonds that link adjacent nucleotides into structures called **base pairs**.

The human genome, which is typical of the genomes of all multicellular animals, consists of two distinct parts (**Figure 1.1**):

- The **nuclear genome** comprises approximately 3,100,000,000 base pairs of DNA, divided into 24 linear molecules, the shortest 47,000,000 base pairs in length and the longest 249,000,000 base pairs, each contained in a different **chromosome**. These 24 chromosomes consist of 22 **autosomes** and the two **sex chromosomes**, X and Y. Altogether, some 44,500 **genes** are present in the human nuclear genome.

- The **mitochondrial genome** is a circular DNA molecule of 16,569 base pairs, up to ten copies of which are present in each of the energy-generating organelles called mitochondria. The human mitochondrial genome contains just 37 genes.

Each of the approximately 3×10^{13} cells in the adult human body has its own copy or copies of the nuclear genome, the only exceptions being those few cell types, such as red blood cells, that lack a **nucleus** in their fully differentiated state. The vast majority of cells are **diploid** and so have two copies of each autosome, plus two sex chromosomes, XX for females or XY for males – 46 chromosomes in all. These are called **somatic cells**, in contrast to **sex cells**, or **gametes**, which are **haploid** and have just 23 chromosomes, comprising one of each autosome and one sex chromosome. Each cell also has multiple copies of the mitochondrial genome, 2000–7000 copies in somatic cells such as those in the liver and heart tissue, and over 100,000 in each female **oocyte**.

DOI: 10.1201/9781003133162-1

**Figure 1.1 The nuclear and mito-
chondrial components of the human
genome.**

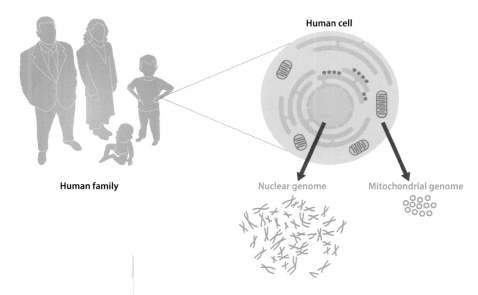

The genome is a store of biological information, but on its own it is unable
to release that information to the cell. Utilization of the biological information
contained in the genome requires the coordinated activity of enzymes and other
proteins, which participate in a complex series of biochemical reactions referred
to as **genome expression** (**Figure 1.2**). The initial product of genome expression
is the **transcriptome**, a collection of RNA molecules derived from those genes
that are active in the cell at a particular time. The transcriptome is maintained
by the process called **transcription**, in which individual genes are copied into
RNA molecules. The second product of genome expression is the **proteome**, the
cell's repertoire of **proteins**, which specifies the nature of the biochemical reactions that the cell is able to carry out. The proteins that make up the proteome
are synthesized by **translation** of some of the individual RNA molecules present
in the transcriptome.

This book is about genomes and genome expression. It explains how
genomes are studied (Part 1), how they are organized (Part 2), how they function (Part 3), and how they replicate and evolve (Part 4). It was not possible to
write this book until quite recently. Since the 1950s, molecular biologists have
studied individual genes or small groups of genes, and from these studies have
built up a wealth of knowledge about how genes work. But only during the last
few years have techniques been available that make it possible to examine
entire genomes. Individual genes are still intensively studied, but information
about individual genes is now interpreted within the context of the genome as
a whole. This new, broader emphasis applies not just to genomes but to all of
biochemistry and cell biology. No longer is it sufficient simply to understand
individual biochemical pathways or subcellular processes. The challenge now
is provided by **systems biology**, which attempts to link together these pathways
and processes into networks that describe the overall functioning of living cells
and living organisms.

This book will lead you through our knowledge of genomes and show you how
this exciting area of research is underpinning our developing understanding of
biological systems. First, however, we must pay attention to the basic principles
of molecular biology by reviewing the key features of the three types of biological
molecule involved in genomes and genome expression: DNA, RNA, and protein.

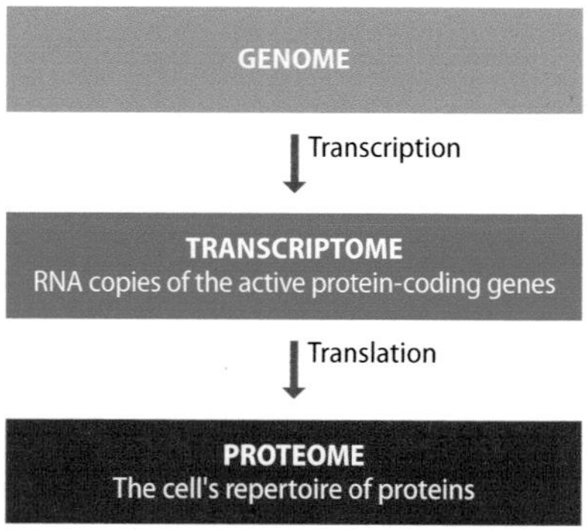

1.1 DNA

DNA was discovered in 1869 by Johann Friedrich Miescher, a Swiss biochemist working in Tübingen, Germany. The first extracts that Miescher made from
human white blood cells were crude mixtures of DNA and chromosomal proteins, but the following year he moved to Basel, Switzerland (where the research

Figure 1.2 Genome expression. The
genome specifies the transcriptome, and
the transcriptome specifies the proteome.

GENOME

↓ Transcription

TRANSCRIPTOME
RNA copies of the active protein-coding genes

↓ Translation

PROTEOME
The cell's repertoire of proteins

institute named after him is now located), and prepared a pure sample of **nucleic acid** from salmon sperm. Miescher's chemical tests showed that DNA is acidic and rich in phosphorus, and also suggested that the individual molecules are very large, although it was not until the 1930s, when biophysical techniques were applied to DNA, that the huge lengths of the polymeric chains were fully appreciated.

Genes are made of DNA

The fact that genes are made of DNA is so well known today that it can be difficult to appreciate that for the first 75 years after its discovery the true role of DNA was unsuspected. As early as 1903, W.S. Sutton had realized that the inheritance patterns of genes parallel the behavior of chromosomes during cell division, an observation that led to the **chromosome theory**, the proposal that genes are located in chromosomes. Examination of cells by **cytochemistry**, which makes use of stains that bind specifically to just one type of biochemical, showed that chromosomes are made of DNA and protein, in roughly equal amounts. Biologists at that time recognized that billions of different genes must exist and the genetic material must therefore be able to take many different forms. But this requirement appeared not to be satisfied by DNA, because in the early part of the twentieth century it was thought that all DNA molecules were the same. On the other hand, it was known, correctly, that proteins are highly variable, polymeric molecules, each one made up of a different combination of 20 chemically distinct, amino acid monomers (Section 1.3). Genes simply had to be made of protein, not DNA.

The errors in understanding DNA structure lingered on, but by the late 1930s it had become accepted that DNA, like protein, has immense variability. The notion that protein was the genetic material initially remained strong, but was eventually overturned by the results of two important experiments:

- Oswald Avery, Colin MacLeod, and Maclyn McCarty showed that DNA is the active component of the **transforming principle**, a bacterial cell extract which, when mixed with a harmless strain of *Streptococcus pneumoniae*, converts these bacteria into a virulent form capable of causing pneumonia when injected into mice (**Figure 1.3A**). In 1944, when the results of this experiment were published, only a few microbiologists appreciated that transformation involves transfer of genes from the cell extract into the living bacteria. However, once this point had been accepted, the true meaning of the 'Avery experiment' became clear: bacterial genes must be made of DNA.

- Alfred Hershey and Martha Chase used **radiolabeling** to show that, when a bacterial culture is infected with **bacteriophages** (also called **phages**, a type of virus), DNA is the major component of the bacteriophages that enters the cells (**Figure 1.3B**). This was a vital observation because it was known that, during the infection cycle, the genes of the infecting bacteriophages are used to direct synthesis of new bacteriophages, and this synthesis occurs within the bacteria. If it is only the DNA of the infecting bacteriophages that enters the cells, then it follows that the genes of these bacteriophages must be made of DNA.

Although from our perspective these two experiments provide the key results that tell us that genes are made of DNA, biologists at the time were not so easily convinced. Both experiments have limitations that leave room for skeptics to argue that protein could still be the genetic material. For example, there were worries about the specificity of the **deoxyribonuclease** enzyme that Avery and colleagues used to inactivate the transforming principle. This result, a central part of the evidence for the transforming principle being DNA, would be invalid if, as seemed possible, the enzyme contained trace amounts of a contaminating **protease** and hence was also able to degrade

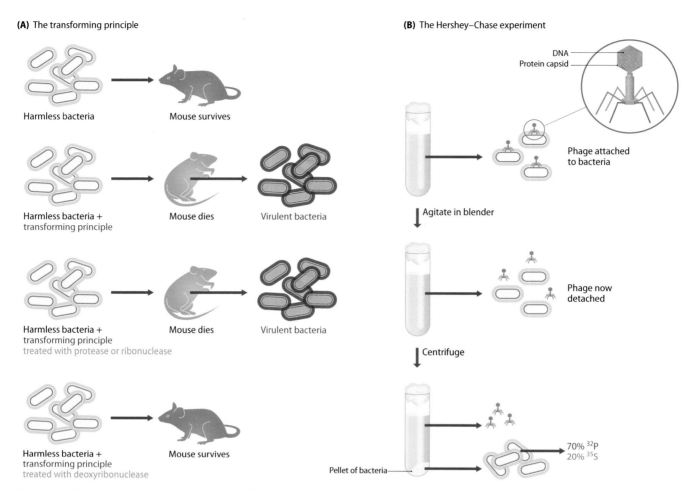

(A) The transforming principle

Harmless bacteria → Mouse survives

Harmless bacteria + transforming principle → Mouse dies → Virulent bacteria

Harmless bacteria + transforming principle treated with protease or ribonuclease → Mouse dies → Virulent bacteria

Harmless bacteria + transforming principle treated with deoxyribonuclease → Mouse survives

(B) The Hershey–Chase experiment

DNA
Protein capsid

Phage attached to bacteria

Agitate in blender

Phage now detached

Centrifuge

Pellet of bacteria

70% ^{32}P
20% ^{35}S

Figure 1.3 The two experiments that suggested that genes are made of DNA. (A) Avery and colleagues showed that the transforming principle is made of DNA. The top two panels show what happens when mice are injected with harmless *Streptococcus pneumoniae* bacteria, with or without addition of the transforming principle, a cell extract obtained from a virulent strain of *S. pneumoniae*. When the transforming principle is present, the mouse dies, because the genes in the transforming principle convert the harmless bacteria into the virulent form, with these virulent bacteria subsequently being recovered from the lungs of the dead mouse. The lower two panels show that treatment with protease or ribonuclease has no effect on the transforming principle, but that the transforming principle is inactivated by deoxyribonuclease. (B) The Hershey–Chase experiment used T2 bacteriophages, each of which comprises a DNA molecule contained in a protein capsid attached to a 'body' and 'legs' that enable the bacteriophage to attach to the surface of a bacterium and inject its genes into the cell. The DNA of the bacteriophages was labeled with ^{32}P, and the protein with ^{35}S. A few minutes after infection, the culture was agitated to detach the empty bacteriophage particles from the cell surface. The culture was then centrifuged, which collects the bacteria plus bacteriophage genes as a pellet at the bottom of the tube, but leaves the lighter bacteriophage particles in suspension. Hershey and Chase found that the bacterial pellet contained 70% of the ^{32}P-labeled component of the bacteriophages (the DNA) but only 20% of the ^{35}S-labeled material (the bacteriophage protein). In a second experiment, Hershey and Chase showed that new bacteriophages produced at the end of the infection cycle contained less than 1% of the protein from the parent bacteriophages. For more details of the bacteriophage infection cycle, see Figure 2.27.

protein. Neither is the bacteriophage experiment conclusive, as Hershey and Chase stressed when they published their results: 'Our experiments show clearly that a physical separation of phage T2 into genetic and non-genetic parts is possible …The chemical identification of the genetic part must wait, however, until some questions … have been answered.' In retrospect, these two experiments are important not because of what they tell us but because they alerted biologists to the fact that DNA might be the genetic material and was therefore worth studying. It was this that influenced Watson and Crick to work on DNA and, as we will see below, it was their discovery of the double-helix structure, which solved the puzzling question of how genes can replicate, which really convinced the scientific world that genes are made of DNA.

DNA is a polymer of nucleotides

The names of James Watson and Francis Crick are so closely linked with DNA that it is easy to forget that when they began their collaboration in October 1951 the detailed structure of the DNA polymer was already known. Their contribution was not to determine the structure of DNA *per se*, but to show that in living cells two DNA chains are intertwined to form the double helix. First, therefore, we should examine what Watson and Crick knew before they began their work.

DNA is a linear, unbranched polymer in which the monomeric subunits are four chemically distinct nucleotides that can be linked together in any order in chains hundreds, thousands, or even millions of units in length. Each nucleotide in a DNA polymer is made up of three components (**Figure 1.4**):

- **2′-Deoxyribose**, which is a **pentose**, a type of sugar composed of five carbon atoms. These five carbons are numbered 1′ (spoken as 'one-prime'), 2′, and so on. The name '2′-deoxyribose' indicates that this particular sugar is a derivative of ribose, in which the hydroxyl (–OH) group attached to the 2′-carbon of ribose has been replaced by a hydrogen (–H) group.

- A **nitrogenous base**, one of **cytosine**, **thymine** (single-ring **pyrimidines**), **adenine**, or **guanine** (double-ring **purines**). The base is attached to the 1′-carbon of the sugar by a **β-*N*-glycosidic bond** attached to nitrogen number one of the pyrimidine or number nine of the purine.

- A **phosphate group** comprising one, two, or three linked phosphate units attached to the 5′-carbon of the sugar. The phosphates are designated α, β, and γ, with the α-phosphate being the one directly attached to the sugar.

A molecule made up of just the sugar and base is called a **nucleoside**; the addition of the phosphates converts this into a nucleotide. Although cells contain nucleotides with one, two, or three phosphate groups, only the nucleoside triphosphates act as substrates for DNA synthesis. The full chemical names of the four nucleotides that polymerize to make DNA are:

- 2′-deoxyadenosine 5′-triphosphate

- 2′-deoxycytidine 5′-triphosphate

- 2′-deoxyguanosine 5-triphosphate

- 2′-deoxythymidine 5-triphosphate

The abbreviations of these four nucleotides are dATP, dCTP, dGTP, and dTTP, respectively, or when referring to a DNA sequence, A, C, G, and T, respectively.

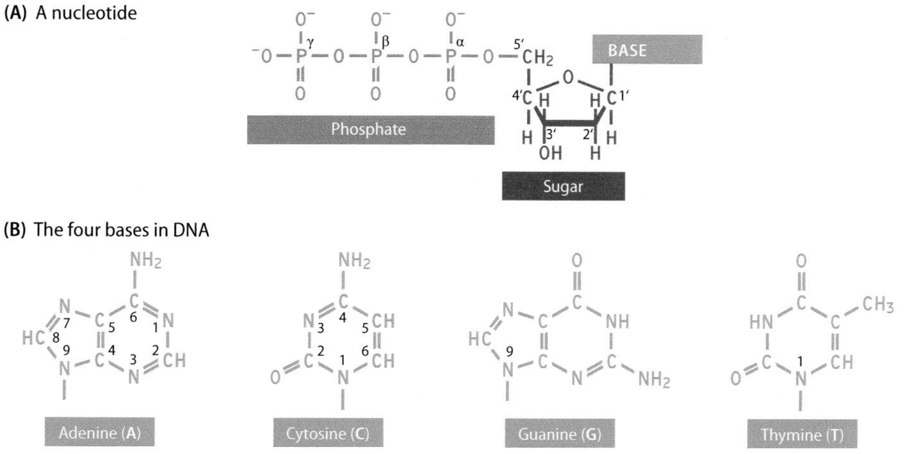

(A) A nucleotide

(B) The four bases in DNA

Figure 1.4 The structure of a nucleotide. (A) The general structure of a deoxyribonucleotide, the type of nucleotide found in DNA. (B) The four bases that occur in deoxyribonucleotides.

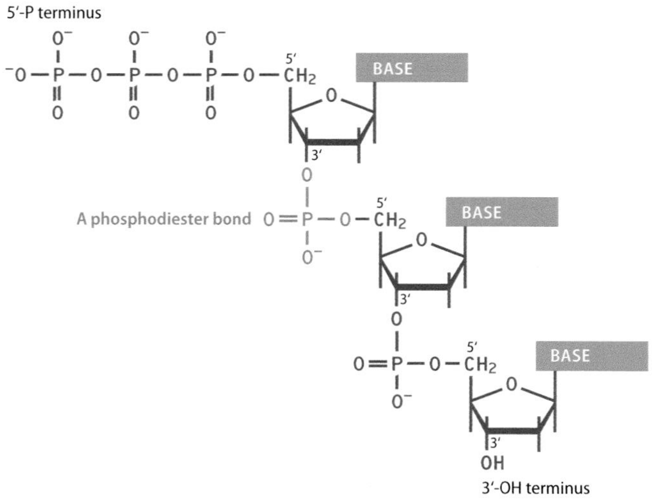

Figure 1.5 A short DNA polynucleo-tide showing the structure of the phosphodiester bond. Note that the two ends of the polynucleotide are chemically distinct.

In a polynucleotide, individual nucleotides are linked together by **phospho-diester bonds** between their 5´- and 3´-carbons (**Figure 1.5**). From the structure of this linkage, we can see that the polymerization reaction (**Figure 1.6**) involves removal of the two outer phosphates (the β- and γ-phosphates) from one nucleotide and replacement of the hydroxyl group attached to the 3´-carbon of the second nucleotide. Note that the two ends of the polynucleotide are chemically distinct, one having an unreacted triphosphate group attached to the 5´-carbon (the **5´** or **5´-P terminus**), and the other having an unreacted hydroxyl attached to the 3´-carbon (the **3´** or **3´-OH terminus**). This means that the polynucleotide has a chemical direction, expressed as either 5´→3´ (down in Figure 1.5) or 3´→5´ (up in Figure 1.5). An important consequence of the polarity of the phosphodiester bond is that the chemical reaction needed to extend a DNA polymer in the 5´→3´ direction is different to that needed to make a 3´→5´ extension. The **DNA polymerase** enzymes present in living organisms are only able to carry out 5´→3´ synthesis, which adds significant complications to the process by which double-stranded DNA is replicated (Section 15.3).

The discovery of the double helix

In the years before 1950, various lines of evidence had shown that cellular DNA molecules are comprised of two or more polynucleotides assembled together in some way. The possibility that unraveling the nature of this assembly might provide insights into how genes work prompted Watson and Crick, among others, to try to solve the structure. According to Watson, in his book *The Double Helix*, their work was a desperate race against the famous American biochemist, Linus Pauling, who initially proposed an incorrect triple helix model, giving Watson and Crick the time they needed to complete the double-helix structure. It is now difficult to separate fact from fiction, especially regarding the part played by Rosalind Franklin, whose **X-ray diffraction studies** provided the bulk of the experimental data in support of the double helix and who was herself very close to solving the structure. The one thing that is clear is that the double helix, discovered by Watson and Crick on Saturday, March 7, 1953, was the single most important breakthrough in biology during the twentieth century.

The discovery of the double helix can be looked on as one of the first multidisciplinary biological research projects. Watson and Crick used four quite different types of information to deduce the double-helix structure:

- Biophysical data of various kinds were used to infer some of the key features of the structure. The water content of DNA fibers was particularly important because it enabled the density of the DNA in a fiber to be estimated. The number of strands in the helix and the spacing between the nucleotides

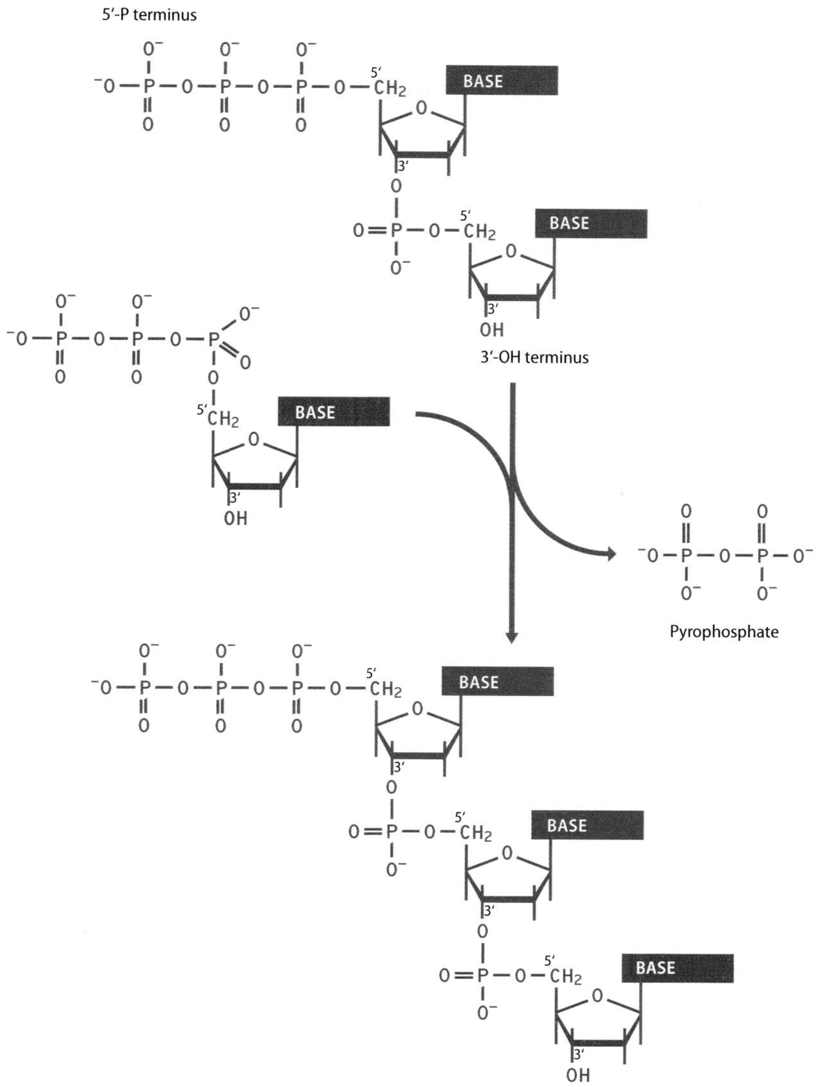

5′-P terminus

3′-OH terminus

Pyrophosphate

Figure 1.6 The polymerization reaction that results in synthesis of a DNA polynucleotide. Synthesis occurs in the 5′→3′ direction, with the new nucleotide being added to the 3′-carbon at the end of the existing polynucleotide. The β- and γ-phosphates of the nucleotide are removed as a pyrophosphate molecule.

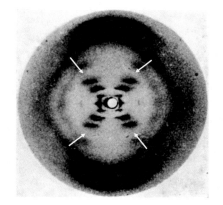

Figure 1.7 Franklin's 'photo 51' showing the X-ray diffraction pattern obtained with a fiber of DNA. The cross shape indicates that DNA has a helical structure, and the extent of the shadowing within the 'diamond' spaces above, below and to either side of the cross show that the sugar–phosphate backbone is on the outside of the helix (see Figure 1.9). The positions of the various smears that make up the arms of the cross enable dimensions such as the diameter, rise per base pair and pitch (see Table 1.1) of the molecule to be calculated. The 'missing smears' (the gap in each arm of the cross, marked by the arrows) indicate the relative positioning of the two polynucleotides. These missing smears enabled Watson and Crick to recognize that there are two grooves of different depths on the outer surface of the helix (see Figure 1.9). (From Franklin R & Gosling RG [1953] Nature 171:740-741. With permission from Springer Nature.)

had to be compatible with the fiber density. Pauling's triple helix model was based on an incorrect density measurement that suggested that the DNA molecule was more closely packed than is actually the case.

- **X-ray diffraction patterns** (Section 11.1), most of which were produced by Rosalind Franklin, revealed the detailed helical structure (**Figure 1.7**).

- The **base ratios**, which had been discovered by Erwin Chargaff of Columbia University, New York, enabled the pairing between the polynucleotides in the helix to be deduced. Chargaff had carried out a lengthy series of chromatographic studies of DNA samples from various sources and showed that, although the values are different in different organisms, the amount of adenine is always the same as the amount of thymine, and the amount of guanine equals the amount of cytosine (**Figure 1.8**). These base ratios led to the base-pairing rules, which were the key to the discovery of the double-helix structure.

- The construction of scale models of possible DNA structures, which was the only major technique that Watson and Crick performed themselves, enabled the relative positioning of the various atoms to be checked, to ensure that pairs that formed bonds were not too far apart, and that other atoms were not so close together as to interfere with one another.

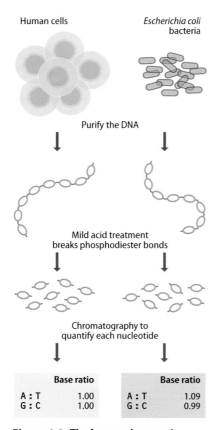

Human cells

Escherichia coli bacteria

Purify the DNA

Mild acid treatment breaks phosphodiester bonds

Chromatography to quantify each nucleotide

	Base ratio
A : T	1.00
G : C	1.00

	Base ratio
A : T	1.09
G : C	0.99

Figure 1.8 The base ratio experiments performed by Chargaff. DNA was extracted from various organisms and treated with acid to hydrolyze the phosphodiester bonds and release the individual nucleotides. Each nucleotide was then quantified by chromatography. The data show some of the actual results obtained by Chargaff. These indicate that, within experimental error, the amount of adenine is the same as that of thymine, and the amount of guanine is the same as that of cytosine.

The double helix is stabilized by base-pairing and base-stacking

The double helix is right-handed, which means that if it were a spiral staircase and you were climbing upwards then the rail on the outside of the staircase would be on your right-hand side. The two strands run in opposite directions (**Figure 1.9A**). The helix is stabilized by two types of chemical interaction:

- **Base-pairing** between the two strands involves the formation of **hydrogen bonds** between an adenine on one strand and a thymine on the other strand, or between a cytosine and a guanine (**Figure 1.9B**). Hydrogen bonds are weak **electrostatic interactions** between an electronegative atom (such as oxygen or nitrogen) and a hydrogen atom attached to a second electronegative atom. Hydrogen bonds are longer than covalent bonds and are much weaker, typical bond energies being 8–29 kJ mol^{-1} at 25°C, compared with up to 348 kJ mol^{-1} for a single covalent bond between a pair of carbon atoms. As well as their role in the DNA double helix, hydrogen bonds stabilize protein secondary structures. The two base-pair combinations – A base-paired with T, and G base-paired with C – explain the base ratios discovered by Chargaff. These are the only pairs that are permissible, partly because of the geometries of the nucleotide bases and the relative positions of the atoms that are able to participate in hydrogen bonds, and partly because the pair must be between a purine

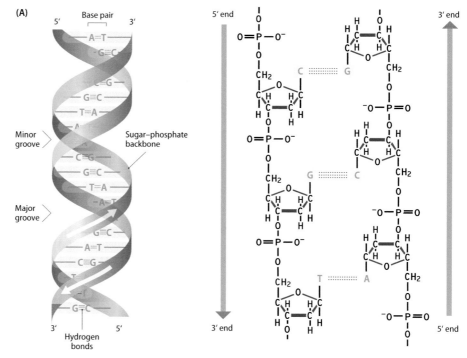

Figure 1.9 The double-helix structure of DNA. (A) Two representations of the double helix. On the left, the structure is shown with the sugar–phosphate 'backbones' of each polynucleotide drawn as a gray ribbon with the base pairs in green. On the right the chemical structure for three base pairs is given. (B) A base-pairs with T, and G base-pairs with C. The bases are drawn in outline, with the hydrogen bonding indicated by dotted lines. Note that a G–C base pair has three hydrogen bonds whereas an A–T base pair has just two.

and a pyrimidine: a purine–purine pair would be too big to fit within the helix, and a pyrimidine–pyrimidine pair would be too small.

- **Base-stacking** involves attractive forces between adjacent base pairs and adds stability to the double helix once the strands have been brought together by base-pairing. Base-stacking is sometimes called **π–π interactions**, because it is thought to involve the p electrons associated with the double bonds of the purine and pyrimidine structures. However, this hypothesis is now being questioned and the possibility that base-stacking involves a type of electrostatic interaction is being explored.

Both base-pairing and base-stacking are important in holding the two polynucleotides together, but base-pairing has added significance because of its biological implications. The limitation that A can only base-pair with T, and G can only base-pair with C, means that **DNA replication** can result in perfect copies of a parent molecule through the simple expedient of using the sequences of the preexisting strands to dictate the sequences of the new strands. This is **template-dependent DNA synthesis**, and it is the system used by all cellular DNA polymerases (Section 2.1). Base-pairing therefore enables DNA molecules to be replicated by a system that is so simple and elegant that as soon as the double-helix structure was publicized by Watson and Crick, every biologist became convinced that genes really are made of DNA.

The double helix has structural flexibility

The double helix described by Watson and Crick, and shown in Figure 1.9A, is called the B-form of DNA or **B-DNA**. Its characteristic features lie in its dimensions: a helical diameter of 2.37 nm, a rise of 0.34 nm per base pair, and a pitch (the distance taken up by a complete turn of the helix) of 3.4 nm, corresponding to ten base pairs per turn. The DNA in living cells is thought to be predominantly in this B-form, but it is now clear that genomic DNA molecules are not entirely uniform in structure. This is mainly because each nucleotide in the helix has the flexibility to take up slightly different molecular shapes. To adopt these different conformations, the relative positions of the atoms in the nucleotide must change slightly. There are a number of possibilities but the most important conformational changes are:

- Rotation around the β-*N*-glycosidic bond, which changes the orientation of the base relative to the sugar: the two possibilities are called the *anti* and *syn* conformations (**Figure 1.10A**). Base rotation influences the positioning of the two polynucleotides.

- **Sugar pucker**, which refers to the three-dimensional shape of the sugar. The ribose component of the nucleotide does not have a planar structure: when viewed from the side, one or two of the carbon atoms are either above or below the plane of the sugar (**Figure 1.10B**). In the C2′-*endo*

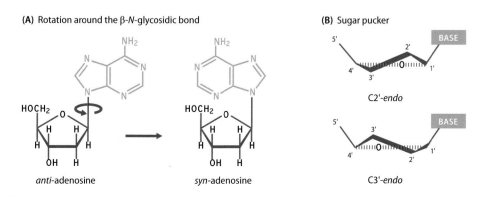

(A) Rotation around the β-*N*-glycosidic bond

anti-adenosine

syn-adenosine

(B) Sugar pucker

C2′-*endo*

C3′-*endo*

Figure 1.10 Changes in nucleotide configuration that can affect the conformation of the double helix. (A) The structures of *anti*- and *syn*-deoxyadenosine. The two structures differ in the orientation of the base relative to the sugar component of the nucleotide, rotation around the β-*N*-glycosidic bond converting one form into the other. The three other nucleotides also have *anti* and *syn* conformations. (B) Sugar pucker, illustrating the positioning of the sugar carbons in the C2′-*endo* and C3′-*endo* configurations.

configuration the 2′-carbon is above the plane and the 3′-carbon slightly below, and in the C3′-*endo* configuration the 3′-carbon is above the plane and the 2′-carbon below. Because the 3′-carbon participates in the phosphodiester bond with the adjacent nucleotide, the two pucker configurations have different effects on the conformation of the sugar–phosphate backbone.

Conformation changes resulting from rotation around the β-*N*-glycosidic bond and sugar pucker can give rise to major changes in the overall structure of the helix. It has been recognized since the 1950s that changes in the dimensions of the double helix occur when fibers containing DNA molecules are exposed to different relative humidities. For example, the modified version of the double helix called **A-DNA** has a diameter of 2.55 nm, a rise of 0.23 nm per base pair, and a pitch of 2.5 nm, corresponding to 11 base pairs per turn (**Table 1.1**). Like the B-form, A-DNA is a right-handed helix and the bases are in the *anti*-conformation relative to the sugar. The main difference lies with the sugar pucker, the sugars in the B-form being in the C2′-*endo* configuration, and those in A-DNA in the C3′-*endo* configuration. Other right-handed variations of the double helix include B′-, C-, C′-, C′′-, D-, E- and T-DNAs.

A more drastic reorganization is also possible, leading to the left-handed **Z-DNA**, in which the sugar–phosphate backbone adopts an irregular zigzag conformation. Z-DNA is a more tightly wound version of the double helix with 12 bp per turn and a diameter of only 1.84 nm (Table 1.1). It is known to occur in regions of a double helix that contain repeats of the motif GC (i.e., the sequence of each strand is ..GCGCGCGC..). In these regions, each G nucleotide has the *syn* and C3′-*endo* conformations, and each C has the *anti* and C2′-*endo* conformations.

The bare dimensions of the various forms of the double helix do not reveal what are probably the most significant differences between them. These relate not to diameter and pitch, but to the extent to which the internal regions of the helix are accessible from the surface of the structure. As shown in Figure 1.9A, the B-form of DNA does not have an entirely smooth surface: instead, two grooves spiral along the length of the helix. One of these grooves is relatively wide and deep and is called the **major groove**; the other is narrow and less deep and is called the **minor groove**. A-DNA also has two grooves (**Figure 1.11**), but with this conformation the major groove is even deeper, and the minor groove shallower compared with B-DNA. Z-DNA is different again, with the major groove virtually nonexistent but the minor groove very narrow and deep. In each form of DNA, part of the internal surface of at least one of the grooves is formed by chemical groups attached to the nucleotide bases. In Chapter 11, we will see that expression of the biological information contained within a genome is mediated by DNA-binding proteins that attach to the double helix and regulate the activity of the genes contained within it. To carry out their

TABLE 1.1 FEATURES OF THE DIFFERENT CONFORMATIONS OF THE DNA DOUBLE HELIX			
Feature	A-form	B-form	Z-DNA
Type of helix	Right-handed	Right-handed	Left-handed
Helical diameter (nm)	2.55	2.37	1.84
Distance between base pairs (nm)	0.23	0.34	0.38
Distance per complete turn (nm)	2.5	3.4	4.6
Number of base pairs per turn	11	10	12
Base orientation	*anti*	*anti*	mixture
Sugar pucker	C3′-*endo*	C2′-*endo*	mixture

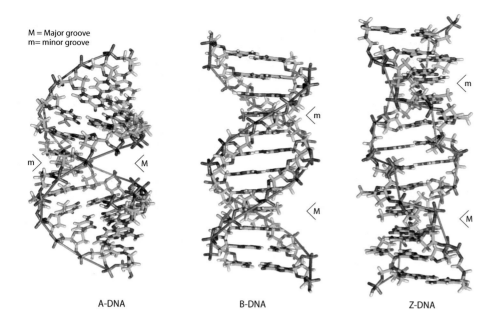

M = Major groove
m = minor groove

A-DNA   B-DNA   Z-DNA

Figure 1.11 The A-, B- and Z-forms of the double helix. The major and minor grooves on each molecule are indicated by 'M' and 'm', respectively. (Courtesy of Richard Wheeler, published under GFDL 1.2.)

function, each DNA-binding protein must attach at a specific position, near to the gene whose activity it will influence. This can be achieved, with at least some degree of accuracy, by the protein reaching down into a groove, within which the DNA sequence can be 'read' without the helix being opened up by breaking the base pairs. A corollary of this is that a DNA-binding protein whose structure enables it to recognize a specific nucleotide sequence within B-DNA, for example, might not be able to recognize that sequence if the DNA has taken up a different conformation. As we will see in Section 11.3, conformational variations along the length of a DNA molecule, together with other structural polymorphisms caused by the nucleotide sequence, could be important in determining the specificity of the interactions between the genome and its DNA-binding proteins.

1.2 RNA AND THE TRANSCRIPTOME

The initial product of genome expression is the transcriptome (see Figure 1.2), the collection of RNA molecules derived from those genes that are active in the cell at a particular time. The RNA molecules of the transcriptome are synthesized by the process called transcription. In this section, we will examine the structure of RNA and then look more closely at the various types of RNA molecule that are present in living cells.

RNA is a second type of polynucleotide

RNA is a polynucleotide similar to DNA but with two important chemical differences (**Figure 1.12**). First, the sugar in an RNA nucleotide is **ribose** and, second, RNA contains **uracil** instead of thymine. The four nucleotide substrates for synthesis of RNA are therefore:

- adenosine 5′-triphosphate
- cytidine 5′-triphosphate
- guanosine 5′-triphosphate
- uridine 5′-triphosphate

These nucleotides are abbreviated to ATP, CTP, GTP, and UTP, or A, C, G, and U, respectively.

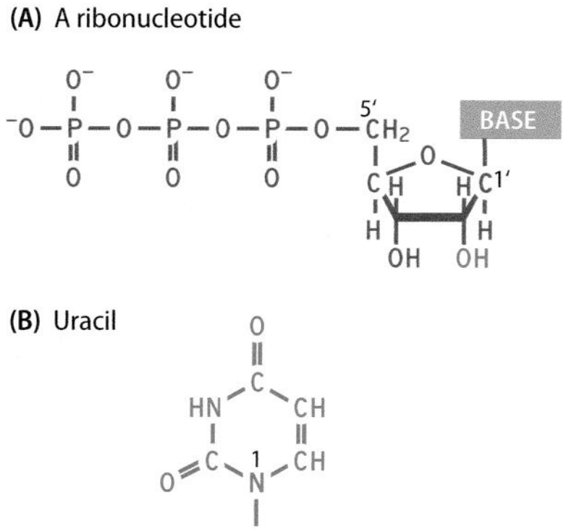

(A) A ribonucleotide

(B) Uracil

Figure 1.12 The chemical differences between DNA and RNA. (A) RNA contains ribonucleotides, in which the sugar is ribose rather than 2′-deoxyribose. The difference is that a hydroxyl group, rather than hydrogen atom, is attached to the 2′-carbon. (B) RNA contains the pyrimidine called uracil instead of thymine.

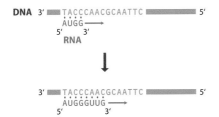

Figure 1.13 Template-dependent RNA synthesis. The RNA transcript is synthesized in the 5´→3´ direction, reading the DNA in the 3´→5´ direction, with the sequence of the transcript determined by base-pairing to the DNA template.

As with DNA, RNA polynucleotides contain 3´–5´ phosphodiester bonds, but these phosphodiester bonds are less stable than those in a DNA polynucleotide because of the indirect effect of the hydroxyl group at the 2´-position of the sugar. RNA molecules are rarely more than a few thousand nucleotides in length, and although many form intramolecular base pairs (for example, see Figure 5.6A), most are **single-** rather than **double-stranded**.

The enzymes responsible for transcription of DNA into RNA are called **DNA-dependent RNA polymerases**. The name indicates that the enzymatic reaction that they catalyze results in polymerization of RNA from ribonucleotides and occurs in a DNA-dependent manner, meaning that the sequence of nucleotides in a DNA template dictates the sequence of nucleotides in the RNA that is made (**Figure 1.13**). It is permissible to shorten the enzyme name to **RNA polymerase**, as the context in which the name is used means that there is rarely confusion with the **RNA-dependent RNA polymerases** that are involved in replication and expression of some virus genomes. The chemical basis of **template-dependent RNA synthesis** is equivalent to that shown for the synthesis of DNA in Figure 1.6. Ribonucleotides are added one after another to the growing 3´ end of the RNA **transcript**, the identity of each nucleotide being specified by the base-pairing rules: A base-pairs with T or U; G base-pairs with C. During each nucleotide addition, the β- and γ-phosphates are removed from the incoming nucleotide, and the hydroxyl group is removed from the 3´-carbon of the nucleotide at the end of the chain, precisely the same as for DNA polymerization.

The RNA content of the cell

A typical bacterium contains 0.05–0.10 pg of RNA, making up about 6% of its total weight. A mammalian cell, being much larger, contains more RNA, 20–30 pg in all, but this represents only 1% of the cell as a whole.

The best way to understand the RNA content of a cell is to divide it into categories and subcategories depending on function. There are several ways of doing this, the most informative scheme being the one shown in **Figure 1.14**. The primary division is between **coding RNA** and **noncoding RNA**. The coding RNA is made up of just one class of molecule, the **messenger RNAs (mRNAs)**, which are transcripts of protein-coding genes and hence are translated into protein in the second stage of genome expression. Messenger RNAs rarely make up more than 4% of the total RNA and are short-lived, being degraded soon after synthesis. Bacterial mRNAs have half-lives of no more than a few minutes, and in **eukaryotes** most mRNAs are degraded a few hours after synthesis. This rapid turnover means that the mRNA composition of the cell is not fixed and can quickly be restructured by changing the rate of synthesis of individual mRNAs.

The second type of RNA is referred to as 'noncoding' as these molecules are not translated into protein. An alternative name is **functional RNA**, which emphasizes that the noncoding RNAs still have essential roles within the cell.

Figure 1.14 The RNA content of a cell. This scheme shows the types of RNA present in all organisms and those categories found only in eukaryotic cells. Precursor RNAs, as described in the next section, are included.

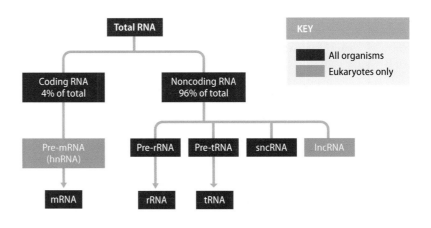

There are several diverse types of noncoding RNA, the two most important being as follows:

- **Ribosomal RNAs (rRNAs)** are present in all organisms and are usually the most abundant RNAs in the cell, making up over 80% of the total RNA in actively dividing bacteria. These molecules are components of **ribosomes**, the structures within which protein synthesis takes place (Section 13.3).

- **Transfer RNAs (tRNAs)** are small molecules that are also involved in protein synthesis and, like rRNA, are found in all organisms. The function of tRNAs is to carry amino acids to the ribosome and ensure that the amino acids are linked together in the order specified by the nucleotide sequence of the mRNA that is being translated (Section 13.3).

The above are the two most important categories of noncoding RNA, but there are several other types with specialist roles in eukaryotic or bacterial cells. In eukaryotes, these RNAs are usually divided into two groups, the **short noncoding RNAs (sncRNAs)**, comprising RNAs less than 200 nucleotides in length, and the **long noncoding RNAs (lncRNAs)**, made up of molecules longer than 200 nucleotides. We will examine the roles of these various types of noncoding RNA in Chapter 12.

Many RNAs are synthesized as precursor molecules

As well as the mature RNAs described above, cells also contain precursor molecules. Many RNAs, especially in eukaryotes, are initially synthesized as a precursor called the **primary transcript** or **pre-RNA**, which has to be processed in order to release the functional molecules.

The most important of these processing events is **splicing**. Some eukaryotic genes are **discontinuous** and contain internal segments that are copied during transcription but then excised from the pre-RNA (**Figure 1.15**). These excised segments are called **introns**, in contrast to the **exons**, which are spliced together to form the mature RNA. Introns are present in some rRNA and tRNA genes, but are particularly common in protein-coding genes. Splicing of **pre-mRNA** is therefore a major part of the process that results in synthesis of the protein-coding component of the transcriptome (Section 12.4). Splicing occurs in the nucleus, with the unspliced pre-mRNA forming the nuclear RNA fraction called **heterogeneous nuclear RNA (hnRNA)**.

Splicing is not the only type of cutting event that occurs during processing of pre-RNA. Many rRNAs and tRNAs are initially synthesized as precursors that contain copies of more than one molecule. The **pre-rRNAs** and **pre-tRNAs** must therefore be cut into pieces to produce the mature RNAs (**Figure 1.16**). This type of processing occurs in both prokaryotes and eukaryotes.

Other processing events result in changes occurring at the ends of RNA molecules. These **end-modifications** occur during the synthesis of eukaryotic mRNAs, most of which have a structure called a **cap** attached at the 5 end and a **poly(A) tail** attached to the 3′ end. The cap structure comprises the modified

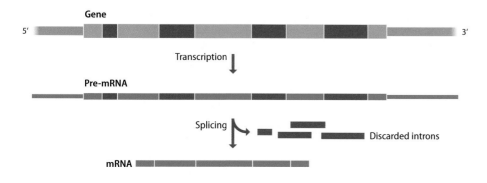

Figure 1.15 Splicing of a eukaryotic pre-mRNA. The introns are cut out of the pre-mRNA and the exons rejoined to give the functional mRNA. In the gene, the exons are shown in green, and in the pre-mRNA and mRNA they are shown in blue.

Figure 1.16 Processing of a bacterial pre-rRNA. The bacterial pre-rRNA contains one copy of each of the three rRNAs that, together with ribosomal proteins, make up the bacterial ribosome. A series of cutting rand trimming reactions releases the mature rRNAs from the precursor molecule.

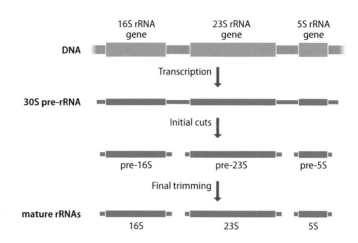

nucleotide called 7-methylguanosine attached to the first nucleotide in the pre-mRNA by an unusual triple-phosphate bond (**Figure 1.17A**). The first and second nucleotides in the pre-mRNA might also be modified by addition of methyl groups. The cap structure is needed to help initiate translation of the mRNA into a protein. The poly(A) tail is a series of up to 250 adenine nucleotides that are present at the 3′ end of the mRNA. The pre-mRNA is cut at a position close to its 3′ end and the adenines are added to this new end by a **template-independent RNA polymerase** called **poly(A) polymerase** (**Figure 1.17B**). The function of this **polyadenylation** process is not fully understood, but we do know that if the poly(A) tail is absent or shorter than usual, then the mRNA is degraded.

The final type of processing event is **chemical modification**. Some nucleotide bases within the rRNAs and tRNAs of all organisms are modified by methylation, deamination (removal of a –NH$_2$ group), and/or thio-substitution (replacement of oxygen with sulfur), and some bases undergo internal rearrangements that change the positions of particular groups and/or convert double bonds to single bonds (**Figure 1.18**). The reasons for many of these modifications are unknown, but functions have been assigned for specific cases. In tRNA, some of the

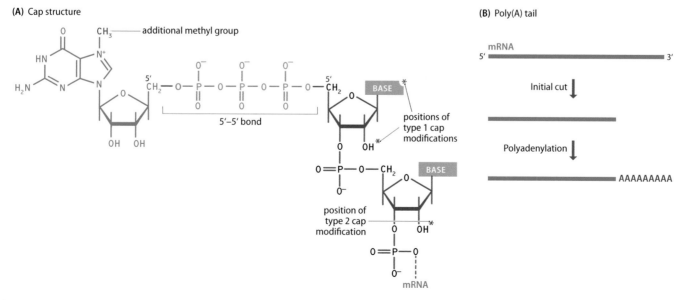

Figure 1.17 Modifications at the 5′ and 3′ ends of a eukaryotic mRNA. (A) The cap structure at the 5′ end of an mRNA. The **type 0** cap comprises the modified nucleotide 7-methylguanosine linked to the first nucleotide in the pre-mRNA by a 5′–5′ triple-phosphate bond. Attachment of additional methyl groups, at the indicated positions, gives rise to the **type 1** and **type 2 cap** structures. (B) Polyadenylation of the 3′ end of an mRNA. The poly(A) tail does not have a counterpart in the DNA sequence and so is not synthesized by RNA polymerase when the gene is transcribed. Instead, the poly(A) tail is added after transcription, by poly(A) polymerase.

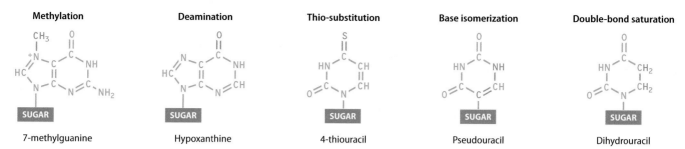

Figure 1.18 Examples of chemically modified bases occurring in rRNA and tRNA molecules. The differences between these modified bases and the standard ones from which they are derived are shown in orange.

modified nucleotides are recognized by the enzymes that attach an amino acid to the 3′ end of the molecule. This reaction is central to the role that tRNA plays during protein synthesis. The correct amino acid has to be attached to the correct tRNA, and the modifications within the tRNA are thought to provide some of the specificity that ensures that this happens.

Eukaryotic mRNAs also undergo chemical modification. Some of these modifications alter the base-pairing properties of a nucleotide and so change the sequence of the mRNA and might also change the amino acid sequence of the encoded protein (**Figure 1.19A**). This type of chemical modification is called **RNA editing**. Other modifications do not change the sequence of the mRNA: for example, N^7-methylguanine still base-pairs with C (**Figure 1.19B**). Although these modifications do not alter the amino acid sequence of the encoded protein, they might influence the way the mRNA is translated (Section 12.5).

There are different definitions of the transcriptome

Although most biologists now define the transcriptome as the total RNA content of a cell, the term, when first introduced in 1997, was initially used to describe just the mRNA component. The mRNA makes up less than 4% of the total cell RNA, but is often looked on as the most significant component because it comprises the coding RNAs that are used in the next stage of genome expression. Even in the simplest organisms, such as bacteria and yeast, many different protein-coding genes are active at any one time. The mRNA content of a cell is therefore complex, containing copies of hundreds, if not thousands, of different genes. By specifying the set of proteins that the cell is able to make, the mRNA content determines the biochemical features of a cell. Many of the early studies of 'transcriptomes' aimed to identify all, or as many as possible, of the mRNAs in a cell, in order to understand the overall pattern of **gene expression**, and how that pattern changes when, for example, a cell becomes cancerous. These types of study are still important today, and because they focus on mRNA there remains a tendency to look on the transcriptome as just referring to the mRNA content of the cell.

The broader definition of 'transcriptome' to include *all* the RNA in a cell reflects our growing awareness of the important roles that noncoding RNAs play in specifying the biochemical properties of a cell. In particular, the sncRNAs called **microRNAs (miRNAs)** regulate gene expression in eukaryotic cells by degrading those mRNAs whose products are no longer needed (Section 12.6). Human cells are able to make about 1000 miRNAs, each one specific for a single or small group of mRNAs. Understanding which miRNAs are synthesized in a particular cell, and how the pattern of miRNA synthesis changes in diseased cells, is an essential complement to the equivalent studies of mRNAs. It is therefore sensible to extend the term 'transcriptome' to include all of the RNA in a cell, because a focus just on mRNA misses the vital role that other parts of the transcriptome play in mediating expression of the biological information contained in the genome.

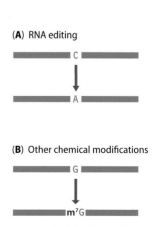

Figure 1.19 Chemical modification of eukaryotic mRNA. (A) RNA editing, which results in a change in the nucleotide sequence of the mRNA. (B) Modifications such as conversion of guanine to N^7-methylguanine (m7G) do not affect base-pairing and so do not change the nucleotide sequence.

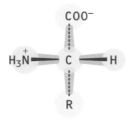

Figure 1.20 The general structure of an amino acid. All amino acids have the same general structure, comprising a central α-carbon attached to a hydrogen atom, a carboxyl group, an amino group, and an R group. The R group is different for each amino acid (see Figure 1.24).

1.3 PROTEINS AND THE PROTEOME

The second product of genome expression is the proteome (see Figure 1.2), the cell's repertoire of proteins, which specifies the nature of the biochemical reactions that the cell is able to carry out. These proteins are synthesized by translation of the mRNA component of the transcriptome.

There are four hierarchical levels of protein structure

A protein, like a DNA molecule, is a linear, unbranched polymer. In proteins, the monomeric subunits are called **amino acids** (**Figure 1.20**) and the resulting polymers, or **polypeptides**, are rarely more than 2000 units in length.

Proteins are traditionally looked upon as having four distinct levels of structure. These levels are hierarchical; the protein is built up stage-by-stage with each level of structure depending on the one below it:

- The **primary structure** of the protein is formed by joining amino acids into a polypeptide. The amino acids are linked by **peptide bonds** that are formed by a condensation reaction between the carboxyl group of one amino acid and the amino group of a second amino acid (**Figure 1.21**). Note that, as with a polynucleotide, the two ends of the polypeptide are chemically distinct: one has a free amino group and is called the **amino, NH$_2$–**, or **N terminus**; the other has a free carboxyl group and is called the **carboxyl, COOH–**, or **C terminus**. The direction of the polypeptide can therefore be expressed as either N→C (left to right in Figure 1.21) or C→N (right to left in Figure 1.21).

- The **secondary structure** refers to the different conformations that can be taken up by the polypeptide. The two main types of secondary structure are the **α-helix** and **β-sheet** (**Figure 1.22**). These are stabilized mainly by hydrogen bonds that form between different amino acids in the polypeptide. Most polypeptides are long enough to be folded into a series of secondary structures, one after another along the molecule.

- The **tertiary structure** results from folding the secondary structural components of the polypeptide into a three-dimensional configuration (**Figure 1.23**). The tertiary structure is stabilized by various chemical forces, notably hydrogen bonding between individual amino acids, electrostatic interactions between the R groups of charged amino acids (see **Figure 1.24**), and **hydrophobic effects**, which dictate that amino acids with nonpolar ('water-hating') side-groups must be shielded from water by embedding within the internal regions of the protein. There may also be covalent linkages called **disulfide bonds** between cysteine amino acid residues at various places in the polypeptide.

- The **quaternary structure** involves the association of two or more polypeptides, each folded into its tertiary structure, into a multisubunit protein. Not all proteins form quaternary structures, but it is a feature of many proteins with complex functions, including several involved in genome expression. Some quaternary structures are held together by disulfide bonds between the different polypeptides, resulting in stable multisubunit proteins that cannot easily be broken down to the component parts. Other quaternary structures comprise looser associations of subunits stabilized by hydrogen bonding and hydrophobic effects, which means that these proteins can revert to their component polypeptides, or change their subunit composition, according to the functional requirements of the cell.

Amino acid diversity underlies protein diversity

Proteins are functionally diverse because the amino acids from which proteins are made are themselves chemically diverse. Different sequences of amino acids

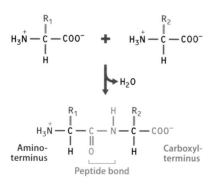

Figure 1.21 In polypeptides, amino acids are linked by peptide bonds. The drawing shows the chemical reaction that results in two amino acids becoming linked together by a peptide bond. The reaction is called a condensation because it results in elimination of water.

(A) α helix
(B) β sheet
H bond

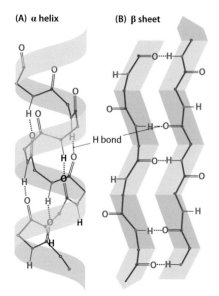

Figure 1.22 The two main secondary structural units found in proteins. (A) the α-helix and (B) the β-sheet. The polypeptide chains are shown in outline. The R groups have been omitted for clarity. Each structure is stabilized by hydrogen (H) bonds between the –C=O and –N–H groups of different peptide bonds. The β-sheet conformation that is shown is antiparallel, the two chains running in opposite directions. Parallel β-sheets also occur.

N-terminus
α-helices
C-terminus
β-sheet
connecting loop

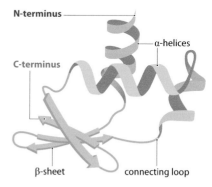

Figure 1.23 The tertiary structure of a protein. This imaginary protein structure comprises three α-helices, shown as coils, and a four-stranded β-sheet, indicated by the arrows.

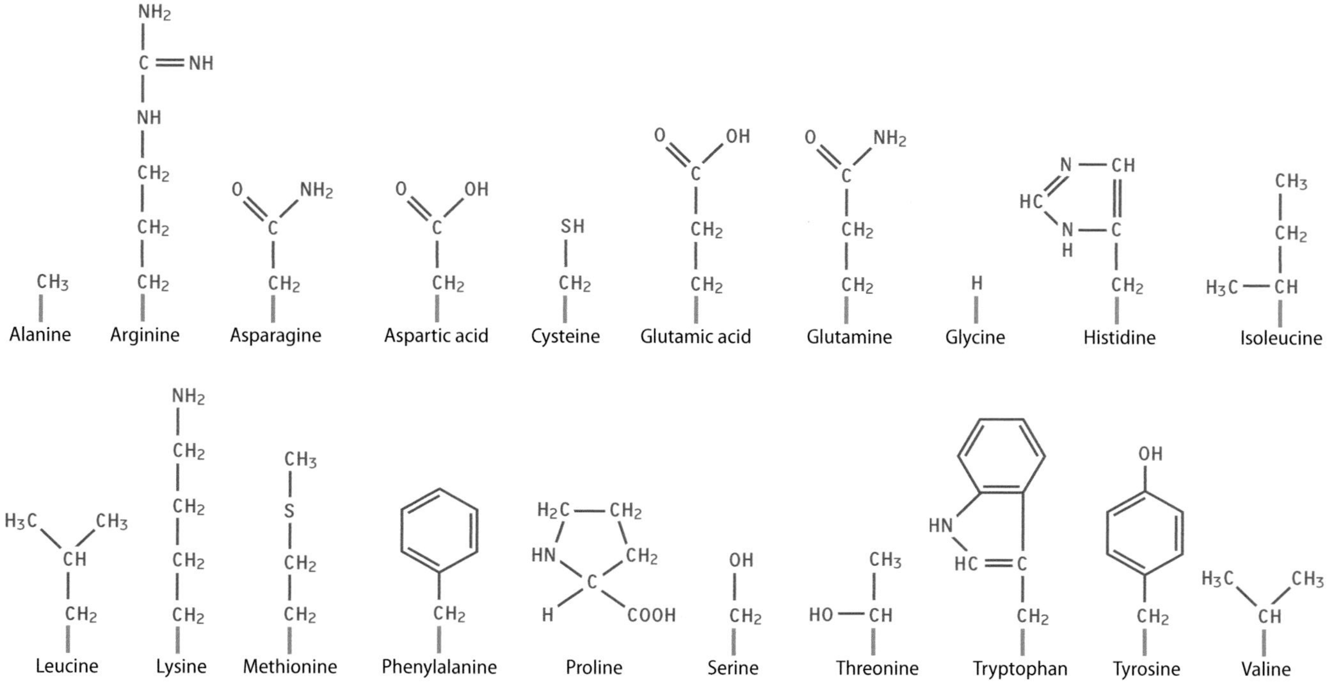

Alanine
Arginine
Asparagine
Aspartic acid
Cysteine
Glutamic acid
Glutamine
Glycine
Histidine
Isoleucine
Leucine
Lysine
Methionine
Phenylalanine
Proline
Serine
Threonine
Tryptophan
Tyrosine
Valine

Figure 1.24 The structures of the amino acid R groups. These 20 amino acids are the ones that are conventionally looked on as being specified by the genetic code. Note that the entire structure of proline is shown, not just its R group. This is because proline has an unusual structure, in which the R group forms a bond not just to the α-carbon but also with the amino group attached to this carbon.

TABLE 1.2 AMINO ACID ABBREVIATIONS

Amino acid	Abbreviation	
	Three-letter	One-letter
Alanine	Ala	A
Arginine	Arg	R
Asparagine	Asn	N
Aspartic acid	Asp	D
Cysteine	Cys	C
Glutamic acid	Glu	E
Glutamine	Gln	Q
Glycine	Gly	G
Histidine	His	H
Isoleucine	Ile	I
Leucine	Leu	L
Lysine	Lys	K
Methionine	Met	M
Phenylalanine	Phe	F
Proline	Pro	P
Serine	Ser	S
Threonine	Thr	T
Tryptophan	Trp	W
Tyrosine	Tyr	Y
Valine	Val	V

therefore result in different combinations of chemical reactivities, these combinations dictating not only the overall structure of the resulting protein but also the positioning on the surface of the structure of reactive groups that determine the chemical properties of the protein.

Amino acid diversity derives from the R group because this part is different in each amino acid and varies greatly in structure. Proteins are made up from a set of 20 amino acids (Figure 1.24; Table 1.2). Some of these have R groups that are small, relatively simple structures, such as a single hydrogen atom (in the amino acid called glycine) or a methyl group (alanine). Other R groups are large, complex aromatic side chains (phenylalanine, tryptophan, and tyrosine). Most amino acids are uncharged at pH 7.4 (the 'physiological pH' of most cells and tissues), but two are negatively charged (aspartic acid and glutamic acid) and three are positively charged (arginine, histidine, and lysine). Some amino acids are **polar** (e.g., serine and threonine), while others are nonpolar (e.g., alanine, leucine, and valine).

The 20 amino acids shown in Figure 1.24 are the ones that are conventionally looked upon as being specified by the genetic code. They are therefore the amino acids that are linked together when mRNA molecules are translated into proteins. However, these 20 amino acids do not, on their own, represent the limit of the chemical diversity of proteins. The diversity is even greater because of two factors:

- At least two additional amino acids – selenocysteine and pyrrolysine (**Figure 1.25**) – can be inserted into a polypeptide chain during protein synthesis, their insertion directed by a modified reading of the genetic code.

- During protein processing, some amino acids are modified by the addition of new chemical groups, for example, by acetylation or phosphorylation, or by attachment of large side chains made up of sugar units (Section 13.4).

Proteins therefore have an immense amount of chemical variability, some of this directly specified by the genome, with the remainder arising from protein processing.

The link between the transcriptome and the proteome

The proteome comprises all the proteins present in a cell at a particular time. A typical mammalian cell, for example, a liver hepatocyte, is thought to contain 10,000–20,000 different proteins, about 8×10^9 individual molecules in all, representing approximately 0.5 ng of protein or 18–20% of the total cell weight. The copy numbers of individual proteins vary enormously, from less than 20,000 molecules per cell for the rarest types to 100 million copies for the commonest ones. Any protein that is present at a copy number of greater than 50,000 per cell is considered to be relatively abundant, and in the average mammalian cell some 2000 proteins fall into this category. When the proteomes of different types of mammalian cell are examined, very few differences are seen among these abundant proteins, suggesting that most of them are **housekeeping** proteins that perform general biochemical activities that occur in all cells. The proteins that provide the cell with its specialized function are often quite rare, although there are exceptions, such as the vast amounts of hemoglobin that are present only in red blood cells.

The proteome is synthesized by translation of the mRNA component of the transcriptome. In the early 1950s, shortly after the double-helix structure of DNA had been discovered, several **molecular biologists** attempted to devise ways in which amino acids could attach directly to mRNAs in an ordered fashion, but in all of these schemes at least some of the bonds had to be shorter or longer than was possible according to the laws of physical chemistry, and each idea was quietly dropped. Eventually, in 1957, Francis Crick cut a way through the confusion by predicting the existence of an adaptor molecule that would form a bridge between the mRNA and the polypeptide being synthesized. Soon afterwards it was realized that the tRNAs are these adaptor molecules. Once this fact had been established, attention turned to the ribosomes, the structures within which proteins are synthesized. Gradually, a detailed understanding of the mechanism by which mRNAs are translated into polypeptides was built up (Section 13.3).

The other aspect of protein synthesis that interested molecular biologists in the 1950s was the **informational problem**. This refers to the second important component of the link between the transcriptome and proteome: the **genetic code**, which specifies how the nucleotide sequence of an mRNA is translated into the amino acid sequence of a protein. It was recognized in the 1950s that a triplet genetic code – one in which each codeword, or **codon**, comprises three nucleotides – is required to account for all 20 amino acids found in proteins. A two-letter code would have only $4^2 = 16$ codons, which is not enough to specify all 20 amino acids, whereas a three-letter code would give $4^3 = 64$ codons. The genetic code was worked out in the 1960s, partly by analysis of polypeptides arising from translation of artificial mRNAs of known or predictable sequence in **cell-free protein synthesizing systems**, and partly by determining which amino acids associated with which RNA sequences in an assay based on purified ribosomes. When this work was completed, it was realized that the 64 codons fall into groups, the members of each group coding for the same amino acid (**Figure 1.26**). Only tryptophan and methionine have just a single codon each: all other amino acids are coded by two, three, four, or six codons. This feature of the code is called **degeneracy**. The code also has four **punctuation codons**, which indicate the points within an mRNA where translation of the nucleotide sequence should start and finish (**Figure 1.27**). The **initiation codon** is usually 5´–AUG–3´, which also specifies methionine (so most newly synthesized polypeptides start with methionine), although other codons such as 5´–GUG–3´ and 5´–UUG–3´ are also used, especially in bacteria. The three **termination codons** are 5´–UAG–3´, 5´–UAA–3´, and 5´–UGA–3´.

The genetic code is not universal

It was originally thought that the genetic code must be the same in all organisms. The argument was that, once established, it would be impossible for the code to change because giving a new meaning to any single codon would result in widespread disruption of the amino acid sequences of proteins. This reasoning seems sound, so it is surprising that, in reality, the genetic code is not universal. The code shown in Figure 1.26 holds for the vast majority of genes in the vast majority of organisms, but deviations are widespread. In particular, mitochondrial genomes often use a nonstandard code (Table 1.3A). This was first discovered in 1979 by Frederick Sanger's group in Cambridge, UK, who found that several human mitochondrial mRNAs contain the sequence UGA, which normally codes for termination, at internal positions where protein synthesis was not expected to stop. Comparisons with the amino acid sequences of the proteins coded by these mRNAs showed that 5´–UGA–3´ is a tryptophan codon in human mitochondria, and that this is just one of four code deviations in this particular genetic system. Mitochondrial genes in other organisms also display code deviations, although at least one of these – the use of 5´–CGG–3´ as a tryptophan codon in plant mitochondria – is probably corrected by RNA editing before translation occurs.

Nonstandard codes are also known for the nuclear genomes of lower eukaryotes. Often a modification is restricted to just a small group of organisms and frequently it involves reassignment of the termination codons (Table 1.3B). Modifications are less common among prokaryotes, but one example is known in *Mycoplasma*. A more important type of code variation is **context-dependent codon reassignment**, which occurs when the protein to be synthesized contains either selenocysteine or pyrrolysine. Proteins containing pyrrolysine are rare, and are probably only present in the group of prokaryotes called the **archaea** (Chapter 8), but selenoproteins are widespread in many organisms, with one example being the enzyme glutathione peroxidase, which helps protect the cells of humans and other mammals against oxidative damage. Selenocysteine is coded by 5´–UGA–3´ and pyrrolysine by 5´–UAG–3´. These codons therefore have a dual meaning because they are still used as termination codons in the

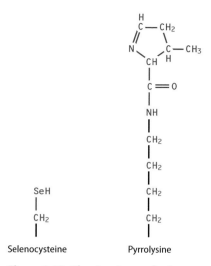

Figure 1.25 The structures of selenocysteine and pyrrolysine. The parts shown in red indicate the differences between these amino acids and cysteine and lysine, respectively.

Figure 1.26 The genetic code. The codons are read in the 5´→3´ direction in an mRNA. Amino acids are designated by the standard, three-letter abbreviations (see Table 1.2).

UUU	Phe	UCU	Ser	UAU	Tyr	UGU	Cys
UUC	Phe	UCC	Ser	UAC	Tyr	UGC	Cys
UUA	Leu	UCA	Ser	UAA	Stop	UGA	Stop
UUG	Leu	UCG	Ser	UAG	Stop	UGG	Trp
CUU	Leu	CCU	Pro	CAU	His	CGU	Arg
CUC	Leu	CCC	Pro	CAC	His	CGC	Arg
CUA	Leu	CCA	Pro	CAA	Gln	CGA	Arg
CUG	Leu	CCG	Pro	CAG	Gln	CGG	Arg
AUU	Ile	ACU	Thr	AAU	Asn	AGU	Ser
AUC	Ile	ACC	Thr	AAC	Asn	AGC	Ser
AUA	Ile	ACA	Thr	AAA	Lys	AGA	Arg
AUG	Met	ACG	Thr	AAG	Lys	AGG	Arg
GUU	Val	GCU	Ala	GAU	Asp	GGU	Gly
GUC	Val	GCC	Ala	GAC	Asp	GGC	Gly
GUA	Val	GCA	Ala	GAA	Glu	GGA	Gly
GUG	Val	GCG	Ala	GAG	Glu	GGG	Gly

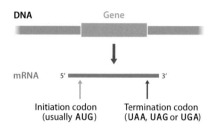

Figure 1.27 The positions of the punctuation codons in an mRNA.

organisms concerned (Table 1.3C). A 5´-UGA-3´ codon that specifies selenocysteine is distinguished from true termination codons by the presence of a **stem-loop** structure in the mRNA, the loop formed by bending the mRNA back on itself, with a short stretch of base-pairing making up the stem that holds the conformation together (Figure 1.28). The stem-loop is positioned just **downstream** of the selenocysteine codon in prokaryotes, and in the **3´ untranslated region** (the part of the mRNA after the termination codon) in eukaryotes. Recognition of the selenocysteine codon requires interaction between the stem-loop structure and a special protein that is involved in translation of these mRNAs. A similar system probably operates for recognition of a pyrrolysine codon.

The link between the proteome and the biochemistry of the cell

The biological information encoded by the genome finds its final expression in a protein whose biological properties are determined by its folded structure and by the spatial arrangement of chemical groups on its surface. By specifying proteins of different types, the genome is able to construct and maintain a proteome whose overall biological properties form the underlying basis of life. The proteome can play this role because of the huge diversity of protein structures that can be formed, the diversity enabling proteins to carry out a variety of biological functions. These functions include the following:

- Biochemical catalysis is the role of the special type of proteins called enzymes. The central metabolic pathways, which provide the cell with energy, are catalyzed by enzymes, as are the biosynthetic processes that result in construction of nucleic acids, proteins, carbohydrates, and lipids. Biochemical catalysis also drives genome expression through the activities of enzymes such as RNA polymerase.

Figure 1.28 Context-dependent reassignment of a 5´-UGA-3´ codon. A 5´-UGA-3´ codon that specifies selenocysteine is distinguished by the stem-loop structure, which is positioned in the mRNA just downstream of the codon in prokaryotes, as shown here, or in the 3´ untranslated region of a eukaryotic mRNA.

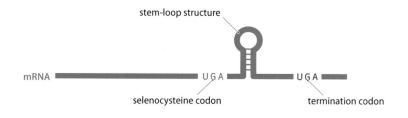

TABLE 1.3 EXAMPLES OF DEVIATIONS FROM THE STANDARD GENETIC CODE

Organism	Codon	Should code for	Actually codes for
(A) Mitochondrial genomes			
Mammals	UGA	Stop	Trp
	AGA, AGG	Arg	Stop
	AUA	Ile	Met
Drosophila	UGA	Stop	Trp
	AGA	Arg	Ser
	AUA	Ile	Met
Saccharomyces cerevisiae	UGA	Stop	Trp
	CUN	Leu	Thr
	AUA	Ile	Met
Fungi	UGA	Stop	Trp
Maize	CGG	Arg	Trp
(B) Nuclear and prokaryotic genomes			
Several protozoa	UAA, UAG	Stop	Gln
Candida cylindracea	CUG	Leu	Ser
Euplotes sp.	UGA	Stop	Cys
Mycoplasma sp.	UGA	Stop	Trp
(C) Context-dependent codon reassignments			
Various	UGA	Stop	Selenocysteine
Archaea	UAG	Stop	Pyrrolysine

Abbreviation: N, any nucleotide.

- Structure, which at the cellular level is determined by the proteins that make up the cytoskeleton, is also the primary function of some extracellular proteins. An example is collagen, which is an important component of bones and tendons.

- Movement is conferred by contractile proteins, of which actin and myosin in cytoskeletal fibers are the best-known examples.

- Transport of materials around the body is an important protein activity. For example, hemoglobin transports oxygen in the bloodstream, and serum albumin transports fatty acids.

- Regulation of cellular processes is mediated by signaling proteins such as **transcription factors** that bind to the genome and influence the expression levels of individual genes and groups of genes (Section 12.3). The activities of groups of cells are regulated and coordinated by extracellular hormones and cytokines, many of which are proteins (e.g., insulin, the hormone that controls blood sugar levels, and the interleukins, a group of cytokines that regulate cell division and differentiation).

- Protection of the body and of individual cells is the function of a range of proteins, including the antibodies and those proteins involved in the blood-clotting response.

- Storage functions are performed by proteins such as ferritin, which acts as an iron store in the liver, and the gliadins, which store amino acids in dormant wheat seeds.

This multiplicity of protein function provides the proteome with its ability to convert the blueprint contained in the genome into the essential features of life.

SUMMARY

- The genome is the store of biological information possessed by every organism on the planet.
- The vast majority of genomes are made of DNA, the few exceptions being those viruses that have RNA genomes.
- Genome expression is the process by which the information contained in the genome is released to the cell.
- The first product of genome expression is the transcriptome, the collection of RNAs derived from those genes that are active at a particular time.
- The second product is the proteome, the cell's repertoire of proteins that specify the nature of the biochemical reactions that the cell is able to carry out.
- Experimental evidence showing that genes are made of DNA was first obtained in 1945–1952, but it was the discovery of the double-helix structure by Watson and Crick in 1953 that convinced biologists that DNA is indeed the genetic material.
- A DNA polynucleotide is an unbranched polymer made up of multiple copies of four chemically different nucleotides.
- In the double helix, two polynucleotides are wound around one another, with the nucleotide bases on the inside of the molecule.
- The polynucleotides are linked by hydrogen bonding between the bases, with A always base-paired to T and G always base-paired to C.
- RNA is also a polynucleotide, but the individual nucleotides have different structures compared with those found in DNA, and RNA is usually single-stranded.
- A cell contains various types of RNA, including mRNAs, which are transcripts of protein-coding genes, and several types of noncoding RNA.
- Many RNAs are initially synthesized as precursor molecules, which are processed by cutting and joining reactions and by chemical modification to give the mature forms.
- Proteins are also unbranched polymers, but in proteins the units are amino acids linked by peptide bonds.
- The amino acid sequence is the primary structure of a protein, the higher levels of structure – secondary, tertiary, and quaternary – being formed by folding of the primary structure into three-dimensional conformations and by association of individual polypeptides into multiprotein structures.

- Proteins are functionally diverse because individual amino acids have different chemical properties that, when combined in different ways, result in proteins with a range of chemical features.

- Proteins are synthesized by translation of mRNAs, with the rules of the genetic code specifying which triplet of nucleotides codes for which amino acid.

- The genetic code is not universal: variations occur in mitochondria and in lower eukaryotes, and some codons can have two different meanings in a single gene.

SHORT ANSWER QUESTIONS

1. Provide a timeline for the discovery of DNA, the discovery that DNA is the genetic material, the discovery of the structure of DNA, and the characterization of the first genome.

2. Which two types of chemical interaction stabilize the double helix?

3. Why does the specific base-pairing between A and T, and G and C, provide a basis for the fidelity of DNA replication?

4. What are the two important chemical differences between RNA and DNA?

5. Why is noncoding RNA also called functional RNA?

6. Outline the various ways in which RNA molecules are processed.

7. Do cells ever lack a transcriptome? Explain the significance of your answer.

8. How do hydrogen bonds, electrostatic interactions, and hydrophobic forces play important roles in the secondary, tertiary, and quaternary structures of proteins?

9. How can proteins have so many diverse structures and functions when they are all synthesized from just 20 amino acids?

10. In addition to the 20 amino acids, proteins have additional chemical diversity because of two factors. What are these two factors, and what is their importance?

11. How can the codon 5′–UGA–3′ function as both a stop codon and as a codon for the modified amino acid selenocysteine?

12. How does the genome direct the biological activity of a cell?

IN-DEPTH PROBLEMS

1. The text (page 6) states that Watson and Crick discovered the double-helix structure of DNA on Saturday, March 7, 1953. Justify this statement.

2. Discuss why the double helix gained immediate universal acceptance as the correct structure for DNA.

3. What experiments led to elucidation of the genetic code in the 1960s?

4. Discuss the reasons why polypeptides can take up a large variety of structures whereas polynucleotides cannot.

5. The transcriptome and proteome are looked on as, respectively, an intermediate and the end-product of genome expression. Evaluate the strengths and limitations of these terms for our understanding of genome expression.

FURTHER READING

Books and articles on the discovery of the double helix and other important landmarks in the study of DNA

Brock, T.D. (1990). *The Emergence of Bacterial Genetics*. Cold Spring Harbor Laboratory Press, New York. *A detailed history that puts into context the work on the transforming principle and the Hershey–Chase experiment.*

Judson, H.F. (1996). *The Eighth Day of Creation: Makers of the Revolution in Biology*. Cold Spring Harbor Laboratory Press, New York. *A highly readable account of the development of molecular biology up to the 1990s.*

Kay, L.E. (1996). *The Molecular Vision of Life*. Oxford University Press, Oxford. *Contains a particularly informative explanation of why genes were once thought to be made of protein.*

Lander, E.S. and Weinberg, R.A. (2000). Genomics: Journey to the center of biology. *Science* 287:1777–1782. *A brief description of genetics and molecular biology from Mendel to the human genome sequence.*

Maddox, B. (2003). *Rosalind Franklin: The Dark Lady of DNA*. HarperCollins, London.

McCarty, M. (1986). *The Transforming Principle: Discovering that Genes are Made of DNA*. Norton, London.

Olby, R. (2012). *The Path to the Double Helix*. Dover Publications, Mineola, New York. *A scholarly account of the research that led to the discovery of the double helix.*

Watson, J.D. (1968). *The Double Helix*. Atheneum, London. *The most important discovery of twentieth-century biology, written as a soap opera.*

Research papers and reviews describing important aspects of DNA, RNA, or proteins

Altona, C. and Sundaralingam, M. (1972) Conformational analysis of the sugar ring in nucleosides and nucleotides: A new description using the concept of pseudorotation. *J. Am. Chem. Soc.* 94:8205–8212. *Information on sugar pucker.*

Eisenberg, D. (2003). The discovery of the α-helix and β-sheet, the principal structural features of proteins. *Proc. Natl Acad. Sci. USA* 100:11207–11210.

Pauling, L. and Corey, R.B. (1951). The pleated sheet, a new layer configuration of polypeptide chains. *Proc. Natl Acad. Sci. USA* 37:251–256. *The first description of the β-sheet.*

Pauling, L., Corey, R.B. and Branson, H.R. (1951). The structure of proteins: two hydrogen-bonded helical configurations of the polypeptide chain. *Proc. Natl Acad. Sci. USA* 37:205–211. *The first description of the α-helix.*

Ravichandran, S., Subramani, V.K. and Kim, K.K. (2019). Z-DNA in the genome: From structure to disease. *Biophys. Rev.* 11:383–387.

Watson, J.D. and Crick, F.H.C. (1953). Molecular structure of nucleic acids: A structure for deoxyribose nucleic acid. *Nature* 171:737–738. *The scientific report of the discovery of the double-helix structure of DNA.*

Yakovchuk, P., Protozanova, E. and Frank-Kamenetskii, M.D. (2006). Base-stacking and base-pairing contributions into thermal stability of the DNA double helix. *Nucleic Acids Res.* 34:564–574.

Zacharias, M. (2020). Base-pairing and base-stacking contributions to double-stranded DNA formation. *J. Phys. Chem B* 124:10345–10352.

Online resources

Protein Data Bank. https://www.wwpdb.org/. *Archive of information on the three-dimensional structures of proteins and nucleic acids.*

The genetic codes. https://www.ncbi.nlm.nih.gov/Taxonomy/Utils/wprintgc.cgi. *Details of nonstandard genetic codes in different species; mirrored at https://en.wikipedia.org/wiki/List_of_genetic_codes.*

STUDYING DNA

Virtually everything we know about genomes and genome expression has been discovered by scientific research: theoretical studies have played very little role in this or any other area of molecular and cell biology. It is possible to learn 'facts' about genomes without knowing very much about how those facts were obtained, but in order to gain a real understanding of the subject we must examine in detail the techniques and scientific approaches that have been used to study genomes. The next five chapters cover these research methods. First, we examine the techniques, centered on the polymerase chain reaction and DNA cloning, that are used to study DNA molecules. These techniques are very effective with short segments of DNA, including individual genes, enabling a wealth of information to be obtained at this level. Chapter 3 then covers the methods that are used to construct maps of genomes, and Chapter 4 describes the methods used to sequence DNA molecules, and to assemble the short sequences that these methods generate into the immensely long sequences that make up individual chromosomes and entire genomes. Finally, in Chapters 5 and 6, we will look at the various approaches that are used to locate the positions of genes in a genome sequence and to identify the functions of those genes. As you read through these chapters, you will begin to appreciate that understanding the structure and function of an individual genome is a major undertaking, and that research is currently in the middle of an exciting discovery phase, with new techniques and new approaches revealing novel and unexpected aspects of genomes almost every week.

The toolkit of techniques used by molecular biologists to study DNA molecules was assembled during the 1970s and 1980s. Before then, the only way in which individual genes could be studied was by classical **genetics**, using techniques that originated with Mendel in the middle part of the 19th century. The development of more direct methods for studying DNA was stimulated by breakthroughs in biochemical research that, in the early 1970s, provided molecular biologists with enzymes that could be used to manipulate DNA molecules in the test tube. These enzymes occur naturally in living cells and are involved in processes such as DNA replication, repair, and recombination, which we will study in Chapters 15–17. In order to determine the functions of these enzymes, many of them were purified, and the reactions that they catalyze were studied. Molecular biologists then adopted the pure enzymes as tools for manipulating DNA molecules in predetermined ways, using them to make copies of DNA molecules, to cut DNA molecules into shorter fragments, and to join them together again in combinations that do not exist in nature (**Figure 2.1**). These manipulations form the basis of **recombinant DNA technology**, in which new or 'recombinant' DNA molecules are constructed from pieces of naturally occurring chromosomes and **plasmids**.

Recombinant DNA methodology led to development of the **polymerase chain reaction** (**PCR**). PCR is a deceptively simple technique – all that it achieves is the repeated copying of a short segment of a DNA molecule (**Figure 2.2**) – but it has become immensely important in many areas of biological research, not least the study of genomes. PCR is covered in detail in Section 2.2. Recombinant DNA techniques also underlie **DNA cloning**, or **gene cloning**, in which a DNA

2.1 ENZYMES FOR DNA MANIPULATION

2.2 THE POLYMERASE CHAIN REACTION

2.3 DNA CLONING

DOI: 10.1201/9781003133162-2

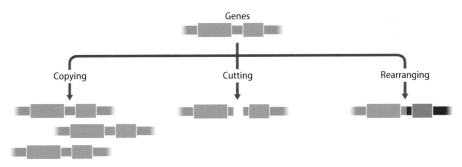

Figure 2.1 Examples of the manipulations that can be carried out with DNA molecules.

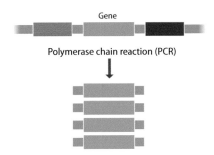

Figure 2.2 The polymerase chain reaction (PCR) is used to make copies of a selected segment of a DNA molecule. In this example, a single gene is copied.

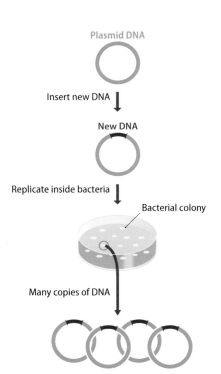

Figure 2.3 DNA cloning. In this example, the fragment of DNA to be cloned is inserted into a plasmid vector, which is subsequently replicated inside a bacterial host.

fragment is inserted into a plasmid or virus chromosome and then replicated in a bacterial or eukaryotic host (**Figure 2.3**). We will examine exactly how DNA cloning is performed, and the reasons why this technique is important in genome research, in Section 2.3.

2.1 ENZYMES FOR DNA MANIPULATION

Recombinant DNA technology was one of the main factors that contributed to the rapid advance in knowledge concerning gene expression that occurred during the 1970s and 1980s. The basis of recombinant DNA technology is the ability to manipulate DNA molecules in the test tube. This, in turn, depends on the availability of purified enzymes whose activities are known and can be controlled, and which can therefore be used to make specified changes to the DNA molecules that are being manipulated. The enzymes available to the molecular biologist fall into four broad categories:

- **DNA polymerases**, which are enzymes that synthesize new polynucleotides **complementary** to an existing DNA or RNA template (**Figure 2.4A**).

- **Nucleases**, which degrade DNA molecules by breaking the phosphodiester bonds that link one nucleotide to the next (**Figure 2.4B**).

- **Ligases**, which join DNA molecules together by synthesizing phosphodiester bonds between nucleotides at the ends of two different molecules, or at the two ends of a single molecule (**Figure 2.4C**).

- **End-modification enzymes**, which make changes to the ends of DNA molecules (**Figure 2.4D**).

We will begin our study of recombinant DNA techniques by examining how each of these types of enzyme are used to make specified changes to DNA molecules.

The mode of action of a template-dependent DNA polymerase

Many of the techniques used to study DNA depend on the synthesis of DNA copies of all or part of existing DNA or RNA molecules. This is an essential requirement for PCR (Section 2.2), **DNA sequencing** (Sections 4.1 and 4.2), and many other procedures that are central to molecular biology research. An enzyme that synthesizes DNA is called a **DNA polymerase**, and one that copies an existing DNA or RNA molecule is called a **template-dependent DNA polymerase**. A template-dependent DNA polymerase makes a new DNA polynucleotide whose sequence is dictated, via the base-pairing rules, by the sequence of nucleotides in the DNA or RNA molecule that is being copied (**Figure 2.5**). The new polynucleotide is always synthesized in the 5′→3′ direction: DNA polymerases that make DNA in the other direction are unknown in nature.

An important feature of template-dependent DNA synthesis is that a DNA polymerase is unable to use an entirely single-stranded molecule as the template.

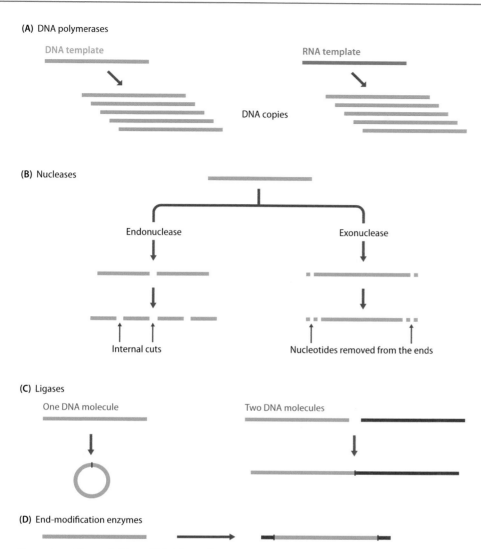

Figure 2.4 The activities of (A) DNA polymerases, (B) nucleases, (C) ligases, and (D) end-modification enzymes. In (A), the activity of a DNA-dependent DNA polymerase is shown on the left, and that of an RNA-dependent DNA polymerase on the right. In (B), the activities of endonucleases and exonucleases are shown. In (C), the green DNA molecule is ligated to itself (left), and to a second DNA molecule (right). In (D) the activity of terminal deoxynucleotidyl transferase is illustrated, this enzyme adding nucleotides to the 3′ ends of a double-stranded DNA molecule.

In order to initiate DNA synthesis, there must be a short, double-stranded region to provide a 3′ end onto which the enzyme will add new nucleotides (**Figure 2.6A**). The way in which this requirement is met in living cells when the genome is replicated is described in Chapter 15. In the test tube, a DNA-copying reaction is initiated by attaching to the template a short, synthetic **oligonucleotide**, usually about 20 nucleotides in length, which acts as a **primer** for DNA synthesis. At first glance, the need for a primer might appear to be an undesired complication in the use of DNA polymerases in recombinant DNA technology, but nothing could be further from the truth. Because **annealing** of the primer to the template depends on complementary base-pairing, the position within the template molecule at which DNA copying is initiated can be specified by synthesizing a primer with the appropriate nucleotide sequence (**Figure 2.6B**). A short, specific segment of a much longer template molecule can therefore be copied, which is much more valuable than the random copying that would occur if DNA synthesis did not need to be primed. You will fully appreciate the importance of priming when we deal with PCR in Section 2.2.

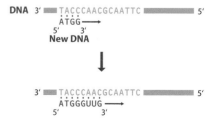

Figure 2.5 The activity of a DNA-dependent DNA polymerase. New nucleotides are added on to the 3′ end of the growing polynucleotide, with the sequence of this new polynucleotide determined by the sequence of the template DNA. Compare this with the process of transcription (DNA-dependent RNA synthesis) shown in Figure 1.13.

Figure 2.6 The role of the primer in template-dependent DNA synthesis. (A) A DNA polymerase requires a primer in order to initiate the synthesis of a new polynucleotide. (B) The sequence of this primer determines the position at which it attaches to the template DNA and hence specifies the region of the template that will be copied. When a DNA polymerase is used to make new DNA *in vitro*, the primer is usually a short oligonucleotide made by chemical synthesis.

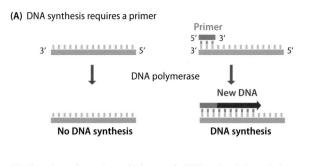

A second general feature of template-dependent DNA polymerases is that many of these enzymes are multifunctional, being able to degrade DNA molecules as well as synthesize them. This is a reflection of the way in which DNA polymerases act in the cell during genome replication (Section 15.3). As well as their 5´→3´ DNA synthesis capability, DNA polymerases can also have one or both of the following exonuclease activities (**Figure 2.7**):

- A **3´→5´ exonuclease** activity enables the enzyme to remove nucleotides from the 3´ end of the strand that it has just synthesized. This is called the **proofreading** activity because it allows the polymerase to correct errors by removing a nucleotide that has been inserted incorrectly.

- A **5´→3´ exonuclease** activity is less common, but is possessed by some DNA polymerases whose natural function in genome replication requires that they must be able to remove at least part of a polynucleotide that is already attached to the template strand that the polymerase is copying.

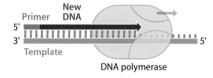

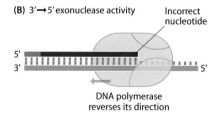

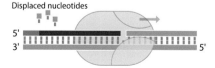

Figure 2.7 The DNA synthesis and exonuclease activities of DNA polymerases. (A) The 5´→3´ DNA synthesis activity enables the polymerase to add nucleotides to the 3´-end of the strand that it is synthesizing. (B) The 3´→5´ exonuclease activity enables the polymerase to remove one or more nucleotides from the 3´-end of the strand that it is making. (C) The 5´→3´ exonuclease activity enables the polymerase to remove one or more nucleotides from the 5´-end of a polynucleotide that is already attached to the template strand.

The types of DNA polymerase used in research

Several of the template-dependent DNA polymerases that are used in molecular biology research (**Table 2.1**) are versions of the *Escherichia coli* DNA polymerase I enzyme, which plays a central role in replication of this bacterium's genome (Section 15.3). This enzyme, sometimes called the **Kornberg polymerase**, after its discoverer Arthur Kornberg, has both the 3´→5´ and 5´→3´ exonuclease activities, which limits its usefulness in DNA manipulation. Its main application is in the synthesis of DNA molecules containing radioactive or fluorescent nucleotides – the process called **DNA labeling**.

Of the two exonuclease activities, it is the 5´→3´ version that causes the most problems when a DNA polymerase is used to manipulate molecules in the test tube. This is because an enzyme that possesses this activity is able to remove nucleotides from the 5´ ends of polynucleotides that have just been synthesized (**Figure 2.8**). It is unlikely that the polynucleotides will be completely degraded, because the polymerase function is usually much more active than the exonuclease function, but some techniques will not work if the 5´ ends of the new polynucleotides are shortened in any way. In particular, some of the older DNA sequencing methods are based on synthesis of new polynucleotides, all of which share exactly the same 5´ end, marked by the primer used to initiate the sequencing reactions. If any 'nibbling' of the 5´ ends occurs, then it is impossible to determine the correct DNA sequence. Because of this problem, when DNA sequencing was first introduced in the late 1970s, a modified version of the Kornberg enzyme called the **Klenow polymerase** was used. The Klenow polymerase was initially prepared by cutting the natural *E. coli* DNA polymerase I enzyme into two segments using a protease. One of these segments retained the polymerase and 3´→5´ exonuclease activities, but lacked the 5´→3´ exonuclease

TABLE 2.1 FEATURES OF TEMPLATE-DEPENDENT DNA POLYMERASES USED IN MOLECULAR BIOLOGY RESEARCH			
Polymerase	**Description**	**Main uses**	**Cross reference**
DNA polymerase I	Unmodified *E. coli* enzyme	DNA labeling	Section 2.1
Klenow polymerase	Modified version of *E. coli* DNA polymerase I	DNA labeling, chain-termination DNA sequencing	Sections 2.1 and 4.1
Taq polymerase	*Thermus aquaticus* DNA polymerase I	PCR	Section 2.2
Reverse transcriptase	RNA-dependent DNA polymerase, obtained from various retroviruses	cDNA synthesis	Sections 3.6, 5.2, and 5.3

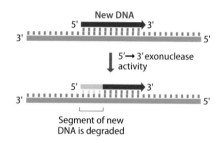

Figure 2.8 The 5′→3′ exonuclease activity of a DNA polymerase can degrade the 5′ end of a polynucleotide that has just been synthesized.

function of the untreated enzyme. Nowadays, the enzyme is almost always prepared from *E. coli* cells whose polymerase gene has been engineered so that the resulting enzyme has the desired properties.

The *E. coli* DNA polymerase I enzyme has an optimum reaction temperature of 37°C, this being the usual temperature of the natural environment of the bacterium, inside the lower intestines of mammals such as humans. Test-tube reactions with either the Kornberg or Klenow polymerases are therefore incubated at 37°C, and terminated by raising the temperature to 75°C or above, which causes the protein to unfold, or **denature**, destroying its enzymatic activity. This regimen is perfectly adequate for most molecular biology techniques, but for reasons that will become clear in Section 2.2, PCR requires a **thermostable** DNA polymerase – one that is able to function at temperatures much higher than 37°C. Suitable enzymes can be obtained from bacteria such as *Thermus aquaticus*, which live in hot springs at temperatures up to 95°C, and whose DNA polymerase I enzyme has an optimum working temperature of 75–80°C. The biochemical basis of protein thermostability is not fully understood, but probably centers on structural features that reduce the amount of protein unfolding that occurs at elevated temperatures.

One additional type of DNA polymerase is important in molecular biology research. This is **reverse transcriptase**, which is an **RNA-dependent DNA polymerase** and so makes DNA copies of RNA rather than DNA templates. Reverse transcriptases are involved in the replication cycles of retroviruses (Section 9.1), including the human immunodeficiency viruses, these viruses having RNA genomes that are copied into DNA after infection of the host. In the test tube, a reverse transcriptase can be used to make DNA copies of mRNA molecules. These copies are called **complementary DNAs (cDNAs)**. Their synthesis is important in some types of gene cloning and in techniques used to map the regions of a genome that specify particular mRNAs (Sections 5.2 and 5.3).

Restriction endonucleases enable DNA molecules to be cut at defined positions

A variety of nucleases have found applications in recombinant DNA technology (Table 2.2). Some nucleases have a broad range of activities, but most are either **exonucleases**, removing nucleotides from the ends of DNA and/or RNA molecules, or **endonucleases**, making cuts at internal phosphodiester bonds. Some nucleases are specific for DNA and some for RNA, some work only on double-stranded DNA and others only on single-stranded DNA, and some are not fussy about what they work on. We will encounter various examples of nucleases in later chapters when we deal with the techniques in which they are used. Only one type of nuclease will be considered in detail here: the

TABLE 2.2 FEATURES OF IMPORTANT NUCLEASES USED IN MOLECULAR BIOLOGY RESEARCH

Nuclease	Description	Main use	Cross reference
Restriction endonucleases	Sequence-specific DNA endonucleases, from many sources	Many applications	Section 2.1
S1 nuclease	Endonuclease specific for single-stranded DNA and RNA, from the fungus *Aspergillus oryzae*	Transcript mapping	Section 5.2
Deoxyribonuclease I	Endonuclease specific for double-stranded DNA and RNA, from *Escherichia coli*	Nuclease footprinting	Section 11.1

restriction endonucleases, which play a central role in all aspects of recombinant DNA technology.

A restriction endonuclease is an enzyme that binds to a DNA molecule at a specific sequence and makes a double-stranded cut at or near that sequence. Because of the sequence specificity, the positions of cuts within a DNA molecule can be predicted, assuming that the DNA sequence is known, enabling defined segments to be excised from a larger molecule. This ability underlies gene cloning and all other aspects of recombinant DNA technology in which DNA fragments of known sequence are required.

There are four main types of restriction endonuclease. With Types I, III, and IV, there is no strict control over the position of the cut relative to the specific sequence in the DNA molecule that is recognized by the enzyme. These enzymes are therefore less useful because the sequences of the resulting fragments are not precisely known. Type II enzymes do not suffer from this disadvantage because the cut is always at the same place, either within the recognition sequence or very close to it (**Figure 2.9**). For example, the Type II enzyme called *Eco*RI (isolated from *E. coli*) cuts DNA only at the hexanucleotide 5′–GAATTC–3′. Digestion of DNA with a Type II enzyme therefore gives a reproducible set of fragments whose sequences are predictable if the sequence of the target DNA molecule is known. Over 60,000 Type II enzymes have been isolated, and more than 600 are available for use in the laboratory. Many enzymes have hexanucleotide target sites, but others recognize shorter or longer sequences (**Table 2.3**). There are also examples of enzymes with degenerate recognition sequences, meaning that they cut DNA at any of a family of related sites. *Hin*fI (from *Haemophilus influenzae*), for example, recognizes 5′–GANTC–3′, where 'N' is any nucleotide, and so cuts at 5′–GAATC–3′, 5′–GATTC–3′, 5′–GAGTC–3′, and 5′–GACTC–3′. Most enzymes cut within the recognition sequence, but a few, such as *Bsr*BI, cut at a specified position outside of this sequence.

Restriction enzymes cut DNA in two different ways. Many make a simple double-stranded cut, giving a **blunt** or **flush end**, but others cut the two DNA strands at different positions, usually two or four nucleotides apart, so that the resulting DNA fragments have short, single-stranded overhangs at each end. These are called **sticky** or **cohesive ends** because base-pairing between them can stick the DNA molecule back together again (**Figure 2.10A**). Some sticky-end cutters give 5′ overhangs (e.g., *Sau*3AI, *Hin*fI), whereas others leave 3′ overhangs (e.g., *Pst*I) (**Figure 2.10B**). One feature that is particularly important in recombinant DNA technology is that some pairs of restriction enzymes have different recognition sequences but give the same sticky ends, examples being *Sau*3AI and *Bam*HI, which both give a 5′–GATC–3′ sticky end, even though *Sau*3AI has a four-base-pair recognition sequence and *Bam*HI recognizes a six-base-pair sequence (**Figure 2.10C**).

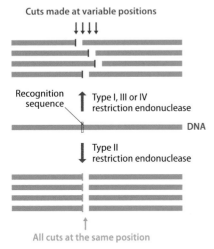

Cuts made at variable positions

Recognition sequence — Type I, III or IV restriction endonuclease

DNA

Type II restriction endonuclease

All cuts at the same position

Figure 2.9 Cuts produced by restriction endonucleases. In the top part of the diagram, the DNA is cut by a Type I, III or IV restriction endonuclease. The cuts are made in slightly different positions relative to the recognition sequence, so the resulting fragments have different lengths. In the lower part of the diagram, a Type II enzyme is used. Each molecule is cut at exactly the same position to give exactly the same pair of fragments. Note that for a Type IV enzyme, the recognition sequence must also be chemically modified, for example, by addition of a methyl group to one or more of the nucleotide bases.

TABLE 2.3 SOME EXAMPLES OF RESTRICTION ENDONUCLEASES

Enzyme	Recognition sequence	Type of ends	End sequences
*Alu*I	5´–AGCT–3´ 3´–TCGA–5´	Blunt	5´–AG CT–3´ 3´–TC GA–5´
*Sau*3AI	5´–GATC–3´ 3´–CTAG–5´	Sticky, 5´ overhang	5´– GATC–3´ 3´–CTAG –5´
*Hin*fI	5´–GANTC–3´ 3´–CTNAG–5´	Sticky, 5´ overhang	5´–G ANTC–3´ 3´–CTNA G–5´
*Bam*HI	5´–GGATCC–3´ 3´–CCTAGG–5´	Sticky, 5´ overhang	5´–G GATCC–3´ 3´–CCTAG G–5´
*Bsr*BI	5´–CCGCTC–3´ 3´–GGCGAG–5´	Blunt	5´– NNNCCGCTC–3´ 3´– NNNGGCGAG–5´
*Eco*RI	5´–GAATTC–3´ 3´–CTTAAG–5´	Sticky, 5´ overhang	5´–G AATTC–3´ 3´–CTTAA G–5´
*Pst*I	5´–CTGCAG–3´ 3´–GACGTC–5´	Sticky, 3´ overhang	5´–CTGCA G–3´ 3´–G ACGTC–5´
*Not*I	5´–GCGGCCGC–3´ 3´–CGCCGGCG–5´	Sticky, 5´ overhang	5´–GC GGCCGC–3´ 3´–CGCCGG CG–5´
*Bgl*I	5´–GCCNNNNNGGC–3´ 3´–CGGNNNNNCCG–5´	Sticky, 3´ overhang	5´–GCCNNNN NGGC–3´ 3´–CGGN NNNNCCG–5´

Abbreviation: N, any nucleotide. Note that most, but not all, recognition sequences have inverted symmetry: when read in the 5´→3´ direction, the sequence is the same in both strands.

(A) Blunt and sticky ends

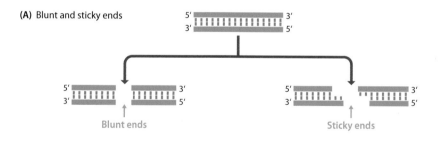

Blunt ends   Sticky ends

(B) 5´ and 3´ overhangs

(C) The same sticky end produced by different enzymes

Figure 2.10 The results of digestion of DNA with different restriction endonucleases. (A) Blunt ends and sticky ends. (B) Different types of sticky end: the 5´ overhangs produced by *Bam*HI and the 3´ overhangs produced by *Pst*I. (C) The same sticky ends produced by two different restriction endonucleases: a 5´ overhang with the sequence 5´–GATC–3´ is produced by both *Bam*HI (recognizes 5´–GGATCC–3´) and *Sau*3AI (recognizes 5´–GATC–3´).

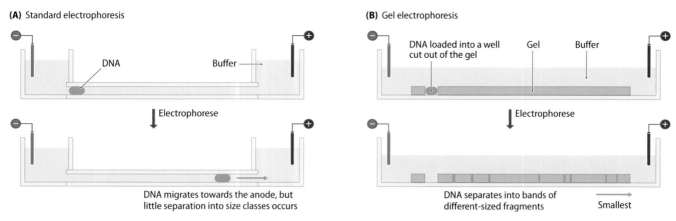

(A) Standard electrophoresis

DNA

Buffer

Electrophorese

DNA migrates towards the anode, but
little separation into size classes occurs

(B) Gel electrophoresis

DNA loaded into a well   Gel   Buffer
cut out of the gel

Electrophorese

DNA separates into bands of
different-sized fragments   Smallest

Figure 2.11 Separation of DNA molecules of different lengths by electrophoresis. (A) Standard electrophoresis does not separate DNA fragments of different sizes, whereas (B) gel electrophoresis does.

Gel electrophoresis is used to examine the results of a restriction digest

Treatment with a restriction endonuclease results in a larger DNA molecule being cut into smaller fragments. How do we measure the sizes of these fragments? The answer is by **gel electrophoresis**. This is the standard method for separating DNA molecules of different lengths; it has many applications in size analysis of DNA fragments and can also be used to separate RNA molecules.

Electrophoresis is the movement of charged molecules in an electric field: negatively charged molecules migrate toward the positive electrode, and positively charged molecules migrate toward the negative electrode. The technique was originally carried out in aqueous solution, in which the predominant factors influencing migration rate are the shape of a molecule and its electric charge. This is not particularly useful for DNA separations because most DNA molecules are the same shape (linear) and although the charge of a DNA molecule is dependent on its length, the differences in charge are not sufficient to result in effective separation (**Figure 2.11A**). The situation is different when electrophoresis is carried out in a gel, because now shape and charge are less important and molecular length is the critical determinant of migration rate. This is because the gel is a network of pores through which the DNA molecules have to travel to reach the positive electrode. Shorter molecules are less impeded by the pores than are longer molecules and so move through the gel more quickly. Molecules of different lengths therefore form bands in the gel (**Figure 2.11B**).

Two types of gel are used in molecular biology: **agarose gels**, as described here, and **polyacrylamide gels**, which are mainly used in DNA sequencing (Section 4.1). Agarose is a polysaccharide that forms gels with pores ranging from 100 nm to 300 nm in diameter, the size depending on the concentration of agarose in the gel. Gel concentration therefore determines the range of DNA fragments that can be separated. The separation range is also affected by the **electroendosmosis** (EEO) value of the agarose, this being a measure of the amount of bound sulfate and pyruvate anions. The greater the EEO, the slower the migration rate for a negatively charged molecule such as DNA.

An agarose gel is prepared by mixing the appropriate amount of agarose powder in a buffer solution, heating it to dissolve the agarose, and then pouring the molten gel onto a Perspex plate with tape around the sides to prevent spillage. A comb is placed in the gel to form wells for the samples. The gel is allowed to set and the electrophoresis is then carried out with the gel submerged under a buffer. In order to follow the progress of the electrophoresis, one or two dyes of known migration rates are added to the DNA samples before loading. The bands of DNA can be visualized by soaking the gel in **ethidium bromide** solution. This compound intercalates between DNA base pairs and fluoresces when activated with ultraviolet radiation (**Figure 2.12**). Unfortunately, the procedure is hazardous because ethidium bromide is a

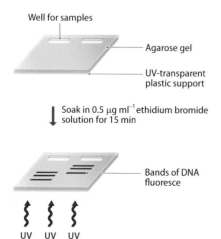

Well for samples

Agarose gel

UV-transparent
plastic support

Soak in 0.5 μg ml^{-1} ethidium bromide
solution for 15 min

Bands of DNA
fluoresce

UV   UV   UV

Figure 2.12 DNA bands in an agarose gel are visualized by staining with ethidium bromide.

powerful mutagen. Nonmutagenic dyes that stain DNA green, red, or blue are therefore now used in many laboratories. The most sensitive dyes are able to detect bands that contain less than 1 ng DNA, compared to a minimum of 10 ng of DNA when ethidium bromide is used.

Depending on the concentration of agarose in the gel, fragments between 100 bp and 50 **kilobase pairs** (**kb**) in length can be separated into sharp bands after electrophoresis (**Figure 2.13**). For example, a 0.5 cm thick slab of 0.5% agarose, which has relatively large pores, would be used for molecules in the size range of 1–30 kb, allowing, for example, molecules of 10 and 12 kb to be clearly distinguished. Alternatively, a 0.3% gel can be used for molecules up to 50 kb, and a 5% gel for molecules 100–500 bp in length.

Interesting DNA fragments can be identified by Southern hybridization

If the DNA that is cut with a restriction endonuclease is a relatively short molecule, and 20 or fewer fragments are produced after restriction, then usually it is possible to select an agarose concentration that results in each fragment being visible as a separate band in the gel. If the starting DNA is long, and so gives rise to many fragments after digestion, then whatever agarose concentration is used the gel may simply show a smear of DNA, because there are fragments of every possible length that all merge together. This is the usual result when genomic DNA is cut with a restriction enzyme.

If the sequence of the starting DNA is known, then the sequences, and hence the sizes, of the fragments resulting from treatment with a particular restriction enzyme can be predicted. The band for a desired fragment (for example, one containing a gene) can then be identified, cut out of the gel, and the DNA purified. Even if its size is unknown, a fragment containing a gene or another segment of DNA of interest can be identified by the technique called **Southern hybridization**. The only requirement is that at least some of the sequence of the gene or DNA segment that we are searching for is known or can be predicted. The first step is to transfer the restriction fragments from the agarose gel to a nitrocellulose or nylon membrane. This is done by placing the membrane on the gel and allowing the buffer to soak through, taking the DNA from the gel to the membrane, where it becomes bound (**Figure 2.14A**). This process results in the DNA bands becoming immobilized in the same relative positions on the surface of the membrane.

The next step is to prepare a **hybridization probe**, which is a labeled DNA molecule whose sequence is complementary to the target DNA that we wish to detect. The label is often a **radioactive marker**. Nucleotides can be synthesized

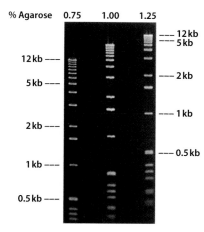

Figure 2.13 The range of fragment sizes that can be resolved depends on the concentration of agarose in the gel. Electrophoresis has been performed with three different concentrations of agarose. The labels indicate the sizes of bands in the left and right lanes. (Courtesy of BioWhittaker Molecular Applications.)

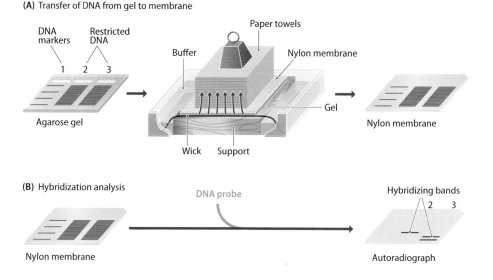

Figure 2.14 Southern hybridization. (A) Transfer of DNA from the gel to the membrane. (B) The membrane is probed with a radioactively labeled DNA molecule. On the resulting autoradiograph, one hybridizing band is seen in lane 2, and two in lane 3.

in which one of the phosphorus atoms is replaced with ^{32}P or ^{33}P, one of the oxygen atoms in the phosphate group is replaced with ^{35}S, or one or more of the hydrogen atoms is replaced with ^{3}H. Radioactive nucleotides still act as substrates for DNA polymerases and so are incorporated into a DNA molecule by any strand-synthesis reaction catalyzed by a DNA polymerase. Alternatively, to avoid the health hazards and disposal issues associated with the use of radioactive chemicals, nucleotides carrying a **fluorescent marker** or one that emits chemiluminescence can be used.

The probe itself could be a synthetic oligonucleotide whose sequence matches part of an interesting gene. Because the probe and target DNAs are complementary, they can base-pair or **hybridize**, the position of the hybridized probe on the membrane being identified by detecting the signal given out by the label attached to the probe. To carry out the hybridization, the membrane is placed in a glass bottle with the labeled probe and some buffer, and the bottle is gently rotated for several hours so that the probe has plenty of opportunity to hybridize with its target DNA. The membrane is then washed to remove any probe that has not become hybridized, and the signal from the label is detected. In the example shown in **Figure 2.14B**, the probe is radioactively labeled and the signal is detected by exposure to an X-ray-sensitive film (**autoradiography**). The band that is seen on the autoradiograph is the one that corresponds to the restriction fragment that hybridizes to the probe and which therefore contains the gene that we are searching for. If a fluorescent marker is used, then the label is detected with a film sensitive to the emission spectrum of the fluorophore. Chemiluminescent markers can be detected in the same way, but these have the disadvantage that the signal is not generated directly by the label, and instead must be developed by treatment of the labeled molecule with chemicals. A popular method involves labeling the DNA with the enzyme alkaline phosphatase, which is detected by applying dioxetane, which the enzyme dephosphorylates to produce the chemiluminescence. All three types of label – radioactive, fluorescent, and chemiluminescent – can also be detected with a digital scanner, which provides an immediate image of the membrane and the position of the label, without the delay incurred with film exposure.

Ligases join DNA fragments together

DNA fragments that have been generated by treatment with a restriction endonuclease can be joined back together again, or attached to a new partner, by a DNA ligase. The reaction requires energy, which is provided by adding either ATP or nicotinamide adenine dinucleotide (NAD) to the reaction mixture, depending on the type of ligase that is being used.

The most widely used DNA ligase is obtained from *E. coli* cells infected with T4 bacteriophage. This enzyme is involved in the replication of the phage DNA and is encoded by the T4 genome. The natural role of this enzyme is to synthesize phosphodiester bonds between unlinked nucleotides present in one polynucleotide of a double-stranded molecule (**Figure 2.15A**). In order to join together two restriction fragments, the ligase has to synthesize two phosphodiester bonds, one in each strand (**Figure 2.15B**). This is by no means beyond the capabilities of the enzyme, but the reaction can occur only if the ends to be joined come close enough to one another by chance – the ligase is not able to catch hold of them and bring them together. If the two

(A) The role of DNA ligase *in vivo*

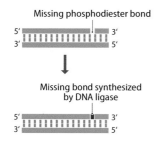

(B) Ligation *in vitro*

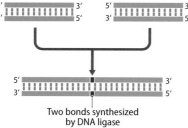

(C) Sticky-end ligation is more efficient

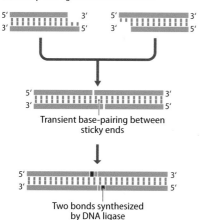

Figure 2.15 Ligation of DNA molecules with DNA ligase. (A) In living cells, DNA ligase synthesizes a missing phosphodiester bond in one strand of a double-stranded DNA molecule. (B) To link two DNA molecules *in vitro*, DNA ligase must make two phosphodiester bonds, one in each strand. (C) Ligation *in vitro* is more efficient when the molecules have compatible sticky ends, because transient base-pairing between these ends holds the molecules together and so increases the opportunity for DNA ligase to attach and synthesize the new phosphodiester bonds.

molecules have complementary sticky ends, and the ends come together by random diffusion events in the ligation mixture, then transient base pairs might form between the two overhangs. These base pairs are not particularly stable, but they may persist for sufficient time for a ligase enzyme to attach to the junction and synthesize phosphodiester bonds to fuse the ends together (**Figure 2.15C**). If the molecules are blunt-ended, then they cannot base-pair to one another, not even temporarily, and ligation is a much less efficient process, even when the DNA concentration is high and pairs of ends are in relatively close proximity.

The greater efficiency of sticky-end ligation has stimulated the development of methods for converting blunt ends into sticky ends. In one method, short double-stranded molecules called **linkers** or **adaptors** are attached to the blunt ends. Linkers and adaptors work in slightly different ways, but both contain a recognition sequence for a restriction endonuclease and so produce a sticky end after treatment with the appropriate enzyme (**Figure 2.16**). Another way to create a sticky end is by **homopolymer tailing**, in which nucleotides are added one after the other to the 3´ terminus at a blunt end (**Figure 2.17**). The enzyme involved is called **terminal deoxynucleotidyl transferase**, which we will meet in the next section. If the reaction mixture contains the DNA, enzyme, and only one of the four nucleotides, then the new stretch of single-stranded DNA that is made consists entirely of just that single nucleotide. It could, for example, be a poly(G) tail, which would enable the molecule to base-pair to other molecules that carry poly(C) tails, created in the same way but with dCTP, rather than dGTP, in the reaction mixture.

End-modification enzymes

Terminal deoxynucleotidyl transferase (see Figure 2.17), obtained from calf thymus tissue, is one example of an end-modification enzyme. It is, in fact, a **template-independent DNA polymerase**, because it is able to extend a DNA polynucleotide without base-pairing of the incoming nucleotides to an existing strand of DNA or RNA. Its main role in recombinant DNA technology is in homopolymer tailing, as described above.

Two other end-modification enzymes are also frequently used. These are **alkaline phosphatase** and **T4 polynucleotide kinase**, which act in complementary ways. Alkaline phosphatase, which is obtained from various sources, including E. coli, calf intestinal tissue, and the arctic shrimp, removes phosphate groups from the 5´ ends of DNA molecules, which prevents these molecules from being ligated to one another. Two ends carrying 5´ phosphates can be ligated to one another, and an end lacking a phosphate group can ligate to one containing a phosphate, but a link cannot be formed between a pair of ends if neither carries a 5´ phosphate. Judicious use of alkaline phosphatase can therefore direct the action of a DNA ligase in a predetermined way so that only desired ligation products are obtained. T4 polynucleotide kinase, obtained from E. coli cells infected with T4 bacteriophage, performs the reverse reaction to alkaline phosphatase, adding phosphates to 5´ ends. Like alkaline phosphatase, the enzyme is used during complicated ligation experiments, but its main application is in the **end-labeling** of DNA molecules.

2.2 THE POLYMERASE CHAIN REACTION

Although methods with similar outcomes were suggested as early as 1971, the invention of PCR is now credited to Kary Mullis, who describes how he experienced a 'Eureka moment' one Friday evening in 1983 as he drove along California State Route 128 from Berkeley to Mendocino. His brainwave was an exquisitely simple technique that results in repeated copying of a selected segment of a longer DNA molecule. The technique is so straightforward that it is sometimes difficult for students encountering it for the first time to appreciate why it has

Figure 2.16 Linkers are used to place sticky ends on to a blunt-ended molecule. In this example, each linker contains the recognition sequence for the restriction endonuclease *Bam*HI. DNA ligase attaches the linkers to the ends of the blunt-ended molecule in a reaction that is made relatively efficient because the linkers are present at a high concentration. The restriction enzyme is then added to cleave the linkers and produce the sticky ends. Note that during the ligation the linkers ligate to one another, so a series of linkers (a concatemer) is attached to each end of the blunt molecule. When the restriction enzyme is added, these linker concatemers are cut into segments, with half of the innermost linker left attached to the DNA molecule. Adaptors are similar to linkers but each one has one blunt end and one sticky end. The blunt-ended DNA is therefore given sticky ends simply by ligating it to the adaptors: there is no need to carry out the restriction step.

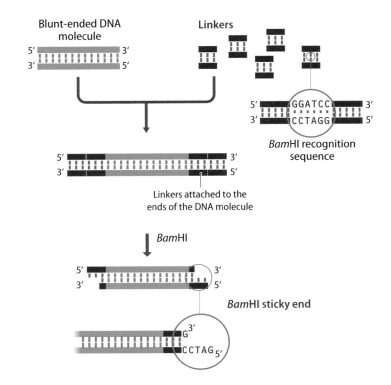

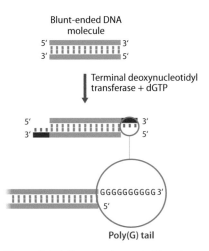

Figure 2.17 Homopolymer tailing. In this example, a poly(G) tail is synthesized at each end of a blunt-ended DNA molecule. Tails comprising other nucleotides are synthesized by including the appropriate dNTP in the reaction mixture.

become so important in modern biology. We will first look at the technique itself, and then explore some of its myriad applications.

Carrying out a PCR

PCR results in the repeated copying of a selected region of a DNA molecule (see Figure 2.2). The reaction is carried out by the purified, thermostable DNA polymerase of *T. aquaticus* (Section 2.1). The reason why a thermostable enzyme is needed will become clear when we look in more detail at the events that occur during a PCR.

To carry out a PCR experiment, the target DNA is mixed with *Taq* DNA polymerase, a pair of oligonucleotide primers, and a supply of nucleotides. The amount of target DNA can be very small because PCR is extremely sensitive and will work with just a single starting molecule. The primers are needed to initiate the DNA synthesis reactions that will be carried out by the *Taq* polymerase (see Figure 2.6). They must attach to the target DNA at either side of the segment that is to be copied. This means that the sequences of these attachment sites must be known so that primers of the appropriate sequences can be synthesized.

The reaction is started by heating the mixture to 94°C. At this temperature the hydrogen bonds that hold together the two polynucleotides of the double helix are broken, so the target DNA becomes denatured into single-stranded molecules (**Figure 2.18**). The temperature is then reduced to 50–60°C, which results in some rejoining of the single strands of the target DNA, but also allows the primers to attach to their annealing positions. DNA synthesis can now begin, so the temperature is raised to 72°C, the optimum for *Taq* polymerase. In this first stage of the PCR, a set of 'long' products is synthesized from each strand of the target DNA. These polynucleotides have identical 5´ ends, but random 3´ ends, the latter representing positions where DNA synthesis terminates by chance.

When the cycle of denaturation–annealing–synthesis is repeated, the long products act as templates for new DNA synthesis, giving rise, in the third cycle, to 'short' products, the 5´ and 3´ ends of which are both set by the primer annealing positions (**Figure 2.19**). In subsequent cycles, the number of short products accumulates in an exponential fashion (doubling during each cycle) until one of the components of the reaction becomes depleted. This means that after 30 cycles, there will be over 130 million short products derived from each starting molecule. In real terms, this equates to several micrograms of PCR product from a few nanograms or less of target DNA.

The rate of product formation can be followed during a PCR

Often a PCR is allowed to reach completion before the outcome is determined. After a pre-set number of cycles, usually 30–40, the reaction is halted and a sample analyzed by agarose gel electrophoresis, which will reveal a single band if the PCR has worked as expected and has amplified a single segment of the target DNA (**Figure 2.20**). Alternatively, the sequence of the product can be determined, using techniques described in Section 4.1.

It is also possible to follow synthesis of the product as the PCR proceeds through its series of cycles. This is called **real-time PCR**, and can be carried out in two different ways. In the simplest method, a dye that gives a fluorescent signal when it binds to double-stranded DNA is included in the PCR mixture. The gradual increase in the fluorescent signal given out by the mixture indicates the rate at which the product is being synthesized. The disadvantage of this approach is that it measures the total amount of double-stranded DNA in the PCR at any particular time, which may overestimate the actual amount of the product. This is because the primers sometimes anneal to themselves in various nonspecific ways, increasing the amount of double-stranded DNA that is present.

The second method for real-time PCR requires a short oligonucleotide called a **reporter probe**, which gives a fluorescent signal when it hybridizes to the PCR product. Because the probe only hybridizes to the PCR product, this method avoids the problems caused by primer–primer annealing. Several systems

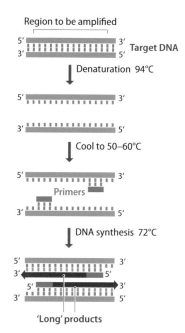

Figure 2.18 The first stage of a PCR.

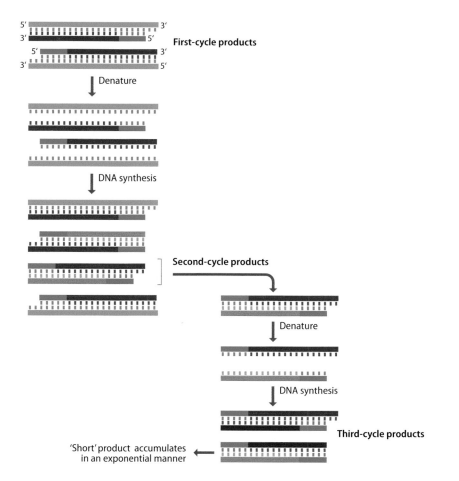

'Short' product accumulates in an exponential manner

Figure 2.19 The synthesis of 'short' products during a PCR. From the first-cycle products shown at the top of the diagram, the next cycle of denaturation–annealing–synthesis leads to four products, two of which are identical to the first-cycle products and two of which are made entirely of new DNA. During the third cycle, the latter give rise to 'short' products that, in subsequent cycles, accumulate in an exponential fashion.

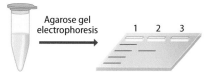

Figure 2.20 Analyzing the results of a PCR by agarose gel electrophoresis. The PCR has been carried out in a microfuge tube. A sample is loaded into lane 2 of an agarose gel. Lane 1 contains DNA size markers, and lane 3 contains a sample of a PCR carried out by a colleague. After electrophoresis, the gel is stained with ethidium bromide. Lane 2 contains a single band of the expected size, showing that the PCR has been successful. In lane 3 there is no band – this PCR has not worked.

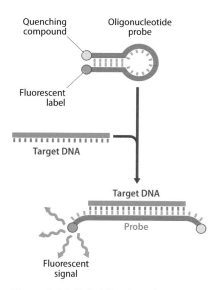

Figure 2.21 Hybridization of a reporter probe to its target DNA. The oligonucleotide reporter probe has two end-labels. One of these is a fluorescent dye, and the other is a quenching compound. The two ends of the oligonucleotide base-pair to one another, so the fluorescent signal is quenched. When the probe hybridizes with its target DNA, the ends of the molecule become separated, enabling the fluorescent dye to emit its signal.

have been developed, one of which makes use of a pair of labels comprising a fluorescent dye and a compound that quenches the fluorescent signal when brought into close proximity to the dye. This quenching is brought about by a process called **Förster resonance energy transfer** (**FRET**). The dye is attached to one end of the reporter probe and the quenching compound to the other end. Normally there is no fluorescence because the probe is designed in such a way that the two ends base-pair to one another, placing the quencher next to the dye (**Figure 2.21**). Hybridization between the probe and PCR product disrupts this base-pairing, moving the quencher away from the dye and enabling the fluorescent signal to be generated.

Both methods can be used as the basis for **quantitative PCR**, which enables the amount of target DNA that was present at the start of the PCR to be measured. The rate of product synthesis during the test PCR is compared with the progress of control PCRs with known amounts of starting DNA. The comparison is usually made by identifying the stage in the PCR at which the amount of fluorescent signal reaches a pre-set threshold (**Figure 2.22**). The more rapidly the threshold is reached, the greater the amount of target DNA in the starting mixture.

PCR has many and diverse applications

Why is PCR so important in modern research? First, we will deal with its limitations. In order to synthesize primers that will anneal at the correct positions, the sequences of the boundary regions of the DNA to be amplified must be known. This means that PCR cannot be used to purify fragments of genes, or other parts of a genome, that have never been studied before. A second constraint is the length of DNA that can be copied. Regions of up to 5 kb can be amplified without too much difficulty, and longer amplifications – up to 40 kb – are possible using modifications of the standard technique. Fragments longer than about 40 kb are unattainable by PCR.

Now we will consider the strengths of PCR. Primary among these is the ease with which products representing a single segment of a genome can be obtained. PCR can therefore be used to screen human DNA samples for mutations associated with genetic diseases such as thalassemia and cystic fibrosis. It also forms the basis of **genetic profiling**, in which natural variations in genome sequences are typed in order to connect samples taken at crime scenes with suspects and to establish paternity in cases where parentage is disputed (Section 7.3).

A second important feature of PCR is its ability to work with minuscule amounts of starting DNA. This means that PCR can be used to obtain sequences from the trace amounts of DNA that are present in hairs, bloodstains, and other forensic specimens, and from bones and other remains preserved at archaeological sites. Our ability to use PCR to amplify DNA from preserved skeletons has led to genome sequences of extinct species such as Neanderthals (Section 4.3). In clinical diagnosis, PCR is able to detect the presence of viral DNA well before the virus has reached the levels needed to initiate a disease response. This is particularly important in the early identification of viral-induced cancers because it means that treatment programs can be initiated before the cancer becomes established.

The above are just a few of the applications of PCR. The technique is now a major component of the molecular biologist's toolkit, and we will discover many more examples of its use as we progress through the chapters of this book.

2.3 DNA CLONING

DNA cloning was the first of the important new research tools that were developed during the early years of the recombinant DNA revolution. Cloning is a logical extension of the ability to manipulate DNA molecules with restriction endonucleases and ligases. First, we will look at the reasons why DNA cloning is a central technique in genomics research, and then we will examine how the technique is carried out.

Figure 2.22 Quantification of the amount of starting DNA by real-time PCR. The graph shows product synthesis during three PCRs, each with a different amount of starting DNA. During a PCR, product accumulates exponentially, the amount present at any particular cycle being proportional to the amount of starting DNA. The red curve is therefore the PCR with the greatest amount of starting DNA, and the blue curve is the one with the least starting DNA. If the amounts of starting DNA in these three PCRs are known, then the amount in a test PCR can be quantified by comparison with these controls. In practice, the comparison is made by identifying the cycle at which product synthesis moves above a threshold amount, indicated by the horizontal line on the graph.

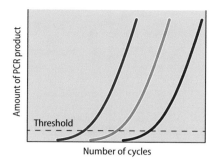

Why is gene cloning important?

Imagine that an animal gene has been obtained as a single restriction fragment after digestion of a larger molecule with the restriction enzyme *Bam*HI, which leaves 5′–GATC–3′ sticky ends (**Figure 2.23**). Imagine also that a plasmid – a small circle of DNA capable of replicating inside a bacterium – has been purified from *E. coli* and treated with *Bam*HI, which cuts the plasmid in a single position. The circular plasmid has therefore been converted into a linear molecule, again with 5′–GATC–3′ sticky ends. Mix the two DNA molecules together and add DNA ligase. Various recombinant ligation products will be obtained, one of which comprises the circularized plasmid with the animal gene inserted into the position originally taken by the *Bam*HI restriction site. If the recombinant plasmid is now reintroduced into *E. coli*, and the inserted gene has not disrupted its replicative ability, then the plasmid plus inserted gene will be replicated and copies passed to the daughter bacteria after cell division. The plasmid therefore acts as a **cloning vector**, providing the replicative ability that enables the cloned gene to be propagated inside the host cell. More rounds of plasmid replication and cell division will result in a colony of recombinant *E. coli* bacteria, each bacterium containing multiple copies of the animal gene. This series of events, as illustrated in Figure 2.23, constitutes the process called DNA or gene cloning.

When DNA cloning was first invented in the early 1970s, it revolutionized molecular biology by making possible experiments that previously had been inconceivable. This is because cloning can provide a pure sample of an individual DNA fragment, separated from all the other fragments that are produced when one or more larger molecules are cut with a restriction enzyme. These larger molecules could be, for example, an entire genome. Each of the fragments resulting from treatment with the endonuclease becomes inserted into a different plasmid molecule to produce a family of recombinant plasmids (**Figure 2.24**). Usually, only one recombinant molecule is transported into any single host cell, so that although the final set of **clones** may contain many different recombinant molecules, each individual clone contains multiple copies of just one. The end result is a **clone library**, whose inserted DNA fragments come from different parts of the starting DNA. If enough clones are obtained, then it is possible to have every part of a genome represented in the library.

Clone libraries are important for two reasons. First, it is often possible to identify from within the library the clone or clones that contain the DNA from a single gene, so that gene can be isolated and studied in detail. Second, a clone library is often the starting point for a genome sequencing project, because by sequencing the individual fragments contained in different clones, the genome sequence can gradually be built up (Section 4.2).

The simplest cloning vectors are based on *E. coli* plasmids

Plasmids replicate efficiently in their bacterial hosts because each plasmid possesses an **origin of replication** that is recognized by the DNA polymerases and other proteins that normally replicate the bacterium's chromosomes. The host cell's replicative machinery will also propagate a plasmid cloning vector, plus any new genes that have been inserted into it, providing that the vector possesses an origin of replication. Cloning vectors based on bacterial plasmids are therefore simple to construct and relatively easy to use.

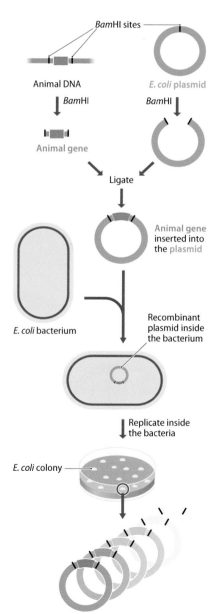

Figure 2.23 An outline of gene cloning.

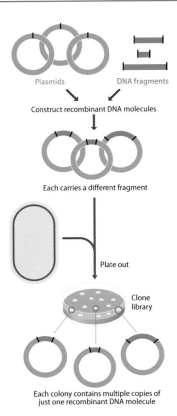

Figure 2.24 Production of a library of cloned fragments. In this example, just three different fragments are cloned. In reality, libraries comprising thousands of fragments, possibly covering an entire genome, are routinely prepared.

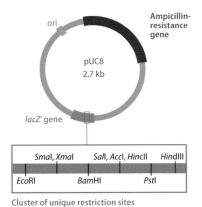

Figure 2.25 pUC8. The map shows the positions of the ampicillin-resistance gene, the *lacZ´* gene, the origin of replication (ori), and the cluster of restriction sites within the *lacZ´* gene. These are unique sites: they do not occur at any other positions in the genome.

One of the most popular plasmid vectors is pUC8, a member of a series of vectors that was first introduced in the early 1980s. The pUC series is derived from an earlier cloning vector, pBR322, which was originally constructed by ligating together restriction fragments from three naturally occurring *E. coli* plasmids: R1, R6.5, and pMB1. pUC8 is a small plasmid, comprising just 2.7 kb. As well as its origin of replication, it carries two genes (**Figure 2.25**):

- A gene for ampicillin resistance. The presence of this gene means that a bacterium containing a pUC8 plasmid is able to synthesize an enzyme, called β-lactamase, that enables the cell to withstand the growth-inhibitory effect of the antibiotic. This means that cells containing pUC8 plasmids can be distinguished from those that do not by plating the bacteria onto an agar medium containing ampicillin. Normal *E. coli* cells are sensitive to ampicillin and cannot grow when the antibiotic is present. Ampicillin resistance is therefore a **selectable marker** for pUC8.

- The *lacZ´* gene, which codes for part of the enzyme β-galactosidase. This enzyme is involved in the breakdown of lactose to glucose plus galactose. It is normally coded by the gene *lacZ*, which resides on the *E. coli* chromosome. Some strains of *E. coli* have a modified *lacZ* gene, one that lacks the segment referred to as *lacZ´*, which codes for the α-peptide portion of β-galactosidase. These mutants can synthesize the enzyme only when they harbor a plasmid, such as pUC8, that carries the missing *lacZ´* segment of the gene.

To carry out a cloning experiment with pUC8, the manipulations shown in Figure 2.23, resulting in construction of a recombinant plasmid, are performed in the test tube with purified DNA. Pure pUC8 DNA can be obtained quite easily from extracts of bacterial cells, and after manipulation the plasmids can be reintroduced into *E. coli* by **transformation**, the process by which 'naked' DNA is taken up by a bacterial cell. This is the system studied by Avery and his colleagues in the experiments that showed that bacterial genes are made of DNA (Section 1.1). Transformation is not a particularly efficient process with many bacteria, including *E. coli*, but the rate of DNA uptake can be enhanced significantly by suspending the cells in calcium chloride before adding the DNA, and then briefly incubating the mixture at 42°C. Even after this enhancement, only a very small proportion of the cells take up a plasmid. This is why the ampicillin-resistance marker is so important – it allows the small number of **transformants** to be selected from the large background of nontransformed cells.

The map of pUC8 shown in Figure 2.25 indicates that the *lacZ´* gene contains a cluster of unique restriction sites. Ligation of new DNA into any one of these sites results in **insertional inactivation** of the gene and hence loss of β-galactosidase activity. This is the key to distinguishing a **recombinant plasmid** – one that contains an inserted piece of DNA – from a nonrecombinant plasmid that has no new DNA. Identifying recombinants is important because the manipulations illustrated in Figures 2.23 and 2.24 result in a variety of ligation products, including plasmids that have recircularized without insertion of new DNA. Screening for β-galactosidase presence or absence is, in fact, quite easy. Rather than assaying for lactose being split to glucose and galactose, the presence of functional β-galactosidase molecules in the cells is checked by a histochemical test with a compound called X-gal (5-bromo-4-chloro-3-indolyl-β-D-galactopyranoside), which the enzyme converts into a blue product. If X-gal (plus an inducer of the enzyme such as isopropylthiogalactoside, IPTG) is added to the agar, along with ampicillin, then nonrecombinant colonies, the cells of which synthesize β-galactosidase, will be colored blue, whereas recombinants with a disrupted *lacZ´* gene, which are unable to make β-galactosidase, will be white (**Figure 2.26**). This system is called **Lac selection**.

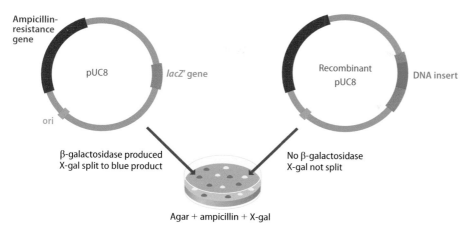

Figure 2.26 Recombinant selection with pUC8.

Bacteriophages can also be used as cloning vectors

Bacteriophage (or 'phage') genomes can also be used as cloning vectors because they too possess origins of replication that enable them to be propagated inside bacteria, either by the bacterial enzymes or by DNA polymerases and other proteins specified by phage genes. *E. coli* bacteriophages were developed as cloning vectors back in the earliest days of the recombinant DNA revolution. The main reason for seeking a different type of vector was the inability of plasmids such as pUC8 to handle DNA fragments greater than about 10 kb in size, larger inserts undergoing rearrangements or interfering with the plasmid replication system in such a way that the recombinant DNA molecules become lost from the host cells. The first attempts to develop vectors able to handle larger fragments of DNA centered on the bacteriophage called **lambda** (λ).

To replicate, a bacteriophage must enter a bacterial cell and subvert the bacterial enzymes into expressing the information contained in the phage genes, so that the bacterium synthesizes new phages. Once replication is complete, the new phages leave the bacterium, usually causing its death as they do so, and move on to infect new cells (**Figure 2.27A**). This is called a **lytic infection cycle**, because it results in **lysis** of the bacterium. As well as the lytic cycle, λ (unlike many other types of bacteriophage) can also follow a **lysogenic infection cycle**, during which the λ genome integrates into the bacterial chromosome, where it can remain quiescent for many generations, being replicated along with the host chromosome every time the cell divides (**Figure 2.27B**).

The size of the λ genome is 48.5 kb, of which some 15 kb or so is 'optional' in that it contains genes that are only needed for integration of the phage DNA into the *E. coli* chromosome (**Figure 2.28A**). These segments can therefore be deleted without impairing the ability of the phage to infect bacteria and direct synthesis of new λ particles by the lytic cycle. Two types of vector have been developed (**Figure 2.29B**):

- **Insertion vectors**, in which part or all of the optional DNA has been removed, and a unique restriction site is introduced at some position within the trimmed-down genome.

- **Replacement vectors**, in which the optional DNA is contained within a **stuffer fragment**, flanked by a pair of restriction sites, that is replaced when the DNA to be cloned is ligated into the vector.

The λ genome is linear, but the two natural ends of the molecule have 12-nucleotide single-stranded overhangs, called **cos** sites, which have complementary sequences and so can base-pair to one another. A λ cloning vector can therefore be obtained as a circular molecule that can be manipulated in the test tube in

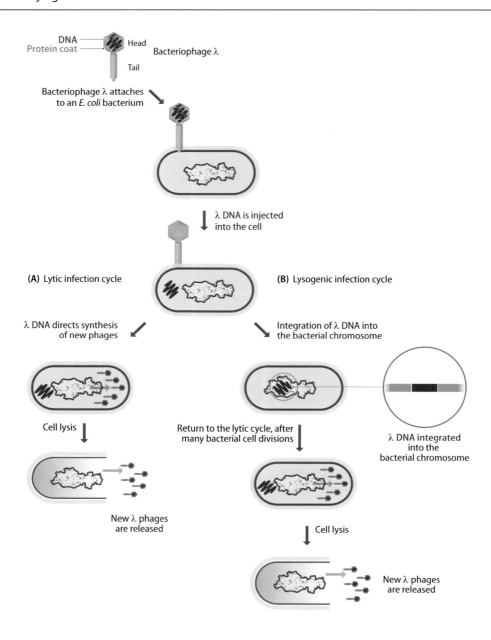

Figure 2.27 The lytic and lysogenic infection cycles of bacteriophage λ. (A) In the lytic cycle, new phages are produced shortly after infection. (B) During the lysogenic cycle, the phage genome becomes inserted into the bacterium's chromosomal DNA, where it can remain quiescent for many generations.

the same way as a plasmid, and reintroduced into *E. coli* by **transfection**, the term used for the uptake of naked phage DNA. Alternatively, a more efficient uptake system called ***in vitro* packaging** can be utilized. This procedure starts with a linear version of the cloning vector, which is cut into two segments, the left and right arms, each with a *cos* site at one end. Ligation is then carried out with carefully measured quantities of each arm and the DNA to be cloned, the aim being to produce concatemers in which the different fragments are linked together in the order left arm–new DNA–right arm, as shown in **Figure 2.29**. The concatemers are then added to an *in vitro* packaging mix, which contains all the proteins needed to make λ phage particles. These proteins form phage particles spontaneously, and will place inside the particles any DNA fragment that is between 37 kb and 52 kb in length and is flanked by *cos* sites. The packaging mix therefore cuts left arm–new DNA–right arm combinations of 37–52 kb out of the concatemers and constructs λ phages around them. The phages are then mixed

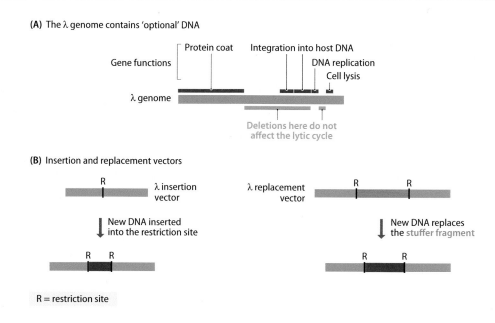

(A) The λ genome contains 'optional' DNA

(B) Insertion and replacement vectors

R = restriction site

Figure 2.28 Cloning vectors based on bacteriophage λ. (A) In the λ genome, the genes are arranged into functional groups. For example, the region marked as 'protein coat' comprises genes coding for proteins that are either components of the phage capsid or are required for capsid assembly, and 'cell lysis' comprises genes involved in lysis of the bacterium at the end of the lytic phase of the infection cycle. The regions of the genome that can be deleted without impairing the ability of the phage to follow the lytic cycle are indicated in green. (B) The differences between a λ insertion vector and a λ replacement vector.

with *E. coli* cells, and the natural infection process transports the vector plus new DNA into the bacteria.

After infection, the cells are spread onto an agar plate. The objective is not to obtain individual colonies but to produce an even layer of bacteria across the entire surface of the agar. Bacteria that were infected with the packaged cloning vector die within about 20 minutes because the λ genes contained in the arms of the vector direct replication of the DNA and synthesis of new phages by the lytic cycle, each of these new phages containing its own copy of the vector plus cloned DNA. Death and lysis of the bacterium releases these phages into the

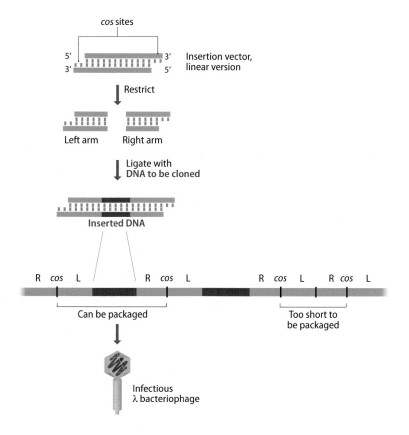

Figure 2.29 Cloning with a λ insertion vector. The linear form of the vector is shown at the top of the diagram. Treatment with the appropriate restriction endonuclease produces the left and right arms, both of which have one blunt end and one end with the 12-nucleotide overhang of the *cos* site. The DNA to be cloned is blunt-ended and so is inserted between the two arms during the ligation step. The arms also ligate to one another via their *cos* sites, forming a concatemer. Some parts of the concatemer comprise left arm–insert DNA–right arm and, assuming this combination is 37–52 kb in length, will be enclosed inside the capsid by the *in vitro* packaging mix. Parts of the concatemer made up of the left arm ligated directly to the right arm, without new DNA, are too short to be packaged.

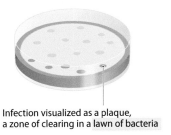

Infection visualized as a plaque, a zone of clearing in a lawn of bacteria

Figure 2.30 Bacteriophage infection is visualized as a plaque on a lawn of bacteria.

surrounding medium, where they infect new cells and begin another round of phage replication and lysis. The end result is a zone of clearing, called a **plaque**, which is visible on the lawn of bacteria that grows on the agar plate (**Figure 2.30**). With some λ vectors, all plaques are made up of recombinant phages because ligation of the two arms without insertion of new DNA results in a molecule that is too short to be packaged. With other vectors it is necessary to distinguish recombinant plaques from nonrecombinant ones. Various methods are used, including the β-galactosidase system described above for the plasmid vector pUC8 (see Figure 2.26), which is also applicable to those λ vectors that carry a fragment of the *lacZ* gene into which the DNA to be cloned is inserted.

Vectors for longer pieces of DNA

The λ phage particle can accommodate up to 52 kb of DNA, so if the genome has 15 kb removed, then up to 18 kb of new DNA can be cloned. This limit is higher than that for plasmid vectors, but is still very small compared with the sizes of intact genomes. The comparison is important if a clone library is to be used as the starting point for a project aimed at determining the sequence of a genome (Section 4.3). If a λ vector is used with human DNA, then over half a million clones are needed for there to be a 95% chance of any particular part of the genome being present in the library (**Table 2.4**). It is possible to prepare a library comprising half a million clones, especially if automated techniques are used, but such a large collection is far from ideal. It would be much better to reduce the number of clones by using a vector that is able to handle fragments of DNA longer than 18 kb. Many of the developments in cloning technology over the last 25 years have been aimed at finding ways of doing this.

One possibility is to use a **cosmid**, which is a special type of plasmid that carries a λ *cos* site (**Figure 2.31**). Concatemers of cosmid molecules, linked at their *cos* sites, act as substrates for *in vitro* packaging because the *cos* site is the only sequence that a DNA molecule needs in order to be recognized as a 'λ genome' by the proteins that package DNA into λ phage particles. Particles containing cosmid DNA are as infective as real λ phages, but once inside the cell the cosmid cannot direct synthesis of new phage particles and instead replicates as a plasmid. Recombinant DNA is therefore obtained from colonies rather than plaques. As with other types of λ vector, the upper limit for the length of the cloned DNA is set by the space available within the λ phage particle. A cosmid can be 8 kb or less in size, so up to 44 kb of new DNA can be inserted before the packaging limit of the λ phage particle is reached. This reduces the size of the

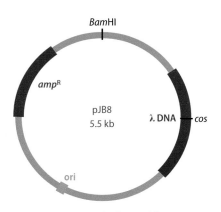

Figure 2.31 A typical cosmid. pJB8 is 5.5 kb in size and carries the ampicillin-resistance gene (*amp*R), a segment of λ DNA containing the *cos* site, and an *E. coli* origin of replication (ori).

		Number of clones*	
TABLE 2.4 SIZES OF HUMAN GENOMIC LIBRARIES PREPARED IN DIFFERENT TYPES OF CLONING VECTOR			
Type of vector	Insert size (kb)	*P* = 95%	*P* = 99%
λ replacement	18	516,000	793,000
Cosmid, fosmid	40	232,000	357,000
P1	100	93,000	143,000
BAC, PAC	300	31,000	47,500

*Calculated from the equation:

$$N = \frac{\ln(1-P)}{\ln\left(1-\dfrac{a}{b}\right)}$$

where *N* is the number of clones required, *P* is the probability that any given segment of the genome is present in the library, *a* is the average size of the DNA fragments inserted into the vector, and *b* is the size of the genome human genomic library to about one-quarter of a million clones, which is an improvement compared with a λ library, but still a massive number of clones to have to work with.

To reduce the size of clone libraries even further, other types of vector with even greater carrying capacities have been developed. The following are the most important of these vectors:

- **Bacterial artificial chromosomes**, or **BACs**, are based on the naturally occurring **F plasmid** of *E. coli*. Unlike the plasmids used to construct the early cloning vectors, the F plasmid is relatively large, and vectors based on it have a higher capacity for inserted DNA. BACs are designed so that recombinants can be identified by Lac selection (see Figure 2.26) and hence are easy to use. They can clone fragments of 300 kb and longer, and the inserts are very stable. BACs were used extensively in the Human Genome Project (Section 4.3), and they are currently the most popular vectors for cloning large pieces of DNA.

- **Bacteriophage P1 vectors** are very similar to λ vectors, being based on a deleted version of a natural phage genome, the capacity of the cloning vector being determined by the size of the deletion and the space within the phage particle. P1 has the advantage over λ that it is able to squeeze 110 kb of DNA into its capsid structure, which results in P1 vectors having a higher capacity than those based on λ. Cosmid-type P1 vectors have been designed and used to clone DNA fragments ranging in size from 75 to 100 kb.

- **P1-derived artificial chromosomes**, or **PACs**, combine features of P1 vectors and BACs, and have a capacity of up to 300 kb.

- **Fosmids** contain the F plasmid origin of replication and a λ *cos* site. They are similar to cosmids in the way they are used and in their capacity for inserted DNA, but have a lower copy number in *E. coli*, which means that they are less prone to instability problems.

The sizes of human genome libraries prepared in these various types of vector are given in Table 2.4.

DNA can be cloned in organisms other than *E. coli*

Cloning is not merely a means of producing DNA for sequencing and other types of analysis. It is also the central part of techniques that are used to identify the function of an unknown gene, and to study its mode of expression and the way in which its expression is regulated. In the wider research world, cloning is also used to carry out genetic engineering experiments aimed at modifying the biological characteristics of the host organism, and to transfer genes for important animal proteins, such as pharmaceuticals, into a new host cell from which the proteins can be obtained in larger quantities than is possible by conventional purification from animal tissue. These multifarious applications demand that genes must frequently be cloned in organisms other than *E. coli*.

Cloning vectors based on plasmids or phages have been developed for most of the well-studied species of bacteria, such as *Bacillus*, *Streptomyces*, and *Pseudomonas*, these vectors being used in exactly the same way as the *E. coli* analogs. Plasmid vectors are also available for yeasts and fungi. Some of these carry the origin of replication from the **2 μm plasmid**, which is present in many strains of *Saccharomyces cerevisiae*, but other plasmid vectors for yeast and fungi only have an *E. coli* origin. An example is YIp5, an *S. cerevisiae* vector that is simply an *E. coli* plasmid that contains a copy of the yeast gene called *URA3* (**Figure 2.32A**). The presence of the *E. coli* origin means that YIp5 is a **shuttle vector** that can be used with either *E. coli* or *S. cerevisiae* as the host. This is a useful feature because cloning in *S. cerevisiae* is a relatively inefficient process, and generating a large number of clones is difficult. If the experiment requires

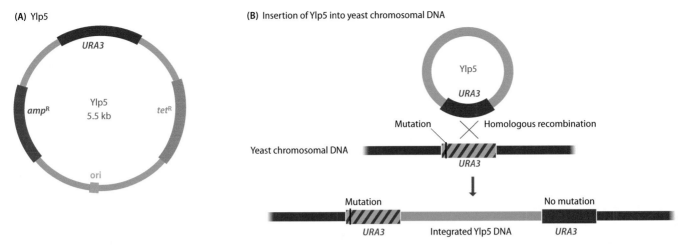

Figure 2.32 Cloning with a YIp. (A) YIp5, a typical yeast integrative plasmid. The plasmid contains the ampicillin-resistance gene (*amp^R*), the tetracycline-resistance gene (*tet^R*), the yeast gene *URA3*, and an *E. coli* origin of replication (ori). The presence of the *E. coli* ori means that recombinant YIp5 molecules can be constructed in *E. coli* before their transfer into yeast cells. (B) YIp5 has no origin of replication that can function inside yeast cells, but can survive if it integrates into the yeast chromosomal DNA by homologous recombination between the plasmid and chromosomal copies of the *URA3* gene. The chromosomal gene carries a small mutation that means that it is nonfunctional so the host cells are *ura3⁻*. One of the pair of *URA3* genes that is formed after integration of the plasmid DNA is mutated, but the other is not. Recombinant cells are therefore *ura3⁺* and can be selected by plating onto minimal medium, which does not contain uracil.

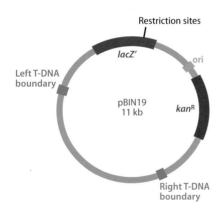

Figure 2.33 The plant cloning vector pBIN19. pBIN19 carries the *lacZ´* gene, the kanamycin-resistance gene (*kan^R*), an *E. coli* origin of replication (ori), and the two boundary sequences from the T-DNA region of the Ti plasmid. These two boundary sequences recombine with plant chromosomal DNA, inserting the segment of DNA between them into the plant DNA. The orientation of the boundary sequences in pBIN19 means that the *lacZ´* and *kan^R* genes, as well as any new DNA ligated into the restriction sites within *lacZ´*, are transferred to the plant DNA. Note that pBIN19 is another example of a shuttle vector, recombinant molecules being constructed in *E. coli*, using the *lacZ´* selection system, before transfer to *A. tumefaciens* and thence to the plant.

that the desired recombinant must be identified from a mixture of clones, then it may not be possible to obtain enough recombinants to find the correct one. To avoid this problem, construction of recombinant DNA molecules and selection of the correct recombinant is carried out with *E. coli* as the host. When the correct clone has been identified, the recombinant YIp5 construct is purified and transferred into *S. cerevisiae*, usually by mixing the DNA with **protoplasts** – yeast cells whose walls have been removed by enzyme treatment. Without an origin of replication, the vector is unable to propagate independently inside yeast cells, but it can survive if it becomes integrated into one of the yeast chromosomes, which can occur by **homologous recombination** (Section 16.1) between the *URA3* gene carried by the vector and the chromosomal copy of this gene (**Figure 2.32B**). 'YIp' in fact stands for 'yeast integrative plasmid.' Once integrated, the YIp, plus any DNA that has been inserted into it, replicates along with the host chromosomes.

Integration into chromosomal DNA is also a feature of many of the cloning systems used with animals and plants, and forms the basis of the construction of **knockout mice**, which have been used to identify the functions of previously unknown genes that are discovered in the human genome (Section 6.2). The vectors are animal equivalents of YIps. A similar range of vectors has been developed for cloning genes in plants. Plasmids can be introduced into plant embryos by bombardment with DNA-coated microprojectiles, a process called **biolistics**. Integration of the plasmid DNA into the plant chromosomes, followed by growth of the embryo, then results in a plant that contains the cloned DNA in most or all of its cells. Some success has also been achieved with plant vectors based on the genomes of caulimoviruses and geminiviruses, but the most interesting types of plant cloning vector are those derived from the **Ti plasmid**, a large bacterial plasmid found in the soil microorganism *Agrobacterium tumefaciens*. Part of the Ti plasmid, the region called the **T-DNA**, becomes integrated into a plant chromosome when the bacterium infects a plant stem and causes crown gall disease. The T-DNA carries a number of genes that are expressed inside the plant cells and induce the various physiological changes that characterize the disease. Vectors such as pBIN19 (**Figure 2.33**) have been designed to make use of this natural genetic engineering system. The recombinant vector is introduced into *A. tumefaciens* cells, which are allowed to infect a cell suspension or plant callus culture, from which mature, transformed plants can be regenerated (**Figure 2.34**).

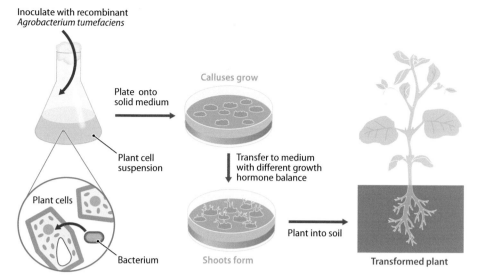

Figure 2.34 Transformation of plant cells by recombinant *A. tumefaciens*. After transformation of a cell suspension, recombinant plant cells are selected by plating onto kanamycin agar, and then regenerated into whole plants.

SUMMARY

- The four main types of enzyme used in recombinant DNA technology are DNA polymerases, nucleases, ligases, and end-modification enzymes.

- DNA polymerases synthesize new DNA polynucleotides and are used in procedures such as DNA sequencing and PCR.

- The most important nucleases are the restriction endonucleases, which cut double-stranded DNA molecules at specific nucleotide sequences, and hence cut a molecule into a predicted set of fragments, the sizes of which can be determined by agarose gel electrophoresis.

- Ligases join molecules together and end-modification enzymes carry out a variety of reactions, including several used to label DNA molecules.

- PCR results in the repeated copying of a selected region of a DNA molecule, but at least part of the DNA sequence of this region must be known.

- Starting with just a single target DNA molecule, over 130 million copies can be made during 30 cycles of a PCR.

- Real-time and quantitative methods enable the dynamics of product synthesis to be followed during a PCR.

- DNA cloning is a means of obtaining a pure sample of an individual gene or other segment of a DNA molecule.

- Many different types of cloning vector have been designed for use with *E. coli* as the host organism, the simplest being based on small plasmids that carry selectable markers such as the *lacZ'* gene.

- Bacteriophage λ has also been used as the basis for a series of *E. coli* cloning vectors, including the plasmid–phage hybrids called cosmids, which are used to clone fragments of DNA up to 44 kb in length.

- Other types of vector, such as bacterial artificial chromosomes, can be used to clone even longer pieces of DNA up to 300 kb.

- Organisms other than *E. coli* can also be used as the hosts for DNA cloning. Several types of vector have been designed for *Saccharomyces cerevisiae*, and specialized techniques are available for cloning DNA in animals and plants.

SHORT ANSWER QUESTIONS

1. Describe how a DNA fragment containing a single gene of interest would be obtained by (A) PCR or (B) DNA cloning.

2. How can a researcher identify a single restriction enzyme fragment containing a gene of interest in a digest of genomic DNA that contains thousands of different restriction fragments?

3. Describe a useful and quick method for increasing the ligation efficiency of blunt-ended DNA molecules.

4. Why are the initial PCR products – produced in the first few cycles of the reaction – long and of varying sizes, and the final PCR products all of a shorter and uniform size?

5. How do the primers determine the specificity of a PCR?

6. Explain how the rate of product formation can be followed during a PCR.

7. Explain why bacterial plasmids are popular cloning vectors.

8. Distinguish between the ways in which antibiotic resistance and Lac selection are used in identification of recombinant cloning vectors.

9. List the features of bacteriophage λ that have led to this phage being used as a cloning vector.

10. Outline the key differences between a λ insertion vector and a λ replacement vector.

11. Why are vectors that can carry larger DNA inserts beneficial for the creation of clone libraries?

12. Describe the important features of the cloning systems used with (A) *Saccharomyces cerevisiae*, (B) animals, and (C) plants.

IN-DEPTH PROBLEMS

1. How might you determine the positions of the restriction sites in a DNA molecule, other than by working out the sequence of the molecule?

2. Calculate the numbers of short and long products that would be present after 20, 25, and 30 cycles of a PCR.

3. When DNA is cloned in pUC8, recombinant bacteria (those containing a circular pUC8 molecule that carries an inserted DNA fragment) are identified by plating onto an agar medium containing ampicillin and the lactose analog called X-gal. An older type of cloning vector, called pBR322, also had the gene for ampicillin resistance but did not carry the *lacZ′* gene. Instead, DNA was inserted into a gene for tetracycline resistance present in pBR322. Describe the procedure that would be needed to distinguish bacteria that had taken up a recombinant pBR322 plasmid from those that had taken up a plasmid that had circularized without insertion of new DNA.

4. What would be the features of an ideal cloning vector? To what extent are these requirements met by any of the existing cloning vectors?

5. Soon after the first gene cloning experiments were carried out in the early 1970s, a number of scientists argued that there should be a temporary moratorium on this type of research. What was the basis of these scientists' fears and to what extent were these fears justified?

FURTHER READING

Textbooks and practical guides on the methods used to study DNA

Brown, T.A. (2021). *Gene Cloning and DNA Analysis: An Introduction*, 8th edition. Wiley-Blackwell, Chichester.

Brown, T.A. (ed.) (2000). *Essential Molecular Biology: A Practical Approach*, 1 and 2, 2nd edition. Oxford University Press, Oxford. *Includes detailed protocols for DNA cloning and PCR.*

Henkin, T.M. and Peters, J.E. (2020). *Molecular Genetics of Bacteria*, 5th edition. ASM Press, Washington. *Provides a detailed description of plasmids and bacteriophages.*

Enzymes for DNA manipulation

Brown, T.A. (1998). *Molecular Biology Labfax. Volume I: Recombinant DNA*, 2nd edition. Academic Press, London. *Contains details of all types of enzymes used to manipulate DNA and RNA.*

Lee, P.Y., Costumbrado, J., Hsu, C.-Y. and Kim, Y.H. (2012). Agarose gel electrophoresis for the separation of DNA fragments. *J. Vis. Exp.* 62:3923. *Takes you through the steps in setting up and running an agarose gel.*

Loenen, W.A.M., Dryden, D.T.F., Raleigh, E.A., et al. (2014). Highlights of the DNA cutters: A short history of the restriction enzymes. *Nucl. Acids Res.* 42:3–19.

Pingoud, A., Fuxreiter, M., Pingoud, V. and Wende, W. (2005). Type II restriction endonucleases: Structure and mechanism. *Cell. Mol. Life Sci.* 62:685–707.

Smith, H.O. and Wilcox, K.W. (1970). A restriction enzyme from *Hemophilus influenzae*: I. general properties. *J. Mol. Biol.* 51:379–391. *One of the first full descriptions of a restriction endonuclease.*

PCR

Higuchi, R., Dollinger, G., Walsh, P.S. and Griffith, R. (1992). Simultaneous amplification and detection of specific DNA sequences. *Biotechnology* 10:413–417. *The first description of real-time PCR.*

Mullis, K.B. (1990). The unusual origin of the polymerase chain reaction. *Sci. Am.* 262(4):56–65.

Rychlik, W., Spencer, W.J. and Rhoads, R.E. (1990). Optimization of the annealing temperature for DNA amplification *in vitro*. *Nucleic Acids Res.* 18:6409–6412.

Saiki, R.K., Gelfand, D.H., Stoffel, S., et al. (1988). Primer-directed enzymatic amplification of DNA with a thermostable DNA polymerase. *Science* 239:487–491.

VanGuilder, H.D., Vrana, K.E. and Freeman, W.M. (2008). Twenty-five years of quantitative PCR for gene expression analysis. *Biotechniques* 44:619–626.

DNA cloning in bacteria

Frischauf, A.-M., Lehrach, H., Poustka, A. and Murray, N. (1983). Lambda replacement vectors carrying polylinker sequences. *J. Mol. Biol.* 170:827–842.

Hohn, B. and Murray, K. (1977). Packaging recombinant DNA molecules into bacteriophage particles *in vitro*. *Proc. Natl Acad. Sci. USA* 74:3259–3263.

Vieira, J. and Messing, J. (1982). The pUC plasmids, an M13mp7-derived system for insertional mutagenesis and sequencing with synthetic universal primers. *Gene* 19:259–268.

High-capacity cloning vectors

Ioannou, P.A., Amemiya, C.T., Garnes, J., et al. (1994). A new bacteriophage P1-derived vector for the propagation of large human DNA fragments. *Nat. Genet.* 6:84–89. *PACs.*

Kim, U.-J., Shizuya, H., de Jong, P.J., et al. (1992). Stable propagation of cosmid sized human DNA inserts in an F factor based vector. *Nucleic Acids Res.* 20:1083–1085. *Fosmids.*

Shizuya, H., Birren, B., Kim, U.-J., et al. (1992). Cloning and stable maintenance of 300-kilobase-pair fragments of human DNA in *Escherichia coli* using an F-factor-based vector. *Proc. Natl Acad. Sci. USA* 89:8794–8797. *The first description of a BAC.*

Sternberg, N. (1990). Bacteriophage P1 cloning system for the isolation, amplification, and recovery of DNA fragments as large as 100 kilobase pairs. *Proc. Natl Acad. Sci. USA* 87:103–107. *Bacteriophage P1 vectors.*

Cloning in plants and animals

Bevan, M. (1984). Binary *Agrobacterium* vectors for plant transformation. *Nucleic Acids Res.* 12:8711–8721.

Colosimo, A., Goncz, K.K., Holmes, A.R., et al. (2000). Transfer and expression of foreign genes in mammalian cells. *Biotechniques* 29:314–324.

Gnügge, R. and Rudolf, F. (2017). *Saccharomyces cerevisiae* shuttle vectors. *Yeast* 34:205–221. *Details of different yeast cloning vectors.*

Hansen, G. and Wright, M.S. (1999). Recent advances in the transformation of plants. *Trends Plant Sci.* 4:226–231.

Kost, T.A. and Condreay, J.P. (2002). Recombinant baculoviruses as mammalian cell gene-delivery vectors. *Trends Biotechnol.* 20:173–180.

Lee, L.-Y. and Gelvin, S.B. (2008). T-DNA binary vectors and systems. *Plant Physiol.* 146:325–332.

Păcurar, D.I., Thordal-Christensen, H., Păcurer, M.L., et al. (2011) *Agrobacterium tumefaciens*: From crown gall tumors to genetic transformation. *Physiol. Mol. Plant Pathol.* 76:76–81.

Online resources

Addgene Vector Database. https://www.addgene.org/vector-database/. *A database of cloning vectors.*

REBASE, The Restriction Enzyme Database. http://rebase.neb.com/rebase/rebase.html. *A comprehensive list of all the known restriction endonucleases and their recognition sequences.*

MAPPING GENOMES

In this Chapter, we will study the various ways in which **genome maps** are constructed. A genome map, like any other type of map, indicates the positions of interesting features and other important landmarks. In a genome map, these features and landmarks are genes and other distinctive DNA sequences. Although a variety of techniques can be used to map genes and other DNA landmarks, the convention is to look on genome mapping as comprising two complementary approaches:

- **Genetic mapping** (Sections 3.2–3.4), also called **linkage analysis**, is based on the use of genetic techniques, including planned breeding experiments or, in the case of humans, the examination of family histories (also called **pedigrees**).
- **Physical mapping** (Sections 3.5 and 3.6) uses molecular biology techniques to examine DNA molecules directly in order to identify the positions of sequence features, including genes.

Before exploring the various techniques involved in genetic and physical mapping, we must first understand why genome maps are important.

3.1 WHY A GENOME MAP IS IMPORTANT

The study of genomes is often looked on as a modern, edgy area of biological research, far divorced from the work of the old era geneticists such as Gregor Mendel. And yet many of the techniques used to construct genome maps are based directly on the discoveries of Mendel and the other early geneticists. We must therefore spend a few minutes understanding why genome mapping, despite being an 'old-fashioned' type of biology, is still important in the fast-paced research of the genomic age.

Genome maps are needed in order to sequence the more complex genomes

During the early days of genome research, it was believed that possession of a detailed map would be an essential prerequisite for assembly of the correct sequence of a genome. This is because DNA sequencing has one major limitation: only with the most sophisticated and recently-introduced technology is it possible to obtain a sequence of more than about 750 bp in a single experiment. This means that the sequence of a long DNA molecule has to be constructed from a series of shorter sequences. This is done by breaking the molecule into fragments, determining the sequence of each one, and using a computer to search for overlaps and build up the master sequence (**Figure 3.1**). This **shotgun method** is the standard approach for genome sequencing, but suffers from two problems. The first of these is that, especially with larger genomes, it might not be possible to obtain sufficient short sequences to produce a contiguous DNA sequence for the entire genome. Instead, the genome sequence might be made up of many short segments separated by gaps that represent parts of the genome that, by chance, are not covered by the sequences that have been obtained (**Figure 3.2**).

DOI: 10.1201/9781003133162-3

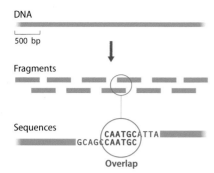

DNA

500 bp

Fragments

Sequences

CAATGCATTA
GCAGCCAATGC

Overlap

Figure 3.1 The shotgun method for sequence assembly. The DNA molecule is broken into small fragments, each of which is sequenced. The master sequence is assembled by searching for overlaps between the sequences of individual fragments.

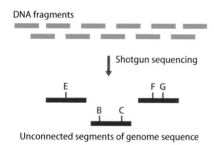

DNA fragments

Shotgun sequencing

E   F G

B   C

Unconnected segments of genome sequence

Identify positions on the genome map

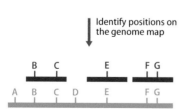

B   C   E   F G

A   B   C   D   E   F G

Positions of mapped sequence features

Figure 3.2 Using a genome map as an aid to sequence assembly. A genome has been broken into short DNA fragments which have been sequenced by the shotgun method. When the sequences are assembled, a series of unconnected genome segments are obtained. The segments contain genes and other sequence features (A, B, C, etc.) whose positions in the genome have been mapped. The map can therefore be used to identify the positions of the segments in the genome sequence.

If these segments are unconnected, then how can they be positioned correctly relative to one another in order to build up the genome sequence? The answer is to identify within those segments features that are located on the genome map. By anchoring the segments onto the map, the correct genome sequence can be obtained, even if that sequence still contains some gaps.

The second problem with the shotgun approach is that it can lead to errors if the genome contains **repetitive DNA** sequences. These are sequences, up to several kilobases in length, that are repeated at two or more places in a genome. When a genome containing repetitive DNA is broken into fragments, some of the resulting pieces will contain the same sequence motifs. It would be very easy to reassemble these sequences so that the DNA between a pair of repeats is left out, or even to connect together two quite separate pieces of the same or different chromosomes (**Figure 3.3A**). Once again, a genome map enables errors of this type to be avoided. If the sequence features either site of a repetitive region match the genome map, then the sequence in that region has been put together correctly. If the sequence and the map do not match, then a mistake has been made and the assembly must be revised (**Figure 3.3B**).

Over the years, sequencing technology has become more powerful, enabling ever-increasing numbers of short sequences to be generated from a single genome, which means that there is less likelihood that the final sequence will contain many gaps. At the same time, the computer algorithms used to assemble those sequences into contiguous segments have become more intelligent, and are now able to recognize when the assembly reaches a region of repetitive DNA, and to take steps to ensure that the sequence around these regions is not put together incorrectly (Section 4.2). Maps have therefore become less important. Many prokaryotic genomes (which are relatively small and have little repetitive DNA) have been sequenced without reference to a map, and an increasing number of eukaryotic genome projects are dispensing with them. But maps are not yet entirely redundant as aids to genome sequencing. One of the greatest challenges today is in obtaining genome sequences for important crop plants, many of which have large genomes with a substantial repetitive DNA content. Barley, for example, has a genome of 5100 Mb (compared to 3100 Mb for the human genome), of which approximately 80% is repetitive DNA. An even greater degree of difficulty is presented by bread wheat, which is a **hexaploid**, meaning that it has three genomes, called A, B, and D. Each is about 5500 Mb (a massive 16,500 Mb in total) with a similar repetitive DNA content to barley. The genome projects for these and other important crops are still ongoing, and because of the complexities of their genomes, comprehensive maps are essential in order to assemble the sequences. This is, of course, a critical area of research: understanding all aspects of the biology of crops is essential for dealing with global hunger over the coming decades.

Genome maps are not just sequencing aids

Maps might have become less generally relevant as aids in assembly of genome sequences, but their value in other aspects of genomics research is undiminished. It is important to recognize that completion of the nucleotide sequence of a genome is not an end in itself. Indeed, every genome is simply a series of As, Cs, Gs, and Ts, and working out the order of these letters does not tell us much, if anything, about the way in which a genome acts as a store of biological information, or how that information is used to specify the characteristics of the species being studied. As we will see in Chapters 5 and 6, the first stage in understanding a genome sequence is to identify the genes that it contains and to assign functions to as many of these as possible. Many of the methods used to assign functions begin with a gene and ask, 'What does this gene do?', but the reverse process, in which we start with a function and ask, 'Which gene is responsible for this characteristic?', is equally important. As we will see in Section 6.4, a genome map is essential in order to answer this second question, because the approach used initially involves identifying the position of the gene being sought relative

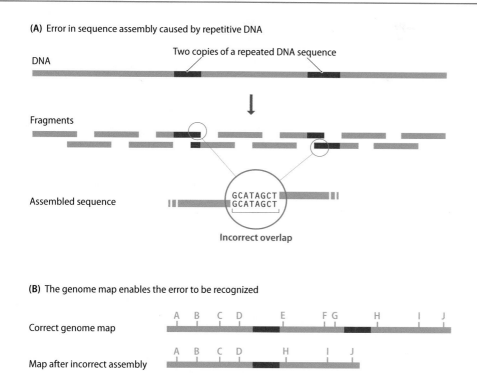

(A) Error in sequence assembly caused by repetitive DNA

DNA

Two copies of a repeated DNA sequence

Fragments

Assembled sequence

GCATAGCT
GCATAGCT

Incorrect overlap

(B) The genome map enables the error to be recognized

Correct genome map

A B C D E F G H I J

Map after incorrect assembly

A B C D H I J

Figure 3.3 A possible error in sequence assembly caused by repetitive DNA. (A) The DNA molecule contains two copies of a repeat sequence. When the shotgun sequences are examined, two fragments appear to overlap, but one fragment contains the left-hand part of one repeat, and the other fragment has the right-hand part of the second repeat. Failure to recognize this assembly error would lead to the segment of DNA between the two repeats being left out of the master sequence. If the two repeats were on different chromosomes, then the sequences of these chromosomes would mistakenly be linked together. (B) The error in sequence assembly is recognized because the relative positions of mapped features (A, B, C, etc.) in the assembled sequence do not correspond with the correct positions of these features in the genome map.

to other genes or sequence features whose locations on the map are already known. This process has been, and continues to be, the key to identification of genes responsible for human diseases such as cystic fibrosis and breast cancer. Similar methods are used to identify groups of genes, possibly spread around the genome, which do not directly cause a disease, but confer differing degrees of susceptibility to that disease. One step further on, are methods used to identify **quantitative trait loci** (**QTLs**), which are regions of a genome, each possibly containing several genes, which control variable traits such as meat productivity in farm animals and pest resistance in crop plants.

The information provided by a genome map on the locations of genes and QTLs controlling commercially important traits in crop plants is also utilized in breeding programs aimed at the development of new varieties with improved agricultural properties. These breeding programs typically generate thousands of seedlings, whose precise biological characteristics are unknown because of the randomness of the inheritance process. A seedling might combine the best features of the two parents, and potentially be an important new crop variety, or it might combine the least useful properties of the parents and be of no commercial value. Many of the traits that crop breeders are interested in are exhibited late in the life cycle of the plant – examples are seed or fruit yield – and can only be assayed by growing each seedling to maturity, which of course takes time and requires large amounts of growing space. We will see in Section 18.4 how the method called **marker-assisted selection** enables DNA screening to be used to identify those seedlings that possess a beneficial characteristic, so these can be retained and other, less interesting seedlings discarded. Marker-associated selection is possible only if a genome map is available. If a map is available, then it can be carried out with success even if the complete genome sequence is unknown, as is the case for crops such as barley and wheat.

3.2 MARKERS FOR GENETIC MAPPING

As with any type of map, a genetic map must show the positions of distinctive features. In a geographic map, these markers are recognizable components of the landscape, such as rivers, roads and buildings. What markers can we use in a genetic landscape?

Genes were the first markers to be used

The first genetic maps, constructed in the early decades of the 20th century for organisms such as the fruit fly, used genes as markers. To be useful as a **genetic marker**, a gene must exist in at least two forms, or **alleles**, each specifying a different **phenotype**, an example being tall or short stems in the pea plants originally studied by Gregor Mendel. To begin with, the only genes that could be studied were those specifying phenotypes that were distinguishable by visual examination. So, for example, the first fruit-fly maps showed the positions of genes for body color, eye color, wing shape, and suchlike, all of these phenotypes being visible simply by looking at the flies with a low-power microscope or the naked eye. This approach was fine in the early days but geneticists soon realized that there were only a limited number of visual phenotypes whose inheritance could be studied, and in many cases their analysis was complicated because a single phenotype could be affected by more than one gene. For example, by 1922, over 50 genes had been mapped onto the four fruit-fly chromosomes, but nine of these genes were for eye color. In later research, geneticists studying fruit flies had to learn to distinguish between fly eyes that were colored red, light red, vermilion, garnet, carnation, cinnabar, ruby, sepia, scarlet, pink, cardinal, claret, purple, or brown. To make gene maps more comprehensive, it was necessary to find characteristics that were more distinctive and less complex than visual ones.

The answer was to use biochemistry to distinguish phenotypes. This has been particularly important with two types of organisms – microbes and humans. Microbes, such as bacteria and yeast, have very few visual characteristics, so gene mapping with these organisms has to rely on biochemical phenotypes such as those listed in Table 3.1. For humans, it is possible to use visual characteristics but, since the 1920s, studies of human genetic variation have been based largely on biochemical phenotypes that can be scored by blood typing. These phenotypes include not only the standard blood groups, such as the ABO series, but also variants of blood serum proteins and of immunological proteins, such as the human leukocyte antigens (the HLA system). A big advantage of these markers is that many of the relevant genes have **multiple alleles**. For example, the gene called *HLA-DRB1* has over 3000 alleles and *HLA-B* has 8200. This is relevant because of the way in which gene mapping is carried out with humans (Section 3.4). Rather than setting up planned breeding experiments, which is the procedure with experimental organisms such as fruit flies or mice, data on the inheritance of human genes have to be gleaned by examining the phenotypes displayed by members of families in which the parents have come together for personal reasons rather than for the convenience of a geneticist. If all the members of a family have the same allele for the gene being studied, then no useful information can be obtained. For gene mapping purposes, it is therefore

TABLE 3.1 TYPICAL BIOCHEMICAL MARKERS USED FOR GENETIC ANALYSIS OF *SACCHAROMYCES CEREVISIAE*

Marker	Phenotype	Method by which cells carrying the marker are identified
ADE2	Requires adenine	Grows only when adenine is present in the medium
CAN1	Resistant to canavanine	Grows in the presence of canavanine
CUP1	Resistant to copper	Grows in the presence of copper
CYH1	Resistant to cycloheximide	Grows in the presence of cycloheximide
LEU2	Requires leucine	Grows only when leucine is present in the medium
SUC2	Able to ferment sucrose	Grows if sucrose is the only carbohydrate in the medium
URA3	Requires uracil	Grows only when uracil is present in the medium

necessary to find families in which the parents, by chance, have different alleles. This is much more likely if the gene being studied has 1800 rather than 2 alleles.

RFLPs and SSLPs are examples of DNA markers

Genes are very useful markers, but they are by no means ideal. One problem, especially with larger genomes, such as those of vertebrates and flowering plants, is that a map based entirely on genes is not very detailed. This would be true even if every gene could be mapped because in most eukaryotic genomes the genes are widely spaced out with large gaps between them. The problem is made worse by the fact that only a fraction of the total number of genes exist in allelic forms that can be distinguished conveniently. Gene maps are therefore not very comprehensive. We need other types of marker.

Mapped features that are not genes are called **DNA markers**. As with gene markers, a DNA marker must have at least two alleles to be useful. Two examples of DNA markers are the sequences called **restriction fragment length polymorphisms** (**RFLPs**) and **simple sequence length polymorphisms** (**SSLPs**).

RFLPs were the first type of DNA marker to be studied. Recall that restriction enzymes cut DNA molecules at specific recognition sequences (Section 2.1). This sequence specificity means that treatment of a DNA molecule with a restriction enzyme should always produce the same set of fragments. This is not always the case with genomic DNA molecules because some restriction sites are polymorphic, existing as two alleles, with one allele displaying the correct sequence for the restriction site and therefore being cut when the DNA is treated with the enzyme, and the second allele having a sequence alteration, so the restriction site is no longer recognized. The result of the sequence alteration is that the two adjacent restriction fragments remain linked together after treatment with the enzyme, leading to a length polymorphism (**Figure 3.4**). This is an RFLP, and its position on a genome map can be worked out by following the inheritance of its alleles, just as is done when genes are used as markers. There are thought to be about 10^5 RFLPs in a mammalian genome.

With small DNA molecules, the two alleles of an RFLP can be distinguished simply by cutting with the appropriate restriction enzyme and identifying the sizes of the resulting fragments in an agarose gel. Typing an RFLP in genomic DNA is more difficult. An enzyme such as *Eco*RI, with a six-nucleotide recognition sequence, should cut approximately once every $4^6 = 4096$ bp and so would give over 750,000 fragments when used with human DNA. After separation by agarose gel electrophoresis, these 750,000 fragments produce a smear of DNA. Southern hybridization, using a probe that spans the polymorphic restriction site, would therefore have to be carried out in order to visualize the fragments relevant to the RFLP (**Figure 3.5A**). This is a lengthy process, and it is difficult to examine more than about 12 DNA samples in a single experiment. RFLP typing is an example of the many procedures that have been made easier since PCR was invented. Using PCR, an RFLP can by typed in a sample of genomic DNA without the need to cut that DNA with the restriction enzyme. Instead, the primers for the PCR are designed so that they anneal on either side of the polymorphic site, and the RFLP is typed by treating the amplified fragment with the restriction enzyme (**Figure 3.5B**). Multiple PCRs can easily be set up in multiwall plates, so up to 96 DNA samples can now be typed in a single run.

SSLPs, are quite different from RFLPs. SSLPs are arrays of repeat sequences that display length variations, with different alleles containing different numbers of repeat units (**Figure 3.6A**). Unlike RFLPs, SSLPs can be multiallelic as each SSLP can have a number of different length variants. There are two types of SSLP:

- **Minisatellites**, also known as **variable number of tandem repeats** (**VNTRs**), in which the repeat unit is up to 25 bp in length.

- **Microsatellites**, or **short tandem repeats** (**STRs**), whose repeats are shorter, usually 13 bp or less.

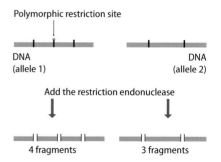

Figure 3.4 A restriction fragment length polymorphism (RFLP). The DNA molecule on the left has a polymorphic restriction site (marked with the asterisk) that is not present in the molecule on the right. The RFLP is revealed after treatment with the restriction enzyme because one of the molecules is cut into four fragments whereas the other is cut into three fragments.

Figure 3.5 Two methods for typing an RFLP. (A) RFLPs can be typed by Southern hybridization. The DNA is digested with the appropriate restriction enzyme and separated in an agarose gel. The smear of restriction fragments is transferred to a nylon membrane and probed with a piece of DNA that spans the polymorphic restriction site. If the site is absent, then a single restriction fragment is detected (lane 2); if the site is present, then two fragments are detected (lane 3). (B) The RFLP can also be typed by PCR, using primers that anneal either side of the polymorphic restriction site. After the PCR, the products are treated with the appropriate restriction enzyme and then analyzed by agarose gel electrophoresis. If the site is absent, then one band is seen on the agarose gel (lane 2); if the site is present, then two bands are seen (lane 3).

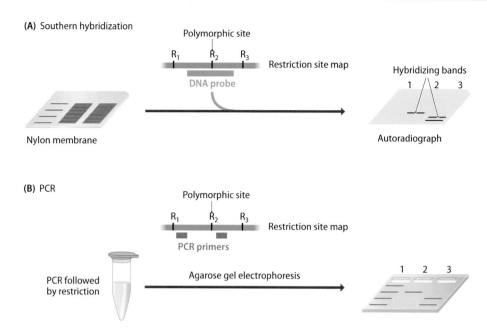

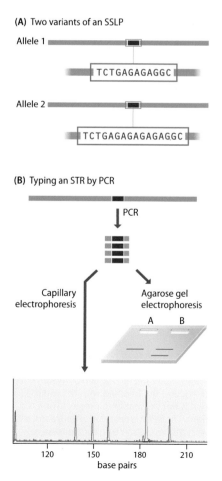

Microsatellites are more popular than minisatellites as DNA markers, for two reasons. First, minisatellites are not spread evenly around the genome but tend to be found more frequently in the telomeric regions at the ends of chromosomes. In geographic terms, this is equivalent to trying to use a map of lighthouses to find one's way around the middle of an island. Microsatellites are more conveniently spaced throughout the genome. Second, the quickest way to type a length polymorphism is by PCR, but PCR typing is much quicker and more accurate with sequences less than 300 bp in length. Most minisatellite alleles are longer than this because the repeat units are relatively large and there tend to be many of them in a single array, so PCR products several kilobases in length are needed to type them. Microsatellites used as DNA markers typically consist of 10–30 copies of a repeat that is no longer than 6 bp in length, and so are much more amenable to analysis by PCR. There are over 4 million microsatellites with repeat units of 2–6 bp in the human genome.

When examined by PCR, the allele present at a microsatellite is revealed by the precise length of the PCR product (**Figure 3.6B**). The length variations can be visualized by agarose gel electrophoresis, but standard gel electrophoresis is a cumbersome procedure that is difficult to automate, and is hence unsuitable for the high-throughput analyses that are demanded by modern genome research. Instead, microsatellites are usually typed by **capillary electrophoresis** in a polyacrylamide gel. Polyacrylamide gels have smaller pore sizes than agarose gels and allow greater precision in the separation of molecules of different lengths. Most capillary electrophoresis systems use fluorescence detection, so a fluorescent label is attached to one or both of the primers before the PCR is carried out. After PCR, the product is loaded into the capillary system and run past a fluorescence detector. A computer attached to the detector correlates the time

Figure 3.6 SSLPs and how they are typed. (A) Two alleles of an SSLP. This particular example is a short tandem repeat (STR), also called a microsatellite. In allele 1, the motif 'GA' is repeated three times, and in allele 2, it is repeated five times. (B) Typing an STR by PCR. The STR and part of the surrounding sequence is amplified, and the size of the product is determined by agarose gel electrophoresis or capillary electrophoresis. In the agarose gel, lane A contains the PCR product, and lane B contains DNA markers that show the sizes of the bands given after PCR of the two alleles. The band in lane A is the same size as the larger of the two DNA markers, showing that the DNA that was tested contained allele 2. The results of capillary electrophoresis are displayed as an electrophoretogram, the position of the blue peak indicating the size of the PCR product. The electrophoretogram is automatically calibrated against size markers (red peaks), so the precise length of the PCR product can be calculated.

of passage of the PCR product with equivalent data for a set of size markers, and hence identifies the precise length of the product.

Single-nucleotide polymorphisms are the most useful type of DNA marker

RFLPs and SSLPs are useful in some types of genomic research, but most modern genetic mapping projects make use of a different type of DNA marker. These are called **single-nucleotide polymorphisms** (**SNPs**). An SNP is a position in a genome where some individuals have one nucleotide (e.g., G), and others have a different nucleotide (e.g., C) (**Figure 3.7**). There are vast numbers of SNPs in the genomes of every species: 8.47×10^7 were discovered when 2504 human genome sequences were compared, and it has been estimated that there are $3\text{-}4 \times 10^8$ SNPs in the human population as a whole. Some SNPs also give rise to RFLPs, but many do not because the sequence in which they lie is not recognized by any restriction enzyme.

Any one of the four nucleotides could be present at any single position in a genome, so it might be imagined that each SNP should have four alleles. Theoretically, this is possible, but in practice most SNPs exist as just two variants. This is because each SNP originates when a **point mutation** (Chapter 17) occurs in a genome, converting one nucleotide into another. If the mutation occurs in the reproductive cells of an individual, then one or more of that individual's offspring might inherit the mutation, and, after many generations, the SNP may eventually become established in the population. But there are just two alleles – the original sequence and the mutated version. For a third allele to arise, a new mutation must occur at the same position in the genome in another individual, and this individual and his or her offspring must reproduce in such a way that the new allele becomes established. This scenario is not impossible, but it is unlikely: consequently, the vast majority of SNPs are biallelic. This disadvantage is more than outweighed by the huge number of SNPs present in each genome – in most eukaryotes, at least one for every 1000 bp of DNA. SNPs therefore enable very detailed genome maps to be constructed.

The frequency of SNPs in a genome means that these markers have assumed considerable importance in projects that utilize a genome map in order to identify genes or QTLs specifying particular characteristics (Section 6.4), as well as in crop breeding programs that use a map as an aid to marker-assisted selection (Section 18.4). These applications have driven the development of methods for rapid typing of individual and large sets of SNPs. Several of these typing methods are based on **oligonucleotide hybridization analysis**. An oligonucleotide is a short, single-stranded DNA molecule, usually less than 50 nucleotides in length, that is synthesized in the test tube. If the conditions are just right, then an oligonucleotide will hybridize with another DNA molecule only if the oligonucleotide forms a completely base-paired structure with the second molecule. If there is a single mismatch – a single position within the oligonucleotide that does not form a base pair – then hybridization does not occur (**Figure 3.8**). Oligonucleotide hybridization can therefore discriminate between the two alleles of an SNP. Various SNP typing strategies based on oligonucleotide hybridization have been devised, including the following:

- **DNA chip** technology makes use of a wafer of glass or silicon, 2 cm² or less in area, carrying many different oligonucleotides in a high-density array. The DNA to be tested is labeled with a fluorescent marker and pipetted onto the surface of the chip. Hybridization is detected by examining the chip with a fluorescence microscope, the positions at which the fluorescent signal is emitted indicating which oligonucleotides have hybridized with the test DNA (**Figure 3.9**). Hybridization requires a complete match between an oligonucleotide and its complementary sequence in the test DNA, and so indicates which of the two versions of a SNP are present in the test. A density of up to 300,000 oligonucleotides per cm² is possible on

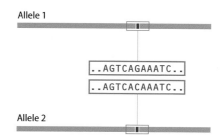

Figure 3.7 A single-nucleotide polymorphism (SNP).

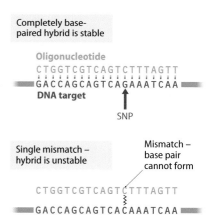

Figure 3.8 The basis for SNP typing by oligonucleotide hybridization analysis. Under highly stringent hybridization conditions, a stable hybrid occurs only if the oligonucleotide is able to form a completely base-paired structure with the target DNA. If there is a single mismatch, then the hybrid does not form. To achieve this level of stringency, the incubation temperature must be just below the **melting temperature**, or T_m, of the oligonucleotide. At temperatures above the T_m, even the fully base-paired hybrid is unstable. At more than 5°C below the T_m, mismatched hybrids might be stable. The T_m for the oligonucleotide shown in the figure would be about 58°C. The T_m in °C is calculated from the formula $T_m = (4 \times$ number of G and C nucleotides) + (2 × number of A and T nucleotides). This formula gives a rough indication of the T_m for oligonucleotides of 15–30 nucleotides in length.

Microarray

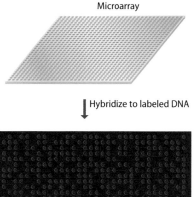

↓ Hybridize to labeled DNA

Figure 3.9 SNP typing with a DNA chip. Oligonucleotides are immobilized in an array on the surface of the chip. Labeled DNA is applied and the positions at which hybridization occurs are determined by laser scanning or fluorescence confocal microscopy.

the surface of the chip, so a chip of 2 cm^2 can type 300,000 SNPs in a single experiment, presuming the chip carries oligonucleotides for both alleles of each SNP.

- **Solution hybridization** techniques are carried out in the wells of a microtiter tray, using a detection system that can discriminate between non-hybridized, single-stranded DNA and the double-stranded product that results when an oligonucleotide hybridizes to the test DNA. The most popular detection system makes use of dye-quenching, which we met in Section 2.2 as the basis for the way in which a reporter probe is used to follow product formation during real-time PCR (see Figure 2.21). In SNP typing, the dye is attached to one end of the oligonucleotide and the quenching compound to the other end. Hybridization between the oligonucleotide and the test DNA is indicated by generation of the fluorescent signal. When used in this context, the dye-quenching technique is sometimes called **molecular beacons**.

Other typing methods make use of an oligonucleotide whose mismatch with the SNP occurs at its extreme 5′ or 3′ end. Under the appropriate conditions, an oligonucleotide of this type will hybridize to the mismatched template DNA with a short, non-base-paired 'tail' (**Figure 3.10A**). This feature is utilized in two different ways:

- The **oligonucleotide ligation assay** (**OLA**) makes use of two oligonucleotides that anneal adjacent to one another, with the 3′ end of one of these oligonucleotides positioned exactly at the SNP. This oligonucleotide will form a completely base-paired structure if one version of the SNP is present in the template DNA, and when this occurs the oligonucleotide can be ligated to its partner (**Figure 3.10B**). If the DNA being examined contains the other allele of the SNP, then the 3′ nucleotide of the test oligonucleotide will not anneal to the template and no ligation occurs. The allele is therefore typed by determining if the ligation product is synthesized. If a single SNP is being assayed, then formation of the ligation product can be identified by running the postreaction mixture in a capillary electrophoresis system, as described above for microsatellite typing.

- In the **amplification refractory mutation system**, or **ARMS test**, the test oligonucleotide is one of a pair of PCR primers. If the 3′ nucleotide of the test primer anneals to the SNP then it can be extended by *Taq* polymerase

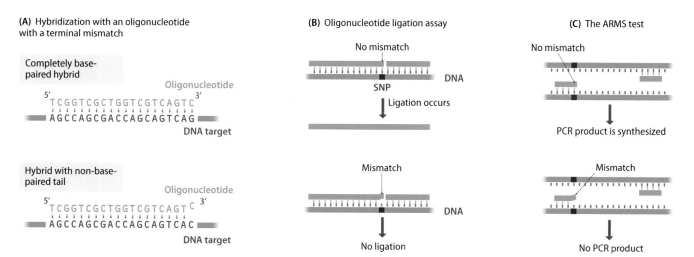

Figure 3.10 Methods for typing SNPs. (A) Under the appropriate conditions, an oligonucleotide whose mismatch with the SNP occurs at its extreme 5′ or 3′ end will hybridize to the mismatched template DNA with a short, non-base-paired tail. (B) SNP typing by the oligonucleotide ligation assay. (C) The ARMS test.

and the PCR can take place, but if it does not anneal, because the alternative version of the SNP is present, then no PCR product is generated (**Figure 3.10C**).

3.3 THE BASIS TO GENETIC MAPPING

Now that we have assembled a set of markers with which to construct a genetic map, we can move on to look at the mapping techniques themselves. The techniques for genetic mapping derive from the seminal discoveries in genetics made in the mid-19th century by Gregor Mendel, so to understand mapping we must take a journey back to Mendel's monastery garden in Brno and look at the experiments that he carried out between 1856 and 1863.

The principles of inheritance and the discovery of linkage

Genetic mapping is based on the principles of inheritance as first described by Gregor Mendel in 1865. From the results of his breeding experiments with peas, Mendel concluded that each pea plant possesses two alleles for each gene, but displays only one phenotype. This is easy to understand if the plant is pure-breeding, or **homozygous**, for a particular characteristic, as it then possesses two identical alleles and displays the appropriate phenotype (**Figure 3.11A**). However, Mendel showed that if two pure-breeding plants with different phenotypes are crossed, then all the progeny (the F$_1$ generation) display the same phenotype. These F$_1$ plants must be **heterozygous**, meaning that they possess two different alleles, one for each phenotype – one allele inherited from the mother and one from the father. Mendel postulated that in this heterozygous condition, one allele overrides the effects of the other allele: he therefore described the phenotype expressed in the F$_1$ plants as being **dominant** over the second, **recessive** phenotype (**Figure 3.11B**).

Mendel's interpretation of the heterozygous condition is perfectly correct for the pairs of alleles that he studied, but we now appreciate that this simple dominant–recessive rule can be complicated by situations that he did not encounter. These include:

- **Incomplete dominance**, where the heterozygous form displays a phenotype intermediate between the two homozygous forms. Flower color in plants such as carnations (but not peas) is an example: when red carnations are crossed with white ones, the F$_1$ heterozygotes are neither red nor white, but pink (**Figure 3.12A**).

- **Codominance**, where the heterozygous form displays both of the homozygous phenotypes. Human blood groups provide several examples of codominance. For example, the two homozygous forms of the MN series are M and N, with these individuals synthesizing the M or N blood glycoproteins, respectively. Heterozygotes, however, synthesize both glycoproteins and hence are designated MN (**Figure 3.12B**).

As well as discovering dominance and recessiveness, Mendel carried out additional experiments that enabled him to establish his two Laws of Genetics. The First Law states that *alleles segregate randomly*. In other words, if the parent's alleles are *A* and *a*, then a member of the F$_1$ generation has the same chance of inheriting *A* as it has of inheriting *a*. The Second Law is that *pairs of alleles segregate independently*, so that inheritance of the alleles of gene *A* is independent of inheritance of the alleles of gene *B*. Because of these laws, the outcomes of **genetic crosses** are predictable (**Figure 3.13**).

When Mendel's work was rediscovered in 1900, his Second Law worried the early geneticists because it was soon established that genes reside on chromosomes, and it was realized that all organisms have many more genes than chromosomes. Chromosomes are inherited as intact units, so it was reasoned that the alleles of some pairs of genes will be inherited together because they are on the same chromosome (**Figure 3.14**). This is the principle of **genetic linkage**, and it

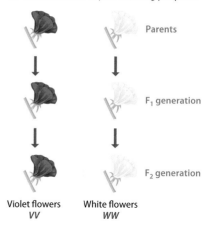

(A) Self-fertilization of pure-breeding pea plants

Parents

F$_1$ generation

F$_2$ generation

Violet flowers   White flowers
VV   *WW*

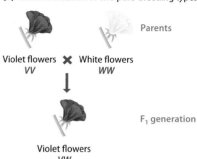

(B) Cross-fertilization of two pure-breeding types

Violet flowers ✕ White flowers
VV   *WW*

Parents

F$_1$ generation

Violet flowers
VW

Figure 3.11 Homozygosity and heterozygosity. Mendel studied seven pairs of contrasting characteristics in his pea plants, one of which was violet and white flower color, as shown here. (A) Pure-breeding plants always give rise to flowers with the parental color. These plants are homozygotes, each possessing a pair of identical alleles, denoted here by *VV* for violet flowers and *WW* for white flowers. (B) When two pure-breeding plants are crossed, only one of the phenotypes is seen in the F$_1$ generation. Mendel deduced that the genotype of the F$_1$ plants was *VW*, so *V* is the dominant allele and *W* is the recessive allele.

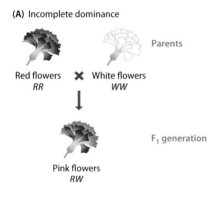

(A) Incomplete dominance

Red flowers   **✕**   White flowers
RR   *WW*

Parents

F₁ generation

Pink flowers
RW

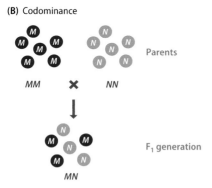

(B) Codominance

MM   **✕**   *NN*

Parents

F₁ generation

MN

Figure 3.12 Two types of allele interaction not encountered by Mendel. (A) Incomplete dominance of flower color in carnations. (B) Codominance of the *M* and *N* blood group alleles.

was quickly shown to be correct, although the results did not turn out exactly as expected. The complete linkage that had been anticipated between many pairs of genes failed to materialize. Pairs of genes were either inherited independently, as expected for genes in different chromosomes, or if they showed linkage, then it was only **partial linkage**: sometimes they were inherited together and sometimes they were not (**Figure 3.15**). The resolution of this contradiction between prediction and observation was the critical step in the development of genetic mapping techniques.

Partial linkage is explained by the behavior of chromosomes during meiosis

The critical breakthrough was achieved by Thomas Hunt Morgan, who made the conceptual leap between partial linkage and the behavior of chromosomes when the nucleus of a cell divides. Cytologists in the late 19th century had distinguished two types of nuclear division: **mitosis** and **meiosis**. Mitosis is more common, being the process by which the diploid nucleus of a somatic cell divides to produce two daughter nuclei, both of which are diploid (**Figure 3.16**). Approximately 10^{17} mitoses are needed to produce all the cells required during a human lifetime. Before mitosis begins, each chromosome in the nucleus is replicated, but the resulting daughter chromosomes do not immediately break away from one another. To begin with, they remain attached at their centromeres, and do not separate until later in mitosis when the chromosomes are distributed between the two new nuclei. Obviously, it is important that each of the new nuclei receives a complete set of chromosomes, and most of the intricacies of mitosis are devoted to achieving this end.

Mitosis illustrates the basic events occurring during nuclear division, but it is the distinctive features of meiosis that interest us. Meiosis occurs only in reproductive cells, and results in a diploid cell giving rise to four haploid **gametes**, each of which can subsequently fuse with a gamete of the opposite sex during sexual reproduction. The fact that meiosis results in four haploid cells, whereas mitosis gives rise to two diploid cells, is easy to explain: meiosis involves two nuclear divisions, one after the other, whereas mitosis is just a single nuclear division. This is an important distinction, but the critical difference between mitosis and meiosis is more subtle. Recall that in a diploid cell there are two, separate copies of each chromosome (Chapter 1). We refer to these as pairs of **homologous**

MONOHYBRID CROSS

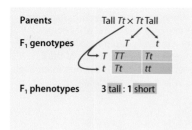

Parents	Tall *Tt* ✕ *Tt* Tall	
F₁ genotypes		
	T	*t*
T	*TT*	*Tt*
t	*Tt*	*tt*
F₁ phenotypes	3 tall : 1 short	

DIHYBRID CROSS

Parents	Round yellow *RrYy* ✕ *RrYy* Round yellow			
F₁ genotypes				
	RY	*Ry*	*rY*	*ry*
RY	*RRYY*	*RRYy*	*RrYY*	*RrYy*
Ry	*RRYy*	*RRyy*	*RrYy*	*Rryy*
rY	*RrYY*	*RrYy*	*rrYY*	*rrYy*
ry	*RrYy*	*Rryy*	*rrYY*	*rryy*
F₁ phenotypes	9 round, yellow : 3 round, green : 3 wrinkled, yellow : 1 wrinkled, green			

Figure 3.13 Mendel's Laws enable the outcome of genetic crosses to be predicted. Two of Mendel's crosses are shown with their predicted outcomes. In a **monohybrid cross**, the alleles of a single gene are followed, in this case allele *T* for tall pea plants and allele *t* for short pea plants. *T* is dominant and *t* is recessive. The grid shows the predicted genotypes and phenotypes of the F₁ generation based on Mendel's First Law, which states that alleles segregate randomly. When Mendel carried out this cross, he obtained 787 tall pea plants and 277 short plants, a ratio of 2.84 : 1. In the **dihybrid cross**, two genes are followed. The first gene determines the shape of the peas, the alleles being *R* (round, the dominant allele) and *r* (wrinkled, which is recessive), and the second specifies the color of the pea, with alleles *Y* (dominant, yellow peas) and *y* (recessive, green peas). The genotypes and phenotypes shown are those predicted by Mendel's First and Second Laws, the latter stating that pairs of alleles segregate independently. Mendel's cross gave 315 round yellow peas, 108 round green, 101 wrinkled yellow and 32 wrinkled green, a ratio of 9.84 : 3.38 : 3.12 : 1.

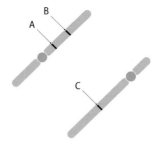

Figure 3.14 Genes on the same chromosome should display linkage. Genes A and B are on the same chromosome and so should be inherited together. Mendel's Second Law should therefore not apply to the inheritance of A and B. Gene C is on a different chromosome, so the Second Law will hold for the inheritance of A and C, or B and C. Mendel did not discover linkage because the seven genes that he studied were each on a different pea chromosome.

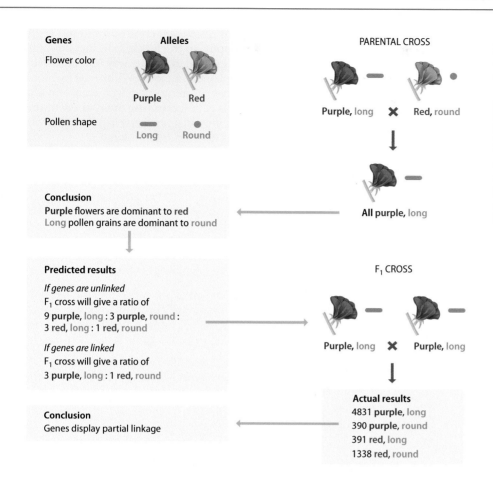

Figure 3.15 **Partial linkage.** Partial linkage was discovered in the early 20th century. The cross shown here was carried out by Bateson, Saunders, and Punnett in 1905 with sweet peas. The parental cross gives the expected result, with all the F_1 plants displaying the same phenotype, indicating that the dominant alleles are purple flowers and long pollen grains. The F_1 cross gives unexpected results, as the progeny show neither a 9:3:3:1 ratio (expected for genes on different chromosomes) nor a 3:1 ratio (expected if the genes are completely linked). An unusual ratio is typical of partial linkage.

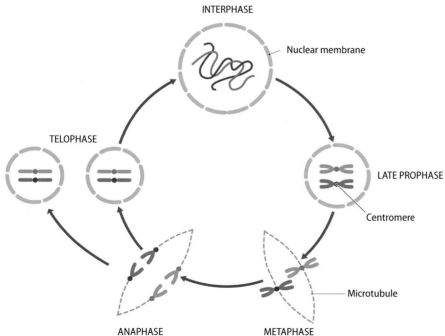

Figure 3.16 **Mitosis.** During interphase (the period between nuclear divisions), the chromosomes are in their extended form (Section 7.1). At the start of mitosis, the chromosomes condense, and by late prophase have formed structures that are visible with the light microscope. Each chromosome has already undergone DNA replication, but the two daughter chromosomes are held together by the centromere. During metaphase, the nuclear membrane breaks down (in most eukaryotes), and the chromosomes line up in the center of the cell. Microtubules now draw the daughter chromosomes toward either end of the cell. In telophase, nuclear membranes re-form around each collection of daughter chromosomes. The result is that the parent nucleus has given rise to two identical daughter nuclei. For simplicity, just one pair of homologous chromosomes is shown; one member of the pair is red, and the other is blue.

chromosomes. During mitosis, homologous chromosomes remain separate from one another, each member of the pair replicating and being passed to a daughter nucleus independently of its homolog. In meiosis, however, the pairs of homologous chromosomes are by no means independent. During the initial stage of meiosis, called prophase I, each chromosome lines up with its homolog to form a **bivalent** (**Figure 3.17**). This occurs after each chromosome has replicated, but before the replicated structures split, so the bivalent in fact contains four chromosome copies, each of which is destined to find its way into one of the four gametes that will be produced at the end of the meiosis. Within the bivalent, the chromosome arms (the **chromatids**) can undergo physical breakage and exchange of segments of DNA. The process is called **crossing over**, or **recombination**, and was discovered by the Belgian cytologist Janssens in 1909. This was just two years before Morgan started to think about partial linkage.

How did the discovery of crossing over help Morgan explain partial linkage? To understand this, we need to think about the effect that crossing over can have on the inheritance of genes. Let us consider two genes, each of which has two alleles. We will call the first gene A and its alleles A and a, and the second gene B with alleles B and b. Imagine that the two genes are located on chromosome number 2 of *Drosophila melanogaster*, the species of fruit fly studied by Morgan. We are going to follow the meiosis of a diploid nucleus in which one copy of chromosome 2 has alleles A and B, and the second has a and b. This situation is illustrated in **Figure 3.18**. Consider the two alternative scenarios:

- A crossover does not occur between genes A and B. If this is what happens, then two of the resulting gametes will contain chromosome copies with alleles A and B, and the other two will contain a and b. In other words, two of the gametes have the genotype AB and two have the genotype ab.

- A crossover does occur between genes A and B. This leads to segments of DNA containing gene A being exchanged between homologous chromosomes. The eventual result is that each gamete has a different genotype: one AB, one aB, one Ab, and one ab.

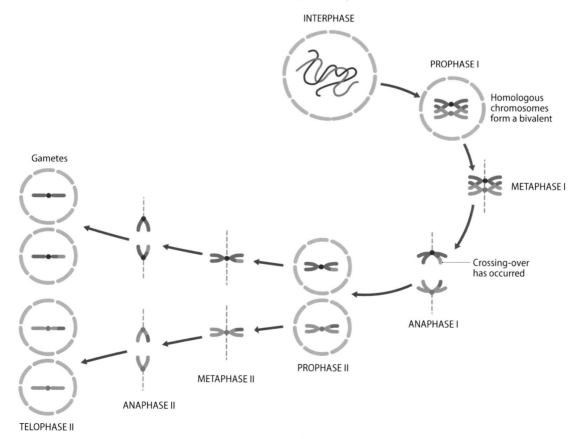

Figure 3.17 Meiosis. The events involving one pair of homologous chromosomes are shown; one member of the pair is red, and the other is blue. At the start of meiosis the chromosomes condense and each homologous pair lines up to form a bivalent. Within the bivalent, crossing over might occur, involving breakage of chromosome arms and exchange of DNA. Meiosis then proceeds by a pair of mitotic nuclear divisions that result initially in two nuclei, each with two copies of each chromosome still attached at their centromeres, and finally in four nuclei, each with a single copy of each chromosome. These final products of meiosis, the gametes, are therefore haploid.

Now think about what would happen if we looked at the results of meiosis in 100 identical cells. If crossovers never occur, then the resulting gametes will have the following genotypes:

200 *AB*
200 *ab*

This is complete linkage: genes A and B behave as a single unit during meiosis. But if (as is more likely) crossovers occur between A and B in some of the nuclei, then the allele pairs will not be inherited as single units. Let us say that crossovers occur during 40 of the 100 meioses. The following gametes will result:

160 *AB*
160 *ab*
40 *Ab*
40 *aB*

The linkage is not complete; it is only partial. As well as the two **parental genotypes** (*AB*, *ab*) we see gametes with **recombinant genotypes** (*Ab*, *aB*).

From partial linkage to genetic mapping

Once Morgan had understood how partial linkage could be explained by crossing over during meiosis, he was able to devise a way of mapping the relative positions of genes on a chromosome. In fact, the most important work was done not by Morgan himself but by an undergraduate in his laboratory, Arthur Sturtevant. Sturtevant assumed that crossing over was a random event, there being an equal chance of it occurring at any position along a pair of lined-up chromatids. If this assumption is correct, then two genes that are close together will be separated by crossovers less frequently than two genes that are more distant from one another. Furthermore, the frequency with which the genes are unlinked by crossovers will be directly proportional to how far apart they are on their chromosome. The **recombination frequency** is therefore a measure of the distance between two genes. If you work out the recombination frequencies for different pairs of genes, you can construct a map of their relative positions on the chromosome.

The first map that Sturtevant constructed showed the positions of four genes on chromosome 1 of the fruit fly (**Figure 3.19**). Morgan's group then set about mapping as many fruit-fly genes as possible and by 1915 had assigned locations for 85 of them. These genes fall into four **linkage groups**, corresponding to the four pairs of chromosomes seen in the fruit-fly nucleus. The distances between genes are expressed in **map units**, with one map unit being the distance between two genes that recombine with a frequency of 1%. According to this notation, the distance between the genes for white eyes and yellow body, which recombine with a frequency of 1.3%, is 1.3 map units (see Figure 3.19). More recently, the name **centiMorgan** (**cM**) has begun to replace the map unit. Each of the 85 genes initially mapped by Morgan specified a phenotype, such as eye color, or wing or body shape, that could be typed simply by examining the fruit flies obtained from the genetic crosses. The technique is equally efficacious with genes that are typed by biochemical tests, and with DNA markers such as RFLPs, SSLPs, and SNPs, whose alleles are identified by PCR or some other type of DNA analysis (Section 3.2). Linkage analysis can therefore be used with many different types of organism, as we will see in the next section, and the resulting maps can show positions of many different types of marker.

Before moving on, there is one final issue relating to the basic principles of linkage analysis that we must consider. It turns out that Sturtevant's assumption about the randomness of crossovers was not entirely justified. Comparisons between genetic maps and the actual positions of markers on DNA molecules, as revealed by physical mapping and DNA sequencing, have shown that some regions of chromosomes, called **recombination hotspots**, are more likely to

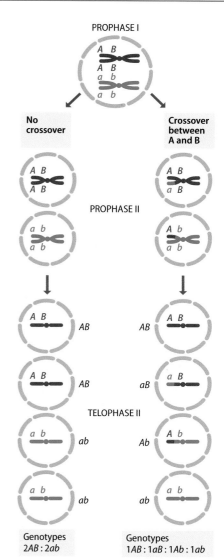

Figure 3.18 The effect of a crossover on linked genes. The drawing shows a pair of homologous chromosomes, one red and the other blue. A and B are linked genes with alleles *A*, *a*, *B*, and *b*. On the left is meiosis with no crossover between A and B: two of the resulting gametes have the genotype *AB* and the other two are *ab*. On the right, a crossover occurs between A and B: the four gametes display all of the possible genotypes – *AB*, *aB*, *Ab*, and *ab*.

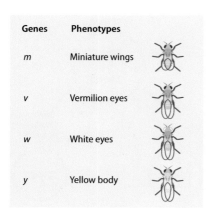

Genes	Phenotypes
m	Miniature wings
v	Vermilion eyes
w	White eyes
y	Yellow body

Recombination frequencies

Between	m	and	v	= 3.0%
Between	m	and	y	= 33.7%
Between	v	and	w	= 29.4%
Between	w	and	y	= 1.3%

Deduced map positions

y	w			v	m
0	1.3			30.7	33.7

Figure 3.19 Working out a genetic map from recombination frequencies. The example is taken from the original experiments carried out with fruit flies by Arthur Sturtevant. All four genes are on chromosome 1 of the fruit fly. Recombination frequencies between the genes are shown, along with their deduced map positions.

be involved in crossovers than others. This means that a genetic map distance does not necessarily indicate the physical distance between two markers (see Figure 3.26). Also, we now realize that a single chromatid can participate in more than one crossover at the same time, but that there are limitations on how close together these crossovers can be, leading to more inaccuracies in the mapping procedure. Despite these qualifications, **linkage analysis** usually makes correct deductions about marker order, and distance estimates are sufficiently accurate to generate genetic maps that are of value as frameworks for genome sequencing projects and for use in techniques such as marker-assisted selection. We will therefore move on to consider how linkage analysis is carried out with different types of organism.

3.4 LINKAGE ANALYSIS WITH DIFFERENT TYPES OF ORGANISM

To see how linkage analysis is actually carried out, we need to consider three quite different situations:

- Linkage analysis with species such as fruit flies and mice, with which we can carry out planned breeding experiments.

- Linkage analysis with humans, with whom we cannot carry out planned experiments but instead can make use of family pedigrees.

- Linkage analysis with bacteria, which do not undergo meiosis.

Linkage analysis when planned breeding experiments are possible

The first type of linkage analysis that we will study is the modern counterpart of the method developed by Morgan and his colleagues. The method is based on analysis of the progeny of experimental crosses set up between parents of known genotypes and is, at least in theory, applicable to all eukaryotes. Ethical considerations preclude this approach with humans, and practical problems such as the length of the gestation period and the time taken for the newborn to reach maturity (and hence to participate in subsequent crosses) limit the effectiveness of the method with some animals and plants.

If we return to Figure 3.18, we see that the key to genetic mapping is being able to determine the genotypes of the gametes resulting from meiosis. In a few situations, this is possible by directly examining the gametes. For example, the gametes produced by some microbial eukaryotes, including the yeast *Saccharomyces cerevisiae*, can be grown into colonies of haploid cells. The genotypes of these haploid colonies can then be identified by biochemical tests and by DNA marker typing. Direct genotyping of gametes is also possible with higher eukaryotes if DNA markers are used, as PCR can be carried out with the DNA from individual spermatozoa, enabling RFLPs, SSLPs, and SNPs to be typed. Unfortunately, sperm typing is laborious. Routine linkage analysis with higher eukaryotes is therefore carried out not by examining the gametes directly but by determining the genotypes of the diploid progeny that result from fusion of two gametes, one from each of a pair of parents. In other words, a **genetic cross** is performed.

The complication with a genetic cross is that the resulting diploid progeny are the product not of one meiosis but of two (one in each parent), and in most organisms crossover events are equally likely to occur during production of the male and female gametes. Somehow we have to be able to disentangle from the genotypes of the diploid progeny the crossover events that occurred in each of these two meioses. This means that the cross has to be set up with care. The standard procedure is to use a **test cross**. This is illustrated in **Figure 3.20**, where we have set up a test cross to map the two markers we met earlier: A (alleles *A* and *a*) and B (alleles *B* and *b*), both on chromosome 2 of the fruit fly. The critical feature of a test cross is the genotypes of the two parents:

- One parent is a **double heterozygote**. This means that all four alleles are present in this parent: its genotype is *AB/ab*. This notation indicates that one of the pair of homologous chromosomes has alleles *A* and *B*, and the other has *a* and *b*. Double heterozygotes can be obtained by crossing two pure-breeding strains, for example, *AB/AB × ab/ab*.

- The second parent is a pure-breeding **double homozygote**. In this parent, both homologous copies of chromosome 2 are the same: in the example shown in Figure 3.20, both have alleles *a* and *b* and the genotype of the parent is *ab/ab*.

The double heterozygote has the same genotype as the cell whose meiosis we followed in Figure 3.18. Our objective is therefore to infer the genotypes of the gametes produced by this parent and to calculate the fraction that are **recombinants**. Note that all the gametes produced by the second parent (the double homozygote) will have the genotype *ab* regardless of whether they are parental or recombinant gametes. Alleles *a* and *b* are both recessive, so meiosis in this parent is, in effect, invisible when the phenotypes of the progeny are examined. This means that, as shown in Figure 3.20, the phenotypes of the diploid progeny can be unambiguously converted into the genotypes of the gametes from the double heterozygous parent. The test cross therefore enables us to make a direct examination of a single meiosis and hence to calculate a recombination frequency and map distance for the two markers being studied.

The power of this type of linkage analysis is enhanced if more than two markers are followed in a single cross. Not only does this generate recombination frequencies more quickly, it also enables the relative order of markers on a chromosome to be determined by simple inspection of the data. This is because two recombination events are required to unlink the central marker from the two outer markers in a series of three, whereas either of the two outer markers can be unlinked by just a single recombination (**Figure 3.21**). A double recombination is less likely than a single one, so unlinking of the central marker will occur relatively infrequently. A set of typical data from a **three-point cross** is shown in Table 3.2. A test cross has been set up between a triple heterozygote (*ABC/abc*) and a triple homozygote (*abc/abc*). The most frequent progeny are those with one of the two parental genotypes, resulting from an absence of recombination events in the region containing the markers A, B, and C. Two other classes of progeny are relatively frequent (51 and 63 progeny in the example shown). Both of these are presumed to arise from a single recombination. Inspection of their genotypes shows that in the first of these two classes, marker A has become unlinked from B and C, and in the second class marker B has become unlinked from A and C. The implication is that A and B are the outer markers. This is confirmed by the number of progeny in which marker C has become unlinked from A and B. There are only two of these, showing that a double recombination is needed to produce this genotype. Marker C is therefore between A and B.

Just one additional point needs to be considered. If, as in Figure 3.20 and Table 3.2, markers whose alleles display dominance and recessiveness are

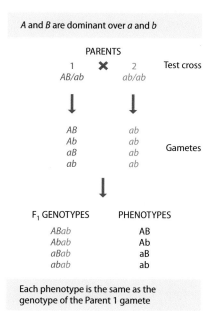

A and B are dominant over a and b

Figure 3.20 **A test cross between alleles displaying dominance and recessiveness.** A and B are markers with alleles *A*, *a*, *B*, and *b*. The resulting progeny are typed by examining their phenotypes. Because the double homozygous parent (Parent 2) has both recessive alleles – *a* and *b* – it effectively makes no contribution to the phenotypes of the progeny. The phenotype of each individual in the F_1 generation is therefore the same as genotype of the gamete from Parent 1 that gave rise to that individual.

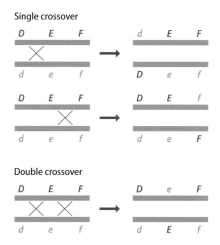

Figure 3.21 **The effects of crossovers during a trihybrid cross.** Either of the two outer markers can be unlinked by just a single recombination event, but two recombinations are required to unlink the central marker from the two outer markers.

TABLE 3.2 SET OF TYPICAL DATA FROM A THREE-POINT TEST CROSS		
Genotypes of progeny	**Number of progeny**	**Inferred recombination events**
ABC/abc or *abc/abc*	987	None (parental genotypes)
aBC/abc or *Abc/abc*	51	One, between A and B/C
AbC/abc or *aBc/abc*	63	One, between B and A/C
ABc/abc or *abC/abc*	2	Two, one between C and A and one between C and B

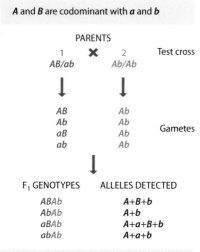

A and *B* are codominant with *a* and *b*

Figure 3.22 A test cross between alleles displaying codominance. A and B are markers whose allele pairs are codominant. In this particular example, the double homozygous parent has the genotype *Ab/Ab*. The alleles present in each F$_1$ individual are directly detected, for example, by PCR. These allele combinations enable the genotype of the Parent 1 gamete that gave rise to each individual to be deduced.

examined in a test cross, then the double or triple homozygous parent must have alleles for the recessive phenotypes. If, on the other hand, codominant markers are used, then the double homozygous parent can have any combination of homozygous alleles (i.e., *AB/AB*, *Ab/Ab*, *aB/aB*, or *ab/ab*). **Figure 3.22**, which gives an example of this type of test cross, shows the reason for this. Note that DNA markers typed by PCR display what is, in effect, codominance: Figure 3.22 therefore shows a typical scenario encountered when linkage analysis is being carried out with DNA markers.

Gene mapping by human pedigree analysis

With humans, it is, of course, impossible to preselect the genotypes of parents and set up crosses designed specifically for mapping purposes. Instead, data for the calculation of recombination frequencies have to be obtained by examining the genotypes of the members of successive generations of existing families. This method is called **pedigree analysis**. Usually, only limited data are available, and their interpretation is often difficult because a human pairing rarely results in a convenient test cross, and often the genotypes of one or more family members are unobtainable because those individuals are dead or unwilling to cooperate.

The problems are illustrated in **Figure 3.23**. In this example, we are studying a genetic disease present in a family of two parents and six children. Genetic diseases are frequently used as gene markers in humans, the disease state being one allele and the healthy state being a second allele. The pedigree in **Figure 3.23A**

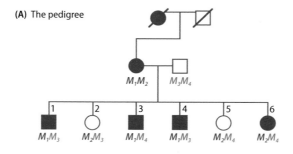

(A) The pedigree

Figure 3.23 An example of human pedigree analysis. (A) The pedigree shows inheritance of a genetic disease in a family of two living parents and six children, with information about the maternal grandparents available from family records. The disease allele (closed symbols) is dominant over the healthy allele (open symbols). The objective is to determine the degree of linkage between the disease gene and the microsatellite M by typing the alleles for this microsatellite (*M$_1$*, *M$_2$*, etc.) in living members of the family. (B) The pedigree can be interpreted in two different ways: Hypothesis 1 gives a low recombination frequency and indicates that the disease gene is tightly linked to microsatellite M. Hypothesis 2 suggests that the disease gene and microsatellite are much less closely linked. In (C), the issue is resolved by the reappearance of the maternal grandmother, whose microsatellite genotype is consistent only with Hypothesis 1.

(B) Possible interpretations of the pedigree

MOTHER'S CHROMOSOMES

		Hypothesis 1	**Hypothesis 2**
		Disease M$_1$	*Healthy M$_1$*
		Healthy M$_2$	*Disease M$_2$*
Child 1	*Disease M$_1$*	Parental	Recombinant
Child 2	*Healthy M$_2$*	Parental	Recombinant
Child 3	*Disease M$_1$*	Parental	Recombinant
Child 4	*Disease M$_1$*	Parental	Recombinant
Child 5	*Healthy M$_2$*	Parental	Recombinant
Child 6	*Disease M$_2$*	Recombinant	Parental
Recombination frequency		1/6 = 16.7%	5/6 = 83.3%

(C) Resurrection of the maternal grandmother

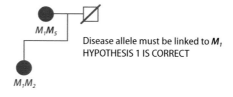

Disease allele must be linked to *M$_1$*
HYPOTHESIS 1 IS CORRECT

KEY

◯ Unaffected female
● Affected female
□ Unaffected male
■ Affected male
╱ Dead

shows us that the mother is affected by the disease, as are four of her children. We know from family accounts that the maternal grandmother also suffered from this disease, but both she and her husband – the maternal grandfather – are now dead. We can include them in the pedigree, with slashes indicating that they are dead, but we cannot obtain any further information on their genotypes. We know that the disease gene is present on the same chromosome as a microsatellite, which we call M, four alleles of which – M_1, M_2, M_3, and M_4 – are present in the living family members. Our aim is to map the position of the disease gene relative to the microsatellite.

To establish a recombination frequency between the disease gene and microsatellite M, we must determine how many of the children are recombinants. If we look at the genotypes of the six children, we see that numbers 1, 3, and 4 have the *Disease* allele and the microsatellite allele M_1. Numbers 2 and 5 have the *Healthy* allele and M_2. We can therefore construct two alternative hypotheses. The first is that the two copies of the relevant homologous chromosomes in the mother have the genotypes *Disease–M_1* and *Healthy–M_2*. Therefore, children 1, 2, 3, 4, and 5 have parental genotypes, and child 6 is the one and only recombinant (**Figure 3.23B**). This would suggest that the disease gene and the microsatellite are relatively closely linked and that crossovers between them occur infrequently. The alternative hypothesis is that the mother's chromosomes have the genotypes *Healthy–M_1* and *Disease–M_2*. This would mean that children 1–5 are recombinants, child 6 has the parental genotype, and the gene and microsatellite are relatively far apart on the chromosome. We cannot determine which of these hypotheses is correct: the data are frustratingly ambiguous.

The most satisfying solution to the problem posed by the pedigree in Figure 3.23 would be to know the genotype of the grandmother. Let us pretend that this is a soap-opera family and that the grandmother is not really dead. To everyone's surprise, she reappears just in time to save the declining audience ratings. Her genotype for microsatellite M turns out to be M_1M_5 (**Figure 3.23C**). This tells us that the chromosome inherited by the mother has the genotype *Disease–M_1*. We can therefore conclude with certainty that Hypothesis 1 is correct and that only child 6 is a recombinant.

Resurrection of key individuals is not usually an option open to real-life geneticists, although DNA can be obtained from old pathology specimens such as slides and Guthrie cards, the latter of which contain blood samples from newborn children. Imperfect pedigrees are analyzed statistically, using a measure called the **lod score**. This stands for *l*ogarithm of the *od*ds that the genes are linked, and is used primarily to determine if the two markers being studied lie on the same chromosome – in other words, if the genes are linked or not. A lod score of three or more corresponds to odds of 1000 : 1 and is usually taken as the minimum for confidently concluding that this is the case. If the lod analysis establishes linkage, then additional lod scores can be calculated for each of a range of recombination frequencies, in order to identify the frequency most likely to have given rise to the data obtained by pedigree analysis. Ideally, the available data will derive from more than one pedigree, increasing confidence in the result. The analysis is less ambiguous for families with larger numbers of children, and as we saw in Figure 3.23, it is important that the members of at least three generations can be genotyped. For this reason, family collections have been established, such as the one maintained by the Centre d'Études du Polymorphisme Humaine (CEPH), Fondation Jean Dausset, in Paris. The CEPH collection contains cultured cell lines from families in which all four grandparents as well as at least eight second-generation children could be sampled. This collection is available for DNA marker mapping by any researcher who agrees to submit the resulting data to the central CEPH database.

Genetic mapping in bacteria

The final type of genetic mapping that we must consider is the strategy used with bacteria. The main difficulty that geneticists faced when trying to develop genetic mapping techniques for bacteria is that these organisms are normally haploid, and so do not undergo meiosis. Some other way therefore had to be devised to induce crossovers between homologous segments of bacterial DNA. The answer was to make use of three natural methods that exist for transferring pieces of DNA from one bacterium to another (**Figure 3.24**):

- In **conjugation**, two bacteria come into physical contact and one bacterium (the donor) transfers DNA to the second bacterium (the recipient). The transferred DNA can be a copy of some or possibly all of the donor cell's chromosome, or it could be a segment of chromosomal DNA – up to 1 Mb in length – integrated into a plasmid. The latter is called **episome transfer**.

- **Transduction** involves transfer of a small segment of DNA – up to 50 kb or so – from donor to recipient via a bacteriophage.

- In **transformation**, the recipient cell takes up from its environment a fragment of DNA, rarely longer than 50 kb, released from a donor cell.

Biochemical markers are often used, the dominant or **wild-type** phenotype being possession of a biochemical characteristic (e.g., ability to synthesize tryptophan) and the recessive phenotype being the complementary characteristic (e.g., inability to synthesize tryptophan). The DNA transfer is usually set up

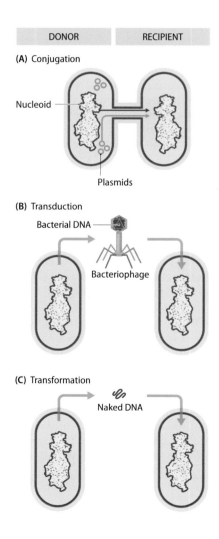

Figure 3.24 Three ways of achieving DNA transfer between bacteria. (A) Conjugation can result in transfer of chromosomal or plasmid DNA from the donor bacterium to the recipient. Conjugation involves physical contact between the two bacteria, with transfer thought to occur through a narrow tube called the **pilus**. (B) Transduction is the transfer of a small segment of the donor cell's DNA via a bacteriophage. (C) Transformation is similar to transduction, but 'naked' DNA is transferred. The events illustrated in (B) and (C) are often accompanied by death of the donor cell. In (B), death occurs when the bacteriophages emerge from the donor cell. In (C), release of DNA from the donor cell is usually a consequence of the cell's death through natural causes.

between a donor strain that possesses the wild-type allele and a recipient with the recessive allele, transfer into the recipient strain being monitored by looking for acquisition of the biochemical function specified by the gene being studied. This is illustrated in **Figure 3.25A**, where we see a functional gene for tryptophan biosynthesis being transferred from a wild-type bacterium (genotype described as *trp*⁺) to a recipient that lacks a functional copy of this gene (*trp*⁻). The recipient is called a tryptophan **auxotroph**, this being the term used to describe a mutant bacterium that can survive only if provided with a nutrient – in this case, trypto-phan – not required by the wild type. After transfer, two crossovers are needed to integrate the transferred gene into the recipient cell's chromosome, converting the recipient from *trp*⁻ to *trp*⁺.

The precise details of the mapping procedure depend on the type of gene transfer that is being used. During conjugation, DNA is transferred from donor to recipient in the same way that a string is pulled through a tube. The relative positions of markers on the DNA molecule can therefore be mapped by determining the times at which the markers appear in the recipi-ent cell. In the example shown in **Figure 3.25B**, markers A, B, and C are transferred 8, 20, and 30 minutes after the beginning of conjugation, respec-tively. The entire *Escherichia coli* chromosome takes approximately 100 minutes to transfer. In contrast, transduction and transformation mapping enable markers that are relatively close together to be mapped, because the transferred DNA segment is short (<50 kb), so the probability of two markers being transferred together depends on how close they are on the bacterial chromosome (**Figure 3.25C**).

The limitations of linkage analysis

As we have seen, the data for linkage analysis are obtained by experimental uti-lization of the natural biological processes that govern inheritance of markers during meiosis and the transfer of DNA between non-meiotic species such as bacteria. It is difficult, if not impossible, to modify these processes to improve the quality of the resulting data. Meiosis, for example, is a complex cellular pathway (see Figure 3.17), and there is a little that a geneticist can do to alter the pathway in a particular organism in order to increase the accuracy or degree of detail in a genetic map. This means that linkage analysis has natural

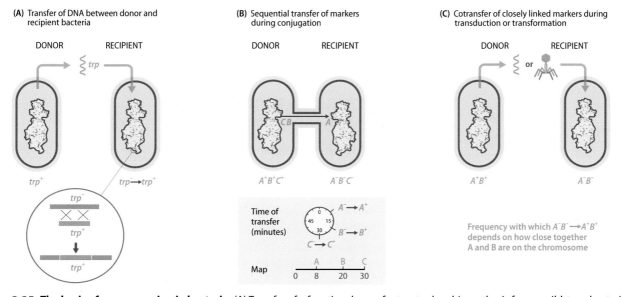

Figure 3.25 The basis of gene mapping in bacteria. (A) Transfer of a functional gene for tryptophan biosynthesis from a wild-type bacterium (genotype described as *trp*⁺) to a recipient that lacks a functional copy of this gene (*trp*⁻). (B) Mapping by conjugation. (C) Mapping by transduc-tion and transformation.

limitations that influence the utility of the resulting map. The two most important limitations are:

- The resolution of a genetic map depends on the number of crossovers that have been scored. This is not a major problem for microorganisms because these can be obtained in huge numbers, enabling many crossovers to be studied, resulting in a highly detailed genetic map in which the markers are just a few kilobases apart. For example, when the *Escherichia coli* genome sequencing project began in 1990, the latest genetic map for this organism comprised over 1400 markers, an average of 1 per 3.3 kb. This was sufficiently detailed to ensure that the genome sequence was assembled correctly. Similarly, the *Saccharomyces cerevisiae* project was supported by a fine-scale genetic map (approximately 1150 markers, on average 1 per 10 kb). The problem with humans and most other eukaryotes is that it is simply not possible to obtain large numbers of progeny, so relatively few meioses can be studied and the resolving power of linkage analysis is restricted. This means that markers that are several tens of kilobases apart may appear at the same position on the genetic map.

- Genetic maps have limited accuracy. We touched on this point in Section 3.3 when we assessed Sturtevant's assumption that crossovers occur at random along chromosomes. This assumption is only partly correct because the presence of recombination hotspots means that crossovers are more likely to occur at some points rather than at others. The effect that this can have on the accuracy of a genetic map was illustrated in 1992 when the sequence of *Saccharomyces cerevisiae* chromosome III was published. This was the first complete eukaryotic chromosome sequence to be obtained, and so enabled the first direct comparison to be made between a genetic map and the actual positions of markers as shown by DNA sequencing (**Figure 3.26**). There were considerable discrepancies, even to the extent that one pair of genes had been ordered incorrectly by genetic analysis. Bear in mind that *S. cerevisiae* is one of the two eukaryotes (fruit fly is the second) whose genomes have been subjected to intensive genetic mapping. If the yeast genetic map is inaccurate, then how precise are the genetic maps of organisms subjected to less-detailed analysis?

Because of these limitations, other ways of mapping markers on to chromosomes have been developed, with the aim of creating maps that have a greater density of markers with those markers more accurately positioned. These alternative methods do not make use of linkage analysis and hence are not based on conventional genetic techniques. The methods, which collectively make the approach called physical mapping, are described in the next two sections.

3.5 PHYSICAL MAPPING BY DIRECT EXAMINATION OF DNA MOLECULES

A plethora of physical mapping techniques have been developed, but these can conveniently be grouped into two categories depending on the approach used to identify marker locations:

- Methods that involve direct examination of DNA molecules or chromosomes.

- Methods that assign markers to DNA fragments whose positions within an intact DNA molecule are known or can be inferred.

The simplest of the direct examination methods are those that enable the positions of restriction sites to be located in a DNA molecule. This process is called **restriction mapping**.

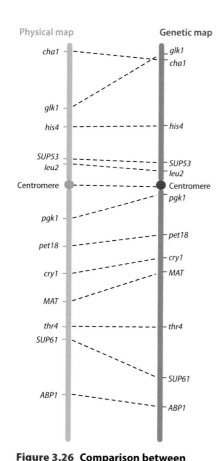

Figure 3.26 Comparison between the genetic and physical maps of *Saccharomyces cerevisiae* chromosome III. The comparison shows the discrepancies between the genetic and physical maps, the latter determined by DNA sequencing. Note that the order of the upper two markers (*glk1* and *cha1*) is incorrect on the genetic map, and that there are also differences in the relative positioning of other pairs of markers.

Conventional restriction mapping is only applicable to small DNA molecules

Genetic mapping using RFLPs as DNA markers can locate the positions of polymorphic restriction sites within a genome (Section 3.2), but very few restriction sites are polymorphic, so many sites are not mapped by this technique (**Figure 3.27**). Could we increase the marker density on a genome map by using an alternative method to locate the positions of some of the nonpolymorphic restriction sites? This is what restriction mapping achieves.

The simplest way to construct a restriction map is to compare the fragment sizes produced when a DNA molecule is digested with two different restriction enzymes that recognize different target sequences. An example using the restriction enzymes *Eco*RI and *Bam*HI is shown in **Figure 3.28**. This example illustrates the conventional approach to restriction mapping of a small DNA molecule, which involves carrying out three types of digest:

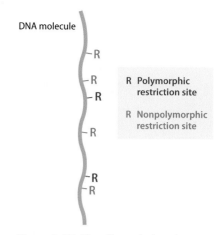

Figure 3.27 Not all restriction sites are polymorphic.

- First, the DNA molecule is digested with just one of the enzymes and the sizes of the resulting fragments are measured by agarose gel electrophoresis. Next, the molecule is digested with the second enzyme and the resulting fragments again sized in an agarose gel. The results so far enable the number of restriction sites for each enzyme to be worked out, but do not allow their relative positions to be determined.

- The DNA molecule is then cut with both enzymes together. In the example shown in Figure 3.28, the fragment sizes produced by this **double restriction** enable three of the four restriction sites to be mapped.

- A **partial restriction** is carried out, in which a single enzyme is used, but the digestion does not to go to completion, because the reaction is incubated for only a short time or a suboptimal incubation temperature is used. Partial restriction leads to a complex set of products, the complete restriction products now being supplemented with partially restricted fragments that still contain one or more uncut restriction sites. The sizes of the partially restricted fragments enable the map to be completed.

The method described above will yield an unambiguous map if there are relatively few cut sites for the enzymes being used. However, as the number of cut sites increases, so also do the numbers of single-, double-, and partial-restriction products whose sizes must be measured and compared in order for the map to be constructed. Computer analysis can be brought into play, but problems still eventually arise. A stage will be reached when a digest contains so many fragments that individual bands merge on the agarose gel, increasing the chances of one or more fragments being measured incorrectly or missed out entirely. If several fragments have similar sizes then even if they can all be identified, it may not be possible to assemble them into an unambiguous map. The conventional approach to restriction mapping is therefore more applicable to small rather than large molecules, with the upper limit depending on the frequency of the restriction sites in the molecule being mapped. In practice, if a DNA molecule is less than 50 kb in length, it is usually possible to construct a restriction map for a selection of enzymes with six-nucleotide recognition sequences. Fifty kilobases is way below the minimum size for bacterial or eukaryotic chromosomes, although it does cover a few viral and organelle genomes, and whole-genome restriction maps constructed in this way have indeed been important in directing sequencing projects with these small molecules. The approach is equally useful after bacterial or eukaryotic genomic DNA has been cloned, if the cloned fragments are less than 50 kb in length.

Optical mapping can locate restriction sites in longer DNA molecules

It is also possible to use methods other than electrophoresis to map restriction sites in DNA molecules. With the technique called **optical mapping**, restriction sites are located simply by observing the cut DNA molecules with a microscope.

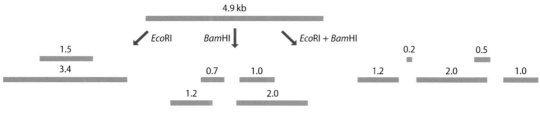

INTERPRETATION OF THE DOUBLE RESTRICTION

Fragments	Conclusions
0.2 kb, 0.5 kb	These must derive from the 0.7 kb *Bam*HI fragment, which therefore has an internal *Eco*RI site:

1.0 kb	This must be a *Bam*HI fragment with no internal *Eco*RI site. We can account for the 1.5 kb *Eco*RI fragment if we place the 1.0 kb fragment thus:

1.2 kb, 2.0 kb	These must also be *Bam*HI fragments with no internal *Eco*RI sites. They must lie within the 3.4 kb *Eco*RI fragment. There are two possibilities:

PREDICTED RESULTS OF A PARTIAL *Bam*HI RESTRICTION

If Map I is correct, then the partial restriction products will include a fragment of 1.2+0.7=1.9 kb
If Map II is correct, then the partial restriction products will include a fragment of 2.0+0.7=2.7 kb

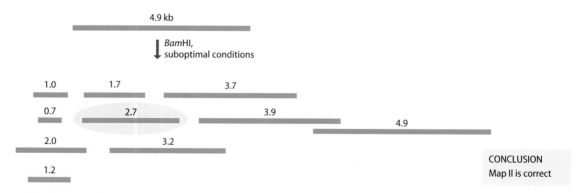

CONCLUSION
Map II is correct

Figure 3.28 The conventional approach to restriction mapping. The objective is to map the *Eco*RI (E) and *Bam*HI (B) sites in a linear DNA molecule of size 4.9 kb. The results of single and double restrictions are shown at the top. The sizes of the fragments given after double restriction enable two alternative maps to be constructed, as explained in the central panel, the unresolved issue being the position of one of the three *Bam*HI sites. The two maps are tested by a partial *Bam*HI restriction (bottom), which shows that Map II is the correct one.

DNA in solution takes up a random coil configuration, with groups of molecules tending to clump together in masses. The key to optical mapping is therefore the ability to extend individual DNA molecules into a linear configuration so that the locations of the cuts made by a restriction enzyme, as visualized by observation of the molecule, accurately reflect the positions of the restriction sites in the DNA sequence. If parts of the DNA molecule remain as a random coil, or are not fully extended, then working out the actual distances between the restriction sites will be much more difficult. In the earliest form of optical mapping, the molecules were extended by a process called **gel stretching**. Chromosomal DNA was suspended in molten agarose and placed on a microscope slide held at a slight angle so that the agarose flowed slowly along the slide as it cooled and solidified. Under these conditions, the DNA molecules contained within the agarose line up and become extended (**Figure 3.29A**). The gel also contains the restriction enzyme, which can be activated by adding magnesium ions – all restriction enzymes require magnesium in order to work. The molecules are then visualized by adding a fluorescent dye, such as DAPI (4,6-diamino-2-phenylindole dihydrochloride), which stains the DNA so that the fibers can be seen when the slide is examined with a high-power fluorescence microscope. The restriction sites in the extended molecules gradually become gaps as the degree of fiber extension is reduced by the natural springiness of the DNA, enabling the relative positions of the cuts to be recorded.

Gel stretching is relatively easy to carry out, but the distortions inherent in viewing the DNA fibers while they are contained in a gel droplet limits the degree of resolution that can be achieved with this approach. An alternative method for stretching out the molecules without the use of a gel is **molecular combing**. A silicone-coated cover slip is dipped into a solution of DNA, left for 5 minutes (during which time the DNA molecules attach to the cover slip by their ends), and then removed from the solution at a constant speed, typically 0.3 mm s^{-1} (**Figure 3.29B**). The force required to pull the DNA molecules through the meniscus causes them to line up. Once in the air, the surface of the cover slip dries, retaining the DNA molecules as an array of parallel fibers. With this method, restriction sites less than 800 bp apart can be visualized.

Optical mapping was first applied to large DNA fragments cloned in BAC vectors (Section 2.3). The feasibility of using this technique with genomic DNA was then established with studies of a 1 Mb chromosome of the malaria parasite

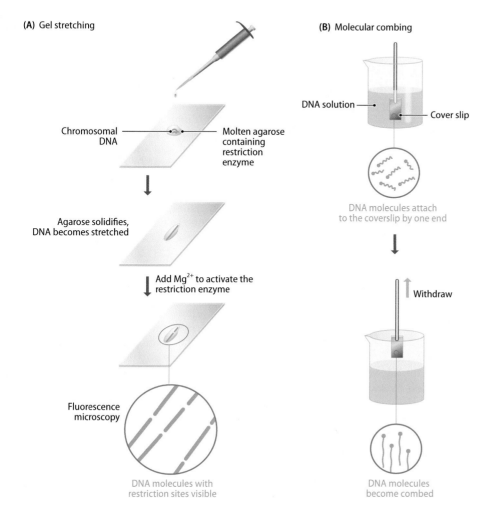

(A) Gel stretching

Chromosomal DNA — Molten agarose containing restriction enzyme

Agarose solidifies, DNA becomes stretched

Add Mg^{2+} to activate the restriction enzyme

Fluorescence microscopy

DNA molecules with restriction sites visible

(B) Molecular combing

DNA solution — Cover slip

DNA molecules attach to the coverslip by one end

Withdraw

DNA molecules become combed

Figure 3.29 Gel stretching and molecular combing. (A) To carry out gel stretching, molten agarose containing chromosomal DNA molecules is pipetted onto a microscope slide held at a slight angle. As the gel flows and solidifies, the DNA molecules become stretched. Addition of magnesium chloride activates the restriction enzyme contained in the gel, which cuts the DNA molecules. As the molecules gradually coil up, the gaps representing the cut sites become visible. (B) In molecular combing, a cover slip is dipped into a solution of DNA. The DNA molecules attach to the cover slip by their ends, and the slip is withdrawn from the solution at a rate of 0.3 mm s^{-1}, which produces a 'comb' of parallel molecules.

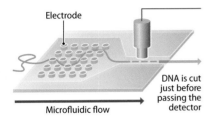

Figure 3.30 A microfluidic device for optical mapping of restriction sites. The DNA molecule becomes partially extended by passage through the grid of electrodes, and is then fully extended when it enters the nanochannel, which is only slightly wider than the double helix. The DNA is cut within the nanochannel, in which there is a magnesium ion gradient.

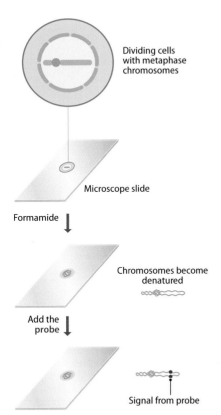

Figure 3.31 Fluorescent *in situ* hybridization. A sample of dividing cells is dried onto a microscope slide and treated with formamide so that the chromosomes become denatured but do not lose their characteristic metaphase morphologies (see Section 7.1). The position at which the probe hybridizes with the chromosomal DNA is visualized by detecting the fluorescent signal emitted by the labeled DNA.

Plasmodium falciparum, and the two chromosomes and the larger of the two plasmids of the bacterium *Deinococcus radiodurans*, which are 2.65, 0.41 and 0.18 Mb, respectively (see Table 8.2). Molecules over 1 Mb in length are difficult to purify and extend without accidental breakage, so most optical maps are built up from the data obtained from a series of overlapping fragments. The 2.65 Mb *D. radiodurans* chromosome, for example, was mapped from 157 fragments. This means that the procedure is labor-intensive as many separate observations have to be made, and the amount of labor increases disproportionately as the length of the starting molecule increases. Recent work has therefore focused on automating the procedure so that restriction sites can be mapped in many fragments in a high-throughput manner. These automated procedures make use of microfluidic devices that extend the molecules and then move them, one by one, past an optical detector. In some systems, a version of molecular combing is used to extend the molecules, but in others the molecules are partially extended by movement through a grid of electrodes and are then fully extended by being pushed, by the solvent flow, into a series of nanochannels that are only just wide enough for the linear molecules to squeeze through (**Figure 3.30**). Of course, this method will only work if cleavage of the DNA fragment by the restriction enzyme is delayed until the fragment enters the nanochannels. One way of achieving this outcome is to design the microfluidic architecture in such a way that a magnesium ion gradient is established within each nanochannel, so that the restriction enzyme is only activated when it enters the channel along with the DNA fragment. The restriction sites are therefore cut within the nanochannels, and the resulting gaps in the DNA fragment are immediately recorded by the detection system. These automated approaches for data generation, along with computer analysis of the resulting data, have greatly extended the scope of optical mapping, with maps of this type now available for a range of plant and animal genomes.

Optical mapping with fluorescent probes

The realization during the 2000s that observation of extended DNA molecules is a feasible method for mapping restriction sites has led to the development of innovative versions of optical mapping, ones that enable markers other than restriction sites to be mapped. These modifications to optical mapping derive, in part, from parallel developments that were occurring in use of a second technique for physical mapping of DNA molecules, called **fluorescent *in situ* hybridization** (**FISH**).

As in optical mapping, FISH enables the position of a marker on a chromosome or extended DNA molecule to be directly visualized. The difference is that, with FISH, the marker is a DNA sequence contained in the DNA molecule, whose location is visualized by hybridization with a fluorescent DNA probe that is complementary to and hence binds to the marker sequence (**Figure 3.31**). The technique was first used in the 1980s with **metaphase chromosomes** (Section 7.1). These chromosomes, prepared from nuclei that are undergoing division, are highly condensed, each chromosome in a set taking up a recognizable appearance, characterized by the position of its centromere and the banding pattern that emerges after the chromosome preparation is stained (see Figure 7.5). This type of FISH can therefore identify the position of a marker relative to the centromere and the chromosome bands, but cannot achieve any degree of high-resolution mapping, with two markers having to be at least 1 Mb apart to be resolved as separate hybridization signals. The main application of metaphase FISH has therefore been in determining on which chromosome a new marker is located, and providing a rough idea of its map position, as a preliminary to finer-scale mapping by other methods.

During the 1990s, modified versions of FISH were developed in which the target material was not the metaphase structures, but instead were mechanically stretched chromosomes or ones prepared from the prophase or interphase stages of nuclear division, when the chromosomes are naturally more extended.

Even with these innovations, markers closer than 25 kb cannot be resolved. To improve the resolution of FISH even further, it was therefore necessary to abandon intact chromosomes and instead use purified DNA. This approach, initially called **fiber-FISH**, is essentially a modified version of optical mapping, and is carried out with stretched DNA fragments in microfluidic devices with architectures similar to those described above. The advantage of fiber-FISH compared with restriction site mapping is that the probe can be designed to target any desired DNA sequence, so there is no limit to the types of marker that can be detected.

The main challenge presented by this type of optical mapping is ensuring that the probe remains attached to its specific position on the DNA fragment as the fragment is extended and passed through the microfluidic channels and past the detector. With a traditional hybridization probe, the target DNA must be at least partially denatured to expose a single-stranded region to which the probe anneals. The second DNA strand will then compete with the probe and possibly displace it, reforming the double-stranded molecule. If this occurs prior to passage of the DNA past the detector, then no data will be obtained. One solution to this problem is to use a **peptide nucleic acid** (**PNA**) as the probe. This is a polynucleotide analog in which the sugar–phosphate backbone is replaced by amide bonds (**Figure 3.32**). The hybridization between a PNA probe and its target on a DNA molecule is more stable than the normal DNA–DNA interaction, for two reasons. First, the stability of a DNA–DNA hybrid is weakened, to some extent, by repulsion between the negatively-charged sugar–phosphate backbones of the two polynucleotides. The amide backbone of a PNA is uncharged, so this repulsion does not occur. Second, there are two different ways in which a PNA can base pair to its target. As well as attaching via standard 'Watson-Crick' base pairs, a PNA with a high pyrimidine content can also form **Hoogsteen base pairs** with the target. Hoogsteen base pairs involve the same combinations (A–T and G–C) as Watson-Crick base pairs, but the hydrogen bonding that holds the pairs together involves different groups on the purine and pyrimidine bases (**Figure 3.33**). This means that a single DNA strand can attach to two PNAs at the same time, one by Watson-Crick pairing and one by Hoogsteen pairing. The resulting **triplex** structure, PNA$_2$DNA, is more stable than a DNA-DNA hybrid and so is unlikely to break down during the optical mapping procedure.

Further innovations extend the scope of optical mapping

The hybrid instability issues that arise when optical mapping is carried out with fluorescent probes would be avoided if the fluorescent marker could be incorporated directly into the DNA molecule that is being examined. This can be achieved by **nick translation**, a long-established method for labeling DNA that involves attachment of a DNA polymerase to a single-stranded nick followed by replacement of a short stretch of the broken polynucleotide, with one or more fluorescent nucleotides added to the reaction mixture to act as substrates for the strand synthesis (**Figure 3.34**). In the standard method, the nicks are generated at random positions by treating the DNA with a single-strand specific endonuclease such as DNase I. For optical mapping, where the objective is to label the DNA at specific positions, variants of restriction endonucleases called **nicking endonucleases** are used. These enzymes recognize specific nucleotide sequences, but rather than making a double-stranded break in the DNA they cut just one strand, resulting in a sequence-specific nick. The polymerase then labels the DNA adjacent to the nick, enabling the positions of the recognition sequences to be identified by optical mapping. Some nicking endonucleases are naturally occurring bacterial or bacteriophage enzymes, whereas others are engineered versions of standard restriction endonucleases. An example of the latter is *Bbv*CI, which comprises two subunits, each cutting a different strand at its 5′–CCTCAGC–3 target sequence. Mutation of the catalytic site of one or other of the subunits results in two modified enzymes, Nb.*Bbv*CI and Nt.*Bbv*CI, which nick rather than cut the DNA at the recognition sequence.

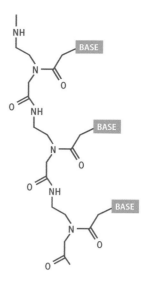

Figure 3.32 A short stretch of peptide nucleic acid. A peptide nucleic acid has an amide backbone instead of the sugar–phosphate structure found in a standard nucleic acid.

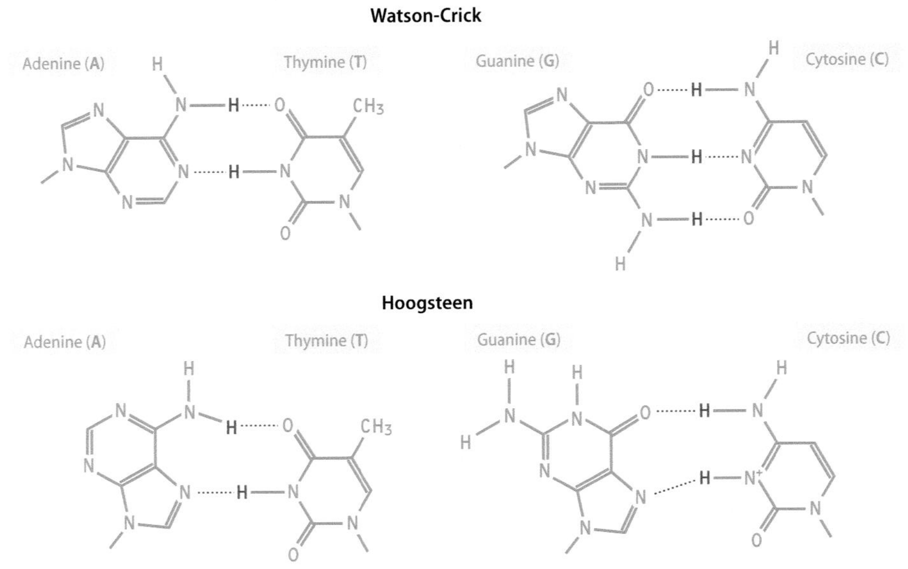

Figure 3.33 Hoogsteen base pairs. This type of base-pairing can form between PNA and DNA strands.

Watson-Crick

Adenine (A)   Thymine (T)   Guanine (G)   Cytosine (C)

Hoogsteen

Adenine (A)   Thymine (T)   Guanine (G)   Cytosine (C)

If a nicking endonuclease is used, then a problem can arise if two recognition sequences for the enzyme are located at positions closer than approximately 400 bp on the target DNA. Two nicks so close together form a 'fragile site' that can affect the integrity of the double-stranded molecule, as the base pairs in the region between the two nicks might not be strong enough to hold the two polynucleotides together. If the base-pairing is disrupted, then the DNA will break into two segments, reducing the lengths of the optical maps that can be obtained. This problem is avoided by using a sequence-specific **DNA methyltransferase** as a **direct labeling enzyme**. This is an enzyme that transfers methyl groups from *S*-adenosylmethionine to adenine or cytosine nucleotides within a double-stranded DNA molecule. If the *S*-adenosylmethionine donor molecules are fluorescently labeled in an appropriate way, then the label is transferred, along with the methyl group, to the DNA. An example is DLE-1, which methylates the second adenine in its 5′–CTTAAG–3′ recognition sequence, without the introduction of nicks or any other type of damage to the DNA molecule (**Figure 3.35**). When used with a microfluidic imaging system (see Figure 3.30), the direct labeling approach enabled 90% of the 730 Mb genome of the important crop plant *Sorghum bicolor* to be mapped.

There are two other innovations of optical mapping that we should briefly consider while we are exploring this method. These innovations are not directly relevant to the identification of marker positions on a genome map, but nonetheless provide important information on the structure and expression of a genome:

- GC-rich regions can be located by partially denaturing the DNA fragments, which can be achieved by raising the temperature or including a chemical denaturant such as formamide in the microfluidic solvent. Because a G–C base pair has three hydrogen bonds, compared to just two for A–T, the GC-rich regions are more likely to remain double-stranded under these conditions. If a double-strand specific dye is added, then the GC-rich regions will be revealed. In some genomes, GC-rich regions indicate the locations of genes, so the data provided by this method can be useful when a genome sequence is being annotated.

- Patterns of DNA methylation can be assessed if optical mapping is carried out with a restriction enzyme that is unable to cut the DNA if its restriction site is methylated. As we will see in Section 10.3, the addition of methyl groups to certain nucleotides is one way in which a gene can be silenced. By identifying whether a particular region of the genome is methylated or unmethylated, it is therefore possible to infer if the genes in that region are inactive or are being expressed.

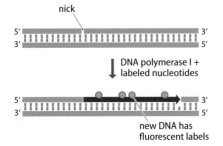

nick

5′
3′

↓ DNA polymerase I + labeled nucleotides

5′
3′

new DNA has fluorescent labels

Figure 3.34 Nick translation. DNA polymerase I attaches to the single-stranded nick and replaces a short stretch of the broken polynucleotide. If one or more fluorescent nucleotides are added to the reaction mixture, then these will be incorporated into the newly-synthesized strand.

3.6 PHYSICAL MAPPING BY ASSIGNING MARKERS TO DNA FRAGMENTS

The second approach to physical mapping involves assigning markers to genome fragments, on the basis that two markers that occur in the same fragment must be located close to one another in the genome. In this method, each marker is called a **sequence-tagged site** (**STS**), an STS simply being a short DNA sequence, generally between 100 bp and 500 bp in length, that is easily recognizable and occurs only once in the chromosome or genome being studied. To map a set of STSs, a collection of overlapping DNA fragments from a single chromosome or from the entire genome is needed. In the example shown in **Figure 3.36**, a fragment collection has been prepared from a single chromosome, with each point along the chromosome represented, on average, five times in the collection. The data from which the map will be derived are obtained by determining which fragments contain which STSs. This can be done by hybridization analysis, but PCR is generally used because it is quicker and has proven to be more amenable to automation. The chances of two STSs being present on the same fragment will, of course, depend on how close together they are in the genome. If they are very close, then there is a good chance that they will always be on the same fragment; if they are further apart, then sometimes they will be on the same fragment and sometimes they will not. The data can therefore be used to calculate the distance between two markers, in a manner analogous to the way in which map distances are determined by linkage analysis (Section 3.4). Remember that in linkage analysis a map distance is calculated from the frequency at which crossovers occur between two markers. **STS mapping** is essentially the same, except that the map distance is based on the frequency at which breaks occur between two markers.

The description of STS mapping given above leaves out some critical questions: What exactly is an STS? How is the DNA fragment collection obtained?

Any unique sequence can be used as an STS

To qualify as an STS, a DNA sequence must satisfy two criteria. The first is that its sequence must be known, so that a PCR assay can be set up to test for the presence or absence of the STS on different DNA fragments. The second requirement is that the STS must have a unique location in the chromosome being studied, or in the genome as a whole if the DNA fragment set covers the entire genome. If the STS sequence occurs at more than one position, then the mapping data will be ambiguous. Care must therefore be taken to ensure that STSs do not include sequences found in repetitive DNA.

These are easy criteria to satisfy and STSs can be obtained in many ways, the most common sources being **expressed sequence tags** (**ESTs**), SSLPs, and **random genomic sequences**.

- Expressed sequence tags are short sequences obtained by analysis of complementary (cDNA) clones. Complementary DNA is prepared by converting an mRNA preparation into double-stranded DNA (**Figure 3.37**).

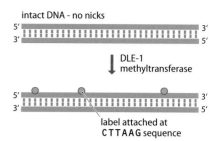

Figure 3.35 Direct labeling. The direct labeling enzyme DLE-1 adds fluorescent methyl groups directly to the second adenine in 5′–CTTAAG–3′ sequences in intact double-stranded DNA molecules.

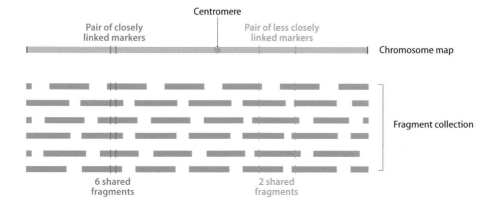

Figure 3.36 A fragment collection suitable for STS mapping. The fragments span the entire length of a chromosome, with each point on the chromosome present on average in five fragments. The two blue markers are close together on the chromosome map and there is a high probability that they will be found on the same fragment. The two green markers are more distant from one another and so are less likely to be found on the same fragment.

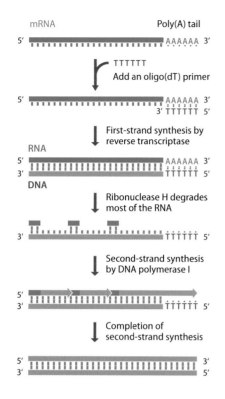

Figure 3.37 One method for preparing cDNA. Most eukaryotic mRNAs have a poly(A) tail at their 3′ end (Section 1.2). This series of A nucleotides is used as the priming site for the first stage of cDNA synthesis, carried out by reverse transcriptase – a DNA polymerase that copies an RNA template (Section 2.1). The primer is a short, synthetic DNA oligonucleotide, typically 20 nucleotides in length, made up entirely of Ts (an 'oligo(dT)' primer). When the first-strand synthesis has been completed, the preparation is treated with ribonuclease H, which specifically degrades the RNA component of an RNA–DNA hybrid. Under the conditions used, the enzyme does not degrade all of the RNA, instead leaving short segments that prime the second DNA strand-synthesis reaction, this one catalyzed by DNA polymerase I. This polymerase possesses a 5′→3′ exonuclease activity (Section 2.1) and so is able to degrade the RNA primers and replace these with DNA, completing synthesis of the second strand of cDNA.

Because the mRNA in a cell is derived from protein-coding genes, cDNAs and the ESTs obtained from them represent the genes that were being expressed in the cell from which the mRNA was prepared. ESTs are looked upon as a rapid means of gaining access to the sequences of important genes, and they are valuable even if their sequences are incomplete. An EST can also be used as an STS, assuming that it comes from a unique gene and not from a member of a gene family in which all the genes have the same or very similar sequences.

- SSLPs, whose use in genetic mapping we examined in Section 3.2, can also act as STSs in physical mapping. SSLPs that have already been mapped by linkage analysis are particularly valuable as they provide a direct connection between the genetic and physical maps.

- Random genomic sequences are obtained by sequencing random pieces of cloned genomic DNA, or simply by downloading sequences that have been deposited in the databases. If these sequences contain SNPs that have already been mapped by linkage analysis, then again a direct connection can be made between the genetic and physical maps.

DNA fragments for STS mapping can be obtained as radiation hybrids

The second component of an STS mapping procedure is the collection of DNA fragments spanning the chromosome or genome being studied. This collection is sometimes called the **mapping reagent** and at present there are two ways in which it can be assembled: as a clone library and as a panel of **radiation hybrids**. We will consider radiation hybrids first.

A radiation hybrid is a cell or organism that contains fragments of chromosomes from a second organism. The technology was initially developed with human chromosomes, starting in the 1970s when it was discovered that exposure of human cells to X-ray doses of 3000–8000 rad causes the chromosomes to break up randomly into fragments, larger X-ray doses producing smaller fragments (**Figure 3.38A**). This treatment is of course lethal for the human cells, but the chromosome fragments can be propagated if the irradiated cells are subsequently fused with nonirradiated cells of a hamster or other rodent. Fusion is stimulated either chemically with polyethylene glycol, or by exposure to the Sendai virus (**Figure 3.38B**). Not all of the hamster cells take up chromosome fragments, so a means of identifying the hybrids is needed. The routine selection process is to use a hamster cell line that is unable to make either thymidine kinase (TK) or hypoxanthine phosphoribosyl transferase (HPRT), deficiencies in either of these two enzymes being lethal when the cells are grown in a medium containing a mixture of hypoxanthine, aminopterin, and thymidine (HAT medium). After fusion, the cells are placed in a HAT medium. Those that grow are hybrid hamster cells that have acquired human DNA fragments that include genes for the human TK and HPRT enzymes, which are synthesized inside the hybrids,

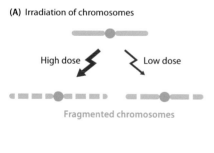

(A) Irradiation of chromosomes

(B) Fusion of cells to produce a radiation hybrid

Figure 3.38 Radiation hybrids. (A) The result of irradiation of human cells: the chromosomes break into fragments, smaller fragments being generated by higher X-ray doses. In (B), a radiation hybrid is produced by fusing an irradiated human cell with an untreated hamster cell. For clarity, only the nuclei are shown.

enabling these cells to grow in the selective medium. The treatment results in hybrid cells that contain a random selection of human DNA fragments inserted into the hamster chromosomes. Typically the fragments are 5–10 Mb in size, with each cell containing fragments equivalent to 15–35% of the human genome. The collection of cells is called a radiation hybrid panel and can be used as a mapping reagent in STS mapping, provided that the PCR assay used to identify the STS does not amplify the equivalent region of DNA from the hamster genome.

Radiation hybrid mapping was important in constructing the first physical maps of the human genome, with a panel of fewer than 200 hybrids enabling 41,000 STSs to be mapped with 100 kb resolution. This means that if two markers are less than 100 kb apart, then they will appear to occupy the same position in the genome. This degree of resolution is much less than for optical mapping, with which it can be possible to distinguish markers less than 500 bp apart, but is still satisfactory for the initial mapping of an unsequenced genome. Following the success of the approach with the human genome, radiation hybrid mapping was applied to other mammals and to non-mammalian species, such as the zebra fish and the chicken. Some progress has also been made in adapting the technique for use with plants. For example, a barley radiation hybrid panel has been created by irradiating barley protoplasts to fragment their chromosomes, and then fusing these cells with tobacco protoplasts. A cotton panel was generated by irradiating pollen of one species of cotton, *Gossypium hirsutum*, and then using this pollen to fertilize the related species *Gossypium barbadense*. Similar approaches are proving successful with wheat, one study with a panel of 115 radiation hybrids enabling 26,299 SNPs to be mapped onto the D genome with 249 kb resolution.

A clone library can be used as the mapping reagent

Sometimes the preliminary task for the sequencing of a large and complex genome is to break the genome or isolated chromosomes into fragments and to clone each one in a high-capacity vector such as a BAC (Section 2.3). This results in a clone library, a collection of DNA fragments, with an average size of several hundred kilobases. The fragments in the various clones form an overlapping series, which means that as well as supporting the sequencing work, the clone library can also be used as a mapping reagent in STS analysis.

A clone library can be prepared from genomic DNA, in which case it represents the entire genome, or a chromosome-specific library can be made if the starting DNA comes from just one type of chromosome. The latter is possible because individual chromosomes can be separated by **flow cytometry**. To carry out this technique, dividing cells (ones with condensed chromosomes) are carefully broken open so that a mixture of intact chromosomes is obtained. The chromosomes are then stained with a fluorescent dye. The amount of dye that a chromosome binds depends on its size, so larger chromosomes bind more dye and fluoresce more brightly than smaller ones. The chromosome preparation is diluted and passed through a fine aperture, producing a stream of droplets, each one containing a single chromosome (**Figure 3.39**). The droplets pass through a detector that measures the amount of fluorescence and hence identifies which droplets contain the particular chromosome being sought. An electric charge is applied to these drops, and no others, enabling the droplets containing the desired chromosome to be deflected and separated from the rest. What if two different chromosomes have similar sizes, as is the case with human chromosomes 21 and 22? These can usually be separated if the dye that is used is not one that binds nonspecifically to DNA, but instead has a preference for AT- or GC-rich regions. Examples of such dyes are Hoechst 33258 and chromomycin A3, respectively. Two chromosomes that are the same size rarely have identical **GC contents**, and so can be distinguished by the amounts of AT- or GC-specific dye that they bind.

Compared with radiation hybrid panels, clone libraries have one important advantage for STS mapping. This is the fact that the assembly of overlapping

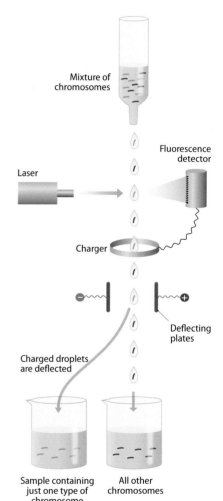

Figure 3.39 Separating chromosomes by flow cytometry. A mixture of fluorescently stained chromosomes is passed through a small aperture so that each drop that emerges contains just one chromosome. The fluorescence detector identifies the signal from drops containing the correct chromosome and applies an electric charge to these drops. When the drops reach the electric plates, the charged ones are deflected into a separate beaker. All other drops fall straight through the deflecting plates and are collected in the waste beaker.

clones can be used as the base material for a lengthy, continuous DNA sequence, and the STS data can then be used to anchor this sequence precisely onto the physical map. If the STSs also include SSLPs or SNPs that have been mapped by linkage analysis, then the DNA sequence, physical map, and genetic map can all be integrated.

SUMMARY

- Genome maps provide the framework for sequencing projects because they indicate the positions of genes and other recognizable features and hence enable the accuracy of an assembled DNA sequence to be checked.

- Genome maps are also used in methods for identifying the functions of genes involved in disease and for identifying QTLs that control traits such as meat productivity in farm animals.

- In the first genetic maps, the markers were genes whose alleles could be distinguished because they gave rise to easily recognized phenotypes, such as different eye colors, or whose alleles could be distinguished by biochemical tests.

- Today, DNA markers are also extensively used for mapping, including restriction fragment length polymorphisms (RFLPs), simple sequence length polymorphisms (SSLPs), and single-nucleotide polymorphisms (SNPs), all of which can be typed quickly and easily by PCR.

- The relative positions of genes and DNA markers on chromosomes are determined by linkage analysis, which enables the recombination frequency between a pair of markers to be calculated, providing the data needed to deduce the relative positions of the markers on the genetic map.

- For many organisms, linkage analysis is carried out by following the inheritance of markers in planned breeding experiments, but this is not possible for humans. Instead, genetic mapping of the human genome depends on examination of marker inheritance in large families, a procedure called pedigree analysis.

- Genetic maps have relatively poor resolution and tend to be inaccurate and must be refined by physical mapping if the map is to be used in a genome sequencing project.

- The positions of restriction sites in a small DNA molecule can be determined by restriction mapping.

- Optical mapping enables the positions of restriction sites and other sequence features to be directly visualized in a long DNA molecule.

- The most detailed physical maps are obtained by sequence-tagged site (STS) content mapping, which makes use of a mapping reagent, a collection of overlapping DNA fragments that span an entire chromosome or genome. The mapping reagent can be a library of clones or a radiation hybrid panel.

SHORT ANSWER QUESTIONS

1. Describe the past and current use of maps in genome sequencing projects.

2. Clearly explain the differences between a genetic and a physical map of a genome.

3. How has PCR made the analysis of RFLPs much faster and easier? What was required to map RFLPs prior to the utilization of PCR?

4. Explain why SNPs are now the most widely used type of DNA marker, and outline the various methods used to type SNPs.

5. How does the linkage between genes provide a critical component of genetic mapping? Describe how genetic maps of individual chromosomes are obtained in (A) fruit flies and (B) humans.

6. What are Mendel's two Laws of Genetics? What component of genetic mapping is not covered by Mendel's Laws?

7. Explain why a double homozygote is used for test crosses in linkage analysis experiments. Why is it preferable that the homozygote alleles be recessive for the traits being tested?

8. Outline the limitations of human pedigree analysis and describe how the impact of these limitations is minimized in actual pedigree studies.

9. Briefly describe the three methods used to obtain maps of bacterial genomes.

10. Describe the basis for optical mapping and explain why optical mapping has become important in genome studies.

11. How are radiation hybrids used in the construction of genome maps?

12. How would a scientist prepare a clone library of DNA from just a single chromosome?

IN-DEPTH PROBLEMS

1. What are the ideal features of a DNA marker that will be used to construct a genetic map? To what extent can RFLPs, SSLPs, or SNPs be considered ideal DNA markers?

2. What features would be desirable for an organism that is to be used for extensive studies of heredity?

3. Will maps ever become entirely unnecessary in genome research?

4. Which is more useful – a genetic or a physical map?

FURTHER READING

Books on the history of genetics

Orel, V. (1996). *Gregor Mendel: The First Geneticist.* Oxford University Press, Oxford.

Shine, I. and Wrobel, S. (2009). *Thomas Hunt Morgan: Pioneer of Genetics.* University Press of Kentucky, Lexington, Kentucky.

Sturtevant, A.H. (2001). *A History of Genetics.* Cold Spring Harbor Laboratory Press, New York. *Describes the early gene mapping work carried out by Morgan and his colleagues.*

Genetic and DNA markers

Sobrino, B., Brión, M. and Carracedo, A. (2005). SNPs in forensic genetics: a review on SNP typing methodologies. *Forensic Sci. Int.* 154:181–194.

Wang, D.G., Fan, J.-B., Siao, C.-J., et al. (1998). Large-scale identification, mapping, and genotyping of single-nucleotide polymorphisms in the human genome. *Science* 280:1077–1082.

Yamamoto, F., Clausen, H., White, T., et al. (1990). Molecular genetic basis of the histo-blood group ABO system. *Nature* 345:229–233.

Linkage analysis

Morton, N.E. (1955). Sequential tests for the detection of linkage. *Am. J. Hum. Genet.* 7:277–318. *The use of lod scores in human pedigree analysis.*

Ott, J., Wang, J. and Leal, S.M. (2015). Genetic linkage analysis in the age of whole-genome sequencing. *Nat. Rev. Genet.* 16:275–284.

Cummings, M. (2015). *Human Heredity: Principles and Issues,* 11th edition. Cengage Learning, Boston. *Chapter 4 covers pedigree analysis.*

Sturtevant, A.H. (1913). The linear arrangement of six sex-linked factors in *Drosophila*, as shown by their mode of association. *J. Exp. Zool.* 14:43–59. *Construction of the first linkage map for the fruit fly.*

Restriction and optical mapping

Deschamps, S., Zhang, Y., Llaca, V., et al. (2018). A chromosome-scale assembly of the sorghum genome using nanopore sequencing and optical mapping. *Nat. Commun.* 9:4844. *Optical mapping by direct labeling.*

Hosoda, F., Arai, Y., Kitamura, E., et al. (1997). A complete *Not*I restriction map covering the entire long arm of human chromosome 11. *Genes Cells* 2:345–357.

Jeffet, J., Margalit, S., Michaeli, Y. and Ebenstein, Y. (2021) Single-molecule optical genome mapping in nanochannels: Multidisciplinarity at the nanoscale. *Essays Biochem.* 65:51–66.

Jing, J.P., Lai, Z.W., Aston, C., et al. (1999). Optical mapping of *Plasmodium falciparum* chromosome 2. *Genome Res.* 9:175–181.

Lin, J., Qi, R., Aston, C., et al. (1999). Whole-genome shotgun optical mapping of *Deinococcus radiodurans*. *Science* 285:1558–1562.

Michalet, X., Ekong, R., Fougerousse, F., et al. (1997). Dynamic molecular combing: Stretching the whole human genome for high-resolution studies. *Science* 277:1518–1523.

Yuan, Y., Chung, C.Y.-L. and Chan, T.-F. (2020). Advances in optical mapping for genomic research. *Comput. Struct. Biotechnol. J.* 18:2051–2062.

Radiation hybrids

Hudson, T.J., Church, D.M., Greenaway, S., et al. (2001). A radiation hybrid map of mouse genes. *Nat. Genet.* 29:201–205.

Itoh, T., Watanabe, T., Ihara, N., et al. (2005). A comprehensive radiation hybrid map of the bovine genome comprising 5593 loci. *Genomics* 85:413–424.

Mazaheri, M., Kianian, P.M.A., Kumar, A., et al. (2015). Radiation hybrid map of barley chromosome 3H. *Plant Genome 8.* doi:10.3835/plantgenome2015.02.0005.

McCarthy, L. (1996). Whole genome radiation hybrid mapping. *Trends Genet.* 12:491–493.

Tiwari, V.K., Heesacker, A., Riera-Lizarazu, O., et al. (2016). A whole-genome, radiation hybrid mapping resource of hexaploid wheat. *Plant J.* 86:195–207.

Walter, M.A., Spillett, D.J., Thomas, P., et al. (1994). A method for constructing radiation hybrid maps of whole genomes. *Nat. Genet.* 7:22–28.

SEQUENCING GENOMES

The ultimate objective of a genome project is the complete DNA sequence for the organism being studied. This chapter describes the techniques and research strategies that are used during the sequencing phase of a genome project, when this ultimate objective is being directly addressed. Techniques for sequencing DNA are clearly of central importance in this context, and we will begin the chapter with a detailed examination of sequencing methodology. This methodology is of little value, however, unless the **sequence reads** that result from individual sequencing experiments can be linked together in the correct order to give the master sequences of the chromosomes that make up the genome. The second part of this chapter therefore describes the strategies used to ensure that the master sequences are assembled correctly. We will then, in the final part of the chapter, examine how the challenges posed by sequence generation and assembly have been and are being met in the series of projects that have resulted in our current understanding of the human genome.

4.1 METHODOLOGY FOR DNA SEQUENCING

Over the years, a number of different methods for DNA sequencing have been developed, and others are likely to become important in the future. The techniques in use today can be divided into three categories:

- The **chain-termination method**, which was first devised by Fred Sanger and colleagues in the mid-1970s and was used in all the genome sequencing projects completed prior to the mid-2000s, including the Human Genome Project and projects for several other eukaryotes and many types of bacteria and archaea. Nowadays, chain-termination sequencing is rarely used in genome projects, but the method is still performed in most molecular biology labs as a means of sequencing short DNA molecules, such as PCR products derived from individual genes.

- **Short-read sequencing**, which utilizes a **massively parallel** strategy in order to generate millions of sequence reads, each less than 600 bp in length, in a single experiment. The technology usually referred to as **Illumina sequencing**, named after the company that markets the necessary equipment, is currently the most popular type of short-read sequencing.

- **Long-read sequencing**, which also uses a parallel strategy to generate multiple reads per experiment, but in this case with individual sequence reads tens or even thousands of kb in length. Because the reads are longer, fewer are needed to assemble a complete genome. At present, most long-read sequencing is performed by either of two methods: **single-molecule real-time (SMRT) sequencing** and **nanopore sequencing.**

Chain-termination sequencing of PCR products

Chain-termination sequencing is based on the principle that single-stranded DNA molecules that differ in length by just a single nucleotide can be separated from one another by **polyacrylamide gel electrophoresis**. If the electrophoresis

DOI: 10.1201/9781003133162-4

is carried out in a capillary tube 50–80 cm in length, with a bore of 0.1 mm, then it is possible to resolve a family of molecules representing all lengths up to 1500 nucleotides, with the single-stranded molecules emerging one after another from the end of the capillary (**Figure 4.1**).

Chain-termination sequencing was originally carried out with fragments of DNA cloned into plasmid or bacteriophage vectors (Section 2.3), but is now more routinely used to sequence PCR products by the method called **thermal cycle sequencing**. The single-stranded molecules are generated by *Taq* polymerase in a similar manner to PCR (Section 2.2), but with the important difference that just one primer is used (**Figure 4.2A**). This means that only one strand of the PCR product is copied, and the new strands accumulate in a linear fashion, not exponentially as is the case in a real PCR. The new strands would normally be full-length copies of the PCR product, but in a chain-termination sequencing experiment each strand synthesis ends prematurely because, as well as the normal deoxynucleotides (dATP, dCTP, dGTP, and dTTP), a small amount of each of four **dideoxynucleotide triphosphates** (ddATP, ddCTP, ddGTP, and ddTTP) is added to the reaction. Each of these dideoxynucleotides is labeled with a different fluorescent marker. *Taq* polymerase enzyme does not discriminate between deoxy- and dideoxynucleotides, but once incorporated a dideoxynucleotide blocks further strand elongation because it lacks the 3′–hydroxyl group needed to form a connection with the next nucleotide (**Figure 4.2B**). Because the normal deoxynucleotides are also present, in larger amounts than the dideoxynucleotides, the strand synthesis does not always terminate close to the primer: in fact, several hundred nucleotides may be polymerized before a dideoxynucleotide is eventually incorporated. The result is a set of new molecules, all of different lengths, and each ending in a dideoxynucleotide whose identity indicates the nucleotide – A, C, G, or T – that is present at the equivalent position in the template DNA (**Figure 4.2C**).

To determine the DNA sequence, all that we have to do is identify the dideoxynucleotide at the end of each chain-terminated molecule. This is where the polyacrylamide gel comes into play. The DNA mixture is loaded onto the capillary gel and electrophoresis carried out to separate the molecules according to their lengths. After separation, the molecules are run past a fluorescence detector capable of discriminating the labels attached to the dideoxynucleotides (**Figure 4.3A**). The detector therefore determines if each molecule ends in an A, C, G, or T. The sequence can be printed out for examination by the operator (**Figure 4.3B**), or entered directly into a storage device for future analysis.

A single sequence read of up to 900 bp can be obtained by a chain-termination experiment. If the PCR product is longer than 900 bp, then two reactions can be carried out, one with each of the two PCR primers, to give **forward** and **reverse sequences** (**Figure 4.4**), the ends of which, if the sequences are long

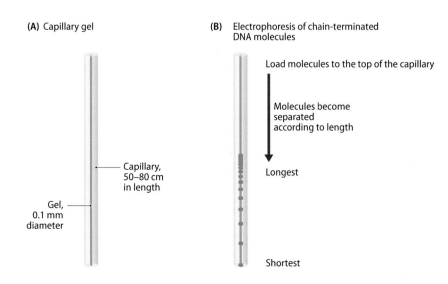

Figure 4.1 Polyacrylamide gel electrophoresis in a capillary system can resolve single-stranded DNA molecules that differ in length by just one nucleotide. (A) The dimensions of a capillary gel used in chain-termination DNA sequencing. (B) Separation of DNA molecules of different lengths during electrophoresis.

(A) Capillary gel

Capillary, 50–80 cm in length

Gel, 0.1 mm diameter

(B) Electrophoresis of chain-terminated DNA molecules

Load molecules to the top of the capillary

Molecules become separated according to length

Longest

Shortest

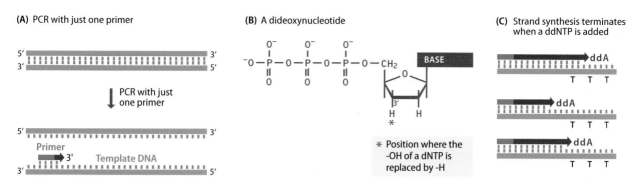

Figure 4.2 Chain-termination DNA sequencing. (A) Chain-termination sequencing can be carried out in a manner similar to PCR but with just one primer. (B) Strand synthesis does not proceed indefinitely because the reaction mixture contains small amounts of each of the four dideoxynucleotides, which block further elongation because they have a hydrogen atom rather than a hydroxyl group attached to the 3´-carbon. (C) Incorporation of ddATP results in chains that are terminated opposite Ts in the template. This generates the 'A' family of terminated molecules. Incorporation of the other dideoxynucleotides generates the 'C,' 'G,' and 'T' families.

enough, will overlap in the middle of the PCR product. Alternatively, it is possible to extend the sequence in one direction by synthesizing a new primer, designed to anneal at a position within the PCR product.

Chain-termination sequencing can be automated with multiple capillary gels working in parallel, so that up to 384 different sequences can be obtained in a one-hour period. If the reads have an average length of 750 bp, then almost 7 Mb of information can be generated per machine per day. This, of course, requires round-the-clock technical support, ideally with robotic devices used to prepare the sequencing reactions and to load the reaction products into the sequencers. If such a factory approach can be established and maintained, then it is possible to use the chain-termination method to obtain enough reads to sequence an entire genome. Chain-termination sequencing carried out in this intensive

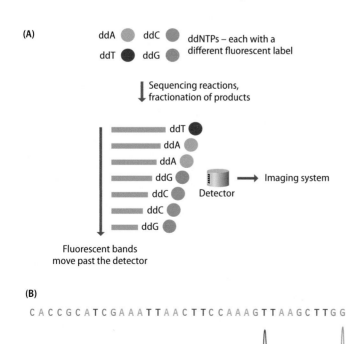

Figure 4.3 Reading the sequence generated by a chain-termination experiment. (A) Each dideoxynucleotide is labeled with a different fluorophore. During electrophoresis, the labeled molecules move past a fluorescence detector, which identifies which dideoxynucleotide is present in each band. The information is passed to the imaging system. (B) A DNA sequencing printout. The sequence is represented by a series of peaks, one for each nucleotide position. In this example, a green peak is an 'A,' blue is 'C,' orange is 'G,' and pink is 'T.'

Figure 4.4 Different types of primer for chain-termination sequencing. Forward, reverse and internal primers enable different parts of a PCR product to be sequenced.

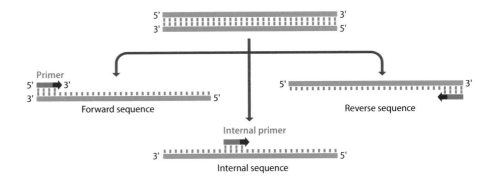

manner was used in the first genome sequencing projects, and proved effective in particular with bacterial genomes, many of which are less than 5 Mb in length. For example, to sequence the *Haemophilus influenzae* genome (Section 4.2), 28,643 chain-termination experiments were carried out, generating 11.6 Mb of sequence. The length of the *H. influenzae* genome is 1.8 Mb, but no sequencing method is entirely accurate, so it is necessary to sequence each region of a genome multiple times, in order to identify errors present in individual sequence reads (**Figure 4.5**). With the chain-termination method, to ensure that errors are identified, at least 5× **sequence depth** or **coverage** is needed, meaning that every nucleotide is present in five different reads. Despite this requirement, it proved possible to use chain-termination sequencing to obtain the human genome sequence, as well as those of other eukaryotes such as the nematode worm *Caenorhabditis elegans* and the fruit fly *Drosophila melanogaster*. The human genome is 3,100 Mb, so for 5× sequence depth, a total of $5 \times 3,100 = 15,500$ Mb of sequence is required, equivalent to over 20 million chain-termination sequences averaging 750 bp in length. In fact, this target was exceeded, with the Human Genome Project generating 23,147 Mb of sequence by the time that the first draft of the genome was published in 2001 (Section 4.3).

Illumina sequencing is the most popular short-read method

Despite the success of the chain-termination method in providing data for the early genome projects, it was realized during the early 2000s that, even when multiple sequencing machines are operated in a factory setting, it would take months or years of effort to obtain enough data to assemble the genome sequence of each new species that was studied. Cost is also an important consideration, especially for **genome resequencing**, in which the sequences of multiple individuals from within a species are obtained. Comparisons between these genome sequences enable variations to be identified that might be associated with characteristics such as inherited disease in humans or adaptation to environmental extremes for a crop plant. One of the goals of **personalized medicine** is to use individual human genome sequences to make accurate diagnoses of a person's risk of developing a disease, and to use that person's genetic characteristics to plan effective therapies and treatment regimes. For personalized medicine and other research programs based on genome resequencing to become a reality, more cost-effective and rapid ways of sequencing individual genomes are needed.

Figure 4.5 Read depth is needed in order to identify sequencing errors. Each region of a genome must be sequenced multiple times, in order to identify errors present in individual sequence reads. In this example, the discrepancy in Read 4 in the highlighted column can be ascribed to a sequencing error, the correct nucleotide at this position being C.

```
                                ACCATCGTAGCTTCAGTATGTATGTACTAG          Read 1
        ATGTTTGTAGCTAGGATCGTAGCTACC                                    Read 2
          TTTGTAGCTAGGATCGTAGCTACCATCGTAGCTT                           Read 3
                                  TTGTAGCTTCAGTATGTATGTACTAG           Read 4
              GGATCGTAGCTACCATCGTAGCTTCAGT                             Read 5
     ATGTTTGTAGCTAGGATCGTA                                            Read 6
     ATGTTTGTAGCTAGGATCGTAGCTACCATCGTAGCTTCAGTATGTATGTACTAG           Deduced sequence
```

Although the chain-termination method is still used to sequence small pieces of DNA, where its ease and speed of use outweigh other issues, for genome projects it has now been superseded by the short- and long-read methods, which are able to generate much larger amounts of sequence data, more quickly and at less cost. When the first short-read methods were developed in the late 2000s, they were called **next-generation sequencing** to emphasize their radical differences compared with the chain-termination technology. The most important of these differences is the initial preparation of a **sequencing library**, a collection of DNA fragments that have been immobilized on a solid support in such a way that multiple sequencing reactions can be carried out side by side (**Figure 4.6**). For Illumina sequencing, which is currently the most popular of the various short-read methods that have been developed, the DNA fragments from which the sequencing library is prepared are 200–500 bp in length. This is much shorter than the lengths of the molecules obtained when genomic DNA is purified from animal or plant tissue, so the initial extract must be treated in some way to break the molecules down to the desired size. One possibility is to use **sonication**, a technique that uses high-frequency sound waves to make random cuts in DNA molecules. Random breakage is important because each fragment will be sequenced from its ends. With short-read methods, it is not possible to direct the sequencing toward the middle of a fragment, as can be done by designing an internal primer for the chain-termination method. The ends must therefore be randomly distributed throughout the starting DNA molecule in order to ensure that all of that molecule is sequenced.

After breakage, the DNA fragments are immobilized within a **flow cell** (**Figure 4.7**), a reaction chamber into which the reagents used in the sequencing experiment can be added and washed away one after the other so the various reactions occur in the correct order. Part of the internal surface of the flow cell is coated with multiple copies of a short oligonucleotide. **Adaptors**, short pieces of double-stranded DNA whose sequences match that of the oligonucleotides, are ligated to the ends of the DNA fragments, which are then denatured. The resulting single-stranded molecules attach to the coated internal surface of the flow cell by base-pairing between their adaptor sequences and the immobilized oligonucleotides (**Figure 4.8**). The immobilized DNA fragments are then amplified by PCR, to produce a sufficient number of identical copies to be sequenced. The adaptors now play a second role as they provide the annealing sites for the primers for this PCR. The same pair of primers can therefore be used to amplify all the fragments, even though the fragments themselves have many different sequences. The PCR products become attached to adjacent oligonucleotides within the flow cell, so each starting fragment is amplified into an immobilized cluster of identical fragments (Figure 4.8).

The Illumina method is based on **reversible terminator sequencing,** which, like the chain-termination method, makes use of modified nucleotides which block strand synthesis when incorporated at the end of a polynucleotide that is being synthesized by a DNA polymerase. The difference is that the termination step is reversible, because the chemical group that is attached to the 3′-carbon of the modified nucleotide can be removed once the identity of this nucleotide has been confirmed (**Figure 4.9**). This removable blocking group is a fluorescent label, a different one for each of the four nucleotides. There are no normal deoxynucleotides present in the reaction mixture, so each step in strand synthesis is accompanied by a pause, during which an optical device detects the fluorescent label, thereby identifying the terminal nucleotide. An enzyme then removes the label, enabling the next terminator nucleotide to be added and the detection process to be repeated. The strand synthesis is initiated by a primer that anneals to the adaptor sequences that were attached to the ends of the DNA fragments during library preparation. Every one of the millions of fragment clusters in the library is therefore sequenced at the same time.

With the current Illumina technology, the maximum read lengths are about 250 bp, although a single fragment can be sequenced from both ends to give **paired-end reads**, which, if they overlap, enable the entire sequence of a <500 bp

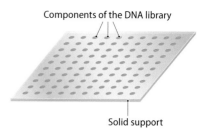

Figure 4.6 A DNA library immobilized on a solid support. In reality, a single massively parallel array will have millions of components.

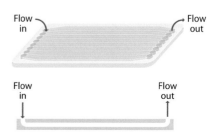

Figure 4.7 A typical flow cell used in DNA sequencing. The sequencing library is immobilized within the channels of the flow cell. To carry out the sequencing reactions, the necessary reagents flow through the cell, one after the other.

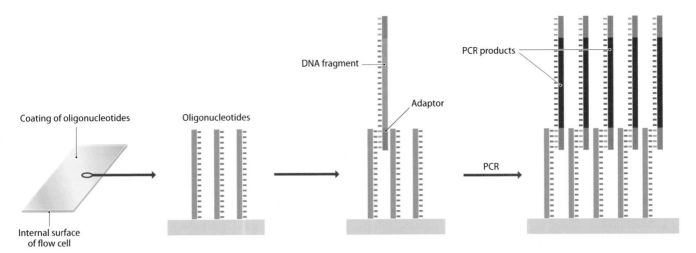

Figure 4.8 Immobilization and amplification of the DNA fragments in a sequencing library by base-pairing to oligonucleotides within a flow cell. Base-pairing between the immobilized oligonucleotides and the adaptors ligated to the ends of the DNA fragments results in attachment of the fragments to the internal surface of the flow cell. After PCR, the fragment copies attach to adjacent oligonucleotides, resulting in a cluster of identical immobilized fragments.

fragment to be obtained. Although these sequences are short, up to 3×10^9 reads can be obtained in a single experiment, giving 7.5×10^{11} bp (750 gigabase pairs or Gb) of sequence data. If the initial fragmentation is entirely random, then this is sufficient to sequence a human genome with almost 250× coverage. Short-read methods therefore enable the vast amounts of data needed to assemble an entire genome sequence to be obtained much more rapidly than was possible with the chain-termination approach.

A variety of other short-read sequencing methods have been devised

Over the last few years, there has been immense competition between different companies striving to develop sequencing platforms that combine the greatest speed in generating accurate sequence data with the lowest cost. Prior to the ascendency of Illumina technology, the most popular short-read technology was based on an approach called **pyrosequencing**. In this method, the reaction mixture contains only deoxynucleotides and template copying proceeds

(A) A reversible terminator nucleotide

(B) Reversible terminator sequencing

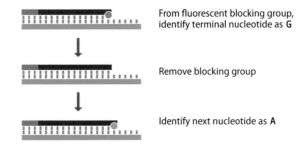

Figure 4.9 Reversible terminator sequencing. (A) The structure of a reversible terminator nucleotide with a removable fluorescent blocking group attached to the 3′-carbon. (B) After each nucleotide addition there is a pause while the fluorescent label is detected and the terminal nucleotide is identified. The blocking group is then removed so that the next nucleotide addition can occur.

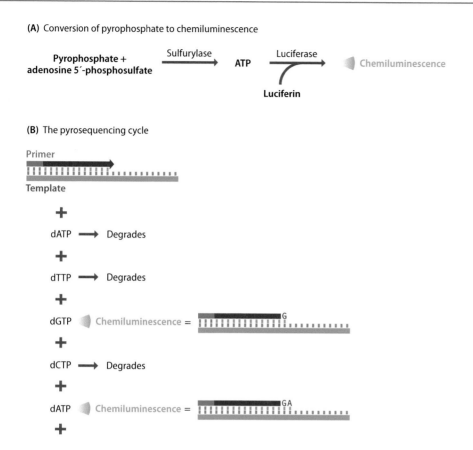

(A) Conversion of pyrophosphate to chemiluminescence

Pyrophosphate + adenosine 5′-phosphosulfate → (Sulfurylase) → **ATP** → (Luciferase) → Chemiluminescence

Luciferin

(B) The pyrosequencing cycle

Primer

Template

+

dATP → Degrades

+

dTTP → Degrades

+

dGTP Chemiluminescence = ...G...

+

dCTP → Degrades

+

dATP Chemiluminescence = ...GA...

+

Figure 4.10 Pyrosequencing. (A) A linked enzymatic reaction involving sulfurylase and luciferase generates chemiluminescence from the molecules of pyrophosphate that are released during DNA synthesis. (B) The strand-synthesis reaction is carried out in the absence of terminating nucleotides. Each deoxy-nucleotide is added individually, along with a nucleotidase enzyme that degrades the deoxynucleotide if it is not incorporated into the strand being synthesized. Incorporation is detected by the flash of chemiluminescence induced by the pyrophosphate released from the deoxynucleotide. The order in which deoxynucleotides are added to the growing strand can therefore be followed.

without artificial termination. Strand synthesis is followed by detecting the molecule of pyrophosphate that is released each time the DNA polymerase adds a deoxynucleotide to the 3′ end of the growing strand. The detection is initiated by the sulfurylase enzyme, which uses the pyrophosphate to convert adenosine 5′-phosphosulfate to ATP. The ATP then acts as the energy source for a second enzyme, luciferase, which oxidizes luciferin in a reaction that is accompanied by release of chemiluminescence (**Figure 4.10A**). A flash of chemiluminescence therefore signals the successful copying of one position in the template molecule. The deoxynucleotides are added individually in a repetitive series (e.g., A, then T, then G, then C, then A, then T, etc.) and the pattern of light emissions used to deduce the order in which nucleotides are incorporated into the growing strand (**Figure 4.10B**). Sequences of up to 1000 bp can be read, and in its most advanced format the technology enabled up to 700 Mb of data to be obtained in a single experiment. However, this type of sequencing is more expensive than the Illumina method, and there is a tendency for homopolymer runs (e.g., TTTTT) to be misread, with nucleotides being added to or omitted from the sequence in these regions.

The **ion torrent** method uses a similar approach to pyrosequencing, with a repetitive series of nucleotides flowing over an immobilized fragment library. However, with this method, the detection system is directed at the hydrogen ions, which, along with pyrophosphate, are released every time a nucleotide is incorporated into the growing strand. The reaction is carried out in wells lined with an **ion-sensitive field effect transistor** (**ISFET**). An ISFET generates an electronic pulse each time it detects hydrogen ions, these pulses being correlated with the flow of nucleotides over the well in order to deduce the sequence of the immobilized fragments. Read lengths of up to 600 bp are possible, and although homopolymer errors can occur, an advantage of this technology is the electronic detection system, which has lower construction and running costs compared with the optical detectors used in the Illumina platform. Ion torrent sequencing is still popular for resequencing projects with small genomes, such

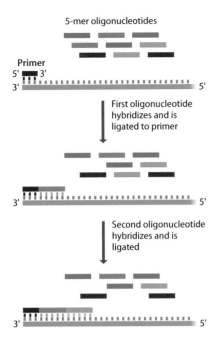

Figure 4.11 The basis of SOLiD sequencing. The mixture of 5-mer oligonucleotides (i.e., oligonucleotides that are five nucleotides in length) includes all 1024 possible sequences. The oligonucleotide that is complementary to the template DNA immediately adjacent to the primer hybridizes and is attached to the primer by DNA ligase. The cycle then repeats, resulting in hybridization and ligation of the second oligonucleotide.

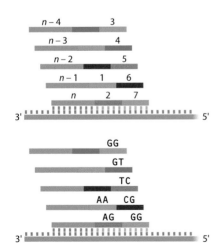

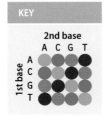

as those of viruses, where the vast numbers of reads provided by the Illumina method are not needed in order to achieve high sequence coverage.

A radically different approach is taken by the technology called **sequencing by oligonucleotide ligation and detection** (**SOLiD**). The sequence is deduced not by polymerase-directed synthesis of a new DNA strand, but by hybridization of a series of oligonucleotides whose sequences are complementary to that of the template. To begin the sequencing process, a primer is attached to the template DNA, using a terminal adaptor sequence as the annealing site, just as in the sequencing-by-synthesis methods. A set of 1024 oligonucleotides, representing each of the possible 5-nucleotide sequences, is then added, along with a DNA ligase. The sequence of one of these oligonucleotides will be complementary to the template DNA immediately adjacent to the primer. This oligonucleotide will hybridize and be attached to the primer by the DNA ligase (**Figure 4.11**). The process of hybridization-ligation continues for a set number of cycles until 50–75 nucleotides of the template have been covered. SOLiD sequencing is therefore quite straightforward in outline. However, it is computationally intensive because of the way in which the sequences of the oligonucleotides that hybridize to the template are deduced. Each of these oligonucleotides is labeled with a fluorescent marker, but only four different markers are used in total. This means that the markers divide the 1024 oligonucleotides into four families, each of 256 sequences. The groupings are not random, and instead each family comprises four 'dibase' sets, each of these sets being the 64 oligonucleotides with the same two initial nucleotides. For example, the AT dibase includes all the oligonucleotides with the sequence ATNNN, where 'N' is any nucleotide. Detection of the marker attached to a hybridizing oligonucleotide therefore assigns a 'color' to the initial pair of nucleotides in that 5-nucleotide sequence. Clearly, this results in an incomplete sequence (only two of every five nucleotides are read) as well as one that is ambiguous (there are 16 dibase sets but only four colors, so each color represents four different dibase sets). The sequencing process is therefore repeated with a second primer that anneals at a position that is offset by one nucleotide compared to the first primer (**Figure 4.12**). This is followed by three further repeats with primers that anneal at the –2, –3, and –4 positions. The result is that every nucleotide in the template is read twice, the combination of colors assigned to that nucleotide enabling its identity to be determined without ambiguity. Because of this double-read, SOLiD sequencing is highly accurate, but the read lengths are shorter than for the other methods, typically 50 or 75 nucleotides depending on the number of hybridization-ligation cycles that are used. Although originally designed for sequencing of purified DNA, the method is now more commonly used for ***in situ*** **sequencing**, which enables the spatial organization of a genome within a single nucleus to be examined (Section 10.1) and can also be used to detect the cellular location of specific RNA molecules (Section 12.2). In these applications, sequencing is carried out within cells that have been treated with a histological fixative (e.g., formaldehyde) or embedded in polyacrylamide gel, and the dibase color of each ligated oligonucleotide is detected with a confocal microscope.

Single-molecule real-time sequencing provides reads up to 200 kb in length

The longer the individual sequencing reads, the fewer that will be needed to cover an entire genome. Long-read sequencing therefore provides a clear benefit for

Figure 4.12 Interpretation of the results of a SOLiD sequencing experiment. A series of hybridization-ligation reactions have been carried out with the initial primer, and then with four other primers offset by one (*n*–1), two (*n*–2), three (*n*–3) and four (*n*–4) nucleotides. The color of the fluorescent markers attached to each of the hybridizing oligonucleotides is shown. In this example, the 3´-terminal nucleotide of the *n* primer is known to be A. The dibase of oligonucleotide 1 must therefore be AN. From the key (the 'color space') this dibase can be identified as AA. This means that the oligonucleotide 2 dibase is AN, its color identifying it as AG. Continuing the process, the sequence is revealed as AAGGTCGG.

genome assembly by reducing the number of overlaps that have to be identified in order to convert the unassembled reads into a complete genome sequence. Long reads also enable the genome sequence to be assembled with greater accuracy by reducing the problems that can be caused by the presence of repetitive DNA in the genome sequence (see Figure 3.3). With longer reads, there is less chance of missing out a segment of DNA between two repeat elements, because the complete sequence of each element, plus its flanking regions, is likely to be contained within at least one of the reads that are obtained (**Figure 4.13**).

The main factor limiting read length when the Illumina method is used is the delay caused by the need to remove the 3′ blocking group after each nucleotide addition. This reduces the **processivity** of the DNA polymerase. Processivity refers to the length of polynucleotide that is synthesized before the polymerase detaches from the template DNA. If there are repeated pauses in the progress of the polymerase along the DNA, then detachment is likely to occur before the new strand is more than a few hundred nucleotides in length. The recent focus has therefore been on methods that avoid a delay at the nucleotide detection step, and enable a sequence to be read during the normal, unimpeded progression of the polymerase along the template. The most successful approach so far developed is single-molecule real-time sequencing, in which a sophisticated optical system called a **zero-mode waveguide** is used to observe the copying of a single DNA template. The templates are circularized molecules and the sequencing takes place in a flow cell that contains several million wells, so that multiple sequences can be generated in parallel. The nucleotide substrates are still labeled with fluorescent markers, but the optical system is so precise that there is no need to use a blocking group to delay the polymerization process in order to allow the detection to take place. Instead, the label is removed immediately after nucleotide incorporation, so strand synthesis progresses without interruption (**Figure 4.14**). Read lengths of up to 200 kb have been reported, with up to 160 Gb of sequence data obtained per flow cell. This technology was initially developed by Pacific Biosciences and is often called **PacBio sequencing**. With the early versions of SMRT sequencing, read accuracy was a problem, with an error rate of 10%, meaning that on average one in ten nucleotides were read incorrectly. High coverage of each nucleotide position was therefore necessary

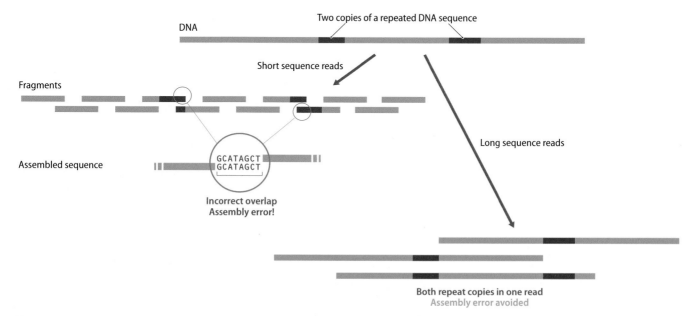

Figure 4.13 Long reads reduce assembly errors due to repetitive DNA. The DNA molecule contains two copies of a repeat sequence. On the left, two of the short reads appear to overlap, but one read contains the left-hand part of one repeat and the other read has the right-hand part of the second repeat. Failure to recognize this problem results in an assembly error, with the segment of DNA between the two repeats left out of the genome sequence. On the right, long reads have been obtained. Both copies of the repeat sequence are present in one of the reads, ensuring that the assembly error is avoided.

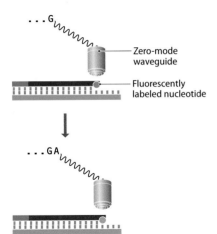

Figure 4.14 Single-molecule real-time sequencing. Each nucleotide addition is detected with a zero-mode waveguide.

in order to ensure that the master sequence was correct. However, with the latest HiFi version of PacBio sequencing an accuracy of >99% can be obtained.

Nanopore sequencing is currently the longest long-read method

The next logical step is to dispense with the strand-synthesis step and read the sequence of a DNA molecule directly without copying the molecule in any way. This is the aim of nanopore sequencing. The original concept behind nanopore sequencing was first proposed in the late 1980s, but it proved challenging to convert these ideas into a functioning methodology and it was not until 2014 that a working sequencer – the **MinION** – became available.

The principle of the method is that the flow of ions through a small pore in a membrane will be impeded when a single-stranded DNA molecule threads its way through the pore. Each of the four nucleotides has a different shape and so occludes the nanopore in a different way, resulting in a slightly different perturbation of the ion flow. These perturbations result in changes in the electrical current across the membrane, which can be measured in order to deduce the sequence of the polynucleotide as it threads its way through the pore. In the initial systems, a biological membrane was used with nanopores provided by embedded pore-forming proteins such as the alpha hemolysin of *Staphylococcus aureus* or the *Mycobacterium smegmatis* MspA protein. More recently, a solid-state technology has been developed with nanopores artificially created in films of materials, such as silicon nitride, aluminum oxide, or graphene.

To initiate the sequencing reaction, an electrical current is set up, positive on one side of the membrane and negative on the other, so electrophoresis causes a DNA molecule to approach one of the nanopores. This molecule is double-stranded, but the presence of a **helicase** enzyme in the vicinity of the nanopore breaks the base pairs, so the DNA unwinds and just one strand passes through the pore (**Figure 4.15**). The accompanying changes in the ion flow through the nanopore are measured and the resulting data are used to deduce the DNA sequence. However, reading the sequence is not a trivial exercise. Even with the most sophisticated nanopores, the changes in ion flow are not caused by individual nucleotides: the nanopores are so large that stretches of adjacent nucleotides are located within the pore at any one time (**Figure 4.16**). With the MinION, four or five nucleotides are present in the pore at the same time, with the central three of these having the greatest effect on the ion flow. Sophisticated computation

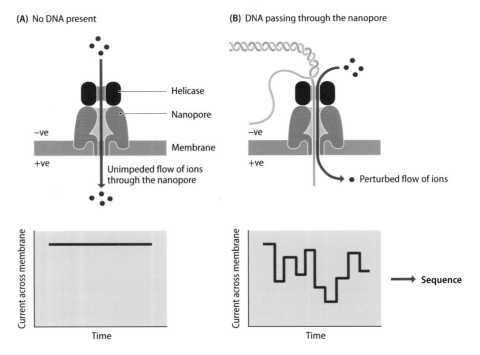

Figure 4.15 Nanopore sequencing. (A) In the absence of DNA, the flow of ions through the nanopore is unimpeded and the electrical current across the membrane is constant. (B) Passage of a polynucleotide through the nanopore perturbs the ion flow. Each nucleotide, or combination of adjacent nucleotides, perturbs the ion flow in a different way, resulting in fluctuations in the current from which the DNA sequence can be deduced.

systems are therefore needed to convert the plot of the electrical changes across the membrane – referred to as a **squiggle** – into a nucleotide sequence. Not surprisingly, read accuracy has always been an issue with nanopore sequencing. Even with the current technology, error rates of 5–10% are expected, compared to >99% accuracy that can be obtained with Illumina sequencing and the latest version of the SMRT methodology. High coverage is therefore needed to identify the errors in individual nanopore reads and construct an accurate master sequence. The great advantage of the method is that no DNA synthesis is involved, so the length of the sequence is not limited by polymerase processivity. In fact, sequence length is determined not by the methodology but by the lengths of the DNA molecules present in the samples being analyzed. If a very gentle extraction method is used, so the chromosomal DNA molecules undergo the minimum amount of fragmentation, then reads of up to 4 Mb can be obtained.

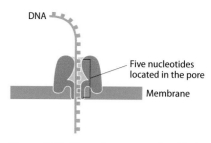

Figure 4.16 More than one nucleotide is present in the nanopore at a single time.

4.2 HOW TO SEQUENCE A GENOME

Having understood the methods available for sequencing DNA, the next question we must address is how the reads generated by these methods are assembled into the sequence of an entire genome. Since the 1990s, when the chain-termination method was first automated, the actual generation of sequence data has not been a limiting factor in genome sequencing projects. Instead, the main challenge lies with **sequence assembly**, the procedure used to convert the thousands or millions of sequence reads into the contiguous genome sequence. The most straightforward approach to sequence assembly is to build up the master sequence directly from the sequences obtained from individual sequencing experiments, simply by examining those sequences for overlaps (see Figure 3.1). This is called the **shotgun method**.

The potential of the shotgun method was proven by the *Haemophilus influenzae* sequence

During the early 1990s, there was extensive debate about whether the shotgun method would work in practice, many molecular biologists being of the opinion that the amount of data handling needed to compare all the sequence reads and identify overlaps, even with the smallest genomes, would be beyond the capabilities of existing computer systems. These doubts were laid to rest in 1995 when the sequence of the 1830 kb genome of the bacterium *Haemophilus influenzae* was published.

The strategy used to obtain the *H. influenzae* genome sequence is shown in **Figure 4.17**. The first step was to sonicate the genomic DNA in order to break it into fragments. The fragments were then electrophoresed, and those in the size range of 1.6–2.0 kb were purified from the agarose gel and ligated into a plasmid vector. From the resulting library, 19,687 clones were taken at random and 28,643 chain-termination sequencing experiments were carried out, the number of sequencing experiments being greater than the number of plasmids because both ends of some inserts were sequenced. Of these sequencing experiments, 16% were considered to be failures because they resulted in less than 400 bp of sequence. The remaining 24,304 reads gave a total of 11,631,485 bp of sequence, which if randomly distributed along the *H. influenzae* genome would give 6.4× coverage, sufficient to enable sequencing errors in individual reads to be recognized (see Figure 4.5). Sequence assembly required 30 hours on a computer with 512 Megabytes of random access memory (RAM), and resulted in 140 lengthy,

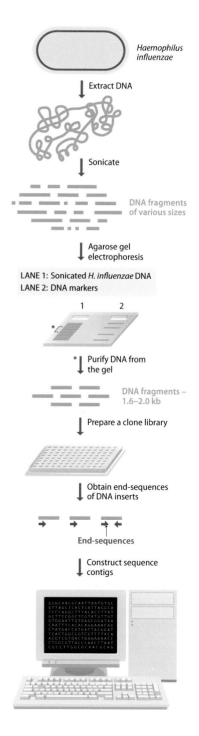

Figure 4.17 Shotgun sequencing of the *Haemophilus influenzae* genome. *H. influenzae* DNA was sonicated, and fragments with sizes between 1.6 kb and 2.0 kb were purified from an agarose gel and ligated into a plasmid vector to produce a clone library. End-sequences were obtained from clones taken from this library, and a computer was used to identify overlaps between sequences. This resulted in 140 sequence contigs.

contiguous sequences, each of these **sequence contigs** representing a different, nonoverlapping portion of the genome.

The next step was to join up pairs of contigs by obtaining sequences from the gaps between them. First, the library was checked to see if there were any clones whose two end-sequences were located in different contigs. If such a clone could be identified, then additional sequencing of its insert would close the 'sequence gap' between the two contigs (**Figure 4.18A**). In fact, there were 99 clones in this category, so 99 of the gaps could be closed without too much difficulty.

This left 42 gaps, which probably consisted of DNA sequences that were unstable in the cloning vector and therefore not present in the library. To close these 'physical gaps' a second clone library was prepared, this one with a different type of vector. Rather than using another plasmid, in which the uncloned sequences would probably still be unstable, the second library was prepared in a bacteriophage λ vector (Section 2.3). This new library was probed with 84 oligonucleotides, one at a time, these 84 oligonucleotides having sequences identical to the sequences at the ends of the unlinked contigs (**Figure 4.18B**). The rationale was that if two oligonucleotides hybridized to the same λ clone, then the ends of the contigs from which they were derived must lie within that clone, and sequencing the DNA in the λ clone would therefore close the gap. Twenty-three of the 42 physical gaps were dealt with in this way. A second strategy for gap closure was to use pairs of oligonucleotides, from the set of 84 described above,

Figure 4.18 Methods used to close the gaps in the initial assembly of the *Haemophilus influenzae* genome sequence. (A) 'Sequence gaps' are ones that can be closed by further sequencing of clones already present in the library. In this example, the end-sequences of contigs 1 and 2 lie within the same plasmid clone, so further sequencing of this DNA insert with internal primers will provide the sequence to close the gap. (B) 'Physical gaps' are stretches of sequence that are not present in the clone library, probably because these regions are unstable in the cloning vector that was used. Two strategies for closing these gaps are shown. On the left, a second clone library, prepared with a bacteriophage λ vector rather than a plasmid vector, is probed with oligonucleotides corresponding to the ends of the contigs. Oligonucleotides 1 and 7 both hybridize to the same clone, whose insert must therefore contain DNA spanning the gap between contigs 1 and 4. On the right, PCRs are carried out with pairs of oligonucleotides. Only numbers 1 and 7 give a PCR product, confirming that the contig ends represented by these two oligonucleotides are close together in the genome. The PCR product or the insert from the λ clone can therefore be sequenced to close the gap between contigs 1 and 4.

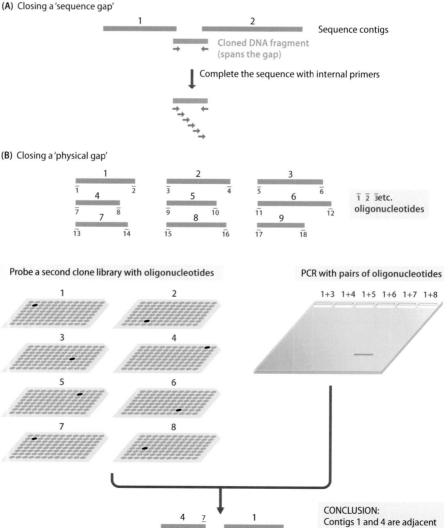

as primers for PCRs of *H. influenzae* genomic DNA. Some oligonucleotide pairs were selected at random, and those located on either side of a gap were identified simply from whether or not they gave a PCR product (see **Figure 4.18B**). Sequencing these PCR products closed the relevant gaps.

Many prokaryotic genomes have been sequenced by the shotgun method

The demonstration that a small genome can be sequenced relatively rapidly by the shotgun method led to a sudden plethora of completed microbial genomes. The 580 kb genome of *Mycoplasma genitalium* was completed in 1995, shortly after publication of the *H. influenzae* sequence. Despite the initial doubts about the feasibility of the approach, it soon became accepted that the shotgun method was capable of assembling the genome sequences of most if not all prokaryotic species, the relatively short lengths of these genomes meaning that the computational requirements for finding sequence overlaps are not too great. With the introduction of short-read sequencing in the 2000s, gap closure became less of an issue, because a single experiment can easily generate enough sequence reads to give several hundred times coverage of the genome.

Many prokaryotic genomes have been assembled by ***de novo*** **sequencing**, as described above for *H. influenzae*, with the genome sequence worked out solely by finding overlaps between individual sequence reads. The genomes of almost 17,000 prokaryotic species have been sequenced in this way. There are, however, more than 450,000 prokaryotic genome sequences in the databases. This is because some species have been sequenced multiple times in order to understand the amount of variation that exists in their genomes (Section 8.2). These resequencing projects make use of a second, less computer-intensive approach to assembly, in which the existing genome sequence is used as the **reference sequence** for assembly of additional genome sequences from the same species. Rather than looking for overlaps among the sequence reads for the genome that is being assembled, individual reads are simply placed on to the reference sequence by looking for regions of sequence identity or similarity (**Figure 4.19**). The reference sequence approach can also be used for *de novo* sequencing if the species whose genome is being sequenced is related to another species whose genome has already been assembled. The rationale is that the two sequences will be sufficiently similar for the known genome to direct assembly of the sequence for the new species. Usually, the reads from the new genome will be assembled into contigs before comparing with the reference, in order to increase the degree of certainty that can be assigned to matches.

Shotgun sequencing of eukaryotic genomes requires sophisticated assembly programs

The shotgun approach is also being applied to eukaryotic genomes, but with these organisms sequence assembly is complicated by two factors. The first of these is the size of the read dataset that must be assembled. Eukaryotic genomes are much longer than prokaryotic ones (e.g., the human genome is 3,100 Mb compared with just 1.83 Mb for *H. influenzae*), so many more sequence reads are required in order to ensure adequate coverage. In fact, the problem as far as assembly is concerned is not the number of reads, but the number of pairs of reads, as we must compare pairs in order to identify overlaps. The required data

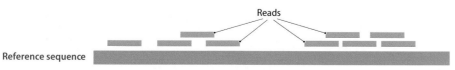

Figure 4.19 Using a reference sequence during a resequencing project. The reference sequence is used to place the reads from the second genome at their correct relative positions.

analysis therefore becomes disproportionately more complex as the number of sequence reads increases – for n reads, the number of pairs is given by $2n^2 - 2n$.

The second problem with the shotgun method is that it can lead to errors if the genome contains repetitive DNA sequences. As we learned in Section 3.1, repetitive DNA causes problems for the shotgun approach because a sequence that lies partly or wholly within one repeat element might accidentally be assigned an overlap with the identical sequence present in a different repeat element (see Figure 3.3A). This could lead to a part of the genome sequence being placed at the incorrect position or left out entirely. Most prokaryotic genomes contain relatively little repetitive DNA, but repeat sequences are common in eukaryotes, and in some species make up more than 50% of the genome.

The first **sequence assemblers** – the software packages that convert sequence reads into contigs – simply compared pairs of reads, merging those with the longest overlaps, repeating the process until no further overlaps were present (**Figure 4.20**). The result is an **overlap graph**, from which a master sequence can be assembled. In computational parlance, this approach uses a **greedy algorithm**, one which makes the most logical choice at each step in an iterative process. However, greedy algorithms are notorious for sometimes ending up with a sub-optimum solution, because each step is taken in isolation without consideration of the nature of the problem as a whole. This is the case with the overlap graph method, as the presence of repetitive DNA is likely to lead to errors in assembly, as described above. Construction of an overlap graph is also computationally intensive. The approach is feasible when the chain-termination method is used to sequence a genome, but is impossible with the much greater number of reads obtained by a short-read method.

More recently developed sequence assemblers take an alternative approach that is less computationally intensive and hence is applicable to short sequence reads. These assemblers make use of a representation called a **de Bruijn graph**, which is a mathematical concept for identifying overlaps between strings of symbols. The strings must be of equal length, so when this method is used in sequence assembly the initial step is to break the sequence reads into smaller segments or **k-mers**, typically of 20–30 nucleotides in length. Duplicate k-mers are discarded, so this step reduces the size of the dataset. Each k-mer is then converted into a 'prefix' sequence (the k-mer minus its last nucleotide) and a 'suffix' sequence (the k-mer minus its first nucleotide) (**Figure 4.21A**). Overlaps between k-mers are then identified by searching for pairs where the suffix of one k-mer is identical to the prefix of the second k-mer. Once all overlaps are identified, the k-mers are linked together as a de Bruijn graph, each k-mer being depicted as a node with lines ('edges') connecting k-mers whose suffix and prefix sequences overlap (**Figure 4.21B**). The master sequence can then be read from the graph. If the sequence contains repetitive DNA, then the de Bruijn graph will be branched rather than linear. If this is the case, then the computer attempts to identify an **Eulerian pathway** through the graph, this being a pathway that visits each edge just once. If an Eulerian pathway can be identified, then the correct sequence assembly will be achieved, despite the presence of the repeated sequences (**Figure 4.21C**). One the other hand, if the de Bruijn graph is so complex that an Eulerian pathway cannot be identified, then the branched structure is still valuable as it indicates the parts of the sequence in which there are unresolved repetitive DNA regions.

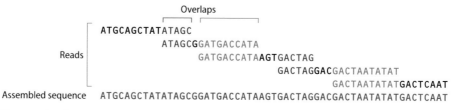

Figure 4.20 Sequence assembly using an overlap graph. Overlaps between pairs of reads are identified in order to build up the master sequence.

(A) The prefix and suffix versions of a *k*-mer

k-mer	ATGCAGCTATATAGCGGATG
Prefix sequence	ATGCAGCTATATAGCGGAT
Suffix sequence	TGCAGCTATATAGCGGATG

(B) Reading a sequence from a de Bruijn graph

TGA → GAC → ACC → CCG → CGC → GCA → CAG → AGT → GTT → TTA

Master sequence TGACCGCAGTTA

(C) A Eulerian pathway through a de Bruijn graph

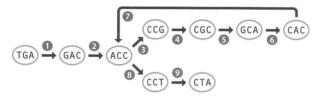

Master sequence TGACCGCACCTA

Figure 4.21 Sequence assembly using a de Bruijn graph. (A) A *k*-mer of 20 nucleotides and its prefix and suffix sequences. (B) An example of sequence assembly using a de Bruijn graph. In this example, the *k*-mers are just three nucleotides in length, which is too short for sequence assembly in the real world, but serves to illustrate the principle of the method. Pairs of *k*-mers with identical suffix and prefix sequences have been identified – 'GA' for the first two *k*-mers in the series, and 'AC' for *k*-mers 2 and 3. The edges, shown here as arrows, link pairs of *k*-mers, resulting in a series from which the master sequence can be read. (C) Sequences containing repetitive DNA regions can be assembled correctly from a de Bruijn graph if an Eulerian pathway through the graph can be identified. This is a pathway that visits every edge just once. In this example, the Eulerian pathway follows the edges labeled 1 to 9 in series, resulting in unambiguous assembly of the master sequence even though that sequence contains a repeat of the motif 'ACC.'

From contigs to scaffolds

Assembly, whether by the overlap or de Bruijn methods, will result in a set of sequence contigs. The next step is therefore to identify contigs that are adjacent to one another in the genome, and hence to establish a series of **scaffolds**, each scaffold comprising a set of sequence contigs separated by gaps which are not covered by the read dataset (**Figure 4.22**). How do we construct scaffolds and how do we close the gaps between them?

If a reference genome is available, then placing the contigs at their appropriate places in the reference sequence will clearly show which contigs are adjacent and how much sequence space lies between them. However, the reference genome approach is not foolproof, and in particular might bias the new genome sequence toward the pattern of **structural variants** in the reference. These are natural variations that result in the genomes of two individuals of the same species differing in the numbers and locations of repeat units, the presence or absence of certain sequences of 50–2000 bp in length, and the positioning of some sequences (**Figure 4.23**). These variants might not be recognized in a resequencing project, because if the individual sequence reads are too short to span the entire sequence containing the variant, then the assembly that is obtained will be biased toward the structure of the reference genome. So even if a reference sequence is used, it is important to check that scaffolds have been constructed correctly. If a reference is not available, and the new sequence is a *de novo* assembly, then these other methods become critical as the only means of placing the contigs into scaffolds.

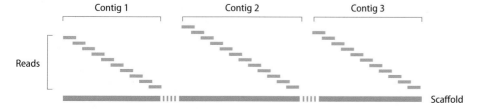

Figure 4.22 A scaffold. The scaffold is made up of sequence contigs that have been shown to lie adjacent to one another.

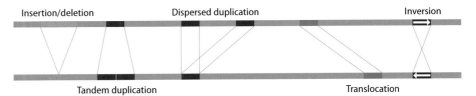

Figure 4.23 Structural variants that can complicate assemble of sequence reads on to a reference genome.

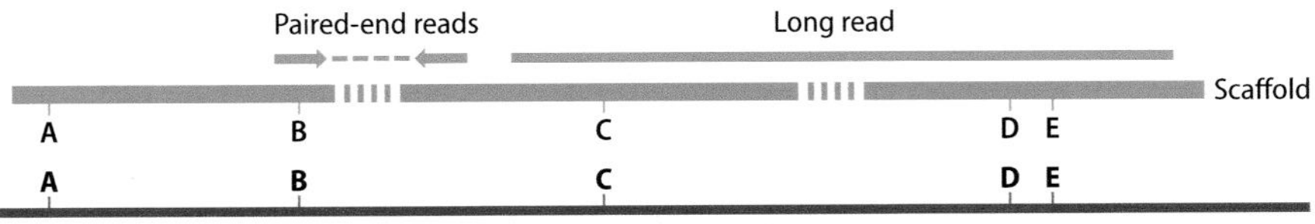

Figure 4.24 Three possible strategies for scaffold construction. Contigs that are adjacent to one another can be identified from paired-end reads, by reference to a genetic map, and from long-read sequences.

We are already familiar with three of the possible strategies for scaffold construction (**Figure 4.24**):

- Paired-end reads might be helpful. If two members of a pair of reads fall in different contigs, then clearly those contigs must be adjacent to one another in the genome sequence, and the gap between them could be filled in by further sequencing of the fragment from which the paired-end reads were obtained.

- If a genetic or physical map is available, then this could be used to anchor contigs on the genome sequence and to identify contigs that are adjacent. Optical maps (Section 3.5) are now frequently used for this purpose because they are relatively easy to generate and have high marker density.

- If the contigs have been assembled from short reads, then SMRT or nanopore sequencing could be used to generate long reads that would show which short-read contigs are adjacent. Until recently, long-read sequencing on its own has not routinely been used to generate data for genome assembly because of the high error rate, which is still a problem with nanopore sequencing and was also an issue with the earlier versions of the SMRT method. However, even if a long sequence read contains a few errors, it should still be possible to identify the location within it of one or more short-read contigs.

As well as the three strategies described above, there are more innovative methods for identifying contigs that are located near to one another in the genome sequence. One of these is **Hi-C sequencing**, which is a version of **chromosome conformation capture**, a method used to identify segments of the genome that are positioned close to one another within the three-dimensional structure of the nucleus. Within the nucleus, the chromosomal DNA molecules are folded into loops (Section 10.1), so two segments of DNA that are tens or hundreds of kb apart in the linear genome sequence might be positioned quite close to one another within the three-dimensional topology of the nucleus (**Figure 4.25**). This three-dimensional topology can be stabilized by treating nuclei with formaldehyde, which forms crosslinks between the $-NH_2$ groups present in DNA and proteins. In the Hi-C method, the resulting network is treated with a restriction endonuclease to break it into pieces, some of which will comprise two crosslinked segments of DNA, one from each of the two regions of the genome that were adjacent in the nucleus (**Figure 4.26**). The cut ends of the molecules are then extended by addition of a nucleotide that is labeled with **biotin**, a small organic molecule that binds strongly to **avidin**, a protein from egg white that has

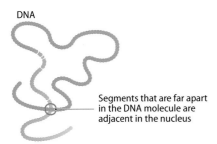

Figure 4.25 Two segments of DNA that are far apart in the genome might be located adjacent to one another in the nucleus.

Figure 4.26 Hi-C sequencing. A nuclear extract is treated with formaldehyde, so crosslinks form between adjacent segments of DNA. The resulting network is cut with a restriction endonuclease and the ends labeled with biotin. Ligation is then carried out in a dilute solution that favors intramolecular ligation, which therefore occurs preferentially between the ends of the crosslinked DNA fragments, which are relatively close to one another. Ligation between the non-crosslinked restriction fragments is less likely as these ends are relatively distant within the reaction solution. Disruption of the crosslinks then gives circular molecules that are broken into fragments, some of which carry a biotin label. Purification and paired-end sequencing of the labeled fragments enables the pairs of genome segments that were linked together by formaldehyde to be identified.

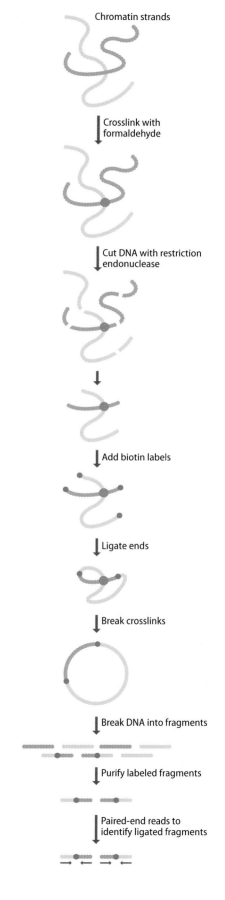

Chromatin strands

Crosslink with formaldehyde

Cut DNA with restriction endonuclease

Add biotin labels

Ligate ends

Break crosslinks

Break DNA into fragments

Purify labeled fragments

Paired-end reads to identify ligated fragments

a high binding affinity for biotin. The ends are then ligated and these molecules are broken into fragments of the appropriate size for short-read sequencing. In a final step, prior to library preparation, the biotin-labeled fragments are purified by attachment to magnetic beads that are coated with streptavidin. Paired-end sequencing is then carried out, enabling the pairs of genome segments that were linked together by formaldehyde to be identified. The rationale is that the closer two segments are in the linear map of the genome, the more likely they are to be linked together: in human nuclei, segments that are 10 kb apart are eight times more likely to be crosslinked than segments 100 kb apart. Analysis of the paired-read data can therefore give statistical information on the relative proximity of contigs in the genome sequence.

A second approach that has been valuable in the construction of scaffolds is **linked-read sequencing**. In this method, the short reads derived from a single, longer DNA molecule are **barcoded** so that they can be identified when their sequences are examined. A barcode is simply a DNA sequence that is unique to a particular DNA fragment or set of fragments, and which can be created by attaching a short adaptor molecule to the end of each fragment (**Figure 4.27**). In one version of linked-read sequencing, the DNA extraction procedure is designed to give DNA molecules of 50–100 kb, which are suspended in oil droplets, ideally one fragment per droplet (**Figure 4.28A**). Each droplet also contains adaptors that provide a barcode specific to that droplet (and hence also to the DNA molecule that it contains), as well as a PCR priming site. The DNA molecule in each droplet is fragmented and the barcoded adaptors are attached. Amplification of the fragments followed by Illumina sequencing results in barcoded reads, which can be sorted into groups to identify those that originate from the same DNA molecule. In an alternative method, the long DNA molecules are attached to magnetic beads that carry bead-specific barcode sequences (**Figure 4.28B**). These are inserted into the DNA molecule by a **transposase**, a type of enzyme that catalyzes **transposition**: the insertion of a shorter DNA molecule into a longer one (Section 9.2). At least some of the short reads obtained from each DNA molecule will therefore contain the barcode sequence.

As well as its importance in construction of scaffolds, the linked-read method also enables the sequences of each of a pair of homologous chromosomes to be distinguished. This is important because the two members of the chromosome pair will not have identical sequences: each will have its own SNP variations inherited from the female or male parent. These **heterozygosities** appear as ambiguities during the assembly process (**Figure 4.29**), with one or other of the variations being chosen to represent that position in the 'genome sequence.' The information that is lost complicates the use of genome sequences in applications where knowledge of the linkage between alleles is important, which include not just genetic mapping (Section 3.4), but also evolutionary studies where linkage is important in understanding the impact of selection on an organism's phenotype. This type of linked-read sequencing has been called **haplotagging**.

What is a 'genome sequence' and do we always need one?

In this chapter, we have examined how genome sequences are obtained. We have not, however, asked ourselves what we mean by the term 'genome sequence.' This is what we must do now.

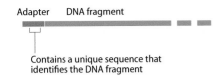

Contains a unique sequence that
identifies the DNA fragment

Figure 4.27 Barcoding. The barcode
is an adaptor that contains a unique
sequence. The adaptor is attached to
a DNA fragment and so 'barcodes' that
fragment.

We have already noted that for most species there is no such thing as *the* genome sequence. The individual members of a species have their own personal variations, the variable positions often making up a substantial fraction of the genome as a whole. The human genome, for example, contains over 10 million SNPs, an average of one SNP for every 300 bp of the genome. This means that a genome sequence is, at best, just the genome of a single representative of a species. In fact, for some of the first species to be studied, the 'genome sequence' was a conglomerate of different genomes, as separate sequencing libraries were prepared for different chromosomes, and the DNA for those libraries did not always come from the same individual.

It is also important to recognize that a 'complete' sequence – one in which every nucleotide is known, with no errors, and every segment placed at its correct position – has only recently become feasible for any genome. The less-complete sequence referred to as the **finished genome** will typically still have some unsequenced gaps between contigs and an average of up to one error per 10^4 nucleotides. In contrast, a **draft genome sequence** has a greater error rate, more gaps, and possibly some ambiguity about the order and/or orientation of some sequence contigs. Moving down the scale, we have partially-assembled genomes, which might be a 'work in progress,' and in extreme cases unassembled collections of sequence reads.

To avoid confusion over the meaning of terms such as 'complete,' 'finished,' and 'draft,' statistical measures have been devised to provide a more accurate indication of the quality of a genome sequence. One of these is the **N50 size**, which can be applied to either contigs or scaffolds. The concept behind the N50 size can be difficult to grasp, but becomes clear when a graphical representation is used (**Figure 4.30A**). Each contig is represented by an oblong, the length of which indicates the length of sequence covered by the contig. The contig oblongs are placed end to end, with the longest contig first, the second contig in second place, and so on, until we reach the shortest contig in the very last place. Now, to work out the N50 size, we find the contig whose position spans

Figure 4.28 Linked-read sequencing.
(A) The oil droplet method. DNA molecules are suspended in oil droplets which also contain barcoded adaptors. The DNA molecule in each droplet is fragmented and the barcoded adaptors attached.
(B) The magnetic bead method. DNA molecules are attached to magnetic beads that carry the barcoded adaptors. These are inserted into the DNA molecule by a transposase.

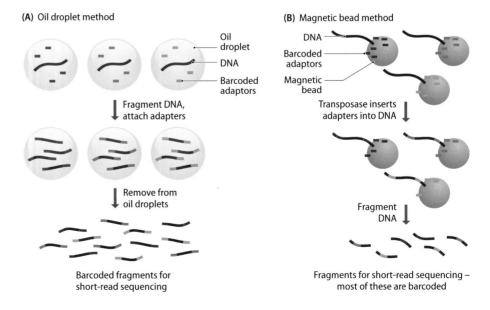

$$\ldots \text{ATGAGCATCGATGCA}^{\text{C}}_{\text{T}}\text{CAGCAGATTGAGCTAC}\ldots$$

Figure 4.29 The effect of heterozygosity on a sequence assembly. In this region of the assembly, some of the reads covering a particular position identify the nucleotide as C, whereas a similar number of reads identify the nucleotide as T. This is because of heterozygosity: one member of the pair of homologous chromosomes has a C at this position, and the other has a T.

the halfway point between the start and end of the linear series. The length of this contig is the N50 size for the genome assembly. In the example shown in Figure 4.30A, this contig is 30 kb in length, so the N50 size is 30 kb. What this means is the 50% of the nucleotides in the assembly are present in contigs that are 30 kb or longer. A higher N50 size, for example, 45 kb, would indicate a more completely assembled genome sequence, in this case with half the nucleotides in contigs greater or equal to 45 kb in length. Exactly the same method can be used to calculate the N50 size for the scaffolds rather than individual contigs.

A disadvantage of the N50 measure is that it takes no account of genome size. This means that two assemblies can have the same N50 sizes but represent quite different degrees of completeness of the genome sequence. A different value, the **NG50 size**, circumvents this problem and enables direct comparisons to be made between genome assemblies for different species. The same graphical method can be used, but the contig or scaffold whose length determines the NG50 size is not the one halfway along linear series: instead, it is the point that marks half the length of the genome that is being studied (**Figure 4.30B**).

Finally, we must appreciate that a genome sequence, however complete, is not always needed. All parts of the genome are fascinating, but the degree of fascination for any component is in the eye of the beholder. For example, for some research questions, the critical issue is the sequence of the **exome**, which is the complete set of exons in the genome, and hence the part that codes for proteins. Sequence variations in the exomes of different individuals might reveal protein polymorphisms that underlie cancer or some other disorder. The human exome comprises approximately 48 Mb of DNA, about 1.5% of the total genome. Sequencing the exome would therefore be a less challenging task than sequencing the entire genome. With short-read methods, sequencing can be directed at the exome, or any other part of the genome, by including **target enrichment** during preparation of the sequencing library. A large set of oligonucleotides, each typically 150 nucleotides in length, is synthesized, with the sequences of these oligonucleotides corresponding to the sequences of the part of the genome that is being targeted (**Figure 4.31**). The oligonucleotides then act as **baits** that hybridize to, and hence capture, the DNA fragments representing the segments of interest. If the baits are attached to magnetic beads, the captured DNA fragments can be collected with a magnet, and non-targeted parts of the genome, which remain in solution, can be discarded. The captured DNA is then released from the baits and used to prepare the sequencing library. Exome sequencing has been particularly important with crop plants, whose large genomes have only recently been sequenced. By focusing on the exome sequences rather than complete genomes, crop breeders were able to study the genetic basis to properties such as environmental adaptation before the full genome sequences became available.

4.3 SEQUENCING THE HUMAN GENOME

To complete this chapter, we will examine how the human genome has been sequenced in order to explore the ways in which the challenges inherent in

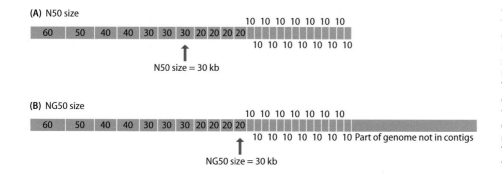

Figure 4.30 Statistical measures of the quality of a genome sequence. The contigs are represented by oblongs, placed end to end in order of size, shown here by numbers from 10 to 60 kb. (A) The N50 size is given by the size of the contig at the halfway point along the linear series. (B) To calculate the NG50 size, an additional oblong is added to represent that part of the genome that is not contained in the contigs. The overall length of the series is now equivalent to the size of the genome. The NG50 size is indicated by the size of the contig at the new halfway point.

Figure 4.31 Target enrichment. (A) Baits are used to capture DNA fragments representing segments of the genome that are of interest. (B) Only the captured DNA fragments are sequenced.

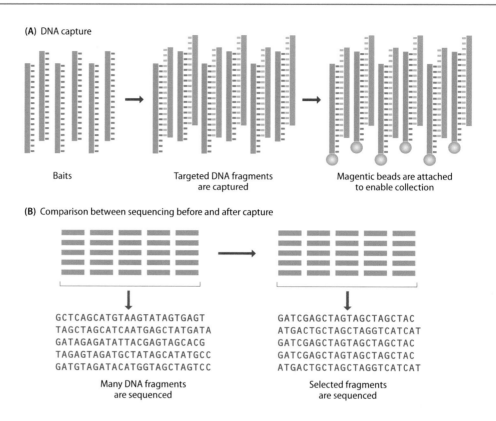

(A) DNA capture

Baits

Targeted DNA fragments are captured

Magentic beads are attached to enable collection

(B) Comparison between sequencing before and after capture

```
GCTCAGCATGTAAGTATAGTGAGT
TAGCTAGCATCAATGAGCTATGATA
GATAGAGATATTACGAGTAGCACG
TAGAGTAGATGCTATAGCATATGCC
GATGTAGATACATGGTAGCTAGTCC
```

Many DNA fragments are sequenced

```
GATCGAGCTAGTAGCTAGCTAC
ATGACTGCTAGCTAGGTCATCAT
GATCGAGCTAGTAGCTAGCTAC
GATCGAGCTAGTAGCTAGCTAC
ATGACTGCTAGCTAGGTCATCAT
```

Selected fragments are sequenced

sequencing and assembling a genome have been dealt with as the technologies have gradually advanced.

The Human Genome Project – genome sequencing in the heroic age

The **Human Genome Project** was established in the late 1980s as a loose but organized collaboration between geneticists in all parts of the world, with the objective of obtaining, by 2005, a finished sequence of at least 95% of the **euchromatin** component of the human genome – the part in which most of the genes are located (Section 10.1). Prior to the beginning of the HGP, complete sequences had only been obtained for viral and organelle genomes, the longest being the chloroplast genomes of tobacco and other plants, which are 155 kb. The massive conceptual leap to the 3,100 Mb human genome was looked on as mad by some biologists. Although comprehensive genetic maps had been constructed for fruit flies and a few other organisms, the problems inherent in analysis of human pedigrees (Section 3.4) and the relative paucity of polymorphic genetic markers, meant that most geneticists doubted whether a map of the human genome could ever be achieved, and it was assumed that without a map it would be impossible to assemble the genome sequence.

The initial goal that the Human Genome Project set itself was therefore a genetic map with a density of one marker per 1 Mb, although it was thought that a density of one per 2–5 Mb might be the realistic limit. An important breakthrough came from the discovery of RFLPs, which were the first highly polymorphic DNA markers to be recognized in animal genomes. In 1987 the first human RFLP map was published, comprising 393 RFLPs and ten additional polymorphic markers, with an average marker density of one per 10 Mb. By 1994, the genetic map had been extended to 7000 markers, most of which were SSLPs, with a density of one marker per 0.7 Mb. Physical mapping did not lag far behind, with publication in 1995 of a radiation hybrid map of 15,088 STS markers, with an average density of one per 199 kb. This map was later supplemented with an additional 20,104 STSs, most of these being ESTs and hence positioning protein-coding genes on the physical map. The combined STS maps included positions for almost 7000 polymorphic SSLPs that had also been mapped onto the genome by genetic means. As a result, the physical and genetic maps could be directly compared, and cloned pieces of DNA that included STS data could be anchored onto both maps. The net result was a comprehensive, integrated map that was then used as the framework for the DNA sequencing phase of the Human Genome Project.

The sequencing phase was begun at a time when it was thought that it would be impossible to assemble an entire eukaryotic genome solely by the shotgun method. A modification of the shotgun approach called **hierarchical shotgun sequencing** was therefore used. This approach involves a pre-sequencing phase during which the genome is broken into large fragments, typically 300 kb in length, that are cloned into a high-capacity vector such as a BAC (Section 2.3). Clones that contain overlapping fragments of DNA are then identified, for example, by searching for ones that share the same STS markers, enabling a contiguous series – a **clone contig** – to be built up (**Figure 4.32**). The sequence of each cloned fragment is then assembled by the shotgun method, and the master sequence is assembled by joining together the fragment sequences in the order dictated by the clone contig. With this approach, repetitive DNA will only cause a problem if two or more copies of the same repeat sequence are present in a single clone insert, and even then it might be possible to identify errors in the assembly by examining the sequences of clones that overlap with the one containing the repeats (**Figure 4.33**). For the Human Genome Project, a library of 300,000 BAC clones with known positions on the genome map was prepared, so the sequence contig from each clone could be anchored onto the genome map.

Just as the sequencing phase of the Human Genome Project was beginning, a second group of researchers started to question the assumption that a eukaryotic genome could not be assembled by shotgun sequencing. They showed that a reasonably accurate sequence of the 140 Mb genome of the fruit fly *Drosophila melanogaster* could be obtained simply by shotgun sequencing, and set out to obtain a draft of the much larger human genome by the same method. The possibility that the official Human Genome Project might not in fact provide the first human genome sequence stimulated the organizers of the Project to bring forward their planned dates for the completion of a working draft. The first draft sequence of an entire human chromosome (number 22) was published in December 1999, and the draft sequence of chromosome 21 appeared a few months later. Finally, on June 26, 2000, accompanied by the President of the United States, Francis Collins and Craig Venter, who were the leaders of the two projects, jointly announced completion of their genome drafts, which appeared in print eight months later.

The draft sequence covered just 90% of the genome, the missing 320 Mb lying predominantly in **constitutive heterochromatin**. This component of the genome is made up of chromosomal regions in which the DNA is very tightly packaged and in which there are few if any genes (Section 10.1). Within the 90% of the genome that was covered, each part had been sequenced at least four times, providing an 'acceptable' level of accuracy, but only 25% had been sequenced

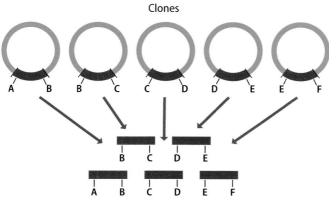

Clones

B C D E

A B C D E F

Contig assembled from shared STS markers

Figure 4.32 A clone contig. The fragments contained in this set of clones form an overlapping series, as indicated by the positions of the shared STS markers B–E.

Figure 4.33 Hierarchical shotgun sequencing can avoid problems with repeat sequences. The fragment in clone 2 contains two copies of a repeat sequence. Assembly of the reads from clone 2 could result in the segment between the two repeats being omitted. This error is avoided because clone 3 contains just one copy of the repeat. The sequence around this repeat can therefore be assembled unambiguously from the clone 3 reads, ensuring that the overall assembly is correct.

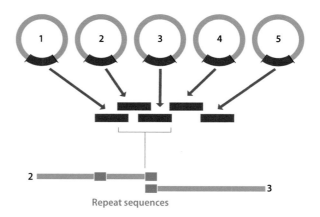

the eight to ten times that is necessary before the work is considered to be 'finished.' Furthermore, this draft sequence had approximately 150,000 gaps, and it was recognized that some segments had probably not been ordered correctly. The International Human Genome Sequencing Consortium, which managed the next phase of the project, set as its goal a 'finished' sequence of at least 95% of the euchromatin, with an error rate of less than one in 10^4 nucleotides, and all except the most refractory gaps filled. Achieving this goal required further chain-termination sequencing of 46,000 BAC, PAC, fosmid, and cosmid clones. The chromosome sequences from this phase of the project began to appear in 2004, with the entire genome sequence being considered complete a year later.

The Human Genome Project was a monumental undertaking that stretched the limits of the mapping and sequencing technologies that were available in the 1980s and 1990s. All of the work up to 2005 made use of chain-termination sequencing, as the first of the high throughput next-generation methods did not become available until the mid-2000s. We will now look at how the 'finished' genome sequence published in 2005 was made even more complete by the application of short- and long-read sequencing.

The human genome – genome sequencing in the modern age

The introduction of high throughput short-read sequencing in the mid-2000s transformed all areas of genome research. Short-read sequencing not only enabled sequence data to be generated much more rapidly than had been possible with the chain-termination method, it also dramatically reduced the cost of sequencing a human genome. The Human Genome Project, resulting in the first human genome sequence, is estimated to have cost over $100,000,000. By 2007, the cost of sequencing a single human genome had decreased, due in part to the availability of the original genome as a reference sequence. Assembly onto a reference sequence is cheaper than *de novo* sequencing because a lower coverage, and hence fewer reads, are needed in order to obtain an accurate sequence. However, in 2007 it still cost in the region of $10,000,000 to sequence a human genome. Today, with the availability of short- and long-read methods, a human genome sequence can be obtained for as little as $1000, taking into account labor, reagents, data processing, and indirect costs, such as depreciation of equipment, and there are indications that this figure will go down as far $100 per genome during the next few years.

Resequencing of human genomes has therefore become a relatively inexpensive exercise. This has prompted projects of ever-increasing ambition aimed at sequencing multiple genomes in order to study issues such as human variation and the association of sequence features with a predisposition to diseases such as breast cancer. The 1000 genomes project, which in fact sequenced 1092 human genomes, was completed in 2013. This was followed by 10,545 genome sequences in 2016, then a 100,000 genomes project carried out primarily in the UK that was completed in 2018, and finally the launch in the USA of a 1,000,000 genomes project scheduled to reach its target in 2028.

With these large-scale resequencing projects, the primary aim is to type as many SNPs and other sequence variations as possible, to provide data from which variants associated with disease can be identified. Despite the huge number of genome sequences that have been obtained, these projects did not significantly improved the quality of the human reference sequence. Although the human genome sequence was considered to be 'finished' in 2005, 15 years later the reference sequence still had unresolved areas, many of these regions containing head-to-tail arrays of repetitive DNA, possibly several Mb in length, in particular in the region of each **centromere**. This is the specialized structure that links together the two chromosome copies that are made prior to nuclear division, and which acts as the assembly point for the **kinetochores**, which attach to the microtubules which pull the chromosome copies apart during the anaphase stages of mitosis of meiosis (Section 7.1). Also lacking was detailed information on **segmental duplications**, regions of 1–400 kb that are repeated at various places in the genome (Section 18.2). The different copies of a duplication have >90% sequence identity and so are difficult to assemble from short reads, because of uncertainties regarding which reads belong to which copy. A growing realization that at least some variations in repetitive DNA arrays and segmental duplications have an important impact on the functioning of the genome made it imperative that more complete versions of the 'finished' genome sequence be obtained.

The development of accurate long-read sequencing methods have made it possible to obtain a genuinely complete human genome sequence. To avoid the problems caused by heterozygosities between two copies of each homologous pair of chromosomes (see Figure 4.29), this new phase of genome sequencing has focused on individual chromosomes, obtained either by flow cytometry (see Figure 3.39) or from **hydatidiform moles**, which sometimes occur when an egg cell that lacks a nucleus is fertilized by a sperm and implants in the uterus wall. The cells resulting from this egg cell form a small growth which contains only the paternal chromosomes. Sequencing typically makes use of the SMRT and nanopore methods to generate long reads with a coverage of >30×, with the accuracy of the assembly, especially in difficult parts of the genome, checked by reference to optical maps and by **polishing**, in which short reads are aligned on to the long-read assemblies to ensure that all sequence errors have been accounted for (**Figure 4.34**). With the X chromosome, this approach provided an accurate sequence of the 3.1 Mb region within which the centromere is located, and also filled in 29 additional gaps that were present in the previous version of the chromosome sequence, including more accurate assemblies of regions containing the CT-X and GAGE genes, which are expressed in some types of cancer. Similarly, with chromosome 8 the 2.08 Mb centromere sequence was resolved, and four other gaps were filled in, including a copy number variation in the β-defensin gene cluster, which codes for antimicrobial proteins synthesized in macrophages and epithelial cells. Additional work, using similar strategies, has also completed the sequences of the other 21 human chromosomes, with accurate coverage of each centromere as well as all the segmental duplications and

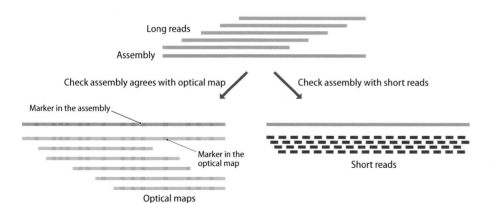

Figure 4.34 Polishing a genome assembly constructed with long reads. Following assembly of the long reads into contigs, the accuracy of the assembly is checked against the optical map (shown here as a series of overlapping maps that together cover the assembled sequence) and by ensuring that short reads can be aligned onto the assembly without ambiguities.

other repetitive sequences. A total of 182 Mb of new sequence has been added to the human reference genome, and another 56 Mb has undergone significant correction, revealing 79 new protein-coding genes whose existence had not previously been recognized.

The Neanderthal genome – assembly of an extinct genome using the human sequence as a reference

Neanderthals are extinct hominins who lived in Europe and parts of Asia between 200,000 and 30,000 years ago. Their preserved skeletons show many features similar to those of our own species, *Homo sapiens*, leading paleontologists to classify Neanderthals as another member of the *Homo* genus, *Homo neanderthalensis*, our closest relatives in the fossil world. Many of the distinctive features of Neanderthals are thought to be adaptations to the cold climates of the Ice Ages, such as an enlarged nasal cavity which might have helped to warm the air before it entered the lungs (**Figure 4.35**).

If the last Neanderthal died out 30,000 years ago, then how can we obtain the sequence of its genome? **Ancient DNA** has provided an answer. It has been known for some years that DNA molecules can survive the death of the organism in which they are contained, being recoverable centuries and possibly millennia later as short, degraded fragments preserved in bones and other biological remains. Unfortunately, the study of ancient DNA has been plagued with controversies. In the early 1990s, there were many reports of ancient human DNA being detected in bones and other archaeological specimens, but often it turned out that what had been sequenced was not ancient DNA at all, but contaminating modern DNA from the archaeologist who had excavated the specimen or the molecular biologist who carried out the DNA extraction. The worldwide success of the film *Jurassic Park* led to reports of DNA in insects preserved in amber and even in dinosaur bones, but all these claims are now known to be incorrect. Many biologists started to wonder if ancient DNA existed at all, but gradually it has become clear that if the work is carried out with extreme care, it is sometimes possible to extract authentic ancient DNA from specimens up to about 1 million years old. This is old enough to include Neanderthals.

Although they are different species, it was anticipated that the Neanderthal and *H. sapiens* genomes would be sufficiently similar for the human genome to act as a reference for assembly of the Neanderthal sequence reads. It would not therefore be necessary to attempt a *de novo* assembly of the Neanderthal sequence. There would, however, be one problem not usually encountered during a genome assembly. DNA degrades over time, and some of these degradation processes result in what ancient DNA researchers call **miscoding lesions**. A miscoding lesion is a chemical change that results in a nucleotide being read incorrectly during a sequencing experiment. The commonest miscoding lesion is removal of the amino group from a cytosine base, which is promoted by the presence of water and gives rise to uracil (**Figure 4.36A**). If this miscoding lesion occurs in an ancient DNA molecule, then the C will mistakenly be read as a T. These sequence errors must be distinguished from genuine positions at which the Neanderthal genome differs from the *H. sapiens* reference. For the Neanderthal genome sequence to be accurate, it would therefore be necessary to achieve a high coverage, so that each nucleotide in the genome is present in as many sequence reads as possible (**Figure 4.36B**). Those reads will be derived from different fragments of ancient DNA, and it would be extremely unlikely for all of those fragments to have a miscoding lesion at the same position. A genuine difference between the Neanderthal and *H. sapiens* sequences would therefore be signaled by a SNP that is present in each of the sequence reads covering a particular nucleotide. A miscoding lesion, on the other hand, would only occur in one of those reads.

The first complete Neanderthal genome sequence was obtained from a toe bone of an adult female, dated to approximately 50,000 years ago, and recovered from a cave site in the Altai mountains of Siberia. Five libraries were prepared

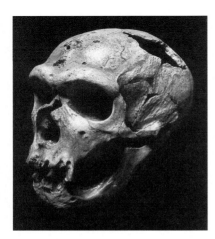

Figure 4.35 Skull of a 40- to 50-year-old male Neanderthal. The specimen is about 50,000 years old and was found at La Chapelle-aux-Saints, France. (From John Reader/Science Photo Library, with permission.)

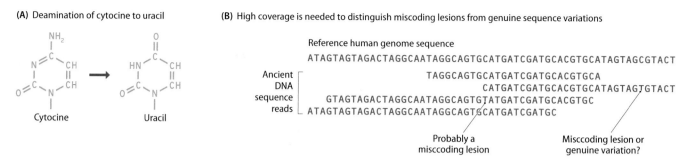

(A) Deamination of cytocine to uracil

Cytocine → Uracil

(B) High coverage is needed to distinguish miscoding lesions from genuine sequence variations

Reference human genome sequence
ATAGTAGTAGACTAGGCAATAGGCAGTGCATGATCGATGCACGTGCATAGTAGCGTACT

Ancient DNA sequence reads
TAGGCAGTGCATGATCGATGCACGTGCA
CATGATCGATGCACGTGCATAGTAGTGTACT
GTAGTAGACTAGGCAATAGGCAGTGTATGATCGATGCACGTGC
ATAGTAGTAGACTAGGCAATAGGCAGTGCATGATCGATGC

Probably a misccoding lesion

Misccoding lesion or genuine variation?

Figure 4.36 Miscoding lesions cause problems when sequencing ancient DNA. (A) Deamination of cytosine gives uracil. When a DNA sequence is read, this miscoding lesion results in the C being read as a T. (B) High coverage is needed to distinguish miscoding lesions from genuine sequence variations. In this example, there are two possible miscoding lesions. One can be identified with confidence as it is present in just one of four sequence reads. The second possible miscoding lesion is present in a part of the genome that is covered by just a single ancient DNA sequence read. It is impossible to tell if this C→T anomaly is a miscoding lesion or a genuine sequence variation.

and sequenced, from which over 2,278,000,000 Neanderthal sequence reads were obtained, using the Illumina method. The average length of these reads was only 75 bp, a second problem with ancient DNA being that the polynucleotides degrade into short fragments over time, limiting the lengths of the individual sequence reads. Nevertheless, it was possible to assemble the reads into a genome sequence with an average of 52× coverage, sufficient for a high degree of sequence accuracy. This work was published in 2014. Since then, high-coverage genome sequences have been obtained for a second Neanderthal specimen from the Altai mountains as well as one from the Vindija cave in Croatia. Ancient DNA sequencing has also provided a high-quality genome sequence for a previously-unknown type of hominin, called the Denisovans, who appear to be related to Neanderthals. Examination of the Neanderthal and Denisovan genome sequences, and comparisons with the *H. sapiens* genome, have shown that all three types of human interbred, and that the *H. sapiens* genome contains some sequences that we have inherited via this interbreeding. Those sequences include novel alleles of genes that may play roles in our immune response and susceptibility to certain diseases. We will return to the Neanderthal and Denisovan genomes in Section 18.3, where we will examine in more detail the fast-moving field of **paleogenomics** – the study of the genomes of extinct species.

The human genome – new challenges

Work on the human genome sequence is far from complete. In fact, the availability of high-quality complete versions of each human chromosome marks the starting point for a new series of sequencing projects aimed at exploring the structural variations present in the genomes of the human population as a whole. The previous 1000, 10,000, and 100,000 genomes projects were largely focused on variations in SNPs and other simple genetic markers, because these projects made use of the 'finished' version of the human genome as the reference sequence, which, as described above, was not sufficiently complete to enable accurate comparison of structural variations. The growing awareness that these structural features have functionality and may affect factors such as predisposition to certain diseases, prompts a new round of multiple genome resequencing aimed at charting the variations in as many genomes as possible.

A second aspect of the human genome that until recently was impossible to explore is the distribution within the sequence of chemical modifications such as methylated nucleotides. In Chapter 10, we will study how DNA methylation, along with chemical modification of some of the proteins associated with DNA within chromosomes, can modulate patterns of gene expression. Methylation, in particular of cytosine to 5′-methylcytosine (**Figure 4.37**) is associated with a repression or absence of gene activity. Methylation of sequences adjacent to genes is usually an indicator that those genes are not expressed in the cells or tissue being studied, and parts of the genome that lack genes, such as the

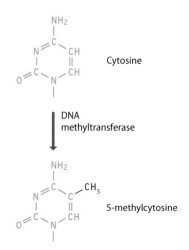

Cytosine

DNA methyltransferase

5-methylcytosine

Figure 4.37 Conversion of cytosine to 5-methylcytosine by a DNA methyltransferase.

centromeres, also have a high degree of methylation. It has been suggested that within the centromere, the pattern of DNA methylation influences the accuracy of chromosome partitioning during cell division, with unusual methylation patterns being seen in some types of tumor. Mapping DNA methylation within a genome is therefore an important requirement in studies of both gene expression patterns and aberrant functioning of the genome. Distinguishing cytosine from 5′-methylcytosine is not possible with any sequencing-by-synthesis method, which includes all the short-read techniques as well as the SMRT long-read method. However, the methylated and nonmethylated versions of a nucleotide can be distinguished during nanopore sequencing, because methylation affects the way in which the nucleotide occludes the nanopore and hence results in a different disruption in the pattern of ion flow, from which the sequence is deduced (see Figure 4.15). This property was exploited during the project that led to the complete version of chromosome 8, during which the methylation pattern in the centromere region was deduced from the nanopore sequence reads that spanned this region. The future challenge is to extend this approach to the human genomes present in different tissues, to chart how methylation dictates genome activity and responds to disease.

Finally, there is an entirely different area of research in which the human genome is looked on as just part of a larger biological system, which can only be understood in completeness if the genomes of other organisms are studied in a similar degree of detail. Humans are vertebrates, as are 71,656 other species that have been named and are believed still to be represented by living members, despite extinctions caused by loss of habitat and human interference. The **Vertebrate Genomes Project** aims to produce at least one high-quality genome sequence for every one of these vertebrate species, using the short- and long-read sequencing approaches, coupled with optical mapping, linked-read and Hi-C sequencing, and the sequencing polishing methods that have been used to complete the human chromosome sequences. The genome sequences resulting from this project will enable many aspects of vertebrate taxonomy and evolution to be studied, including the evolution of genomes themselves (Chapter 18). Even more ambitious is the **Earth BioGenome Project**, whose aim is to sequence the genome of every eukaryotic species. There is estimated to be 1.35 million of these, but with the increasing efficiency of the methodology for genome sequencing and assembly it is hoped that the project can be completed by 2028.

SUMMARY

- Chain-termination sequencing is used to sequence short DNA molecules such as PCR products.

- Next-generation sequencing is a collection of methods that enable thousands or millions of DNA fragments to be sequenced in parallel in a single experiment.

- The Illumina method begins with preparation of a library of DNA fragments that have been immobilized within a flow cell, followed by generation of up to 3×10^9 short reads corresponding to 750 Gb of sequence data.

- Other short-read methods have made use of pyrosequencing, the ion torrent method, and sequencing by ligation.

- Long reads can be obtained either by single-molecule real-time sequencing or by the nanopore method.

- When a genome is being sequenced, the major challenge is assembling all the sequence reads, obtained from the multiple sequencing experiments, in the correct order.

- With a small bacterial genome, sequence assembly is possible by the shotgun method, which simply involves examining sequence reads for overlaps.

- Shotgun sequencing of eukaryotic genomes requires more complex assembly procedures, such as the use of de Bruijn graphs.

- The initial product of sequence assembly is a set of contigs that can be assembled into scaffolds by various means, including the use of long-range information on the structure of the genome, such as HiC sequencing and linked-read sequencing.

- The completeness of a genome sequence is described by its N50 size.

- A draft of the human genome sequence was completed in 2000. An improved sequence, described as 'finished,' was published in 2005.

- More accurate sequences of the human chromosomes were obtained by a combination of long- and short-read sequencing, with the accuracy of assemblies checked against optical maps.

- The Neanderthal genome has been sequenced from ancient DNA.

- Nanopore sequencing enables the methylated and nonmethylated versions of cytosine to be distinguished, so patterns of DNA methylation can be studied.

- Ambitious projects have begun to sequence the genomes of all eukaryotic species by 2028.

SHORT ANSWER QUESTIONS

1. Explain how the inclusion of dideoxynucleotides in a strand-synthesis reaction enables a DNA sequence to be read.

2. Compare the strengths and weaknesses of the chain-termination and next-generation sequencing methods. What are the applications of chain-termination sequencing in modern genomics research?

3. How is a sequencing library prepared during an Illumina sequencing project?

4. Describe how the DNA sequence is obtained by (A) reversible terminator sequencing, (B) pyrosequencing, (C) the ion torrent method, and (D) SOLiD sequencing.

5. Outline the methodology used in single-molecule real-time sequencing and in the nanopore method.

6. Outline the approach that was used in order to sequence the *Haemophilus influenzae* genome.

7. What factors complicate the application of shotgun sequencing to a eukaryotic genome?

8. Describe how sequence reads are assembled into contiguous sequences.

9. What methods are available to assemble sequence contigs into scaffolds?

10. Explain how the NG50 size of a genome assembly is calculated.

11. Describe how the hierarchical shotgun method was used in the Human Genome Project.

12. What methods have been used to obtain complete sequences of human chromosomes?

13. How can errors resulting from DNA degradation be identified when ancient DNA is being sequenced?

14. What is DNA methylation, why is it important, and how can it be studied by DNA sequencing?

IN-DEPTH PROBLEMS

1. In the late 1970s, two methods for DNA sequencing – the chain-termination and chemical-degradation methods – were developed. Initially, both methods were used extensively, but gradually the chain-termination method became more popular. Why did the chemical-degradation method go out of favor?

2. You have isolated a new species of bacterium whose genome is a single DNA molecule of approximately 2.6 Mb. Write a detailed project plan to show how you would obtain the genome sequence for this bacterium.

3. A 122 bp DNA molecule is randomly broken into overlapping fragments and those fragments are sequenced. The resulting sequence reads are as follows:

    CGTAGCTAGCTAGCGATT
    GATTTAGTTCGCCCATTCG
    GCTGTAGCATGTTTTCGC
    TTCGCTCAGCATCGGATTT
    AGCTAGCTAGCGATTTAGT
    TAGCATGTTTTCGCTCAGC
    TTTCGCTCAGCATCGGATT
    ATTTAGTTTAGCTGTAGCA
    CATTCGCGATGCTATCTCT
    GTTGACGCATACGGCGGG
    TCGTAGCTAGCTAGCGAT
    ATGCTATCTCATCTGATTT
    ATTTAGTTCGCCCATTCGC
    ATTTAGTTGACGCATACGG
    ATGCATCGTAGCTAGCTAG
    CTCAGCATCGGATTTAGTT
    CGATGCTATCTCATCTGAT
    CGCATACGGCGGGGGGAT

    Is it possible to reconstruct the sequence of the original molecule by searching for overlaps between pairs of reads? If not, then what problem has arisen and how might this problem be solved?

4. A pharmaceutical company has invested a great deal of time and money to sequence the gene for a genetic disease. The company is now studying the gene and its protein product and is working to develop drugs to treat the disease. Does the company have the right (in your opinion) to patent the gene sequence? Justify your answer.

FURTHER READING

DNA sequencing methods

Chen, F., Dong, M., Ge, M., et al. (2013). The history and advances of reversible terminators used in new generations of sequencing technology. *Genom. Proteom. Bioinf.* 11:34–40.

Giani, A.M., Gallo, G.R., Gianfranceshi, L., et al. (2020). Long walk to genomics: History and current approaches to genome sequencing and assembly. *Comput. Struct. Biotechnol. J.* 18:9–19.

Kono, N. and Arakawa, K. (2019). Nanopore sequencing: Review of potential applications in functional genomics. *Dev. Growth Differ.* 61:316–326.

Logsdon, G., Vollger, M.R. and Eichler, E.E. (2020). Long-read human genome sequencing and its applications. *Nat. Rev. Genet.* 21:597–614.

Prober, J.M., Trainor, G.L., Dam, R.J., et al. (1987). A system for rapid DNA sequencing with fluorescent chain-terminating dideoxynucleotides. *Science* 238:336–341.

Rang, F.J., Kloosterman, W.P. and de Ridder, J. (2018). From squiggle to basepair: Computational approaches for improving nanopore sequencing read accuracy. *Genome Biol.* 19:90.

Ronaghi, M., Uhlén, M. and Nyrén, P. (1998). A sequencing method based on real-time pyrophosphate. *Science* 281:363–365. *Pyrosequencing.*

Sanger, F., Nicklen, S. and Coulson, A.R. (1977). DNA sequencing with chain-terminating inhibitors. *Proc. Natl Acad. Sci. USA* 74:5463–5467. *The first description of chain-termination sequencing.*

Shenure, J., Balasubramanian, S., Church, G.M., et al. (2017). DNA sequencing at 40: Past, present and future. *Nature* 550:345–353.

Wang, Y., Zhao, Y., Bollas, A., et al. (2021). Nanopore sequencing technology, bioinformatics and applications. *Nat. Biotechnol.* 39:1348–1365.

Shotgun sequencing of microbial genomes
Fleischmann, R.D., Adams, M.D., White, O., et al. (1995). Whole-genome random sequencing and assembly of *Haemophilus influenzae* Rd. *Science* 269:496–512.

Loman, N.J., Constantinidou, C., Chan, J.Z.M., et al. (2012). High-throughput bacterial genome sequencing: An embarrassment of choice, a world of opportunity. *Nat. Rev. Microbiol.* 10:599–606.

Sequence assembly
Ekblom, R. and Wolf, J.B.W. (2014). A field guide to whole-genome sequencing, assembly and annotation. *Evol. Appl.* 7:1026–1042.

Li, Z., Chen, Y., Mu, D., et al. (2011). Comparison of the two major classes of assembly algorithms: Overlap-layout-consensus and de-bruijn-graph. *Brief. Funct. Genomics* 11:25–37.

Luo, J., Wei, Y., Lyu, M., et al. (2021). A comprehensive review of scaffolding methods in genome assembly. *Brief. Bioinformatics* 22:1–19.

Miller, J.R., Koren, S. and Sutton, G. (2010). Assembly algorithms for next-generation sequencing data. *Genomics* 95:315–327.

Neier, J.I., Salazar, P.A., Kučka, M., et al. (2021). Haplotype tagging reveals parallel formation of hybrid races in two butterfly species. *Proc. Natl. Acad. Sci USA* 118:e2015005118. *Haplotagging.*

Ott, A., Schnable, J.C., Yeh, C.-T., et al. (2018). Linked read technology for assembling large complex and polyploid genomes. *BMC Genom.* 19:651.

Pal, K., Forcato, M. and Ferrari, F. (2019). Hi-C analysis: From data generation to integration. *Biophys. Rev.* 11:67–78.

Sequencing the human genome
International Human Genome Sequencing Consortium (2001). Initial sequencing and analysis of the human genome. *Nature* 409:860–921. *The draft sequence obtained by the Human Genome Project.*

International Human Genome Sequencing Consortium (2004). Finishing the euchromatic sequence of the human genome. *Nature* 431:931–945.

Logsdon, G.A., Vollger, M.R., Hsieh, P.H., et al. (2021). The structure, function and evolution of a complete human chromosome 8. *Nature* 593:101–107.

Miga, K.H., Koren, S., Rhie, A., et al. (2020). Telomere-to-telomere assembly of a complete human X chromosome. *Nature* 585:79–84.

Nurk, S., Koren, S., Rhie, A., et al. (2022). The complete sequence of the human genome. *Science* 376:44–53. *End-to-end sequences of the 22 human autosomes and the X chromosome.*

Prüfer, K., Racimo, F., Patterson, N., et al. (2014). The complete genome sequence of a Neanderthal from the Altai Mountains. *Nature* 505:43–49.

Rhie, A., McCarthy, S.A., Fedrigo, O., et al. (2021). Towards complete and error-free genome assemblies of all vertebrate species. *Nature* 592:737–746.

Simpson, J.T., Workman, R.E., Zuzarte, P.C., et al. (2017). Detecting DNA cytosine methylation using nanopore sequencing. *Nat. Methods.* 14:407–410.

Telenti, A., Pierce, L.C.T., Biggs, W.H., et al. (2016). Deep sequencing of 10,000 human genomes. *Proc. Natl. Acad. Sci. USA* 113:11901–11906.

Venter, J.C., Adams, M.D., Myers, E.W., et al. (2001). The sequence of the human genome. *Science* 291:1304–1351. *The draft sequence obtained by the shotgun method.*

GENOME ANNOTATION

A genome sequence is not an end in itself. Major challenges still have to be met in locating the genes and other interesting features within the genome sequence and assigning functions to those genes whose roles are unknown. These challenges can be addressed by a combination of computer analysis and experimentation, but a complete description of a genome is rarely possible. Even though the finished version of the human genome sequence has been available since 2005, there are still uncertainties regarding the number of genes in the genome, and many of the identified genes have unknown functions. The development of new methods for understanding genome sequences is one of the central goals of genomics research.

This chapter describes the methods used in **genome annotation**, the process by which genes are located in a genome sequence. In the following chapter, we will explore the various ways in which functions can be assigned to unknown genes.

5.1 GENOME ANNOTATION BY COMPUTER ANALYSIS OF THE DNA SEQUENCE

Once an assembled genome sequence has been obtained, various methods can be employed to locate the genes that are present. These methods can be divided into those that involve simply inspecting the sequence, by eye or, more frequently, by computer, and those methods that locate genes by experimental analysis.

Sequence inspection can be used to locate genes because genes are not random series of nucleotides but instead have distinctive features. At present, we do not fully understand the nature of all of these specific features, and sequence inspection is therefore not a foolproof way of locating genes, but it is still a powerful tool and is usually the first method that is applied to analysis of a new genome sequence. The computer techniques used in sequence inspection form part of the methodology called **bioinformatics**, and it is with these that we begin.

The coding regions of genes are open reading frames

Genes that code for proteins comprise **open reading frames (ORFs)** consisting of a series of codons that specify the amino acid sequence of the protein that the gene codes for (**Figure 5.1**). The ORF begins with an initiation codon – usually (but not always) ATG – and ends with a termination codon: TAA, TAG, or TGA (Section 1.3). **ORF scanning** or *ab initio* **gene prediction**, which involves searching a DNA sequence for ORFs that begin with an ATG and end with a termination triplet, is therefore one way of looking for genes. The analysis is complicated by the fact that each DNA sequence has six **reading frames**, three in one direction and three in the reverse direction on the complementary strand (**Figure 5.2**), but computers are quite capable of scanning all six reading frames for ORFs. How effective is this as a means of gene location?

DOI: 10.1201/9781003133162-5

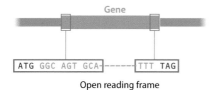

Figure 5.1 A protein-coding gene is an open reading frame of triplet codons. The first four and last two codons of the gene are shown. The first four codons specify methionine/initiation–glycine–serine–alanine, and the last two specify phenylalanine–termination.

The key to the success of ORF scanning is the frequency with which termination codons appear in the DNA sequence. If the DNA has a random sequence and a GC content of 50%, then each of the three termination codons – TAA, TAG, and TGA – will appear, on average, once every $4^3 = 64$ bp. If the GC content is greater than 50%, then the termination codons, being AT-rich, will occur less frequently, but one will still be expected every 100–200 bp. This means that random DNA should not show many ORFs longer than 50 codons in length, especially if the presence of a starting ATG triplet is used as part of the definition of an ORF. Most genes, on the other hand, are longer than 50 codons: the average lengths are 300–350 codons for bacterial genes and approximately 450 codons for humans. ORF scanning, in its simplest form, therefore takes a figure of, say, 100 codons as the shortest length of a putative gene and records positive hits for all ORFs longer than this.

How well does this strategy work in practice? With bacterial genomes, simple ORF scanning is an effective way of locating most of the genes in a DNA sequence. This is illustrated by **Figure 5.3**, which shows a segment of the *E. coli* genome with all ORFs longer than 50 codons highlighted. The real genes in the sequence cannot be mistaken because they are much longer than 50 codons in length. With bacteria the analysis is further simplified by the fact that the genes are very closely spaced and hence there is relatively little **intergenic DNA** in the genome (only 11% for *E. coli*; see Section 8.2). If we assume that the real genes do not overlap, which is true for most bacterial genes, then a short, spurious ORF would have to be located in an intergenic region in order for it to be mistaken for a real gene. This means that if the intergenic component of a genome is small, then there is a reduced chance of making mistakes in interpreting the results of a simple ORF scan.

Simple ORF scans are less effective with genomes of higher eukaryotes

Although ORF scans work well for bacterial genomes, they are less effective for locating genes in DNA sequences from higher eukaryotes. This is partly because there is substantially more space between the real genes in a eukaryotic genome, increasing the chances of finding spurious ORFs. But the main problem with the human genome and the genomes of higher eukaryotes in general is that their genes are often split by introns (Section 1.2), and so do not appear as continuous ORFs in the DNA sequence. Many exons are shorter than 100 codons, some consisting of fewer than 50 codons, and continuing the reading frame into an intron usually leads to a termination sequence that appears to close the ORF (**Figure 5.4**). In other words, the genes of a higher eukaryote do not appear in the genome sequence as long ORFs, and simple ORF scanning cannot locate them.

Solving the problem posed by introns is the main challenge for bioinformaticians writing new software programs for ORF location. Three modifications to the basic procedure for ORF scanning have been adopted:

- **Codon bias** is taken into account. 'Codon bias' refers to the fact that not all codons are used equally frequently in the genes of a particular organism. For example, leucine is specified by six codons in the genetic code (TTA, TTG, CTT, CTC, CTA, and CTG; see Figure 1.26), but in human genes leucine is most frequently coded by CTG and is only rarely specified by TTA or CTA. Similarly, of the four valine codons, human genes use GTG four times more frequently than GTA. The biological reason for codon bias is not understood, but all organisms have a bias, which is different in different species. Real exons are expected to display the codon bias, whereas chance series of triplets do not. The codon bias of the organism being studied is therefore written into the ORF-scanning software.

- **Exon–intron boundaries** can be searched for as these have distinctive sequence features, although unfortunately the distinctiveness of these

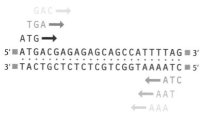

Figure 5.2 A double-stranded DNA molecule has six reading frames. Both strands are read in the 5′→3′ direction. Each strand has three reading frames, depending on which nucleotide is chosen as the starting position.

```
GCGCAACGCAATTAATGTGCGTTAGCTCACTCATTAGGCACCCCAGGCTTTACACTTTATGCTTCCGGCTCGTATGTTGTGTGGAATTGTGAGCGGATAACAATTTCACACAGGAAACAGCTATGACCATGATTACGGATTCACTGGCCGTCGTTTTACAACGTCGTGACTGGGAAAACCCTGGCGTTACCCAACTTAATCGCCTTGCAGCACATCCCCCTTTCGCCAGCTGGCGTAA

TAGCGAAGAGGCCCGCACCGATCGCCCTTCCCAACAGTTGCGCAGCCTGAATGGCGAATGGCGCTTTGCCTGGTTTCCGGCACCAGAAGCGGTGCCGGAAAGCTGGCTGGAGTGCGATCTTCCTGAGGCCGATACTGTCGTCGTCCCCTCAAACTGGCAGATGCACGGTTACGATGCGCCCATCTACACCAACGTAACCTATCCCATTACGGTCAATCCGCCGTTTGTTCCCACGGAG

AATCCGACGGGTTGTTACTCGCTCACATTTAATGTTGATGAAAGCTGGCTACAGGAAGGCCAGACGCGAATTATTTTTGATGGCGTTAACTCGGCGTTTCATCTGTGGTGCAACGGGCGCTGGGTCGGTTACGGCCAGGACAGTCGTTTGCCGTCTGAATTTGACCTGAGCGCATTTTTACGCGCCGGAGAAAACCGCCTCGCGGTGATGGTGCTGCGTTGGAGTGACGGCAGTTATC

TGGAAGATCAGGATATGTGGCGGATGAGCGGCATTTTCCGTGACGTCTCGTTGCTGCATAAACCGACTACACAAATCAGCGATTTCCATGTTGCCACTCGCTTTAATGATGATTTCAGCCGCGCTGTACTGGAGGCTGAAGTTCAGATGTGCGGCGAGTTGCGTGACTACCTACGGGTAACAGTTTCTTTATGGCAGGGTGAAACGCAGGTCGCCAGCGGCACCGCGCCTTTCGGCGG

TGAAATTATCGATGAGCGTGGTGGTTATGCCGATCGCGTCACACTACGTCTGAACGTCGAAAACCCGAAACTGTGGAGCGCCGAAATCCCGAATCTCTATCGTGCGGTGGTTGAACTGCACACCGCCGACGGCACGCTGATTGAAGCAGAAGCCTGCGATGTCGGTTTCCGCGAGGTGCGGATTGAAAATGGTCTGCTGCTGCTGAACGGCAAGCCGTTGCTGATTCGAGGCGTTAAC

CGTCACGAGCATCATCCTCTGCATGGTCAGGTCATGGATGAGCAGACGATGGTGCAGGATATCCTGCTGATGAAGCAGAACAACTTTAACGCCGTGCGCTGTTCGCATTATCCGAACCATCCGCTGTGGTACACGCTGTGCGACCGGCTACGGCCTGTATGTGGTGGATGAAGCCAATATTGAAACCCACGGCATGGTGCCAATGAATCGTCTGACCGATGATCCGCGCTGGCTACCGG

CGATGAGCGAACGCGTAACGCGAATGGTGCAGCGCGATCGTAATCACCCGAGTGTGATCATCTGGTCGCTGGGGAATGAATCAGGCCACGGCGCTAATCACGACGCGCTGTATCGCTGGATCAAATCTGTCGATCCTTCCCGCCCGGTGCAGTATGAAGGCGGCGGAGCCGACACCACGGCCACCGATATTATTTGCCCGATGTACGCGCGCGTGGATGAAGACCAGCCCTTCCCGGC

TGTGCCGAAATGGTCCATCAAAAAATGGCTTTCGCTACCTGGAGAGACGCGCCCGCTGATCCTTTGCGAATAGCGCCACGCGATGGGTAACAGTCTTGGCGGTTTCGCTAAATACTGGCAGGCGTTTCGTCAGTATCCCCGTTTACAGGGCGGCTTCGTCTGGGACTGGGTGGATCAGCTGCTGATTAAATATGATGAAAACGGCAACCCGTGGTCGGCTTACGGCGGTGATTTTGGC

GATACGCCGAACGATCGCCAGTTCTGTATGAACGGTCTGGTCTTTGCCGACCGCACGCCGCATCCAGCGCTGACGGAAGCAAAACACCAGCAGCAGTTTTTCCAGTTCCGTTTATCCGGGCAAACCATCGAAGTGACCAGCGAATACCTGTTCCGTCATAGCGATAACGAGCTCCTGCACTGGATGGTGGCGCTGGATGGTAAGCCGCTGGCAAGCGGTGAAGTGCCTCTGGATGTCG

CTCCACAAGGTAAACAGTTGATTGAACTGCCTGAACTACCGCAGCCGGAGAGCGCCGGGCAACTCTGGCTCACAGTACGCGTAGTGCAACCGAACGCGACCGCATGGTCAGAAGCCGGGCACATCAGCGCCTGGCAGCAGTGGCGTCTGGCGGAAAACCTCAGTGTGACGCTCCCCGCCGCGTCCCACGCCATCCCGCATCTGACCACCAGCGAAATGGATTTTTGCATCGAGCTGGG

TAATAAGCGTTGGCAATTTAACCGCCAGTCAGGCTTTCTTTCACAGATGTGGATTGGCGATAAAAAACAACTGCTGACGCCGCTGCGCGATCAGTTCACCCGTGCACCGCTGGATAACGACATTGGCGTAAGTGAAGCGACCCGCATTGACCCTAACGCCTGGGTCGAACGCTGGAAGGCGGCGGGCCATTACCAGGCCGAAGCAGCGTTGTTGCAGTGCACGGCAGATACACTTGCT

GATGCGGTGCTGATTACGACCGCTCACGCGTGGCAGCATCAGGGGAAAACCTTATTTATCAGCCGGAAAACCTACCGGATTGATGGTAGTGGTCAAATGGCGATTACCGTTGATGTTGAAGTGGCGAGCGAATACACCGCATCCGGCGGATTGGCCTGAACTGCCAGCTGGCGCAGGTAGCAGAGCGGGTAAACTGGCTCGGATTAGGGCCGCAAGAAAACTATCCCGACCGCCTTA

CTGCCGCCTGTTTTGACCGCTGGGATCTGCCATTGTCAGACATGTATACCCCGTACGTCTTCCCGAGCGAAAACGGTCTGCGCTGCGGGACGCGCGAATTGAATTATGGCCCACACCAGTGGCGCGGCGACTTCCAGTTCAACATCAGCCGCTACAGTCAACAGCAACTGATGGAAACCAGCCATCGCCATCTGCTGCACGCGGAAGAAGGCACATGGCTGAATATCGACGGTTTCCA

TATGGGGATTGGTGGCGACGACTCCTGGAGCCCGTCAGTATCGGCGGAATTCCAGCTGAGCGCCGGTCGCTACCATTACCAGTTGGTCTGGTGTCAAAAATAATAATAACCGGGCAGGCCATGTCTGCCCGTATTTCGCGTAAGGAAATCCATTATGTACTATTTAAAAAACACAAACTTTTGGATGTTCGGTTTATTCTTTTTCTTTTACTTTTTTATCATGGGAGCCTACTTCCCG

TTTTTCCCGATTTGGCTACATGACATCAACCATATCAGCAAAAGTGATACGGGTATTATTTTTGCCGCTATTTCTCTGTTCTCGCTATTATTCCAACCGCTGTTTGGTCTGCTTTCTGACAAACTCGGGCTGCGCAAATACCTGCTGGTGGATTATTACCGGATGTTAGTGATGTTTGCGCCGTCTTCATTTATTTTATCTTCGGGCCACTGTTACAATACAACACATTTTAGTAGGATCGA

TTGTTGGTGGTTATTATCTAGGCTTTTGTTTTTAACGCCGGTGCGCCAGCAGTAGAGGCATTTATTGAGAAAGTCAGCCGTCGCAGTAAATTTCGAATTTGGTCGCGCGGCGGATGTTTGGCTGTCGTTGGCTGGGCGCTGTGTGCCTCGATTGTCGGCATCATGTTCACCATCAATAATCAGTTTGTTTTCTGGCTGGGCTCTGGCTGCACTCATCCTCGCCGTTTACTCTTTTCGC

CAAAACGGATGCGCCCTCTTCTGCCACGGTTGCCAATGCGGTAGGTGCCAACCATTCGGCATTTAGCCTTAAGCTGGCACTGGAACTGTTCAGACAGCCAAAACTGTGGTTTTTGTCACTGTATGTTATTGGCGTTTCCTGCACCTACGATGTTTTTGACCAACAGTTTGCTAATTTCTTTACTTCGTTCTTTGCTACCGGTGAACAGGGTACGCGGGTATTTGGCTACGTAACGACA

ATGGGCGAATTACTTAACGCCTCGATTATGTCTTTGCGCCACTGATCATTAATCGCATCGGTGGGAAAAACGCCCTGCTGCTGGCTGGCACTATTATGTCTGTACGTATTATTGGCTCATCGTTCGCCACCTCAGCGCTGGAAGTGGTTATTCTGAAAACGCTGCATATGTTTGAAGTACCGTTCCTGCTGGTGGGCTGCTTTAAATATATTACCAGCCAGTTTGAAGTGCGTTTTT

CAGCGACGATTTATCTGGCTCGTTTCTGCGTTCTTTAAGCAACTGGCGATGATTTTTTATGTCGTACTGGCGGGCAATATGTATGAAAGCATCGGTTTCCAGGGCGCTTATCCTGGTGCTGGGTCTGGTGGCGCTGGGCTTCACCTTAATTTCCGTGTTCACGCTTAGCGGCCCCGGCCCGCTTTCCCTGCGTGCGTCGTCAGGTGAATGAAGTCGCTTAAGCAATCAATGTCGGATGCGG
```

Figure 5.3 **ORF scanning is an effective way of locating genes in a bacterial genome.** The diagram shows 4522 bp of the lactose operon of *Escherichia coli* with all ORFs longer than 50 codons marked. The sequence contains two real genes – *lacZ* and *lacY* – indicated by the red lines. These real genes cannot be mistaken because they are much longer than the spurious ORFs, shown in yellow.

sequences is not so great as to make their location a trivial task. The sequence of the upstream exon–intron boundary is usually described as:

5′–AG↓GTAAGT–3′

the arrow indicating the precise boundary point. However, only the 'GT' immediately after the arrow is invariable: elsewhere in the sequence, nucleotides other than the ones shown are quite often found. In other words, the sequence is a **consensus**, by which we mean that the sequence shows the most frequent nucleotide at each position in all of the upstream exon–intron boundaries that are known, but that in any particular boundary sequence one or more of these positions might have a different nucleotide (**Figure 5.5**). The downstream intron–exon boundary is even less well-defined:

5′–PyPyPyPyPyPyNCAG↓–3′

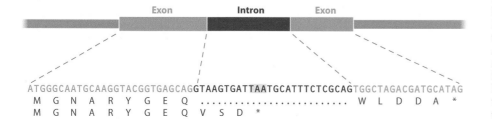

Figure 5.4 ORF scans are complicated by introns. The nucleotide sequence of a short gene containing a single intron is shown. The correct amino acid sequence of the protein translated from the gene is given immediately below the nucleotide sequence: in this sequence, the intron has been left out because it is removed from the transcript before the mRNA is translated into protein. In the lower line, the sequence has been translated without realizing that an intron is present. As a result of this error, the amino acid sequence appears to terminate within the intron. The amino acid sequences have been written using the one-letter abbreviations (see Table 1.2). Asterisks indicate the positions of termination codons.

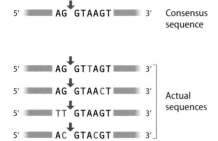

Figure 5.5 The relationship between a consensus sequence for an upstream exon–intron boundary and the actual sequences found in real genes. Differences from the consensus sequence are shown in red. At upstream exon–intron boundaries, only the 'GT' immediately after the splice site (shown by the arrow) is invariant

where 'Py' means one of the pyrimidine nucleotides (T or C) and 'N' is any nucleotide.

- **Upstream regulatory sequences** can be used to locate the regions where genes begin. This is because these regulatory sequences, like exon–intron boundaries, have distinctive sequence features that they possess in order to carry out their role as recognition signals for the DNA-binding proteins involved in gene expression (Section 12.3). Unfortunately, as with exon–intron boundaries, the regulatory sequences are variable, more so in eukaryotes than in prokaryotes, and in eukaryotes not all genes have the same collection of regulatory sequences. Using these to locate genes is therefore problematic.

These three extensions of simple ORF scanning, despite their limitations, are generally applicable to the genomes of all higher eukaryotes. Additional strategies are also possible with individual organisms, based on the special features of their genomes. For example, vertebrate genomes contain **CpG islands** upstream of many genes, these being sequences of approximately 1 kb in which the GC content is greater than the average for the genome as a whole. Some 40–50% of human genes are associated with an upstream CpG island. These sequences are distinctive and when one is located in vertebrate DNA, a strong assumption can be made that a gene begins in the region immediately downstream.

Ab initio gene prediction with eukaryotic genomes remains an inefficient process, despite the increasing sophistication of the computer programs that have been developed for this task. For most genomes, the starts and ends of genes can be predicted with almost 100% accuracy, but the accuracy in identifying exon–intron boundaries is much lower, usually only 60–70%. These figures assume that there is some *a priori* knowledge of parameters such as codon bias. If the genome is completely unstudied, then the accuracy of gene prediction will be lower, even though most gene prediction software includes a machine learning function, so the computer becomes 'trained' to recognize appropriate patterns of codon usage as it gradually builds up the genome annotation.

Locating genes for noncoding RNA

ORF scanning is appropriate for protein-coding genes, but what about those genes for noncoding RNAs such as rRNA and tRNA (Section 1.2)? These genes do not comprise open reading frames and hence will not be located by the methods described above. Noncoding RNA molecules do, however, have their own distinctive features, which can be used to aid their discovery in a genome sequence. The most important of these features is the ability to fold into a secondary structure, such as the **cloverleaf** adopted by tRNA molecules (**Figure 5.6A**). These secondary structures are held together by base-pairing not between two separate polynucleotides, as in the DNA double helix, but between different parts of the same polynucleotide – what we call **intramolecular base-pairing**. In order for intramolecular base pairs to form, the nucleotide sequences in the two parts of the molecule must be complementary, and to produce a complex structure such as the cloverleaf, the components of these pairs of complementary sequences must be arranged in a characteristic order within the RNA sequence (**Figure 5.6B**). These features provide a wealth of information that can be used to locate tRNA genes in a genome sequence, and programs designed for this specific purpose are usually very successful.

Ribosomal RNAs and some of the short noncoding RNAs (Section 1.2) also adopt secondary structures that have sufficient complexity to enable their genes to be identified without too much difficulty. Other noncoding RNA genes are less easy to locate because the RNAs take up structures that involve relatively little base-pairing or the base-pairing is not in a regular pattern. Three approaches are being used for location of the genes for these RNAs:

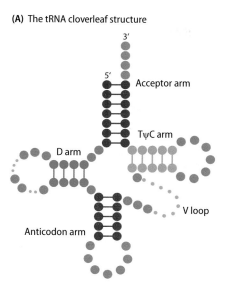

(A) The tRNA cloverleaf structure

(B) Sequence of one of the *Escherichia coli* tRNA^{leu} genes

5′ GCCGAAGTGGCGA**AATCGGTAG**TCGCAGTTGATTCAAAATCAACCGTAGAAATACGT**GCCGG**TTCGAGT**CCGGC**CTTCGGC**ACCA** 3′

Figure 5.6 The distinctive features of tRNAs aid location of the genes for these noncoding RNAs. (A) All tRNAs fold into the cloverleaf structure, which is held together by intramolecular base-pairing in the four highlighted regions. (B) The DNA sequence of the gene for one of the *Escherichia coli* tRNAs specific for the amino acid leucine is shown. The highlighted segments correspond to the regions of intramolecular base-pairing shown in part A. The sequence constraints imposed by the need for these segments to be able to base-pair to one another provide features that can be searched for by computer programs designed to locate tRNA genes.

- Although some noncoding RNAs do not adopt complex secondary structures, most contain one or more **stem-loops** (or **hairpins**), which result from the simplest type of intramolecular base-pairing (**Figure 5.7**). Programs that scan DNA sequences for such structures therefore identify regions where noncoding RNA genes might be present. These programs incorporate thermodynamic rules that enable the stability of a stem-loop to be estimated, taking into account features such as the size of the loop, the number of base pairs in the stem, and the proportion of G–C base pairs (these being more stable than A–T pairs as they are held together by three rather than two hydrogen bonds; see Figure 1.9). A putative stem-loop structure with an estimated stability above a chosen limit is considered a possible indicator of the presence of a noncoding RNA gene.

- As with protein-coding genes, a search can be made for regulatory sequences associated with genes for noncoding RNAs. These regulatory sequences are different to those for protein-coding genes, and may be present within a noncoding RNA gene as well as upstream of it.

- In compact genomes, attention is directed toward regions that remain after a comprehensive search for protein-coding genes. Often these 'empty spaces' are not empty at all and a careful examination will reveal the presence of one or more noncoding RNA genes.

Homology searches and comparative genomics give an extra dimension to gene prediction

The limitations of *ab initio* gene prediction can be offset to a certain extent by the use of a **homology search** to test whether a series of triplets is a real exon or a chance sequence. In this analysis the DNA databases are searched to compare the test sequence with genes that have already been sequenced. Obviously, if the test sequence is part of a gene that has already been sequenced by someone else, then an identical match will be found, but this is not the point of a homology search. Instead, the intention is to determine if an entirely new sequence is similar to

Figure 5.7 A typical RNA stem-loop structure.

any known genes because, if it is, then there is a chance that the test and match sequences are **homologous**, meaning that they represent genes that are evolutionarily related. The main use of homology searching is to assign functions to newly discovered genes, and we will therefore return to it when we deal with this aspect of genome analysis in the next chapter (Section 6.1). The technique is also central to gene prediction because it enables the authenticity of tentative exon sequences located by ORF scanning to be tested. If the tentative exon sequence gives one or more positive matches after a homology search, then it is probably a real exon, but if it gives no match then its authenticity must remain in doubt until it is assessed by one or other of the experiment-based genome annotation techniques.

A more precise version of homology searching is possible when genome sequences are available for two or more related species. Related species have genomes that share similarities inherited from their common ancestor, overlaid with species-specific differences that have arisen since the species began to evolve independently (**Figure 5.8**). Because of natural selection, the sequence similarities between related genomes are greatest within the genes and lowest in the intergenic regions. Therefore, when related genomes are compared, homologous genes are easily identified because they have high sequence similarity, and any ORF that does not have a clear homolog in the second genome can be discounted as almost certainly being a chance sequence and not a genuine gene. The value of this type of analysis – called **comparative genomics** – was illustrated during annotation of the genome of the yeast *Saccharomyces cerevisiae*. Complete or partial sequences were available not only for this yeast but also for various other members of the Saccharomycetes, including *Saccharomyces paradoxus*, *Saccharomyces mikatae*, and *Saccharomyces bayanus*, the species most closely related to *S. cerevisiae*. Comparisons between these genomes confirmed the authenticity of a number of *S. cerevisiae* ORFs, and also enabled almost 500 putative ORFs to be removed from the *S. cerevisiae* catalog on the grounds that they have no equivalents in the related genomes. The analysis is made even more powerful by the **synteny** – conservation of gene order – displayed by the genomes of these related yeasts. Although each genome has undergone its own species-specific rearrangements, there are still substantial regions where the gene order in the *S. cerevisiae* genome is the same as in one or more of the related genomes. This makes it very easy to identify homologous genes but, more importantly, enables a spurious ORF, especially a short one, to be discarded with great confidence, because its expected location in a related genome can be searched in detail to ensure that no equivalent is present (**Figure 5.9**).

Synteny is also displayed by other related groups of species. A second important example occurs among the grasses, including commercially important cereals such as barley, maize, sorghum, sugar cane, foxtail millet, and rice. The rice genome, which at 430 Mb is the smallest of this group, was sequenced in 2005. The wild grass *Brachypodium distachyon* has an even smaller genome, of 270 Mb, and this also has been sequenced and shown to be syntenic with the larger cereal genomes. The synteny within this family has been particularly important in annotating the genomes of those cereals, such as barley, that have very large and complex genomes.

Figure 5.8 Related species have similar genomes. (A) An illustration of how gene organization might change as two species diverge from their common ancestor. The common ancestor has five genes labeled A to E. In one of the derived species, gene C is no longer present, and in the other species, gene A has become truncated. (B) Related species display DNA sequence similarities. The diagram shows a short section of a gene sequence from an ancestral organism, along with the homologous sequences for this gene segment in the derived species.

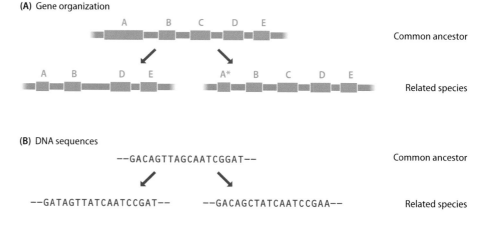

(A) Gene organization

A   B   C   D   E   Common ancestor

A   B   D   E   A*   B   C   D   E   Related species

(B) DNA sequences

--GACAGTTAGCAATCGGAT--   Common ancestor

--GATAGTTATCAATCCGAT--   --GACAGCTATCAATCCGAA--   Related species

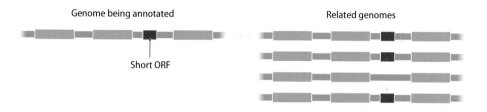

Figure 5.9 Using comparisons with syntenic genomes to test the authenticity of a short ORF. In this example, the ORF is present in three of the four related genomes, and hence is likely to be a genuine gene.

5.2 GENOME ANNOTATION BY ANALYSIS OF GENE TRANSCRIPTS

The second approach to genome annotation makes use of experimental techniques to locate genes within a genome sequence. These methods are not usually based on direct examination of DNA molecules but instead rely on detection of the RNA molecules that are transcribed from genes. All genes are transcribed into RNA, and if the gene is discontinuous then the primary transcript is subsequently processed to remove the introns and link up the exons (Section 12.5). Techniques that map the positions of transcribed sequences in a DNA fragment can therefore be used to locate exons and entire genes. The only problem to be kept in mind is that for protein-coding genes the transcript is usually longer than the open reading frame: the transcript begins several tens of nucleotides upstream of the initiation codon and continues several tens or hundreds of nucleotides downstream of the termination codon (**Figure 5.10**). Because of these upstream and downstream **untranslated regions** (**UTRs**), transcript analysis does not give a precise definition of the start and end of a protein-coding gene, but it does tell you that a gene is present in a particular region and it can locate the exon–intron boundaries. Often this is sufficient information to enable the coding region to be delineated.

Hybridization tests can determine if a fragment contains one or more genes

The simplest procedures for studying transcribed sequences are based on hybridization analysis. RNA molecules can be separated by specialized forms of agarose gel electrophoresis, transferred to a nitrocellulose or nylon membrane, and examined by the process called **northern hybridization**. This differs from Southern hybridization (Section 2.1) only in the precise conditions under which the transfer is carried out, and the fact that it was not invented by a Dr. Northern and so does not have a capital 'N.' If a northern blot of cellular RNA is probed with a labeled fragment of the genome, then RNAs transcribed from genes within that fragment will be detected (**Figure 5.11**). Northern hybridization is therefore, theoretically, a means of determining the number of genes present in a DNA fragment and the approximate size of each coding region. There are two weaknesses to this approach:

- Some individual genes give rise to two or more transcripts of different lengths because of **alternative** splicing pathways, which result in exons being joined together in different combinations (Section 7.2). If this is the case, then a fragment that contains just one gene could detect two or more

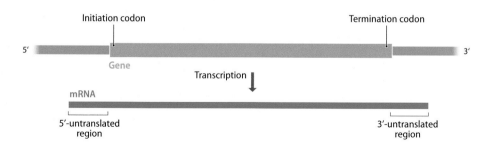

Figure 5.10 The transcript of a protein-coding gene is longer than the open reading frame.

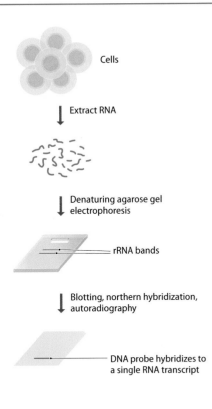

Cells

Extract RNA

Denaturing agarose gel electrophoresis

rRNA bands

Blotting, northern hybridization, autoradiography

DNA probe hybridizes to a single RNA transcript

Figure 5.11 Northern hybridization. An RNA extract is electrophoresed under denaturing conditions in an agarose gel. After ethidium bromide staining, two bands are seen. These are the two largest rRNA molecules (Section 1.2), which are abundant in most cells. The smaller tRNAs, which are also abundant, are not seen because they are so short that they run out the bottom of the gel and, in most cells, none of the mRNAs are abundant enough to form a band visible after ethidium bromide staining. The gel is blotted onto a nylon membrane and, in this example, probed with a radioactively labeled DNA fragment. A single band is visible on the autoradiograph, showing that the DNA fragment used as the probe contains part or all of one transcribed sequence.

hybridizing bands in the northern blot. A similar problem can occur if the gene is a member of a multigene family (Section 7.2).

- With many species, it is not practical to make an mRNA preparation from an entire organism so the extract is obtained from a single organ or tissue. Consequently, any genes not expressed in that organ or tissue will not be represented in the RNA population, and so will not be detected when the RNA is probed with the DNA fragment being studied. Even if the whole organism is used, not all genes will give hybridization signals because many are expressed only at a particular developmental stage, and others are weakly expressed, meaning that their RNA products are present in amounts too low to be detected by hybridization analysis.

A second and quite different type of hybridization analysis avoids the problems with poorly expressed and tissue-specific genes by searching not for RNAs but for related sequences in the DNAs of other organisms. This approach, like homology searching, is based on the fact that homologous genes in related organisms have similar sequences, whereas the intergenic DNA is usually quite different. If a DNA fragment from one species is used to probe a Southern transfer of DNAs from related species, and one or more hybridization signals are obtained, then it is likely that the probe contains one or more genes (**Figure 5.12**). This is called **zoo-blotting**.

Methods are available for precise mapping of the ends of transcripts

Northern hybridization and zoo-blotting can identify a DNA fragment that contains a gene, but those methods do not enable the gene to be positioned within that fragment with any degree of accuracy. More precise mapping of the transcript onto the DNA sequence requires a different approach. One possibility is a special type of PCR that uses RNA rather than DNA as the starting material. The first step in this type of PCR is to convert the RNA into single-stranded cDNA with reverse transcriptase (see **Figure 3.37**), after which the cDNA is amplified with *Taq* polymerase in the same way as in a normal PCR. These methods go under the collective name of **reverse-transcriptase PCR** (**RT-PCR**), but the particular version that interests us at present is **rapid amplification of cDNA ends** (**RACE**). In the simplest form of this method, one of the primers is specific for an internal region of the gene being studied, close to its start site. This primer attaches to the transcript for the gene and directs the first reverse-transcriptase-catalyzed stage of the process, during which a cDNA copy of the start of the RNA is made, the 3´ end of this cDNA corresponding exactly with the 5´ end of the RNA (**Figure 5.13**). The 3´ end of the cDNA is then extended by treatment with terminal deoxynucleotidyl transferase, to give a short poly(A) tail. The second primer anneals to this

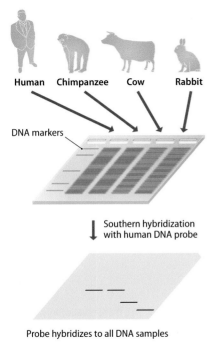

Human   Chimpanzee   Cow   Rabbit

DNA markers

Southern hybridization with human DNA probe

Probe hybridizes to all DNA samples

Figure 5.12 Zoo-blotting. The objective is to determine if a fragment of human DNA hybridizes with DNAs from related species. Samples of human, chimpanzee, cow and rabbit DNAs are therefore prepared, restricted, and electrophoresed in an agarose gel. Southern hybridization is then carried out with a human DNA fragment as the probe. A positive hybridization signal is seen with each of the animal DNAs, suggesting that the human DNA fragment contains an expressed gene. Note that the hybridizing restriction fragments from the cow and rabbit DNAs are smaller than the hybridizing fragments in the human and chimpanzee samples. This indicates that the restriction map around the transcribed sequence is different in cows and rabbits, but does not affect the conclusion that a homologous gene is present in all four species.

Figure 5.13 RACE – rapid amplification of cDNA ends. The RNA being studied is converted into a partial cDNA by extension of a DNA primer that anneals at an internal position close to the 5′ end of the molecule. The 3′ end of the cDNA is further extended by treatment with terminal deoxynucleotidyl transferase in the presence of dATP, which results in a series of As being added to the cDNA. This series of As acts as the annealing site for the anchor primer. Extension of the anchor primer leads to a double-stranded DNA molecule, which can now be amplified by a standard PCR. This is 5′-RACE, so-called because it results in amplification of the 5′ end of the starting RNA. A similar method – 3′-RACE – can be used if the 3′ end-sequence is desired.

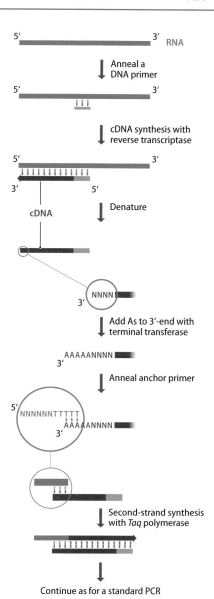

poly(A) sequence and, during the first round of the normal PCR, converts the single-stranded cDNA into a double-stranded molecule, which is subsequently amplified as the PCR proceeds. The sequence of this amplified molecule will reveal the position of the start of the transcript.

Other methods for precise transcript mapping involve **heteroduplex analysis**. This method requires a single-stranded version of the gene being studied, which can be obtained by cloning in a vector based on **M13 bacteriophage**. The replication process of M13 involves synthesis of phage particles that contain a single-stranded version of the phage genome. DNA that has been cloned in an M13 vector can therefore be obtained as single-stranded DNA by purification from the recombinant phages. When mixed with an appropriate RNA preparation, the transcribed sequence in the single-stranded DNA hybridizes with the equivalent RNA, forming a double-stranded heteroduplex. In the example shown in **Figure 5.14**, a restriction fragment spanning the start of the RNA has been cloned. Some of the cloned fragment participates in the heteroduplex, but the rest does not. The single-stranded regions can be digested by treatment with a single-strand-specific nuclease such as S1. The size of the heteroduplex is determined by degrading the RNA component with alkali and electrophoresing the resulting single-stranded DNA in an agarose or polyacrylamide gel. This size measurement is then used to position the start of the transcript relative to the restriction site at the end of the cloned fragment.

Exon–intron boundaries can also be located with precision

Heteroduplex analysis can also be used to locate exon–intron boundaries. The method is almost the same as that shown in Figure 5.14, with the exception that the cloned restriction fragment spans the exon–intron boundary being mapped rather than the start of the transcript.

A second method for finding exons in a genome sequence is called **exon trapping**. This requires a special type of vector that contains a **minigene** consisting of two exons flanking an intron sequence, the first exon being preceded by the sequence signals needed to initiate transcription in a eukaryotic cell (**Figure 5.15**). To use the vector, the piece of DNA to be studied is inserted into a restriction site located within the vector's intron region. The vector is then introduced into a suitable eukaryotic cell line, where it is transcribed and the RNA produced from it is spliced. The result is that any exon contained in the genomic fragment becomes attached between the upstream and downstream exons from the minigene. RT-PCR with primers annealing within the two minigene exons is now used to amplify a DNA fragment, which is sequenced. As the minigene sequence is already known, the nucleotide positions at which the inserted exon starts and ends can be determined, precisely delineating this exon.

5.3 ANNOTATION BY GENOME-WIDE RNA MAPPING

The transcript analysis methods that we have studied so far are designed for the mapping of individual genes onto short sequences of DNA. As there are tens of thousands of genes in the average vertebrate or plant genome, reliance on these methods for genome annotation would make RNA mapping a lengthy and highly tedious process. We therefore need to explore alternative methods that enable multiple transcripts to be mapped simultaneously.

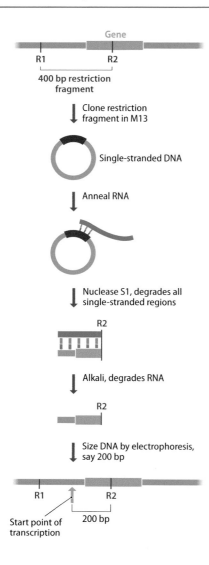

Figure 5.14 Nuclease S1 mapping. This method of transcript mapping makes use of nuclease S1, an enzyme that degrades single-stranded DNA or RNA polynucleotides, including single-stranded regions in predominantly double-stranded molecules, but has no effect on double-stranded DNA nor on DNA–RNA hybrids. In the example shown, a restriction fragment that spans the start of a transcription unit has been cloned in an M13 vector and the resulting single-stranded DNA hybridized with an RNA preparation. After nuclease S1 treatment, the resulting heteroduplex has one end marked by the start of the transcript and the other by the downstream restriction site (R2). The size of the DNA fragment in the heteroduplex is therefore measured by gel electrophoresis in order to determine the position of the start of the transcription unit relative to the downstream restriction site.

Tiling arrays enable transcripts to be mapped on to chromosomes or entire genomes

The first methods for multiple, parallel transcript mapping made use of **tiling arrays**. A tiling array is a special type of DNA chip. We learned in Section 3.2 that a DNA chip is a small piece of glass or silicon onto which many different oligonucleotides have been immobilized in an ordered array, and we saw how, by hybridizing a DNA sample to the chip, the SNPs within that DNA could be typed (see Figure 3.9). In a tiling array, the oligonucleotides form a series that covers the length of a chromosome sequence, or the sequence of an entire genome, the series either overlapping or with small gaps between adjacent oligonucleotides (**Figure 5.16**). To squeeze all of the necessary oligonucleotides on to the chip a high-density array is needed. In the traditional technology, each oligonucleotide was synthesized separately and then spotted onto the chip at its appropriate position. This method is suitable for preparing low-density arrays for typing a relatively small number of SNPs, but it does not enable a high-density tiling array to be prepared. To achieve high density, the oligonucleotides must be synthesized directly on the surface of the chip. This presents a challenge as the normal synthesis method involves adding nucleotides one-by-one to the growing end of an oligonucleotide, the sequence determined by the order in which the nucleotide substrates are added to the reaction mixture. If used for synthesis on a chip, this method would result in every oligonucleotide having the same sequence. Instead, modified nucleotide substrates are used – ones that have to be light-activated before the growing oligonucleotide can be extended. The nucleotides are added one after another to the chip surface, **photolithography** being used to direct pulses of light onto individual positions in the array. Only those oligonucleotides that are light-activated will be extended by the nucleotide that is present at any particular step (**Figure 5.17**). This method enables the highest density arrays, with up to 300,000 oligonucleotides per cm^2, to be prepared.

The tiling array is hybridized not to DNA, as would be the case with SNP typing, but to a labeled sample of RNA from the organism whose genome is being

Figure 5.15 Exon trapping. The exon-trap vector consists of two exon sequences preceded by a promoter sequence, the latter containing the signals required for gene expression in a eukaryotic host (Section 12.3). New DNA containing an unmapped exon is ligated into the vector and the recombinant molecule is introduced into the host cell. Splicing of the primary transcript joins the unmapped exon to the upstream and downstream minigene exons. The resulting RNA molecule is then amplified by RT-PCR and sequenced in order to identify the boundaries of the unmapped exon.

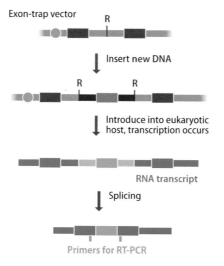

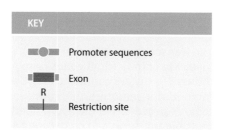

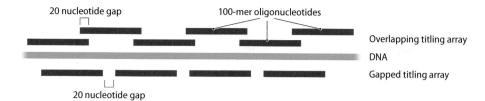

20 nucleotide gap

100-mer oligonucleotides

Overlapping titling array

DNA

Gapped titling array

20 nucleotide gap

Figure 5.16 Tiling arrays. Two arrays are shown, both comprising oligonucleotides of 100 nucleotides in length ('100-mers'). The top array has overlapping oligonucleotides, and in the lower array there are short gaps between the oligonucleotides.

annotated. Those positions on the array that hybridize will be ones containing oligonucleotides that hybridize to molecules in the RNA samples, and hence will reveal the positions in the genome of transcribed sequences. The hybridization data will not accurately locate each gene, for two reasons. First, as described above, the transcript might be longer than the gene, so oligonucleotides whose positions lie in the upstream and downstream UTRs will also give signals. Second, the accuracy of mapping depends on the lengths of the oligonucleotides and of the overlaps or gaps between them in the array. In the examples of array design shown in Figure 5.16, the accuracy is ±30 for the overlapping array and ±70 nucleotides for the gapped array.

The source of the RNA preparation also has to be chosen with care to ensure that the sample contains transcripts of as many genes as possible. With a higher eukaryote such as humans, it is very difficult, if not impossible, to obtain a sample that contains transcripts of every gene in the genome, because not all genes are expressed in a single organ or cell type, and even in one cell type the gene expression pattern varies over time. In fact, the major use of tiling arrays is in understanding these cell-specific and time-dependent patterns of gene expression. It is therefore necessary to test RNA samples from many different tissues in order to build up a complete genome annotation by this, or any other, genome-wide method for transcript mapping. The RNA that is prepared can also be fractionated in various ways so that only certain types of gene are targeted by the tiling array. The most commonly used fractionation procedure is to pre-select those RNAs that carry a poly(A) tail at their 3′ ends. The tail is a series of up to 250 adenine nucleotides that is added to the end of a eukaryotic mRNA after transcription (Section 1.2). An **affinity chromatography** column containing oligo(dT)–cellulose (cellulose beads to which short oligonucleotides of deoxythymidine have been attached) can therefore be used to purify the polyadenylated mRNA fraction from a eukaryotic RNA sample (**Figure 5.18**).

Transcript sequences can be directly mapped onto a genome

The most direct way of using transcript analysis in genome annotation is to sequence the transcripts and then to use the sequence data to search the genome for the genes from which the RNAs are transcribed. Since the development of next-generation sequencing, this approach to genome annotation has become increasingly attractive. **RNA-seq** is simply the application of a high-throughput sequencing method to a library that has been prepared from RNA rather than directly from DNA. For the sequencing-by-synthesis methods (e.g., Illumina and SMRT sequencing), the RNA preparation is converted into cDNA prior to library preparation, or if the nanopore method is being used then the RNA can be directly sequenced. RNA-seq reads therefore correspond to segments of the transcripts in the original RNA sample. Those reads can be mapped directly onto a genome sequence, in a manner identical to the way in which DNA sequence reads are mapped onto a reference genome during a genome resequencing project (Section 4.2). The only difference is, of course, that the RNA reads do not give lengthy scaffolds, but instead form clusters that map specifically on to the transcribed parts of the genome (**Figure 5.19A**). An alternative strategy that achieves the same result is to apply a *de novo* assembly method to the collection of RNA-seq reads, and then map the assembled contigs onto the reference genome (**Figure 5.19B**). The advantage of the latter approach is that many genes form **multigene families**, the members of which display sequence similarity (Section 7.2). If individual, short RNA reads are mapped directly onto

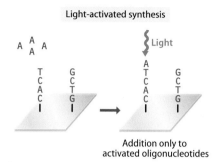

Light-activated synthesis

Light

Addition only to activated oligonucleotides

Figure 5.17 Light-activated oligonucleotide synthesis on the surface of a DNA chip.

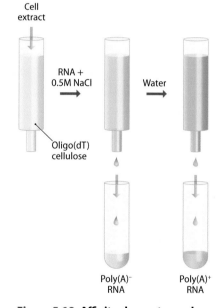

Cell extract

RNA + 0.5M NaCl

Water

Oligo(dT) cellulose

Poly(A)⁻ RNA

Poly(A)⁺ RNA

Figure 5.18 Affinity chromatography for the purification of polyadenylated RNA. The chromatography column contains oligo(dT)-cellulose. The RNA fraction is applied in a high salt solution, which promotes hybridization between the oligo(dT) molecules in the column and the poly(A) tails that are present on most eukaryotic mRNAs. The polyadenylated RNA is therefore retained in the column while the nonpolyadenylated fraction, which mainly comprises noncoding RNA, passes directly through. The poly(A) fraction is then eluted with water, which disrupts the oligo(dT)-poly(A) base-pairing.

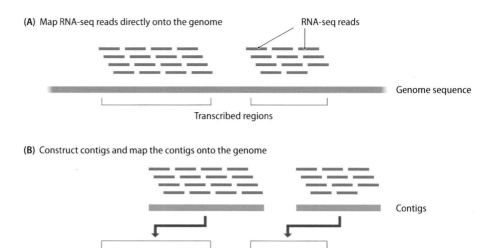

Figure 5.19 Two approaches to mapping RNA-seq reads onto a reference genome. (A) Direct mapping of reads onto the genome sequence. (B) Initial assembly of RNA-seq contigs, followed by mapping of the contigs onto the genome.

the reference genome, then some might be identical to segments of two or more members of a multigene family, complicating the mapping process. If, on the other hand, the complete transcript sequence is determined prior to mapping, then the members of a gene family are more easily distinguished.

RNA-seq might appear to be the simplest and most foolproof way to identify transcripts and map their positions on a genome sequence, but in practice there are several problems. The most important of these is the differential abundance of certain transcripts in any particular cell or tissue. If RNA is simply extracted and sequenced, then the reads will be dominated by ribosomal RNA, which make up the major component of the transcriptomes of most cells and tissues. As with the use of tiling arrays, it is necessary to fractionate the RNA preparation prior to sequencing, for example, with oligo(dT)–cellulose if protein-coding genes are being mapped. Alternatively, genes for short noncoding RNAs can be sequenced after size fractionation of the RNA sample so that only transcripts with lengths less than 200 nucleotides are retained.

Even if the RNA preparation is carefully fractionated before sequencing, some transcripts might be missed because they form such a small part of the total RNA that they still do not give rise to any reads in the resulting sequence dataset. This is a greater problem with long-read sequencing because a single SMRT or nanopore flow cell gives many fewer reads than is obtained in a single Illumina sequencing run, resulting in a lower sequence coverage and hence a greater chance that rare transcripts are not sequenced. However, long-read RNA-seq, in particular by the nanopore method, is increasingly popular because it has a number of benefits compared to short-read methods (**Figure 5.20**):

- Errors in distinguishing reads from different members of a multigene family are less likely to occur with long as opposed to short reads.

- The entire sequence of a transcript might be contained in a single long read, especially if the nanopore method is used. If the sequence is complete, then it will include precise information on the locations of each of the two ends of the transcript on the genome sequence. With short reads, the end sequences are often missing or present at low coverage, so precise annotation of the start and end points of a transcript might be impossible.

- Long-read RNA-seq is able to identify the products of alternative splicing pathways and hence provide a more comprehensive annotation of the genome than is possible with a short-read method.

One way of increasing the sensitivity of long-read RNA-seq, so that rare transcripts are less likely to be missed, is to enrich the sample by **cDNA capture** or **cDNA selection**. In its original format, this method was used to capture transcripts

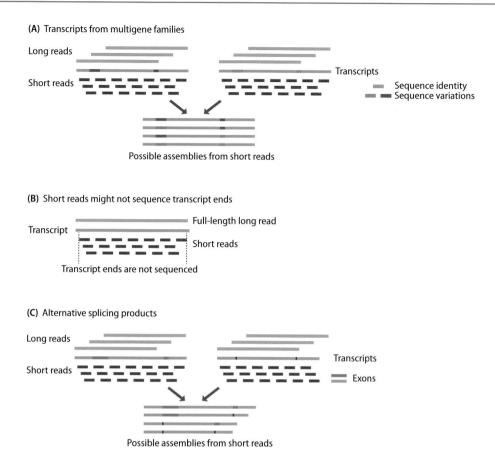

(A) Transcripts from multigene families

Long reads

Short reads

Transcripts

▬▬ Sequence identity
▬▬ Sequence variations

Possible assemblies from short reads

(B) Short reads might not sequence transcript ends

Transcript

Full-length long read

Short reads

Transcript ends are not sequenced

(C) Alternative splicing products

Long reads

Short reads

Transcripts

▬▬ Exons

Possible assemblies from short reads

Figure 5.20 Long reads have benefits compared to short reads in transcript sequencing. (A) Long reads enable the transcripts from multigene families to be distinguished. In this example, the two transcripts have identical sequences except for two short regions, shown in orange and brown. The long reads span both variable regions, and hence show that there are two different transcripts. If only the short-read sequences are used, then four transcripts can be assembled, but two of these do not actually exist. (B) A full-length long read reveals the exact start and end points of a transcript. Short reads might miss one or both ends, because when the cDNA obtained from the RNA is fragmented prior to sequencing, the end fragments might be lost because they are too short for the sequencing method. (C) Long reads enable the products of alternative splicing pathways to be distinguished. These two transcripts are derived from the same gene by two alternative splicing pathways, the first linking all five exons together, and the second omitting exons 2 and 4. Long reads reveal that there are just two transcripts, but short reads erroneously identify four transcripts.

corresponding to genes located in a relatively short piece of DNA in, for example, a single BAC clone. The RNA preparation is converted into cDNA and this pool of cDNAs is repeatedly hybridized to the BAC clone, with nonhybridized cDNAs washed away and discarded. As an adjunct to long-read RNA-seq, cDNA capture makes use of an oligonucleotide tiling array that covers the genome segment that is being studied. Because the cDNA pool contains so many different sequences, it is generally not possible to discard all the irrelevant cDNAs by these repeated hybridizations, but it is possible to significantly increase the frequency of those cDNAs that specifically hybridize to the oligonucleotide array. This increases the likelihood that reads corresponding to rare transcripts mapping to the targeted region will be obtained.

Obtaining transcript sequences by SAGE and CAGE

RNA-seq mapping is becoming increasingly important in genome annotation and in studies of gene expression patterns in different tissues. The analysis is, however, computationally intensive and methods that still make use of RNA-seq data but take shortcuts to gene mapping are constantly being sought. Two related methods – **serial analysis of gene expression (SAGE)** and **cap analysis gene expression (CAGE)** – enable more rapid characterization of the mRNA content of a transcriptome by providing short sequences, each of which represents a single RNA. These sequences, despite their shortness, are sufficient in many cases to enable the gene that codes for the mRNA to be identified.

To carry out SAGE, the mRNAs are first immobilized in an oligo(dT) cellulose column (**Figure 5.21**). The mRNA is then converted into double-stranded cDNA and treated with a restriction enzyme that recognizes a 4 bp target site, such as *Alu*I, and so cuts frequently in each cDNA. The terminal restriction fragment of each cDNA remains attached to the cellulose beads, enabling all the other fragments to be eluted and discarded. A short linker is now attached to the free end of each of the terminal cDNA fragments; this linker contains a recognition

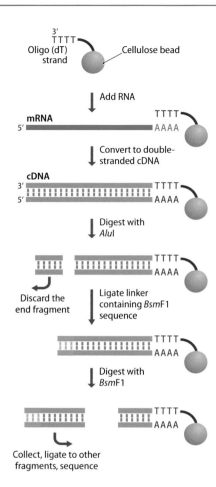

Oligo (dT) strand
Cellulose bead

Add RNA

mRNA

Convert to double-stranded cDNA

cDNA

Digest with *Alu*I

Discard the end fragment

Ligate linker containing *Bsm*F1 sequence

Digest with *Bsm*F1

Collect, ligate to other fragments, sequence

Figure 5.21 SAGE: serial analysis of gene expression. Messenger RNAs are immobilized by attachment to oligo(dT) cellulose, converted into double-stranded cDNA and treated with *Alu*I. A short linker containing a *Bsm*FI recognition sequence is attached to the free end of each terminal cDNA fragment. Treatment with *Bsm*FI then removes fragments with an average length of 12 bp, which are ligated into concatemers that are long enough to be sequenced.

sequence for *Bsm*FI, a Type III restriction endonuclease (Section 2.1) that makes a cut not within the recognition site but 10–14 nucleotides downstream. Treatment with *Bsm*FI therefore removes a fragment with an average length of 12 bp from the end of each cDNA. The fragments are collected and ligated head-to-tail to produce concatemers that are long enough to be sequenced. The individual mRNA sequences can be identified within each concatemer because they are separated by *Bsm*FI sites.

The disadvantage with SAGE is that the sequenced fragments represent internal regions of a transcript, close to but a variable distance away from the start of the poly(A) tail. The second method, CAGE, can be more useful in genome annotation because it identifies the exact start point of each mRNA. The basis of this method is the presence of the **cap structure** at the 5′ end of a eukaryotic mRNA. The cap is a 7-methylguanine nucleotide that is added to the start of the mRNA after transcription, becoming attached to the first nucleotide of the transcript by an unusual 5′– 5′ phosphodiester bond (Section 1.2). The cap is utilized in the CAGE method in the following way. The first stage of the standard cDNA synthesis process is carried out, creating an RNA–DNA hybrid (**Figure 5.22**). The hybrid is then treated with RNase I, which degrades the single-stranded RNA that has not been copied into cDNA. Those cDNAs that retain the cap structure must therefore be full-length copies or truncated fragments that include the start of the mRNA. Because the 7-methylguanine within the cap structure is attached to the next nucleotide by a 5′– 5′ bond, both its 2′ and 3′ carbons retain their hydroxyl groups. These two hydroxyls form a **diol** structure. Addition of an oxidizing agent specifically breaks the bond between the 2′ and 3′ carbons of the diol, creating an open-chain sugar structure that can be covalently attached to a molecule of biotin. This enables the capped cDNAs to be captured, through use of magnetic beads that are coated with avidin. Following capture, the RNA component of the hybrids is degraded and a short oligonucleotide containing a recognition sequence for the restriction enzyme *Eco*P15I is ligated to the start of the cDNA. *Eco*P15I is another Type III restriction endonuclease (Section 2.1), this one making a cut 27 base pairs away from its recognition sequence. After synthesis of the second cDNA strand, the enzyme is added, releasing a 27 bp tag fragment representing the 5′ end of the original mRNA. All of the tags from all of the cDNAs in the sample are then collected, sequenced and mapped onto the reference genome.

Although SAGE and CAGE are methodologically complex, the computational phase of the procedures is less intensive than the mapping of reads from complete RNA-seq libraries. The two methods therefore hold promise as a means of rapidly identifying the genes in a genome sequence.

5.4 GENOME BROWSERS

The most convenient way of displaying the results of a genome annotation project is in a graphical format, with the DNA sequence forming the *x*-axis and the positions of genes and other interesting features marked at their appropriate map positions. A **genome browser** is a software package that enables genome annotation data to be displayed in this way. The display might be quite complex, because as well as indicating the positions of ORFs that have been identified with high confidence, the browser might also display short ORFs of questionable authenticity in all six reading frames, noncoding RNA genes, the starts and ends of transcripts, positions of mapped DNA markers, locations and identities of repetitive DNA sequences, etc. The software will enable the map to be displayed at different levels of magnification, so the entire length of a chromosome can be

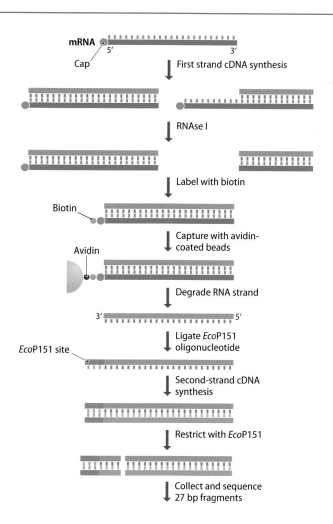

Figure 5.22 CAGE: cap analysis of gene expression. First-strand cDNA is synthesized and the cDNA-RNA hybrids are treated with RNase I to degrade the uncopied RNA. Those hybrids still carrying a cap structure are labeled with biotin and purified by capture on avidin-coated magnetic beads. The RNA component of each hybrid is degraded and a short oligonucleotide containing an *Eco*P151 recognition sequence is ligated to the start of the cDNA. The second cDNA strand is synthesized and *Eco*P151 is added, releasing a 27 bp fragment representing the 5′ end of the original mRNA. The fragments are then sequenced.

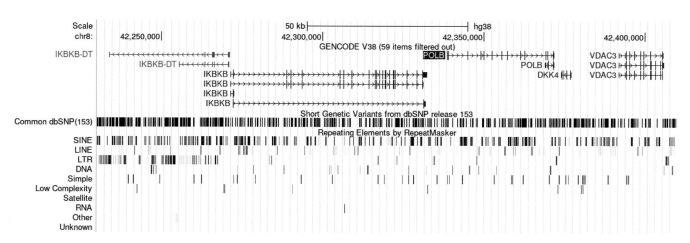

Figure 5.23 An example of the information provided by an online genome browser. This example shows the annotation for the 180 kb region of human chromosome 8 between nucleotide positions 42,250,000 and 42,410,000, as displayed by the UCSC Genome Browser. The user can configure the browser to show many different features of the genome annotation. This configuration shows the locations of genes (according to the GENCODE version 38 annotation), short genetic variants (which include SNPs and short insertions or deletions), and various types of repetitive DNA elements. Protein-coding genes are in blue and noncoding RNA genes are in green, with introns shown as boxes or vertical lines, and arrows indicating the DNA strand on which the gene is located. The different entries for an individual gene represent different transcription units with variable start or end points and/or different splicing patterns. The genes are: IRBKB-DT, long noncoding RNA gene active mainly in testes; IRBKB, inhibitor of NF-κB kinase subunit β; POLB, DNA polymerase β; DKK4, WNT signaling pathway inhibitor; VDAC3, mitochondrial porin. The repeats shown are classified as short interspersed nuclear elements (SINEs), long interspersed nuclear elements (LINEs), long terminal repeat (LTR) retrotransposons, DNA transposons, simple repeats (microsatellites), low complexity repeats, satellite DNA, RNA repeats (various short noncoding RNA genes such as transfer RNAs and short nuclear RNAs), various other repeats, and ones of unknown category.

viewed, or the operator can zoom in to a level at which individual nucleotides are distinguished. Most browsers also incorporate a search facility so that particular genes, markers or map positions can be quickly located.

Many genome browsers are used online, with the draft and final annotations of a newly sequenced genome accessible to other researchers. One of the imperatives of genomics research is that data must be made publically available. Databases for the curation of DNA sequences have been established for many years, the most important being **GenBank**, which is maintained by the National Center for Biotechnology Information (NCBI), part of the US National Institutes of Health. Online genome browsers fulfill the same purpose for genome annotations. Two of the most widely used are **Ensembl**, which is maintained by the European Bioinformatics Institute and the UK Sanger Institute, and the **UCSC Genome Browser** of the University of California, Santa Cruz (**Figure 5.23**). Both Ensembl and the UCSC Genome Browser hold annotations of the human genome along with those of several other vertebrates and invertebrates. There are also more specialized online genome browsers, such as the **Plant GDB**, which as its name indicates holds plant genome annotations.

SUMMARY

- Genome annotation is the process by which genes are located in a genome sequence.

- Protein-coding genes can be identified by searching for open reading frames, though this is complicated in eukaryotes by the presence of introns, whose boundary sequences are variable and which therefore cannot be identified accurately.

- Genes for noncoding RNAs can be located by searching for their characteristic features, primarily the ability of the RNAs to fold into secondary structures based on the formation of base-paired stem-loops.

- Genes can also be located by homology analysis, which uses the presence of an equivalent gene in a second genome as evidence that a putative gene in the test genome is genuine.

- Experimental methods for gene location are based on detection of RNA molecules transcribed from the genome. These techniques include transcript mapping by reverse-transcriptase PCR (RT-PCR) or heteroduplex analysis.

- Exon–intron boundaries can be identified experimentally by the procedure called exon trapping.

- Tiling arrays are used for the genome-wide mapping of transcript positions.

- RNA sequencing using short- or long-read methods is becoming increasingly important in the identification of the transcribed parts of a genome.

- Serial analysis of gene expression and cap analysis gene expression are two methods for sequencing tags representing mRNAs.

- Genome annotations are displayed using a genome browser.

SHORT ANSWER QUESTIONS

1. Why is it relatively easy to identify ORFs in prokaryotic genomes by computer analysis?

2. Describe how ORF scans are used to search for genes in eukaryotic genome sequences.

3. What is meant by the term 'codon bias'?

4. Outline the structural features of functional RNA molecules, such as tRNA and rRNA, that can be searched for in a genome sequence to identify the genes encoding these RNA molecules.

5. Define the term 'homologous,' as used when comparing gene sequences.

6. Give examples of the use of comparative genomics in genome annotation.

7. What are the two limitations that arise when northern analysis is used to determine the number of genes present in a DNA fragment?

8. Describe how rapid amplification of cDNA ends is used to map the transcription initiation site of a gene.

9. How is heteroduplex analysis used to map transcripts?

10. Describe a method for the experimental identification of exon–intron boundaries.

11. What is a tiling array, and how are tiling arrays used in genome annotation?

12. Describe the advantages and disadvantages of the use of long-read methods in transcript identification and mapping.

13. Explain how the computational challenges of RNA-seq are reduced by the methods called serial analysis of gene expression (SAGE) and cap analysis gene expression (CAGE).

IN-DEPTH PROBLEMS

1. To what extent do you believe it will be possible in future years to use bioinformatics to obtain a complete description of the locations and functions of the protein-coding genes in a eukaryotic genome sequence?

2. Devise a hypothesis to explain the codon biases that occur in the genomes of various organisms. Can your hypothesis be tested?

3. 'Comparative genomics has an important role to play in the study of disease genes.' Evaluate this statement.

4. Studies with tiling arrays have revealed that many transcripts are synthesized from the sequences in the intergenic spaces lying between the open reading frames in the human genome. Use the internet (and possibly other chapters of this book) to explore the nature and possible functions of these RNAs.

5. Use the Ensembl Bacteria genome browser (http://bacteria.ensembl .org/index.html) to locate the position of the β-galactosidase gene in the *Escherichia coli* genome. Draw (or export) a map of the genes in the 30 kb region centered on the β-galactosidase gene.

FURTHER READING

Gene location by computer analysis

Ejigu, G.F. and Jung, J. (2020). Review on the computational genome annotation of sequences obtained by next-generation sequencing. *Biology* 9:295.

Guigó, R., Flicek, P., Abril, J.F., et al. (2006). EGASP: The human ENCODE Genome Annotation Assessment Project. *Genome Biol.* 7(Suppl 1):S2, 1–31. *Comparison of the accuracy of different computer programs for gene location.*

Pavesi, G., Mauri, G., Stefani, M. and Pesole, G. (2004). RNAProfile: An algorithm for finding conserved secondary structure motifs in unaligned RNA sequences. *Nucleic Acids Res.* 32:3258–3269. *Locating noncoding RNA genes.*

Quax, T.E.F., Claassens, N.J., Söll, D. and van de Oost, J. (2015). Codon bias as a means to fine-tune gene expression. *Mol. Cell* 59:149–161.

Salzberg, S.L. (2019). Next-generation genome annotation: We still struggle to get it right. *Genome Biol.* 20:92.

Yandell, M. and Ence, D. (2012). A beginner's guide to eukaryotic genome annotation. *Nat. Rev. Genet.* 13:329–342.

Yella, V.R., Kumar, A. and Bansal, M. (2018). Identification of putative promoters in 48 eukaryotic genomes on the basis of DNA free energy. *Sci. Rep.* 8:4520.

Comparative genomics

Alföldi, J. and Lindblad-Toh, K. (2013). Comparative genomics as a tool to understand evolution and disease. *Genome Res.* 23:1063–1068.

Kellis, M., Patterson, N., Endrizzi, M., et al. (2003). Sequencing and comparison of yeast species to identify genes and regulatory elements. *Nature* 423:241–254.

König, S., Romoth, L. and Stanke, M. (2018). Comparative genome annotation. *Methods Mol. Biol.* 1704:189–212.

Paterson, A.H., Bowers, J.E., Feltus, F.A., et al. (2009). Comparative genomics of grasses promises a bountiful harvest. *Plant Physiol.* 149:125–131.

Experimental methods for gene location

Berk, A.J. (1989). Characterization of RNA molecules by S1 nuclease analysis. *Methods Enzymol.* 180:334–347.

Church, D.M., Stotler, C.J., Rutter, J.L., et al. (1994). Isolation of genes from complex sources of mammalian genomic DNA using exon amplification. *Nat. Genet.* 6:98–105. *Exon trapping.*

Frohman, M.A., Dush, M.K. and Martin, G.R. (1988). Rapid production of full-length cDNAs from rare transcripts: Amplification using a single gene-specific oligonucleotide primer. *Proc. Natl. Acad. Sci. USA* 85:8998–9002. *An example of RACE.*

Pellé, R. and Murphy, N.B. (1993). Northern hybridization: Rapid and simple electrophoretic conditions. *Nucleic Acids Res.* 21:2783–2784.

Genome-wide RNA mapping

Lagarde, J. and Johnson, R. (2018). Capturing a long look at our genetic library. *Cell Syst.* 6:153–155. *Long-read sequencing of RNA transcripts.*

Lemetre, C. and Zhang, Z.D. (2013). A brief introduction to tiling microarrays: Principles, concepts, and applications. *Methods Mol. Biol.* 1067:3–19.

Lovett, M. (1994). Fishing for complements: Finding genes by direct selection. *Trends Genet.* 10:352–357. *cDNA capture.*

Lowe, R., Shirley, N., Bleackley, M., et al. (2017). Transcriptomics technologies. *PLoS Comput. Biol.* 13:e1005457. *Reviews all aspects of RNA mapping.*

Stark, R., Grzelak, M., and Hadfield, J. (2019). RNA sequencing: The teenage years. *Nat. Rev. Genet.* 20:631–656.

Takahashi, H., Kato, S., Murata, M. and Carninci, P. (2012). CAGE (cap analysis of gene expression): A protocol for the detection of promoter and transcriptional networks. *Methods Mol. Biol.* 786:181–200.

Velculescu, V.E., Vogelstein, B., and Kinzler, K.W. (2000). Analysing uncharted transcriptomes with SAGE. *Trends Genet.* 16:423–425.

Yazaki, J., Gregory, B.D. and Ecker, J.R. (2007). Mapping the genome landscape using tiling array technology. *Curr. Opin. Plant Biol.* 10:534–542.

Genome browsers

Duvick, J., Fu, A., Muppirala, U., et al. (2007). PlantGDB: A resource for comparative plant genomics. *Nucl Acids Res.* 36:D959–D965.

Fernandes JD, Hinrichs AS, Clawson H, et al. (2020). The UCSC SARS-CoV-2 Genome Browser. *Nat Genet.* 52:991–998.

Howe, K.L., Achuthan, P., Allen, J., et al. (2021). Ensembl 2021. *Nucl. Acids Res.* 49:D884–D891.

Wang, J., Kong, L., Gao, G., et al. (2012). A brief introduction to web-based genome browsers. *Brief. Bioinformatics* 14:131–143.

URLs for genome browsers

Ensembl. http://www.ensembl.org/index.html

Ensembl Bacteria. http://bacteria.ensembl.org/index.html

UCSC Genome Browser. https://genome.ucsc.edu/

Plant GDB. http://www.plantgdb.org/

IDENTIFYING GENE FUNCTIONS

Once a new gene has been located in a genome sequence, the question of its function has to be addressed. This is turning out to be an important area of genomics research, because completed sequencing projects have revealed that we knew rather less than we thought about the content of individual genomes. *Escherichia coli* and *Saccharomyces cerevisiae*, for example, were studied intensively by conventional genetic analysis before the advent of sequencing projects, and geneticists were at one time fairly confident that most of the genes in these species had been identified. The genome sequences revealed that in fact there are large gaps in our knowledge. Of the 4288 protein-coding genes in the initial annotation of the *E. coli* genome sequence, only one-third were described as 'well characterized' and 38% had no attributed function. The figures were very similar for *S. cerevisiae*. Methods that enable functions to be assigned to genes are therefore of critical importance in understanding a genome sequence.

As with genome annotation, attempts to determine the functions of unknown genes are made by computer analysis and by experimental studies. We begin with the computer methods.

6.1 COMPUTER ANALYSIS OF GENE FUNCTION

We have already seen that computer analysis plays an important role in locating genes in DNA sequences, and that one of the most powerful tools available for this purpose is homology searching, which locates genes by comparing the DNA sequence under study with all the other DNA sequences in the databases. The basis of homology searching is that related genes have similar sequences and so a new gene can be discovered by virtue of its similarity to an equivalent, already-sequenced gene from a different organism. Now we will look more closely at homology analysis and see how it can be used to assign a function to a new gene.

Homology reflects evolutionary relationships

Homologous genes are ones that share a common evolutionary ancestor, revealed by sequence similarities between the genes. Homologous genes fall into two categories (**Figure 6.1**):

- **Orthologous** genes are those homologs that are present in different organisms and whose common ancestor predates the split between the species. Orthologous genes usually have the same, or very similar, functions. For example, the myoglobin genes of humans and chimpanzees are orthologs.

- **Paralogous** genes are present in the same organism, often as members of a recognized multigene family (Section 7.2). Their common ancestor may or may not predate the species in which the genes are now found. For example, the myoglobin and β-globin genes of humans are paralogs: they originated by duplication of an ancestral gene some 550 million years ago (Section 18.2).

DOI: 10.1201/9781003133162-6

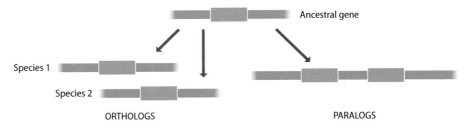

Figure 6.1 Orthologous and paralogous genes.

Sequence 1   GGTGAGGGTATCATCCCATCTGACTACACCTCATCGGGAGACGGAGCAGT
Sequence 2   GGTCAGGATATGATTCCATCACACTACACCTTATCCCGAGTCGGAGCAGT
Identities   * * *   * * *   * * *   * *   * * * * *   * * * * * * * *   * * *   * * *   * * * * * * * *

Figure 6.2 Two DNA sequences with 80% sequence identity.

A pair of homologous genes do not usually have identical nucleotide sequences, because the two genes undergo different random changes by mutation, but they have similar sequences because these random changes have operated on the same starting sequence, the common ancestral gene. Homology searching makes use of these sequence similarities. The basis of the analysis is that if a newly sequenced gene turns out to be similar to a previously sequenced gene, then an evolutionary relationship can be inferred and the function of the new gene is likely to be the same, or at least similar, to the function of the known gene.

It is important not to confuse the words homology and similarity. It is incorrect to describe a pair of related genes as '80% homologous' if their sequences have 80% nucleotide identity (**Figure 6.2**). A pair of genes are either evolutionarily related or they are not; there are no in-between situations and it is therefore meaningless to ascribe a percentage value to homology.

Homology analysis can provide information on the function of a gene

A homology search can be conducted with a DNA sequence but usually a tentative gene sequence is converted into an amino acid sequence before the search is carried out. One reason for this is that there are 20 different amino acids in proteins but only four nucleotides in DNA, so genes that are unrelated usually appear to be more different from one another when their amino acid sequences are compared (**Figure 6.3**). A homology search is therefore less likely to give spurious results if the amino acid sequence is used.

A homology search program begins by making alignments between the query sequence and sequences from the databases. For each alignment, a score is calculated from which the operator can gauge the likelihood that the query and test sequences are homologs. There are two ways of generating the score:

- The simplest programs count the number of positions at which the same amino acid is present in both sequences. This number, when converted into a percentage, gives the degree of identity between two sequences.

- More sophisticated programs use the chemical relatedness between nonidentical amino acids to assign a score to each position in the alignment, a higher score for identical or closely related amino acids (e.g., leucine and isoleucine, or aspartic acid and asparagine) and a lower score for less-related amino acids (e.g., cysteine and tyrosine, or phenylalanine and serine). This analysis determines the degree of similarity between a pair of sequences.

To achieve the highest possible score, the algorithm introduces gaps at various positions in one or both sequences, up to limits set by the operator, paralleling processes thought to occur during the evolution of genes, when blocks of

```
              G   A   P   G   M   W   L   R   L   A   A   G   S   F   Q   H   A   G
Sequence 1   GGT GCA CCC GGT ATG TGG CTG CGA TTA GCA GCG GGA TCG TTT CAG CAT GCA GGG
              *   *  **** ****  ****  **  *** ****  ***** *** ** **** ** *
Sequence 2   GAT ACA CCC CGT ATT TGG CAG CAA TTT GCA GGG GGA TGG TTG CAC CAT GGA GCG
              D   T   P   R   I   W   Q   Q   F   A   G   G   W   L   H   H   G   A
```

Figure 6.3 Lack of homology between two sequences is often more apparent when comparisons are made at the amino acid level. Two nucleotide sequences are shown, with nucleotides that are identical in the two sequences given in green and nonidentities given in red. The two nucleotide sequences are 76% identical, as indicated by the asterisks. This might be taken as evidence that the sequences are homologous. However, when the sequences are translated into amino acids, the identity decreases to 28%. Identical amino acids are shown in gold, and nonidentities are in red. The comparison between the amino acid sequences suggests that the genes are not homologous, and that the similarity at the nucleotide level was fortuitous. The amino acid sequences have been written using the one-letter abbreviations (see Table 1.2).

nucleotides coding for individual or adjacent amino acids might be inserted into or deleted from a gene.

The practicalities of homology searching are not at all daunting. Several software programs exist for this type of analysis, the most popular being **BLAST** (**Basic Local Alignment Search Tool**). The analysis can be carried out simply by logging on to the web site for one of the DNA databases and entering the sequence into the online search tool. The standard BLAST program is efficient at identifying homologous genes that have more than 40% sequence similarity, but is less effective at recognizing evolutionary relationships if the similarity is lower than this amount. The modified version called **PSI-BLAST** (**position-specific iterated BLAST**) identifies more distantly related sequences, by combining the homologous sequences from a standard BLAST search into a profile, the features of which are used to identify additional homologous sequences that were not detected in the initial search.

Homology searching with BLAST and similar programs has gained immense importance in genomics research, but its limitations must be recognized. A growing problem is the presence in the databases of genes whose stated functions are incorrect. If one of these genes is identified as a homolog of the query sequence, then the incorrect function will be passed on to this new sequence, adding to the problem. There are also several cases where homologous genes have quite different biological functions, an example being the crystallins of the eye lens, some of which are homologous to metabolic enzymes. Homology between a query sequence and a crystallin therefore does not mean that the query sequence is a crystallin and, similarly, an apparently clear homology between a query sequence and a metabolic enzyme might not mean that the query sequence is a metabolic enzyme.

Identification of protein domains can help to assign function to an unknown gene

What if a homology search with the DNA or amino acid sequence of an unknown gene fails to reveal any matches in the databases? All is not lost, as it may be possible to deduce at least some part of the function of the gene by searching the amino acid sequence for motifs that encode protein **domains** of known function. A protein domain is a protein segment that possesses a characteristic tertiary structure that provides the protein with a particular biochemical function. An example of a protein domain is the **Cys$_2$His$_2$ zinc finger**, which comprises a series of 12 or so amino acids, including two cysteines and two histidines, which form a segment of β-sheet followed by an α-helix. These two structures, which form the 'finger' projecting from the surface of the protein, hold between them a bound zinc atom, coordinated with the two cysteines and two histidines (**Figure 6.4**). Zinc fingers are DNA-binding structures (Section 11.2), so identification in an unknown gene of an amino acid sequence that can encode a zinc finger indicates that the gene codes for a DNA-binding protein. As well as function, the amino acid sequence can also provide information on the subcellular location of a protein. This can be inferred by searching for motifs called **sorting sequences**, which direct proteins to organelles such as the nucleus or mitochondria, or might specify that the protein is secreted from the cell.

As with homology searching, identification of conserved sequence motifs in an unknown gene can be performed online, using search tools at protein structure databases such as **Pfam**, hosted by the European Bioinformatics Institute, and **PROSITE**, which is maintained by the Swiss Institute of Bioinformatics. These databases contain collections of amino acid sequences for different types of domain, which are compared with the sequence of the test protein to identify the

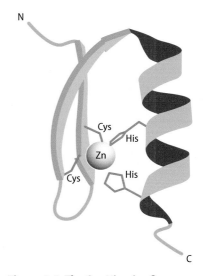

Figure 6.4 The Cys$_2$His$_2$ zinc finger. This particular zinc finger is from the yeast SWI5 protein. The zinc atom is held between two cysteines within the β-sheet of the motif and two histidines in the α-helix. The orange lines indicate the R groups of these amino acids. 'N' and 'C' indicate the N- and C-termini of the motif, respectively.

domains present in that protein. The results of such an analysis must, however, be interpreted with care. The presence of a shared domain indicates that two proteins can perform a similar biochemical activity, but that does not necessarily mean that the proteins have similar overall functions. This point is illustrated by the **Tudor domain** family of proteins. As the name implies, these proteins are related in that each possesses one or more copies of the Tudor domain, a five-stranded β-sheet structure encoded by an approximately 60-amino acid sequence. The Tudor domain binds to methylated arginine and/or lysine amino acids contained in other proteins. This is a specific biochemical activity, but one that is associated with a variety of different protein functions. The first Tudor domain protein to be discovered, coded by the *Drosophila melanogaster* gene called *tudor*, is involved in synthesis of **piwi-interacting RNAs** (**piRNAs**) in the developing oocyte. These are short noncoding RNAs that bind to **piwi proteins**, forming complexes that regulate gene expression during various developmental processes (Section 12.1). Other Tudor domain proteins, in a range of species, are also involved in piRNA synthesis, but other members of the family are associated with RNA splicing (Section 12.4), the RNA interference pathway (Section 12.6), the response to DNA damage (Section 17.2), and histone modification (Section 10.2). In all of these processes, the Tudor domain protein is thought to exert its effect through attachment to methylated arginine and/or lysine amino acids in target proteins. Identification of a Tudor domain sequence in an unknown gene therefore enables a specific biochemical activity to be identified, but on its own does not enable the actual function of the gene to be assigned, beyond placing that function among the variety of roles performed by known Tudor domain proteins.

Annotation of gene function requires a common terminology

An easily overlooked but nonetheless vital aspect of genome annotation is the need to have an agreed and consistent terminology with which to describe the functions of different genes. Consistency is needed for two reasons. First, a rigorous comparison can only be made between the genome annotations of two different species if the same terminology has been used to name the genes in those two genomes. If the annotations use different terms for the same gene function, then attempts to identify the similarities between the two genomes will give inaccurate results, because the computer making the comparison will fail to recognize the co-identity of the genes. If the terminology is based on a hierarchical classification system, then a second advantage is that a gene whose function has only been partially deduced – for example, by identification of one or more protein domains – can still be given a meaningful descriptor that indicates what is known about its function.

The first comprehensive protein classification system predates the advent of genome sequencing but provides the hierarchical structure that is needed for genome annotation. This system, which applies only to enzymes, was first agreed by the International Union of Biochemistry and Molecular Biology in 1961. In this classification, enzymes are initially divided into six broad groups:

- EC (Enzyme Commission) 1, the oxidoreductases.
- EC 2, transferases.
- EC 3, hydrolases.
- EC 4, lyases.
- EC 5, isomerases.
- EC 6, ligases.

Each of these groups is further subdivided in such a way that every individual enzyme has its own four-part **EC number**. For example, EC 3.2.1.2 is a hydrolase (EC 3) that breaks glycosidic bonds (EC 3.2) in which the linkage between the sugar and the second molecule includes an oxygen or sulfur atom (EC 3.2.1), specifically the (α1→4) *O*-glycosidic bonds between glucose units in starch,

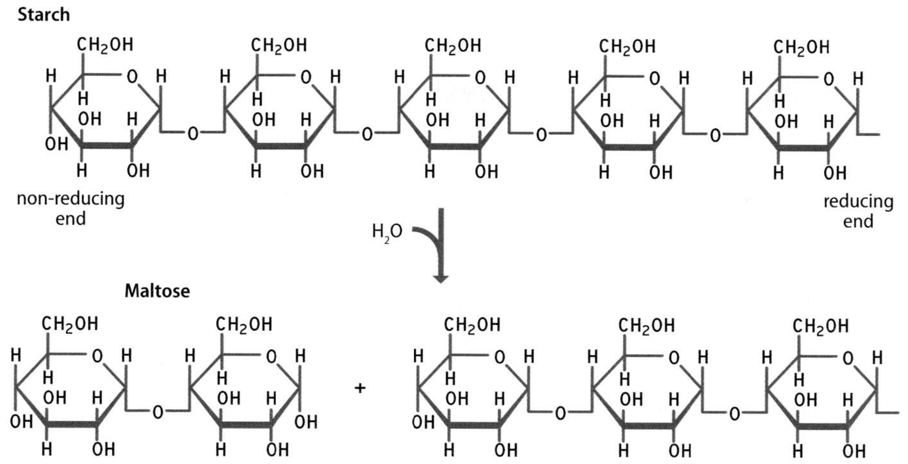

Starch

non-reducing
end

reducing
end

H_2O

Maltose

+

Figure 6.5 The mode of action of the enzyme specified by EC 3.2.1.2. The first three digits of the EC number indicate that this enzyme is a hydrolase, which means that water is involved in the enzymatic reaction (EC 3), and that it breaks glycosidic bonds, which are any bonds that link two sugar units or a sugar and second molecule (EC 3.2), and that the linkage with the second molecule contains an oxygen (or sulfur) atom (EC 3.2.1). The fourth digit indicates that the enzyme specifically cuts the ($\alpha1 \rightarrow 4$) O-glycosidic bonds between glucose units in starch, glycogen and related polysaccharides, in such a way as to release maltose disaccharide units from the non-reducing ends of the polymers.

glycogen, and related polysaccharides, in such a way as to release maltose disaccharide units from the non-reducing ends of the polymers (**Figure 6.5**). The enzyme activity is therefore described in a specific and hierarchical fashion, one that is much more informative than 'β-amylase,' which is the common name for this enzyme. It has been estimated that if a homology search reveals 40% amino acid sequence similarity between an unknown gene and a database entry, then the function of the unknown gene can be assigned as far as the first three parts of the EC number (e.g., EC 3.2.1). This degree of similarity indicates that the catalytic mechanism of the protein specified by the unknown gene is the same as that of the database match. If the similarity is 60% or higher, then both the query and match are likely to use the same substrate, so the fourth digit of the EC number can be assigned.

A second scheme for describing gene functions, the **gene ontology** (**GO**) nomenclature, was initially designed for annotation of the *Drosophila* genome but subsequently has been applied to many other species. The GO system, which can be applied to any protein, not just to enzymes, is not so much a classification scheme as a detailed set of standardized words and phrases that are used to describe a protein's molecular function, the biological process to which it contributes, and its location in the cell. For example, the molecular function of β-amylase is described as:

"Catalysis of the reaction (1,4-alpha-D-glucosyl)(n+1) + H2O = (1,4-alpha-D-glucosyl)(n–1) + alpha-maltose. This reaction is the hydrolysis of 1,4-alpha-glucosidic linkages in polysaccharides so as to remove successive maltose units from the non-reducing ends of the chains."

This description is accompanied by an information modeling device called a **directed acyclic graph** (**DAG**), which gives a hierarchical categorization of the molecular function, similar to the way an enzyme function is categorized within an EC number (**Figure 6.6**).

Because the GO vocabulary is standardized, GO descriptions can be searched using a computer, which not only enables homologous genes in different genomes to be identified, but also makes it possible to identify groups of genes with similar functions in a single genome or across a set of genomes. The DAG enables the search to be conducted at different levels, so in the case of β-amylase the search can be for 'beta-amylase activity' and hence directed specifically at β-amylase homologs, or the search term could be 'hydrolase activity, hydrolyzing O-glycosyl' so that a broader group of related hydrolase enzymes are also identified.

6.2 ASSIGNING FUNCTION BY GENE INACTIVATION AND OVEREXPRESSION

Computer methods for assigning functions to unknown genes are becoming increasingly sophisticated, but the bioinformatics approach has limitations and cannot identify the function of every new gene that is discovered in a genome.

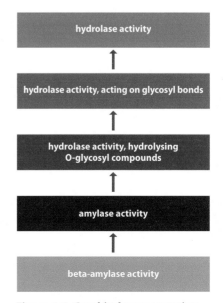

hydrolase activity

hydrolase activity, acting on glycosyl bonds

hydrolase activity, hydrolysing O-glycosyl compounds

amylase activity

beta-amylase activity

Figure 6.6 Graphical representation of gene ontology for β-amylase. The representation is hierarchical from most specific at the bottom to least specific at the top.

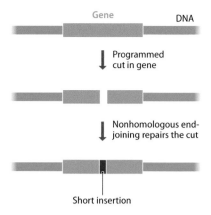

Figure 6.7 Gene inactivation with a programmable nuclease. The cut made by the nuclease is repaired by nonhomologous end-joining, which is error-prone and likely to insert or delete a few base pairs of DNA at the repair site, disrupting the target gene.

Experimental methods are therefore needed to complement and extend the results of computer analysis.

Functional analysis by gene inactivation

Devising experimental methods for functional analysis of new genes is proving to be one of the biggest challenges in genomics research. Most molecular biologists would agree that the methodologies and strategies currently in use are not entirely adequate for assigning functions to the vast numbers of unknown genes being discovered by sequencing projects. The problem is that the objective – to plot a course from gene to function – is the reverse of the route conventionally taken by genetic analysis, in which the starting point is a phenotype and the objective is to identify the underlying gene or genes. The problem we are currently addressing takes us in the opposite direction: starting with a new gene and hopefully leading to identification of the associated phenotype.

In conventional genetic analysis, the genetic basis of a phenotype is usually studied by searching for mutant organisms in which the phenotype has become altered. The mutants might be obtained experimentally, for example, by treating a population of organisms (e.g., a culture of bacteria) with ultraviolet radiation or a mutagenic chemical, or the mutants might be present in a natural population. The gene or genes that have been altered in the mutant organism are then studied by genetic crosses (Section 3.4), which can locate the position of a gene in a genome and also determine if the gene is the same as one that has already been characterized. The gene can then be studied further by molecular biology techniques such as cloning and sequencing.

The general principle of this conventional analysis is that the genes responsible for a phenotype can be identified by determining which genes are inactivated in organisms that display a mutant version of the phenotype. If the starting point is the gene, rather than the phenotype, then the equivalent strategy would be to mutate the gene and identify the phenotypic change that results. This is the basis of most of the techniques used to assign functions to unknown genes.

Gene inactivation by genome editing

The most efficient way to inactivate a specific gene is by **genome editing** with a **programmable nuclease**. This is a nuclease that can be directed to a specific site in a genome where it makes a double-stranded cut (**Figure 6.7**). The cut stimulates a natural repair process, called **nonhomologous end-joining** (**NHEJ**) in eukaryotes (Section 17.2), which joins the DNA strands together again. However, NHEJ is error-prone, and usually results in a short insertion or deletion occurring at the repair site. If the repair is within a gene, then the change in nucleotide sequence will inactivate the gene. This is called **error-prone gene editing**.

Several genome editing systems have been explored, but the one that has gained prominence makes use of the **Cas9 endonuclease**, which is a component of the prokaryotic immune system called **clustered regularly interspaced short palindromic repeats** (**CRISPR**; Section 8.2). The endonuclease is directed to its target site by a 20-nucleotide guide RNA whose binding site must be immediately upstream of a **protospacer adjacent motif (PAM)**, which for Cas9 is a 5´–NGG–3´ or 5´–NAG–3´ sequence (where 'N' is any nucleotide), giving a 23-bp target that is cleaved by the endonuclease (**Figure 6.8**). Carrying out the experiment is straightforward, simply requiring that a cloning vector be used to introduce the *Cas9* gene and a DNA sequence specifying the guide RNA into the cells whose genomes are being edited. To ensure that the endonuclease is active in a eukaryotic host, an artificial gene can be used, based on the sequence of a naturally occurring bacterial *Cas9* gene but utilizing the preferred codons for the eukaryotic host (Section 5.1), and including sequences specifying nuclear localization signals. These are short amino acid sequences that ensure that the protein, following synthesis by a ribosome in the cytoplasm, is imported into the nucleus. It is also possible to edit a genome simply by introducing a ribonucleoprotein complex comprising the Cas9 endonuclease protein and the guide RNA.

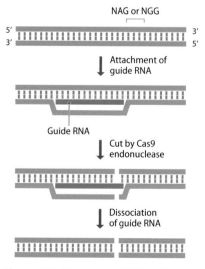

Figure 6.8 Cleavage of DNA by the Cas9 endonuclease. The cut position is specified by the 20-nucleotide guide RNA, which must be designed to base pair to a target site immediately upstream of a 5´–NGG–3´ or 5´–NAG–3´ sequence (where 'N' is any nucleotide).

This DNA-free system has been trialed with plants, where gene inactivation is important not only in genome annotation but also in engineering projects aimed at developing new varieties of crops with improved properties.

A critical issue with the use of a programmable endonuclease is the specificity of the editing process for the target sequence in the genome. The expected frequency of a 20-nucleotide motif (the length of the Cas9 guide RNA) in a DNA sequence is once every $4^{20} = 1.1 \times 10^{12}$ bp, which is 350 times the length of the human genome. This means that it is unlikely that there will be a second exact version of the target sequence in the genome, unless the guide RNA has been designed poorly and hybridizes to a repeat sequence. However, the Cas9 system does not require complete base-pairing between the guide RNA and genomic DNA, and in some early experiments **off-target editing** was detected at sites where the DNA–RNA heteroduplex had as many as five mismatched positions (**Figure 6.9**). Various ways of improving specificity have been explored, including the use of modified endonucleases that bind less strongly to the strand of the double helix that does not form the heteroduplex with the guide RNA. This might seem a counter-intuitive way to increase specificity, but the rationale is that the weaker protein-DNA interaction encourages base-pairing between the two DNA strands, which now out-competes attachment of the guide RNA unless the RNA is able to form a perfectly matched, and hence stable, heteroduplex.

Gene inactivation by homologous recombination

Before the development of genome editing, the most popular way to inactivate a specific gene was to disrupt it with an unrelated segment of DNA by **homologous recombination** between the chromosomal copy of the gene and a second piece of DNA that shares some sequence identity with the target gene (**Figure 6.10**). Homologous and other types of recombination are complex events, which we will deal with in detail in Chapter 16. For present purposes it is enough to know that if two DNA molecules have similar sequences, then recombination can result in segments of the molecules being exchanged.

How is gene inactivation carried out in practice? We will consider two examples, the first with *S. cerevisiae*. After completing the genome sequence in 1996, yeast molecular biologists embarked on a coordinated, international effort to determine the functions of as many of the unknown genes as possible. Much of this work made use of the technique shown in **Figure 6.11**. The central component is the **deletion cassette**, which carries a gene for antibiotic resistance. This gene is not a normal component of the yeast genome but it will work if transferred into a yeast chromosome, giving rise to a transformed yeast cell that is resistant to the antibiotic geneticin. Before using the deletion cassette, new segments of DNA are attached as tails to either end. These segments have sequences identical to parts of the yeast gene that is going to be inactivated. After the modified cassette is introduced into a yeast cell, homologous recombination occurs between the DNA tails and the chromosomal copy of the yeast gene, replacing the latter with the antibiotic-resistance gene. Cells which have undergone the replacement are therefore selected by plating the culture onto agar medium containing geneticin. The resulting colonies lack the target gene activity and their phenotypes can be examined to gain some insight into the function of the gene.

This method of gene inactivation is straightforward to carry out, but time-consuming if it needs to be applied independently to each gene being studied. In the yeast project, this was an important consideration as only 60% of the 6274 ORFs greater than 100 codons in length could be assigned functions based either on the results of previous genetic analysis of yeast or by homology searching. The remaining 40%, over 2500 genes in total, had to be subjected to experimental analysis in order to assign functions. A high throughput version of the gene inactivation method was therefore devised, called the **barcode deletion strategy**. This strategy uses a modified version of the basic deletion cassette system, the difference being that the cassette also includes two 20-nucleotide barcode sequences, different for each deletion, which act as tags for that particular

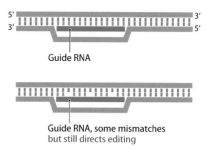

Figure 6.9 Off-target editing. The guide RNA can still initiate editing even if it is not completely base-paired to the target DNA.

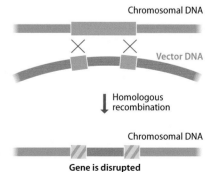

Figure 6.10 Gene inactivation by homologous recombination. The vector caries two segments of DNA matching the ends of the gene that we wish to inactivate. These end segments recombine with the chromosomal copy of the target gene. As a result, the target gene becomes inactivated.

Deletion cassette

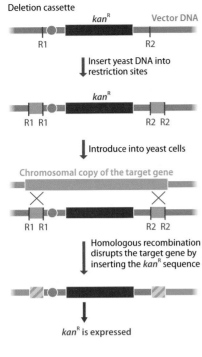

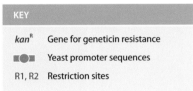

KEY

kan^R   Gene for geneticin resistance

    Yeast promoter sequences

R1, R2   Restriction sites

Figure 6.11 The use of a yeast deletion cassette. The deletion cassette consists of an antibiotic-resistance gene preceded by the promoter sequences needed for expression in yeast, and flanked by two restriction sites. The start and end segments of the target gene are inserted into the restriction sites, and the vector is introduced into yeast cells. Recombination between the gene segments in the vector and the chromosomal copy of the target gene results in disruption of the latter. Cells in which the disruption has occurred are identifiable because they now express the antibiotic-resistance gene and so will grow on an agar medium containing geneticin. The gene designation 'kan' is an abbreviation for 'kanamycin resistance,' kanamycin being the family name of the group of antibiotics that include geneticin.

mutant (**Figure 6.12**). Each barcode is flanked by the same pair of sequences and so can be amplified by a single PCR. This means that groups of mutated yeast strains, each with a different inactivated gene, can be mixed together and their phenotypes screened in a single experiment. For example, to identify genes required for growth in a glucose-rich medium, a collection of mutants would be mixed together and cultured under these conditions. After incubation, DNA is prepared from the culture and the barcoding PCR carried out. The result is a mixture of PCR products, each representing a different barcode, the relative abundance of each barcode indicating the abundance of each mutant after growth in glucose-rich medium. Those barcodes that are absent or present only at low abundance indicate the mutants whose inactivated genes were needed for growth under these conditions.

A similar approach to gene inactivation can also be used with mice. The mouse is a popular **model organism** for humans because the mouse genome is similar to the human genome, containing many of the same genes. Functional analysis of unknown human genes can therefore be carried out by inactivating the equivalent genes in the mouse. The procedure makes use of homologous recombination in a manner identical to that described for yeast, and once again results in a cell in which the target gene has been inactivated. The problem is that we do not want just one mutated cell, we want a whole mutant mouse, as only with the complete organism can we make a full assessment of the effect of the gene inactivation on the phenotype. To achieve this it is necessary to use a special type of mouse cell: an **embryonic stem** or **ES cell**. Unlike most mouse cells, ES cells are **totipotent**, meaning that they are not committed to a single developmental pathway and can therefore give rise to all types of differentiated cell. The engineered ES cell is therefore injected into a mouse embryo, which continues to develop and eventually gives rise to a **chimera**, a mouse whose cells are a mixture of mutant ones, derived from the engineered ES cells, and nonmutant ones, derived from all the other cells in the embryo. This is still not quite what we want, so the chimeric mice are allowed to mate with one another. Some of the offspring result from fusion of two mutant gametes, and will therefore be nonchimeric, as every one of their cells will carry the inactivated gene. These are **knockout mice**, and with luck their phenotypes will provide the desired information on the function of the gene being studied. This works well for many gene inactivations but some are lethal and so cannot be studied in a homozygous knockout mouse. Instead, a heterozygous mouse is obtained, the product of fusion between one normal and one mutant gamete, in the hope that the phenotypic effect of the gene inactivation will be apparent even though the mouse still has one correct copy of the gene being studied.

Gene inactivation by transposon tagging and RNA interference

A third method that can be used to disrupt a gene in order to study its function is **transposon tagging**, in which inactivation is achieved by the insertion of a transposable element, or transposon, into the gene. Most genomes contain transposable elements (Section 9.2) and although the bulk of these are inactive, there are usually a few that retain their ability to move to new positions in the genome. Under normal circumstances, transposition is a relatively rare event, but it is sometimes possible to use recombinant DNA techniques to make modified transposons that change their position in response to an external stimulus. One way of doing this, involving the yeast transposon *Ty1*, is shown in **Figure 6.13**. Transposon tagging is also important in analysis of the fruit-fly genome, using the endogenous *Drosophila* transposon called the **P element**. The weakness of transposon tagging is that it is difficult to target individual genes, because transposition is more or less a random event and it is impossible to predict where a transposon will end up after it has jumped. If the intention is to inactivate a particular gene, then it is necessary to induce a substantial number of transpositions and then to screen all the resulting organisms to find one with the correct insertion. Transposon tagging is therefore more applicable to global studies of genome function, in which genes are inactivated at random and groups of genes with similar functions are identified by examining the progeny for interesting phenotype changes.

A completely different approach to gene inactivation is provided by **RNA interference**, or **RNAi**, one of a series of natural processes by which short RNA molecules influence gene expression in living cells (Section 12.6). When used in genomics research, RNAi provides a means of silencing the expression of a target gene, not by disrupting the gene itself, but by destroying its mRNA. This is accomplished by introducing into the cell short, double-stranded RNA molecules whose sequences match that of the mRNA being targeted. The double-stranded RNAs are broken down into shorter molecules, which induce degradation of the mRNA (**Figure 6.14**).

RNA interference was initially shown to work effectively in the worm *Caenorhabditis elegans*, the genome of which has been completely sequenced and which is looked on as an important model organism for higher eukaryotes (Section 14.3). Virtually all of the 20,000 predicted genes in the *C. elegans* genome were individually silenced by RNA interference. The key step in any RNAi experiment is introducing into the test organism the double-stranded RNA molecule that will give rise to the single-stranded interfering RNAs. With *C. elegans* this can be achieved by feeding the RNA to the worms. *C. elegans* eats bacteria, including *Escherichia coli*, and is often grown on a lawn of bacteria on an agar plate. If the bacteria contain a cloned gene that directs expression of a double-stranded RNA with the same sequence as a *C. elegans* gene, then, after ingestion, the RNAi pathway begins to operate. Alternatively, the double-stranded RNA can be directly microinjected into the worm, but this is more time-consuming.

RNA interference occurs naturally in most eukaryotes, but its general application as a means of studying gene function has been hindered by three issues:

- RNAi does not always result in complete silencing of the target gene; often, the silencing is incomplete, and is referred to as 'knockdown' rather than 'knockout.' Depending on the degree of silencing, it might or might not be possible to assess the effect of a gene knockdown on phenotype.

- The interfering RNAs are so short that off-target effects are possible. These occur when the interfering RNAs bind to mRNAs other than the targets, resulting in silencing of more than one gene.

- In mammals, the artificial introduction of double-stranded RNA often results in activation of signaling proteins called **interferons**, which stimulate an antiviral defense process that is displayed by both cultured cells and by whole organisms. This interferon response results in phenotypic changes that can mask the specific change occurring due to silencing of the target gene. Some mammalian cells, such as mouse oocytes, lack the interferon response, but with most mammalian systems special strategies have to be devised when RNAi is being used for gene knockdown.

Gene overexpression can also be used to assess function

So far, we have concentrated on techniques that result in inactivation of the gene being studied ('loss of function'). The complementary approach is to engineer an organism in which the test gene is much more active than normal ('gain of function') and to determine what changes, if any, this has on the phenotype. The results of these experiments must be treated with caution because of the need to distinguish between a phenotype change that is due to the specific function of an overexpressed gene, and a less specific phenotype change that reflects the abnormality of the situation where a single gene product is being synthesized in excessive amounts, possibly in tissues in which the gene is normally inactive. Despite this qualification, overexpression has provided some important information on gene function.

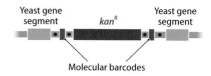

Figure 6.12 A deletion cassette used in the barcode strategy. The two molecular barcodes are 20 bp sequences, different for each cassette, which can be amplified by PCR. During homologous recombination, the barcodes are inserted into the yeast genome along with the kanamycin-resistance gene. The barcodes therefore provide specific tags for each individual gene deletion.

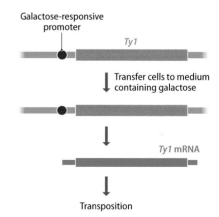

Figure 6.13 Artificial induction of transposition. Recombinant DNA techniques have been used to place a promoter sequence that is responsive to galactose upstream of a *Ty1* element in the yeast genome. When galactose is absent, the *Ty1* element is not transcribed and so remains quiescent. When the cells are transferred to a culture medium containing galactose, the promoter is activated and the *Ty1* element is transcribed, initiating the transposition process.

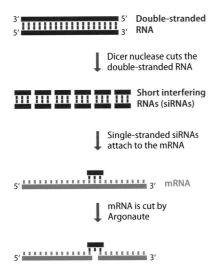

Figure 6.14 RNA interference. The double-stranded RNA molecule is broken down by the Dicer ribonuclease into 'short interfering RNAs' (siRNAs) of length 21–25 bp. One strand of each siRNA base-pairs to a target mRNA, which is then cleaved by an endoribonuclease called the Argonaute protein.

To overexpress a gene, a special type of cloning vector must be used, one designed to ensure that the cloned gene directs the synthesis of as much protein as possible. The vector is therefore **multicopy**, meaning that it multiplies inside the host organism to 40–200 copies per cell, so there are many copies of the test gene. The vector must also contain a highly active promoter sequence (Section 12.3) so that each copy of the test gene is converted into large quantities of mRNA, again ensuring that as much protein as possible is made. In the example shown in **Figure 6.15**, the cloning vector contains a highly active promoter that is expressed only in the liver, so each **transgenic mouse** overexpresses the test gene in its liver. This approach has been used with genes whose sequences suggest that they code for proteins that are secreted into the bloodstream. After synthesis in the liver, the test protein is secreted and the phenotype of the transgenic mouse examined in the search for clues regarding the function of the cloned gene. An interesting discovery was made when it was realized that one mouse containing a human transgene had bones that were significantly more dense than those of normal mice. This was important for two reasons: first, it enabled the relevant gene to be identified as one involved in bone synthesis; second, the discovery of a protein that increases bone density has implications for the development of treatments for human osteoporosis, a fragile-bone disease.

The phenotypic effect of gene inactivation or overexpression may be difficult to discern

The critical aspect of a gene inactivation or overexpression experiment is the need to identify a phenotypic change, the nature of which gives a clue to the function of the manipulated gene. This can be much more difficult than it sounds. Even with a unicellular organism such as yeast, the **phenotype ontology** is quite lengthy. This is a list that describes the characteristics of an organism that can be distinguished by visual inspection or by biochemical tests. For yeast there are 148 characteristics describing variations in 32 different phenotypes (Table 6.1). With multicellular eukaryotes the ontologies are much more complex, with over 13,000 characteristics to consider when assigning a function to a mammalian gene. Furthermore, the effect of gene inactivation can be very subtle and may not be recognized when the phenotype is examined. A good example of the problems that occur was provided by the longest gene on yeast chromosome III, which, at 2167 codons and with typical yeast codon bias, simply had to be a functional gene rather than a spurious ORF. Inactivation of this gene had no apparent effect, with the mutant yeast cells appearing to have an identical phenotype to normal yeast. For some time it was thought that perhaps this gene is dispensable, its protein product either involved in some completely nonessential function, or having a function that is duplicated by a second gene. Eventually it was shown that the mutants die when they are grown at low pH in the presence of glucose and acetic acid, which normal yeasts can tolerate. This enabled the gene to be linked with the processes by which yeast cells secrete unwanted compounds,

Figure 6.15 Functional analysis by gene overexpression. The objective is to determine if overexpression of the gene being studied has an effect on the phenotype of a transgenic mouse. A cDNA of the gene is therefore inserted into a multicopy cloning vector carrying a highly active promoter sequence that directs expression of the cloned gene in mouse liver cells. The cDNA is used rather than the genomic copy of the gene because the former does not contain introns and so is shorter and easier to manipulate in the test tube.

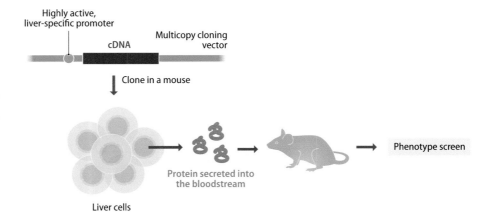

TABLE 6.1 PHENOTYPE ONTOLOGY FOR *SACCHAROMYCES CEREVISIAE*		
Phenotype group	**Phenotype**	**Number of characteristics**
Metabolism and growth	RNA modification	1
	Protein activity	1
	Chemical compound accumulation	1
	Nutrient utilization	12
	Vegetative growth	3
	Protein/peptide accumulation	1
	Protein/peptide distribution	1
	Anaerobic metabolism	1
	Chemical compound excretion	1
	RNA accumulation	1
	Protein/peptide modification	1
	Redox state	1
Morphology	Cellular morphology	19
	Culture appearance	6
Development	Sexual cycle	13
	Budding	3
	Filamentous growth	2
	Lifespan	2
Cellular processes	Mitotic cell cycle	14
	Intracellular transport	23
	Stress resistance	20
	Chromosome/plasmid maintenance	8
	Prion state	3
	Cell death	2
Essentiality	Inviable	1
	Viable	1
Interaction with host/environment	Adhesion	1
	Virulence	1
Fitness	Competitive fitness	1
	Viability	1
	Haploinsufficient	1
	Haploproficient	1

such as acetate, out of the cell. This is definitely an essential function, but this essentiality was difficult to track down from the phenotype tests.

Even when the most careful screens are carried out, many gene inactivations appear to give no discernible phenotypic change. Almost 5000 of the over 6000 genes in the yeast genome can be individually inactivated without causing the cells to die, and inactivation of many of these 5000 genes has no detectable effect on the metabolic properties of the cell under normal growth conditions. The phenotypic effects of these genes only become apparent, if at all, when the cells are grown under a range of different conditions, or when groups of genes that contribute to the same phenotype are co-inactivated. In the human genome, 166 genes have been identified that appear to be nonessential, as both copies can be inactivated, due to natural mutation, without any discernible effect on the health of the individual. These observations suggest that a complete functional annotation of the genomes of many species will not be achievable by approaches that are based solely on gene inactivation or overexpression.

6.3 UNDERSTANDING GENE FUNCTION BY STUDIES OF ITS EXPRESSION PATTERN AND PROTEIN PRODUCT

Gene inactivation and overexpression are the primary techniques used by genome researchers to determine the function of a new gene, but they are not the only procedures that can be used to provide information on gene activity. Additional insights into gene function can be obtained by identifying in which tissues, and at what times, a gene is expressed, and by direct examination of the protein coded by the gene.

Reporter genes and immunocytochemistry can be used to locate where and when genes are expressed

Clues to the function of a gene can often be obtained by determining where and when the gene is active. If gene expression is restricted to a particular organ or tissue of a multicellular organism, or to a single set of cells within an organ or tissue, then this positional information can be used to infer the general role of the gene product. The same is true of information relating to the developmental stage at which a gene is expressed. This type of analysis has proved particularly useful in understanding the activities of genes involved in the earliest stages of development in *Drosophila* (Section 14.3) and is increasingly being used to unravel the genetics of mammalian development. It is also applicable to those unicellular organisms, such as yeast, that have distinctive developmental stages in their life cycle.

Determining the pattern of gene expression within an organism is possible with a **reporter gene**. This is a gene whose expression can be monitored in a convenient way, ideally by visual examination (**Table 6.2**), with cells that express the reporter gene becoming blue, fluorescing, or giving off some other visible signal. For the reporter gene to give a reliable indication of where and when a test gene is expressed, the reporter must be subject to the same regulatory signals as the test gene. This is achieved by replacing the ORF of the test gene with the ORF

TABLE 6.2 EXAMPLES OF REPORTER GENES

Gene	Gene product	Assay
lacZ	β-galactosidase	Histochemical test
uidA	β-glucuronidase	Histochemical test
lux	Luciferase	Bioluminescence
GFP	Green fluorescent protein	Fluorescence

of the reporter gene (**Figure 6.16**). Most of the regulatory signals that control gene expression are contained in the region of DNA upstream of the ORF, so the reporter gene should now display the same expression pattern as the test gene. The expression pattern can therefore be determined by examining the organism for the reporter signal.

As well as knowing in which cells a gene is expressed, it is often useful to locate the position within the cell where the protein coded by the gene is found. For example, key data regarding gene function can be obtained by showing that the protein product is located in mitochondria, in the nucleus, or on the cell surface. Reporter genes cannot help here because the DNA sequence upstream of the gene – the sequence to which the reporter gene is attached – is not involved in targeting the protein product to its correct intracellular location. Instead, it is the amino acid sequence of the protein itself that is important. Therefore the only way to determine where the protein is located is to search for it directly. This can be done by **immunocytochemistry**, which makes use of an antibody that is specific for the protein of interest and so binds to this protein and no other. The antibody is labeled so that its position in the cell, and hence the position of the target protein, can be visualized (**Figure 6.17**). Fluorescent labeling and confocal microscopy are used for low-resolution studies; alternatively, high-resolution immunocytochemistry can be carried out by electron microscopy using an electron-dense label such as colloidal gold.

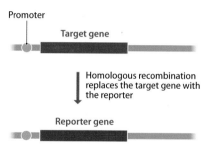

Figure 6.16 A reporter gene. The open reading frame of the reporter gene replaces the open reading frame of the gene being studied. The result is that the reporter gene is placed under control of the regulatory sequences that usually dictate the expression pattern of the test gene.

CRISPR can be used to make specific changes in a gene and the protein it encodes

Inactivation and overexpression can determine the general function of a gene, but they cannot provide detailed information on the activity of a protein coded by a gene. For example, it might be suspected that part of a gene specifies an amino acid sequence that directs its protein product to a particular compartment in the cell, or is responsible for the ability of the protein to respond to a chemical or physical signal. To test these hypotheses, it would be necessary to alter the relevant part of the gene sequence, but to leave the bulk unmodified so that the protein is still synthesized and retains the major part of its activity. A variety of procedures referred to as **site-directed** or *in vitro* **mutagenesis** can be used to make these subtle changes. These are important techniques whose applications lie not only with the study of gene activity but also in **genetic modification**, where the intention is to alter the phenotype of an organism in a defined way, and **protein engineering**, which aims to create novel proteins with properties that are better suited for use in industrial or clinical settings.

In Section 6.2, we studied how the CRISPR system can be used to inactivate specific genes. In this process a guide RNA directs the Cas9 endonuclease to a target site in the genome, where it makes a double-stranded cut that is repaired by the nonhomologous end-joining pathway in an error-prone fashion that usually results in the random insertion or deletion of a few nucleotides. In a second version of CRISPR genome editing, the repair is directed by a short template DNA molecule that is introduced into the cell along with the endonuclease gene and the guide RNA sequence. When the template DNA is present the nonhomologous repair process does not occur, and instead a version of homologous recombination replaces the two broken DNA ends with the intact template sequence. If the sequence of the template DNA is identical to that of the target site then the break is repaired, without insertion or deletion, and the original sequence is restored. However, the template DNA is able to participate in the recombination event even if its sequence is slightly different to that of the target site. If this is the case, then after repair the target sequence will contain the variations that were originally in the template DNA (**Figure 6.18**). This type of gene editing, called **homology-directed repair**, is therefore able to introduce directed mutations into the target gene.

Two other modifications of genome editing have taken the power and adaptability of the CRISPR system to even higher levels. In the first of these methods, a

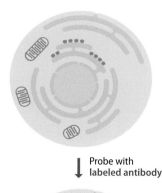

Probe with labeled antibody

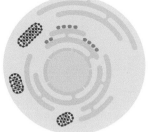

Labeling appears within mitochondria

Figure 6.17 Immunocytochemistry. The cell is treated with an antibody that is labeled with a red fluorescent marker. Examination of the cell shows that the fluorescent signal is associated with the inner mitochondrial membrane. A working hypothesis would therefore be that the target protein is involved in electron transport and oxidative phosphorylation, as these are the main biochemical functions of the inner mitochondrial membrane.

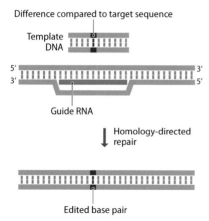

Figure 6.18 Homology-directed repair. When a short template DNA is present, the Cas9 endonuclease directs a homologous recombination event that inserts the template into the target DNA. If the sequence of the template is different to that of the target, then the target DNA becomes edited.

fusion is made between the endonuclease and a **base editor**, an enzyme capable of changing one nucleotide to another within a DNA molecule. An example is the adenine deaminase of *Escherichia coli*, which converts adenine nucleotides to inosine during the synthesis of tRNAs (**Figure 6.19A**). When fused to the Cas9 endonuclease, adenine deaminase carries out the adenine to inosine conversion within the region specified by the guide RNA. The endonuclease is itself modified so that it can no longer cut the target DNA, and in effect is simply a vehicle for transporting the base editor to the position where the guide RNA is bound. Editing occurs within the displaced strand of DNA – the strand that is not base-paired to the guide RNA – primarily in the region furthest from the PAM (**Figure 6.19B**). This editing window could contain two or more adenines and hence multiple sites will be mutated, which might or might not be desirable depending on the project design. Modified versions of the editing enzyme have been developed to reduce the editing window to as little as two nucleotides, thereby avoiding the possibility of unwanted **bystander mutations** in the sequence adjacent to the targeted nucleotide. After editing, the inosine base-pairs with cytosine when the gene is transcribed or replicated, so in effect an A-to-G edit has been made. Base editors that convert C to T have also been designed but these involve a fusion between three proteins: the endonuclease, a cytosine deaminase editing enzyme and a uracil glycosylase inhibitor. The latter is needed because the initial product of cytosine deamination is uracil (**Figure 6.20**), which is rapidly cut out of the DNA by the cell's natural repair mechanism (Section 17.2). To prevent this happening during base editing, the fusion includes an inhibitor of the glycosylase repair enzyme.

The second modification of genome editing, called **prime editing**, holds promise as the most specific of the systems that have been developed, with the lowest frequency of off-target events. The RNA component, which is called a **prime editing guide RNA (pegRNA)**, is longer than a standard guide RNA and contains sequences that are able to base-pair to both of the DNA strands at the target site (**Figure 6.21**). The editing enzyme is a fusion between an endonuclease that has been engineered so that it cuts just one strand of the DNA, and a reverse transcriptase. To initiate the editing process, the guide RNA component of the pegRNA attaches to the target site, displacing a strand of DNA that is cut by the endonuclease at a position three nucleotides upstream of the PAM. The other end of the pegDNA then base-pairs with the cut strand, enabling the 3′ end of the cut strand to be extended by the reverse transcriptase, using the pegRNA as the template. The part of the pegRNA that is copied contains the edited positions that we wish to introduce into the target site, which therefore appear in the new segment of DNA. Following dissociation of the guide RNA, the DNA forms one of two possible a flap structures. In one of these (shown on the left in Figure 6.21), the displaced strand rehybridizes with its partner, and the new DNA is detached as the flap; in the alternative structure (on the right in Figure 6.21), the new DNA has hybridized and the original DNA forms the flap. Both versions mimic a structure that arises during DNA replication (Section 15.3) and which can therefore be repaired by the enzyme called the **flap endonuclease (FEN1)**, working in conjunction with a DNA ligase. In the first example, the repair restores the DNA

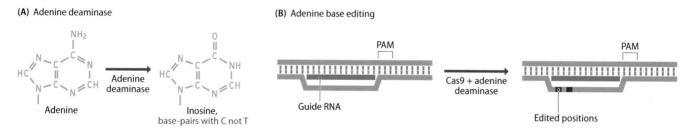

Figure 6.19 Adenine base editing. (A) Adenine deaminase converts adenine to inosine, which base-pairs with cytosine rather than thymine. (B) Fusion between the Cas9 endonuclease and adenine deaminase creates the base editor, which carries out the adenine to inosine conversion in the region of the displaced strand furthest from the PAM.

to its original, unedited sequence. However, if the original DNA forms the flap, then this is cut off and the edits are incorporated into the target site.

Other methods for site-directed mutagenesis

Prior to the development of CRISPR genome editing by Jennifer Doudna and Emmanuelle Charpentier in the early 2010s, genome researchers were reliant on a variety of more complicated methods for site-directed mutagenesis of target genes. These methods could not be applied directly to genomic DNA, and instead the target gene had to be cloned and then reintroduced into the host cell after mutagenesis. Although more cumbersome than CRISPR editing, these versions of site-directed mutagenesis are still important in some projects. The three most important of these methods are:

- **Oligonucleotide-directed mutagenesis**, which requires a single-stranded version of the relevant gene, the latter usually obtained by cloning with a **M13 bacteriophage** vector. M13 vectors are circular double-stranded DNA molecules that can be manipulated in the same way as a plasmid vector, but which give rise to bacteriophage particles that contain single-stranded DNA. An oligonucleotide containing a single base-pair mismatch, corresponding to the desired mutation, is annealed to the single-stranded DNA and a DNA polymerase added, so that the oligonucleotide primes a strand-synthesis reaction that continues all the way around the circular template molecule (**Figure 6.22A**). After introduction into *E. coli*, DNA replication produces numerous copies of this recombinant DNA molecule, half of these being copies of the original strand of

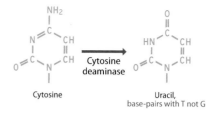

Figure 6.20 Cytosine deaminase converts cytosine to uracil, which base-pairs with thymine rather than guanine.

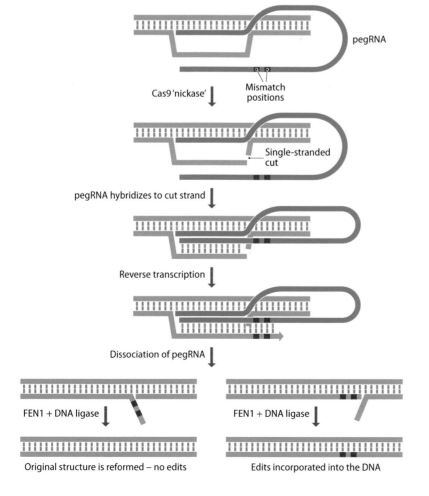

Figure 6.21 Prime editing. The guide component of the pegRNA attaches to the target site and the modified Cas9 endonuclease (now a 'nickase') cuts the displaced strand. The 3′ end of the cut strand is then extended by reverse transcriptase, using the pegRNA as the template. Following dissociation of the guide RNA, the DNA forms one of two possible flap structures. On the left, the displaced strand has rehybridized with its partner, and the new DNA is detached as a flap structure. The flap endonuclease FEN1 and DNA ligase cut off the flap and repair the nick, leaving the DNA with its original sequence. However, on the right, the new DNA has hybridized and the original DNA forms the flap. Repair of this structure incorporates the edits into the target site.

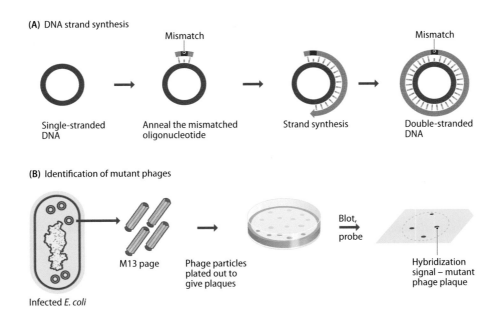

(A) DNA strand synthesis

Mismatch

Mismatch

Single-stranded
DNA

Anneal the mismatched
oligonucleotide

Strand synthesis

Double-stranded
DNA

(B) Identification of mutant phages

Infected *E. coli*

M13 page

Phage particles
plated out to
give plaques

Blot,
probe

Hybridization
signal – mutant
phage plaque

Figure 6.22 Oligonucleotide-directed mutagenesis. (A) A short oligonucleotide, containing a mismatch to the template DNA, is used to prime synthesis of the second DNA strand. (B) Replication in *E. coli* produces M13 phages containing a mixture of mutated and unmutated molecules. Plaques are blotted onto a membrane and the ones containing the mutation are identified by hybridization probing with the original oligonucleotide.

DNA, and half copies of the strand that contains the mutated sequence. All of these double-stranded molecules direct synthesis of M13 phage particles, so about half the phages released from the infected bacteria carry a copy of the mutated molecule. The phages are plated onto solid agar so that plaques are produced, and the mutant ones identified by hybridization probing with the original oligonucleotide (**Figure 6.22B**).

- **Artificial gene synthesis** involves constructing the gene in the test tube, placing mutations at all the desired positions. The gene is constructed by synthesizing a series of partially overlapping oligonucleotides, each typically about 150 nucleotides in length. The gene is assembled by filling in the gaps between the overlaps with DNA polymerase.

- PCR can also be used to create mutations in cloned genes, though like oligonucleotide-directed mutagenesis, only one mutation can be created per experiment. The method shown in **Figure 6.23** involves two PCRs, each with one normal primer (which forms a fully base-paired hybrid with the template DNA), and one mutagenic primer (which contains a single base-pair mismatch). The mutation is therefore initially present in two PCR products, each corresponding to one-half of the starting DNA molecule. The two PCR products are then mixed together and a final PCR cycle is carried out to construct the full-length, mutated DNA molecule.

After mutagenesis, the mutated gene can be placed back in its original host by homologous recombination, as described in Section 6.2, or transferred to an *E. coli* vector designed for synthesis of protein from cloned DNA, so that a sample of the mutated protein can be obtained. If homologous recombination is used, then we must have a way of knowing which cells have taken up a copy of the mutated gene. Even with yeast this will only be a fraction of the total. Normally we would solve this problem by placing a marker gene (e.g., one coding for antibiotic resistance) next to the mutated gene and looking for cells that take on the phenotype conferred by this marker. In most cases, cells that insert the marker gene into their genome also insert the closely attached mutated gene and so are the ones we want. The problem is that in a site-directed mutagenesis experiment we must be sure that any change in the activity of the gene being studied is the

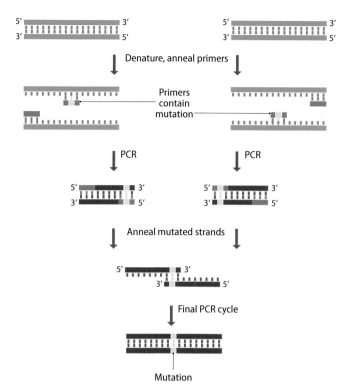

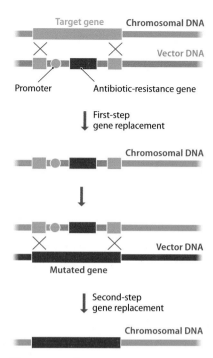

Figure 6.24 Two-step gene replacement.

Figure 6.23 One method for site-directed mutagenesis by PCR.

result of the specific mutation that was introduced into the gene, rather than the indirect result of changing its environment in the genome by inserting a marker gene next to it. The answer is to use a more complex, two-step gene replacement (**Figure 6.24**). In this procedure, the target gene is first replaced with the marker gene on its own, with the cells in which this recombination takes place being identified by selecting for the marker gene phenotype. These cells are then used in the second stage of the gene replacement, when the marker gene is replaced by the mutated gene; success is monitored by looking for cells that have lost the marker gene phenotype. These cells contain the mutated gene and their pheno-types can be examined to determine the effect of the directed mutation on the activity of the protein product.

6.4 USING CONVENTIONAL GENETIC ANALYSIS TO IDENTIFY GENE FUNCTION

We noted at the start of Section 6.2 that functional annotation of a genome rep-resents the reverse of the conventional approach to genetics, because functional annotation starts with a gene and attempts to discover its function, whereas con-ventional genetics starts with a phenotype and attempts to discover the gene or genes responsible for that phenotype. The conventional approach is sometimes called **forward genetics**, with **reverse genetics** comprising the methods that start with the gene. So far, we have only studied the use of reverse genetics in functional annotation of a genome, but that does not mean that forward genetics is no longer important. Quite the opposite is true. Forward genetics is still cen-tral to many areas of genomics research, in particular as a means of identifying human genes responsible for **inherited diseases**.

Identification of human genes responsible for inherited diseases

An inherited disease is one that is caused by a defect in the genome and which can be passed from parents to offspring. There are over 6000 **monogenic** inher-ited diseases known in the human population, each resulting from a defect in a

TABLE 6.3 SOME OF THE COMMONEST GENETIC DISEASES IN THE UK

Disease	Symptoms	Frequency (births per year)
Inherited breast cancer	Cancer	1 in 300 females
Cystic fibrosis	Lung disease	1 in 2000
Huntington's chorea	Neurodegeneration	1 in 2000
Duchenne muscular dystrophy	Progressive muscle weakness	1 in 3000 males
Hemophilia A	Blood disorder	1 in 4000 males
Sickle cell anemia	Blood disorder	1 in 10,000
Phenylketonuria	Mental retardation	1 in 12,000
β-Thalassemia	Blood disorder	1 in 20,000
Retinoblastoma	Cancer of the eye	1 in 20,000
Hemophilia B	Blood disorder	1 in 25,000 males
Tay–Sachs disease	Blindness, loss of motor control	1 in 200,000

single gene. The frequencies of these diseases vary enormously: the commonest types, such as inherited breast cancer and cystic fibrosis, occur once every few hundred or few thousand births, but there are also very rare diseases with just a few births every year (**Table 6.3**). Inherited diseases also affect other animals, in particular ones with low genetic diversity due to artificial breeding, such as some types of pedigree dog.

Forward genetics is able to identify the gene that confers a phenotype even if very little is known about that phenotype. In fact, all that is necessary is to establish that inheritance of the phenotype follows a simple Mendelian pattern (Section 3.3) and hence is specified by a single gene. If this is the case for an inherited disease, then we simply need DNA samples from members of affected families, so that pedigree analysis can be carried out to determine the linkage between the disease gene and mapped DNA markers (Section 3.4). To illustrate the methodology in greater detail we will examine the way in which pedigree analysis, coupled with reverse genetics, was used to identify a gene that confers susceptibility to inherited breast cancer.

The first pedigree studies of inherited breast cancer aimed to locate the position of the causative gene relative to RFLPs (Section 3.2) that had already been mapped onto the human genome. This work showed that in families with a high incidence of breast cancer, a significant number of the women who suffered from the disease possessed the same allele of the RFLP called *D17S74*. This observation suggested that the breast cancer gene must lie close to *D17S74* on the human genome. The rationale is that recombination is unlikely between two markers that are close together, so the alleles of those two markers will be inherited together (**Figure 6.25**). This is called **linkage disequilibrium**. In our example, the defective allele of the disease gene is linked to one allele of *D17S74*, and the nondefective allele of the gene is linked to the second allele of *D17S74*. As the RFLP had previously been mapped to the long arm of chromosome 17, it could

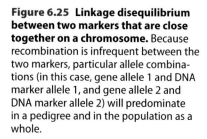

Figure 6.25 Linkage disequilibrium between two markers that are close together on a chromosome. Because recombination is infrequent between the two markers, particular allele combinations (in this case, gene allele 1 and DNA marker allele 1, and gene allele 2 and DNA marker allele 2) will predominate in a pedigree and in the population as a whole.

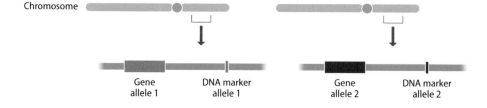

be concluded that the breast cancer gene must also be located in this part of the genome, probably within the chromosomal region designated q21 (**Figure 6.26**). A more specific map position was then deduced from additional pedigree analyses, these examining linkage between the breast cancer gene and STRs known to be present in the q21 region. These analyses placed the breast cancer gene between two STRs called *D17S1321* and *D17S1325*, which are approximately 600 kb apart.

Further pedigree analysis, with increasing numbers of families, could conceivably have resulted in an even more specific location for the breast cancer gene, but once a human gene has been placed within a genomic region of 1 Mb or less, it is usually possible for reverse genetics methods to complete the identification. The genome annotation shows that the region between *D17S1321* and *D17S1325* contains over 60 genes, any of which could be the breast cancer gene. The expression patterns of these **candidate genes** were studied, the expectation being that the breast cancer gene would be expressed in breast and ovarian tissue, as ovarian cancer is frequently associated with inherited breast cancer. Those genes that had the expected expression profile were then used in BLAST searches of other mammalian genomes, on the basis that a human gene that is important enough to cause disease when mutated would be expected to have homologs in a range of other mammals. Finally, the sequences of those genes still looked on as candidates were examined in women with and without inherited breast cancer to see if the genes from affected individuals contained mutations that might explain why they have the disease. When these analyses were complete, the most likely candidate was an approximately 100 kb gene, comprising 22 exons and coding for a 1863-amino acid protein. This gene, subsequently named *BRCA1*, is expressed in breast and ovary tissue, and has homologs in mice, rats, rabbits, sheep and pigs, but not in chickens. Critically, the alleles of this gene in five susceptible families contained mutations likely to lead to a nonfunctioning protein. Subsequent studies have shown that the protein specified by *BRCA1* is involved in transcription regulation and DNA repair, and also acts as a tumor suppressor gene, inhibiting abnormal cell division.

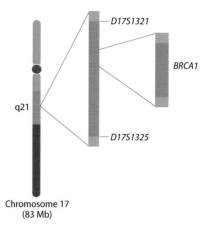

Figure 6.26 Mapping the breast cancer susceptibility gene *BRCA1*. Initially, the gene was mapped to a segment of the long arm of chromosome 17, within the region designated q21 (highlighted in the left-hand drawing). Additional mapping experiments narrowed this down to a 600 kb region flanked by two STR loci, *D17S1321* and *D17S1325* (middle drawing). After examination of expressed sequences, a strong candidate for *BRCA1* was eventually identified (right-hand drawing).

Genome-wide association studies can also identify genes for diseases and other traits

Although pedigree analysis can be used to identify the gene responsible for a monogenic disease, it is less successful with the many human diseases that have more complex genetic backgrounds. Several types of cancer, as well as disorders such as coronary heart disease and osteoporosis, are polygenic, meaning that they are controlled not just by one gene but by many genes working together. Most polygenic traits are quantitative, so affected individuals display different degrees of susceptibility, depending on the particular combination of alleles that they possess. This means that two members of a family that display the disease can have different genotypes. Under these circumstances, the disease phenotype will not follow a simple Mendelian inheritance pattern, and it becomes impossible to use pedigree data to link the disease to DNA markers with any degree of certainty.

A **genome-wide association study** (**GWAS**) is an alternative approach to gene identification that can work with polygenic traits. Rather than linking DNA markers with individual genes, a GWAS attempts to identify all of the markers, from all over the genome, that are associated with the disease. The positions of these markers will reveal the positions of candidate genes that might form part of the polygenic trait. A GWAS therefore requires both a large cohort of individuals, including people with and without the disease, and a large panel of DNA markers that will be typed in DNA samples from those individuals. The DNA samples are often obtained from **biobanks**, which are collections of biological material such as blood samples, donated following informed consent by patients and volunteers, with each sample accompanied by a detailed description of that individual's disease status. The DNA markers are invariably SNPs, because of the huge number of SNPs whose precise positions in the human genome are known, and because of the ease with which multiple SNPs can be typed using chip technology (Section 3.2). One of the first GWAS projects studied age-related macular degeneration, which causes vision problems in elderly people. By typing 226,204 SNPs in 96 affected individuals and 50 controls, two SNPs displaying strong association with the disease were identified. Both SNPs were located in an intron within the gene for complement factor B, a protein involved in control of the inflammatory response. This is one of five genes, on three different chromosomes, now thought to be involved in age-related macular degeneration.

GWAS projects have gradually become more ambitious, with highly complex traits such as hypertension (high blood pressure) now being addressed. These studies use cohort sizes of hundreds of thousands and may type a million or more SNPs, the larger numbers enabling a greater degree of resolution so that associations with multiple genetic loci can be identified in a single screen. Over 30 genes have been associated with hypertension as a result of a series of GWAS projects carried out in different parts of the world. The GWAS approach has also been applied to species other than humans, including crop plants, where it is proving valuable in identifying genes responsible for complex traits such as heading date (the time taken for the plant to reach the flowering stage) and the number and weight of the seeds that are produced.

SUMMARY

- Gene functions can tentatively be assigned by homology analysis, because homologous genes are evolutionarily related and often, but not always, have similar functions.

- Identification of conserved sequence motifs can also help to determine the function of a gene.

- Most experimental techniques for the functional analysis of genes involve examining the effect of gene inactivation on the phenotype of the organism.

- Gene inactivation can be performed by the error-prone repair mechanism of the CRISPR procedure.

- Inactivation is also achieved by homologous recombination with a defective version of the target gene, by insertion of a transposon into the gene, and by RNA interference.

- Gene overexpression can also be used to assess function.

- With both inactivation and overexpression experiments, it may be difficult to discern a phenotypic change, and the precise function of the gene may remain elusive.

- The cellular location of a protein can be determined by expression of a reporter gene or by immunocytochemistry.

- More detailed studies of gene function can be carried out by site-directed mutagenesis using the homology-directed repair version of CRISPR and by base or prime editing.

- Other site-directed mutagenesis methods can be used with cloned genes.

- Candidates for human disease genes can be identified by pedigree analysis and by genome-wide association studies.

SHORT ANSWER QUESTIONS

1. What is the difference between orthologous and paralogous genes?

2. Describe how a BLAST search is carried out and explain why errors are sometimes made when assigning gene function using this approach.

3. How is identification of protein domains used in the functional analysis of a genome sequence?

4. Outline the key features of the EC and GO systems for the classification of gene functions.

5. Describe how a gene can be inactivated by the CRISPR system.

6. Discuss the role of homologous recombination in studies of gene function.

7. Distinguish between the ways in which genes are inactivated by transposon tagging and by RNA interference.

8. Outline the applications of gene overexpression in studies of gene function.

9. Compare the strengths and weaknesses of gene inactivation and overexpression experiments in the functional annotation of a genome sequence.

10. Describe how reporter genes and immunocytochemistry are used to understand gene functions.

11. Describe the key features of the homology-directed repair, base editing and prime editing versions of CRISPR.

12. Outline the methods available for site-directed mutagenesis of cloned genes.

13. Describe the possible approaches that can be used to identify the gene or genes responsible for an inherited disease.

IN-DEPTH PROBLEMS

1. Perform a BLAST search (https://blast.ncbi.nlm.nih.gov/Blast.cgi) with the following amino acid sequence:

   GLSDGEWQLVLNVWGKVEADLAGHGQEVLIRLFKGHPETLEKFDKFKHLK SEKGSEDLKKHGNTVETALEGILKKKALELFKNDIAAKTKELGFLG

   What protein has this amino acid sequence? Are the homologous sequences identified by this search mostly orthologs or paralogs?

2. An important protein domain has the following amino acid sequence:

   KRARTAYTRYQTLELEKEFHFNRYLTRRRRIEIAHALCLSERQIKIWFQN RRMKWKKDN

   Identify the domain and describe its function.

3. Gene inactivation studies have suggested that at least some genes in a genome are redundant, meaning that they have the same function as a second gene and so can be inactivated without affecting the phenotype of the organism. What evolutionary questions are raised by genetic redundancy? What are the possible answers to these questions?

4. Explore the natural role of RNA interference in living organisms.

5. Gene overexpression has so far provided limited but important information on the function of unknown genes. Assess the overall potential of this approach in functional analysis.

FURTHER READING

Assigning function by computer analysis

Altschul, S.F., Gish, W., Miller, W., et al. (1990). Basic local alignment search tool. *J. Mol. Biol.* 215:403–410. *The BLAST program.*

Ejigu, G.F. and Jung, J. (2020). Review on the computational genome annotation of sequences obtained by next-generation sequencing. *Biology* 9:295.

Henikoff, S. and Henikoff, J.G. (1992). Amino acid substitution matrices from protein blocks. *Proc. Natl Acad. Sci. USA* 89:10915–10919. *Describes the chemical relationships between amino acids, from which sequence similarity scores are calculated.*

Hill, D.P., Smith, B., McAndrews-Hill, M.S., et al. (2008). Gene Ontology annotations: What they mean and where they come from. *BMC Bioinform.* 9:S2.

Lee, D., Redfern, O. and Orengo, C. (2007). Predicting protein function from sequence and structure. *Nat. Rev. Mol. Cell Biol.* 8:995–1005.

Mistry, J., Chuguransky, S., Williams, L., et al. (2021). Pfam: The protein families database in 2021. *Nucl. Acids Res.* 49:D412–D419.

Pek, J.W., Anand, A. and Kai, T. (2012). Tudor domain proteins in development. *Development* 139:2255–2266.

Wang, Y., Zhang, H., Zhong, H., et al. (2021). Protein domain identification methods and online resources. *Comp. Struct. Biotechnol. J.* 19:1145–1153.

CRISPR genome editing

Adli, M. (2018). The CRISPR tool kit for genome editing and beyond. *Nat. Comm.* 9:1911.

Anzalone, A.V., Koblan, L.W. and Liu, D.R. (2020). Genome editing with CRISPR-Cas nucleases, base editors, transposases and prime editors. *Nat. Biotechnol.* 38:824–844.

Kantor, A., McClements M.E. and MacLaren, R.E. (2020). CRISPR-Cas9 DNA base-editing and prime-editing. *Int. J. Mol. Sci.* 21:6240.

Porto, E.M., Komor, A.C., Slaymaker, I.M., et al. (2020). Base editing: Advances and therapeutic opportunities. *Nat. Rev. Drug. Discov.* 19:839–859.

Scholefield, J. and Harrison, P.T. (2021). Prime editing – An update on the field. *Gene Ther.* 28:396–401.

Other methods for gene inactivation

Evans, M.J., Carlton, M.B.L. and Russ, A.P. (1997). Gene trapping and functional genomics. *Trends Genet.* 13:370–374. *The use of ES cells.*

Fraser, A.G., Kamath, R.S., Zipperlen, P., et al. (2000). Functional genomic analysis of *C. elegans* chromosome I by systematic RNA interference. *Nature* 408:325–330.

Ross-Macdonald, P., Coelho, P.S.R., Roemer, T., et al. (1999). Large-scale analysis of the yeast genome by transposon tagging and gene disruption. *Nature* 402:413–418.

Wach, A., Brachat, A., Pöhlmann, R. and Philippsen, P. (1994). New heterologous modules for classical or PCR-based gene disruptions in *Saccharomyces cerevisiae*. *Yeast* 10:1793–1808. *Gene inactivation by homologous recombination.*

Overexpression, immunocytochemistry, and directed mutagenesis of cloned genes

Carrigan, P.E., Ballar, P. and Tuzmen, S. (2011). Site-directed mutagenesis. *Methods Mol. Biol.* 700:107–124.

Kunkel, T.A. (1985). Rapid and efficient site-specific mutagenesis without phenotypic selection. *Proc. Natl Acad. Sci. USA* 82:488–492. *Oligonucleotide-directed mutagenesis.*

Ramos-Vara, J.A. (2005). Technical aspects of immunohistochemistry. *Vet. Pathol.* 42:405–426.

Tsien, R. (1998). The green fluorescent protein. *Annu. Rev. Biochem.* 67:509–544. *A reporter gene system.*

Identifying gene function by conventional genetics

Bush, W.S. and Moore, J.H. (2012). Genome-wide association studies. *PLoS Comput. Biol.* 8:el002822.

Hall, J.M., Lee, M.K., Newman, B., et al. (1990). Linkage of early-onset familial breast cancer to chromosome 17q21. *Science* 250:1684–1689.

Huang, X. and Han, B. (2014). Natural variations and genome-wide association studies in crop plants. *Annu. Rev. Plant Biol.* 65:531–551.

Miki, Y., Swensen, J., Shattuck-Eidens, D., et al. (1994). A strong candidate for the breast and ovarian cancer susceptibility gene *BRCA1*. *Science* 266:66–71.

Padmanabhan, S. and Dominiczak, A.F. (2020). Genomics of hypertension: The road to precision medicine. *Nat. Rev. Cardiol.* 18:235–250. *Describes the results of GWAS studies of hypertension.*

Online resources

BLAST. https://blast.ncbi.nlm.nih.gov/Blast.cgi. *Online tool for conducting homology searches of nucleotide and amino acid sequences.*

ExPASy Enzyme Nomenclature Database. http://enzyme.expasy.org/. *Access to enzyme EC numbers.*

Pfam. http://pfam.xfam.org/. *Database of protein families and domains.*

PROSITE. https://prosite.expasy.org/. *Another database of protein families and domains.*

The Gene Ontology Resource. http://geneontology.org/. *The unified representation of gene functions and other biological characteristics.*

The Mammalian Phenotype Ontology. https://www.ebi.ac.uk/ols/ontologies/mp. *Terms and definitions for describing mammalian phenotypes.*

PART 2

GENOME ANATOMIES

EUKARYOTIC NUCLEAR GENOMES

In the next three chapters, we will survey the anatomies of the various types of genome that are found on our planet. There are three chapters because there are three types of genome to consider:

- **Eukaryotic nuclear genomes** (this chapter), of which the human genome is the one of greatest interest to us.

- **The genomes of prokaryotes and of eukaryotic organelles** (Chapter 8), which we will consider together because eukaryotic organelles are descended from ancient prokaryotes.

- **Virus genomes and mobile genetic elements** (Chapter 9), grouped together because some mobile elements are related to virus genomes.

7.1 NUCLEAR GENOMES ARE CONTAINED IN CHROMOSOMES

The nuclear genome is split into a set of linear DNA molecules, each contained in a chromosome. No exceptions to this pattern are known: all eukaryotes that have been studied have at least one chromosome and the DNA molecules are always linear. The only variability lies with chromosome number, which appears to be unrelated to the biological features of the organism. For example, yeast has 16 chromosomes, four times as many as the fruit fly. The ant *Myrmecia croslandi* has just one chromosome, and the Indian muntjac deer has only four. Nor is chromosome number linked to genome size: some salamanders have genomes 30 times bigger than the human version but split into half the number of chromosomes. These comparisons are interesting but at present do not tell us anything useful about the genomes themselves; they are more a reflection of the nonuniformity of the evolutionary events that have shaped genome architecture in different organisms.

Chromosomes are made of DNA and protein

Chromosomes are made up of both DNA and protein. The complex between the two is referred to as **chromatin**. This term was first used by the German cytologist Walther Flemming in 1879 to describe the fibrous material in the nuclei of cells that becomes visible by light microscopy after staining with dyes such as haematoxylin. A few years later, it became clear that this fibrous material forms discrete structures, which became known as the chromosomes. The term chromatin then fell out of favor, but was revived with its more precise meaning in the 20th century.

DOI: 10.1201/9781003133162-7

Figure 7.1 Nuclease protection analysis of chromatin from human nuclei. Chromatin is gently purified from nuclei and treated with a nuclease enzyme. On the left, the nuclease treatment is carried out under limiting conditions so that the DNA is cut, on average, just once in each of the linker regions between the bound proteins. After removal of the protein, the DNA fragments are analyzed by agarose gel electrophoresis and found to be 200 bp in length, or multiples thereof. On the right, the nuclease treatment proceeds to completion, so all the DNA in the linker regions is digested. The remaining DNA fragments are all 146 bp in length. The results show that in this form of chromatin, protein complexes are spaced along the DNA at regular intervals, one for each 200 bp, with 146 bp of DNA closely attached to each protein complex.

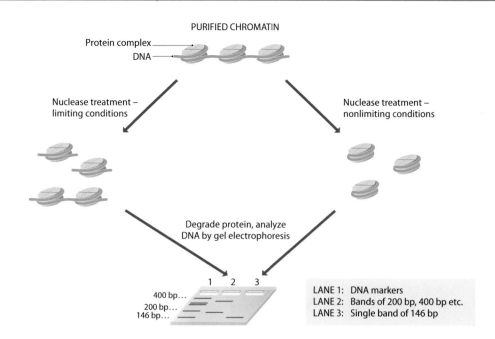

Figure 7.2 Nucleosomes. The model for the 'beads-on-a-string' structure, in which each bead is a barrel-shaped nucleosome with the DNA wound twice around the outside. Each nucleosome is made up of eight proteins: a central tetramer of two histone H3 and two histone H4 subunits, plus a pair of H2A–H2B dimers, one above and one below the central tetramer (see Figure 10.21).

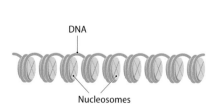

The protein component of chromatin is largely made up of DNA-binding proteins called **histones**. These are very basic proteins (they contain a high amount of lysine and arginine, which are two of the amino acids with basic chemical properties) and display a high degree of sequence similarity between different species. For example, the H4 histones of pea and cow are 102 amino acids in length and differ at only two positions. This is much more similarity than we expect for proteins of equivalent function in such divergent species, and indicates that histones have undergone little evolutionary change over millions of years.

The first important breakthroughs in understanding the exact nature of the association between DNA and histones were made in the early 1970s by a combination of biochemical analysis and electron microscopy. Several groups carried out **nuclease protection experiments** on chromatin that had been gently extracted from mammalian nuclei by methods designed to retain as much of the chromatin structure as possible. In a nuclease protection experiment, the complex is treated with an enzyme that cuts the DNA at positions that are not 'protected' by attachment to a protein. The sizes of the resulting DNA fragments indicate the positioning of the protein complexes on the original DNA molecule (**Figure 7.1**). After limited nuclease treatment of purified chromatin, the bulk of the DNA fragments have lengths of approximately 200 bp and multiples thereof. This indicates that the histone proteins are located at regular intervals along the DNA molecule, one protein or protein complex every 200 bp. After complete nuclease digestion, fragments of 146 bp remain, suggesting that this amount of DNA is closely associated with, and hence protected by, each protein complex.

In 1974 these biochemical results were supplemented by electron micrographs of purified chromatin, which enabled the regular spacing inferred by the protection experiments to be visualized as beads of protein on the string of DNA. This structure has a width of approximately 10 nm and is called the **10 nm fiber**. Further biochemical analysis indicated that each bead, or **nucleosome**, contains eight histone protein molecules, these being two each of histones H2A, H2B, H3, and H4. Structural studies have shown that these eight proteins form a barrel-shaped **core octamer** with the DNA wound twice around the outside (**Figure 7.2**). Between 140 bp and 150 bp of DNA (depending on the species) is associated with the nucleosome particle, and each nucleosome is separated by 50–70 bp of **linker DNA**, giving the repeat length of 190–220 bp previously shown by the nuclease protection experiments.

As well as the proteins of the core octamer, there is a group of additional histones, all closely related to one another and collectively called **linker histones**. In humans, there are 11 of these, 6 of which (H1.1–H1.5 and H1x) are present in all somatic cells, with the remaining five only found in sex cells (H1t, H1T2, H1oo, and H1LS1) or differentiated cells (H1.0). A single linker histone is attached to each nucleosome, to form the **chromatosome**. For several years there was uncertainty regarding the precise positioning of the linker histone, but recent cryo-electron microscopic studies support a model in which the linker histone is located on the surface of the chromatosome, possibly acting as a clamp, preventing the coiled DNA from detaching from the nucleosome (**Figure 7.3**). The different types of linker histone appear to adopt slightly different positions relative to the dyad axis of the nucleosome, and there is also evidence that an individual linker histone can shift from an on- to off-dyad position or vice versa.

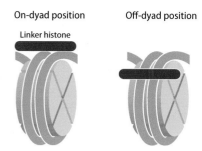

Figure 7.3 The possible positions of the linker histone in the chromatosome. The on- and off-dyad positions are shown.

The special features of metaphase chromosomes

When the nucleus divides, the DNA adopts a compact form of packaging, resulting in the highly condensed **metaphase chromosomes** that can be seen with the light microscope and which have the appearance generally associated with the word 'chromosome' (**Figure 7.4**). The metaphase chromosomes form at a stage in the **cell cycle** after DNA replication has taken place, and so each one contains two copies of its chromosomal DNA molecule. The two copies are held together at the **centromere**, which has a specific position within each chromosome. The arms of the chromosome, which are called **chromatids** and have terminal structures called **telomeres**, are of different lengths in different chromosomes. Individual chromosomes can therefore be recognized because of the lengths of their chromatids and the location of the centromere relative to the telomeres. Further distinguishing features are revealed when chromosomes are stained. There are a number of different staining techniques (**Table 7.1**), each resulting in a banding pattern that is characteristic of a particular chromosome. This means that the set of chromosomes possessed by an organism can be represented as a **karyogram**, in which the banded appearance of each one is depicted. The human karyogram is shown in **Figure 7.5**.

The human karyogram is typical of that of the great majority of eukaryotes, but some organisms have unusual features not displayed by the human genome. These include the following:

- **Microchromosomes** are found in birds and some fish, reptiles, and amphibians. They are relatively short in length, less than 20 Mb, but often they are rich in genes. The chicken genome, for example, is split into 38 autosomes and two sex chromosomes, the latter called Z and W. Of the 38 autosomes, only five are classified as **macrochromosomes**, these being longer than 50 Mb and hence comparable in size to human chromosomes.

TABLE 7.1 STAINING TECHNIQUES USED TO PRODUCE CHROMOSOME BANDING PATTERNS

Technique	Procedure	Banding pattern
G-banding	Mild proteolysis followed by staining with Giemsa	Dark bands are AT-rich Pale bands are GC-rich
R-banding	Heat denaturation followed by staining with Giemsa	Dark bands are GC-rich Pale bands are AT-rich
Q-banding	Stain with quinacrine	Dark bands are AT-rich Pale bands are GC-rich
C-banding	Denature with barium hydroxide and then stain with Giemsa	Dark bands contain constitutive heterochromatin (see Section 10.1)

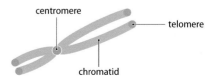

Figure 7.4 The typical appearance of a metaphase chromosome. Metaphase chromosomes are formed after DNA replication has taken place, so each one is, in effect, two chromosomes linked together at the centromere. The arms are called the chromatids. A telomere is the extreme end of a chromatid.

Figure 7.5 The human karyogram.
The chromosomes are shown with their G-banding patterns. Chromosome numbers are given below each structure. 'rDNA' is a region containing a cluster of repeat units for the ribosomal RNA genes (Section 1.2). Constitutive heterochromatin is very compact chromatin that has few or no active genes (Section 10.1)

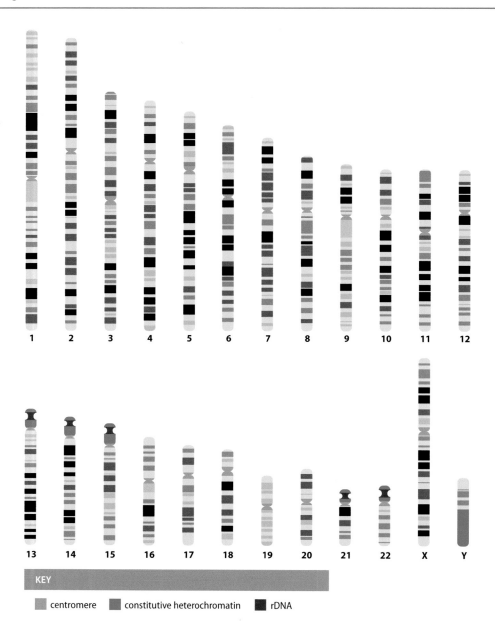

The other 33 autosomes include five of 20–50 Mb, usually referred to as 'intermediate' chromosomes, and 28 microchromosomes, the majority of these less than 10 Mb and several less than 1 Mb. Although the microchromosomes make up only 18% of the genome length, they contain approximately 31% of all of the genes.

- **B chromosomes** are additional chromosomes possessed by some individuals in a population, but not all. They are common in plants and also known in fungi, insects, and animals. B chromosomes appear to be fragmentary versions of normal chromosomes that result from unusual events during nuclear division. Some contain genes, often for rRNAs, but it is not clear if these genes are active. The presence of B chromosomes can affect the biological characteristics of the organism, particularly in plants, where they are associated with reduced viability. It is presumed that B chromosomes are gradually lost from cell lineages as a result of irregularities in their inheritance pattern.

- **Holocentric chromosomes** do not have a single centromere but instead have multiple structures that act as centromeres spread along their length. The nematode worm *Caenorhabditis elegans* has holocentric chromosomes.

Centromeres and telomeres have distinctive DNA sequences

Centromeres and telomeres are specialized structures that play important functions within the nucleus. The centromere, as well as holding together the two daughter chromosomes during the early stages of cell division, is the assembly position for the **kinetochore**. This is the structure that attaches to the microtubules that draw the divided chromosomes into the daughter nuclei (**Figure 7.6**). Telomeres are also important because they mark the ends of each chromosome and therefore enable the cell to distinguish a real end from an unnatural end caused by chromosome breakage – an essential requirement because the cell must repair the latter but not the former. The DNA contained within centromeres and telomeres, and the proteins attached to this DNA, have particular features related to the functions of these structures.

It has been known since the 1960s that the centromere contains extensive amounts of repetitive DNA, but it was not until the advent of long-read sequencing that it became possible to obtain an accurate understanding of how these repeat elements are organized. In humans, the monomeric repeat unit in the centromeres is a 171-bp sequence called **alphoid DNA**. These monomeric units have sequence variability and are arranged in head-to-tail arrays to form higher-order repeats (HORs). For example, the centromere of the human X chromosome is made up predominantly of 1475 copies of a 12-monomer HOR, with much smaller numbers of other HORs interspersed between some of the 12-monomer units. The centromere of human chromosome 8 has a more mixed structure, made up of multiple copies of 4-, 7-, 8, and 11-monomer HORs (**Figure 7.7**). These HOR arrays make up the central regions of the centromeric DNA, 3.1 Mb in length for the X chromosome and 2.1 Mb for chromosome 8, and are surrounded on either side by pericentromeric regions of 300–600 kb, which, as well as alphoid DNA units, also contain other repeat sequences, including some that are common in other parts of the chromosomes. Altogether, the centromeric and pericentromeric repeats make up 6.2% of the human reference genome. This figure may, however, be different in other examples of the human genome, as the amount of repetitive DNA in the centromeric regions, and the exact arrangement of the HOR arrays, is thought to be variable between individuals. The centromeres of other animals and of higher plants have similar structures to the human centromere: in *Arabidopsis thaliana*, a type of cress that has been used as a model organism for plant genomics, the centromeres span 0.4–3.0 Mb of DNA, with each one made up largely of 178–180 bp repeat sequences.

Humans and *Arabidopsis* display the basic pattern for centromeric DNA, as seen in virtually all eukaryotes. We call these **regional centromeres**, to indicate that each centromere covers a region of the chromosomal DNA. The centromeric DNA is attached to nucleosomes, though in some of these nucleosomes, histone H3 is replaced by a centromeric protein called CENP-A, which interacts with other proteins to form the kinetochore. An interesting variation occurs in the yeast *Saccharomyces cerevisiae*, which has short **point centromeres**, which do not contain repetitive DNA and instead are defined by a single-copy sequence, approximately 120 bp in length. This sequence is made up of two short elements, called CDEI and CDEIII, which flank a longer element called CDEII (**Figure 7.8A**). The sequence of CDEII is variable, though always very rich in A

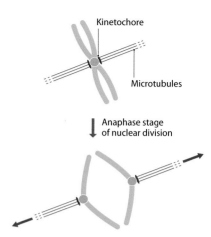

Figure 7.6 The role of the kinetochore during nuclear division. During the anaphase period of nuclear division, individual chromosomes are drawn apart by the contraction of microtubules attached to the kinetochores.

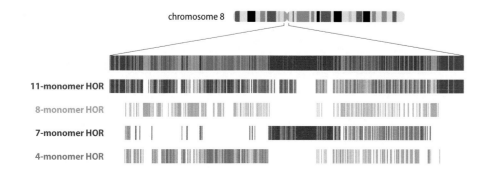

Figure 7.7 The repetitive structure of the centromeric DNA of human chromosome 8. The sequence is made up of multiple copies of 4-, 7-, 8-, and 11-monomer HORs, shown here in different colors. (From Logsdon GA, Vollger MR, Hsieh PH, et al. [2021] *Nature* 593:101–107. With permission from Springer Nature.)

(A) Yeast centromeric DNA

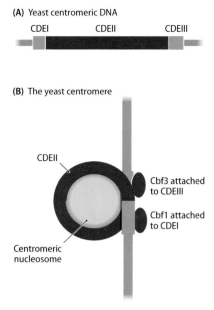

(B) The yeast centromere

Figure 7.8 The point centromere of *Saccharomyces cerevisiae*. (A) *Saccharomyces cerevisiae* centromeric DNA. CDEI is 9 bp in length, CDEII is 80–90 bp, and CDEIII is 11 bp. Additional sequences flanking the region shown here are looked on as part of the centromeric DNA, whose full length is approximately 120 bp. (B) DNA-protein interactions in the yeast centromere.

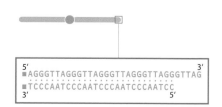

Figure 7.9 Telomeres. The sequence at the end of a human telomere is shown. The length of the 3′ extension is different in each telomere but is usually in the range of 50–400 nucleotides.

and T nucleotides, whereas both CDEI and CDEIII are highly conserved, meaning that their sequences are very similar in all 16 yeast chromosomes. Mutations in CDEII rarely affect the function of the centromere, but a mutation in CDEI or CDEIII usually prevents the centromere from forming. A key role is played by a special chromosomal protein called Cse4, which is similar in structure to histone H3 and replaces the latter in a special centromeric nucleosome that is a hemisome (one molecule each of Cse4 and histones H2A, H2B, and H4) rather than a standard octamer. Two other proteins, Cbf1 and Cbf3, form interactions with the CDEI and CDEIII sequences, respectively, and also bind to at least some of the 20 or so additional proteins that form the kinetochore (**Figure 7.8B**).

The second important part of the chromosome is the terminal region or **telomere**. Telomeric DNA is also made up of hundreds of repeat units, but the repeated motif, 5′–TTAGGG–3′ in humans, is much shorter than alphoid DNA and is invariant in a particular species. The repeats form a short extension of the 3′ terminus of the double-stranded DNA molecule (**Figure 7.9**). A series of proteins, including TRF1, TRF2, and POT1, bind to the telomere repeat sequence and together with other proteins make a structure called a **shelterin**. This structure protects the telomeres from degradation by nuclease enzymes, and mediates the enzymatic activity that maintains the length of each telomere during DNA replication (Section 15.4).

7.2 THE GENETIC FEATURES OF NUCLEAR GENOMES

Now that we understand the physical structure of the eukaryotic nuclear genome, we can move on to consider the genetic features – the genes that are present and the way in which those genes are arranged.

Gene numbers can be misleading

The current annotation of the human genome recognizes 20,442 protein-coding genes and 23,982 genes for noncoding RNAs. During recent years, the trend has been for the number of protein-coding genes to decrease as questionable ORFs are gradually discarded, with some estimates suggesting that there might be as few as 19,000 in the human genome. The number of noncoding genes, on the other hand, has increased substantially in recent years as different types of noncoding RNA have been discovered. This area of genomics research is very fluid at the moment, as we will discover when we study the composition of transcriptomes in Section 12.1. Future changes in the accepted numbers of noncoding genes, in any organism, are likely to be a balance between the discovery of new genes whose products are genuinely functional, and the removal of sequences initially identified as genes but whose RNA transcripts are subsequently interpreted as 'junk.'

Our expectation might be that humans, being the most sophisticated species on the planet, would have more genes than any other organism. An initial comparison between the number of protein-coding genes in different species supports this prejudice (**Table 7.2**). Yeast has just 6600 genes, the fruit fly has a few fewer than 14,000, and chickens have 16,878. But the correlation begins to go awry when we look more carefully at the figures. Humans and other primates are by no means the most complex organisms in terms of gene counts, *Arabidopsis thaliana* having over 27,000 protein-coding genes and rice having almost 38,000. We might ascribe the higher gene numbers of plants to the need for plants to code for proteins involved in photosynthesis, but this would be a misassumption, the photosynthetic capability of plants being outweighed in this regard by the many unique specializations of mammals and other higher vertebrates. In any case, the higher gene content of plant genomes is not the only nor most striking anomaly revealed by the figures in Table 7.2. The genome of the microscopic worm called *Caenorhabditis elegans*, whose adult body comprises just over 1000 cells, contains 20,191 protein-coding genes, which is about the same number as in the human genome.

TABLE 7.2 NUMBERS OF PROTEIN-CODING GENES FOR VARIOUS EUKARYOTES

Species	Protein-coding genes
Saccharomyces cerevisiae (budding yeast)	6600
Schizosaccharomyces pombe (fission yeast)	5145
Caenorhabditis elegans (nematode worm)	20,191
Arabidopsis thaliana (plant)	27,655
Drosophila melanogaster (fruit fly)	13,968
Oryza sativa (rice)	37,960
Gallus gallus (chicken)	16,878
Homo sapiens (human)	20,442

Data taken from Ensembl release 104, Ensembl Plants release 51, and Ensembl Fungi release 51.

These gene number comparisons lead us into an important aspect of genome biology. Before the human genome was sequenced, it was anticipated that there would be 80,000–100,000 protein-coding genes, this number remaining in vogue up to a few months before the draft sequence was completed in 2000. This early estimate was high because it was based on the supposition that, in most cases, a single gene specifies a single mRNA and a single protein. According to this model, the number of genes in the human genome should be similar to the number of proteins in human cells, leading to the estimates of 80,000–100,000. The discovery that the actual number of protein-coding genes is much lower than this indicates that it is possible for an individual gene to specify more than one protein. This is the case for many of the discontinuous genes in the human genome (Section 1.2). When introns were first discovered, it was thought that a discontinuous gene would have just one **splicing pathway**, in which all of the exons are joined together to give a single mRNA. We now know that many discontinuous genes have **alternative splicing** pathways, which means that their pre-mRNAs can be processed in a variety of ways, to give a series of mRNAs made up of different combinations of exons. Each of these genes can therefore direct synthesis of related but different proteins. An example of a human gene with two splicing pathways, one followed in the thyroid and a second in nervous tissue, is shown in **Figure 7.10**. Alternative splicing is relatively common in vertebrates, with 75% of all human protein-coding genes, representing 95% of those with two or more introns, undergoing alternative splicing, giving rise to an average of four different spliced mRNAs per gene and enabling the 20,442 human genes to specify a total of 78,120 proteins. Alternative splicing also occurs in lower eukaryotes, but it is less prevalent. In *C. elegans*, for example, only about 25% of the protein-coding genes have alternative splicing pathways, with an average of 2.2 variants per gene.

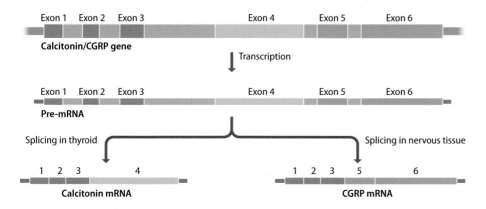

Figure 7.10 An example of a human gene with two splicing pathways. The human calcitonin/CGRP gene has two splicing pathways that give rise to different proteins. In the thyroid, exons 1–2–3–4 are spliced together to give the mRNA for calcitonin, a short peptide hormone that regulates the calcium ion concentration in the bloodstream. In nervous tissue, exons 1–2–3–5–6 are joined to give the calcitonin gene-related peptide (CGRP), which is a neurotransmitter active in sensory neurons and involved in the pain response.

Because of alternative splicing, the question 'How many genes are there?' has no real biological significance, as the number of genes does not indicate the number of proteins that can be synthesized and hence is not a measure of the coding capacity of a genome. A better measure would be provided by categorizing the genes according to function, but now the problem becomes a lack of completeness: because of the difficulties in identifying functions (Section 6.2) some 25% of the human gene catalog comprises genes whose roles are, at best, only partly understood. However, it is possible to sidestep this difficulty by focusing not on entire proteins but on protein domains, the structural components that provide a protein with its particular biochemical capabilities. Examples are the **zinc finger**, which is one of several domains that enable a protein to bind to a DNA molecule (Section 11.2), and the 'death domain,' made up of six α-helices, which is present in many proteins involved in apoptosis. Each domain has a characteristic amino acid sequence, usually not exactly the same sequence in every example of the domain, but close enough for the presence of a particular domain to be recognizable by examining the nucleotide sequence of a gene. The genes in a genome can therefore be categorized according to the protein domains that they specify, even if in some cases the biological function of the encoded protein is not known. An analysis of the domains encoded by human, fruit-fly, *C. elegans*, yeast, and *A. thaliana* genomes reveals that the genomes of the more complex eukaryotes encode a more extensive repertoire of domain types. The human genome, for example, contains information for a number of protein domains that are absent from the genomes of one or more of the other four organisms; these domains include several involved in activities such as cell adhesion, electrical coupling between cells, and growth of nerve cells (Table 7.3). These functions are interesting because they are ones that we look on as conferring the distinctive features of vertebrates compared with other types of eukaryote. Deep mining of the information content of genomes can therefore provide indications of the genetic features that characterize different types of organism.

Genes are not evenly distributed within a genome

Much of what we know about the identities and relative positions of the genes in eukaryotic genomes has been obtained by annotation of genome sequences using the bioinformatic and experimental methods that we studied in Chapters 5 and 6. Prior to the genome sequencing era, linkage analysis had enabled some genes to be mapped and their functions assigned, at least for well-studied species such as yeast, fruit flies, and humans. DNA sequences were also available for many individual genes and for short segments of individual chromosomes, such as the 65 kb stretch of human chromosome 11 that contains the β-globin gene cluster (see Figure 7.16). Geneticists at that time therefore had a broad understanding of the distribution of genes within, for example, the human genome,

TABLE 7.3 EXAMPLES OF PROTEIN DOMAINS SPECIFIED BY DIFFERENT GENOMES

Domain	Function	Number of genes in the genome containing the domain				
		Human	*D. melanogaster*	*C. elegans*	*A. thaliana*	*S. cerevisiae*
Zinc finger, Cys$_2$His$_2$ type	DNA binding	c.3000	850	282	693	50
Zinc finger, GATA type	DNA binding	45	27	42	117	10
Homeobox	Gene regulation during development	889	294	151	474	9
Death	Programmed cell death	187	29	25	0	0
Connexin	Electrical coupling between cells	102	0	0	0	0
Ephrin	Nerve cell growth	14	1	5	0	0

Data taken from InterPro 86.0.

along with a detailed knowledge of the structures of those individual genes that had been cloned and sequenced. This information led to a growing awareness that genes are not arranged evenly along the length of a eukaryotic chromosome, a possibility that had previously been suggested by the banding patterns that are produced when chromosomes are stained. The dyes used in these procedures (see Table 7.1) bind to DNA molecules, but in most cases with preferences for certain base pairs. Giemsa, for example, has a greater affinity for DNA regions that are rich in A and T nucleotides. The dark G-bands in the human karyogram (see Figure 7.5) were therefore thought to be AT-rich regions of the genome. The base composition of the human genome as a whole is 59.7% A + T, so the dark G-bands must have AT contents substantially greater than 60%. Cytogeneticists therefore predicted that there would be fewer genes in dark G-bands because protein-coding genes generally have AT contents of 45–50%.

When the first eukaryotic genome sequences became available, it was possible for the first time to obtain a detailed picture of the genetic organization of individual chromosomes. The first fact that became evident is that genes are not divided evenly among chromosomes: some chromosomes in a karyogram are relatively gene-rich, and others are gene-poor. For example, the gene density for the human chromosomes ranges from 41.2 genes per Mb for chromosome 19 down to 6.3 genes per Mb for chromosome 13 and as few as 3.0 genes for Mb for the poor little Y chromosome. The hypothesis that genes are not spread evenly along the length of a chromosome was also found to be correct, with the gene-rich areas interspersed with relatively empty stretches of DNA, the latter including **gene deserts** in which the density is very low over regions as long as several Mb (**Figure 7.11**). Finally, to a certain extent, the interpretation of G-bands as regions of low gene content has been confirmed; these bands often contain just a few long genes with multiple introns, occasionally with a single gene spanning an entire band.

A segment of the human genome

The variations in gene density that occur along the length of a eukaryotic chromosome mean that it is difficult to identify regions in which the organization of the genes can be looked on as 'typical' of the genome as a whole. Despite this difficulty, it is clear that the overall pattern of gene organization varies greatly between different eukaryotes, and we need to understand these differences because they reflect important distinctions between the genetic features of these genomes. To begin to address this issue, we will look in detail at a small part of the human genome.

The segment that we will examine (**Figure 7.12**) comes from midway along the long arm of human chromosome 1. It is 200 kb in length and runs from nucleotide position 55,000,000 to position 55,200,000. The segment contains:

- All or part of three protein-coding genes. These are:

  - The end of the *BSND* gene, which starts at position 54,998,933. *BSND* codes for a chloride channel protein. This is a membrane-bound protein that forms a pore through which various ions, including chloride, can enter and leave the cell.

  - *PCSK9*, which codes for proprotein convertase subtilisin/kexin type 9, a protein made in liver, intestine, and kidney tissues and which is involved in breakdown of low-density lipoproteins, thereby playing an important role in the metabolism of cholesterol.

Figure 7.11 Gene density along human chromosome 1.

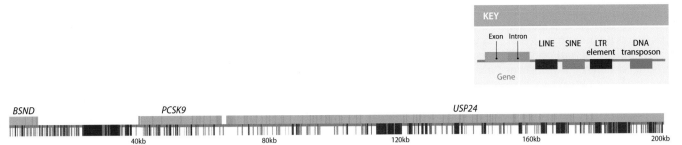

Figure 7.12 A 200 kb segment of the human genome. The segment runs from nucleotide position 55,000,000 to position 55,200,000 of chromosome 1. Within the genes, exons are shown as green boxes and introns as gray boxes. Data taken from the UCSC Genome Browser hg38 assembly.

- The start of *USP24*, specifying ubiquitin-specific peptidase 24, a protease that removes **ubiquitin** sidechains from proteins that have been modified by **ubiquitination**. Ubiquitin is a small regulatory protein whose addition to or removal from a protein controls that protein's location in the cell and eventual degradation (Section 13.3). *USP24* ends at position 55,215,364, so most of it is contained in the segment shown in Figure 7.12.

  Note that each of these three genes is discontinuous: there are three introns in *BSND*, 11 in *PCSK9*, and 67 in *USP24*.

- A vast number of **interspersed repeat** sequences. These are sequences that recur at many places in the genome. There are four main types of interspersed repeat, called **SINEs (short interspersed nuclear elements)**, **LINEs (long interspersed nuclear elements)**, **LTR (long terminal repeat) elements**, and **DNA transposons** (Section 9.2). Multiple copies of each type are seen in this short segment of the genome, in both the intergenic regions and the introns of the protein-coding genes.

The most striking feature of this 200 kb segment of the human genome is the relatively small amount of space taken up by the coding parts of the genes. When added together, the total length of the exons present in this segment (the parts of the genes that contain the biological information) is 10,664 bp, equivalent to 5.33% of the 200 kb segment. In fact, this segment is rather rich in genes: all the exons in the protein-coding genes of the human genome make up only 48 Mb, just 1.5% of the total. In contrast, 48.5% of the genome is taken up by interspersed repeats (**Figure 7.13**).

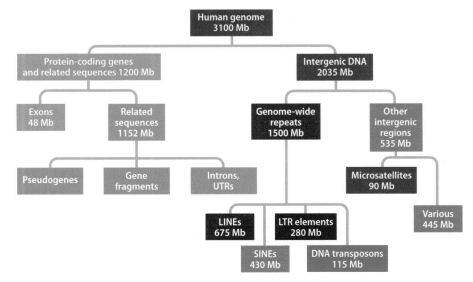

Figure 7.13 The composition of the human genome.

The yeast genome is very compact

How extensive are the differences in gene organization among eukaryotes? There are certainly very substantial differences in genome size, with the smallest eukaryotic genomes being less than 10 Mb in length, and the largest over 100,000 Mb. As can be seen in **Figure 7.14**, this size range coincides to a certain extent with the complexity of the organism, with the simplest eukaryotes, such as fungi, having the smallest genomes, and higher eukaryotes, such as vertebrates and flowering plants, having the largest ones. This might appear to make sense, as one would expect the complexity of an organism to be related to the number of genes in its genome – higher eukaryotes need larger genomes to accommodate the extra genes. However, the correlation is far from precise. The human genome is 3100 Mb and contains 20,442 protein-coding genes. This means that the genome of the yeast *Saccharomyces cerevisiae*, which at 12.2 Mb is 0.004 times the size of the human nuclear genome, would therefore be expected to contain 0.004 × 20,442 genes, which is just 82. In fact, the *S. cerevisiae* genome contains 6600 protein-coding genes.

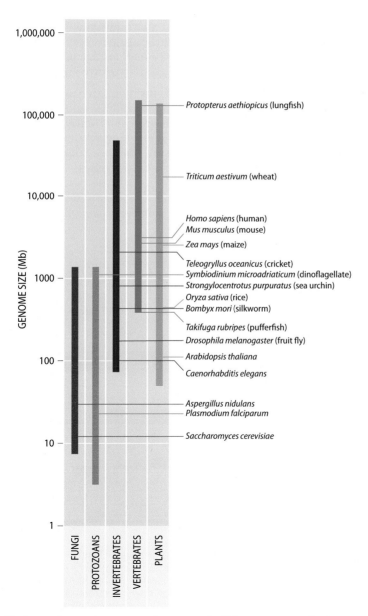

Figure 7.14 Approximate size ranges of genomes in different groups of eukaryotes.

For many years the lack of precise correlation between the complexity of an organism and the size of its genome was looked on as a bit of a puzzle, the so-called **C-value paradox**. In fact, the answer is quite simple: space is saved in the genomes of less complex organisms because the genes are more closely packed together. The *S. cerevisiae* genome illustrates this point, as we can see in **Figure 7.15B**, where a 'typical' 200 kb segment of the yeast genome is displayed. This segment comes from chromosome IV, which is the largest of the 16 yeast chromosomes but still only 1.53 Mb in length, reflecting the much smaller size of the yeast genome as a whole. The 200 kb segment, running from nucleotide position 250,000 to position 450,000, therefore comprises 13% of the length of chromosome IV, and in fact ends adjacent to the centromere of this chromosome. When we compare this segment with the 200 kb stretch of the human genome that we previously examined, three differences immediately become apparent:

- The gene density in the yeast genome is much higher than that of humans. This segment of chromosome IV contains 106 genes thought to code for proteins, four that specify transfer RNAs and one small nucleolar RNA gene.

- Relatively few of the yeast genes are discontinuous. In this segment of chromosome IV, there are seven introns, one in each of seven protein-coding genes. In the entire yeast genome, there are only 287 discontinuous genes and the vast majority of these genes have just one intron each.

- This part of chromosome IV contains just two interspersed repeats. Both are truncated LTR elements; one is called a delta sequence and the other a tau sequence (Section 9.2). Interspersed repeats make up only 3.3% of the yeast genome, the most prevalent types being full-length LTR elements (about 50 copies in total, depending on the particular strain of *S. cerevisiae*) and truncated LTR sequences, (300–400 copies).

The genetic organization of the yeast genome is clearly much more compact than that of the human version. The genes themselves are much shorter, having

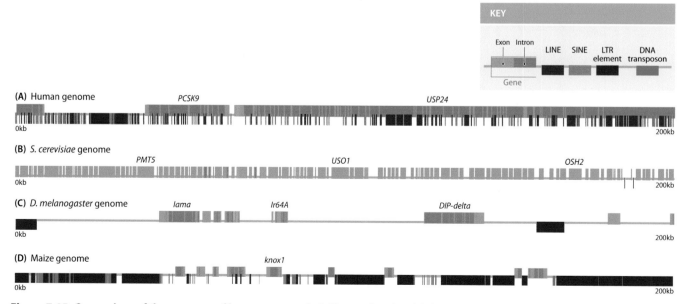

Figure 7.15 Comparison of the genomes of humans, yeast, fruit flies, and maize. (A) The 200 kb segment of human chromosome 1 from Figure 7.12. (B) A 200 kb segment from nucleotide positions 250,000–450,000 of *Saccharomyces cerevisiae* chromosome IV. Data from the UCSC Genome Browser sacCer3 assembly. (C) A 200 kb segment from nucleotide positions 5,300,000–5,500,000 of *Drosophila melanogaster* chromosome 3. Data from the UCSC Genome Browser dm6 assembly. (D) A 200 kb segment from nucleotide positions 5,000,000–5,200,000 of *Zea mays* chromosome 1. Data from the EnsemblPlants AGPv4 assembly. Possible genes for long noncoding RNAs are omitted from the *D. melanogaster* display.

fewer introns, and the spaces between the genes are relatively short, with much less space taken up by interspersed repeats and other noncoding sequences.

Gene organization in other eukaryotes

The hypothesis that less complex eukaryotes have more compact genomes holds when other species are examined. Next, we will examine a 200 kb segment of the fruit-fly genome. If we agree that a fruit fly is more complex than a yeast cell but less complex than a human, then we would expect the organization of the fruit-fly genome to be intermediate between that of yeast and humans. This is what we see in **Figure 7.15C**. Again we have chosen the largest of the chromosomes in the karyogram, our segment running from positions 5,300,000–5,500,000 on the left arm of chromosome 3. There are nine genes in this region, eight of which are discontinuous, with some of the introns similar in length to those in human genes. There are only two interspersed repeat sequences, both of these being LTR elements. The fruit-fly genome also contains SINEs, LINEs and DNA transposons, but there are none in this particular segment. The picture is similar when the entire genome sequences of the three organisms are compared (Table 7.4). The gene density in the fruit-fly genome is intermediate between that of yeast and humans, and the average fruit-fly gene has many more introns than the average yeast gene, but fewer introns than the average human gene.

The comparison between the yeast, fruit-fly, and human genomes also holds true when we consider the interspersed repeats (see Table 7.4). These make up 3.3% of the yeast genome, about 20.5% of the fruit-fly genome, and 48.5% of the human genome. It is beginning to become clear that the interspersed repeats play an intriguing role in dictating the compactness or otherwise of a genome. This is strikingly illustrated by the maize genome, which at 2500 Mb is relatively small for a flowering plant. The 200 kb segment shown in **Figure 7.15D** has nine genes, eight of which contain one or more short introns. Instead of the genes, the dominant feature of this genome segment is the interspersed repeats, which have been described as forming a sea within which islands of genes are located. The interspersed repeats are mainly of the LTR element type, which comprise a large part of the intergenic part of the segment, and on their own are estimated to make up over 50% of the maize genome. It is becoming clear that one or more families of interspersed repeats have undergone a massive proliferation in the genomes of certain species. This may provide an explanation for the most puzzling aspect of the C-value paradox, which is not the general increase in genome size that is seen in increasingly complex organisms, but the fact that similar organisms can differ greatly in genome size. A good example is provided by *Teleogryllus oceanicus*, the oceanic field cricket, which might be expected to have a genome of around 150 Mb, similar to other insects such as *D. melanogaster*. In fact, the *T. oceanicus* genome is over ten times larger than this, at 2050 Mb, and other species of cricket have similarly large genomes. The annotations for those crickets whose genomes have been sequenced show that 45% of their DNA is made up of repetitive sequences, much more than is present in the fruit-fly genome and comparable to the amount of repetitive DNA in the human genome.

TABLE 7.4 COMPACTNESS OF THE YEAST, FRUIT-FLY AND HUMAN GENOMES			
Feature	**Yeast**	**Fruit fly**	**Human**
Gene density (average number of protein-coding genes per Mb)	543	97	7
Introns per protein-coding gene (average)	0.05	3.5	10
Amount of the genome that is taken up by interspersed repeats	3.3%	20.5%	48.5%

Families of genes

Since the earliest days of DNA sequencing, it has been known that **multigene families** –groups of genes of identical or similar sequence – are common features of many genomes. For example, every eukaryote that has been studied (as well as all but the simplest bacteria) has multiple copies of the genes for the ribosomal RNAs. This is illustrated by the human genome, which contains several thousand genes for the 5S rRNA, many of these located in a single cluster on chromosome 1. There are also several hundred copies of a repeat unit containing the 28S, 5.8S, and 18S rRNA genes, with major groups of this unit on each of chromosomes 13, 14, 15, 21, and 22 (see Figure 7.5). Ribosomal RNAs are components of the protein-synthesizing particles called ribosomes, and it is presumed that their genes are present in multiple copies because there is a heavy demand for rRNA synthesis during cell division, when several tens of thousands of new ribosomes must be assembled.

The rRNA genes are examples of 'simple' or 'classical' multigene families, in which all the members have identical or nearly identical sequences. These families are believed to have arisen by gene duplication, with the sequences of the individual members kept identical by an evolutionary process that, as yet, has not been fully described (Section 18.2). Other multigene families, more common in higher eukaryotes than in lower eukaryotes, are called 'complex' because the individual members, although similar in sequence, are sufficiently different for the gene products to have distinctive properties. The mammalian globin genes are one of the best examples of this type of multigene family. The globins are the blood proteins that combine to make hemoglobin, each molecule of hemoglobin being made up of two α-type and two β-type globins. In humans, the α-type globins are coded by a small multigene family on chromosome 16, and the β-type globins by a second family on chromosome 11 (**Figure 7.16**). These genes were among the first to be sequenced, back in the late 1970s. The sequence data showed that the genes in each family are similar to one another, but by no means identical. In fact, the nucleotide sequences of the two most different genes in the β-type cluster, coding for the β- and ε-globins, display only 79.1% identity. Although this is similar enough for both proteins to be β-type globins, it is sufficiently different for them to have distinctive biochemical properties. Similar variations are seen in the α-cluster.

Why are the members of the globin gene families so different from one another? The answer was revealed when the expression patterns of the individual genes were studied. It was discovered that the genes are expressed at different stages in human development; for example, in the β-type cluster, ε is expressed in the early embryo, γ_G and γ_A (whose protein products differ by just one amino acid) in the fetus, and δ and β in the adult (see Figure 7.16). The biochemical properties of the different β-globin proteins reflect the slight changes in the physiological role that hemoglobin plays during the course of human development. For example, the developing fetus must obtain oxygen from its mother. This means that oxygen must be transferred from the mother's

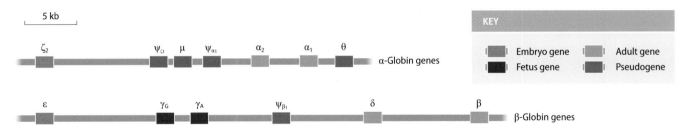

Figure 7.16 The human α- and β-globin gene clusters. The α-globin cluster is located on chromosome 16, and the β-cluster is on chromosome 11. Both clusters contain genes that are expressed at different developmental stages and each includes at least one pseudogene. Note that expression of the α-type gene ζ_2 begins in the embryo and continues during the fetal stage; there is no fetal-specific α-type globin. The θ pseudogene is expressed, but its protein product is inactive. The μ pseudogene is transcribed, but its protein product has never been detected. None of the other pseudogenes is expressed.

hemoglobin molecules to those of the fetus. For this to be possible, fetal hemoglobin must have a greater affinity for oxygen compared to adult hemoglobin, so when the two types of hemoglobin mix in the placenta, the fetus is able to 'steal' oxygen from its mother. The nucleotide sequences of the γ_G and γ_A genes specify globin proteins with the higher degree of oxygen affinity that is needed to allow this to happen.

In some multigene families, the individual members are clustered, as with the globin genes, but in others the genes are dispersed around the genome. An example of a dispersed family is provided by the five human genes for aldolase, an enzyme involved in energy generation, which are located on chromosomes 3, 9, 10, 16, and 17. The important point is that, even though dispersed, the members of the multigene family have sequence similarities that point to a common evolutionary origin. When these sequence comparisons are made it is sometimes possible to see relationships not only within a single gene family but also between different families. All of the genes in the α- and β-globin families, for example, have some sequence similarity and are thought to have evolved from a single ancestral globin gene. We therefore refer to these two multigene families as comprising a single globin **gene superfamily**, and from the similarities between the individual genes we can chart the duplication events that have given rise to the series of genes that we see today (Section 18.2).

Pseudogenes and other evolutionary relics

As well as the functional genes that are expressed at different developmental stages, the human globin gene clusters also contain four **pseudogenes**. These are the sequences labeled $\Psi_{\zeta 1}$, $\Psi_{\alpha 1}$, and θ in the α-globin cluster and $\Psi_{\beta 1}$ among the β-globins (see Figure 7.16). Two of the five genes in the aldolase family are pseudogenes, these being the copies on chromosomes 3 and 10. What are these pseudogenes?

A pseudogene is a sequence of nucleotides that resembles a genuine gene but which does not specify a functional RNA or protein. Pseudogenes are derived from genuine genes and can therefore be looked on as an evolutionary relic, an indication that genomes are continually undergoing change. In some cases, a gene loses its function and becomes a pseudogene simply because its nucleotide sequence changes by mutation. Many mutations have only minor effects on the activity of a gene, but some are more important, and it is quite possible for a single nucleotide change to result in a gene becoming completely nonfunctional. Once a pseudogene has become nonfunctional, it will degrade through accumulation of more mutations, and eventually will no longer be recognizable as a gene relic. Pseudogenes that arise in this way are called **conventional** or **nonprocessed pseudogenes**. They fall into two broad classes:

- A **duplicated pseudogene** arises when a member of a multigene family becomes inactivated by mutation. This event is usually not deleterious to the organism because the other genes in the family are still active and the function specified by the pseudogene is not lost. Comparisons between different genomes have revealed many instances where a pseudogene in one species is the homolog of a functional gene in a second species. For example, the δ-globin gene, which is active in humans, is a pseudogene in mice. The implication is that the δ-globin gene became inactivated by a mutation that occurred at some point during the evolutionary lineage leading to mice, after this lineage diverged from the one leading to humans.

- **Unitary pseudogenes** also arise from mutation, but in this case the gene is not a member of a family, so the resulting loss of function is not compensated by the activity of other genes. Unitary pseudogenes are rare, because the loss of function will usually be lethal, which means that cells that

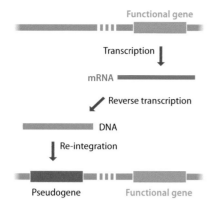

Figure 7.17 The origin of a processed pseudogene. A processed pseudogene arises by integration into the genome of a copy of the mRNA transcribed from a functional gene. The mRNA is reverse transcribed into a cDNA copy, which might integrate into the same chromosome as its functional parent, or possibly into a different chromosome.

experience such a mutation will die and not contribute to the subsequent evolutionary lineage. Those unitary pseudogenes that appear in a genome are therefore ones whose loss of function could be tolerated. There are probably less than 50 unitary pseudogenes in the human genome, the best-known example being the L-gulono-γ-lactone oxidase pseudogene. The functional version of this gene enables many mammals to synthesize ascorbic acid, but in the Haplorhini group of primates, the gene is a pseudogene, which means that haplorhines, including humans, must obtain ascorbic acid (otherwise known as vitamin C) from their diets.

Other pseudogenes arise by a process not involving mutation. These are called **processed pseudogenes** and they result from an abnormal adjunct to gene expression. A processed pseudogene is derived from the mRNA of a gene by synthesis of a cDNA copy, which subsequently reinserts into the genome (**Figure 7.17**). Because a processed pseudogene is a copy of an mRNA molecule, it does not contain any introns that were present in its parent gene. It also lacks the nucleotide sequences immediately upstream of the parent gene, which is the region in which the signals used to switch on expression of the parent gene are located. The absence of these signals means that a processed pseudogene is inactive. Additionally, genomes also contain other evolutionary relics in the form of **truncated genes**, which lack a greater or lesser stretch from one end of the complete gene, and **gene fragments**, which are short, isolated regions from within a gene (**Figure 7.18**).

There are different ways of defining if a sequence is a genuine pseudogene or not, and depending on the annotation methodology that is used between 11,000 and 18,000 are recognized in the human genome, about 70% of which are processed pseudogenes. In recent years, there has been increasing debate about the possibility that some of these sequences do in fact have a functional role. Over 1000 human pseudogenes are transcribed into RNA and a smaller number, about 140, also direct synthesis of a protein. Expression is not, in itself, evidence of a function, because it is conceivable that a nonprocessed pseudogene could be transcribed and/or translated simply because its upstream signals and open reading frame have not yet decayed to the stage where expression is impossible. Before reassigning a pseudogene as a functional sequence, it is necessary to prove that the expression product plays some active role in the cell. There are suggestions that this might be the case for at least a few pseudogenes, an example in humans being *PTENP1*, which is a nonprocessed pseudogene derived from the gene for the PTEN phosphatase, an enzyme involved in one of the signal transduction pathways that control cell division. Expression of the *PTEN* gene is regulated in part by miRNAs that attach to the *PTEN* RNA and promote its degradation (Section 12.6). The transcripts from the *PTENP1* pseudogene act as a **competing endogenous RNA** (**ceRNA**) and also bind some of these miRNAs, reducing their abundance in the cell and ensuring that the *PTEN* gene is not completely silenced (**Figure 7.19A**). In experimental systems, reducing the level of *PTENP1* transcription leads to silencing of *PTEN*, which in turn results in an increased cell division rate. These results correlate with the observation that the *PTENP1* pseudogene is deleted in some types of colon cancer, implying that in these cells the absence of *PTENP1* transcripts results in *PTEN* silencing and the uncontrolled division that gives rise to the cancerous state (**Figure 7.19B**). The system as whole would appear to provide strong evidence that the *PTENP1* sequence is not a genuine pseudogene but instead plays an important regulatory function. However, if a gene is providing a useful function, then we expect natural selection to be acting in a positive way on that gene, and evolutionary studies have failed to find evidence for this being the case with *PTENP1*. Similar evolutionary studies have also failed to reveal indicators of positive selection for most of the human pseudogenes that give rise to protein products. Questions remain, therefore, about the importance of the apparent roles played by those pseudogenes that are still expressed.

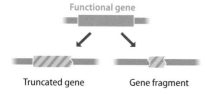

Figure 7.18 A truncated gene and a gene fragment.

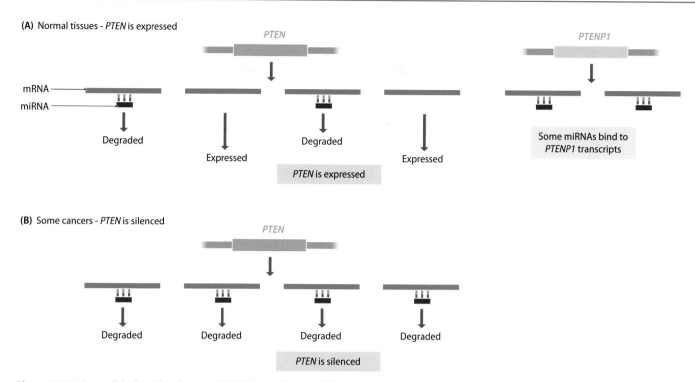

Figure 7.19 **A possible function for the PTENP1 pseudogene.** (A) In normal tissue, the binding of miRNAs to *PTENP1* transcripts is thought to prevent complete silencing of *PTEN*. (B) In some cancers, *PTENP1* is deleted. The absence of the miRNA binding sites normally provided by the *PTENP1* transcripts might lead to additional silencing of *PTEN* mRNAs, resulting in a loss of control over cell division.

7.3 THE REPETITIVE DNA CONTENT OF EUKARYOTIC NUCLEAR GENOMES

Our examination of the genomes of humans and other eukaryotes showed us that large parts of these DNA sequences are made up of repetitive elements (see **Figures 7.12** and **7.15**). To complete our study of the anatomies of eukaryotic genomes, we must look more closely at these different types of repeat sequence.

Repetitive DNA can be divided into two categories (**Figure 7.20**): **interspersed repeats**, whose individual repeat units are distributed around the genome in an apparently random fashion, and **tandemly repeated DNA**, whose repeat units are placed next to each other in an array.

Tandemly repeated DNA is found at centromeres and elsewhere in eukaryotic chromosomes

Tandemly repeated DNA is also called **satellite DNA** because DNA fragments containing tandemly repeated sequences form 'satellite' bands when genomic DNA is fractionated by **isopycnic** or **density gradient centrifugation**. In this method, the DNA fragments are pipetted on to the surface of a high-density

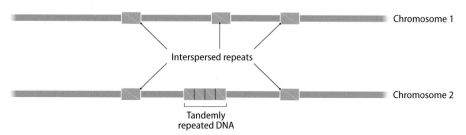

Figure 7.20 **The two types of repetitive DNA: interspersed repeats and tandemly repeated DNA.**

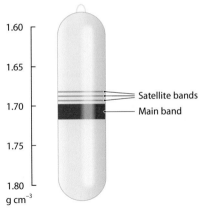

Figure 7.21 Satellite DNA from the human genome. Human DNA has an average GC content of 40.3% and average buoyant density of 1.701 g cm⁻³. Fragments made up mainly of single-copy DNA have a GC content close to this average and are contained in the main band in the density gradient. The satellite bands at 1.687, 1.693, and 1.697 g cm⁻³ consist of fragments containing repetitive DNA. The GC contents of these fragments depend on their repeat motif sequences and are different from the genome average, meaning that these fragments have different buoyant densities to single-copy DNA and migrate to different positions in the density gradient.

solution, such as 8 M cesium chloride, which is then centrifuged at a very high speed: at least 450,000 × g for several hours. The centrifugal force carries some of the CsCl molecules toward the bottom of the centrifuge tube, forming a density gradient. Each DNA fragment migrates to the position in the tube where the density of the CsCl solution equals its own **buoyant density**. The latter is influenced by a number of factors, including conformation (linear, circular, and supercoiled versions of the same DNA molecule have different buoyant densities), but if all the molecules being studied are linear then their buoyant densities are determined mainly by their GC contents, according to the formula:

$$\%\text{GC content} = \frac{\text{bouyant density}\left(\text{g cm}^{-3}\right) - 1.660}{0.098} \times 100$$

For example, when broken into fragments 50–100 kb in length, human DNA forms a main band (buoyant density 1.701 g cm⁻³) and three satellite bands (1.687, 1.693, and 1.697 g cm⁻³) (**Figure 7.21**). The main band contains DNA fragments made up mostly of single-copy sequences with GC compositions close to 40.3%, the average value for the human genome. The satellite bands contain fragments of repetitive DNA, and hence have GC contents and buoyant densities that are atypical of the genome as a whole. This repetitive DNA is made up of long series of tandem repeats, possibly hundreds of kilobases in length. A single genome can contain several different types of satellite DNA, each with a different repeat unit, these units being anything from less than 5 bp to more than 200 bp in length. The three satellite bands in human DNA include at least four different repeat types.

We have already encountered one type of human satellite DNA, the alphoid DNA repeats found in the centromere regions of chromosomes (Section 7.1). Although some satellite DNA is scattered around the genome, most is located in the centromeres, where it may play a structural role, possibly as binding sites for one or more of the special centromeric proteins.

Minisatellites and microsatellites

Although not appearing in satellite bands on density gradients, two other types of tandemly repeated DNA are also classed as 'satellite' DNA. These are **minisatellites** and **microsatellites**. Minisatellites form clusters up to 20 kb in length, with repeat units up to 25 bp in length; microsatellite clusters are shorter, usually less than 150 bp, and the repeat unit is usually 13 bp or less.

Minisatellite DNA is a second type of repetitive DNA that we are already familiar with because of its association with the structural features of chromosomes. Telomeric DNA, which in humans comprises hundreds of copies of the motif 5′–TTAGGG–3′ (see Figure 7.9), is an example of a minisatellite. We know a certain amount about how telomeric DNA is formed, and we know that it has an important function in DNA replication (Section 15.4). In addition to telomeric minisatellites, some eukaryotic genomes contain various other clusters of minisatellite DNA, many, although not all, near the ends of chromosomes. The functions of these other minisatellite sequences have not been identified.

Microsatellites are also examples of tandemly repeated DNA. The commonest type of human microsatellite are dinucleotide repeats, with approximately 325,000 copies in the genome as a whole, the most frequent of these being repeats of the motif 'AC.' There are also over 200,000 trinucleotide repeats. It is not clear if microsatellites have a function. It is known that they arise through an error in the process responsible for copying the genome during cell division, called 'slippage' (Section 17.1), and they might simply be unavoidable products of genome replication.

Although their function, if any, is unknown, microsatellites have proved very useful to geneticists. Many microsatellites are variable, meaning that the number of repeat units in the array is not the same in all members of a species. This is because additional slippage sometimes occurs when a microsatellite is copied during DNA replication, leading to insertion or, less frequently, deletion of one

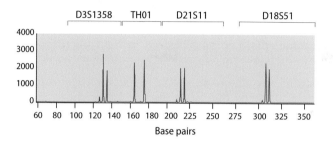

Figure 7.22 Part of a DNA profile. In DNA profiling, a series of PCRs is carried out to identify the lengths of the alleles present at 20 microsatellite loci, referred to as the CODIS (Combined DNA Index System) set. This diagram shows the results for four of those microsatellites, D3S1358, TH01, D21S11, and D18S51. The four sets of PCR products have been run together in a single capillary electrophoresis gel. There are two peaks for each microsatellite because this individual is heterozygous at each of the four loci. (Courtesy of Promega Corporation.)

or more of the repeat units. In the population as a whole, there might be as many as ten different versions of a particular microsatellite, each of these alleles characterized by a different number of repeats. No two humans alive today, except monozygotic twins, triplets, etc., have exactly the same combination of microsatellite length variants. This means that if enough microsatellites are examined, then a unique **DNA profile** can be established for every person. The current methodology for DNA profiling, called CODIS (Combined DNA Index System), makes use of 20 microsatellites with sufficient variability to give only a one in 10^{18} chance that two individuals, other than identical twins, have the same profile. As the world population is around 7.7×10^9, the statistical likelihood of two individuals on the planet sharing the same profile is so low as to be considered implausible when DNA evidence is presented in a court of law. Each microsatellite is typed by PCRs with primers that are fluorescently labeled and anneal on either side of the variable repeat region. The alleles present at the microsatellite are then typed by determining the sizes of the amplified fragments by capillary gel electrophoresis (**Figure 7.22**). DNA profiling is well known as a tool in forensic science, but identification of criminals is a fairly trivial application of microsatellite variability. More sophisticated methodology makes use of the fact that a person's DNA profile is inherited partly from the mother and partly from the father. This means that microsatellites can be used to establish kinship relationships and population affinities, not only for humans but also for other animals, and for plants.

Interspersed repeats

Tandemly repeated DNA sequences are thought to have arisen by expansion of a progenitor sequence, either by replication slippage, as described for microsatellites, or by DNA recombination processes. Both of these events are likely to result in a series of linked repeats, rather than individual repeat units scattered around the genome. Interspersed repeats must therefore have arisen by a different mechanism, one that can result in a copy of a repeat unit appearing in the genome at a position distant from the location of the original sequence. The most frequent way in which this occurs is by **transposition**, and most interspersed repeats have inherent transpositional activity. Transposition is also a feature of some viral genomes, which are able to insert into the genome of the infected cell and then move from place to place within that genome. Some interspersed repeats are clearly descended from transposable viruses, and because of this relationship we will postpone discussion of these and the other types of interspersed repeat until Chapter 9, after we have looked in detail at the features of virus genomes.

SUMMARY

- A eukaryotic nuclear genome is split into a set of linear DNA molecules, each of which is contained in a chromosome.

- Within a chromosome, the DNA is packaged by association with histone proteins to form nucleosomes.

- The most compact organization of chromatin results in the metaphase chromosomes that can be observed by light microscopy of dividing cells and which take up characteristic banding patterns after staining.

- The centromeres, which are visible in metaphase chromosomes, contain special proteins that make up the kinetochore, the attachment point for the microtubules that draw the divided chromosomes into the daughter nuclei.

- Telomeres, the structures that maintain the chromosome ends, contain repetitive DNA and special binding proteins.

- Humans have 20,442 protein-coding genes, about the same number as the nematode worm *Caenorhabditis elegans*. However, the human genome specifies more proteins than that of *C. elegans* because of alternative splicing.

- Genes are not evenly spread along vertebrate chromosomes; some chromosomes have gene deserts where the gene density is very low.

- The coding parts of genes make up only a small part of the human genome, less than 1.5%, with 44% of the genome made up of various types of repetitive DNA sequence. In contrast, the *S. cerevisiae* genome is much more compact, with only 3.4% of the DNA taken up by repeat sequences. In general, larger genomes are less compact, explaining why organisms with similar numbers of genes can have genomes of very different sizes.

- Many genes are organized into multigene families whose members have similar or identical sequences, and in some families, such as the vertebrate globin genes, the members are expressed at different developmental stages.

- Eukaryotic nuclear genomes also contain evolutionary relics such as pseudogenes and gene fragments.

- The repetitive DNA content of a eukaryotic nuclear genome comprises interspersed repeats, much of which has transpositional activity, and tandemly repeated DNA, which includes the satellite DNA found at centromeres, minisatellites such as telomeric DNA, and microsatellites.

SHORT ANSWER QUESTIONS

1. What does the treatment of eukaryotic chromatin with nucleases reveal about the packaging of eukaryotic DNA?

2. Describe how the nucleosomes are arranged in the 10 nm fiber.

3. List the special features of (A) microchromosomes and (B) B chromosomes.

4. What has long-read sequencing revealed about the organization of DNA in the centromere regions of human chromosomes?

5. Explain why it is important that chromosomes have telomeres at their ends.

6. The human genome contains many fewer genes than were predicted by many researchers. Why were these initial predictions so high?

7. How are gene distributed within the human chromosomes?

8. What differences in gene distribution and repetitive DNA content are seen when yeast and human chromosomes are compared?

9. Describe the organization of the human globin gene families and indicate the functions of each of the genes in these families.

10. Distinguish between the two types of nonprocessed pseudogene.

11. Describe the events that give rise to a processed pseudogene.

12. What types of repetitive DNA are present in the human genome?

13. How are microsatellites used in DNA profiling?

IN-DEPTH PROBLEMS

1. What impact is DNA packaging likely to have on the expression of individual genes?

2. Discuss possible functions for the intergenic component of the human genome.

3. To what extent is it possible to describe the typical features of a eukaryotic genome?

4. What would be the implications for genome evolution if some pseudogenes retained their functions or acquired new functions?

FURTHER READING

Chromosome structure

Altemose, N., Logsdon, G.A., Bzikadze, A.V., et al. (2022). Complete genomic and epigenetic maps of human centromeres. *Science* 376:eabl4178.

Cutter, A.R. and Hayes, J.J. (2015). A brief review of nucleosome structure. *FEBS Lett.* 589:2022–2914.

de Lange, T. (2005). Shelterin: The protein complex that shapes and safeguards human telomeres. *Genes Dev.* 19:2100–2110.

Fierz, B. (2019). Revealing chromatin organization in metaphase chromosomes. *EMBO J.* 38:e101699.

Harshman, S.W., Young, N.L., Parthun, M.R. and Freitas, M.A. (2013). H1 histones: Current perspectives and challenges. *Nucl. Acids Res.* 41:9593–9609.

Henikoff, S., Ramachandran, S., Krassovsky, K., et al. (2014). The budding yeast centromere DNA element II wraps a stable Cse4 hemisome in either orientation in vivo. *eLife* 3:e01861.

Logsdon, G.A., Vollger, M.R., Hsieh, P.H., et al. (2021). The structure, function and evolution of a complete human chromosome 8. *Nature* 593:101–107. *Includes details of the organization of DNA in the centromere.*

Miga, K.H., Koren, S., Rhie, A., et al. (2020). Telomere-to-telomere assembly of a complete human X chromosome. *Nature* 585:79–84. *Includes the centromere DNA organization.*

Öztürk, M.A., Cojocaru, V. and Wade, R.C. (2018). Toward an ensemble view of chromatosome structure: A paradigm shift from one to many. *Structure* 26:1050–1057. *Details of the possible location of the linker histone in the chromatosome.*

Willard, H.F. and Waye, J.S. (1987). Hierarchical order in chromosome-specific human alpha satellite DNA. *Trends Genet.* 3:192–198.

Gene numbers and distribution

Hatje, K., Mühlahausen, S., Simm, D., et al. (2019). The protein-coding human genome: Annotating high-hanging fruits. *BioEssays* 41:1900066. *Describes changes in the estimated numbers of human genes and the difficulties in identifying the exact number.*

Ovcharenko, I., Loots, G.G., Nobrega, M.A., et al. (2005). Evolution and functional classification of vertebrate gene deserts. *Genome Res.* 15:137–145.

Piovesan, A., Antonaros, F., Vitale, L., et al. (2019). Human protein-coding genes and gene features statistics in 2019. *BMC Res. Notes* 12:315.

Salzberg, S. (2018). Open questions: How many genes do we have? *BMC Biol.* 16:94.

Key papers on eukaryotic genome structure and content

Adams, M.D., Celniker, S.E., Holt, R.A., et al. (2000). The genome sequence of *Drosophila melanogaster*. *Science* 287:2185–2195.

Arabidopsis Genome Initiative (2000). Analysis of the genome sequence of the flowering plant *Arabidopsis thaliana*. *Nature* 408:796–815.

C. elegans Sequencing Consortium (1998). Genome sequence of the nematode *C. elegans*: A platform for investigating biology. *Science* 282:2012–2018.

Dujon, B. (1996). The yeast genome project: What did we learn? *Trends Genet.* 12:263–270.

International Human Genome Sequencing Consortium (2001). Initial sequencing and analysis of the human genome. *Nature* 409:860–921.

Venter, J.C., Adams, M.D., Myers, E.W., et al. (2001). The sequence of the human genome. *Science* 291:1304–1351.

Genetic features

Cheetham, S.W., Faulkner, G.J. and Dinger, M.E. (2020). Overcoming challenges and dogmas to understand the functions of pseudogenes. *Nat. Rev. Genet.* 21:191–201.

Csink, A.K. and Henikoff, S. (1998). Something from nothing: The evolution and utility of satellite repeats. *Trends Genet.* 14:200–204.

Elliott, T.A. and Gregory, T.R. (2015). What's in a genome? The C-value enigma and the evolution of eukaryotic genome content. *Phil. Trans. R. Soc. B* 370:20140331.

Fritsch, E.F., Lawn, R.M. and Maniatis, T. (1980). Molecular cloning and characterization of the human β-like globin gene cluster. Cell 19:959–972.

Goh, S.H., Lee, Y.T., Bhanu, N.V., et al. (2005). A newly discovered human alpha-globin gene. *Blood.* 106:1466–1472. *Discovery of the μ globin subunit.*

Hoyt, S.J., Storer, J.M., Hartley, G.A., et al. (2022). From telomere to telomere: The transcriptional and epigenetic state of human repeat elements. *Science* 376:eabk3112.

Payseur, B.A., Jing, P. and Haasl, R.J. (2011). A genomic portrait of human microsatellite variation. *Mol. Biol. Evol.* 28:303–312.

Poliseno, L., Salmena, L., Zhang, J., et al. (2010). A coding-independent function of gene and pseudogene mRNAs regulates tumour biology. *Nature* 465:1033–1038. *The PTENP1 pseudogene.*

Online resources

Ensembl. http://www.ensembl.org/index.html. *Genome browser enabling detailed examination of genome organization in different species.*

InterPro. https://www.ebi.ac.uk/interpro/. *Protein database including details of protein domains encoded by different genomes.*

KEGG (Kyoto Encyclopedia of Genes and Genomes). http://www .genome.jp/kegg/. *A collection of databases including details of the structures and contents of all sequenced genomes.*

MicroSatellite DataBase. http://tdb.ccmb.res.in/msdb. *Database of the microsatellite content of different genomes.*

UCSC Genome Browser. https://genome.ucsc.edu/. *Genome browser enabling detailed examination of genome organization in different species.*

GENOMES OF PROKARYOTES AND EUKARYOTIC ORGANELLES

Prokaryotes are organisms whose cells lack extensive internal compartments. There are two very different groups of prokaryotes, distinguished from one another by characteristic genetic and biochemical features:

- The **bacteria**, which include most of the commonly encountered prokaryotes such as the gram-negatives (e.g., *Escherichia coli*), the gram-positives (e.g., *Bacillus subtilis*), the cyanobacteria (e.g., *Anabaena*), and many more.

- The **archaea**, which are less well studied, and were once thought to be **extremophiles**, only living in inhospitable environments such as hot springs and acidic streams. Now we know that they are much more common, being present in many non-extreme environments, including the human gut.

In this chapter, we will examine the genomes of prokaryotes, and also of eukaryotic **mitochondria** and **chloroplasts**, which, as they are descended from bacteria, have genomes that display many prokaryotic features. Because of the relatively small sizes of prokaryotic genomes, over 450,000 complete sequences, representing 17,000 species of bacteria and archaea, have been obtained. As a result, we are beginning to understand a great deal about the anatomies of prokaryotic genomes, and in some respects we know more about these organisms than we do about eukaryotes. The picture that is emerging is one of immense variability among the prokaryotes as a whole and in some cases even between closely related species.

8.1 THE PHYSICAL FEATURES OF PROKARYOTIC GENOMES

Prokaryotic genomes are very different from eukaryotic ones, in particular with regard to the physical organization of the genome within the cell. Although the word 'chromosome' is used to describe the DNA–protein structures present in prokaryotic cells, this is a misnomer as this structure has few similarities with a eukaryotic chromosome.

The traditional view has been that in a typical prokaryote the genome is contained in a single, circular DNA molecule, localized within the **nucleoid** – the lightly staining region of the otherwise featureless prokaryotic cell (**Figure 8.1**). This is certainly true for *E. coli* and many of the other commonly studied bacteria. However, as we will see, our growing knowledge of prokaryotic genomes is leading us to question several of the preconceptions that became established during the pre-genome era of microbiology. These preconceptions relate both to the physical structure of the prokaryotic genome and its genetic organization.

The traditional view of the prokaryotic chromosome

As with eukaryotic chromosomes, a prokaryotic genome has to squeeze into a relatively tiny space (the circular *E. coli* chromosome has a circumference of 1.6 mm, whereas an *E. coli* cell is just $1.0 \times 2.0 \, \mu m$) and, as with eukaryotes, this

8.1 THE PHYSICAL FEATURES OF PROKARYOTIC GENOMES

8.2 THE GENETIC FEATURES OF PROKARYOTIC GENOMES

8.3 EUKARYOTIC ORGANELLE GENOMES

DOI: 10.1201/9781003133162-8

Figure 8.1 The *Escherichia coli* nucleoid. This transmission electron micrograph shows the cross-section of a rapidly growing, dividing *E. coli* cell. The nucleoid is the lightly staining area in the center of the cell. (Courtesy of Conrad Woldringh, University of Amsterdam.)

is achieved with the help of DNA-binding proteins that package the genome in an organized fashion.

Most of what we know about the organization of DNA in the nucleoid comes from studies of *E. coli*. The first feature to be recognized was that the circular *E. coli* genome is **supercoiled**. Supercoiling occurs when additional turns are introduced into the DNA double helix (positive supercoiling) or if turns are removed (negative supercoiling). With a linear molecule, the torsional stress introduced by over- or underwinding is immediately released by rotation of the ends of the DNA molecule, but a circular molecule, having no ends, cannot reduce the strain in this way. Instead, the circular molecule responds by winding around itself to form a more compact structure (**Figure 8.2**). Supercoiling is therefore an ideal way to package a circular molecule into a small space. Evidence that supercoiling is involved in packaging the circular *E. coli* genome was first obtained in the 1970s from examination of isolated nucleoids, and subsequently confirmed as a feature of DNA in living cells in 1981. In *E. coli*, the supercoiling is thought to be generated and controlled by the DNA topoisomerase called DNA gyrase, which we will look at in more detail in Section 15.1 when we examine the roles of these enzymes in DNA replication.

Studies of isolated nucleoids suggest that the *E. coli* DNA molecule does not have unlimited freedom to rotate once a break is introduced. The most likely explanation is that the bacterial DNA is attached to proteins that restrict its ability to relax, so rotation at a break site results in loss of supercoiling from only a small segment of the molecule (**Figure 8.3**). The strongest evidence for this domain model has come from experiments that exploit the ability of trimethylpsoralen to distinguish between supercoiled and relaxed DNA. When photoactivated by a pulse of light of wavelength 360 nm, trimethylpsoralen binds to double-stranded DNA at a rate that is directly proportional to the degree of torsional stress possessed by the molecule. The degree of supercoiling can therefore be assayed by measuring the amount of trimethylpsoralen that binds to a molecule in unit time. After *E. coli* cells have been irradiated to introduce single-strand breaks into their DNA molecules, the amount of trimethylpsoralen binding is indirectly proportional to the radiation dose (**Figure 8.4**). This is the response predicted by the domain model, in which the overall supercoiling of the molecule is gradually relaxed as greater doses of radiation cause breaks within an increasing number of domains. In contrast, if the *E. coli* nucleoid was not organized into domains, then a single break in the DNA molecule would lead to complete loss of supercoiling: irradiation would therefore have an all-or-nothing effect on trimethylpsoralen binding.

The current model has the *E. coli* DNA attached to a protein core from which loops, called **chromosomal interaction domains** (**CIDs**) radiate out into the cell. Each CID contains 40–300 kb of supercoiled DNA, the amount of DNA that becomes unwound after a single break. During active growth there are 31 CIDs, but the number is variable with fewer (and hence longer) CIDs when there are

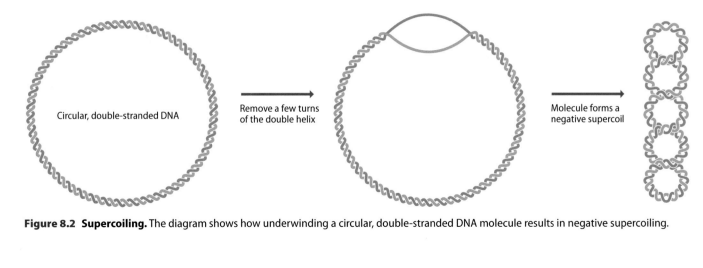

Figure 8.2 Supercoiling. The diagram shows how underwinding a circular, double-stranded DNA molecule results in negative supercoiling.

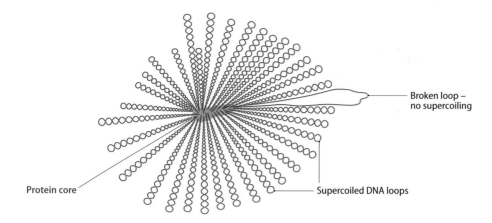

Figure 8.3 A model for the structure of the *Escherichia coli* nucleoid. Supercoiled loops of DNA radiate from the central protein core. One of the loops is shown in circular form, indicating that a break has occurred in this segment of DNA, resulting in a loss of the supercoiling.

lower amounts of available nutrients. The CIDs associate with one another to form four distinct nucleoid regions or **macrodomains**, one associated with the origin of replication, one with the position at the opposite side of the circular genome where replication terminates, and the remaining two formed from the 'left' and 'right' parts of the genome (**Figure 8.5**). The protein component of the nucleoid is made up of a variety of **nucleoid-associated proteins**, which are presumed to include those that are involved in packaging of the chromosome. Early research in this area was influenced by the discovery that the **HU family** of nucleoid-associated proteins, which are present in most bacteria that have been studied, have some amino acid sequence similarity with the eukaryotic histone H2B. Each HU protein is a dimer of two subunits, either two HUα, two HUβ, or a heterodimer comprising one of each type of monomer. HU proteins induce bends in the DNA in order to facilitate formation of the supercoiled loops: in cells that are HU-deficient, the nucleoid is less compact than normal. A similar bending function is provided by two other types of nucleoid-associated proteins, called **integration host factor** (IHF) and **factor for inversion stimulation** (Fis). A packaging role has also been proposed for the **histone-like nucleoid structuring protein** (H-NS), which binds specifically to AT-rich regions, which are thought to be present at the boundaries of the supercoiled loops. H-NS proteins can form filaments, with a single filament able to bind to two different regions of the genome and hence forming a DNA loop.

The above discussion refers specifically to the *E. coli* chromosome, which we look on as typical of bacterial chromosomes in general. But we must be careful to make a distinction between the bacterial chromosome and that of the second group of prokaryotes, the archaea. Although one group of archaea, the Crenarchaeota, which include many marine species, have nucleoid proteins with DNA bending properties similar to the bacterial ones that we have discussed above, another group, the Euryarchaeota, possess proteins that are much more similar to histones. In some species, these form a tetramer that associates with approximately 60 bp of DNA to form a structure similar to a eukaryotic nucleosome. In other species, the histone-like proteins form larger multimers of various sizes able to bind longer segments of DNA. Currently, we have very little information on the archaeal nucleoid, but the assumption is that in the Euryarchaeota these histone-like proteins play a central role in DNA packaging.

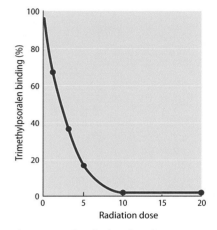

Figure 8.4 Graph showing the relationship between radiation dose and trimethylpsoralen binding. Trimethylpsoralen binding decreases as the radiation dose increases, supporting the domain model, which predicts that supercoiling of the nucleoid DNA will gradually be lost as more radiation hits accumulate in the molecule.

Some bacteria have linear or multipartite genomes

The *E. coli* genome, as described above, is a single, circular DNA molecule. This is also the case with the vast majority of bacterial and archaeal chromosomes that have been studied, but an increasing number of linear versions are being found. The first of these, for *Borrelia burgdorferi*, the organism that causes Lyme disease, was described in 1989, and during the following years similar discoveries were made for *Streptomyces coelicolor* and *Agrobacterium tumefaciens*. Linear molecules have free ends, which must be distinguishable from DNA breaks, so

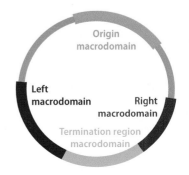

Figure 8.5 Locations of the four macrodomains in the *E. coli* genome.

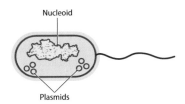

Figure 8.6 Plasmids are small, circular DNA molecules that are found inside some prokaryotic cells.

these chromosomes require terminal structures equivalent to the telomeres of eukaryotic chromosomes (Section 7.1). In *Borrelia* and *Agrobacterium*, the real chromosome ends are distinguishable because a covalent linkage is formed between the 5´ and 3´ ends of the polynucleotides in the DNA double helix, and in *Streptomyces* the ends are capped by special binding proteins.

A second and more widespread variation on the *E. coli* theme is the presence in some prokaryotes of multipartite genomes – genomes that are divided into two or more DNA molecules. With these multipartite genomes, a problem often arises in distinguishing a genuine part of the genome from a plasmid. A plasmid is a small piece of DNA, often but not always circular, that coexists with the main chromosome in a bacterial cell (**Figure 8.6**). Some types of plasmid are able to integrate into the main genome, but others are thought to be permanently independent. Their replication process is distinct from that of the main chromosome, and some can reach copy numbers of a thousand or more in a single cell. When the bacterium divides, the plasmids are partitioned between the daughter cells by a process different from the one that results in each daughter receiving its copy of the main chromosome. Plasmids carry genes that are not usually present in the main chromosome, but in many cases these genes are nonessential to the bacterium, coding for characteristics such as antibiotic resistance, which the bacterium does not need if the environmental conditions are amenable (Table 8.1). As well as this apparent dispensability, many plasmids are able to transfer from one cell to another, and the same plasmids are sometimes found in bacteria that belong to different species. These various features of plasmids suggest that they are independent entities and that in most cases the plasmid content of a prokaryotic cell should not be included in the definition of its genome.

With a bacterium such as *E. coli* K12, which has a 4.64 Mb chromosome and can harbor various combinations of plasmids, none of which is more than a few kilobases in size and all of which are dispensable, it is acceptable to define the main chromosome as the 'genome.' With other prokaryotes, it is not so easy (Table 8.2). *Vibrio cholerae* MS6, the pathogenic bacterium that causes cholera, has two circular DNA molecules, one of 2.94 Mb and the other of 1.09 Mb, with 72% of the organism's 3720 genes on the larger of these. It would appear obvious that these two DNA molecules together constitute the *Vibrio* genome, but closer examination reveals that most of the genes for the central cellular activities, such as genome expression and energy generation, as well as the genes that confer pathogenicity, are located on the larger molecule. The smaller molecule contains many essential genes but also has certain features that are considered characteristic of plasmids, notably the presence of an **integron** – a set of genes and other DNA sequences that enable plasmids to capture genes from bacteriophages and other plasmids. It therefore appears possible that the smaller genome is a 'megaplasmid' that was acquired by the ancestor to *Vibrio* at some period in the bacterium's evolutionary past. *Deinococcus radiodurans* R1, whose genome is of particular interest because it contains many genes that help this bacterium resist the harmful effects of radiation, is constructed on similar lines, with essential genes distributed among two circular chromosomes and two plasmids. However, the *Vibrio* and *Deinococcus* genomes are relatively noncomplex compared with

TABLE 8.1 FEATURES OF TYPICAL PLASMIDS		
Type of plasmid	**Gene functions**	**Examples**
Resistance	Antibiotic resistance	Rbk of *Escherichia coli* and other bacteria
Fertility	Conjugation and DNA transfer between bacteria	F of *E. coli*
Killer	Synthesis of toxins that kill other bacteria	Col of *E. coli*, for colicin production
Degradative	Enzymes for metabolism of unusual molecules	TOL of *Pseudomonas putida*, for toluene metabolism
Virulence	Pathogenicity	Ti of *Agrobacterium tumefaciens*, conferring the ability to cause crown gall disease on dicotyledonous plants

TABLE 8.2 EXAMPLES OF GENOME ORGANIZATION IN PROKARYOTES

Species	Genome organization		
	DNA molecules	Size (Mb)	Number of genes
Escherichia coli K-12	One circular molecule	4.642	4609
Vibrio cholerae MS6	Two circular molecules		
	Chromosome	2.937	2664
	Megaplasmid	1.094	1056
Deinococcus radiodurans R1	Four circular molecules		
	Chromosome 1	2.649	2682
	Chromosome 2	0.412	372
	Circular plasmid	0.177	143
	Circular plasmid	0.046	42
Borrelia burgdorferi B331	Nine circular molecules, nine linear molecules		
	Linear chromosome	0.904	850
	Circular plasmid cp9	0.009	9
	Circular plasmid cp26	0.027	26
	Circular plasmid cp32–1	0.030	41
	Circular plasmid cp32–4	0.031	42
	Circular plasmid cp32–5	0.018	28
	Circular plasmid cp32–6	0.030	41
	Circular plasmid cp32–7	0.030	41
	Circular plasmid cp32–10	0.030	42
	Circular plasmid cp32–11	0.030	41
	Linear plasmid lp17	0.016	6
	Linear plasmid lp25	0.025	14
	Linear plasmid lp28–3	0.028	10
	Linear plasmid lp28–5	0.023	9
	Linear plasmid lp28–6	0.029	16
	Linear plasmid lp36	0.030	20
	Linear plasmid lp38	0.039	19
	Linear plasmid lp54	0.054	19
	Linear plasmid lp56	0.028	28

Data taken from NCBI Genome.

Borrelia burgdorferi B331, whose linear chromosome of 904 kb, carrying 850 genes, is accompanied by 18 linear and circular plasmids, which together contribute another 507 kb and another 452 genes. Although the functions of most of these genes are unknown, those that have been identified include several that would not normally be considered dispensable, such as genes for membrane

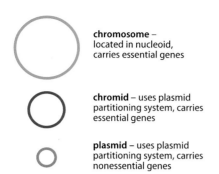

chromosome – located in nucleoid, carries essential genes

chromid – uses plasmid partitioning system, carries essential genes

plasmid – uses plasmid partitioning system, carries nonessential genes

Figure 8.7 The differences between prokaryotic chromosomes, chromids, and plasmids.

proteins and purine biosynthesis. The implication is that at least some of the *Borrelia* plasmids are essential components of the genome, leading to the possibility that some prokaryotes have highly multipartite genomes, comprising a number of separate DNA molecules, more akin to what we see in the eukaryotic nucleus rather than the 'typical' prokaryotic arrangement.

The complications posed by bacteria such as *Vibrio* and *Deinococcus* have prompted microbial geneticists to invent a new term – **chromid** – to describe a plasmid that carries essential genes. This means that we now distinguish between three, rather than just two, types of DNA molecule that might be found in a bacterium (**Figure 8.7**):

- One or more bacterial chromosomes, carrying essential genes and located in the nucleoid.

- Genuine plasmids, which are distinct from a bacterial chromosome because of their special plasmid partitioning system and whose genes are nonessential to the bacterium.

- Chromids, which use a plasmid partitioning system but which carry genes that the bacterium needs to survive.

According to this nomenclature, *Vibrio cholerae* has one chromosome and one chromid, and *Deinococcus radiodurans* has two chromosomes and two chromids.

8.2 THE GENETIC FEATURES OF PROKARYOTIC GENOMES

Genome annotation by sequence inspection is much easier for prokaryotes compared with eukaryotes (Section 5.1), and for most of the prokaryotic genomes that have been sequenced we have reasonably accurate estimates of gene number and fairly comprehensive lists of gene functions. The results of these studies have been surprising, and have forced microbiologists to reconsider the meaning of 'species' when applied to prokaryotes. We will examine these evolutionary issues later in this chapter. First, we must look at the way in which the genes are organized in a prokaryotic genome.

Gene organization in the *E. coli* K12 genome

We are already familiar with the notion that bacterial genomes have compact genetic organizations with very little space between genes, as this was an important part of our discussion of the strengths and weaknesses of open reading frame (ORF) scanning as a means of identifying the genes in a genome sequence (see Figure 5.3). To reemphasize this point, the complete circular gene map of the *E. coli* K12 genome is shown in **Figure 8.8**. There is intergenic DNA in the *E. coli* genome, but it accounts for only 11% of the total and it is distributed round the genome in small segments that do not show up when the map is drawn at this scale. In this regard, *E. coli* is typical of all prokaryotes whose genomes have so far been sequenced – prokaryotic genomes have very little wasted space. There are theories that this compact organization is beneficial to prokaryotes, for example, by enabling the genome to be replicated relatively quickly, but these ideas have never been supported by hard experimental evidence.

Let us now look more closely at the *E. coli* K12 genome. A typical 50 kb segment is shown in **Figure 8.9**. When we compare this segment with a typical part of the human genome (see Figure 7.12), it is immediately obvious that in the *E. coli* segment there are more genes and much less space between them, with 43 genes taking up 85.9% of the segment. Some genes have virtually no space between them: *thrA* and *thrB*, for example, are separated by a single nucleotide, and *thrC* begins at the nucleotide immediately following the last nucleotide of *thrB*. In general, prokaryotic genes are shorter than their eukaryotic counterparts, the average length of a bacterial gene being about two-thirds that of a eukaryotic

Origin of replication

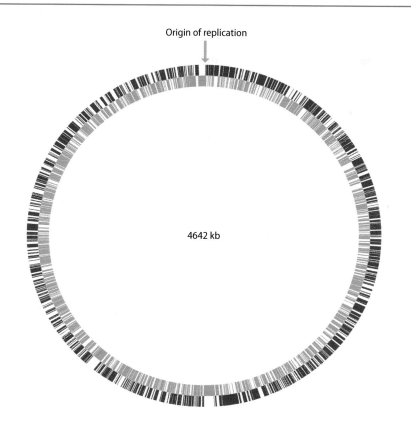

4642 kb

Figure 8.8 The genome of *Escherichia coli* K12. The map is shown with the origin of replication positioned at the top. Genes on the outside of the circle are transcribed in the clockwise direction and those on the inside are transcribed in the anticlockwise direction. (Courtesy of Frederick Blattner, University of Wisconsin-Madison.)

gene, even after the introns have been removed from the latter. Bacterial genes appear to be slightly longer than archaeal ones.

Two other features of prokaryotic genomes can be deduced from **Figure 8.9**. The first of these is the infrequency of repetitive sequences. Most prokaryotic genomes do not have anything equivalent to the high-copy-number interspersed repeat families found in eukaryotic genomes. They do, however, possess certain sequences that might be repeated elsewhere in the genome, examples being the **insertion sequences** IS1 and IS186 that can be seen in the 50 kb segment shown in Figure 8.9. These are examples of transposable elements, sequences that have the ability to move around the genome and, in the case of insertion elements, to transfer from one organism to another, even sometimes between two different species. The positions of the IS1 and IS186 elements shown in Figure 8.9 refer only to the particular *E. coli* isolate from which this sequence was obtained: if a different isolate is examined, then the insertion sequences could well be in different positions or might be entirely absent from the genome. Several families of transposable elements are known in prokaryotic genomes, and we will examine their structures when we study mobile genetic elements in more detail in Section 9.2. Many prokaryotic genomes also contain at least a few non-transposable repeat sequences, falling into various classes, the two most important being:

- **Repetitive extragenic palindromic (REP) sequences**, most of which are 20–50 bp in length and occur singly or in arrays. If one or more REP sequences are located immediately downstream of a gene, then they can be transcribed as an extension of the mRNA, with the REP sequences then

Figure 8.9 A 50 kb segment of the *Escherichia coli* genome. The segment runs between nucleotide positions 377–50,377. Data taken from the UCSC Microbial Genome Browser.

folding into stem-loop structures, which might play a role in gene regulation. There are approximately 600 REP sequences in the *E. coli* K12 genome, and several thousand in some other bacteria, such as *Pseudomonas.*

- **Clustered regularly interspaced short palindromic repeats** (**CRISPRs**), which we have already met as the source of the programmable nuclease that forms the basis for one of the gene inactivation procedures used to assign functions to eukaryotic genes (Section 6.2). CRISPRs are 20–50 bp sequences found in tandem arrays, with each repeat separated from its neighbor by a spacer of similar length but with a variable sequence. Some spacer sequences resemble segments of bacteriophage genomes, leading to the suggestion that CRISPRs represent a prokaryotic immune system, transcripts of the spacers acting as guide RNAs that bind to invading phage genomes, enabling a Cas endonuclease (whose gene is usually located adjacent to a CRISPR array) to cut and hence inactivate the phage DNA.

The number of transposable and non-transposable repeat sequences present in different prokaryotic genomes varies enormously. Usually, they take up less than 1% of the genome sequence, but there are exceptions. The genome of the meningitis bacterium *Neisseria meningitidis* Z2491 has over 3700 copies of 15 different types of repeat sequence, collectively making up almost 11% of the 2.18 Mb genome.

The second feature of prokaryotic genomes that can be deduced from Figure 8.9 is the scarcity of introns. *E. coli* K12 has no discontinuous genes at all, and introns are uncommon among other bacteria and archaea. Those that have been discovered belong to the **Group I** and **Group II** types, which are quite different from the introns present in eukaryotic pre-mRNA. Unlike, pre-mRNA introns, the Group I and II types can fold into complex base-paired structures which have the ability to self-splice, meaning that they can remove themselves from RNA transcripts without the aid of catalytic proteins. At least some are also able to move from one position to another in a genome. Because they are autocatalytic, the insertion of one of these introns into a gene does not affect the ability of that gene to be expressed. Once the gene is transcribed, the intron self-splices, leaving a functional copy of the mRNA. Prokaryotic introns might therefore be looked on as a special type of transposable element, one that targets gene sequences rather than intergenic regions as its insertion sites.

Operons are characteristic features of prokaryotic genomes

One characteristic feature of prokaryotic genomes illustrated by *E. coli* K12 is the presence of **operons**. An operon is a group of genes that are located adjacent to one another in the genome, with perhaps just one or two nucleotides between the end of one gene and the start of the next. Examples within our 50-kb segment of the *E. coli* genome are the three *thr* genes, which form the threonine operon, and the *dna* operon comprising *dnaK* and *dnaJ*. The similarity in the gene names within each operon indicates that the encoded proteins have related functions. For example, the three *thr* genes code for enzymes that catalyze different steps in the biochemical pathway that converts aspartate 4-semialdehyde into threonine amino acid, and the proteins coded by *dnaK* and *dnaJ* work together to prevent other proteins from being damaged when the bacterium is exposed to heat or osmotic shock.

The placement of related genes in an operon means that the bacterium can ensure the coordinated synthesis of the proteins needed for the biochemical function encoded by the operon. This principle is illustrated by the *E. coli* **lactose operon**, the first operon to be discovered, which contains three genes involved in conversion of the disaccharide sugar lactose into its monosaccharide units – glucose and galactose (**Figure 8.10A**). The monosaccharides are substrates for the energy-generating glycolytic pathway, so the function of the genes in the

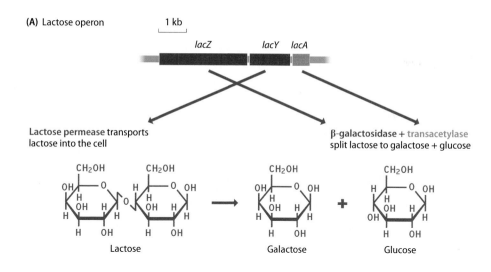

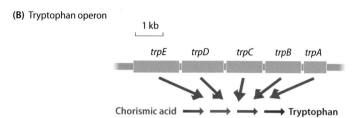

Figure 8.10 Two operons of *Escherichia coli*. (A) The lactose operon. The three genes are called *lacZ*, *lacY*, and *lacA*, the first two separated by 52 bp and the second two by 64 bp. All three genes are expressed together, *lacY* coding for the lactose permease that transports lactose into the cell, and *lacZ* and *lacA* coding for enzymes that split lactose into its component sugars – galactose and glucose. (B) The tryptophan operon, which contains five genes coding for enzymes involved in the multistep biochemical pathway that converts chorismic acid into the amino acid tryptophan. The genes in the tryptophan operon are closer together than those in the lactose operon: *trpE* and *trpD* overlap by 1 bp, as do *trpB* and *trpA*; *trpD* and *trpC* are separated by 4 bp, and *trpC* and *trpB* by 12 bp.

lactose operon is to convert lactose into a form that can be utilized by *E. coli* as an energy source. Lactose is not a common component of *E. coli*'s natural environment, so most of the time the operon is not expressed and the enzymes for lactose utilization are not made by the bacterium. When lactose becomes available, the operon is switched on; all three genes are expressed together, resulting in coordinated synthesis of the lactose-utilizing enzymes. This is the classic example of gene regulation in bacteria.

The lactose operon is an example of an **inducible operon**, one that is switched on by a substrate for the enzymes coded by the genes in the operon. Other operons are **repressible**, being controlled by a product of the pathway catalyzed by the gene products. An example is the tryptophan operon, which contains five genes that specify the set of enzymes needed to synthesize this amino acid from a precursor called chorismic acid (**Figure 8.10B**). The regulatory molecule for this operon is tryptophan: when tryptophan levels are low, the operon is expressed so more is made. When the levels of tryptophan have been replenished, the operon is switched off.

Some operons have been studied in detail, with co-expression of the adjacent genes confirmed by identification of their mRNAs. The vast majority, however, are predicted during genome annotation by searching for adjacent genes with related biochemical functions, possibly supported by comparative genomics (Section 5.1), on the basis that a genuine operon will be present not just in the genome being studied but also in the genomes of similar species. This means that the exact number of operons in a genome is uncertain, as some predictions might be incorrect, but it is clear that they are a common feature of most prokaryotes. For example, according to one analysis, the *E. coli* genome contains 239 operons, comprising a total of 2620 genes, 64% of the entire complement. Operons were once thought to be exclusively a feature of prokaryotic genomes, but we now know that they are not entirely absent in eukaryotes. Closely spaced clusters of genes that are transcribed as a single unit are relatively common in the *Caenorhabditis elegans* genome, and some examples are also known in *Drosophila melanogaster*.

Prokaryotic genome sizes and gene numbers vary according to biological complexity

There is some overlap in size between the largest prokaryotic and smallest eukaryotic genomes, but on the whole prokaryotic genomes are much smaller (Table 8.3). For example, the *E. coli* K12 genome is just 4.64 Mb, two-fifths the size of the yeast genome, and has only 4419 genes. Most prokaryotic genomes are less than 5 Mb in size, but the overall range among sequenced genomes is from 100 kb to over 25 Mb.

The compact organization of the *E. coli* K12 genome, with the genes making up 89% of the genome sequence, is typical of other prokaryotic genomes, the average gene density being 87%, with most genomes in the range of 85–90%. This means that genome size is proportional to gene number. Gene numbers therefore vary over an extensive range, with these numbers reflecting the nature of the ecological niches within which different species of prokaryote live. The largest genomes tend to belong to free-living species that are found in the soil, the environment which is generally looked on as providing the broadest range of physical and biological conditions, to which the genomes of these species must be able to respond. The 14.8 Mb genome of *Sorangium cellulosum* So0157–2 provides a good example. Its 10,400 protein-coding genes include some specifying enzymes that enable this bacterium to break down cellulose into sugars, and others coding for enzymes that synthesize antibacterial and antifungal compounds that help it to compete in the complex soil ecosystem. There are also genes for proteins involved in cell-to-cell communication, these enabling the bacteria to migrate together in swarms and to associate into a multicellular fruiting body that produces resistant spores. At the other end of the scale, many of the smallest genomes belong to species that are obligate parasites. *Nasuia deltocephalinicola* NAS-ALF, for example, is an endosymbiont of leafhoppers, living inside specialized structures within the insect's abdomen. The bacteria provide the leafhoppers with two amino acids that *Nasuia* can synthesize but which insects must obtain from their diet. In return, the bacteria receive various nutrients from the insect, which means that *Nasuia* is able to dispense with many of the enzymes needed by free-living bacteria for the synthesis of metabolites and for energy generation. As a consequence, the *Nasuia* genome is just 112 kb and contains

TABLE 8.3 GENOME SIZES AND GENE NUMBERS FOR VARIOUS PROKARYOTES

Species	Size of genome (Mb)	Number of genes
Bacteria		
Nasuia deltocephalinicola NAS-ALF	0.12	169
Mycoplasma genitalium G37	0.58	519
Streptococcus pneumoniae R6	2.04	2116
Vibrio cholerae O1 El Tor	4.03	3950
Mycobacterium tuberculosis H37Rv	4.41	3976
Escherichia coli K12	4.64	4419
Pseudomonas aeruginosa PA01	6.26	5678
Sorangium cellulosum So0157–2	14.78	10,473
Archaea		
Methanocaldococcus jannaschii DSM2661	1.74	1813
Archaeoglobus fulgidus DSM4304	2.18	2456

Data taken from the *Kyoto Encyclopedia of Genes and Genomes* (KEGG) release 100.0.

only 169 genes, of which 137 are protein-coding, the majority of these involved in essential functions such as DNA replication, transcription and translation.

Comparisons between the genomes of different prokaryotes have led to speculation about the smallest number of genes needed to specify a free-living cell. One of the first genomes to be sequenced was that of *Mycoplasma genitalium* G37, a genuinely free-living organism with just 476 protein-coding genes. Experiments in which increasing numbers of *Mycoplasma* genes were inactivated by mutation suggested that 382 of these genes were essential. However, this is the minimal gene set needed for an *M. genitalium* bacterium, and similar inactivation studies with other genomes have shown that the number of essential genes that are identified in this way is species-specific. With some species, the minimal gene set is greater than 382, and with a few the number is smaller. In one project, a set of just 230 genes was shown to be sufficient for growth of *Salmonella typhimurium* LT2, providing that the mutated bacteria were cultured in a rich culture medium from which they could obtain nutrients such as amino acids. By providing many compounds that in its natural environment a bacterium would synthesize for itself, the required gene set can be reduced to a number similar to that of *Nasuia* and other symbiotic species.

Genome sizes and gene numbers vary within individual species

Genome projects have confused our understanding of what constitutes a 'species' in the prokaryotic world. This has always been a problem in microbiology because the standard biological definitions of species have been difficult to apply to microorganisms. The early taxonomists such as Linneaus described species in morphological terms, all members of one species having the same or very similar structural features. This form of classification was in vogue until the early 20th century and was first applied to microorganisms in the 1880s by Robert Koch and others, who used staining and biochemical tests to distinguish between bacterial species. However, it was recognized that this type of classification was imprecise because many of the resulting species were made up of a variety of types with quite different properties. An example is provided by *E. coli* which, like many bacterial species, includes strains with distinctive pathogenic characteristics, ranging from harmless through to lethal. During the 20th century, biologists redefined the species concept in evolutionary terms, and we now look on a species as a group of organisms that can interbreed with one another. If anything, this is more problematic with microorganisms because there are a variety of methods by which genes can be exchanged between prokaryotes that, according to their biochemical and physiological properties, are different species (see Figure 3.24). The barrier to **gene flow** that is central to the species concept as applied to higher organisms therefore does not hold with prokaryotes.

Genome sequencing has emphasized even further the difficulties in applying the species concept to prokaryotes. It quickly became clear that different strains of a single species can have very different genome sequences, and often have individual sets of strain-specific genes. This was first shown by a comparison between two strains of *Helicobacter pylori*, which causes gastric ulcers and other diseases of the human digestive tract. The two strains were isolated in the United Kingdom and the United States and had genomes of 1.67 Mb and 1.64 Mb, respectively. The initial annotations identified 1552 genes in the larger genome and 1495 in the smaller one, 1406 of these genes being present in both strains. In other words, some 6–9% of the gene content of each genome was unique to that strain. A much more extreme distinction between strains was revealed when the genome sequence of the common laboratory strain of *E. coli*, K12, was compared with that of one of the most pathogenic strains, O157:H7. The lengths of the two genomes are significantly different – 4.64 Mb for K12 and 5.59 Mb for O157:H7 – with the extra DNA in the pathogenic strain scattered around the genome at almost 200 separate positions. These 'O-islands' contain over 1300 genes not present in *E. coli* K12, many of these genes coding for toxins and other proteins that are clearly involved in the pathogenic properties of O157:H7. But it is not simply a case of

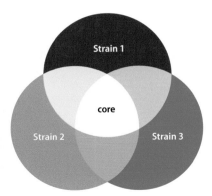

Figure 8.11 The pan-genome concept.
The gene contents of the genomes of three strains of a prokaryote are depicted. Each gene set is represented by a circle, the overlap between the three circles (shown in white) being the core genome. The accessory genome comprises those genes that lie outside of the core set. These genes can be further subdivided into those which are present in just one genome (shown in red, dark blue, and green), and genes shared by two genomes (yellow, pink, and light blue).

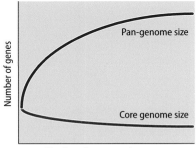

Figure 8.12 The predicted changes in the sizes of the core and pan-genomes as more genomes are sequenced. The expectation is that the size of the core genome (red line) will go down as strains are discovered that lack one or more of the genes previously assigned to the core set. Conversely, the pan-genome (blue line) is expected to increase in size as the strain-specific genes from each new genome that is sequenced are added to this set.

O157:H7 containing extra genes that make it pathogenic. K12 also has 234 segments of its own unique DNA, and although these 'K-islands' are, on average, smaller than the O-islands, they still contain over 500 genes that are absent from O157:H7. The situation, therefore, is that *E. coli* O157:H7 and *E. coli* K12 each has a set of strain-specific genes, which make up approximately 25% and 12% of the gene catalogs, respectively. This is substantially more variation than can be tolerated by the species concept as applied to higher organisms, and is difficult to reconcile with any definition of species yet devised for microorganisms.

The differences in genome sizes and gene contents that occur within a prokaryotic species has led to the **pan-genome concept**. According to this concept, the genome of a species is divided into two components (**Figure 8.11**):

- The **core genome**, which contains the set of genes possessed by all members of the species.

- The **accessory genome**, which is the entire collection of additional genes present in different strains and isolates of that species.

The core genome can therefore be looked on as specifying the basic biochemical and cellular activities that define a particular species, whereas the accessory genome describes the complete biological capability of the species as a whole, components of which are expressed by individual strains. The pan-genome concept therefore takes us away from our conventional view of a genome as one or more DNA molecules possessed by a single cell, and instead redefines the genome in terms of the gene content of the species as a whole.

One of the first pan-genomes to be described was that of *Streptococcus agalactiae*, a bacterium that inhabits the human gastrointestinal and genitourinary tracts and exists as both harmless and pathogenic strains, the latter causing urinary infections in adults and potentially life-threatening infections in newborn children. A comparison of the genome sequences of eight isolates of *S. agalactiae* revealed a pan-genome of 2700 genes, of which 1800 genes made up the core and the remaining 900 were accessory genes. Of the latter, 260 were singleton genes found in just one strain, and the remainder were present in two or more strains, but not, by definition, in each of the eight strains.

It was clear from the analysis of the *S. agalactiae* pan-genome that the numbers of genes in the core and accessory genomes would change as additional strains are added to the dataset. The size of the core genome would be expected to go down as strains were discovered that lacked one or more genes previously assigned to the core set. Conversely, the accessory gene number would increase as the strain-specific genes from each new genome were added to this set (**Figure 8.12**). This prediction has turned out to be correct for some, but not all, species. The *E. coli* pan-genome, for example, continues to grow even though almost 200,000 strains have now been sequenced, and may contain as many as 60,000 genes. As additional strains are added to the catalog, the size of the *E. coli* core genome has gradually decreased; in one study of 1324 high-quality genome sequences a core set of just 420 genes was identified. Because its gene numbers are still increasing, *E. coli* is looked on as having an **open pan-genome** (**Figure 8.13**). In contrast, some species have **closed pan-genomes**, ones that are no longer increasing in size as new strains are sequenced. An example is *Bacillus anthracis*, whose pan-genome has been estimated to contain 2978 genes, of which 2893 form the core. It has been suggested that the relatively small number of accessory genes possessed by a species with a closed pan-genome reflects a more limited ecological range compared to a species with an open pan-genome, whose vast array of accessory genes presumably enables the species to colonize a larger variety of ecological niches.

Distinctions between prokaryotic species are further blurred by horizontal gene transfer

It has been known since the 1940s that plasmids and occasionally chromosomal genes can move between bacteria by conjugation, transduction via a bacteriophage, or by the simple uptake of DNA fragments from the environment. These **horizontal** or **lateral gene transfer** processes have been extensively studied, at least with model species such as *E. coli*, because they form the basis for techniques for gene mapping in bacteria (Section 3.4). This early work revealed that, under some circumstances, genes could be transferred between bacteria of different species, for example, between *E. coli* and *S. typhimurium*. A stark illustration that horizontal gene transfer is not simply an abstract concept was then provided in the 1960s by the accelerating spread of antibiotic-resistance genes through bacterial metapopulations, in hospitals and more broadly in the environment. The notion that the same genes might be present in different prokaryotic species had therefore been established before the beginning of the genomics era. The extent of horizontal gene transfer, as revealed by comparisons between the thousands of prokaryotic genomes that are now available, has nonetheless been a major surprise.

There are three ways of detecting genes that might have participated in horizontal gene transfer (**Figure 8.14**):

- A phylogenetic approach can be used, searching for an inconsistency between the apparent evolutionary relationship of a pair of species as inferred by studies of individual genes compared with studies of their genomes as a whole. This is because the transfer of a gene from one species to another will have occurred more recently than the original divergence of the two species from a common ancestor. The copies of the transferred gene in the two species will therefore have relatively similar sequences, as there has been insufficient time for the sequences to diverge greatly by mutation. Comparisons based on the entire genome sequences will therefore place the two species a different positions on the prokaryotic evolutionary tree, whereas comparisons between the gene sequences will place them close together.

- Compositional differences within a genome can be used to identify regions that might have been obtained by horizontal gene transfer. Compositional features such as GC content, codon usage, and amino acid content of the encoded proteins tend to be consistent throughout a genome sequence,

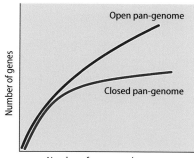

Figure 8.13 Open and closed pan-genomes. An open pan-genome continues to grow as more genome sequences are obtained. A closed pan-genome has reached its maximum size and does not increase any further when more genomes are added.

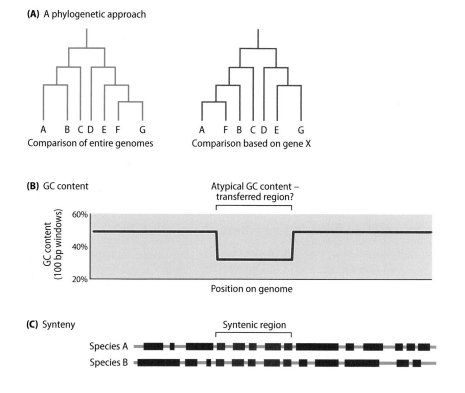

(A) A phylogenetic approach

A   B C D E   F   G
Comparison of entire genomes

A   F B C D E   G
Comparison based on gene X

(B) GC content

Atypical GC content – transferred region?

60%
40%
20%

GC content (100 bp windows)

Position on genome

(C) Synteny

Syntenic region

Species A
Species B

Figure 8.14 Three ways of detecting genes that might have participated in horizontal gene transfer. (A) A phylogenetic approach. In this example, gene X has recently been transferred between species A and F. As a result, a phylogenetic tree constructed from the sequence of gene X shows species A and F to be closely related, whereas when their entire genome sequences are compared, these two species are shown to be unrelated. (B) The GC content for species A is shown, measured in 100 bp windows along the length of the genome. A region where the GC content is atypical of the genome as a whole indicates a segment that might have been transferred from a different species. (C) Alignment between the genome sequences of species A and B reveal a region of synteny, which might indicate a horizontal gene transfer event between these two species.

but variable when different genomes are compared. For example, The *E. coli* genome has a GC content of 51.3%, a bias toward CCG as the codon for proline, and uses leucine, alanine and glycine as the three most frequent amino acids in its encoded proteins. In contrast, the GC content of the *Staphylococcus aureus* genome is 33.3%, CCG is rarely used as a proline codon, and the commonest amino acids are leucine, isoleucine and lysine. If a segment of a genome sequence has atypical compositional features, then it is possible that it has been obtained by horizontal gene transfer, and comparisons with the compositional data for other genomes might indicate the donor species.

- Synteny between segments of two genomes might indicate regions that have participated in horizontal gene transfer. Synteny is the conservation of gene order between two species (Section 5.1). The genomes of two unrelated prokaryotes are expected to have little if any synteny, so if a syntenic region is detected then it is likely that horizontal gene transfer has occurred between those species.

Using these analyses, many possible examples of horizontal gene transfer have been identified, in both bacteria and archaea. Based on the compositional approach, it has been estimated that almost 10% of *E. coli* genes have been acquired in this way, and in other species the number is substantially higher: the plant pathogen *Xylella fastidiosa* appears to have obtained one-fifth of its genes by transfer from other species. One outcome of these discoveries has been a greater focus on the processes by which bacteria take up DNA from the environment, and the realization that DNA uptake is not only more widespread than originally thought, but that many species have proteins in their cell membranes whose specific role is to capture DNA fragments from the environment and to transport those fragments into the cell.

Although extensive horizontal gene transfer has been detected among prokaryotes, including transfers between bacteria and archaea, the impact of these events on the evolution of individual species is still not understood. It has been suggested that almost 25% of the 1895 genes of the thermophilic bacterium *Thermotoga maritima* were obtained from archaea, these genes possibly forming the basis for this bacterium's ability to tolerate high temperatures. Transfer of approximately 1000 genes from bacteria to an anaerobic ancestor of the Haloarchaea might have enabled this genus to evolve a tolerance to oxygen and adopt an aerobic lifestyle, albeit as extremophiles in brine pools and other high salt environments. Other possible examples of metabolic innovation via horizontal gene transfer include a variation of the acetyl-CoA synthesis pathway in *Methanosarcina*, by transfer of two genes from a cellulose-degrading *Clostridium* species, and the ability of *Thermosipho* to synthesize vitamin B12 from glutamate, resulting from transfer of 31 genes from members of the Fimicutes. However, there are debates about the timing of these gene transfer events, especially in those cases where the new capability of the recipient species is dependent on the simultaneous acquisition of multiple genes. In many cases, the available data make it equally or more likely that the genes were transferred individually or in small numbers over a long period of time, indicating that horizontal gene transfer was not directly responsible for the biochemical innovation.

Metagenomes describe the members of a community

In the conventional approach to prokaryotic genomics, a sequencing project is directed at the genome of a single species. A pure culture of that species is prepared, the DNA extracted, and the resulting sequencing reads assembled to give the genome sequence. This approach has one major limitation. For many years, microbiologists have been aware that the artificial culture conditions used to isolate bacteria and archaea from their natural habitats do not suit all species, and that many will not grow under these conditions and hence will remain undetected. If a species cannot be grown in culture, then its genome cannot be sequenced, at least not by the conventional approach.

Metagenomics addresses this problem by obtaining DNA sequences from all the genomes in a particular habitat, for example, from seawater or from soil. DNA is prepared directly from the environmental sample without any attempt to isolate individual species. The resulting sequence reads therefore derive from many different genomes, including those of species that cannot be cultured. Assembling the vast mixture of reads into individual genome sequences is a challenge, but is possible providing that a sufficiently large number of reads are obtained, unless the sample is extremely complex in terms of the numbers of species that are present. As well as enabling the species present in a particular habitat to be identified, the relative abundances of each one can be assessed from the relative numbers of reads that are obtained for each genome. The resulting genome assemblies include species that no microbiologist has ever seen, and which are known to science only from their genome sequences. The metabolic capabilities of an unknown species can, however, be inferred from its genome sequence, and an assessment made of the contribution that the species makes to its ecosystem, for example, in cycling of nutrients.

In one of the first metagenomic studies, over a megabase of sequence was obtained from bacterial DNA from 1500 liters of surface water from the Sargasso Sea. The sequence included segments of the genomes of over 1800 species, of which 148 were totally new. Similar studies have been carried out with samples from sites that have become contaminated with petroleum or acid mine drainage, to assess how the microbial community responds to and helps remediate a polluted environment, and from agriculture soils with the aim of understanding how microbial activity influences the growth and productivity of crops. But the greatest efforts in metagenomics are being applied to studies of the human **microbiome**. These are the microorganisms that live on or within the human body. It is estimated that the entire microbiome of a healthy adult includes at least 1000 different species, with many of these present in the gut. Most of the species are harmless, pathogens making a significant contribution to the microbiome only when an individual has a specific infection. For many years, the microbiome has been looked on as unimportant, but increasing evidence is suggesting that at least some of the species carry out useful activities. In the digestive tract, it appears that bacteria break down some types of carbohydrate into metabolites that can be further digested by the intestinal cells. Without the bacterial activity, the human host could not use these carbohydrates as nutrients. The aims of the various metagenomic studies that are being made of the human microbiome are to catalog the genera that are present in the different parts of the human ecosystem (e.g., the gut, the respiratory tract, the genitourinary tract, the skin), to establish the extent to which these catalogs vary in different people and in different parts of the world, and to understand how the microbiome influences human health and changes in response to disease.

8.3 EUKARYOTIC ORGANELLE GENOMES

Now we return to the eukaryotic world to examine the genomes present in mitochondria and chloroplasts. The possibility that some genes might be located outside the nucleus – **extrachromosomal genes** as they were initially called – was first raised in the 1950s as a means of explaining the unusual inheritance patterns of certain genes in the fungus *Neurospora crassa*, the yeast *Saccharomyces cerevisiae*, and the photosynthetic alga *Chlamydomonas reinhardtii*. Electron microscopy and biochemical studies at about the same time provided hints that DNA molecules might be present in mitochondria and chloroplasts. Eventually, in the early 1960s, these various lines of evidence were brought together and the existence of **mitochondrial** and **chloroplast genomes**, independent of and distinct from the eukaryotic nuclear genome, was accepted.

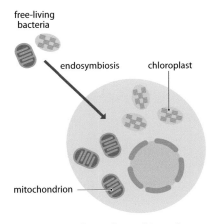

Figure 8.15 The endosymbiont theory. According to this theory, mitochondria and chloroplasts are the relics of free-living bacteria that formed a symbiotic association with the precursor of the eukaryotic cell.

The endosymbiont theory explains the origin of organelle genomes

The discovery of organelle genomes led to many speculations about their origins. Today most biologists subscribe to the **endosymbiont theory**, at least in outline, even though it was considered quite unorthodox when first proposed in the 1960s. The endosymbiont theory is based on the observation that the gene expression processes occurring in organelles are similar in many respects to equivalent processes in bacteria. In addition, when nucleotide sequences are compared, organelle genes are found to be more similar to equivalent genes from bacteria than they are to eukaryotic nuclear genes. The endosymbiont theory therefore holds that mitochondria and chloroplasts are the relics of free-living bacteria that formed a symbiotic association with the precursor of the eukaryotic cell, way back at the very earliest stages of evolution, at least 1.1 billion years ago (**Figure 8.15**). Detailed comparisons between chloroplasts and the photosynthetic prokaryotes alive today indicate that the chloroplast precursor was a type of cyanobacterium, with the closest living relative being *Gloeomargarita lithophora*, which inhabits low-salt environments. The living species most similar to mitochondria are members of two related orders: the Rickettsiales, which are obligate intracellular parasites, and the Pelagibacterales, which are free-living marine bacteria. However, the apparent similarity between mitochondrial genomes and those of the two bacterial groups might be an artifact due to the small sizes of these genomes. Both bacterial groups have undergone **reductive evolution**, a process that involves extensive gene loss and a decrease in the lengths of intergenic regions so that the genome becomes as compact as possible. For the Rickettsiales, this genome reduction reflects their parasitic lifestyle and ability to obtain some nutrients from the host cells. For the free-living Pelagibacterales, the reduction results in **genome streamlining**, which may confer a reproductive advantage compared to the non-streamlined competitors in the marine environment. Mitochondrial genomes are also small and compact, and may resemble those of the Rickettsiales and Pelagibacterales simply because the latter have undergone reductive evolution, rather than because of any evolutionary relationship.

Support for the endosymbiont theory has come from the discovery of organisms which appear to exhibit stages of endosymbiosis that are less advanced than seen with mitochondria and chloroplasts. For example, the group of algae called the glaucophytes possess photosynthetic structures, called **cyanelles**, that are different from chloroplasts and instead resemble ingested cyanobacteria (**Figure 8.16**). Each cyanelle has an external layer of peptidoglycan, thought to be the remnants of the cyanobacterial cell wall, and their light-harvesting proteins resemble those used in free-living cyanobacteria rather than the equivalent structures present in chloroplasts. Modern-day precursors of the mitochondrial endosymbiosis have been more difficult to find, but one possibility is *Pelomyxa*, a type of anaerobic amoeba that lacks mitochondria but instead contains at least three types of symbiotic bacteria. However, the analogy between these bacteria and mitochondrial precursors is questionable as it is by no means certain that the bacteria provide the amoeba with energy, and instead their role might be to convert oxygen to methane and thereby create a low-oxygen microenvironment within which the anaerobic amoeba can survive. Also, phylogenetic studies suggest that *Pelomyxa* is descended from species that did possess mitochondria, so rather than being a primitive type of eukaryotic cell *Pelomyxa* should be looked on as a eukaryote that has lost its mitochondria.

If mitochondria and chloroplasts were once free-living bacteria, then since the endosymbiosis was set up, there must have been a transfer of genes from the organelle into the nucleus. We do not understand how this occurred, or indeed whether there was a mass transfer of many genes at once, or a gradual trickle from one site to the other. But we do know that DNA transfer from organelle to nucleus, and indeed between organelles, still occurs. This was discovered in the early 1980s, when the first partial sequences of chloroplast genomes were obtained. It was found that in some plants the chloroplast genome contains

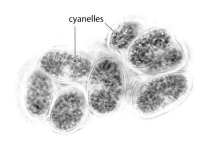

Figure 8.16 Cyanelles in cells of the glaucophyte alga *Cyanophora paradoxa*.

segments of DNA, often including entire genes, that are copies of parts of the mitochondrial genome. The implication is that this so-called **promiscuous DNA** has been transferred from one organelle to the other. We now know that this is not the only type of transfer that can occur. The *Arabidopsis* mitochondrial genome contains various segments of nuclear DNA as well as 16 fragments of the chloroplast genome, including six tRNA genes that have retained their activity after transfer to the mitochondrion. The nuclear genome of this plant includes several short segments of the chloroplast and mitochondrial genomes as well as a 270 kb piece of mitochondrial DNA located within the centromeric region of chromosome 2. The transfer of mitochondrial DNA to vertebrate nuclear genomes has also been documented.

As well as the indications that promiscuous DNA provides for the possible transfer of DNA between genomes, there is also one striking example of an endosymbiosis in which the relationship between host and organelle is less developed than is the case with mitochondria and chloroplasts. *Paulinella* is an amoeba with photosynthetic organelles called **chromatophores**. As with cyanelles, photosynthesis in chromatophores resembles the cyanobacterial processes more closely than the equivalent events in chloroplasts. However, chromatophores, unlike cyanelles, retain a miniature version of a cyanobacterium genome, 1.02 Mb in size and specifying 867 protein-coding genes. This is substantially larger than the typical chloroplast or cyanelle genome, which is less than 0.2 Mb and codes for only 200 genes. The process of genome reduction therefore appears to have reached only an intermediate stage in *Paulinella*. Examination of the chromatophore gene catalog shows that the endosymbiont has lost the ability to make amino acids and some other metabolites, the genes for these entire pathways being absent from the chromatophore genome. In contrast, the chromatophore retains genes for synthesis of all of the proteins and enzymes needed to carry out photosynthesis, as well as genes for DNA replication, transcription and translation. The chromatophore therefore has the typical features of a symbiont, being dependent on its host for provision of metabolites that it can no longer make, but remaining autonomous with regards to energy generation and replication and expression of its genome. In contrast, the genome of an organelle has become so reduced in size and gene content that the organelle is unable to generate energy or replicate and express its genome without the aid of proteins and enzymes coded by nuclear genes.

The physical and genetic features of organelle genomes

Almost all eukaryotes have mitochondrial genomes, and most photosynthetic eukaryotes have chloroplast genomes. Initially, it was thought that virtually all organelle genomes were circular DNA molecules. Electron microscopy had revealed both circular and linear DNA in some organelles, but it was assumed that the linear molecules were simply fragments of circular genomes that had become broken during preparation for electron microscopy. We still believe that most mitochondrial and chloroplast genomes are circular, but we now recognize that there is a great deal of variability in different organisms. In many eukaryotes, the circular genomes coexist in the organelles with linear versions and, in the case of chloroplasts, with smaller circles that contain subcomponents of the genome as a whole. The latter pattern reaches its extreme in the marine algae called dinoflagellates, whose chloroplast genomes are split into many small circles, each containing just a single gene. We also now realize that the mitochondrial genomes of some microbial eukaryotes, such as *Paramecium*, *Chlamydomonas* and several yeasts, are always linear.

Copy numbers for organelle genomes are not particularly well understood. Each human mitochondrion contains about ten identical molecules, which means that there are about 8000 per cell, but in *S. cerevisiae* the total number is probably smaller, perhaps less than 100 per cell. Photosynthetic microorganisms such as *Chlamydomonas* have 80–90 chloroplast genomes per organelle and approximately 1000 genomes per cell, about one-tenth the number of

TABLE 8.4 SIZES OF MITOCHONDRIAL AND CHLOROPLAST GENOMES

Species	Type of organism	Genome size (kb)
Mitochondrial genomes		
Plasmodium falciparum	Protozoan (malaria parasite)	6
Chlamydomonas leiostraca	Green alga	14
Mus musculus	Vertebrate (mouse)	16
Homo sapiens	Vertebrate (human)	17
Drosophila melanogaster	Insect (fruit fly)	19
Chondrus crispus	Red alga	26
Aspergillus nidulans	Ascomycete fungus	33
Reclinomonas americana	Protozoa	69
Saccharomyces cerevisiae	Yeast	86
Brassica oleracea	Flowering plant (cabbage)	360
Arabidopsis thaliana	Flowering plant (vetch)	367
Zea mays	Flowering plant (maize)	681
Cucumis sativus	Flowering plant (cucumber)	1556
Chloroplast genomes		
Bigelowiella natans	Chlorarachniophytes alga	69
Marchantia polymorpha	Liverwort	121
Pisum sativum	Flowering plant (pea)	122
Oryza sativa	Flowering plant (rice)	135
Nicotiana tabacum	Flowering plant (tobacco)	156
Pelargonium spinosum	Flowering plant	217
Floydiella terrestris	Green alga	521
Haematococcus lacustris	Green alga	1350

Data taken from NCBI Genome.

chloroplast genomes present in a leaf cell from a higher plant. One mystery, which dates back to the 1950s and has never been satisfactorily solved, is that when organelle genes are studied in genetic crosses the results suggest that there is just one copy of a mitochondrial or chloroplast genome per cell. This is clearly not the case but indicates that our understanding of the transmission of organelle genomes from parent to offspring is less than perfect.

Mitochondrial genome sizes are variable (**Table 8.4**) and are unrelated to the complexity of the organism. Most multicellular animals have small mitochondrial genomes with a compact genetic organization, the genes being close together with little space between them. The human mitochondrial genome (**Figure 8.17**), at 16,569 bp, is typical of this type. Most lower eukaryotes, such as *S. cerevisiae* (**Figure 8.18**), as well as flowering plants, have larger and less compact mitochondrial genomes, with a number of the genes containing introns. Chloroplast genomes are less variable in size (Table 8.4) and most have a structure similar to that shown in **Figure 8.19** for the rice chloroplast genome.

Because organelle genomes are much smaller than their nuclear counterparts, we anticipate that their gene contents are much more limited, which is indeed the case. Again, mitochondrial genomes display greater variability, with gene contents ranging from three for the malaria parasite *P. falciparum* to 93 for the protozoan *Reclinomonas americana* (Table 8.5). All but the smallest mitochondrial genomes contain genes for the noncoding rRNAs, and all specify at least some of the protein components of the respiratory chain, the latter being the main biochemical feature of the mitochondrion. The more gene-rich genomes also code for tRNAs, ribosomal proteins, and proteins involved in transcription, translation, and transport of other proteins into the mitochondrion from the surrounding cytoplasm. Most chloroplast genomes appear to possess the same set of 200 or so genes, again coding for rRNAs and tRNAs, as well as ribosomal proteins and proteins involved in photosynthesis (**Figure 8.19**).

A general feature of organelle genomes emerges from Table 8.5. These genomes specify some of the proteins found in the organelle, but not all of them. The other proteins are coded by nuclear genes, synthesized in the cytoplasm, and transported into the organelle. If the cell has mechanisms for transporting proteins into mitochondria and chloroplasts, then why not have all the organelle proteins specified by the nuclear genome? We do not yet have a convincing answer to this question, although it has been suggested that at least some of the proteins coded by organelle genomes are extremely hydrophobic and cannot be transported through the membranes that surround mitochondria and chloroplasts, and so simply cannot be moved into the organelle from the cytoplasm. The only way in which the cell can get them into the organelle is to make them there in the first place.

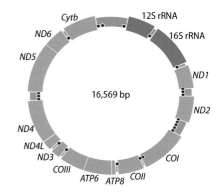

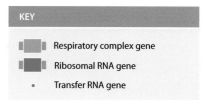

Figure 8.17 The human mitochondrial genome. The human mitochondrial genome is small and compact, with little wasted space – so much so that the *ATP6* and *ATP8* genes overlap. Abbreviations: *ATP6, ATP8*, genes for ATPase subunits 6 and 8; *COI, COII, COIII*, genes for cytochrome *c* oxidase subunits I, II, and III; *Cytb*, gene for apocytochrome *b*; *ND1–ND6*, genes for NADH dehydrogenase subunits 1–6.

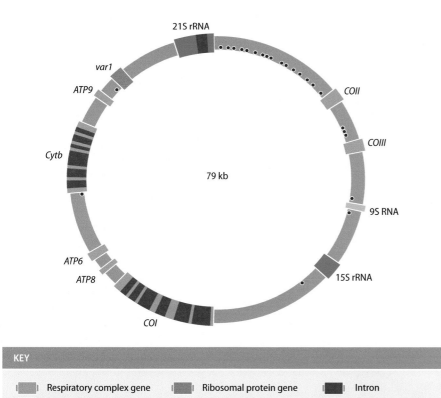

Figure 8.18 The *Saccharomyces cerevisiae* mitochondrial genome. In the yeast mitochondrial genome, the genes are more widely spaced than in the human mitochondrial genome and some of the genes have introns. This type of organization is typical of many lower eukaryotes and plants. Abbreviations: *ATP6, ATP8, ATP9*, genes for ATPase subunits 6, 8, and 9; *COI, COII, COIII*, genes for cytochrome *c* oxidase subunits I, II, and III; *Cytb*, gene for apocytochrome *b*; *var1*, gene for a ribosome-associated protein. The 9S RNA gene specifies the RNA component of the enzyme ribonuclease P. There are also open reading frames located within some of the introns; these ORFs code for proteins involved in the splicing pathway that removes these introns from the pre-mRNAs.

TABLE 8.5 GENE CONTENTS OF MITOCHONDRIAL GENOMES

Feature	Plasmodium falciparum	Chlamydomonas leiostraca	Homo sapiens	Saccharomyces cerevisiae	Arabidopsis thaliana	Reclinomonas americana
Protein-coding genes	3	7	13	19	33	67
Respiratory complex	3	7	13	7	19	25
Ribosomal proteins	0	0	0	1	7	27
Transport proteins	0	0	0	0	1	6
RNA polymerase	0	0	0	0	0	4
Translation factor	0	0	0	0	0	1
Other	0	1	0	11	6	4
Noncoding RNA genes	0	12	24	27	25	30
Ribosomal RNA genes	0	9	2	2	3	4
Transfer RNA genes	0	3	22	24	22	26
Other RNA genes	0	0	0	1	0	0
Total number of genes	3	19	37	46	58	93
Genome size	6 kb	14 kb	17 kb	86 kb	367 kb	69 kb

Data taken from NCBI Genome.

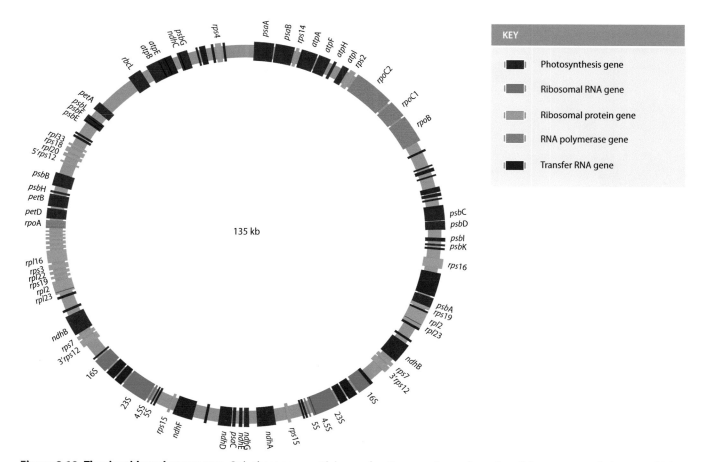

Figure 8.19 The rice chloroplast genome. Only those genes with known functions are shown. A number of the genes contain introns, which are not indicated on this map. These discontinuous genes include several of those for tRNAs, which is why the tRNA genes are of different lengths even though the tRNAs that they specify are all of similar size.

SUMMARY

- Prokaryotes comprise two distinct types of organism, the bacteria and the archaea.

- The bacterial genome is localized within the nucleoid, the lightly staining region of the otherwise featureless prokaryotic cell. The DNA is attached to a core of binding proteins from which supercoiled loops radiate out into the cell.

- The *E. coli* genome is a single, circular DNA molecule, but some prokaryotes have linear genomes, and some have multipartite genomes made up of two or more circular and/or linear molecules. In the more complex cases, it can be difficult to distinguish which molecules are genuine parts of the genome and which are dispensable plasmids.

- Prokaryotic genomes are very compact, with little repetitive DNA.

- Many genes are organized into operons, the members of which are expressed together and usually have a functional relationship.

- Gene number is related to biological complexity. The largest genomes belong to free-living species found in the soil, and the smallest genomes belong to species that are obligate parasites.

- Genome sizes and gene numbers vary within individual prokaryotic species. The core genome is the set of genes possessed by all members of the species, and the accessory genome is the collection of additional genes present in different strains and isolates of that species.

- Many examples of horizontal gene transfer between different prokaryotic species have been identified, including some between bacteria and archaea.

- Metagenomics, the study of all the genomes in a habitat such as seawater, shows that a substantial proportion of the species that are present have never been identified.

- The genomes in the mitochondria and chloroplasts of eukaryotic cells are descended from free-living bacteria that formed a symbiotic association with the precursor of the eukaryotic cell, and hence these genomes have prokaryotic features.

- Most mitochondrial and chloroplast genomes are circular, possibly multipartite, with copy numbers of several thousand per cell.

- Mitochondrial genomes vary in size from 5 to 1500 kb and contain 3–100 genes, including genes for mitochondrial rRNAs, tRNAs, and proteins such as components of the respiratory complex.

- Most chloroplast genomes are 60–525 kb with a similar set of some 200 genes, the majority coding for functional RNAs and photosynthetic proteins.

SHORT ANSWER QUESTIONS

1. Outline the differences between a eukaryotic chromosome and the *E. coli* chromosome.

2. What experimental evidence suggests that the *E. coli* chromosome is organized into supercoiled domains and is attached to proteins that restrict its ability to relax?

3. What similarities, if any, are there between *E. coli* nucleoid-associated proteins and eukaryotic histone proteins?

4. The *E. coli* genome is a single, circular DNA molecule. What other types of genome structure are found among prokaryotes?

5. Describe how the genes and other sequence features are organized in a typical prokaryotic genome. When prokaryotic and mammalian genomes are compared, what differences are seen in the gene density, number of introns, and repetitive DNA content?

6. List the key features of operons and assess the overall importance of the operon as a component of prokaryotic genomic organization.

7. Discuss the factors that influence the number of genes possessed by a prokaryote.

8. Distinguish between the terms core genome and accessory genome.

9. What impact has horizontal gene transfer had on the gene content of prokaryotic genomes?

10. Describe the novel information on biology that has been obtained by metagenomic studies of environments such as seawater.

11. Outline the key features of the endosymbiont theory for the origin of mitochondria and chloroplasts.

12. Compare the gene contents of the mitochondria and chloroplasts of different species.

IN-DEPTH PROBLEMS

1. Should the traditional view of the prokaryotic genome as a single, circular DNA molecule be abandoned? If so, what new definition of the prokaryotic genome should be adopted?

2. Speculate on the identities of the approximately 230 genes that constitute the minimum set for a free-living cell.

3. Can the concept of prokaryotic species survive the discoveries that are being made by genome sequencing?

4. Is a definitive test of the endosymbiont theory possible?

5. Why do organelle genomes exist?

FURTHER READING

Prokaryotic nucleoids

Anuchin, A.M., Goncharenko, A.V., Demidenok, O.I. and Kaprelyants, A.S. (2011). Histone-like proteins of bacteria. *Appl. Biochem. Microbiol.* 47:580–585.

Dame, R.T., Rashid, F.-Z.M., and Grainger, D.C. (2020). Chromosome organization in bacteria: Mechanistic insights into genome structure and function. *Nat. Rev. Genet.* 21:227–242.

Dillon, S. and Dorman, C.J. (2010). Bacterial nucleoid-associated proteins, nucleoid structure and gene expression. *Nat. Rev. Microbiol.* 8:185–195.

Laursen, S.P., Bowerman, S. and Luger, K. (2021. Archaea: The final frontier of chromatin. *J. Mol. Biol.* 433:166791.

Sinden, R.R. and Pettijohn, D.E. (1981). Chromosomes in living *Escherichia coli* cells are segregated into domains of supercoiling. *Proc. Natl Acad. Sci. USA* 78:224–228. *Probing the structure of the nucleoid with trimethylpsoralen.*

Iconic prokaryotic genome sequences

Blattner, F.R., Plunkett, G., Bloch, C.A., et al. (1997). The complete genome sequence of *Escherichia coli* K-12. *Science* 277:1453–1462.

Bult, C.J., White, O., Olsen, G.J., et al. (1996). Complete genome sequence of the methanogenic archaeon *Methanococcus jannaschii. Science* 273:1058–1073.

Fraser, C.M., Casjens, S., Huang, W.M., et al. (1997). Genomic sequence of a Lyme disease spirochaete, *Borrelia burgdorferi. Nature* 390:580–586.

Heidelberg, J.F., Eisen, J.A., Nelson, W.C., et al. (2000). DNA sequence of both chromosomes of the cholera pathogen *Vibrio cholerae*. *Nature* 406:477–483.

Parkhill, J., Achtman, M., James, K.D., et al. (2000). Complete DNA sequence of a serogroup A strain of *Neisseria meningitidis* Z2491. *Nature* 404:502–506.

White, O., Eisen, J.A., Heidelberg, J.F., et al. (1999). Genome sequence of the radioresistant bacterium *Deinococcus radiodurans* R1. *Science* 286:1571–1577.

Prokaryotic gene numbers

Alm, R.A., Ling, L.-S.L., Moir, D.T., et al. (1999). Genomic sequence comparison of two unrelated isolates of the human gastric pathogen *Helicobacter pylori*. *Nature* 397:176–180.

Hutchison, C.A., Chuang, R.-Y., Noskov, V.N., et al. (2016). Design and synthesis of a minimal bacterial genome. *Science* 351:aad6253.

Perna, N.T., Plunkett, G., Burland, V., et al. (2001). Genome sequence of enterohaemorrhagic *Escherichia coli* O157:H7. *Nature* 409:529–533.

Rouli, L., Merhej, V., Fournier, P.E., et al. (2015). The bacterial pangenome as a new tool for analysing pathogenic bacteria. *New Microbes New Infect.* 7:72–85.

Tantaso, E., Eisenhaber, B., Kirsch, M., et al. (2022). To kill or to be killed: Pangenome analysis of *Escherichia coli* strains reveals a tailocin specific for pandemic ST131. *BMC Biol.* 20:146. *Pangenome analysis of 1324 E. coli genomes.*

Tettelin, H., Masignani, V., Cieslewicz, M.J., et al. (2005). Genome analysis of multiple pathogenic isolates of *Streptococcus agalactiae*: Implications for the microbial 'pan-genome'. *Proc. Natl Acad. Sci. USA* 102:13950–13955.

Horizontal gene transfer

Garcia-Vallve, S., Guzman, E., Montero, M.A., et al. (2003). HGT-DB: A database of putative horizontally transferred genes in prokaryotic complete genomes. *Nucl. Acids Res.* 31:187–189.

Groussin, M., Boussau, B., Szöllõsi, G., et al. (2016). Gene acquisitions from bacteria at the origins of major archaeal clades are vastly overestimated. *Mol. Biol. Evol.* 33:305–310. *Discusses some of the difficulties involved in estimating when genes were transferred.*

Mell. J.C. and Redfield, R.J. (2014). Natural competence and the evolution of DNA uptake specificity. *J. Bacteriol.* 196:1471–1483.

Ochman, H., Lawrence, J.G. and Groisman, E.A. (2000). Lateral gene transfer and the nature of bacterial innovation. *Nature* 405:299–304.

Devillya, G., Adato, O. and Snir, S. (2020). Detecting horizontal gene transfer: A probabilistic approach. *BMC Genom.* 21(Suppl 1):106.

Soucy, S.M., Huang, J. and Gogarten, J.P. (2015). Horizontal gene transfer: Building the web of life. *Nat. Rev. Genet.* 16:472–482.

Metagenomics

Chiu, C.Y. and Miller, S.A. (2019). Clinical metagenomics. *Nat. Rev. Genet.* 20:341–355.

Gilbert, J.A., Blaser, M.J., Caporaso, J.G., et al. (2018). Current understanding of the human microbiome. *Nat. Med.* 24:392–400.

Huang, L.-N., Kuang, J.-L. and Shu, W.-S. (2016). Microbial ecology and evolution in the acid mine drainage model system. *Trends. Microbiol.* 24:581–593.

Venter, J.C., Remington, K., Heidelberg, J.F., et al. (2004). Environmental genome shotgun sequencing of the Sargasso Sea. *Science* 304:66–74.

Zhong, C., Chen, C., Wang, L., et al. (2021). Integrating pangenome with metagenome for microbial community profiling. *Comput. Struct. Biotechnol. J.* 19:1458–1466.

Organelle genomes

Boguszewska, K., Szewczuk, M., Kaźmierczak-Barańska, J., et al. (2020). The similarities between human mitochondria and bacteria in the context of structure, genome, and base excision repair system. *Molecule* 25:2857.

Gutiérrez, G., Chistyakova, L.V., Villalobo, E., et al. (2017). Identification of *Pelomyxa palustris* endosymbionts. *Protist* 168:408–424.

Keeling, P.J. and Archibald, J.M. (2008). Organelle evolution: What's in a name? *Curr. Biol.* 18:R345–R347. *Description of Paulinella.*

Margulis, L. (1970). *Origin of Eukaryotic Cells.* Yale University Press, New Haven, Connecticut. *The first description of the endosymbiont theory for the origin of mitochondria and chloroplasts.*

Roger, A.J., Muñoz-Gómez, S.A. and Kamikawa, R. (2017). The origin and diversification of mitochondria. *Curr. Biol.* 27:R1177–R1192.

Sánchez-Baracaldo, P., Raven, J.A., Pisani, D., et al. (2017). Early photosynthetic eukaryotes inhabited low-salinity habitats. *Proc. Natl. Acad. Sci. USA* 114:E7727–E7745. *Identifying the cyanobacterial ancestor of chloroplasts.*

Online resources

Horizontal Gene Transfer Database (HGT-DB). https://usuaris.tinet.cat/debb/HGT/welcomeOLD.html

NCBI Organelle Genomes Resources. https://www.ncbi.nlm.nih.gov/genome/organelle/

ODB (Operon DataBase). http://operondb.jp/

VIRUS GENOMES AND MOBILE GENETIC ELEMENTS

The viruses are the last and simplest form of life whose genomes we will investigate. In fact, viruses are so simple in biological terms that we have to ask ourselves if they can really be thought of as living organisms. Doubts arise partly because viruses are constructed along lines different from all other forms of life – viruses are not cells – and partly because of the nature of the virus life cycle. Viruses are obligate parasites of the most extreme kind: they reproduce only within a host cell, and in order to replicate and express their genomes they must subvert at least part of the host's genetic machinery to their own ends. Some viruses possess genes coding for their own DNA polymerase and RNA polymerase enzymes, but many depend on the host enzymes for genome replication and transcription. All viruses make use of the host's ribosomes and translation apparatus for synthesis of the polypeptides that make up the protein coats of their progeny. This means that virus genes must be matched to the host genetic system. Viruses are therefore quite specific for particular organisms, and individual types cannot infect a broad spectrum of species.

In this chapter, we will also consider the mobile genetic elements that make up a substantial part of the repetitive component of eukaryotic and prokaryotic genomes. We link these elements with virus genomes because it has become clear in recent years that at least some of these repetitive sequences are derived from viruses, and are, in effect, viral genomes that have lost the ability to escape from their host cell.

9.1 THE GENOMES OF BACTERIOPHAGES AND EUKARYOTIC VIRUSES

There are a multitude of different types of virus, but the ones that have received most attention from geneticists are those that infect bacteria. These are called bacteriophages or phages and have been studied in great detail since the 1930s, when the early molecular biologists, notably Max Delbrück, chose phages as convenient model organisms with which to study genes. We will follow the lead taken by Delbrück and use bacteriophages as the starting point for our investigation of viral genomes.

Bacteriophage genomes have diverse structures and organizations

Bacteriophages are constructed from two basic components: protein and nucleic acid. The protein forms a coat, or **capsid**, within which the nucleic acid genome is contained. There are three basic capsid structures (**Figure 9.1**):

- **Icosahedral**, in which the individual polypeptide subunits (**protomers**) are arranged into a three-dimensional geometric structure that surrounds the nucleic acid. Examples are MS2 phage, which infects *Escherichia coli*, and PM2, which infects *Pseudomonas aeruginosa*.

- **Filamentous**, or helical, in which the protomers are arranged in a helix, producing a rod-shaped structure. The *E. coli* phage called M13 is an example.

- **Head-and-tail**, a combination of an icosahedral head, containing the nucleic acid, attached to a filamentous tail and possibly additional structures that facilitate entry of the nucleic acid into the host cell. This is a common structure possessed by, for example, the *E. coli* phages T4 and λ, and phage SPO1 of *Bacillus subtilis*.

DOI: 10.1201/9781003133162-9

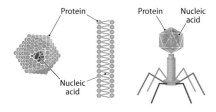

Figure 9.1 The three types of capsid structure commonly displayed by bacteriophages.

The term 'nucleic acid' has to be used when referring to phage genomes because in some cases these molecules are made of RNA. Viruses are the one form of 'life' that contradict the conclusion of Avery and his colleagues and of Hershey and Chase that the genetic material is DNA (Section 1.1). Phages and other viruses also break another rule: their genomes, whether of DNA or RNA, can be single-stranded as well as double-stranded. A whole range of different genome structures is known among the phages, as summarized in **Table 9.1**. With most types of phage there is a single DNA or RNA molecule that comprises the entire genome. However, this is not always the case and a few RNA phages have **segmented genomes**, meaning that their genes are carried by a number of different RNA molecules. The sizes of phage genomes vary enormously, from about 1.6 kb for the smallest phages to over 150 kb for large ones such as T2, T4, and T6.

Bacteriophage genomes, being relatively small, were among the first to be studied comprehensively by the rapid and efficient DNA sequencing methods that were developed in the late 1970s. Gene numbers vary from just four in the case of MS2, to over 200 for the more complex head-and-tail phages (see Table 9.1). The smaller phage genomes of course contain relatively few genes, but these can be organized in a very complex manner. Phage φX174, for example, manages to pack into its genome 'extra' biological information, as several of its genes overlap (**Figure 9.2**). These **overlapping genes** share nucleotide sequences (gene *B*, for example, is contained entirely within gene *A*) but code

TABLE 9.1 FEATURES OF SOME TYPICAL BACTERIOPHAGES AND THEIR GENOMES

Phage	Host	Capsid structure	Genome structure	Genome size (kb)	Number of genes
λ	Enterobacteria	Head-and-tail	Double-stranded linear DNA	48.5	73
M13	Enterobacteria	Filamentous	Single-stranded circular DNA	6.4	10
MS2	Enterobacteria	Icosahedral	Single-stranded linear RNA	3.6	4
φ6	*Pseudomonas*	Icosahedral	Double-stranded segmented linear RNA	2.9, 4.0, 6.4	13
φX174	Enterobacteria	Icosahedral	Single-stranded circular DNA	5.4	11
PM2	*Pseudoalteromonas*	Icosahedral	Double-stranded circular DNA	10.0	22
SPO1	*Bacillus*	Head-and-tail	Double-stranded linear DNA	133	204
T4	Enterobacteria	Head-and-tail	Double-stranded linear DNA	169	278
T7	Enterobacteria	Head-and-tail	Double-stranded linear DNA	39.9	60

Data taken from NCBI Genome. The genome structure is that in the phage capsid; some genomes exist in different forms within the host cell.

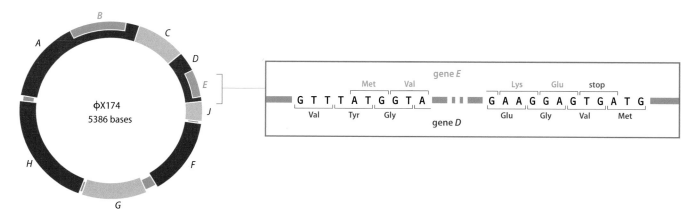

Figure 9.2 The φX174 genome contains overlapping genes. The genome is made of single-stranded DNA. The expanded region shows the start and end of the overlap between genes *E* and *D*. Two other overlapping genes – *A** and *K* – are not shown on this map.

for different gene products, as the transcripts are translated from different start positions and, in most case, in different reading frames. Overlapping genes are not uncommon in viruses. The larger phage genomes contain more genes, reflecting the more complex capsid structures of these phages and a dependence on a greater number of phage-encoded enzymes during the infection cycle. The T4 genome, for example, includes some 40 genes involved solely in construction of the phage capsid. Despite their complexity, even these large phages still require at least some host-encoded proteins and RNAs in order to carry through their infection cycles.

Replication strategies for bacteriophage genomes

Bacteriophages are classified into two groups according to their life cycle: **lytic** and **lysogenic**. The fundamental difference between these groups is that a lytic phage kills its host bacterium very soon after the initial infection, whereas a lysogenic phage can remain quiescent within its host for a substantial period of time, even throughout numerous generations of the host cell. These two life cycles are typified by two *E. coli* phages: the lytic (or **virulent**) T4 and the lysogenic (or **temperate**) λ.

The T series of *E. coli* phages (T1 to T7) were the first to become available to molecular geneticists and have been the subject of much study. Their lytic infection cycle was first investigated in 1939 by Emory Ellis and Max Delbrück, who added T4 phages to a culture of *E. coli*, waited three minutes for the phages to attach to the bacteria, and then measured the number of infected cells over a period of 60 minutes. Their results (**Figure 9.3A**) showed that there is no change in the number of infected cells during the first 22 minutes of infection, this **latent period** being the time needed for the phages to reproduce within their hosts. After 22 minutes, the number of infected cells started to increase, showing that lysis of the original hosts had occurred and that the new phages that had been produced were now infecting other cells in the culture. The molecular events occurring at the different stages of this **one-step growth curve** are shown in **Figure 9.3B**. The initial event is attachment of the phage particle to a receptor protein on the outside of the bacterium. Different types of phage have different receptors: for T4, the receptor is a protein called OmpC ('Omp' stands for 'outer membrane protein'), which is a type of porin, the proteins that form channels through the outer cell membrane and facilitate the uptake of nutrients. After attachment, the phage injects its DNA genome into the cell through its tail structure. Immediately after entry of the phage DNA, the synthesis of host DNA, RNA, and protein stops and transcription of the phage genome begins. Within five minutes, the bacterial DNA molecule has depolymerized, and the resulting nucleotides are being utilized in replication of the T4 genome. After 12 minutes, new phage capsid proteins start to appear and the first complete phage particles are assembled. Finally, at the end of the latent period, the cell bursts and the new phages are released. A typical infection cycle produces 200 to 300 T4 phages per cell, all of which can go on to infect other bacteria.

Most phages can follow the lytic infection cycle, but some, such as λ, can also pursue a lysogenic cycle. In Section 2.3, when we looked at the use of λ phages as cloning vectors, we discovered that during a lysogenic cycle the phage genome becomes integrated into the host DNA. This occurs immediately after entry of the phage DNA into the cell, and results in a quiescent form of the bacteriophage, called the **prophage** (**Figure 9.4A**). Integration occurs by **site-specific recombination** (Section 16.2) between identical 15 bp sequences present in the λ and *E. coli* genomes. Note that this means that the λ genome always integrates at the same position within the *E. coli* DNA molecule. The integrated prophage can be retained in the host DNA molecule for many cell generations, being replicated along with the bacterial genome and passed with it to the daughter cells. However, the switch to the lytic mode of infection occurs if the prophage is **induced** by any one of several chemical or physical stimuli. Each of these appears to be linked to DNA damage, and possibly therefore signals the imminent death of the host by natural causes. In response to these stimuli, a second recombination event excises the phage genome from the host DNA, phage DNA replication

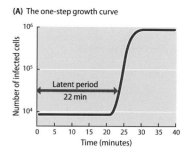

(A) The one-step growth curve

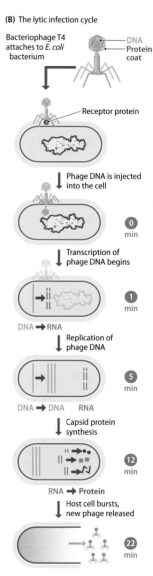

(B) The lytic infection cycle

Figure 9.3 The lytic infection cycle. (A) The one-step growth curve, as revealed by the experiment conducted by Ellis and Delbrück. (B) The molecular events occurring during the lytic infection cycle.

Figure 9.4   The lysogenic infection cycle, as followed by bacteriophage λ. After induction, the infection cycle is similar to the lytic mode.

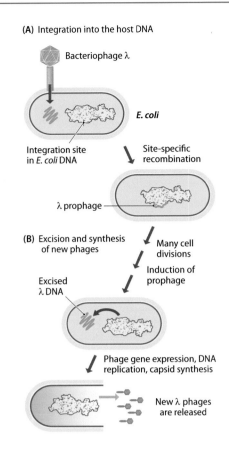

(A) Integration into the host DNA

Bacteriophage λ

E. coli

Integration site in *E. coli* DNA

Site-specific recombination

λ prophage

(B) Excision and synthesis of new phages

Many cell divisions

Induction of prophage

Excised λ DNA

Phage gene expression, DNA replication, capsid synthesis

New λ phages are released

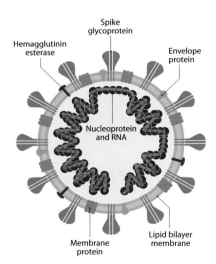

Spike glycoprotein

Hemagglutinin esterase

Envelope protein

Nucleoprotein and RNA

Membrane protein

Lipid bilayer membrane

Figure 9.5   The structure of the COVID-19 virus. COVID-19 is caused by the severe acute respiratory syndrome coronavirus 2 (SARS-CoV-2), which displays many of the typical features of a eukaryotic virus. The capsid is surrounded by a lipid membrane to which additional viral proteins are attached, including the spike protein, which attaches to an ACE2 cell surface protein in order to mediate entry of the virus into a host cell.

begins, and phage coat proteins are synthesized (**Figure 9.4B**). Eventually, the cell bursts and new λ phages are released. Lysogeny adds an additional level of complexity to the phage life cycle, and ensures that the phage is able to adopt the particular infection strategy best suited to the prevailing conditions.

Structures and replication strategies for eukaryotic viral genomes

Now we move on to consider those viruses that infect eukaryotic cells. The capsids of eukaryotic viruses are either icosahedral or filamentous: the head-and-tail structure is unique to bacteriophages. One distinct feature of eukaryotic viruses, especially those with animal hosts, is that the capsid may be surrounded by a lipid membrane, forming an additional component of the virus structure (**Figure 9.5**). This membrane is derived from the host when the new virus particle leaves the cell, and may subsequently be modified by insertion of virus-specific proteins.

Eukaryotic viral genomes display a great variety of structures (**Table 9.2**). They may be DNA or RNA, single- or double-stranded (or partly double-stranded with single-stranded regions), linear or circular, segmented or nonsegmented. For reasons that no one has ever understood, the vast majority of plant viruses have RNA genomes. Genome sizes cover approximately the same range as seen with phages, although the largest viral genomes (e.g., vaccinia virus at 195 kb) are rather bigger than the largest phage genomes.

Although most eukaryotic viruses follow only the lytic infection cycle, few take over the host cell's genetic machinery to the extent that a bacteriophage does. Many viruses coexist with their host cells for long periods, possibly years, with the host cell functions ceasing only toward the end of the infection cycle, when the virus progeny that have been stored in the cell are released. Other viruses continually synthesize new virus particles that are extruded from the cell. These long-term infections can occur even if the viral genome does not integrate

TABLE 9.2 FEATURES OF SOME TYPICAL EUKARYOTIC VIRUSES AND THEIR GENOMES

Virus	Host	Genome structure
Adenovirus	Vertebrates	Double-stranded linear DNA
Anellovirus	Vertebrates	Single-stranded circular DNA
Coronavirus	Mammals, birds	Single-stranded linear RNA
Hepatovirus A	Vertebrates	Single-stranded linear RNA
Influenza A virus	Mammals, birds	Single-stranded segmented linear RNA
Parvovirus	Mammals	Single-stranded linear DNA
Poliovirus	Humans	Single-stranded linear RNA
Reovirus	Wide host range	Double-stranded segmented linear RNA
Retroviruses	Mammals, birds	Single-stranded linear RNA
Tobacco mosaic virus	Plants	Single-stranded linear RNA
Vaccinia virus	Mammals	Double-stranded linear DNA

The genome structure is that in the virus capsid; some genomes exist in different forms within the host cell.

into the host DNA. Other eukaryotic viruses have life cycles that are more similar to those of lysogenic bacteriophages. A number of DNA and RNA viruses are able to integrate into the genomes of their hosts, sometimes with drastic effects on the host cell.

The **viral retroelements** are examples of integrative eukaryotic viruses. Their replication pathways include a novel step in which an RNA version of the genome is converted into DNA. There are two kinds of viral retroelement: the **retroviruses**, whose capsids contain the RNA version of the genome, and the **pararetroviruses**, whose encapsidated genome is made of DNA. The ability of viral retroelements to convert RNA into DNA was confirmed independently in 1970 by Howard Temin and by David Baltimore. Working with cells infected with retroviruses, both Temin and Baltimore isolated the enzyme, now called **reverse transcriptase**, which is capable of making a DNA copy of an RNA template (and is of immense utility in the experimental study of genomes – Section 2.1). The typical retroviral genome is a single-stranded RNA molecule, 7–12 kb in length. After entry into the cell, the genome is copied into double-stranded DNA by a few molecules of reverse transcriptase that the virus carries in its capsid. The double-stranded version of the genome then integrates into the host DNA (**Figure 9.6**). Unlike λ, the retroviral genome has no sequence similarity with its insertion site

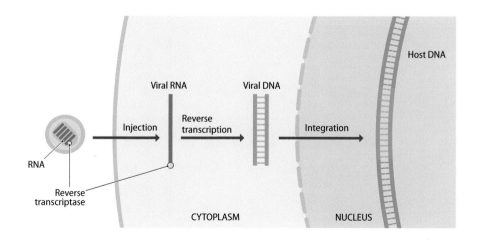

Figure 9.6 Insertion of a retroviral genome into a host chromosome.

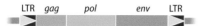

Figure 9.7 A retrovirus genome. Each LTR is a long terminal repeat of 250–1400 bp, which plays an important role in replication of the genome.

and so can integrate at any position in the host chromosomes. Integration is a prerequisite for expression of three retrovirus genes, called *gag*, *pol*, and *env* (**Figure 9.7**), each of which codes for a **polyprotein** that is cleaved, after translation, into two or more functional gene products. These products include the viral coat proteins (from *env*) and the reverse transcriptase (from *pol*). The protein products combine with full-length RNA transcripts of the retroviral genome to produce new virus particles.

The causative agents of HIV/AIDS (human immunodeficiency virus infection and acquired immune deficiency syndrome) were shown to be retroviruses in 1983–1984. The first human immunodeficiency virus was isolated independently by two groups, led by Luc Montagnier and Robert Gallo. This virus is called HIV-1, and is responsible for the most prevalent and pathogenic form of HIV/AIDS. A related virus, HIV-2, discovered by Montagnier in 1985, is less widespread and causes a milder form of the disease. The HIVs attack certain types of lymphocyte in the bloodstream, thereby depressing the immune response of the host. These lymphocytes carry on their surfaces multiple copies of a protein called CD4, which acts as a receptor for the virus. An HIV particle binds to a CD4 protein and then enters the lymphocyte after fusion between its lipid envelope and the cell membrane.

Some retroviruses cause cancer

The human immunodeficiency viruses are not the only retroviruses capable of causing diseases. Several retroviruses induce **cell transformation**, which involves changes in cell morphology and physiology that can lead to cancer. In cell cultures, transformation results in a loss of control over growth, so that transformed cells grow as a disorganized mass, rather than as a monolayer (**Figure 9.8**). In whole animals, cell transformation is thought to underlie the development of tumors.

There appear to be two distinct ways in which retroviruses can cause cell transformation. With some retroviruses, such as the leukemia viruses, cell transformation is a natural consequence of infection, although it may be induced only after a long latent period during which the integrated provirus lies quiescent within the host genome. Other retroviruses cause cell transformation because of abnormalities in their genome structures. These viruses carry cellular genes that they have captured by some undefined process. With at least one transforming retrovirus (Rous sarcoma virus), this cellular gene is in addition to the standard retroviral genes (**Figure 9.9A**). With others, the cellular gene replaces part of the retroviral gene complement (**Figure 9.9B**). In the latter case, the retrovirus may be **defective**, meaning that it is unable to replicate and produce new viruses, as it has lost genes coding for essential replication enzymes and/or capsid proteins. These defective retroviruses are not always inactive, as they can make use of proteins provided by other retroviruses in the same cell (**Figure 9.10**).

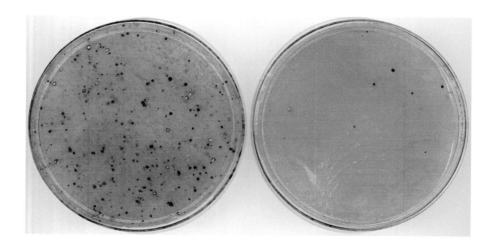

Figure 9.8 Transformation of cultured avian cells. In the Petri dish on the right, normal avian fibroblasts are growing in a monolayer. On the left, cells have been transformed. Piled-up cells forming micro-tumors can be seen, showing that some of the processes that normally control cell growth have been disrupted. (Courtesy of Klaus Bister and Markus Hartl, University of Innsbruck.)

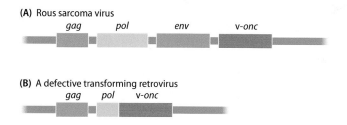

Figure 9.9 Transforming retrovirus genomes. (A) Rous sarcoma virus, which has a full complement of virus genes. (B) A defective transforming retrovirus.

The ability of a transforming retrovirus to cause cell transformation lies with the nature of the cellular gene that has been captured. Often this captured gene (called a v-*onc*, with 'onc' standing for **oncogene**) codes for a protein involved in cell proliferation. The normal cellular version of the gene is subject to strict regulation and is expressed only in limited quantities when needed. It is thought that expression of the v-*onc* follows a different, less controlled pattern, either because of changes in the structure of the gene or because of the influence of expression signals within the retrovirus. One result of this altered expression pattern could be a loss of control over cell division, leading to the transformed state.

Genomes at the edge of life

Viruses occupy the boundary between the living and nonliving worlds. At the very edge of this boundary – or perhaps beyond it – reside a variety of nucleic acid molecules that might or might not be classified as genomes. The **satellite RNAs** or **virusoids** are examples. These are RNA molecules, some 320–400 nucleotides in length, which do not encode any proteins, instead moving from cell to cell within the capsids of helper viruses. The distinction between the two groups is that a satellite virus shares the capsid with the genome of the helper virus whereas a virusoid RNA molecule becomes encapsidated on its own. They are generally looked on as parasites of their helper viruses, although there appear to be at least a few cases where the helper cannot replicate without the satellite RNA or virusoid, suggesting that at least some of the relationships are symbiotic. Satellite RNAs and virusoids are both found predominantly in plants, as is a more extreme group called the **viroids**. These are RNA molecules, 240–475 nucleotides in length, which contain no genes and never become encapsidated, spreading from cell to cell as naked RNA. They include some economically important pathogens, such as the citrus exocortis viroid, which reduces the growth of citrus fruit trees. Viroid and virusoid molecules are circular and single-stranded and are replicated by enzymes coded by the host or helper virus genomes. The replication process results in a series of RNAs joined head to tail which, with some

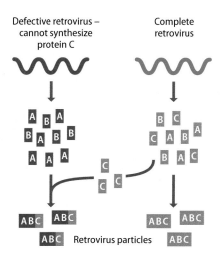

Figure 9.10 A defective retrovirus may be able to give rise to infective virus particles if it shares the cell with a nondefective retrovirus. The nondefective retrovirus acts as a 'helper', providing the proteins that the defective virus is unable to synthesize.

(A) Self-catalyzed cleavage of viroid and virusoid RNAs

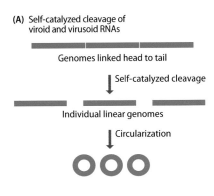

Genomes linked head to tail

Self-catalyzed cleavage

Individual linear genomes

Circularization

(B) The cleavage structure

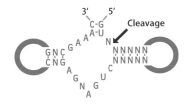

Cleavage

Figure 9.11 Self-catalyzed cleavage of linked genomes during replication of viroids and virusoids. (A) The replication pathway. (B) The **hammerhead** structure, which forms at each cleavage site and which has enzymatic activity. N, any nucleotide.

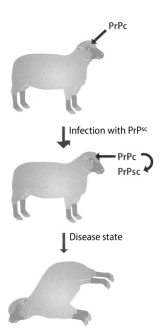

PrPc

Infection with PrP^{SC}

PrPc
PrPsc

Disease state

Figure 9.12 The mode of action of a prion. A normal, healthy sheep has PrP^C proteins in its brain. Infection with PrP^{SC} molecules leads to conversion of newly synthesized PrP^C proteins into PrP^{SC}, leading to the disease state – scrapie in sheep.

viroids and virusoids, are cleaved by a self-catalyzed reaction in which the RNA molecule acts as an RNA enzyme or **ribozyme** (**Figure 9.11**).

Nucleic acid molecules that replicate within plant cells can perhaps be looked on as genomes even if they contain no genes. The same cannot be said for **prions**, as these infectious, disease-causing particles contain no nucleic acid. Prions are responsible for scrapie in sheep and goats, and their transmission to cattle has led to the new disease called BSE – bovine spongiform encephalopathy. Whether their further transmission to humans causes a variant form of Creutzfeldt–Jakob disease (CJD) is controversial but accepted by many biologists. At first, prions were thought to be viruses, but it is now clear that they are made solely of protein. The normal version of the prion protein, called PrP^C, is coded by a mammalian nuclear gene and synthesized in the brain, although its function is unknown. PrP^C is easily digested by proteases, whereas the infectious version, PrP^{SC}, has a more highly β-sheeted structure that is resistant to proteases and forms fibrillar aggregates that are seen in infected tissues. Once inside a cell, PrP^{SC} molecules are able to convert newly synthesized PrP^C proteins into the infectious form, by a mechanism that is not yet understood, resulting in the disease state. Transfer of one or more of these PrP^{SC} proteins to a new animal results in accumulation of new PrP^{SC} proteins in the brain of that animal, transmitting the disease (**Figure 9.12**). Infectious proteins with similar properties are known in lower eukaryotes, examples being the Ure3 and Psi$^+$ prions of *Saccharomyces cerevisiae*. It is clear, however, that prions are gene products rather than genetic material and despite their infectious properties, which led to the initial confusion regarding their status, they are unrelated to viruses or to subviral particles such as viroids and virusoids.

9.2 MOBILE GENETIC ELEMENTS

In Chapters 7 and 8, we learned that eukaryotic genomes, and to a lesser extent those of prokaryotes, contain interspersed repeats, some with copy numbers of several thousand per genome, with the individual repeat units distributed in an apparently random fashion. For many interspersed repeats, a genome-wide distribution pattern is set up by **transposition**, the process by which a segment of DNA can move from one position to another in a genome. These movable segments are called **transposable elements**, or **transposons**. Some types move by a **conservative** process, which involves the excision of the sequence from its original position followed by its reinsertion elsewhere. Conservative transposition therefore results in the transposon simply changing its position in the genome without increasing its copy number (**Figure 9.13**). **Replicative transposition**, on the other hand, results in an increase in copy number, because during this process the original element remains in place while a copy is inserted at the new position. This replicative process can therefore lead to a proliferation of the transposon at interspersed positions around the genome.

Both types of transposition involve recombination, and we will therefore deal with the details of the processes when we study recombination and related types of genome rearrangement in Section 16.3. What interests us here is the variety of structures displayed by the transposable elements found in eukaryotic and prokaryotic genomes, and the link that exists between these elements and viral genomes.

RNA transposons with long terminal repeats are related to viral retroelements

Replicative transposons can be subdivided into those that transpose via an RNA intermediate and those that do not. The process that involves an RNA intermediate, which is called **retrotransposition**, begins with synthesis of an RNA copy of the **retrotransposon** by the normal process of transcription (**Figure 9.14**). The transcript is then copied into double-stranded DNA, which initially exists as an independent molecule outside of the genome. Finally, the DNA copy of

Figure 9.13 Conservative and replicative transposition.

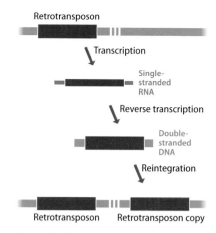

Figure 9.14 Retrotransposition.

the transposon integrates into the genome, possibly back into the same chromosome occupied by the original unit, or possibly into a different chromosome. The end result is that there are now two copies of the transposon, at different points in the genome. If we compare the mechanism for retrotransposition with that for replication of a viral retroelement, as shown in Figure 9.6, then we see that the two processes are very similar, the one significant difference being that the RNA molecule that initiates the process is transcribed from an endogenous genomic sequence during retrotransposition and an exogenous viral genome during replication of a viral retroelement. This close similarity alerts us to the relationships that exist between these two types of element.

RNA transposons, or **retroelements**, are common features of eukaryotic genomes but are much less common, and less well studied, in prokaryotes. The eukaryotic versions can be broadly classified into two types: those that possess **long terminal repeats** (**LTRs**) and those that do not. Long terminal repeats, which play a central role in the process by which the RNA copy of an LTR element is reverse transcribed into double-stranded DNA (Section 16.3), are also possessed by viral retroelements (see Figure 9.7). It is now clear that these viruses are one member of a superfamily of elements that also includes endogenous LTR transposons. The first of the endogenous elements to be discovered was the *Ty* sequence of yeast, which is 6.3 kb in length and has a copy number of about 50 in most *Saccharomyces cerevisiae* genomes.

There are several types of *Ty* element in yeast genomes. The most abundant of these, *Ty1*, is similar to the *copia* retroelement of the fruit fly. These elements are therefore now called the *Ty1/copia* family. If we compare the structure of a *Ty1/copia* retroelement with that of a viral retroelement, then we see clear family relationships (**Figure 9.15A and B**). Each *Ty1/copia* element contains two genes, called *TyA* and *TyB* in yeast, which are similar to the *gag* and *pol* genes of a viral retroelement. In particular, *TyB* codes for a polyprotein that includes the reverse transcriptase that plays the central role in transposition of a *Ty1/copia* element. Note, however, that the *Ty1/copia* element lacks an equivalent of the viral *env* gene, the one that codes for the viral coat proteins. This means that *Ty1/copia* retroelements cannot form infectious virus particles and therefore cannot escape from their host cell. They do, however, form virus-like particles (VLPs) consisting of the RNA and DNA copies of the retroelement attached to core proteins derived from the TyA polyprotein. In contrast, the members of a second family of LTR retroelements, called *Ty3/gypsy* (again after the yeast and fruit-fly versions), do have an equivalent of the *env* gene (**Figure 9.15C**), and at least some of these can form infectious viruses. Although classed as endogenous transposons, these infectious versions should be looked upon as viral retroelements.

Yeast genomes also contain 300–400 additional copies of the 330 bp LTRs of *Ty* elements; these solo sequences probably arise from homologous recombination between the two LTRs of a *Ty* element, which could excise the bulk of the element leaving a single LTR (**Figure 9.16**). This excision event is probably unrelated to transposition of a *Ty* element, which occurs by the RNA-mediated process shown in Figure 9.14. The most common of these solo LTRs are called delta sequences, which derive from *Ty1/copia* elements. Sigma elements, which are solo LTRs from *Ty3/gypsy* retrotransposons, have copy numbers of 20–30 per genome.

LTR retroelements make up substantial parts of many eukaryotic genomes, and are particularly abundant in larger plant genomes, especially those of grasses such as maize (see Figure 7.15D). They also make up an important component of

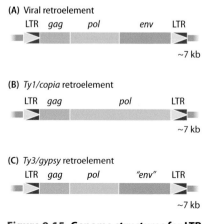

Figure 9.15 Genome structures for LTR retroelements.

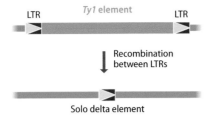

Figure 9.16 Homologous recombination between the LTRs at either end of a *Ty1* element could give rise to a delta sequence.

invertebrate and some vertebrate genomes, but in the genomes of humans and other mammals most of the LTR elements appear to be decayed viral retroelements rather than true transposons. These sequences are called **endogenous retroviruses** (**ERVs**), and they make up 9% of the human genome (Table 9.3). Human ERVs are 6–11 kb in length and have copies of the *gag*, *pol*, and *env* genes. Although most contain mutations or deletions that inactivate one or more of these genes, a few members of the human ERV group HERV-K have functional sequences. By comparing the positions of the HERV-K elements in the genomes of different individuals, it has been inferred that at least some of the HERV-K family are active retrotransposons. There is also evidence that the RNA copies of some HERV-K elements can be packaged into virus-like particles with the ability to move from cell to cell. These discoveries have prompted studies into the possible roles of HERV-K elements in human diseases. HERV-K transcripts and protein products are detectable in the brains of patients suffering from amylotrophic lateral sclerosis (ALS), a neurodegenerative disorder also called Lou Gehrig's disease, after the famous American baseball player who died of ALS in 1941.

Detection of HERV-K products does not prove that these are responsible for ALS, but a possible association is suggested by the demonstration that expression of the HERV-K *env* gene in mouse brains leads to the breakdown of motor neuron function and symptoms similar to ALS. It is also possible that HERV-K elements are involved in susceptibility to autoimmune diseases such as rheumatoid arthritis and in the development of some types of cancer.

Some RNA transposons lack LTRs

Not all types of RNA transposon have LTR elements. In mammals, the most important types of non-LTR retroelements, or **retroposons**, are the **SINEs** (**short interspersed nuclear elements**) and **LINEs** (**long interspersed nuclear elements**). SINEs have the highest copy number for any type of interspersed repetitive DNA in the human genome, with over 1.7 million copies comprising almost 14% of the genome as a whole (Table 9.3). LINEs are less frequent, with about 850,000 copies, but as they are longer they make up a larger fraction of the genome – over 20%. The abundance of SINEs and LINEs in the human genome is underlined by their frequency in the 200 kb segment that we looked at in Section 7.2 (see Figure 7.12).

There are several families of LINEs in the human genome, of which one group, LINE-1, is both the most frequent and the only type that is able to transpose, the

TABLE 9.3 TRANSPOSABLE ELEMENTS IN THE HUMAN GENOME		
Class	**Family**	**Fraction of genome (%)**
LTR retroelements	ERVL	5.8
	ERV1	2.8
	ERVK	0.3
SINE	Alu	10.5
	MIR	2.7
LINE	L1	17.4
	L2	3.5
	CR1	0.4
	RTE	0.1
DNA transposons	hAT	2.1
	TcMar	1.4

Data taken from Hoyt et al. (2022) – see *Further Reading*.

other families being made up of inactive relics. A full-length LINE-1 element is 6.1 kb and has two genes, one of which codes for a polyprotein similar to the product of the viral *pol* gene (**Figure 9.17A**). There are no LTRs, but the 3′ end of the LINE is marked by a series of A–T base pairs, giving what is usually referred to as a poly(A) sequence (though of course it is a poly(T) sequence on the other strand of the DNA). Not all copies of LINE-1 are full length, because the reverse transcriptase coded by LINEs does not always make a complete DNA copy of the initial RNA transcript, meaning that part of the 3′ end of the LINE may be lost. This truncation event is so common that only 80–100 of the 500,000 LINE-1 elements in the human genome are full-length versions, with the average size of all the copies being just 900 bp.

SINEs are much shorter than LINEs, being just 100–400 bp and not containing any genes, which means that SINEs do not make their own reverse transcriptase enzymes (**Figure 9.17B**). Instead, they 'borrow' reverse transcriptases that have been synthesized by LINEs. The commonest SINE in primate genomes is **Alu**, which has a copy number of approximately 1.2 million in humans. An Alu element comprises two halves, each half made up of a similar 120 bp sequence, with a 31–32 bp insertion in the right half (**Figure 9.18**). The mouse genome has a related element, called B1, which is 130 bp in length and equivalent to one half of an Alu sequence. Some Alu elements are actively copied into RNA, providing the opportunity for proliferation of the element. Alu is derived from the gene for the 7SL RNA, a noncoding RNA involved in movement of proteins around the cell. The first Alu element may have arisen by the accidental reverse transcription of a 7SL RNA molecule and integration of the DNA copy into the human genome. Other SINEs are derived from tRNA genes which, like the gene for the 7SL RNA, are transcribed by RNA polymerase III in eukaryotic cells, suggesting that some features of the transcripts synthesized by this polymerase make these molecules prone to occasional conversion into retroposons.

Although transposition of LINEs and SINEs are rare events, LINE-1 transposition has been observed in cultured human and mouse cells and the recent insertions of LINE-1, Alu and other SINEs into protein-coding sequences is thought to have led to gene inactivations that have given rise to inherited human disorders. This was recognized with a small number of hemophilia patients, whose factor VIII gene was disrupted by a LINE-1 sequence that prevented synthesis of this important blood clotting protein. Since this initial discovery, LINE-1 insertion has been implicated as the causative factor in examples of over 25 diseases, including several types of cancer, and increasing numbers of examples involving SINEs are being discovered.

A few non-LTR retroelements are known in prokaryotic genomes, but these have much lower copy numbers than eukaryotic RNA transposons. The prokaryotic versions have a broad distribution among bacteria and archaea, but the distribution is uneven – some strains of *E. coli* possess retroelements and others do not. The commonest type of bacterial retroelement is the **retron**, which is a 2 kb sequence that includes a reverse transcriptase gene. A second part of the retron specifies a 70–80 nucleotide RNA which is copied into DNA by the reverse transcriptase. The 5′ end of this single-stranded DNA then forms a 2′–5′ phosphodiester bond with a guanine nucleotide within the RNA, giving rise to an RNA-DNA hybrid that adopts a base-paired secondary structure. Whether this structure has any function is still being debated, but there is recent evidence that a complex made up of the RNA-DNA hybrid, the reverse transcriptase and an additional, effector protein is able to disrupt the bacteriophage replication cycle, thereby aiding in the defense against phage infection.

DNA transposons are common in prokaryotic genomes

Not all transposons require an RNA intermediate. Those called **DNA transposons** are able to transpose in a more direct DNA-to-DNA manner. DNA transposons are an important component of many prokaryotic genomes. The insertion sequences, IS1 and IS186, present in the 50 kb segment of *E. coli* DNA that we

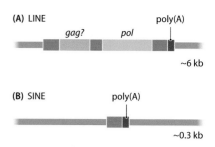

Figure 9.17 Non-LTR retroelements. Both LINEs and SINEs have poly(A) sequences at their 3′ ends.

Figure 9.18 The structure of an Alu element. The element consists of two halves, each of 120 bp, with a 31–32 bp insertion in the right half, and a poly(A) tail at the 3′ end. The two halves (excluding the insertion) have about 85% sequence identity.

Figure 9.19 DNA transposons of prokaryotes. Four types are shown. Insertion sequences, Tn3-type transposons, and transposable phages are flanked by short (<50 bp) inverted terminal repeat (ITR) sequences. The resolvase gene of the Tn3-type transposon codes for a protein involved in the transposition process.

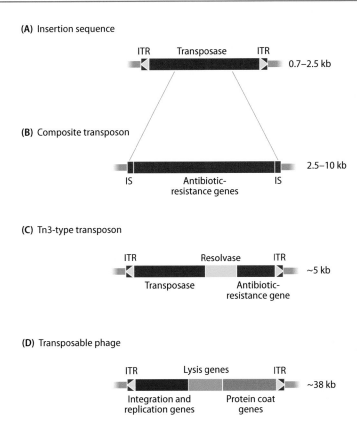

(A) Insertion sequence

ITR   Transposase   ITR

0.7–2.5 kb

(B) Composite transposon

2.5–10 kb

IS   Antibiotic-resistance genes   IS

(C) Tn3-type transposon

ITR   Resolvase   ITR

~5 kb

Transposase   Antibiotic-resistance gene

(D) Transposable phage

ITR   Lysis genes   ITR

~38 kb

Integration and replication genes   Protein coat genes

examined in Section 8.2 (see Figure 8.9), are examples of DNA transposons. Their copy numbers vary in different species and different strains, but a single *E. coli* genome will usually contain 30–50 insertion sequences of various types. An insertion sequence is 0.7–2.5 kb in length, with most of its sequence taken up by one or two genes that specify the **transposase** enzyme that catalyzes its transposition (**Figure 9.19A**). There are a pair of inverted repeats at either end of each IS element, up to about 50 bp in length depending on the type of IS, and insertion of the element into the target DNA creates a pair of short (usually 4–15 bp) direct repeats in the host genome. IS elements can transpose either replicatively or conservatively.

IS elements are also components of a second type of DNA transposon first characterized in *E. coli* and now known to be common in many prokaryotes. These **composite transposons** are made up of a pair of IS elements flanking a segment of DNA, usually containing one or more genes, often ones coding for antibiotic resistance (**Figure 9.19B**). Tn10, for example, carries a gene for tetracycline resistance, and Tn5 and Tn903 both carry a gene for resistance to kanamycin. Some composite transposons have identical IS elements at either end, and others have one element of one type and one of another. In some cases the IS elements are orientated as direct repeats and sometimes as inverted repeats. These variations do not appear to affect the transposition mechanism for a composite transposon, which is conservative in nature and catalyzed by the transposase coded by one or both of the IS elements.

Various other classes of DNA transposon are known in prokaryotes. Two additional important types found in *E. coli* are:

- **Tn3-type** or **unit transposons**, which have their own transposase gene and so do not require flanking IS elements in order to transpose (**Figure 9.19C**). Tn3 elements transpose replicatively.

- **Transposable phages**, which are bacterial viruses that transpose replicatively as part of their normal infection cycle (**Figure 9.19D**).

Prokaryotic genomes also contain IS fragments that have lost the ability to transpose. These are called **miniature inverted-repeat transposable elements** (**MITEs**). MITE is a general term for the truncated relic of a DNA transposon, which were first discovered in plants.

DNA transposons are less common in eukaryotic genomes

About 3.7% of the human genome is made up of DNA transposons of various types (see Table 9.3), all with terminal inverted repeats and all containing a gene for a transposase enzyme that catalyzes the transposition event. However, the vast majority of these elements are inactive, either because the transposase gene is nonfunctional or because sequences at the ends of the transposon, which are essential for active transposition, are missing or mutated.

Active DNA transposons are more common in plants, and include the **Ac/Ds transposon** and the **Spm element**, both of which are found in maize. The Ac/Ds elements were the first transposons to be discovered, by Barbara McClintock in the 1950s. Her conclusions – that some genes are mobile and can move from one position to another in a chromosome – were based on exquisite genetic experiments, the molecular basis of transposition not being understood until the late 1970s. An interesting feature of these plant transposons is that they work together in family groups. For example, the Ac element codes for an active transposase that recognizes both Ac elements and Ds sequences. The latter are versions of Ac that have internal deletions that remove part of the transposase gene, meaning that a Ds element cannot make its own transposase and can move only through the activity of the transposase synthesized by a full-length Ac element (**Figure 9.20**). Similarly, full-length Spm elements are accompanied by deleted versions which transpose through use of the transposase enzymes coded by the intact elements. The activity of Ac elements is apparent during the normal life cycle of a maize plant, transposition in somatic cells resulting in changes in gene expression which are manifested in, for example, variegated pigmentation in maize kernels (**Figure 9.21**).

McClintock's realization that the maize genome contains transposable elements resulted from her studies into the genetic basis of the different color patterns displayed by kernels. The P element, a DNA transposon in *Drosophila melanogaster*, was similarly discovered from studies of an unusual genetic event which, as it turns out, arises from transposition. This event is called **hybrid dysgenesis** and occurs when females from laboratory strains of *D. melanogaster* are crossed with males from wild populations. The offspring resulting from such crosses are sterile and have chromosomal abnormalities along with a variety of other genetic malfunctions. The explanation is that the genomes of wild fruit flies contain inactive versions of **P elements**, which are typical DNA transposons

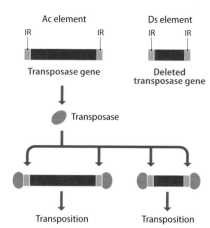

Figure 9.20 The Ac/Ds transposon family of maize. The full-length Ac element is 4.2 kb and contains a functional transposase gene. The transposase recognizes the 11 bp inverted repeats (IRs) at either end of the Ac sequence and catalyzes its transposition. The Ds element has an internal deletion and so does not synthesize its own transposase. It still has the IR sequences, which are recognized by the transposase made by the Ac element. Hence the Ds element is also able to transpose. There are approximately ten different types of Ds element in the maize genome, with deletions ranging in size from 194 bp to several kilobases.

Figure 9.21 Variegated pigmentation in maize kernels caused by transposition in somatic cells. The highly colored forms of *Zea mays* are popularly known as 'Indian corn'. (Courtesy of Lena Struwe, Rutgers University.)

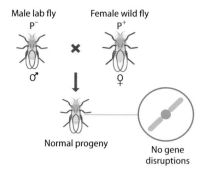

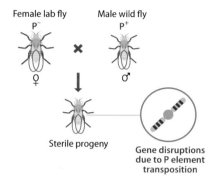

Figure 9.22   Hybrid dysgenesis. Crosses between male lab flies and female wild flies give normal progeny, but when the male partner is a wild fly the offspring are sterile. One possible explanation of hybrid dysgenesis is that the cytoplasm of flies with P elements (P⁺ in this diagram) contains a repressor that prevents P element transposition. The fertilized egg resulting from a cross between a female P⁺ fly and male P⁻ fly will contain this repressor and so the progeny are normal. However, the repressor will not be carried in the sperm from a male P⁺ fly, so the fertilized egg from a cross between a male P⁺ and a female P⁻ fly lacks the repressor, allowing P element transposition to occur and resulting in progeny displaying hybrid dysgenesis.

comprising a transposase gene flanked by inverted terminal repeats. Laboratory strains lack these elements. After crossing, the elements inherited from the wild flies become active in the fertilized eggs, transposing into various new positions and causing the gene disruptions that characterize hybrid dysgenesis (**Figure 9.22**). Exactly why this activation occurs is not known, but a more interesting question is why the genomes of wild populations of *D. melanogaster* contain P elements, whereas laboratory strains do not. Most of the laboratory strains are descended from flies collected by Thomas Hunt Morgan some 90 years ago, and used by Morgan and his colleagues in the first gene mapping experiments (Section 3.3). It appears that wild populations at that time lacked P elements, which have somehow proliferated in wild genomes during the last 90 years. The inability of wild and laboratory flies to produce viable offspring means that these two populations fail one of the main criteria used to identify biological species – the ability of all individuals to mate productively. This raises the intriguing possibility that speciation might, at least in some organisms, be driven by differential proliferation of transposable elements within the genomes of members of different populations.

SUMMARY

- Early studies of viruses focused largely on the bacteriophages, the viruses that infect bacteria.

- Bacteriophages are constructed of protein and nucleic acid, the protein forming a capsid that encloses the genome.

- There are three basic types of capsid structure but many types of genome organization. Different phages have single- or double-stranded DNA or RNA genomes, and the genome may be contained in a single molecule or divided into two or more segments.

- Bacteriophages follow two distinct infection cycles. All phages can infect via the lytic cycle, which results in the immediate synthesis of new bacteriophages, usually accompanied by death of the host cell. Some phages can also follow the lysogenic cycle, during which a copy of the phage genome becomes inserted into the host DNA, where it may remain in quiescent form for many generations.

- Eukaryotic viruses are equally diverse in terms of genome organization but display just two capsid structures.

- Most eukaryotic viruses follow a lytic infection cycle, but this does not always result in the immediate death of the host cell. A number of DNA and RNA viruses can integrate their genomes into eukaryotic chromosomes in a manner similar to a lysogenic bacteriophage.

- The viral retroelements, which include HIV, the causative agent of HIV/AIDS, are examples of integrative RNA viruses.

- Satellite RNAs and virusoids are different types of infective RNA molecule that contain no genes and depend on other viruses for their transmission. Viroids are small infective RNA molecules that self-replicate, and prions are infective proteins.

- Some mobile genetic elements, which are DNA sequences that can transpose within a genome but cannot escape from the cell, are related to RNA viruses. These elements transpose via an RNA intermediate in a pathway similar to the infection process of viral retroelements.

- The *Ty1/copia* and *Ty3/gypsy* retrotransposons, and the endogenous retroviruses of mammals, are the mobile elements most closely related to RNA viruses.

- Mammalian genomes also contain other types of retrotransposon, called LINEs and SINEs, most of which have lost their ability to transpose.

- DNA transposons do not make use of an RNA intermediate in their transposition pathway. These transposons are common in bacteria.

- DNA transposons are less widespread in eukaryotes but include some important examples, such as the Ac/Ds transposon of maize, the first transposon of any kind to be studied in detail, and the P element of *Drosophila melanogaster*, which is responsible for the hybrid dysgenesis that occurs when female laboratory fruit flies are crossed with wild male flies.

SHORT ANSWER QUESTIONS

1. How are viruses different from cells? Is it appropriate to describe viruses as living organisms?

2. Outline the key differences between viral and cellular genomes.

3. Using examples, explain what is meant by overlapping genes, as found in some viral genomes.

4. How long does it take a lytic bacteriophage to lyse a host cell following the initial infection? What is the timeline for the lytic infection cycle of T4 phage?

5. Describe the differences between the capsids of bacteriophages and eukaryotic viruses.

6. List the key stages in the infection cycle of a retrovirus.

7. What is a transposon?

8. Describe the characteristic features of the LTR retroelements present in the human genome.

9. Discuss the properties and types of retroposons present in the human genome.

10. What are the general properties of composite transposons?

11. What are the important features of the DNA transposons found in plants?

12. Describe the basis of hybrid dysgenesis in fruit flies.

IN-DEPTH PROBLEMS

1. To what extent can viruses be considered a form of life?

2. Bacteriophages with small genomes (for example, φX174) are able to replicate very successfully in their hosts. Why then should other bacteriophages, such as T4, have large and complicated genomes?

3. Some bacteriophages, such as T4, modify the host RNA polymerase after infection so that this polymerase no longer recognizes *E. coli* genes but transcribes bacteriophage genes instead. How might this modification be carried out?

4. Genetic elements that reproduce within or along with a host genome, but confer no benefit on the host, are sometimes called selfish DNA. Discuss this concept, in particular, as it applies to transposons.

5. Why do LTR retroelements have long terminal repeats?

FURTHER READING

Classic papers on bacteriophage genetics

Delbrück, M. (1940) The growth of bacteriophage and lysis of the host. *J. Gen. Physiol.* 23:643–660.

Doermann, A.H. (1952) The intracellular growth of bacteriophages. *J. Gen. Physiol.* 35:645–656.

Ellis, E.L. and Delbrück, M. (1939) The growth of bacteriophage. *J. Gen. Physiol.* 22:365–384.

Lwoff, A. (1953) Lysogeny. *Bacteriol. Rev.* 17:269–337.

Bacteriophage genome sequences

Dunn, J.J. and Studier, F.W. (1983) Complete nucleotide sequence of bacteriophage T7 DNA and the locations of T7 genetic elements. *J. Mol. Biol.* 166:477–535.

Sanger, F., Air, G.M., Barrell, B.G., et al. (1977) Nucleotide sequence of bacteriophage ΦX174 DNA. *Nature* 265:687–695.

Sanger, F., Coulson, A.R., Hong, G.F., et al. (1982) Nucleotide sequence of bacteriophage λ DNA. *J. Mol. Biol.* 162:729–773.

Eukaryotic viruses

Baltimore, D. (1970) RNA-dependent DNA polymerase in virions of RNA tumour viruses. *Nature* 226:1209–1211.

Dimmock, N.J., Easton, A.J. and Leppard, K.N. (2016) *An Introduction to Modern Virology*, 7th ed. Blackwell Scientific Publishers, Oxford. *The best general text on viruses.*

Lesbats, P., Engelman, A.N. and Cherepanov, P. (2016) Retroviral DNA integration. *Chem. Rev.* 116:12730–12757.

Temin, H.M. and Mizutani, S. (1970) RNA-dependent DNA polymerase in virions of Rous sarcoma virus. *Nature* 226:1211–1213.

Edge of life

Flores, R., Gas, M.-E., Molina-Serrano, D., et al. (2009) Viroid replication: Rolling-circles, enzymes and ribozymes. *Viruses* 1:317–334.

Mastrianni, J.A. (2010) The genetics of prion diseases. *Genet. Med.* 12:187–195.

Palukaitis, P. (2016) Satellite RNAs and satellite viruses. *Mol. Plant-Microbe Interact.* 29:181–186.

Prusiner, S.B. (1996) Molecular biology and pathogenesis of prion diseases. *Trends Biochem. Sci.* 21:482–487.

Terry, C. and Wadsworth, J.D.F. (2019) Recent advances in understanding mammalian prion structure: A mini review. *Front. Mol. Neurosci.* 12:169.

RNA transposons

Gifford, R. and Tristem, M. (2003) The evolution, distribution and diversity of endogenous retroviruses. *Virus Genes* 26:291–315.

Grandi, N. and Tramontano, E. (2018) Human endogenous retroviruses are ancient acquired elements still shaping innate immune responses. *Front. Immunol.* 12:169.

Hoyt, S.J., Storer, J.M., Hartley, G.A., et al. (2022) From telomere to telomere: The transcriptional and epigenetic state of human repeat elements. *Science* 376:eabk3112.

Krastanova, O., Hadzhitodorov, M. and Pesheva, M. (2005) Ty elements of the yeast *Saccharomyces cerevisiae. Biotechnol. Biotechnol. Equip.* 19(Suppl 2):19–26.

Li, W., Lee, M.-H., Henderson, L., et al. (2015) Human endogenous retrovirus-K contributes to motor neuron disease. *Sci. Transl. Med.* 7:307ra153.

Millman, A., Bernheim, A., Stokar-Avihail, A., et al. (2020) Bacterial retrons function in anti-phage defense. *Cell* 183:1551–1561.

Richardson, S.R., Doucet, A.J., Kopera, H., et al. (2015) The influence of LINE-1 and SINE retrotransposons on mammalian genomes. *Microbiol. Spectrum* 3:MDNA3-0061-2014.

Song, S.U., Gerasimova, T., Kurkulos, M., et al. (1994) An Env-like protein encoded by a *Drosophila* retroelement: Evidence that gypsy is an infectious retrovirus. *Genes Dev.* 8:2046–2057.

Zhang, X., Zhang, R. and Yu, J. (2020) New understanding of the relevant role of LINE-1 retrotransposition in human disease and immune modulation. *Front. Cell Dev. Biol.* 8:657.

DNA transposons

Comfort, N.C. (2001) *The Tangled Field: Barbara McClintock's Search for the Patterns of Genetic Control.* Harvard University Press, Cambridge, MA. *A biography of the geneticist who discovered transposable elements; for a highly condensed version, see Comfort, N.C. (2001) Trends Genet.* 17:475–478.

Gierl, A., Saedler, H. and Peterson, P.A. (1989) Maize transposable elements. *Annu. Rev. Genet.* 23:71–85.

Majumdar, S. and Rio, D.C. (2015) P transposable elements in *Drosophila* and other eukaryotic organisms. *Microbiol. Spectr.* 3:MDNA3-0004-2014.

Siguier, P., Gourbeyre, E. and Chandler, M. (2015) Bacterial insertion sequences: Their genomic impact and diversity. *FEMS Microbiol. Rev.* 38:865–891.

Siguier, P., Gourbeyre, E., Varani, A., et al. (2015) Everyman's guide to bacterial insertion sequences. *Microbiol. Spectrum* 3:MDNA3-0030-2014.

Online resource

RepeatMasker. http://www.repeatmasker.org/ *The 'Genome Analysis and Downloads' component enables the repetitive DNA content of various genomes to be viewed.*

PART 3

HOW GENOMES ARE EXPRESSED

ACCESSING THE GENOME

In order for the cell to utilize the biological information contained within its genome, groups of genes, each gene representing a single unit of information, have to be expressed in a coordinated manner. This coordinated gene expression determines the makeup of the transcriptome, which in turn specifies the nature of the proteome and defines the activities that the cell is able to carry out. In Part 3 of *Genomes*, we will examine the events that result in the transfer of biological information from genome to proteome. Our knowledge of these events was initially gained through studies of individual genes, often as 'naked' DNA in test-tube experiments. These experiments provided an interpretation of gene expression that in recent years has been embellished by more sophisticated studies that have taken greater account of the fact that, in reality, it is the genome that is expressed, not individual genes, and this expression occurs in living cells rather than in a test tube.

We begin our investigation of genome expression, here in Chapter 10, by examining the substantial and important impact that the nuclear environment has on utilization of the biological information contained in the genomes of eukaryotes, the accessibility of that information being dependent on the way in which the DNA is packaged into chromatin, and being responsive to processes that can silence or inactivate part or all of a chromosome. Chapter 11 then describes the critical role played by DNA-binding proteins in genome expression, and explains how the structures of these proteins enable them to recognize their specific binding sites. The composition of the transcriptome, and the role of the transcriptome in the overall process of genome expression, is dealt with in Chapter 12, and Chapter 13 covers the equivalent issues regarding the composition and role of the proteome. As you read Chapters 10–13, you will discover that control over the composition of the transcriptome and of the proteome can be exerted at various stages during the overall chain of events that make up genome expression. These regulatory threads will be drawn together in Chapter 14, where we examine how the genome acts within the context of cell and organism by responding to extracellular signals and by driving the biochemical changes that underlie differentiation and development.

10.1 INSIDE THE NUCLEUS

10.2 NUCLEOSOME MODIFICATIONS AND GENOME EXPRESSION

10.3 DNA MODIFICATION AND GENOME EXPRESSION

10.1 INSIDE THE NUCLEUS

When we look at a genome sequence written out as a series of As, Cs, Gs, and Ts, or use a genome browser to examine a segment of a chromosome (as in Figure 5.23, for example), there is a tendency to imagine that all parts of the genome are readily accessible to the DNA-binding proteins that are responsible for its expression. In reality, the situation is very different. The DNA in the nucleus of a eukaryotic

DOI: 10.1201/9781003133162-10

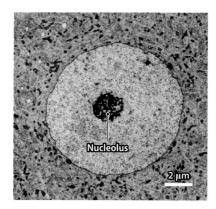

Figure 10.1   The nucleolus. This electron micrograph shows a nucleus from a nerve cell in a spinal ganglion. The nucleolus is the only distinct structure that is visible when nuclei are examined by conventional electron microscopy. (From Martinelli C., Sartori P., Ledda M., Pannese E. [2003] Age-related quantitative changes in mitochondria of satellite cell sheaths enveloping spinal ganglion neurons in the rabbit. *Brain Research Bulletin* 61: 147-151.)

cell or the nucleoid of a prokaryote is attached to a variety of proteins that are not directly involved in genome expression and which must be displaced in order for the RNA polymerase and other expression proteins to gain access to the genes. We know very little about these events in prokaryotes, a reflection of our generally poor knowledge about the physical organization of the prokaryotic genome (Section 8.1), but we are beginning to understand how the packaging of DNA into chromatin (Section 7.1) influences genome expression in eukaryotes. This is an exciting area of molecular biology, with recent research indicating that histones and other packaging proteins are not simply inert structures around which the DNA is wound, but instead are active participants in the processes that determine which parts of the genome are expressed in an individual cell. Many of the discoveries in this area have been driven by new insights into the substructure of the nucleus, and it is with this topic that we begin.

The nucleus has an ordered internal structure

The internal architecture of the nucleus was first examined by light and electron microscopy. Using conventional techniques, the inside of the nucleus appears to be relatively unstructured, made up of lighter and darker regions with just one distinct feature, the **nucleolus**, which is the center for synthesis and processing of rRNA molecules, and which appears as a dark area when nuclei are observed with the electron microscope (**Figure 10.1**). These early microscopy studies therefore suggested that the nucleus has little internal organization, a typical 'black box' in common parlance. In recent years this interpretation has been overthrown, and we now appreciate that the nucleus has an ordered structure that is related to the variety of biochemical activities that it must carry out. Indeed, the inside of the nucleus is just as complex as the cytoplasm of the cell, the only difference being that, in contrast to the cytoplasm, the functional compartments within the nucleus are not individually enclosed by membranes, and so are not visible when the cell is observed using conventional light or electron microscopy techniques.

The revised picture of nuclear structure first began to emerge when techniques were developed for labeling different types of nuclear protein with fluorescent markers. This can be achieved by ligating the coding sequence for the **green fluorescent protein** (**GFP**) to the gene for the protein being studied. Standard cloning techniques are then used to insert the modified gene into the host genome, leading to a recombinant cell that synthesizes a fluorescent version of the protein. Observation of the cell using a fluorescence microscope now reveals the distribution of the labeled protein within the nucleus. For example, nucleoli can be visualized by labeling the fibrillarin protein with GFP (**Figure 10.2A**), fibrillarin being a component of the small nucleolar ribonucleoproteins that are involved in rRNA processing. Labeling of proteins involved in mRNA splicing (Section 12.4) has shown that this activity is also localized into distinct regions, called **speckles** (**Figure 10.2B**), although these are more widely distributed and less well defined than the nucleoli. Other structures, such as **Cajal bodies** (visible in Figure 10.2A), which are probably involved in synthesis

Figure 10.2   The internal architecture of the eukaryotic nucleus. Images of living nuclei obtained after labeling of proteins with GFP. In (A), the nucleolus is shown in blue and the Cajal bodies in yellow. In (B), the colored areas are speckles, which contain proteins involved in RNA splicing. (From Misteli T. [2001] *Science* 291: 843–847. With permission from the American Association for the Advancement of Science.)

(B)

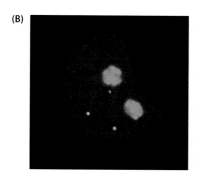

(C)

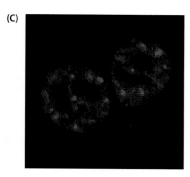

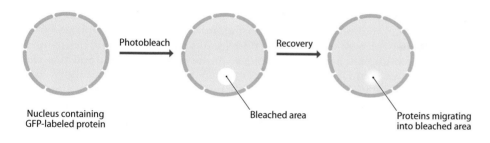

Nucleus containing
GFP-labeled protein

Photobleach

Bleached area

Recovery

Proteins migrating
into bleached area

Figure 10.3 Fluorescence recovery after photobleaching (FRAP). A small area of a nucleus containing a fluorescently labeled protein is bleached by exposure to a laser beam. The migration of proteins into the bleached region from the unexposed area of the nucleus can then be observed.

of small nuclear and small nucleolar RNAs (Section 12.1), can also be seen after fluorescent labeling.

Another new microscopy technique, called **fluorescence recovery after photobleaching** (**FRAP**), enables the movement of fluorescently labeled proteins within the nucleus to be visualized. A small area of the nucleus is **photobleached** by exposure to a tightly focused pulse from a high-energy laser. The laser pulse inactivates the fluorescent signal in the exposed area, leaving a region that appears bleached in the microscopic image. This bleached area gradually retrieves its fluorescent signal, not by a reversal of the bleaching effect, but by migration into the bleached region of fluorescent proteins from the unexposed area of the nucleus (**Figure 10.3**). Rapid reappearance of the fluorescent signal in the bleached area therefore indicates that the tagged proteins are highly mobile, whereas a slow recovery indicates that the proteins are relatively static. Studies of this type have shown that the migration of nuclear proteins does not occur as rapidly as would be expected if their movement were totally unhindered, which is entirely expected in view of the large amounts of DNA and RNA in the nucleus, but that it is still possible for a protein to traverse the entire diameter of a nucleus in a matter of minutes. Proteins involved in genome expression therefore have the freedom needed to move from one activity site to another, as dictated by the changing requirements of the cell. In particular, the linker histones (Section 7.1) continually detach and reattach to their binding sites on the genome. This discovery is important because it emphasizes that the DNA-protein complexes that make up chromatin are dynamic, an observation that has considerable relevance to genome expression, as we will see later in this chapter.

Chromosomal DNA displays different degrees of packaging

When nondividing nuclei are examined by conventional electron microscopy, all that can be seen, other than the nucleolus, is a mixture of light and dark areas (**Figure 10.4**). The light areas are called **euchromatin** and comprise those parts of the chromosomal DNA that contain active genes – ones that are being transcribed into RNA. These regions adopt an open configuration so that RNA polymerase and other proteins involved in transcription can access these genes. In contrast, the dark areas, called **heterochromatin**, contain DNA that is transcriptionally silent, containing few if any active genes, and which therefore adopts a more compact organization. Two types of heterochromatin are recognized:

- **Constitutive heterochromatin** is a permanent feature of all cells and represents DNA that contains no genes and so can always be retained in a compact organization. This fraction includes centromeric and telomeric DNA, as well as certain regions of some other chromosomes. For example, most of the human Y chromosome is made of constitutive heterochromatin.

- **Facultative heterochromatin** is not a permanent feature but is seen in some cells some of the time. Facultative heterochromatin is thought to contain genes that are inactive in some cells or at some periods of the cell cycle. When these genes are inactive, their DNA regions are compacted into heterochromatin.

Heterochromatin   Euchromatin

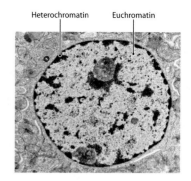

Figure 10.4 Euchromatin and heterochromatin. Electron micrograph of a plasma cell from guinea pig. The light areas in the nucleus are euchromatin, and the dark areas are heterochromatin. (Courtesy of Don Fawcett, Science Photo Library.)

Figure 10.5   The 30 nm chromatin fiber. The solenoid model for the 30 nm fiber held that the 10 nm fiber forms a coil or solenoid structure. The side-view of the model is shown on the left, and the view from the top is on the right.

Understanding the different types of DNA packaging that give rise to euchromatin, heterochromatin and, in its most compact form, the metaphase chromosomes that are seen when the cell divides, has proved to be one of the most challenging problems in genome biology. In Section 7.1, we learned that the association between DNA and nucleosomes gives rise to the 10 nm fiber. This fiber has frequently been observed in living nuclei and is thought to be the configuration adopted by DNA in the euchromatin regions. For several years, it was thought that the 10 nm fiber could form a coiled structure called the **30 nm fiber** (**Figure 10.5**), which represented the next level in a hierarchical series of increasing compact configurations that eventually enable a chromosomal DNA molecule with an average length of 4 cm to be packaged into a metaphase chromosome of just a few μm. However, the 30 nm fiber has only been identified in cell extracts and never directly observed in living cells, even with the most sophisticated imaging systems now available. It is now looked on as an artifact that can form in low-salt conditions *in vitro*, but is not a significant structure within the nucleus. Instead, within the nucleus the inactive parts of the genome comprise 10 nm fibers that have been packaged into globular structures, with a less regular internal organization than that displayed by the 30 nm fiber (**Figure 10.6**). These globular structures are responsible for the darker staining of the heterochromatin parts of the nucleus.

The metaphase chromosomes that form when a nucleus divides are just a few μm in length, but each contains a DNA molecule that, if laid out as a naked double helix, would be a few cm from end to end. Compaction into the 10 nm fiber only decreases the length of the DNA molecule by about one-sixth. Clearly, higher levels of packaging are needed to fit a DNA molecule into its metaphase chromosome. In one model, this compaction is achieved by stacking nucleosomes into plate-like structures that can be packed side by side and in multiple layers (**Figure 10.7A**). An alternative model envisages a protein core from which loops of 10 nm fiber radiate, the loops condensing by interactions between the nucleosomes to form a disordered but compact structure similar to the globules thought to be present in heterochromatin (**Figure 10.7B**).

The nuclear matrix is a dynamic structure

The existence within the nucleus of a network of fibers, equivalent to the cytoskeleton that extends throughout the cytoplasm, was first proposed in the 1940s. However, the role and very existence of this **nuclear matrix** remains a controversial area of cell biology. It is known that there is a thin network of filaments on the internal side of the nuclear membrane, forming a **nuclear lamina** that is attached to proteins that are embedded in the nuclear membrane. What is questioned is whether this network extends into the internal regions of the nucleus, forming a matrix that permeates the entire nucleoplasm.

The existence of a nuclear matrix was suggested by examination of mammalian nuclei that had been prepared in a special way. After dissolution of membranes by soaking in a mild, nonionic detergent such as one of the Tween compounds, followed by treatment with a deoxyribonuclease to degrade the nuclear DNA, and salt extraction to remove the chemically basic histone proteins, a complex network of protein and RNA fibrils was revealed, apparently extending throughout the entire nucleus (**Figure 10.8**). These experiments might appear to give clear evidence for the existence of the nuclear matrix, but could the structures seen in these electron micrographs be an artifact resulting from the preparation method? Doubts were first raised when intact nuclei were examined by **immunofluorescence microscopy**. Like the methods based on green fluorescent protein that we discussed above, immunofluorescence microscopy utilizes a fluorescent marker to visualize the location of particular proteins within a cell. The difference is that visualization is achieved not by engineering the cell to synthesize a GFP fusion of the protein of interest, but instead by treating a tissue section with fluorescently labeled antibodies that bind specifically to that protein. We might anticipate that when this technique is used with antibodies specific to proteins of the nuclear matrix, the latter will be revealed as a network of fluorescence. This

10 nm fiber

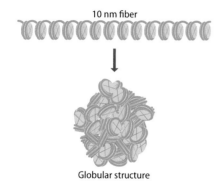

Globular structure

Figure 10.6   Packaging of 10 nm fibers into globular chromatin structures.

(A) Plate model **(B)** Disordered model

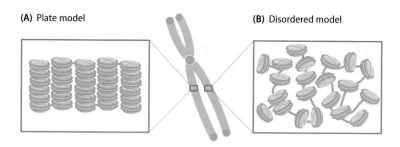

Figure 10.7 Two models for the higher order packaging of the 10 nm fiber into a metaphase chromosome. (A) The plate model, in which nucleosomes are stacked into plate-like structures that can be packed side by side and in multiple layers. (B) A disordered model, in which the nucleosomes form a disordered but compact structure.

is not what has been seen: instead, the matrix proteins appear to form distinct spots rather than fibers.

Additional evidence for the existence of a structural matrix of some kind in the nuclei of most cells is the presence within chromosomal DNA of nucleotide sequences that bind to the matrix proteins. These segments have been given various names in the past but are now commonly called **scaffold/matrix attachment regions (S/MARs)**. Individual S/MARs are 100–4000 bp in length, and although they do not contain any diagnostic sequences they are AT-rich and can be recognized in genome sequences. There are over 7350 S/MARs in the *Drosophila melanogaster* genome and over 20,000 in the *Arabidopsis thaliana* genome. Originally it was suggested that there are 60,000–70,000 in the human genome, but identification of binding sites for proteins that specifically attach to S/MARs indicates that there could be as many as 285,000. The positions of S/MARs often coincide with, or are close to, replication origins and to the transcription start points for protein-coding genes, in particular ones with a high rate of transcription. Pairs of S/MARs are thought to act as anchor sequences, with the intervening DNA forming a loop between 20 kb and 200 kb in length. Some S/MARs appear to be used in all cells, and so may contribute to the underlying structural organization of the nucleus. Other S/MARs are only used in particular cell types, and some form or break their attachments with the matrix proteins in response to extracellular signals such as the presence of a hormone or growth factor. The implication is that the attachments made by these S/MARs can change, so that adjacent loops can be merged into larger ones when the genes contained in those loops are active, and active loops can be condensed into smaller ones when their genes are switched off.

The exact nature of the nuclear matrix is still being debated, but it is now generally accepted that the rigid fibrous network that was originally identified does not exist in the living nucleus, and instead results from the self-assembly of nuclear proteins into filaments when the cells are being prepared for examination with the electron microscope. The nuclear matrix is now viewed as a more dynamic entity with a biochemical composition that is variable within a nucleus and also changes at different stages of the cell cycle. The biochemical composition of the matrix is complex, with over 250 proteins forming a core set present in both fruit-fly and mouse nuclei, and additional proteins that are unique to different groups of organism. The matrix proteins include some with structural roles, such as matrins and lamins, but also many enzymes and other proteins involved in genome replication, repair and transcription, as well as splicing and

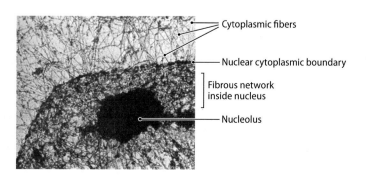

Cytoplasmic fibers

Nuclear cytoplasmic boundary

Fibrous network inside nucleus

Nucleolus

Figure 10.8 Evidence for the existence of a nuclear matrix. Transmission electron micrograph showing the nuclear matrix of a cultured human HeLa cell. Cells were treated with a nonionic detergent to remove membranes, digested with a deoxyribonuclease to degrade most of the DNA, and extracted with ammonium sulfate to remove histones and other chromatin-associated proteins. (From Penman S., Fulton A., Capco D. et al. [1982] *Symp. Quant. Biol.* 46: 1013–1028. With permission from Cold Spring Harbor Laboratory Press.)

Figure 10.9 Chromosome territories revealed by chromosome painting. (A) Human chromosomes in an interphase nucleus, visualized by hybridization labeling with probes of different colors. The black areas are nucleoli. (B) Interpretation of the image in (A), giving the identity of each chromosome. (From Speicher M. & Carter N. [2005] *Nat Rev Genet* 6: 782–792. With permission from Springer Nature.)

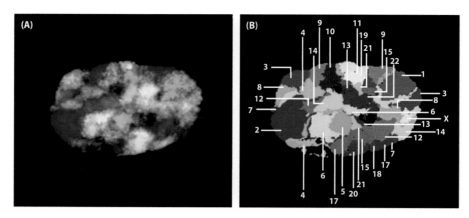

other RNA processing events. The composition of the matrix is therefore consistent with both a structural role and active participation in the events occurring within the nucleus.

Each chromosome has its own territory within the nucleus

Initially, it was thought that chromosomes are distributed randomly within a eukaryotic nucleus. We now know that this view is incorrect and that each chromosome occupies its own space, or **territory**. These were first visualized by **chromosome painting**, which is a version of fluorescent *in situ* hybridization (FISH; Section 3.5). Cells are first treated with a fixative so that the nuclear substructure is stabilized, and then probed with a mixture of DNA molecules, with those specific for different regions of a single chromosome all carrying the same fluorescent label. Each chromosome can therefore be 'painted' a different color. When applied to interphase nuclei, chromosome painting reveals territories occupied by individual chromosomes (**Figure 10.9**). These territories take up the majority of the space within the nucleus, but are separated from one another by **nonchromatin regions**, within which the enzymes and other proteins involved in expression of the genome are located.

The images obtained by chromosome painting have recently been supplemented by ***in situ* sequencing**, which is able to provide greater definition to the territories occupied by different chromosomes and can even identify the positions within those territories of individual genes. In this method, Tn5 transposase is used to insert PCR primers and unique barcodes into genomic DNA at random positions in fixed nuclei (**Figure 10.10**). Following amplification, the barcodes are read *in situ* by SOLiD sequencing (Section 4.1) with labeled oligonucleotides whose incorporation can be followed by fluorescent imaging. The positions of the different barcodes in the nucleus are therefore located. The amplified fragments are then dissociated from the fixed nuclei and paired-end reads obtained. These reads will include both the barcode and the adjacent DNA sequences, enabling the barcodes to be mapped on to the genome sequence. By combining the positional information from the SOLiD sequencing with the genome sequences from the paired-end reads, the positions of chromosomes and individual genes can be identified within the nuclear substructure (**Figure 10.11**)

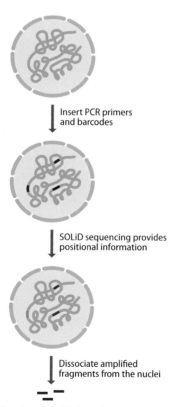

Figure 10.10 *In situ* **sequencing.** Tn5 transposase is used to insert PCR primers and unique barcodes into genomic DNA in fixed nuclei. Following amplification, *in situ* SOLiD sequencing locates the positions of the barcodes and hence provides positional information. The amplified fragments are then dissociated from the fixed nuclei and paired-end reads obtained, enabling the barcodes to be mapped on to the genome sequence.

Chromosome territories appear to be fairly static within an individual nucleus. This has been concluded from experiments in which CENP-B proteins, components of the centromeres (Section 7.1), are labeled with green fluorescent protein and the locations of these proteins, and hence of the centromeres, observed over a period of time. On the whole, individual centromeres remain stationary throughout the cell cycle, though there are occasional bursts of relatively slow movement. Although fairly static during the lifetime of a cell, most studies have suggested that the relative positioning of territories is not retained after cell division, as different patterns are observed in the nuclei of daughter cells. However, during early embryo development there are similarities in territory structure at the two- and four-cell stages, suggesting that positional information can, at least under some circumstances, be passed from parent to daughter nuclei. Some constraint on territory location would help to explain the apparently nonrandom nature of chromosome **translocations**, which result in a segment of one chromosome becoming attached to another chromosome. For example, a translocation between human chromosomes 9 and 22, resulting in the abnormal product called the **Philadelphia chromosome**, is a common cause of chronic myeloid leukemia (**Figure 10.12**). The repeated occurrence of the same translocation suggests that the territories of the interacting pair of chromosomes are frequently close to one another in the nucleus. There is also evidence that, at least in some organisms, certain chromosomes preferentially occupy territories close to the periphery of the nucleus. Relatively little genome expression occurs in this region, and it is often here that those chromosomes that contain few active genes are found, examples being the macrochromosomes of the chicken genome (Section 7.1).

The positioning of active genes within individual chromosome territories is a further topic of debate. At one time, it was thought that the active genes were located on the surface of a territory, adjacent to the nonchromatin region and hence within easy reach of the enzymes and proteins involved in gene transcription. This view has been questioned, partly as a result of experiments that have shown that RNA transcripts are distributed within territories as well as on their surfaces. More refined microscopic examination has suggested that channels run through chromosome territories, linking different parts of the nonchromatin regions, and providing a means by which the transcription machinery can penetrate into the internal parts of these territories (**Figure 10.13**).

Chromosomal DNA is organized into topologically associating domains

Some of the most powerful methods that are being used to study the internal nuclear architecture are versions of **chromosome conformation capture**, which we met in Section 4.2 as the basis of the Hi-C sequencing approach that can be used to identify contigs that are adjacent in a genome sequence. Chromosome conformation capture involves treating nuclei with a cross-linking agent, such as formaldehyde, so that covalent bonds are formed between the DNA and proteins in strands of chromatin that are close to one another (**Figure 10.14**). The resulting network is treated with a restriction endonuclease to break it into fragments, each of which contains two linked pieces of DNA, one from each of the two regions of the genome that were adjacent in the nucleus. DNA ligase is then added to join the ends of the fragments, and the crosslinks are disrupted by heating to 70°C. The end-product of these manipulations is a circle of DNA, comprising the two DNA fragments that originally were located close together in the nucleus. In its original version, called **3C**, chromosome conformation capture was used to determine if particular interactions which were thought to occur between different parts of a DNA molecule had actually taken place. The pair of fragments that would be ligated to form the circular end-product were therefore predicted, and a PCR was designed to detect if that circle was present in the mixture resulting from the 3C experiment. More sophisticated versions of chromosome conformation capture, such as the Hi-C method (see Figure 4.26), utilize short-read sequencing to identify all of the interactions occurring in a nucleus, by mapping the positions of each pair of linked fragments on to the genome sequence.

The first chromosome conformation capture experiments that were carried out with crosslinked chromatin showed that the vast majority of interactions occur between DNA sequences present in the same chromosome, with relatively few interactions between different chromosomes. These results helped establish that chromosomes occupy individual territories within the nucleus, rather than being intertwined in some way. More detailed projects then revealed that the contacts formed within a chromosome were not random. Instead, a single chromosome is made up of a series of **topologically associating domains** (**TADs**), each comprising a contiguous segment of chromatin folded into coils and loops;

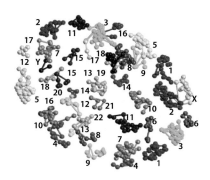

Figure 10.11 Positions of chromosomes in a human fibroblast nucleus revealed by *in situ* sequencing. The chromosomes are numbered and homologous pairs depicted in the same color. (From Payne A.C., Chiang Z.D., Reginato P.L. [2021] *Science* 371. Doi: 10.1126/science.aay3446. With permission from the American Association for the Advancement of Science.)

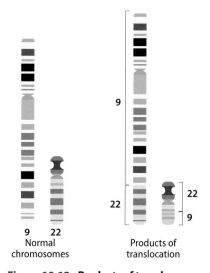

9   22
Normal chromosomes

Products of translocation

Figure 10.12 Products of translocation between human chromosomes 9 and 22. The normal human chromosomes 9 and 22 are shown on the left and the translocation products are on the right. The Philadelphia chromosome is the smaller of the two translocation products. Chromosomes 9 and 22 commonly break at the positions indicated. Often the breaks are correctly repaired, but occasionally misrepair creates the hybrid products. It is thought that the relatively high frequency with which the Philadelphia chromosome arises indicates that chromosomes 9 and 22 occupy adjacent territories in the human nucleus. The chromosome 9 breakpoint lies within the *ABL* gene, the product of which is involved in cell signaling (Section 14.1). The translocation attaches a new coding sequence to the start of this gene, resulting in an abnormal protein which causes cell transformation and gives rise to chronic myeloid leukemia.

Figure 10.13 Channels in chromosome territories. The view on the left shows the original model with each chromosome territory forming a block, implying that active genes are located on the surface of a territory. The view on the right shows an alternative model, with channels running through the territories.

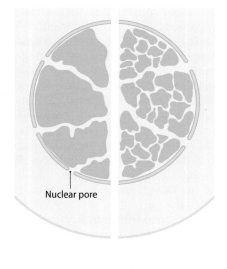

Nuclear pore

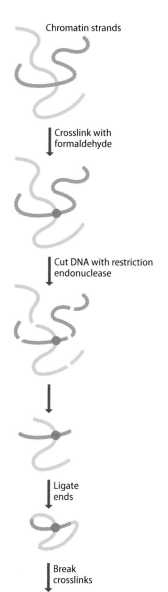

Chromatin strands

Crosslink with formaldehyde

Cut DNA with restriction endonuclease

Ligate ends

Break crosslinks

Circle can be analyzed by PCR or sequenced

this architecture gives rise to many intradomain interactions but few interactions between domains (**Figure 10.15**). In mammalian genomes, TADs have an average size of 880 kb though some are much longer than this, with the largest up to 2 Mb. They make up about 90% of the genome, with the remainder comprising unstructured regions between adjacent TADs. At least some TADs have an internal structure comprising 'sub-TADs' with an average size of 185 kb. The *Drosophila melanogaster* genome contains at least 1000 TADs, ranging in size from 10 to 1000 kb with an average of 100 kb. TADs are present in all metazoan genomes and in at least some plants: they are not obvious components of the *Arabidopsis* genome but have been identified in rice, cotton and various other plant genomes.

Despite the increasing amount of information about the presence of TADs in chromosomal DNA, there are still debates about their importance as functional units within a genome. Initial reports suggested that the same TADs are present in all cells of a single species, and that there are similarities in the domain organization of related species: for example, over 50% of human TAD boundaries are also present in the mouse genome. The degree of conservation within and between species was later questioned when chromosome conformation capture methods were refined to the stage where single cells could be examined, as these revealed more variability between individual cells than had been inferred from earlier studies. The most convincing evidence that TADs do in fact have a functional role comes from experiments in which TAD boundary sequences are deleted, which can lead to aberrant patterns of gene expression, and from the discovery that TAD boundaries are disrupted in some types of cancerous tissue. There is also some evidence that regulatory sequences called **enhancers** (Section 12.3), which can be located up to 1 Mb away from their target genes, only influence expression of genes that lie in the same TAD, and cannot work across TAD boundaries. The relationship between TAD boundaries and nuclear matrix attachment points is also being explored. The chromatin loops thought to be formed by pairs of S/MARs are, at 20–200 kb, smaller than most TADs, though comparable in size to mammalian sub-TADs. In the regions of heterochromatin adjacent to centromeres in *D. melanogaster* chromosomes, there is

Figure 10.14 Chromosome conformation capture. A simplified version of the 3C technique is shown. After crosslinking with formaldehyde, the DNA in the chromatin strands is cut with a restriction endonuclease. Addition of DNA ligase then joins together those DNA ends that are close to one another, giving a circle made up of the two segments of DNA that were adjacent in the nucleus. If the identity of these segments can be predicted, then formation of the circle can be confirmed by PCR, using primers that anneal either side of one of the ligation points. Alternatively, if there is no *a priori* information regarding the likely interactions occurring in the chromatin, then the Hi-C method can be used to sequence the segments of DNA that have been ligated together (see Figure 4.26).

good correspondence between TAD boundaries and S/MAR sequences, and in mammalian cells several of the proteins that bind to S/MAR sequences also bind to TAD boundaries. At present, it seems likely that chromatin loops formed by S/MARs are part of the substructure of individual TADs, but further research is needed to elucidate the exact relationship.

Insulators prevent crosstalk between segments of chromosomal DNA

The inability of regulatory sequences such as enhancers to affect expression of genes outside of their TAD is mediated by sequences, 300 bp to 2 kb in length, called **insulators**. Insulator sequences were first discovered in *Drosophila* and have now been identified in a range of species. They are present within regions identified as TADs as well as at TAD borders. Examples are the pair of sequences called scs and scs´ (scs stands for 'specialized chromatin structure'), which are located on either side of the two *hsp70* genes in the fruit-fly genome (**Figure 10.16**).

Insulators maintain the independence of segments of chromosomal DNA, preventing crosstalk between those segments. If scs or scs´ is excised from its normal location and reinserted between a gene and the upstream regulatory modules that control expression of that gene, then the gene no longer responds to its regulatory modules: it becomes insulated from their effects (**Figure 10.17A**). This observation suggests that, in their normal positions, insulators prevent the genes within one segment of DNA from being influenced by the regulatory modules present in an adjacent segment (**Figure 10.17B**). This ability also enables insulators to overcome the **positional effect** that occurs during a gene cloning experiment with a eukaryotic host. The positional effect refers to the variability in gene expression that occurs after a new gene has been inserted into a eukaryotic chromosome. It is thought to result from the random nature of the insertion event, which could deliver the gene to a region of highly packaged chromatin, where it will be inactive, or into an area of open chromatin, where it will be expressed (**Figure 10.18A**). The ability of scs and scs´ to overcome the positional effect was demonstrated by placing them on either side of a fruit-fly gene for eye color. When flanked by the insulators, this gene was always highly expressed when inserted back into the *Drosophila* genome, in contrast to the

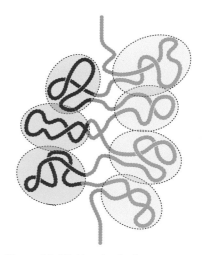

Figure 10.15 Topologically associating domains. Each domain is a contiguous segment of chromatin.

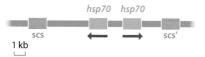

Figure 10.16 Insulator sequences in the fruit-fly genome. The diagram shows the region of the *Drosophila* genome containing the two *hsp70* genes. The insulator sequences scs and scs´ are on either side of the gene pair. The arrows below the two genes indicate that they lie on different strands of the double helix and so are transcribed in opposite directions.

(A) Insulators block the regulatory signals that control gene expression

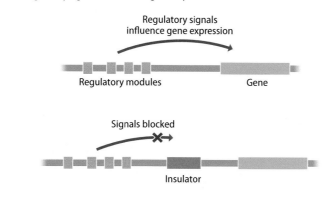

(B) Insulators prevent crosstalk between segments

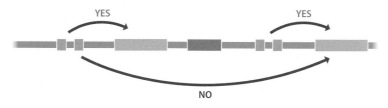

Figure 10.17 Insulators maintain the independence of DNA segments. (A) When placed between a gene and its upstream regulatory modules, an insulator sequence prevents the regulatory signals from reaching the gene. (B) In their normal positions, insulators prevent crosstalk between DNA segments, so the regulatory modules of one gene do not influence expression of a gene in a different segment.

(A) The positional effect

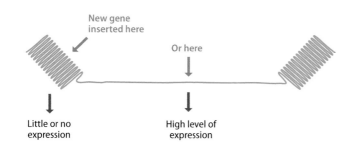

(B) Insulators overcome the positional effect

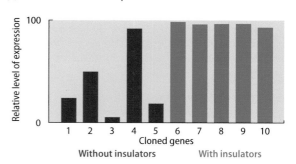

Figure 10.18 The positional effect. (A) A cloned gene that is inserted into a region of highly packaged chromatin will be inactive, but one inserted into open chromatin will be expressed. (B) The results of cloning experiments without (red) and with (blue) insulator sequences. When insulators are absent, the expression level of the cloned gene is variable, depending on whether it is inserted into packaged or open chromatin. When flanked by insulators, the expression level is consistently high because the insulators influence the chromatin structure at the insertion site.

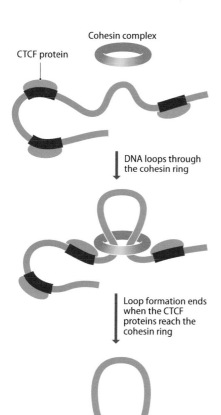

Figure 10.19 Formation of a loop of DNA. In this model, the loop is generated by the ring-shaped cohesin complex. Extrusion of the loop through the complex terminates when the CTCF proteins that are bound to the DNA meet at the cohesin ring. (From Martinelli C, Sartori P, Ledda M & Pannese E [2003] *Brain Res. Bull.* 61: 147–151. With permission from Elsevier.)

variable expression that was seen when the gene was cloned without the insulators (**Figure 10.18B**). The deduction from this and related experiments is that insulators can bring about modifications to chromatin packaging when inserted into a new site in the genome.

How insulators carry out their roles is not yet known, but it is presumed that the functional component is not the insulating sequence itself but the **insulator-binding proteins**, such as Su(Hw) in *Drosophila* and **CCCTC-binding factor** (**CTCF**) in mammals, that attach specifically to insulators. CTCF is thought to work in conjunction with ring-shaped protein complexes called **cohesins** to generate loops of DNA. The loop itself is generated by the cohesin complex, with the size of the loop specified by CTCF proteins, which bind to the DNA and prevent further loop formation when a pair of proteins meet as the DNA is threaded through the cohesin complex (**Figure 10.19**). This process could underlie the formation of TADs, as chromosome conformation capture reveals particularly strong interactions between the DNA at the two boundaries of a TAD, even though these segments might be separated by 1 Mb or more of DNA. The implication is that the boundaries are adjacent to one another in the nucleus. The presence of CTCF binding sites at TAD boundaries is therefore consistent with TAD formation by the cohesin-CTCF system. However, there are more than 10,000 CTCF binding sites in a mammalian genome, many of which are located within TADs and some in sequences that do not have insulator activity. This suggests that insulator-binding proteins play diverse roles in genome organization, and formation of TADs and sub-TADs is just one of these functions.

10.2 NUCLEOSOME MODIFICATIONS AND GENOME EXPRESSION

The previous sections have introduced us to the notion that chromatin structure influences genome expression, with the degree of chromatin packaging displayed by a particular segment of a chromosome determining whether RNA polymerase and the other transcription proteins can gain access to the genes within that segment. But so far we have simply examined the structure of chromatin and not asked how changes in that structure are brought about. This is the issue that we must now explore. We will discover that the key to chromatin structure is the pattern of chemical modifications that are made to the histone proteins, these modifications influencing the degree of packaging taken up by the nucleosomes. We will also discover that the DNA itself can be chemically modified by methylation, and that methylation results in parts of the genome becoming silenced.

The various chemical modifications that are made to the DNA and to the nucleosomes attached to the DNA are referred to as the **epigenome**, a term that is derived from the Greek word *epi*, and means *upon the genome*. We therefore make an important distinction between the nucleotide sequence of the genome, which is the same in all cells of a multicellular organism, and the chemical modifications of the epigenome, which are specific to a particular cell type. The epigenome can therefore be looked on as providing the flexibility that enables a single genome sequence to specify a range of cell phenotypes.

Acetylation of histones influences many nuclear activities, including genome expression

Nucleosomes appear to be the primary determinants of chromatin structure in eukaryotes, the precise chemical structure of the histone proteins contained within nucleosomes being the major factor influencing the degree of packaging displayed by a segment of chromatin. Histones can undergo various types of modification, the best studied of these being **histone acetylation** – the attachment of acetyl groups to lysine amino acids in the N-terminal regions of each of the core molecules (**Figure 10.20**). These N termini form tails that protrude from the nucleosome core octamer (**Figure 10.21**). Their acetylation reduces the affinity of the histones for DNA and possibly also reduces the interaction between individual nucleosomes, destabilizing nucleosome globules so that the chromatin takes up the more open 10 nm fiber structure. The histones in

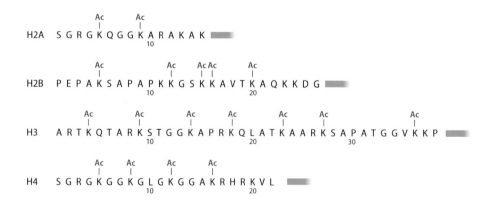

Figure 10.20 The positions at which acetyl groups are attached to lysines within the N-terminal regions of the four core histones. The sequences shown are those of the human histones. Each sequence begins with the N-terminal amino acid. Data taken from *HIST*ome2.

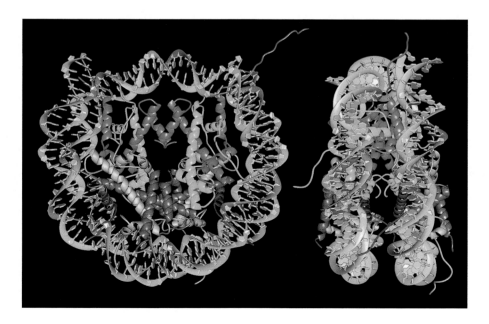

Figure 10.21 Two views of the nucleosome core octamer. The view on the left is downward from the top of the barrel-shaped octamer; the view on the right is from the side. The two strands of the DNA double helix wrapped around the octamer are shown in brown and green. The octamer comprises a central tetramer of two histone H3 (blue) and two histone H4 (bright green) subunits plus a pair of H2A (yellow)–H2B (red) dimers, one above and one below the central tetramer. The N-terminal tails of the histone proteins can be seen protruding from the core octamer. (From Luger K., Mäder A.W., Richmond R.K. et al [1997] *Nature* 389: 251–260. With permission from Springer Nature.)

heterochromatin are generally unacetylated, whereas those in active parts of the chromosomes are acetylated, a clear indication that this type of modification is linked to DNA packaging.

The relevance of histone acetylation to genome expression was underlined in 1996 when, after several years of trying, the first examples of **histone acetyltransferases** (**HATs**) – the enzymes that add acetyl groups to histones – were identified. It was then realized that some proteins that had already been shown to have important influences on genome expression had HAT activity. For example, one of the first HATs to be discovered, the *Tetrahymena* protein called p55, was shown to be a homolog of a yeast protein, GCN5, which was known to activate assembly of the transcription initiation complex (Section 12.3). Similarly, the mammalian protein called **p300/CBP**, which had been ascribed a clearly defined role in activation of a variety of genes, was found to be a HAT. These observations, plus the demonstration that different types of cell display different patterns of histone acetylation, underline the prominent role that histone acetylation plays in regulating genome expression.

Individual HATs can acetylate histones in the test tube, but most have negligible activity on intact nucleosomes, indicating that, in the nucleus, HATs almost certainly do not work independently, but instead form multiprotein complexes, such as the SAGA and ADA complexes of yeast and the TFTC complex of humans. These complexes are typical of the large multiprotein structures that catalyze and regulate the various steps in genome expression, many examples of which we will meet as we progress through the next few chapters. SAGA, for example, comprises 19 proteins with a combined molecular mass of 1.8 million. The complex forms a particle with dimensions 27 × 17 × 13 nm, which means that it is larger than the nucleosome core octamer, which with its associated DNA has a diameter of 11 nm. As well as GCN5 – the protein with HAT activity – the SAGA complex contains a set of proteins that interact with the **TATA-binding protein** (**TBP**), which initiates the process by which a gene is transcribed (Section 12.3), as well as five of the **TBP-associated factors** (**TAFs**), which help TBP fulfill its role. The complexity of SAGA and the other HAT complexes, and the presence within these complexes of proteins with distinct roles in the initiation of gene expression, indicates that the individual events that result in a gene becoming active are intimately linked, with histone acetylation being an integral part, but just one part, of the overall process.

There are at least five different families of HAT proteins. The GCN5-related acetyltransferases, or GNATs, which are components of SAGA, ADA, and TFTC, are clearly associated with activation of gene transcription, but are also involved in the repair of some types of damaged DNA, in particular double-strand breaks and lesions resulting from ultraviolet irradiation (Section 17.1). A second family of HATs, called MYST after the initial letters of four of the proteins in this family, is similarly involved in transcription activation and DNA repair, and has also been implicated in control of the cell cycle, though this may simply be another aspect of the DNA repair function, as the cell cycle stalls if the genome is extensively damaged (Section 15.5). Different complexes appear to acetylate different histones and some can also acetylate other proteins involved in genome expression, such as the general transcription factors TFIIE and TFIIF, which we will meet in Section 12.3. HATs are therefore emerging as versatile proteins that may have diverse functions in expression, replication, and maintenance of the genome.

Histone deacetylation represses active regions of the genome

Gene activation must be reversible; otherwise, genes that become switched on will remain permanently active. Hence it is no surprise that there is a set of enzymes that can remove acetyl groups from histone tails, overturning the transcription-activating effects of the HATs described above. This is the role of the **histone deacetylases** (**HDACs**). The link between HDAC activity and gene silencing was established in 1996, when mammalian HDAC1, the first of these enzymes to be discovered, was shown to be related to the yeast protein called Rpd3, which was known to be a repressor of transcription. The link between histone deacetylation and repression of transcription was therefore established in the same way as the link between acetylation and activation – by showing that two proteins that were initially thought to have different activities are in fact related. These are good examples of the value of homology analysis in studies of gene and protein function (Section 6.1).

HDACs, like HATs, are contained in multiprotein complexes. One of these is the mammalian Sin3 complex, which comprises at least seven proteins, including HDAC1 and HDAC2, along with others that do not have deacetylase activity but which provide ancillary functions essential to the process. Examples of ancillary proteins are RBBP4 and RBBP7, which are members of the Sin3 complex and are thought to contribute to the histone-binding capability. RBBP4 and RBBP7 were first recognized through their association with the retinoblastoma protein, which controls cell proliferation by inhibiting expression of various genes until their activities are required and which, when mutated, leads to cancer. This link between Sin3 and a protein implicated in cancer provides a powerful argument for the importance of histone deacetylation in gene silencing. Other deacetylation complexes include NuRD in mammals, which combines HDAC1, HDAC2, RBBP4, and RBBP7 with a different set of ancillary proteins, and yeast Sir2, which is different from other HDACs in that it has an energy requirement. The distinctive features of Sir2 show that HDACs are more diverse than originally realized, possibly indicating that novel roles for histone deacetylation are waiting to be discovered.

Studies of HDAC complexes are beginning to reveal links between the different mechanisms for genome activation and silencing. NuRD contains proteins that bind to methylated DNA (Section 10.3) and others that are very similar to components of the nucleosome remodeling complex Swi/Snf (see below). NuRD does in fact act as a typical nucleosome remodeling machine *in vitro*. Further research will almost certainly unveil additional links between what we currently look on as different types of chromatin modification system, but which in reality may simply be different facets of a single grand design.

Acetylation is not the only type of histone modification

Lysine acetylation/deacetylation is the best-studied form of histone modification, but it is by no means the only type. Various other kinds of covalent modification are known to occur:

- Methylation of lysine, arginine, and glutamine residues. Methylation was originally thought to be irreversible and hence responsible for permanent changes to chromatin structure. This view has been challenged by the discovery of enzymes that demethylate lysine and arginine residues, but it is still accepted that the effects of methylation are relatively long term.

- Phosphorylation of serine, threonine, and tyrosine residues.

- Ubiquitination of lysine residues at the C terminal regions of H2A, H2B, and H4. This modification involves addition of the small, common ('ubiquitous') protein called **ubiquitin**, or a related protein called **small ubiquitin-like modifier** or **SUMO**.

- Citrullination in the N-terminal regions of H2A, H3, and H4. Citrullination is the conversion of arginine to the related amino acid called citrulline, by replacement of the terminal =NH group of the arginine side chain with an =O group.

- Biotinylation or ribosylation of various lysine amino acids. Biotinylation is attachment of the organic compound biotin, and ribosylation involves attachment of a poly(ADP-ribose) unit.

As with acetylation, these other types of modification influence chromatin structure and have a significant impact on cellular activity. For example, ubiquitination of histone H2B is part of a general role that ubiquitin plays in control of the cell cycle. The effects of methylation of a pair of lysine amino acids at the fourth and ninth positions from the N terminus of histone H3 are particularly interesting. Methylation of lysine-9 forms a binding site for the HP1 protein, which induces chromatin packaging and silences gene expression, but this event is blocked by the presence of two or three methyl groups attached to lysine-4. Methylation of lysine-4 therefore promotes an open chromatin structure and is associated with active genes. In the part of human chromosome 11 that contains the β-globin genes, and probably elsewhere, lysine-4 methylation also prevents binding of the NuRD deacetylase to histone H3, ensuring that this histone remains acetylated. Lysine-4 methylation may therefore work hand-in-hand with histone acetylation to activate regions of chromatin.

Altogether, at least 80 sites in the four core histones are known to be subject to covalent modification (**Figure 10.22**). Our growing awareness of the variety of histone modifications that occur, and of the way in which different modifications work together, has led to the suggestion that there is a **histone code**, by which the pattern of chemical modifications specifies which regions of the genome are expressed at a particular time and dictates other aspects of genome biology, such as the repair of damaged sites and coordination of genome replication with the cell cycle (Table 10.1). This idea is still unproven, but it is clear that the pattern of specific histone modifications within the genome is linked closely to gene activity. Studies of human chromosomes 21 and 22, for example, have shown that regions within these chromosomes where lysine-4 of histone H3 is trimethylated and lysine-9 and lysine-14 are acetylated correspond to the transcription start points for active genes, and that dimethylated lysine-4 is also sometimes found in these regions (**Figure 10.23**). As with all aspects of chromatin modification, the key question is to distinguish cause and effect: are these patterns of histone modification the reason why these particular genes are active, or merely a by-product of the processes responsible for their activation?

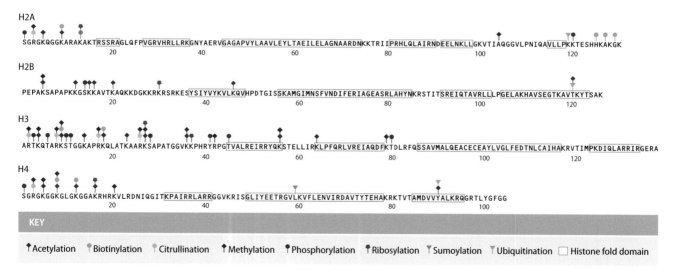

Figure 10.22   Positions where modifications occur in the four core histones. The sequences shown are those of the human histones. The histone fold domains are the parts of the histone proteins that bind directly to the DNA associated with the nucleosome. Data taken from *HISTome2*.

TABLE 10.1 FUNCTIONS OF HISTONE MODIFICATIONS

Histone	Amino acid	Modification	Function
H2A	serine-1	phosphorylation	transcription repression
	lysine-5	acetylation	transcription activation
	lysine-119	ubiquitination	spermatogenesis
H2B	lysine-5	monomethylation	transcription activation
	lysine-5	trimethylation	transcription repression
	lysine-12	acetylation	transcription activation
	serine-14	phosphorylation	DNA repair
	lysine-15	acetylation	transcription activation
	lysine-20	acetylation	transcription activation
	lysine-120	ubiquitination	transcription activation, meiosis
H3	threonine-3	phosphorylation	mitosis
	lysine-4	methylation	transcription activation
	lysine-9	methylation	transcription repression
	lysine-9	acetylation	transcription activation
	serine-10	phosphorylation	transcription activation
	threonine-11	phosphorylation	mitosis
	lysine-14	acetylation	transcription activation, DNA repair
	arginine-17	methylation	transcription activation
	lysine-18	acetylation	transcription activation, DNA repair
	lysine-27	monomethylation	transcription activation
	lysine-27	di-, trimethylation	transcription repression
	lysine-27	acetylation	transcription activation
	serine-28	phosphorylation	mitosis
	lysine-36	trimethylation	transcription activation
	lysine-79	mono-, dimethylation	transcription activation
H4	serine-1	phosphorylation	mitosis
	arginine-3	methylation	transcription activation
	lysine-5	acetylation	transcription activation
	lysine-8	acetylation	transcription activation, DNA repair
	lysine-12	acetylation	transcription activation, DNA repair
	lysine-16	acetylation	transcription activation, DNA repair
	lysine-20	mono-, trimethylation	transcription repression

Figure 10.23 The pattern of histone H3 modification is linked to gene activity. Segments of human chromosomes 21 and 22 are shown, each segment 100 kb in length. Regions that are enriched for dimethylated lysine-4, trimethylated lysine- 4, and acetylated lysine-9 and lysine-14 in lung fibroblasts are shown relative to the locations of known genes. The arrows indicate the directions in which the genes are transcribed.

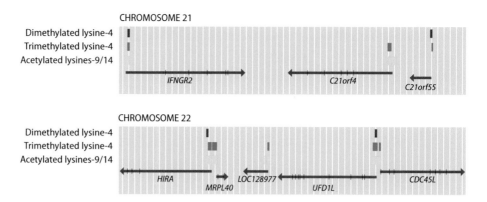

Nucleosome repositioning also influences gene expression

A second type of chromatin modification that can influence genome expression is **nucleosome remodeling**. This term refers to the repositioning of nucleosomes within a short region of the genome, so that DNA-binding proteins can gain access to their attachment sites. This does not appear to be an essential requirement for transcription of all genes, and in at least a few cases it is possible for a protein that switches on gene expression to achieve its effect either by binding to the surfaces of nucleosomes or interacting with the linker DNA without affecting the nucleosome positions. In other examples, repositioning of nucleosomes has been clearly shown to be a prerequisite for gene activation. This is apparent from the detection of **deoxyribonuclease (DNase) I hypersensitive sites** in the vicinity of genes that are being actively transcribed. For example, the DNA just upstream of the human β-globin gene cluster contains five DNase I hypersensitive sites (**Figure 10.24**). Each of these is a short region of DNA that is cleaved by DNase I more easily than other parts of the gene cluster. These sites are thought to coincide with positions where nucleosomes have been moved or are absent and which are therefore accessible to binding proteins that attach to the DNA. Similarly, activation of the *hsp70* gene of *Drosophila melanogaster*, which codes for a protein involved in folding other proteins (Section 13.4), is associated with the creation of a DNase I hypersensitive region upstream of the gene. Altogether there are almost 2.9 million DNase I hypersensitive sites in the human genome, the vast majority of these only detectable in those tissues within which the genes whose positions they mark are active.

Unlike acetylation and the other chemical modifications described in the previous section, nucleosome remodeling does not involve covalent alterations to histone molecules. Instead, remodeling is induced by an energy-dependent process that weakens the contact between the nucleosome and the DNA with which it is associated. Three distinct types of change can occur (**Figure 10.25**):

- Remodeling, in the strict sense, involves a change in the structure of the nucleosome, but no change in its position. The structural change may involve disruption of the core octamer and/or a weakening of the attachment between the DNA and the nucleosome. Exactly how this is achieved is unknown, but there is increasing evidence that remodeling is accompanied by replacement of one or more of the core histones with a variant such as H2AZ.

- Sliding, or *cis*-displacement, physically moves the nucleosome along the DNA.

- One or more adjacent nucleosomes can be removed.

Figure 10.24 DNase I hypersensitive sites upstream of the human β-globin gene cluster. A series of hypersensitive sites are located in the 20 kb of DNA upstream of the start of the β-globin gene cluster. Additional hypersensitive sites are seen immediately upstream of each gene, at the position where RNA polymerase attaches to the DNA. These hypersensitive sites are specific to different developmental stages, being seen only during the phase of development when the adjacent gene is active.

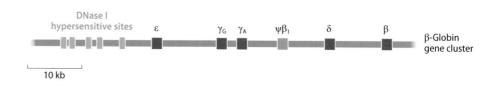

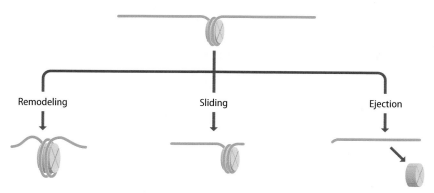

Figure 10.25 Nucleosome remodeling, sliding, and ejection.

As with histone acetyltransferases, the proteins responsible for nucleosome remodeling work together in large complexes. One of these is SWI/SNF, which is made up of at least 10–12 proteins and is present in many eukaryotes. The subunit proteins include some that have DNA-binding capability and others that detect the presence of histone acetylations and which presumably direct the complex to active parts of the genome. One subunit of the mammalian SWI/SNF complex is able to add ubiquitin tags to lysine-120 of histone H2B (see Figure 10.22), a modification that is associated with highly-transcribed regions. SWI/SNF therefore combines the two activities – nucleosome repositioning and histone modification – that are currently looked on as central to genome activation.

10.3 DNA MODIFICATION AND GENOME EXPRESSION

Important alterations in genome activity can also be achieved by making chemical changes to the DNA itself. These changes are associated with the semipermanent silencing of regions of the genome, possibly entire chromosomes, and often the modified state is inherited by the progeny arising from cell division. The modifications are brought about by **DNA methylation**.

Genome silencing by DNA methylation

In eukaryotes, cytosine bases in chromosomal DNA molecules are sometimes changed to 5-methylcytosine by the addition of methyl groups by enzymes called **DNA methyltransferases** (**Figure 10.26**). Cytosine methylation is relatively rare in lower eukaryotes, but in vertebrates up to 10% of the total number of cytosine bases in a genome are methylated, and in plants the figure can be as high as 30%. The methylation pattern is not random, instead being limited to the cytosine in some copies of the sequences 5′–CG–3′ and, in plants, 5′–CNG–3′. Two types of methylation activity have been distinguished (**Figure 10.27**). The first is **maintenance methylation** which, following genome replication, is responsible for adding methyl groups to the newly synthesized strand of DNA at positions opposite methylated sites on the parent strand. The maintenance activity therefore ensures that the two daughter DNA molecules retain the methylation pattern of the parent molecule, which means that the pattern can be inherited after cell division. The second activity is ***de novo* methylation**, which adds methyl groups at totally new positions and so changes the pattern of methylation in a localized region of the genome.

Both maintenance and *de novo* methylation result in repression of gene activity. This has been shown by experiments in which methylated or unmethylated genes have been introduced into cells by cloning and their expression levels measured: expression does not occur if the DNA sequence is methylated. The link with gene expression is also apparent when the methylation patterns in chromosomal DNAs are examined, these showing that active genes are located in unmethylated regions. For example, in humans, 40–50% of all genes

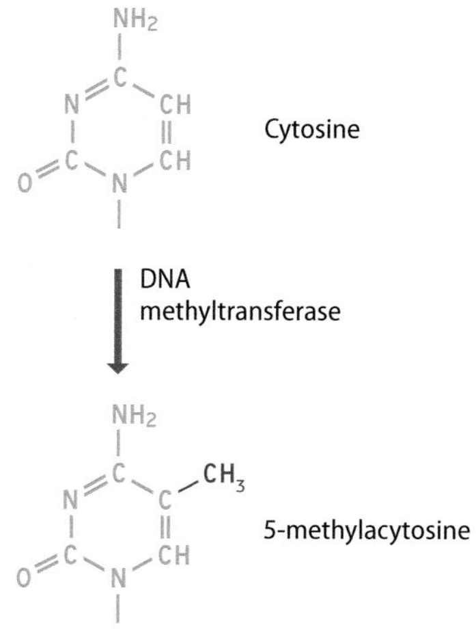

Cytosine

DNA methyltransferase

5-methylacytosine

Figure 10.26 Conversion of cytosine to 5-methylcytosine by a DNA methyltransferase.

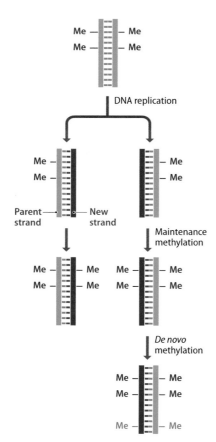

Figure 10.27 Maintenance methylation and de novo methylation.

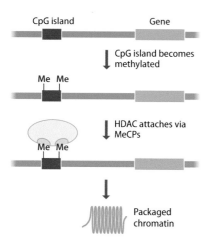

Figure 10.28 A model for the link between DNA methylation and genome expression. Methylation of the CpG island upstream of a gene provides recognition signals for the methyl-CpG-binding protein (MeCP) components of a histone deacetylase complex (HDAC). The HDAC modifies the chromatin in the region of the CpG island and hence inactivates the gene. Note that the relative positions and sizes of the CpG island and the gene are not drawn to scale.

are located close to CpG islands (Section 5.1), with the methylation status of the CpG island reflecting the expression pattern of the adjacent gene. Housekeeping genes – those that are expressed in all tissues – have unmethylated CpG islands, whereas the CpG islands associated with tissue-specific genes are unmethylated only in those tissues in which the gene is expressed. Note that because the methylation pattern is maintained after cell division, information specifying which genes should be expressed is inherited by the daughter cells, ensuring that in a differentiated tissue the appropriate pattern of gene expression is retained even though the cells in the tissue are being replaced and/or added to by new cells.

The importance of DNA methylation is underlined by studies of human diseases. The syndrome called ICF (immunodeficiency, centromere instability, and facial anomalies) which, as the name suggests, has wide-ranging phenotypic effects, is associated with undermethylation of various genomic regions, and is caused by a mutation in the gene for DNA methyltransferase 3b, one of the enzymes responsible for *de novo* methylation. The opposing situation – hypermethylation – is seen within the CpG islands of genes that exhibit altered expression patterns in certain types of cancer, although in these cases the abnormal methylation could equally well be a result rather than the cause of the disease state.

How methylation influences genome expression was a puzzle for many years. Now it is known that **methyl-CpG-binding proteins** (**MeCPs**) are components of both the Sin3 and NuRD histone deacetylase complexes. This discovery has led to a model in which methylated CpG islands are the target sites for attachment of HDAC complexes that modify the surrounding chromatin in order to silence the adjacent genes (**Figure 10.28**).

Methylation is involved in genomic imprinting and X inactivation

Further evidence, if it is needed, of the link between DNA methylation and genome silencing is provided by two intriguing phenomena called **genomic imprinting** and **X inactivation**.

Genomic imprinting is a relatively uncommon but important feature of mammalian genomes in which only one of a pair of genes, present on homologous chromosomes in a diploid nucleus, is expressed, the second being silenced by methylation. It also occurs in some insects (though apparently not *Drosophila melanogaster*) and some plants. It is always the same member of a pair of genes that is imprinted and hence inactive: for some genes this is the version inherited from the mother, and for other genes it is the paternal version. Almost 200 genes in humans and mice have been shown to display imprinting, including both protein-coding genes and genes specifying noncoding RNAs. Imprinted genes are distributed around the genome but tend to occur in clusters. For example, in humans there is a 2.2 Mb segment of chromosome 15, within which there are at least ten imprinted genes, and a smaller, 1 Mb region of chromosome 11 which contains at least eight imprinted genes.

An example of an imprinted gene in humans is *Igf2*, which codes for a growth factor, a protein involved in signaling between cells (Section 14.1). Only the paternal gene is active (**Figure 10.29**), because on the chromosome inherited from the mother, various segments of DNA in the region of *Igf2* are methylated, preventing expression of this copy of the gene. A second imprinted gene, *H19*, is located some 90 kb away from *Igf2*, but the imprinting is the other way round: the maternal version of *H19* is active and the paternal version is silent. Imprinting is controlled by **imprint control elements**, DNA sequences that are found within a few kilobases of clusters of imprinted genes. These centers mediate the methylation of the imprinted regions, but the mechanism by which they do this has not yet been described in detail. There is also uncertainty regarding the function of imprinting. One possibility is that it has a role in development, because artificially created parthenogenetic mice, which have two copies of the maternal genome, fail to develop properly. In addition, several of the genetic diseases that are associated with dysfunctional imprinting, such as Prader-Willi syndrome and Angelman syndrome, are characterized by developmental abnormalities. Individual genes displaying imprinting have been implicated in physiological functions as diverse as body temperature maintenance and behaviors such as sleep and maternal care. More subtle explanations of the role of imprinting, based on the evolutionary conflicts between the males and females of a species, have also been proposed.

X inactivation is less enigmatic. This is a special form of imprinting that leads to almost total inactivation of one of the X chromosomes in a female mammalian cell. It occurs because females have two X chromosomes, whereas males have only one. If both of the female X chromosomes were active then proteins coded by genes on the X chromosome might be synthesized at twice the rate in females compared with males. To avoid this undesirable state of affairs, one of the female X chromosomes is silenced and is seen in the nucleus as a condensed structure called the **Barr body**, which is comprised entirely of heterochromatin. Most of the genes on the inactivated X chromosome become silenced but, for reasons that are unknown, some 20% escape the process and remain functional.

Silencing occurs early in embryo development and is controlled by the X inactivation center (*Xic*), a discrete region present on each X chromosome. In a cell undergoing X inactivation, the inactivation center on one of the X chromosomes initiates the formation of heterochromatin, which spreads out from the nucleation point until the entire chromosome is affected, with the exception of a few short segments containing those genes that remain active. The process takes several days to complete. The exact mechanism is not understood, but it involves, although is not entirely dependent upon, a gene called *Xist*, located in the inactivation center, which is transcribed into a 17 kb noncoding RNA, copies of which coat the chromosome as heterochromatin is formed. At the same time, various histone modifications occur. Lysine-9 of histone H3 becomes methylated (recall that this modification is associated with genome inactivation – see Table 10.1), histone H4 becomes deacetylated (as usually occurs in heterochromatin), and the histone H2A molecules are replaced with a special histone, macroH2A1. Certain DNA sequences become hypermethylated by DNA methyltransferase 3a, although this appears to occur after the inactive state has been set up. X inactivation is heritable and is displayed by all cells descended from the initial one within which the inactivation took place.

In a normal diploid female, one X chromosome is inactivated and the other remains active. Remarkably, in females with unusual sex chromosome constitutions, the process still results in just a single X chromosome remaining active. For example, in those rare individuals with an XXX karyotype, two of the three X chromosomes are inactivated (**Figure 10.30A**). This means that there must be a mechanism by which the X chromosomes in the nucleus are counted and the appropriate number suppressed. In fact, this mechanism does not simply count the X chromosomes; it also counts the autosomes and compares the two numbers. This is evident because if the cell has four X chromosomes but is otherwise diploid, then three X chromosomes are inactivated, but if it is tetraploid (i.e., has four X chromosomes and also four copies of each autosome), then only two X chromosomes are inactivated (**Figure 10.30B**).

How the cell counts its chromosomes has puzzled cytogeneticists for many years, and continues to puzzle us, but it has been suggested that X inactivation is initiated by a regulatory molecule that is specified by an X chromosome gene and whose amount in the nucleus is therefore proportional to the number of X chromosomes that are present. The amount of regulator synthesized by a single X chromosome is insufficient to initiate X inactivation, so in male nuclei the single X chromosome remains active. If two X chromosomes are present, as in a female nucleus, then a greater amount of regulator is made, sufficient to reach a threshold concentration at which the regulator initiates X inactivation. Inactivation of a single X chromosome reduces the regulator below the threshold level and stabilizes the X-inactivated state. This model explains how the additional X chromosomes in an XXX nucleus are inactivated: in this case, reduction of the regulator below the threshold concentration requires inactivation of two

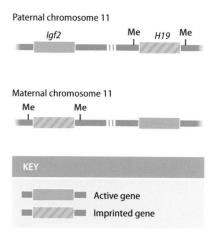

Paternal chromosome 11

Maternal chromosome 11

KEY

Active gene

Imprinted gene

Figure 10.29 A pair of imprinted genes on human chromosome 11. *Igf2* is imprinted on the chromosome inherited from the mother, and *H19* is imprinted on the paternal chromosome. The drawing is not to scale: the two genes are approximately 90 kb apart.

(A) Inactivation of unusual karyotypes

(B) Inactivation involves chromosome counting

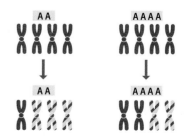

Figure 10.30 X inactivation. (A) If a single X chromosome is present, then no inactivation occurs; if there are three X chromosomes, then two are inactivated. (B) Three X chromosomes are inactivated in a cell that has a diploid complement of autosomes (AA) but four X chromosomes. In contrast, only two X chromosomes are inactivated if the cell is tetraploid (AAAA).

(A) One X chromosome

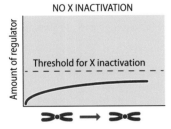

(B) Two X chromosomes

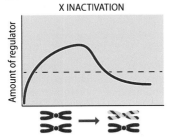

(C) Three X chromosomes

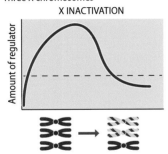

Figure 10.31   A model to explain how the cell counts the number of X chromosomes, so one remains active. According to this model, a regulatory molecule is coded by a gene on the X chromosome. The amount of regulator in the nucleus is proportional to the number of X chromosomes. (A) When there is only one X chromosome (e.g., in a male nucleus), insufficient regulator is synthesized to initiate X inactivation, so the single X chromosome remains active. (B) If there are two X chromosomes, the amount of regulator rises above the threshold and X inactivation occurs. Inactivation of one of the pair of X chromosomes reduces the regulator below the threshold level and stabilizes the X-inactivated state. (C) With three X chromosomes, a greater amount of regulator is made, again initiating X inactivation. Now it is necessary to inactivate two of the X chromosomes in order to reduce the level of the activator below the threshold.

X chromosomes (**Figure 10.31**). One candidate for the regulatory molecule is a protein called Rnf12, which plays a central role in initiating X inactivation and is specified by a gene located on the X chromosome close to *Xist*. Rnf12 is a ubiquitin ligase, a type of enzyme that adds ubiquitin to other proteins, which results in those proteins being degraded (Section 13.3). The targets for Rnf12 include at least two autosomal proteins that inhibit X inactivation, and by degrading these proteins Rnf12 switches on the X inactivation process. The Rnf12 concentration in the nucleus could therefore control X inactivation in a dosage-dependent manner as outlined in Figure 10.31. As the proteins targeted by Rnf12 are coded by autosomal genes their concentration will be dependent on the overall ploidy of the nucleus, explaining why only two X chromosomes are inactivated in a tetraploid cell.

SUMMARY

- The nuclear environment has a substantial and important impact on expression of the genome.

- A eukaryotic nucleus has a highly ordered internal architecture that includes structures associated with rRNA processing, mRNA splicing, and synthesis of small nuclear and small nucleolar RNAs.

- The most compact form of chromatin is heterochromatin, within which genes are inaccessible and cannot be expressed.

- The nucleus is thought to contain a fibrous matrix to which chromosomal DNA is attached, although the existence of this matrix has been questioned.

- Each chromosome has its own territory within the nucleus, these territories being separated from one another by nonchromatin regions within which the enzymes and other proteins involved in genome expression are located.

- A chromosome is made up of a series of topologically associating domains, each comprising a contiguous segment of chromatin folded into coils and loops.

- Each domain is delimited by a pair of insulators, which maintain the independence of the domain.

- Nucleosomes appear to be the primary determinants of genome activity in eukaryotes, not only by virtue of their positioning on a strand of DNA, but also because the precise chemical structure of the histone proteins contained within nucleosomes is the major factor determining the degree of packaging displayed by a segment of chromatin.

- Acetylation of lysine amino acids in the N-terminal regions of each of the core histones is associated with activation of a region of the genome, and deacetylation leads to genome silencing. Histones can also be modified by methylation, phosphorylation, ubiquitination, citrullination, biotinylation, and ribosylation.

- There may be a histone code that specifies how particular combinations of nucleosome modifications should be interpreted by the genome.

- Nucleosome repositioning is required for the expression of some, but not all, genes.

- Regions of the genome can also be silenced by DNA methylation, the relevant enzymes possibly working in conjunction with histone deacetylases.

- Methylation is responsible for genomic imprinting, which results in one of a pair of genes on homologous chromosomes becoming silenced, and X inactivation, which leads to the almost complete inactivation of one of the X chromosomes in a female nucleus.

SHORT ANSWER QUESTIONS

1. Describe the methods that have been used to examine the structural organization of the nucleus.

2. Distinguish between the terms constitutive heterochromatin and facultative heterochromatin.

3. What has chromosome painting and *in situ* sequencing revealed about the location of chromosomes within the nucleus?

4. Translocations occur at a higher frequency between certain pairs of chromosomes. What does this tell us about the distribution of chromosomes in the nucleus?

5. How are topologically associating domains identified in a eukaryotic chromosome?

6. What is the explanation of the positional effect that sometimes occurs when a gene is cloned in a eukaryotic host?

7. What are insulator sequences and what unique properties do they possess?

8. Give examples of histone acetyltransferases and describe the role of these enzymes in nucleosome modification.

9. What is the role of histone deacetylases in the regulation of genome expression?

10. Explain what is meant by the term 'histone code.'

11. Why is DNase I used to study changes in chromatin structure? What does the susceptibility of DNA to cleavage by DNase I tell us about gene expression?

12. How does DNA methylation influence genome activity?

IN-DEPTH PROBLEMS

1. To what extent can it be assumed that the picture of nuclear architecture built up by modern electron microscopy is an accurate depiction of the actual structure of the nucleus, as opposed to an artifact of the methods used to prepare cells for examination?

2. Explore and assess the histone code hypothesis.

3. In many areas of biology it is difficult to distinguish between cause and effect. Evaluate this issue with regard to nucleosome remodeling and genome expression: does nucleosome remodeling cause changes in genome expression or is remodeling the effect of these expression changes?

4. Maintenance methylation ensures that the pattern of DNA methylation displayed by two daughter DNA molecules is the same as the pattern on the parent molecule. In other words, the methylation pattern, and the information on gene expression that it conveys, is inherited. Other aspects of chromatin structure might also be inherited in a similar way. How do these phenomena affect the Mendelian view that inheritance is specified by genes?

5. What might be the means by which the numbers of X chromosomes and autosomes in a nucleus are counted so that the appropriate number of X chromosomes can be inactivated?

FURTHER READING

The internal structure of the nucleus and the matrix controversy

Fierz, B. (2019) Revealing chromatin organization in metaphase chromosomes. *EMBO J.* 38:e101699.

Galganski, L., Urbanek, M.O. and Krzyzosiak, W.J. (2017) Nuclear speckles: Molecular organization, biological function and role in disease. *Nucl. Acids Res.* 45:10350–10368.

Lippincott-Schwartz, J., Snapp, E.L. and Phair, R.D. (2018) The development and enhancement of FRAP as a key tool for investigating protein dynamics. *Biophys. J.* 115:1146–1155.

Maeshima, K., Ide, S. and Babokhov, M. (2019) Dynamic chromatin organization without the 30-nm fiber. *Curr. Opin. Cell Biol.* 58: 95–104.

Narwade, N., Patel, S., Alam, A., et al. (2019) Mapping of scaffold/matrix attachment regions in human genome: A data mining exercise. *Nucl Acids Res.* 47:7247–7261.

Ou, H.D., Phan, S., Deerinck, T.J., et al. (2017) ChromEMT: Visualizing 3D chromatin structure and compaction in interphase and mitotic cells. *Science* 357(6349):eaag0025.

Razin, A.V., Iarovaia, O.V. and Vassetzky, Y.S. (2014) A requiem for the nuclear matrix: From a controversial concept to 3D organization of the nucleus. *Chromosoma* 123:217–224.

Sawyer, I.A., Sturgill, D., Sung, M.-H., et al. (2016) Cajal body function in genome organization and transcriptome diversity. *Bioessays* 38:1197–1208.

Sureka, R. and Mishra, R. (2020) Identification of evolutionarily conserved matrix proteins and their prokaryotic origins. *J. Proteome Res.* 20:518–530.

Methods for studying DNA organization in the nucleus

De Wit, E. and de Laat, W. (2012) A decade of 3C technologies: Insights into nuclear organization. *Genes Dev.* 26:11–24.

Denker, A. and de Laat, W. (2016) The second decade of 3C technologies: Detailed insights into nuclear organization. *Genes Dev.* 30:1357–1382.

McCord, R.P., Kaplan, N. and Giorgetti, L. (2020) Chromosome conformation capture and beyond: Towards an integrative view of chromosome structure and function. *Mol. Cell* 77:688–708.

Payne, A.C., Chiang, Z.D., Reginato, P.L., et al. (2021) In situ genome sequence resolves DNA sequence and structure in intact biological samples. *Science* 371(6532):eaay3446.

Chromosome territories

Cremer, T. and Cremer, M. (2010) Chromosome territories. *Cold Spring Harb. Perspect. Biol.* 2:a003889.

Fritz, A.J., Sehgal, N., Pliss, A., et al. (2019) Chromosome territories and the global regulation of the genome. *Genes Chromosom. Cancer* 58:407–426.

Gerlich, D., Beaudouin, J., Kalbfuss, B., et al. (2003) Global chromosome positions are transmitted through mitosis in mammalian cells. *Cell* 112:751–764.

Szczepińska, T., Rusek, A.M. and Plewczynski, D. (2019) Intermingling of chromosome territories. *Genes Chromosom. Cancer* 58:500–506.

Topologically associating domains

Beagan, J.A. and Phillips-Cremins, J.E. (2020 On the existence and functionality of topologically associating domains. *Nat. Genet.* 52:8–16.

Bell, A.C., West, A.G. and Felsenfeld, G. (2001) Insulators and boundaries: Versatile regulatory elements in the eukaryotic genome. *Science* 291:447–450.

Fujioka, M., Mistry, H., Schedl, P., et al. (2016) Determinants of chromosome architecture: Insulator pairing in *cis* and in *trans*. *PLoS Genet.* 12:el005889.

Hansen, A.S., Cattoglio, C., Darzacq, X., et al. (2018) Recent evidence that TADs and chromatin loops are dynamic structures. *Nucleus* 9:20–32.

Matharu, N.K. and Ahanger, S.H. (2015) Chromatin insulators and topological domains: Adding new dimensions to 3D genome architecture. *Genes* 6:790–811.

Özdemir, I. and Gambetta, M.C. (2019) The role of insulation in patterning gene expression. *Genes* 10:767.

Pirrotta, V. (2014) Binding the boundaries of chromatin domains. *Genome Biol.* 15:121. *Insulator binding proteins.*

Sexton, T. and Cavalli, G. (2015) The role of chromosome domains in shaping the functional genome. *Cell* 160:1049–1059.

Sexton, T., Yaffe, E., Kenigsberg, E., et al. (2012) Three-dimensional folding and functional organization principles of the *Drosophila* genome. *Cell* 148:458–472.

Szabo, Q., Bantignies, F. and Cavalli, G. (2019) Principles of genome folding into topologically associating domains. *Sci. Adv.* 5:eaaw1668.

Covalent modification of histones

Ahringer, J. (2000) NuRD and SIN3: Histone deacetylase complexes in development. *Trends Genet.* 16:351–356.

Bannister, A.J. and Kouzarides, T. (2011) Regulation of chromatin by histone modifications. *Cell Res.* 21:381–395.

Bernstein, B.E., Kamal, M., Lindblad-Toh, K., et al. (2005) Genomic maps and comparative analysis of histone modifications in human and mouse. *Cell* 120:169–181. *Correlates the positions of histone modifications in chromosomes 21 and 22 with gene activity.*

Carrozza, M.J., Utley, R.T., Workman, J.L., et al. (2003) The diverse functions of histone acetyltransferase complexes. *Trends Genet.* 19:321–329.

Khorasanizadeh, S. (2004) The nucleosome: From genomic organization to genomic regulation. *Cell* 116:259–272. *Review of histone modification, nucleosome remodeling, and DNA methylation.*

Lachner, M., O'Carroll, D., Rea, S., et al. (2001) Methylation of histone H3 lysine 9 creates a binding site for *HP1* proteins. *Nature* 410:116–120.

Lawrence, M., Daujat, S. and Schneider, R. (2016) Lateral thinking: How histone modifications regulate gene expression. *Trends Genet.* 32:42–56.

Sneppen, K. and Dodd, I.B. (2012) A simple histone code opens many paths to epigenetics. *PLoS Comput. Biol.* 8:el002643. *Models the possible ways in which a histone code might operate.*

Timmers, H.T. and Tora, L. (2005) SAGA unveiled. *Trends Biochem. Sci.* 30:7–10.

Verdin, E., Dequiedt, F. and Kasler, H.G. (2003) Class II histone deacetylases: Versatile regulators. *Trends Genet.* 19:286–293.

Zhang, T., Cooper, S. and Brockdorff, N. (2015) The interplay of histone modifications – Writers that read. *EMBO Rep.* 16:1467–1481.

Nucleosome remodeling

Becker, P.B. and Workman, J.L. (2013) Nucleosome remodeling and epigenetics. *Cold Spring Harb. Perspect. Biol.* 5:a017905.

Euskirchen, G., Auerbach, R.K. and Snyder, M. (2012) SWI/SNF chromatin- remodeling factors: Multiscale analyses and diverse functions. *J. Biol. Chem.* 287:30897–30905.

Li, X.S., Trojer, P., Matsumura, T., et al. (2010) Mammalian SWI/SNF-A subunit BAF20/ARID1 is an E3 ubiquitin ligase that targets histone H2B. *Mol. Cell. Biol.* 30:1673–1688.

Tang, L., Nogales, E. and Ciferri, C. (2010) Structure and function of SWI/SNF chromatin remodeling complexes and mechanistic implications for transcription. *Prog. Biophys. Mol. Biol.* 102:122–128.

DNA methylation, imprinting, and X inactivation

Ballabio, A. and Willard, H.F. (1992) Mammalian X-chromosome inactivation and the *XIST* gene. *Curr. Opin. Genet. Dev.* 2:439–447.

Barlow, D.P. and Bartolomei, M.S. (2016) Genomic imprinting in mammals. *Cold Spring Harb. Perspect. Biol.* 6:a018382.

Brown, C.J. and Greally, J.M. (2003) A stain upon the silence: Genes escaping X inactivation. *Trends Genet.* 19:432–438.

Costanzi, C. and Pehrson, J.R. (1998) Histone macroH2A1 is concentrated in the inactive X chromosome of female mammals. *Nature* 393:599–601.

Du, Q., Luu, P.-L., Stirzaker, C., et al. (2015) Methyl-CpG-binding domain proteins: Readers of the epigenome. *Epigenomics* 7:1051–1073.

Mutzel, V. and Schulz, E.G. (2020) Dosage sensing, threshold responses, and epigenetic memory: A systems biology perspective on random X-chromosome inactivation. *Bioessays* 42:1900163.

Rauluseviciute, I., Drabløs, F. and Rye, M.B. (2020) DNA hypermethylation associated with upregulated gene expression in prostate cancer demonstrates the diversity of epigenetic regulation. *BMC Med. Genet.* 13:6.

Smith, Z.D. and Meissner, A. (2013) DNA methylation: Roles in mammalian development. *Nat. Rev. Genet.* 14:204–220.

Online resource

HISTome2: The HISTone Infobase. http://www.actrec.gov.in/histome2/. *Data on histones, modifications and modifying enzymes.*

THE ROLE OF DNA–BINDING PROTEINS IN GENOME EXPRESSION

In Chapter 10, we learned that the activity of individual genes is influenced by the degree of packaging exhibited by the chromatin domain in which those genes are contained, and also by the precise positioning of the nucleosomes in the vicinity of the genes. Although our focus was on higher-order structures such as chromatin and nucleosomes, we recognized that, at a more basic level, the accessibility of the genome is controlled by the interactions between histone proteins and the DNA to which they are attached. These interactions are an example of the central role that **DNA-binding proteins** play in genome expression. Histones are DNA-binding proteins, as are several of the proteins responsible for transcription of individual genes. There are also DNA-binding proteins that are involved in DNA replication, repair, and recombination, as well as a large group of related proteins that bind to RNA rather than DNA. Many DNA-binding proteins recognize specific nucleotide sequences and bind predominantly to these target sites, whereas others, such as histones, lack sequence specificity and attach at various positions in the genome.

In this chapter, we will examine the special structural features of DNA-binding proteins and explore how these features enable a DNA-binding protein to attach to the genome, focusing in particular on the way in which a sequence-specific binding protein recognizes its attachment sites. We will therefore begin by studying the methods used to elucidate the structures of DNA-binding proteins and to identify their binding positions on a DNA molecule.

11.1 METHODS FOR STUDYING DNA-BINDING PROTEINS AND THEIR ATTACHMENT SITES

The methods that are used to study the interactions between DNA-binding proteins and the genome fall into two categories:

- Various technologies, most importantly **X-ray crystallography** and **nuclear magnetic resonance (NMR) spectroscopy**, which are used to study the structures of DNA-binding proteins, and in particular to identify the structural features that enable a protein to make sequence-specific attachments to a DNA molecule.

- Methods which identify, with varying degrees of accuracy, the positions on a DNA molecule at which a DNA-binding protein attaches.

X-ray crystallography provides structural data for any protein that can be crystallized

Once a DNA-binding protein has been purified, attempts can be made to determine its structure, in isolation or attached to its binding site. This enables the precise structure of the DNA-binding part of the protein to be studied, and allows the identity and nature of the contacts with the DNA molecule to be elucidated. Two techniques – X-ray crystallography and NMR spectroscopy – are central to this area of research.

DOI: 10.1201/9781003133162-11

X-ray crystallography, which is a long-established technique whose pedigree stretches back to the late 19th century, is based on **X-ray diffraction**. X-rays have very short wavelengths, between 0.01 and 10 nm, which is comparable with the spacings between atoms in chemical structures. When a beam of X-rays is directed onto a crystal, some of the X-rays pass straight through, but others are diffracted and emerge from the crystal at a different angle from which they entered. If the crystal is made up of many copies of the same molecule, all positioned in a regular array, then different X-rays are diffracted in similar ways, resulting in overlapping circles of diffracted waves which interfere with one another. An X-ray-sensitive photographic film or electronic detector placed across the beam reveals a series of spots, an **X-ray diffraction pattern**, from which the structure of the molecule in the crystal can be deduced (**Figure 11.1**). The relative positioning of the spots indicates the arrangement of the molecules in the crystal, and their relative intensities provide information on the structure of the molecule. The more complex the molecule, the greater the number of spots and the larger the number of comparisons that must be made between them. Computational help is therefore required for all but the simplest molecules.

If successful, analysis of an X-ray diffraction pattern enables an **electron density map** to be constructed (**Figure 11.2**), which, with a protein, provides a chart of the folded polypeptide from which the positioning of structural features such as α-helices and β-sheets can be determined. If sufficiently detailed, the R groups of the individual amino acids in the polypeptide can be identified, and their orientations relative to one another established, allowing deductions to be made about the hydrogen bonding and other chemical interactions occurring within the protein structure. In the most successful projects, a resolution of 0.1 nm is possible, which means that structures just 0.1 nm apart can be distinguished. In proteins, most carbon–carbon bonds are 0.1–0.2 nm in length, and carbon–hydrogen bonds are 0.08–0.12 nm. This means that at 0.1 nm resolution, a very detailed three-dimensional model of the protein can be constructed. The one limitation of X-ray crystallography is that the protein must be crystallized before its structure can be studied by this method. For many proteins this is not a problem, as good-quality crystals can be obtained from a supersaturated solution. Other proteins, especially membrane proteins which have external hydrophobic regions, are less easy or even impossible to crystallize.

NMR spectroscopy is used to study the structures of small proteins

Like X-ray crystallography, NMR is a long-established technique that traces its origins to the early part of the 20th century, first being described in 1936. The principle of the technique is that rotation of an atomic nucleus generates a magnetic moment. When placed in an applied electromagnetic field, the spinning nucleus orientates in one of two ways, called α and β (**Figure 11.3**), the α-orientation (which is aligned with the magnetic field) having a slightly lower energy. In NMR spectroscopy the magnitude of this energy separation is determined by

(A) Production of a diffraction pattern

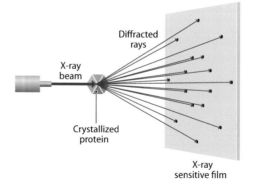

Diffracted rays

X-ray beam

Crystallized protein

X-ray sensitive film

(B) X-ray diffraction pattern for ribonuclease

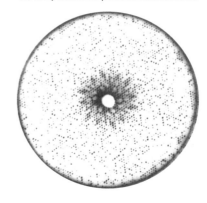

Figure 11.1 X-ray crystallography. (A) An X-ray diffraction pattern is obtained by passing a beam of X-rays through a crystal of the molecule being studied. (B) The diffraction pattern obtained with crystals of ribonuclease.

(A) Section of the ribonuclease electron density map

(B) Interpretation of an electron density map at 0.2 nm resolution revealing a tyrosine side-chain

Figure 11.2 Electron density maps. (A) Part of the electron density map derived from the X-ray diffraction pattern for ribonuclease. (B) Interpretation of an electron density map at 0.2 nm resolution revealing a tyrosine R group.

measuring the frequency of the electromagnetic radiation needed to induce the transition from α to β, this value being described as the **resonance frequency** of the nucleus being studied. The critical point is that although each type of nucleus has its own specific resonance frequency, the measured frequency is often slightly different from the standard value (typically by less than ten parts per million) because electrons in the vicinity of the rotating nucleus shield it to a certain extent from the applied magnetic field. This **chemical shift** (the difference between the observed resonance energy and the standard value for the nucleus being studied) enables the chemical environment of the nucleus to be inferred, and hence provides structural information. Particular types of analysis (called COSY and TOCSY) enable atoms linked by chemical bonds to the spinning nucleus to be identified; other analyses (e.g., NOESY) identify atoms that are close to the spinning nucleus in space but not directly connected to it.

To be suitable for NMR, a chemical nucleus must have an odd number of protons and/or neutrons; otherwise, it will not spin when placed in an electromagnetic field. Most protein NMR projects are ^{1}H studies, with the aim being to identify the chemical environments and covalent linkages of every hydrogen atom, and from this information to infer the overall structure of the protein. These studies are frequently supplemented by analyses of substituted proteins in which at least some of the carbon and/or nitrogen atoms have been replaced with the rare **isotopes** ^{13}C and ^{15}N, these also giving good results with NMR.

When successful, NMR results in the same level of resolution as X-ray crystallography and so provides very detailed information on protein structure. The main advantage of NMR is that it works with molecules in solution and so avoids the problems that sometimes occur when attempting to obtain crystals of a protein for X-ray analysis. Solution studies also offer greater flexibility if the aim is to examine changes in protein structure, as occur during protein folding or in response to addition of a substrate. The disadvantage of NMR is that it is only suitable for relatively small proteins. There are several reasons for this, one being the need to identify the resonance frequencies for each, or as many as possible, of the ^{1}H or other nuclei being studied. This depends on the various nuclei having different chemical shifts so that their frequencies do not overlap. The larger the protein, the greater the number of nuclei and the greater the chances that frequencies overlap and structural information is lost. Although this limits the applicability of NMR, the technique is still very valuable. There are many interesting proteins that are small enough to be studied by NMR, and important information can also be obtained by structural analysis of peptides which, although not complete proteins, can act as models for aspects of protein activity such as nucleic acid binding.

Gel retardation identifies DNA fragments that bind to proteins

Many DNA-binding proteins, such as histones, attach to DNA molecules of any nucleotide sequence, but others display sequence specificity and form stable attachments only at certain positions in the genome. A complement to structural

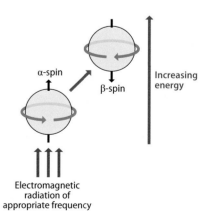

α-spin

β-spin

Increasing energy

Electromagnetic radiation of appropriate frequency

Figure 11.3 The basis of NMR. A rotating nucleus can take up either of two orientations in an applied electromagnetic field. The energy separation between the α and β spin states is determined by measuring the frequency of electromagnetic radiation needed to induce an α→β transition.

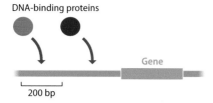

DNA-binding proteins

Gene

200 bp

Figure 11.4 Attachment sites for DNA-binding proteins are located immediately upstream of a gene.

studies of DNA-binding proteins is therefore provided by methods that enable the attachment sites for these sequence-specific proteins to be identified. Many of these methods were invented during the pre-genome era, and are designed to identify protein-binding sites in cloned fragments of DNA up to about 2 kb in length. The stimulus to development of these methods was the discovery that many **transcription factors** – proteins that control gene transcription – exert their effect by binding to sequence-specific attachment sites located immediately upstream of their target genes (**Figure 11.4**). This means that the sequence of a newly discovered gene, assuming that it includes the upstream region, provides immediate access to the binding sites for at least some of the proteins that control its expression.

The various methods that are available enable the protein-binding sites in a cloned DNA fragment to be located with different degrees of accuracy. The least accurate of these methods, but the easiest to carry out, makes use of the substantial difference between the electrophoresis properties of a 'naked' DNA fragment and one that carries a bound protein. Recall that DNA fragments are separated by agarose gel electrophoresis because smaller fragments migrate through the pore-like structure of the gel more quickly than larger fragments (Section 2.1). If a DNA fragment has a protein bound to it, then its mobility through the gel will be impeded, and the DNA-protein complex will form a band at a position nearer to the starting point (**Figure 11.5**). This is called **gel retardation**. In practice the technique is carried out with a collection of restriction fragments that span the region thought to contain a protein-binding site. The digest is mixed with an extract of nuclear proteins (assuming that a eukaryote is being studied) and retarded fragments are identified by comparing the banding pattern obtained after electrophoresis with the pattern for restricted fragments that have not been mixed with proteins. A nuclear extract is used because at this stage of the project the DNA-binding protein has not usually been purified. If, however, the protein is available, then the experiment can be carried out just as easily with the pure protein.

Protection assays pinpoint binding sites with greater accuracy

Gel retardation gives a general indication of the location of a protein-binding site in a DNA sequence, but does not pinpoint the site with great accuracy. Often the retarded fragment is several hundred base pairs in length, compared with the expected length of the binding site of a few tens of base pairs at most, and there is no indication of where in the retarded fragment the binding site lies. Also, if the retarded fragment is long, then it might contain separate binding sites for several proteins, or if it is quite small then there is the possibility that the binding site also includes nucleotides on adjacent fragments, ones that on their own do not form a stable complex with the protein and so do not lead to gel retardation. Gel retardation studies are therefore a starting point, but other techniques are needed to provide more accurate information.

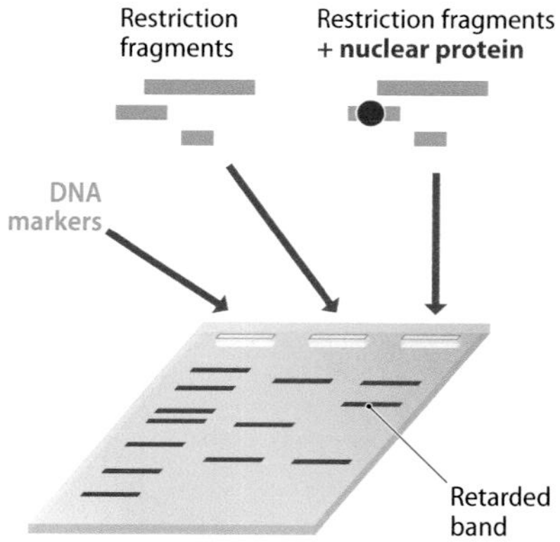

Modification protection assays can take over where gel retardation leaves off. The basis of these techniques is that if a DNA molecule carries a bound protein, then part of its nucleotide sequence will be protected from modification. There are two ways of carrying out the modification:

- By treatment with a nuclease, which cleaves all phosphodiester bonds except those protected by the bound protein.

- By exposure to a methylating agent, such as dimethyl sulfate, which adds methyl groups to guanine nucleotides. Any guanines protected by the bound protein will not be methylated.

Both methods utilize an experimental approach called **footprinting**. In nuclease footprinting, the DNA fragment being examined is labeled at one end, complexed with binding protein (as a nuclear extract or as pure protein), and treated with deoxyribonuclease I (DNase I). Normally, DNase I cleaves every phosphodiester

Restriction
fragments

Restriction fragments
+ **nuclear protein**

DNA
markers

Retarded
band

Figure 11.5 Gel retardation analysis.
A nuclear extract has been mixed with a DNA restriction digest, and a DNA-binding protein in the extract has attached to one of the restriction fragments. The DNA-protein complex has a larger molecular mass than the 'naked' DNA and so runs more slowly during gel electrophoresis. As a result, the band for this fragment is retarded and can be recognized by comparison with the banding pattern produced by restriction fragments that have not been mixed with the nuclear extract.

bond, leaving only the DNA segment protected by the binding protein. This is not very useful because it can be difficult to sequence such a small fragment. It is quicker to use the more subtle approach shown in **Figure 11.6**. The nuclease treatment is carried out under limiting conditions, such as a low temperature and/or very little enzyme, so that on average each copy of the DNA fragment suffers a single 'hit' – meaning that it is cleaved at just one position along its length. Although each fragment is cut just once, in the entire population of fragments all bonds are cleaved except those protected by the bound protein. The protein is now removed, the mixture electrophoresed, and the labeled fragments visualized. Each of these fragments has the label at one end and a cleavage site at the other. The result is a ladder of bands corresponding to fragments that differ in length by one nucleotide, with the ladder broken by a blank area in which no labeled bands occur. This blank area, or 'footprint,' corresponds to the positions of the protected phosphodiester bonds, and hence of the bound protein, in the starting DNA.

In the second modification protection method, instead of DNase I digestion, the fragments are treated with limited amounts of dimethyl sulfate so that on average a single guanine base is methylated in each fragment (**Figure 11.7**). Guanines that are protected by the bound protein cannot be modified. After removal of the protein, the DNA is treated with piperidine, which makes single-stranded cuts at the modified nucleotide positions. Electrophoresis is carried out in the presence of a denaturant, such as urea, so that the double-stranded molecules are separated into single strands, some of which have one labeled end and one end created by piperidine nicking. After electrophoresis, the control DNA strands – those not incubated with the nuclear extract – give a banding pattern that indicates the positions of all of the guanines in the restriction fragment, and the footprint seen in the banding pattern for the test sample shows which guanines were protected.

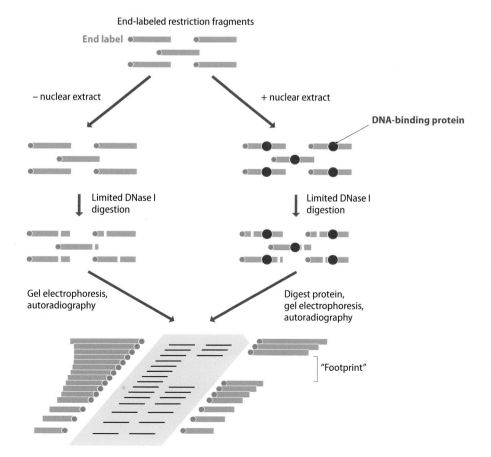

Figure 11.6 DNase I footprinting. Two samples of DNA, with and without binding protein, are treated with DNase I under limiting conditions so that, on average, each DNA fragment is cleaved just once. The binding protein is removed, and both samples are then separated by electrophoresis, and the labeled fragments are visualized. In the entire population of fragments, all the bonds are cleaved except for those protected by the bound protein; this protected area is visible as a 'footprint' in the ladder of bands resulting from electrophoresis. The restriction fragments used at the start of the procedure must be labeled at just one end. This is usually achieved by treating a set of longer restriction fragments with an enzyme that attaches labels at both ends, then cutting these labeled molecules with a second restriction enzyme, and purifying one of the sets of end fragments. The DNase I treatment is carried out in the presence of a manganese salt, which induces the enzyme to make random double-stranded cuts in the target molecules, leaving blunt-ended fragments. In this example, the fragments are radioactively labeled, and the banding patterns in the electrophoresis gel are revealed by autoradiography.

Figure 11.7   The dimethyl sulfate (DMS) modification protection assay. The technique is similar to DNase I footprinting. Instead of DNase I digestion, the fragments are treated with limited amounts of DMS so that a single guanine base is methylated in each fragment. Guanines that are protected by the bound protein cannot be modified. After removal of the protein, the DNA is treated with piperidine, which cuts at the modified nucleotide positions. For simplicity, the diagram shows the double-stranded molecules being cleaved at this stage. In fact, they are only nicked, as piperidine only cuts the strand that is modified, rather than making a double-stranded cut across the entire molecule. The samples are therefore examined by *denaturing* gel electrophoresis so that the two strands are separated. The resulting autoradiograph shows the sizes of the strands that have one labeled end and one end created by piperidine nicking. The banding pattern for the control DNA strands – those not incubated with the nuclear extract – indicates the positions of guanines in the restriction fragment, and the footprint seen in the banding pattern for the test sample shows which guanines were protected.

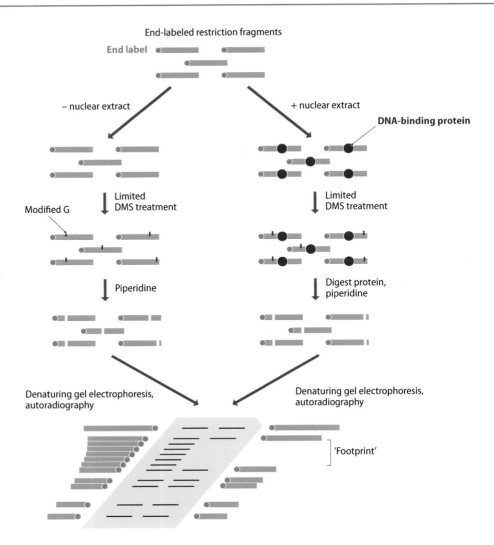

Modification interference identifies nucleotides central to protein binding

Modification protection should not be confused with **modification interference**, a different technique that provides an extra dimension to the study of protein binding. Modification interference works on the basis that if a nucleotide critical for protein binding is altered, for example, by addition of a methyl group, then binding may be prevented. One of this family of techniques is illustrated in **Figure 11.8**. The DNA fragment, labeled at one end, is treated with the modification reagent, in this case dimethyl sulfate, under limiting conditions so that just one guanine per fragment is methylated. Now the binding protein or nuclear extract is added, and the fragments are electrophoresed. Two bands are seen, one corresponding to the DNA-protein complex and one containing DNA without bound protein. The latter contains molecules that have been prevented from attaching to the protein because the methylation treatment has modified one or more guanines that are crucial for the binding. To identify which guanines are modified, the fragment is purified from the gel and treated with piperidine. As in the modification protection assay, the results of cleavage are visualized by denaturing gel electrophoresis. The length(s) of the labeled fragment(s) reveal which nucleotide(s) in the original fragment were methylated and hence identifies the positions in the DNA sequence of guanines that participated in the binding reaction. Equivalent techniques can be used to identify the A, C, and T nucleotides involved in binding.

Genome-wide scans for protein attachment sites

Gel retardation and modification assays have been used to map the attachment sites for DNA-binding proteins in the regions adjacent to many genes in different species. One outcome of this work has been the identification of consensus sequences for the binding sites of several important transcription factors. An example in animals is the **cyclic AMP response element** (**CRE**), which has the consensus sequence 5´–TGACGTCA–3´, and is the recognition site for the **cAMP response element-binding (CREB) protein**. This protein responds to elevated levels of **cyclic AMP** by binding to CRE sequences and activating the genes adjacent to those sequences. The cyclic AMP level in a cell is influenced by the presence of hormones that regulate physiological functions such as blood sugar level. The CREB protein is therefore the final link in the pathway that transduces the extracellular signal provided by the hormone into a change in the pattern of gene expression within the cell. The consensus sequences for most transcription factor binding sites are relatively unambiguous (compare the CRE consensus with, for example, the downstream intron-exon boundary sequence – Section 5.1) and we might therefore imagine that they could be located in a genome sequence simply by scanning for these motifs. In practice, this is not a regular component of a genome annotation project, because sequence scans usually result in a high proportion of false-positive identifications. These include sequences that have all the canonical features of a binding site but appear never to be used, and others that have the appropriate sequence but are methylated in most cells, the additional methyl groups blocking protein binding.

Even if sequence scanning was a reliable way of identifying protein-binding sites in a genome, the resulting catalog would not be particularly interesting. What we really wish to know is which sites are occupied in a particular tissue and how that occupancy pattern changes in response to external stimuli and during differentiation or development. That information enables genome annotation to move beyond the simple mapping of sequence motifs within a genome to a description of how the genome is expressed in different tissues and under different physiological conditions.

The most useful method for genome-wide detection of occupied protein-binding sites is **chromatin immunoprecipitation sequencing** or **ChIP-seq**. The initial step in a ChIP-seq experiment is treatment of the cells with formaldehyde to form DNA-protein crosslinks, similar to the use of formaldehyde in the chromosome conformation capture method (Section 10.1). The DNA-protein complexes are extracted and sonicated to break the DNA into fragments (**Figure 11.9**). Because of the crosslinking, transcription factors and other DNA-binding proteins that are attached to the genome are not displaced during these extraction and fragmentation steps. The next step is to separate out of the mixture just those DNA fragments that are attached to the transcription factor of interest. This can be achieved by immunoprecipitation of the desired DNA-protein complexes with an antibody that is specific for that transcription factor. The DNA-protein crosslinks are now broken by heating to 70°C and the DNA fragments that are released sequenced. The reads are mapped onto the genome sequence, revealing the location of those binding sites that were occupied by the transcription factor in the cells from which the extract was prepared. A related method called **ChIP-on-chip** or **ChIP-chip** achieves the same end, but instead of sequencing uses hybridization to a microarray to identify the DNA fragments that are released from the purified DNA-protein complexes.

11.2 THE SPECIAL FEATURES OF DNA-BINDING PROTEINS

The structures of many DNA-binding proteins and DNA-protein complexes have been determined by methods such as X-ray crystallography and NMR. When the structures of sequence-specific DNA-binding proteins are compared, it is immediately evident that the family as a whole can be divided into a limited number

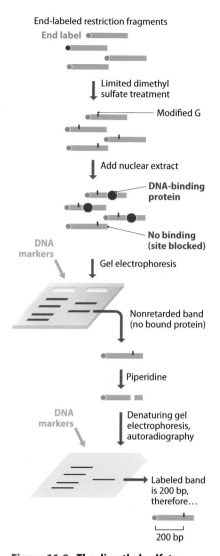

Figure 11.8 The dimethyl sulfate (DMS) modification interference assay. The technique is similar to the DMS protection assay in that DMS is used to modify guanines in the DNA fragments. However, DMS treatment is carried out before the binding protein is introduced, so the effects of modification on binding can be assayed. As in Figure 11.7, the diagram is simplified by showing double-stranded molecules being cleaved by piperidine, when in fact piperidine only cuts the strand that is modified.

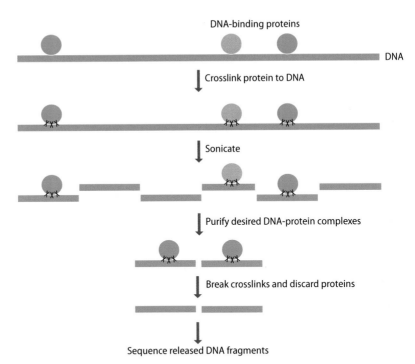

Figure 11.9 Chromatin immunoprecipitation sequencing (ChIP-seq).

of different groups according to the structure of the segment of the protein that interacts with the DNA molecule (**Table 11.1**). Each of these **DNA-binding motifs** is present in a range of proteins, often from very different organisms, and at least some of them probably evolved more than once. We will look at two in detail – the **helix–turn–helix (HTH) motif** and the **zinc finger** – and then briefly survey the others.

The helix–turn–helix motif is present in prokaryotic and eukaryotic proteins

The helix–turn–helix (HTH) motif was the first DNA-binding structure to be identified. As the name suggests, the motif is made up of two α-helices separated by a turn (**Figure 11.10**). The latter is not a random conformation but a specific structure, referred to as a **β-turn**, made up of four amino acids, the second of which is usually glycine. This turn, in conjunction with the first α-helix, positions the second α-helix on the surface of the protein in an orientation that enables it to fit inside the major groove of a DNA molecule. This second α-helix is therefore the **recognition helix** that makes the vital contacts which enable the DNA sequence to be read. The HTH structure is usually 20 or so amino acids in length and so is just a small part of the protein as a whole. Some of the other parts of the protein might also form attachments with the surface of the DNA molecule, primarily to aid the correct positioning of the recognition helix within the major groove.

Many prokaryotic and eukaryotic DNA-binding proteins utilize an HTH motif. In bacteria, HTH motifs are present in some of the best-studied regulatory proteins, which switch on and off the expression of individual genes. An example is the **lactose repressor**, which regulates expression of the lactose operon in *Escherichia coli* (Section 12.3). The various eukaryotic HTH proteins also include many whose DNA-binding properties are important in the developmental regulation of genome expression, such as the **homeodomain** proteins, whose roles we will examine in Section 14.3. The homeodomain is an extended HTH motif made up of 60 amino acids which form four α-helices, numbers 2 and 3

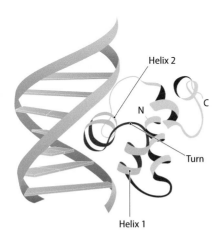

Figure 11.10 The helix–turn–helix motif. The illustration shows the orientation of the helix–turn–helix motif (in red) of a DNA-binding protein in the major groove of a DNA double helix. 'N' and 'C' indicate the N- and C-termini of the motif, respectively.

TABLE 11.1 DNA-BINDING MOTIFS

Motif	Examples of proteins with this motif
Sequence-specific DNA-binding motifs	
Helix–turn–helix family	
Standard helix–turn–helix	*Escherichia coli* lactose repressor
Homeodomain	*Drosophila* Antennapedia protein
Paired homeodomain	Metazoan Pax transcription factors
POU domain	Metazoan transcription factors PIT-1, OCT-1, OCT-2
Winged helix–turn–helix	Eukaryotic transcription factors TFIIE, TFIIF
Zinc-finger family	
Cys_2His_2 finger	Eukaryotic transcription factor TFIIIA
GATA zinc finger	GATA family of eukaryotic transcription factors
Treble clef finger	Metazoan nuclear receptor transcription factors
Basic helix–loop–helix	Eukaryotic MYC transcription factors
Ribbon–helix–helix	Bacterial MetJ, Arc, and Mnt repressors
High mobility group (HMG) box	Mammalian sex determination protein SRY
TBP domain	Eukaryotic TATA-binding protein
β-Barrel dimer	Papillomavirus E2 protein
Rel homology domain (RHB)	Mammalian transcription factor NF-κB
Nonspecific DNA-binding motifs	
Histone fold	Eukaryotic histones
HU/IHF motif*	Bacterial HU and IHF proteins
Polymerase cleft	DNA and RNA polymerases

*The HU/IHF motif is a nonspecific DNA-binding motif in bacterial HU proteins (Section 8.1) but directs sequence-specific binding of the IHF (integration host factor) protein.

separated by a β-turn, with number 3 acting as the recognition helix and number 1 stabilizing the structure (**Figure 11.11**).

Other versions of the HTH motif found in eukaryotes include:

- The **POU domain**, which is usually found in proteins that also have a homeodomain, the two motifs probably working together by binding different regions of a double helix. The name 'POU' comes from the initial letters of the names of the first proteins found to contain this motif.

- The **winged helix–turn–helix** motif, which is another extended version of the basic HTH structure, this one with a third α-helix on one side of the HTH motif and a β-sheet on the other side.

Many proteins, prokaryotic and eukaryotic, possess an HTH motif, but the details of the interaction of the recognition helix with the major groove are not exactly the same in all cases. The length of the recognition helix varies, generally being longer in eukaryotic proteins, the orientation of the helix in the major groove is not always the same, and the positions within the recognition helix of those amino acids that make contacts with the nucleotides are different.

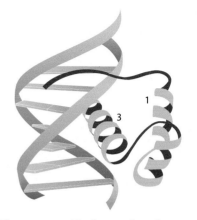

Figure 11.11 The homeodomain motif. The first three helices of a typical homeodomain are shown with helix 3 orientated in the major groove. Helices 1–3 run in the N→C direction along the motif.

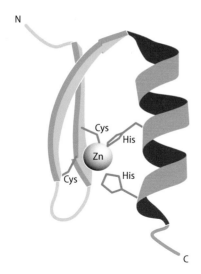

Figure 11.12 The Cys₂His₂ zinc finger. This particular zinc finger is from the yeast SWI5 protein. The zinc atom is held between two cysteines within the β-sheet of the motif and two histidines in the α-helix. The orange lines indicate the R groups of these amino acids. 'N' and 'C' indicate the N- and C-termini of the motif, respectively.

Figure 11.13 The treble clef zinc finger. Each zinc atom is shown coordinated between four cysteines. 'N' and 'C' indicate the N- and C-termini of the motif, respectively.

Zinc fingers are common in eukaryotic proteins

The second type of DNA-binding motif that we will look at in detail is the zinc finger, which is rare in prokaryotic proteins but very common in eukaryotes (see Table 7.3). There are more than 40 different versions of the zinc-finger motif. The first to be studied in detail was the **Cys₂His₂ finger**, which comprises a series of 12 or so amino acids, including two cysteines and two histidines, which form a segment of β-sheet followed by an α-helix. These two structures, which form the 'finger' projecting from the surface of the protein, hold between them a bound zinc atom, coordinated with the two cysteines and two histidines (**Figure 11.12**). The α-helix is the part of the motif that makes the critical contacts within the major groove, its positioning within the groove being determined by the β-sheet, which interacts with the sugar-phosphate backbone of the DNA, and the zinc atom, which holds the β-sheet and α-helix in the appropriate positions relative to one another. The α-helix of a Cys₂His₂ finger is therefore a recognition helix, similar to the second helix of the helix–turn–helix structure. Other versions of the zinc finger differ in the structure of the finger, some lacking the β-sheet component and consisting simply of one or more α-helices, and the precise way in which the zinc atom is held in place also varies. For example, the **multicysteine zinc fingers** lack histidines, the zinc atom being coordinated between four cysteines. The multicysteine group includes the **GATA zinc fingers**, which are found in the GATA family of transcription factors.

An interesting feature of the zinc finger is that multiple copies of the finger are sometimes found on a single protein. Several proteins have two, three, or four fingers, but there are examples with many more than this – there are 36 Cys₂His₂ zinc fingers in the human ZNF91 transcription factor. In most cases, the individual zinc fingers are thought to make independent contact with the DNA molecule, but in some cases the relationship between different fingers is more complex. In one particular group of transcription factors – the nuclear receptor family – two α-helices containing six cysteines combine to coordinate two zinc atoms in a single DNA-binding domain, larger than a standard zinc finger, called a **treble clef finger** (**Figure 11.13**). Within this motif, it appears that one of the α-helices enters the major groove, whereas the second makes contact with other proteins.

Other nucleic acid-binding motifs

The various other DNA-binding motifs that have been discovered in different proteins include:

- **The basic helix–loop–helix** motif, which is distinct from the HTH family, and is found in a number of eukaryotic transcription factors. The first α-helix includes a region containing a high proportion of basic amino acids (e.g., arginine, histidine and lysine), this region binding to the major groove of the DNA (**Figure 11.14**). The remainder of the helix–loop–helix structure facilitates dimerization, so the active transcription factor consists of two subunits, which might have identical (homodimer) or different (heterodimer) structures.

- The **ribbon–helix–helix** motif, which is one of the few motifs that achieves sequence-specific DNA binding without making use of an α-helix as the recognition structure. Instead, the ribbon (i.e., two strands of a β-sheet) makes contact with the major groove (**Figure 11.15**). Ribbon–helix–helix motifs are found in some gene regulation proteins in bacteria.

- The **high mobility group (HMG) box** domain is approximately 75 amino acids in length and comprises three α-helices that form an L shape. The high mobility group proteins are a large and diverse collection of chromatin proteins that were originally classified together because of their electrophoretic properties. They include several sequence-specific transcription factors, which typically have a single HMG box, and other DNA-binding proteins that lack sequence specificity. Members of the latter group, which are involved in processes such as DNA replication and repair, usually have more than one HMG box, and recognize distorted regions of a DNA molecule, such as occur when a nucleotide has undergone chemical modification due to the action of a mutagen (Section 17.2).

RNA-binding proteins also have specific motifs that form attachments with RNA molecules, most of these acting in a sequence-independent manner. The most important are as follows:

- The **RNA recognition domain** comprises four β-strands and two α-helices in the order β–α–β–β–α–β. The two central β-strands make the critical attachments with a single-stranded RNA molecule. The RNA recognition domain is the commonest RNA-binding motif and a similar domain is also present in a few proteins that bind to single-stranded DNA.

- The **double-stranded RNA-binding domain** (**dsRBD**) is similar to the RNA recognition domain but has the structure α–β–β–β–α. The RNA-binding function lies between the β-strand and the α-helix at the end of the structure. As the name implies, the motif is found in proteins that bind double-stranded RNA.

- The **K homology** (**KH**) **domain** has the structure β–α–α–β–β–α, with the binding function between the pair of α-helices. There are two groups of KH domains: in one group, the three β-strands form an antiparallel β-sheet, and in the other group two of the strands are in the parallel conformation. Proteins of the first group are found mainly in eukaryotes, and those in the second group are found mainly in prokaryotes. KH domains are also found in some single-strand DNA-binding proteins.

- The **LSm fold** is found in the LSm family of RNA-binding proteins, which are components of eukaryotic small nuclear ribonucleoproteins and also include the prokaryotic Hfq protein (Section 12.1). The LSm fold comprises a five-stranded β-sheet with an N-terminal α-helix.

Additionally, the DNA-binding homeodomain may also have RNA-binding activity in some proteins. One ribosomal protein uses a structure similar to a homeodomain to attach to rRNA, and some homeodomain proteins, such as Bicoid of *Drosophila melanogaster* (Section 14.3), can bind both DNA and RNA.

11.3 THE INTERACTION BETWEEN DNA AND ITS BINDING PROTEINS

In recent years, our understanding of the interaction between the genome and DNA-binding proteins has begun to change. It has always been accepted that proteins that recognize a specific sequence as their binding site can locate this site by forming contacts with chemical groups attached to the bases that are exposed within the major and minor grooves that spiral around the double helix (see Figure 1.9). This is called **direct readout** and is still looked on as the predominant component of the interaction between a DNA-binding protein and its attachment site. What we are now beginning to recognize is that direct readout can be aided by the influence that the nucleotide sequence has on the precise conformation of the helix. These conformational features represent a second, less direct way in which the DNA sequence can influence protein binding.

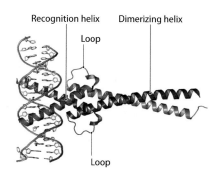

Figure 11.14 The basic helix–loop–helix motif. A dimer of two basic helix–loop–helix proteins is shown, the two subunits held together by interactions between the dimerizing helices. 'N' and 'C' indicate the N- and C-termini of the motif, respectively.

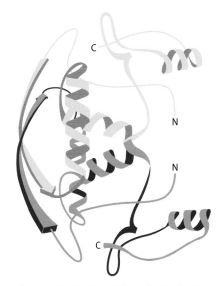

Figure 11.15 The ribbon–helix–helix motif. The drawing is of the ribbon–helix–helix motif of the *Escherichia coli* MetJ repressor, which consists of a dimer of two identical proteins, one shown in orange and the other in purple. The β-strands at the left of the structure make contact with the major groove of the double helix. 'N' and 'C' indicate the N- and C-termini of the motif, respectively.

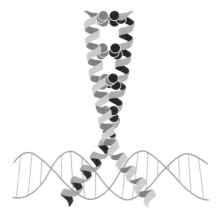

Figure 11.16   A leucine zipper. This is a bZIP type of leucine zipper. The red and orange structures are parts of different proteins. Each set of spheres represents the R group of a leucine amino acid. Leucines in the two helices associate with one another via hydrophobic interactions to hold the two proteins together in a dimer. In a bZIP structure, each dimerization helix is extended to form a basic domain DNA-binding motif. This is an α-helix containing a high proportion of basic amino acids, similar to the recognition helix of the basic helix–loop–helix motif. In this drawing, the basic domain helices are shown making contacts in the major groove.

Contacts between DNA and proteins

The contacts formed between DNA and its binding proteins are noncovalent. Within the major groove, hydrogen bonds form between the nucleotide bases and the R groups of amino acids in the recognition structure of the protein, whereas in the minor groove hydrophobic interactions are more important. On the surface of the DNA helix, the major interactions are electrostatic, between the negative charges on the phosphate component of each nucleotide and the positive charges on the R groups of amino acids such as lysine and arginine, although some hydrogen bonding also occurs. In some cases, hydrogen bonding on the surface of the helix or in the major groove is direct between DNA and protein, or alternatively the hydrogen bonding might be mediated by water molecules. Few generalizations can be made: at this level of DNA-protein interaction, each example has its own unique features and the details of the bonding have to be worked out by structural studies rather than by comparisons with other proteins.

Most proteins that recognize specific nucleotide sequences are also able to bind nonspecifically to other parts of a DNA molecule. In fact, it has been suggested that the amount of DNA in a cell is so large, and the numbers of each DNA-binding protein so small, that the proteins spend most of their time attached nonspecifically to DNA. The distinction between the nonspecific and specific forms of binding is that the latter is more favorable in thermodynamic terms. As a result, there is a high likelihood that a protein will form an attachment with its specific site even though there are literally millions of other sites to which it could bind nonspecifically. To achieve this thermodynamic favorability, the specific binding process must involve the greatest possible number of DNA-protein contacts, which explains in part why the recognition structures of many DNA-binding motifs have evolved to fit snugly into the major groove of the helix, where the opportunity for DNA-protein contacts is greatest. It also explains why some DNA-protein interactions result in conformational changes to one or other partner, increasing still further the complementarity of the interacting surfaces, and allowing additional bonding to occur.

The need to maximize contacts in order to ensure specificity is also one of the reasons why many DNA-binding proteins are dimers, consisting of two proteins attached to one another. This is the case for most HTH proteins and many of the zinc-finger type. Dimerization occurs in such a way that the DNA-binding motifs of the two proteins are both able to access the helix, possibly with some degree of cooperativity between them, so that the resulting number of contacts is greater than twice the number achievable by a monomer. As well as their DNA-binding motifs, many proteins contain additional characteristic domains that participate in the protein–protein contacts that result in dimer formation. One of these is the **leucine zipper**, which is an α-helix that coils more tightly than normal and presents a series of leucines on one of its faces. These can form contacts with the leucines on the zipper of a second protein, forming the dimer (**Figure 11.16**). As mentioned above, the basic helix–loop–helix motif is also able to form protein–protein contacts that result in homo- or heterodimer formation.

Direct readout of the nucleotide sequence

It was clear from the double-helix structure described by Watson and Crick (see Figure 1.9) that although the nucleotide bases are on the inside of the DNA molecule, they are not entirely buried, and some of the chemical groups attached to the purine and pyrimidine bases are accessible from outside the helix. Direct readout of the nucleotide sequence should therefore be possible without breaking the base pairs and opening up the molecule.

In order to form chemical bonds with groups attached to the nucleotide bases, a binding protein must make contact within one or both of the grooves on the surface of the helix. With the B-form of DNA, the identity and orientation of the exposed parts of the bases within the major groove is such that most sequences can be read unambiguously, whereas within the minor groove the identification is less precise (**Figure 11.17**). Direct readout of the B-form therefore predominantly involves contacts in the major groove. With other DNA types, there is much less information on the contacts formed with binding proteins, but the picture is likely to be quite different. In the A-form, for example, the major groove is deep and narrow and less easily penetrated by any part of a protein molecule (see Figure 1.11). The shallower minor groove is therefore likely to play a more substantial role in direct readout. With Z-DNA, the major groove is virtually non-existent and direct readout is possible to a certain extent without moving beyond the surface of the helix.

An intriguing question is whether the specificity of DNA binding can be understood in sufficient detail for the nucleotide sequence of a protein's target site to be predicted from examination of the structure of the recognition helix of a DNA-binding motif. To date, this objective has largely eluded us, but it has been possible to deduce some rules for the interaction involving certain types of zinc finger. In these proteins, four amino acids – three in the recognition helix and one immediately adjacent to it – form critical attachments with the nucleotide bases of the target site. By comparing the sequences of amino acids in the recognition helices of different zinc fingers with the sequences of nucleotides at the binding sites, it has been possible to identify a set of rules governing the interaction. These enable the nucleotide sequence specificity of a new zinc-finger protein to be predicted, admittedly with the possibility of some ambiguity, once the amino acid composition of its recognition helix is known.

The conformation of the helix also influences protein binding

Originally it was thought that cellular DNA molecules have fairly uniform structures, made up mainly of the B-form of the double helix, perhaps with some Z-DNA near the ends of a molecule. We now recognize that DNA is much more polymorphic, and that it is possible for the A-, B-, and Z-DNA configurations, and intermediates between them, to coexist within a single DNA molecule, with different parts of the molecule having different structures. There may also be regions that are identified as B-DNA but in which the major and/or minor grooves have atypical dimensions. These conformational variations are sequence-dependent, being in part the result of the base-stacking interactions that occur between adjacent base pairs. As well as being responsible, along with base-pairing, for the stability of the helix, the base-stacking also influences the amount of rotation that occurs around the covalent bonds within individual nucleotides and hence determines the conformation of the helix at a particular position. The rotational possibilities in one base pair are influenced, via the base-stacking interactions, by the identities of the neighboring base pairs. This means that the nucleotide sequence indirectly affects the overall conformation of the helix, possibly providing structural information that a binding protein can use to help it locate its appropriate attachment site on a DNA molecule.

Although the vast majority of DNA-binding proteins make attachments only with the B-form of the double helix, one type of binding structure is known to be specific for Z-DNA. This is the **Z-binding** or **Zα domain**, which is present in four proteins that are able to bind both to Z-DNA and to double-stranded RNAs that have adopted a left-handed helical conformation. An example is the adenosine deaminase acting on RNA (ADAR1) protein which, as its name implies, acts mainly on RNA, converting adenosines to inosines, possibly as an antiviral activity. The ADAR1 protein also binds to Z-DNA, although the biological role of this function is not understood. The Z-binding domain is, in structural terms, a version of the winged helix–turn–helix motif, but it lacks extensive amino acid similarity with those winged helix–turn–helix motifs that make sequence-specific contacts in the major groove of B-DNA. Instead, the Z-binding domain is thought to recognize the configuration of the phosphate groups in a short stretch

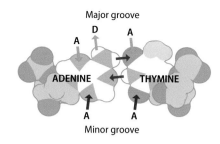

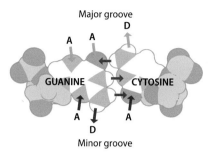

Figure 11.17 Recognition of base pairs by direct readout. The A–T and G–C base pairs are shown as space-filling models with the sugar-phosphate components in green, nitrogen atoms in blue, oxygen atoms in red, and the methyl group of thymine in gray. Green arrows indicate the positions of hydrogen bond acceptors (arrows labeled 'A') and hydrogen bond donors (arrows labeled 'D') in the major groove, and purple arrows indicate these groups in the minor groove. In the major groove, the arrangement of the hydrogen bond donors and acceptors, and the presence of the bulge caused by the methyl group, enable the two base pairs to be distinguished. In the minor groove, the geometries of the hydrogen bond acceptors are similar for both base pairs, and the only distinctive feature is the hydrogen bond donor from the amino group of guanine. It is therefore less easy to distinguish the two base pairs by direct readout of the minor groove. This is why the recognition structures of most sequence-specific DNA-binding proteins make contact with the major rather than minor groove.

of the Z-DNA sugar-phosphate backbone. As well as binding directly to Z-DNA, ADAR1 is also able to induce some segments of B-DNA to adopt the Z conformation, specifically in regions where the nucleotide sequence gives some predisposition toward formation of Z-DNA. Conversion of one DNA form into another is also a feature of several transcription factors, including the **TATA-binding protein** (**TBP**), which plays a central role in the initiation of gene transcription (Section 12.3). The **TBP domain** possessed by this protein forms a saddle-like structure that encloses part of the double helix (see Figure 12.27). When attaching to the DNA, TBP induces a conformational change at its target site from B-DNA to the A-form.

The shape of the helix is also affected by smaller-scale changes that do not induce a complete conversion of the B-form into A- or Z-DNA. For example, certain DNA sequences result in a narrowing of the minor groove of the B-form, this narrowing being one of the features involved in the sequence specificity of some homeodomain proteins. An abrupt alteration in the linearity of the helix, referred to as a **DNA kink**, occurs when the stacking between two adjacent bases is disrupted. This is more prevalent at certain dinucleotide positions, such as TA, and is utilized by the *Eco*RV restriction endonuclease and a variety of other proteins in the process by which they bind to their DNA attachment sites. A related conformational change is **DNA bending**, which is induced by various nucleotide sequences, for example, two or more groups of repeated adenines, each group comprising 3–5 As with the individual groups separated by 10 or 11 nucleotides. When this sequence is present, the DNA will bend at the 3′ end of the adenine-rich region. DNA bending may aid attachment of the *Drosophila* homeodomain protein called fushi tarazu to its target site, and has also been implicated in the DNA-binding mechanism for several viral and bacteriophage proteins.

SUMMARY

- The central players in genome expression and other aspects of genome activity are DNA-binding proteins that attach to the genome in order to perform their biochemical functions.

- The structures of these proteins have been studied by X-ray crystallography and nuclear magnetic resonance spectroscopy.

- The binding positions of these proteins on DNA molecules can be identified by gel retardation analysis and delineated in greater detail by modification assays. Genome-wide scans for the binding sites of a protein can be carried out by chromatin immunoprecipitation sequencing.

- DNA-binding proteins are able to attach to specific DNA sequences by virtue of special domains that form interactions with the double helix.

- The helix–turn–helix motif is a common domain in prokaryotic and eukaryotic DNA-binding proteins.

- Zinc fingers are common DNA-binding domains in eukaryotic proteins.

- There are also RNA-binding proteins with domains specific for attachment to RNA polynucleotides.

- Many DNA-binding proteins act as dimers, contacting the helix at two positions simultaneously. Special structures on the protein surface, such as leucine zippers, aid dimerization.

- Within the major groove of the B-form of the double helix, the identity of nucleotides can be determined from the positions of the chemical groups attached to the purine and pyrimidine nucleotides.

- Some proteins specifically recognize the Z-form of DNA or may induce a conformational change from B-DNA to Z-DNA when they bind. The TATA-binding protein induces a B- to A-DNA change.

- Specific attachment of a binding protein can be influenced by various indirect effects that the nucleotide sequence has on the conformation of the helix, including narrowing of the minor groove and the formation of kinks and bends.

SHORT ANSWER QUESTIONS

1. Describe how X-ray crystallography is used to study protein structures.

2. Outline the strengths and weaknesses of nuclear magnetic resonance spectroscopy as a means of studying protein structure.

3. Explain how gel retardation analysis is used to identify protein-binding sites in a DNA molecule.

4. How are modification assays used to delineate the positions at which proteins bind to DNA molecules?

5. Describe the techniques that are available for scanning a genome to locate the positions of the attachment sites of a particular DNA-binding protein.

6. Describe the different types of helix–turn–helix motif that are known in prokaryotic and/or eukaryotic DNA-binding proteins.

7. What are the general properties of the Cys_2His_2 zinc-finger motif and how does this motif bind to DNA?

8. Describe three examples of RNA-binding domains.

9. What is a leucine zipper?

10. Explain how direct readout of a nucleotide sequence is possible without detaching the two polynucleotides of a double helix.

11. What indirect effects does nucleotide sequence have on the structure of the double helix, and how might these effects influence the attachment of DNA-binding proteins?

IN-DEPTH PROBLEMS

1. DNA does not form crystals but X-ray diffraction analysis was very important in the work that led to discovery of the double-helix structure. Explain how X-ray diffraction analysis can be used with DNA.

2. The resolution achievable by NMR is directly related to the field strength of the magnet that is used. Explore how this relationship has affected development of NMR over the last 30 years, and speculate about the future potential of the procedure.

3. Describe how a DNA chip or microarray might be used in a genome-wide scan for protein-binding sites in a DNA molecule.

4. Why are there so many different types of DNA-binding motif?

5. To what extent is it possible to use the amino acid sequence of a zinc finger in order to deduce the nucleotide sequence of the binding site for a protein containing one or more copies of that finger?

FURTHER READING

X-ray crystallography and NMR spectroscopy

Cavalli, A., Salvatella, X., Dobson, C.M., et al. (2007) Protein structure determination from NMR chemical shifts. *Proc. Natl. Acad. Sci. USA* 104:9615–9620.

Cavanagh, J., Fairbrother, W.J., Palmer, A.G., et al. (2006) *Protein NMR Spectroscopy: Principles and Practice*, 2nd ed. Academic Press, London.

Garman, E.F. (2014) Developments in X-ray crystallographic structure determination of biological macromolecules. *Science* 343:1102–1108.

Su, X.-D., Zhang, H., Terwilliger, T.C., et al. (2015) Protein crystallography from the perspective of technology developments. *Crystallogr. Rev.* 21:122–153.

Methodology for identifying protein-binding sites

Galas, D. and Schmitz, A. (1978) DNAse footprinting: A simple method for the detection of protein-DNA binding specificity. *Nucl. Acids Res.* 5:3157–3170.

Garner, M.M. and Revzin, A. (1981) A gel electrophoresis method for quantifying the binding of proteins to specific DNA regions: Application to components of the *Escherichia coli* lactose operon regulatory system. *Nucl. Acids Res.* 9:3047–3060. *Gel retardation.*

Mundade, R., Ozer, H.G., Wei, H., et al. (2014) Role of ChIP-seq in the discovery of transcription factor binding sites, differential gene regulation mechanism, epigenetic marks and beyond. *Cell Cycle* 13:2847–2852.

Park, P.J. (2009) ChIP-seq: Advantages and challenges of a maturing technology. *Nat. Rev. Genet.* 10:669–680.

DNA- and RNA-binding motifs

Fierro-Monti, I. and Mathews, M.B. (2000) Proteins binding to duplexed RNA: One motif, multiple functions. *Trends Biochem. Sci.* 25:241–246.

Gangloff, Y.G., Romier, C., Thuault, S., et al. (2001) The histone fold is a key structural motif of transcription factor TFIID. *Trends Biochem. Sci.* 26:250–257.

Harrison, S.C. and Aggarwal, A.K. (1990) DNA recognition by proteins with the helix-turn-helix motif. *Annu. Rev. Biochem.* 59:933–969.

Herr, W., Sturm, R.A., Clerc, R.G., et al. (1988) The POU domain: A large conserved region in the mammalian *pit*-1, oct-1, oct-2, and *Caenorhabditis elegans unc-86* gene products. *Genes Dev.* 2:1513–1516.

Hudson, W.H. and Ortlund, E.A. (2014) The structure, function and evolution of proteins that bind DNA and RNA. *Nat. Rev. Mol. Cell Biol.* 15:749–760.

Kluska, K., Adamczyk, J., and Krężel, A. (2018) Metal binding properties, stability and reactivity of zinc fingers. *Coord. Chem. Rev.* 367:18–64.

Mackay, J.P. and Crossley, M. (1998) Zinc fingers are sticking together. *Trends Biochem. Sci.* 23:1–4.

Malhotra, S. and Sowdhamini, R. (2015) Collation and analyses of DNA-binding protein domain families from sequence and structural databanks. *Mol. Biosyst.* 11:1110–1118.

Najafabadi, H.S., Mnaimneh, S., Schmitges, F.W. et al. (2015). C_2H_2 zinc finger proteins greatly expand the human regulatory lexicon. *Nat. Biotechnol.* 33:555–562.

Schreiter, E.R. and Drennan, C.L. (2007) Ribbon–helix–helix transcription factors: Variations on a theme. *Nat. Rev. Microbiol.* 5:710–720.

Stros, M., Launholt, D. and Grasser, K.D. (2007) The HMG-box: A versatile protein domain occurring in a wide variety of DNA-binding proteins. *Cell. Mol. Life Sci.* 64:2590–2606.

Teichmann, M., Dumay-Odelot, H., and Fribourg, S. (2012) Structural and functional aspects of winged-helix domains at the core of transcription initiation complexes. *Transcription* 3:2–7.

Valverde, R., Edwards, L. and Regan, L. (2008) Structure and function of KH domains. *FEBS J.* 275:2712–2726.

Interactions between DNA and DNA-binding proteins

Dutta, S., Madan, S., and Sundar, D. (2016) Exploiting the recognition code for elucidating the mechanism of zinc finger protein-DNA interactions. *BMC Genom.* 17:1037.

Halford, S.E. and Marko, J.F. (2004) How do site-specific DNA-binding proteins find their targets? *Nucl. Acids Res.* 32:3040–3052.

Kielkopf, C.L., White, S., Szewczyk, J.W., et al. (1998) A structural basis for recognition of A•T and T•A base pairs in the minor groove of B-DNA. *Science* 282:111–115.

Rohs, R., Jin, X., West, S.M., et al. (2010) Origins of specificity in protein-DNA recognition. *Annu. Rev. Biochem.* 79:233–269.

Siggers, T. and Gordân, R. (2013) Protein-DNA binding: Complexities and multi-protein codes. *Nucl. Acids Res.* 42:2099–2111.

Van Quyen, D., Ha, S.C., Lowenhaupt, K., et al. (2007) Characterization of DNA-binding activity of Z alpha domains from poxviruses and the importance of the beta-wing regions in converting B-DNA to Z-DNA. *Nucl. Acids Res.* 35:7714–7720.

TRANSCRIPTOMES

The transcriptome is the collection of RNA molecules present in a cell. These RNA molecules are the transcripts of the genes that are active in that particular cell, or whose recent activity gave rise to transcripts that have not yet been degraded. In the past, the transcriptome has been defined as simply the mRNA content of the cell, and hence a reflection of the capacity of the cell to make proteins. In recent years, the term has been expanded to encompass all of a cell's RNA, acknowledging the greater perception that we now have of the importance of noncoding RNA in diverse cellular activities.

We will begin this chapter by examining the components of prokaryotic and eukaryotic transcriptomes and surveying the methods used to catalog the RNAs present in individual transcriptomes. We are already familiar with those methods as they are the same ones that are used to map RNAs onto genomes during annotation projects (Section 5.3). We will then study the events that result in synthesis, processing, and degradation of individual components of a transcriptome, and how those events are controlled, which will enable us to understand how the composition of a transcriptome is maintained by the cell and altered in response to changing environmental or physiological conditions.

12.1 THE COMPONENTS OF THE TRANSCRIPTOME

In most prokaryotes and eukaryotes, the coding component of the transcriptome – the mRNAs – makes up less than 5% of the total RNA content of the cell. The remainder of the transcriptome consists of noncoding RNA molecules that are not translated into protein but nonetheless have important functional roles in the cell (Section 1.2). Two types of noncoding RNA have been known since the 1950s. These are ribosomal RNAs (rRNAs), which are components of ribosomes, and transfer RNAs (tRNAs), which carry amino acids to the ribosome and ensure that these amino acids are linked together in the order specified by the nucleotide sequence of the mRNA that is being translated (Section 13.3).

In addition to rRNA and tRNA, the cells of most organisms also contain a variety of other noncoding RNAs, several of which have been discovered very recently and whose functions are still not clear. In eukaryotes, these RNAs are divided into two groups based on size. Those shorter than 200 nucleotides are called **short noncoding RNAs (sncRNAs)**, and the longer ones, unsurprisingly, are called **long noncoding RNAs (lncRNAs)**. This division is not generally used with prokaryotic noncoding RNAs, because other than the rRNAs, most of these are shorter than 400 nucleotides.

The mRNA fraction of a transcriptome is small but complex

Although only a small part of a transcriptome, rarely more than 5% of the total RNA, the mRNA fraction is complex, containing copies of many different genes. Studies of human transcriptomes, for example, have shown that between 10,000 and 15,000 genes are expressed in a single tissue, with the cerebellum and testes being the most complex in this regard, and skeletal muscle and liver least complex. These gene numbers do not, however, reveal the true complexity of the mRNA content of a transcriptome. As we will begin to appreciate, as we progress through this chapter, alternative splicing (Section 7.2), as well as the presence

DOI: 10.1201/9781003133162-12

Figure 12.1   A single gene can give rise to many mRNAs. In this example, the gene has four exons, three alternative transcription start-points, and three alternative end-points. Five of the transcripts that could be obtained from the gene are shown. If every combination of start- and end-points is allowed, and every possible combination of exons is used, then 135 different transcripts could be synthesized from this gene. This assumes that exons can be omitted from the mRNA, but that the order of exons cannot be changed (e.g., 1–3–4 is possible but not 1–4–3). This assumption is in accord with what we know about alternative splicing.

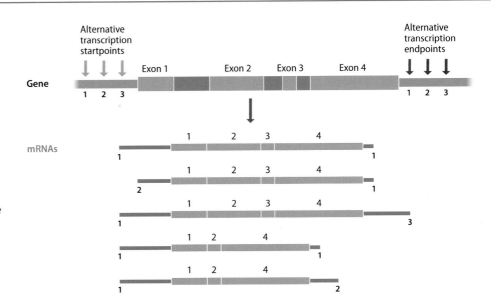

of multiple start-points (**alternative promoters** – Section 12.3) and end-points (**alternative polyadenylation** sites – Section 1.2) for some transcripts, means that an individual gene can give rise to many different mRNAs (**Figure 12.1**). Because of these alternative synthesis and processing pathways, the 10,000–15,000 active genes in a cell can give rise to a transcriptome that contains over 100,000 different mRNAs.

Although the identities of the different mRNAs in a transcriptome can be cataloged with some degree of accuracy, measuring the copy numbers of individual mRNAs is much more difficult. The relative abundances of different mRNAs can be assessed from the intensity of hybridization occurring at each position in a tiling array, or from the numbers of reads mapping to each gene in an RNA-seq database (Section 5.3), but converting these data into absolute numbers for individual transcripts is problematic. It would require controls containing known mRNA copy numbers from which a calibration curve could be constructed, and this is rarely attempted in a microarray or RNA-seq experiment. It is possible to estimate copy numbers of individual mRNAs by quantitative PCR (Section 2.2) or fluorescent *in situ* hybridization (Section 3.5), but neither of these methods is capable of a rigorous distinction between all possible splice variants. Bearing in mind these limitations, the available evidence suggests that a typical mammalian cell contains approximately 200,000 mRNA molecules in total, implying a mean copy number of about 15 for the mRNAs derived from a single gene. Undoubtedly, there is considerable variation around this mean. For example, in cerebral cells the CREB transcription factor, which binds to cyclic AMP response element (CRE) sites (Section 11.1), has been estimated to have an mRNA copy number of fewer than 25, but a repressor protein that also binds to CRE sites has an mRNA copy number of up to 240. There are also some tissues that have highly specialized biochemistries, which are reflected by transcriptomes in which one or a few mRNAs predominate. Wheat seeds are an example. The cells in the endosperm of wheat seeds synthesize large amounts of the gliadin proteins, which accumulate in the dormant grain and provide a source of amino acids for the germinating seedling. Within the developing endosperm, the gliadin mRNAs can make up as much as 30% of the total mRNA content of certain cells.

Short noncoding RNAs have diverse functions

Although tRNAs were discovered in 1958, the existence of other types of sncRNA was not suspected until the 1960s, when gel electrophoresis was first applied to RNA extracts. The gels revealed the presence in eukaryotes of small RNAs, with a relatively high uracil content, located primarily in the nucleus. To reflect the uracil content, these molecules are sometimes called **U-RNAs**, but the more commonly used term is **small nuclear RNA** (**snRNA**). The individual types are called U1-snRNA, U2-snRNA, etc., and their typical sizes are in the range of 105–190 nucleotides, though longer examples are known. Most snRNAs (the Sm subgroup) are transported to the cytoplasm where they attach to proteins to form complexes called **small nuclear ribonucleoproteins** (**snRNPs**) which then return to the nucleus, but a minor set (the Lsm subgroup) are converted into snRNPs in the nucleus, where they remain throughout their functional life. Most snRNPs are found within the nuclear regions called speckles (see Figure 10.2B), where they associate together to form **spliceosomes**, the structures that splice (i.e., remove introns from) pre-mRNA transcripts (Section 12.4). Splicing is the major function of snRNAs, but some other roles are also known, in particular for U7-snRNA, which is not present in spliceosomes and instead is involved specifically in the processing of histone pre-mRNAs.

Eukaryotic nuclei also contain a second major group of small noncoding RNAs, called **small nucleolar RNAs** (**snoRNAs**). As the name indicates, these molecules are located in the nucleoli (Section 10.1), where they participate in the chemical modification of rRNAs (Section 1.2). As with snRNAs, snoRNAs are divided into two subgroups, in this case depending on whether they aid in methylation or pseudouridylation reactions (see Figure 1.18). Other snoRNAs are responsible for chemical modification of snRNAs, which takes place in the Cajal bodies (see Figure 10.2A). This group of snoRNAs is therefore sometimes called **small Cajal body-specific RNAs** (**scaRNAs**).

Moving out of the nucleus, we encounter additional groups of small noncoding RNAs, which have a variety of roles, including the regulation of genome expression. These RNAs were first discovered in the late 1990s when processes responsible for **RNA silencing** in *Caenorhabditis elegans* and various species of plants were first described at the molecular level (Section 12.6). Since 2000, we have begun to realize that there are several types of these small regulatory RNAs, and that they play central roles in the control of expression of individual genes. The main groups are:

- **Short interfering RNAs** (**siRNAs**) are 20–25 nucleotides in length and are involved in the **RNA interference** pathway, which results in 'silencing' of particular mRNAs. Silencing occurs because an siRNA base-pairs with a region in the target mRNA, forming a double-stranded structure that is cut by a nuclease. We will examine the details of the process later in this chapter (Section 12.6).

- **MicroRNAs** (**miRNAs**) are similar to siRNAs and also participate in RNA interference. The difference is that miRNAs are synthesized by cleavage of precursor molecules that form stem-loop structures, whereas the precursors of siRNAs are linear molecules (**Figure 12.2**).

- **piwi-interacting RNAs** (**piRNAs**) are 25–30 nucleotides and hence slightly longer than siRNAs or miRNAs. They associate with piwi proteins, which were first discovered in *Drosophila melanogaster* but are now known to

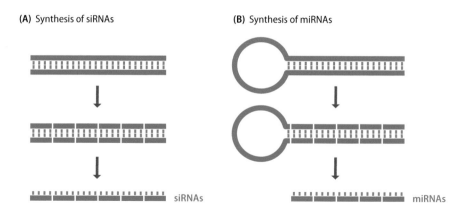

(A) Synthesis of siRNAs

(B) Synthesis of miRNAs

siRNAs

miRNAs

Figure 12.2 Synthesis of siRNAs and miRNAs. (A) siRNAs are obtained by cleavage of a linear double-stranded RNA precursor. (B) miRNAs are obtained by cleavage of the double-stranded part of a single-stranded RNA that has formed a stem-loop.

be present in many metazoans. Although piRNAs are the largest group of small regulatory RNAs in animal cells, their roles are not fully understood. They participate in RNA interference, but are also involved in other processes that result in repression of gene activity, and in particular prevent expression of the genes present in retrotransposons (Section 9.2), both by mRNA silencing and also by methylation of the DNA copies of these elements. Hybrid dysgenesis is thought to be caused by the absence of a particular type of piRNA in laboratory strains of fruit flies (see Figure 9.22).

Additional types of small noncoding RNAs have more specialist roles. **Vault RNAs** are present in protein–RNA complexes called vaults, which are found in most eukaryotic cells but whose functions are not known. Vaults tend to be associated with the nuclear pore complexes and so may aid transport of molecules into and out of the nucleus. Other protein–RNA structures include the eukaryotic signal recognition particle, which is involved in movement of newly translated polypeptides into the endoplasmic reticulum. This particle is a complex of six proteins and one noncoding RNA, the latter called the **7SL RNA**. Another eukaryotic noncoding RNA, the **7SK RNA**, is a component of a protein–RNA complex that controls the activity of a second protein complex, called P-TEFb (positive transcriptional elongation factor b). P-TEFb regulates the rate at which protein-coding genes are transcribed into mRNA. Various other noncoding RNAs are components of enzymes, such as ribonuclease P, which is involved in tRNA processing, and telomerase, the enzyme that prevents a chromosomal DNA molecule from decreasing in length every time the chromosome replicates (Section 15.4).

Bacterial and archaeal transcriptomes also contain a variety of noncoding RNAs, but these appear to be unique to prokaryotes. They have diverse roles in the control of gene expression, influencing biochemical activities such as the stress response and antibiotic resistance. Examples are the **transfer-messenger RNA**, which forms part of a recovery system that enables damaged mRNAs to be translated, and **Qrr RNAs**, which are involved in quorum sensing in *Vibrio* species, the process by which bacteria are able to sense and adapt to their local population density. Interestingly, Qrr is one of several bacterial RNAs that interacts with the Hfq protein, whose RNA-binding domain is an Lsm fold similar to that found in eukaryotic snRNPs (Section 11.2).

Long noncoding RNAs are enigmatic transcripts

One of the more remarkable discoveries arising from genome annotation projects has been the realization that eukaryotic transcriptomes contain many long noncoding RNAs that collectively cover a large proportion of the intergenic space within a genome. The latest annotation of the human genome, for example, includes 18,000 lncRNA genes, which give rise to over 50,000 transcripts due to variable spicing pathways and the use of alternative start-points and end-points. Many of these are located entirely within an intergenic region and are called **long intergenic noncoding RNAs** (**lincRNAs**); others map within the introns of mRNA transcription units, and others overlap with protein-coding exons, often being transcribed from the other strand of the double helix and hence giving rise to an antisense version of the mRNA (**Figure 12.3**). In

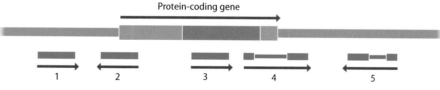

Figure 12.3 Examples of possible map positions for long noncoding RNAs (lncRNAs) and long intergenic noncoding RNAs (lincRNAs).

1 lincRNA, sense strand
2 lncRNA overlapping with protein-coding exon, antisense strand
3 lncRNA located in intron, sense strand
4 lncRNA with intron, sense strand
5 lincRNA with intron, antisense strand

some parts of a eukaryotic genome, multiple lncRNAs form overlapping arrays of sense and antisense transcripts covering tens of kilobases of sequence. Most lncRNAs are synthesized only in certain tissues or at certain developmental stages. Their individual copy numbers are, on average, much lower than mRNA copy numbers, and in many cases there may be as few as two or three copies of a particular lncRNA per cell.

Since their discovery, there has been extensive debate about the possible functionality of lncRNAs. Some are transcripts of pseudogenes and others contain open reading frames up to 100 codons in length, but by definition a lncRNA is noncoding and does not give rise to a functional protein: a few that have been shown to code for short peptides have been reassigned as mRNAs. Difficulties in assigning functions to lncRNAs led to the suggestion that many of these are **transcriptional noise**. According to this hypothesis, within the vast intergenic regions of a eukaryotic genome, there will inevitably be sequences that arise by chance mutation and resemble the promoters and other sequences that direct transcription of a functional gene. The cell might be unable to prevent these sequences from acting as transcription signals and will therefore make RNAs that it does not need.

Although the transcriptional noise hypothesis might account for some lncRNAs, it is becoming clear that many of these transcripts do in fact have roles in the cell. The group includes the *Xist* RNA, which mediates the X inactivation process in female mammals, as well as other RNAs that induce heterochromatin formation to bring about genomic imprinting (Section 10.3) and to silence genes not needed in differentiated tissues (Section 14.2). Other lncRNAs appear to be involved in the formation of speckles and other nuclear substructures (Section 10.1) and may participate in the RNA processing events occurring in these structures. A particularly interesting group influence gene expression by acting as **decoy** or **sponge RNAs** for regulatory proteins and small RNAs. The human lncRNA called *PANDA* provides an example. This 1.5 kb transcript contains binding sites for the NF-YA transcription factor, and so can act as a 'decoy' for these regulatory proteins. NF-YA, when not decoyed by binding to *PANDA*, activates a variety of target genes, including ones that induce **apoptosis** – programmed cell death. When *PANDA* lncRNAs are present, NF-YA binds to the decoy sites, reducing occupancy of the sites upstream of the target genes. These genes are therefore switched off and apoptosis does not occur (**Figure 12.4**). It has been proposed that *PANDA* mediates the balance between cell-cycle arrest and apoptosis in cells that have damaged genomes. DNA damage activates a second transcription factor, p53, which switches on genes that bring about cell-cycle arrest and, via NF-YA, other genes that induce apoptosis. The *PANDA* gene

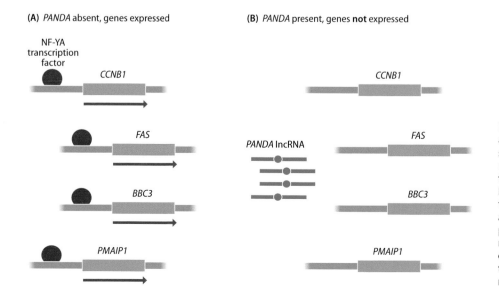

(A) *PANDA* absent, genes expressed

(B) *PANDA* present, genes **not** expressed

NF-YA transcription factor

CCNB1

CCNB1

PANDA lncRNA

FAS

FAS

BBC3

BBC3

PMAIP1

PMAIP1

Figure 12.4 *PANDA*, an example of a decoy lncRNA in the human transcriptome. *PANDA*, although an RNA molecule, is able to bind the transcription factor NF-YA. (A) When *PANDA* is absent, NF-YA attaches upstream of various target genes, including ones that induce apoptosis. (B) When *PANDA* lncRNAs are present, NF-YA binds to the RNA sites, reducing occupancy of the sites upstream of the target genes. These genes are therefore switched off and apoptosis does not occur.

is itself switched on by p53, resulting in synthesis of the lncRNA, which then prevents or delays activation of the apoptosis genes, thereby providing the cell with an opportunity to repair the damage and continue its active life, rather than immediately dying.

Defects in *PANDA* activity have been implicated in the development of both lung and esophageal cancer. More generally, up- or down-regulation of lncRNA expression has been associated with several different types of cancer, and levels of particular lncRNAs are increasingly being used as biomarkers in cancer diagnosis. These lncRNAs include prostate cancer-associated transcript 6 (PCAT6), which is a 1.2 kb lncRNA transcribed from human chromosome 1. Copy numbers of PCAT6 are higher in the transcriptomes of cancerous prostate tissues compared with normal prostates, and have a significant correlation with the potential of the cancer to undergo metastasis. Similar results have been reported for the association of PCAT6 with lung cancer, where there is evidence that the transcript plays an active role in cancer progression. Long noncoding RNAs are also being assigned important functions in other types of eukaryote: for example, in *Arabidopsis* the initiation of flowering is controlled, in part, by lncRNA activity, and *Drosophila* lncRNAs are involved in processes as diverse as aging and sleep behavior.

12.2 TRANSCRIPTOMICS: CATALOGING THE TRANSCRIPTOMES OF CELLS AND TISSUES

Transcriptomes are highly variable, reflecting the different patterns of genome expression that occur in different cell types. Cataloging the individual transcripts that are present in a transcriptome, and their relative amounts, is therefore one of the keys to understanding how the genome specifies the biochemical capabilities of different cell types and how these capabilities are modulated in response to changes in the environment and to external stimuli such as the presence of hormones. The methods used to catalog a transcriptome are called **transcriptomics**.

Microarray analysis and RNA sequencing are used to study the contents of transcriptomes

We are already familiar with the methods used to study transcriptomes as these are the same methods that are used for genome-wide mapping of RNA transcripts during genome annotation (Section 5.3). Transcriptomics initially depended almost entirely on hybridization of RNA or complementary DNA preparations to DNA chips, either tiling arrays or **microarrays**, the latter usually representing the coding sequences in a genome. Microarray analysis is still used routinely today, but largely in applied settings such as clinical diagnosis, where detection of an altered gene expression pattern might identify a diseased tissue or the onset of cancer. More recently, with the growing ease of short- and long-read sequencing, RNA-seq methods have become increasingly important, especially in research projects.

The one difference between the use of microarray analysis and RNA-seq in transcript mapping and in transcriptome analysis is that the latter often requires quantitative data. With RNA mapping, the only important information is that a transcript is present in the transcriptome. To understand the content of the transcriptome in detail, we also wish to know the relative abundances of individual RNAs, and to be able to compare these abundances in different transcriptomes. Often the key distinction between two tissues, or between the normal and diseased versions of a single tissue, is not the presence or absence of particular transcripts, but their relative amounts, which reflect the patterns of up- or down-regulation of gene expression that are occurring in those tissues.

If a microarray is being used, then a number of experimental variables must be controlled in order to be certain that a difference between the hybridization intensities for the same gene in two RNA samples is due to a genuine difference in the amount of the transcript. Errors can arise if the probe is not labeled to precisely the same extent in the two experiments or if there are variations in the effectiveness of the hybridization process. Even in a single laboratory, these factors can rarely be controlled with absolute precision, and exact reproducibility between different laboratories is more or less impossible. This means that the data analysis must include normalization procedures that enable results from different array experiments to be accurately compared. The arrays must therefore include negative controls so that the background can be determined in each experiment, as well as positive controls that should always give identical signals. For vertebrate transcriptomes, the actin gene is often used as a positive control as its expression level tends to be fairly constant in a particular tissue, regardless of the developmental stage or disease state. A more satisfactory alternative is to design the experiment so that the two transcriptomes can be directly compared, in a single analysis using a single array. This is done by labeling the cDNA preparations with different fluorescent probes, and then scanning the array at the appropriate wavelengths to determine the relative intensities of the two fluorescent signals at each position, and hence to determine the differences between the RNA contents of the two transcriptomes (**Figure 12.5**).

If RNA-seq is being used, then the relative amounts of transcripts in different transcriptomes are assessed from the number of reads that map to the gene of interest. The rationale is, of course, that if an RNA is relatively abundant in transcriptome A but less abundant in transcriptome B, then there will be more reads derived from the RNA in the RNA-seq dataset for transcriptome A compared with that for transcriptome B. As with microarray analysis, genes whose expression levels are expected to be the same in the two tissues are used as controls to enable the RNA-seq data to be normalized.

Presuming accurate comparisons can be made between two or more transcriptomes, quite complex differences in gene expression patterns can be distinguished. Genes that display similar expression profiles are likely to be ones with related functions, so as well as cataloging transcriptome components it is also possible to address questions relating to the functional annotation of a genome. To do this, a rigorous method is needed for comparing the expression profiles of different genes. The standard method, which can be applied both to both microarray and RNA-seq data, is called **hierarchical clustering**. This

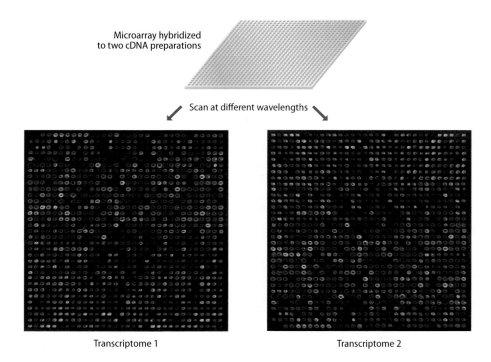

Microarray hybridized
to two cDNA preparations

Scan at different wavelengths

Transcriptome 1   Transcriptome 2

Figure 12.5 Comparing two transcriptomes in a single experiment. (From Strachan T, Abitbol M, Davidson D & Beckmann JS [1997] *Nat. Genet.* 16: 126–132. With permission from Macmillan Publishers Ltd.)

Figure 12.6   Comparing the expression profiles of five genes in seven transcriptomes. Seven transcriptomes have been prepared from cells at different periods after addition of an energy-rich nutrient to the growth medium. The expression profiles for five genes are shown, with red indicating high expression and green indicating low expression. After analysis of the data by hierarchical clustering, a dendrogram is constructed, showing the degree of relationship between the expression profiles of the five genes.

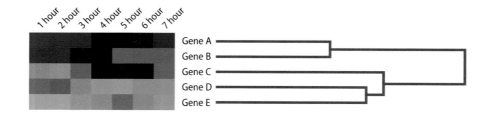

involves comparing the expression levels of every pair of genes in the transcriptomes that have been analyzed, and assigning a value that indicates the degree of relatedness between those expression levels. These data can then be expressed as a **dendrogram**, in which genes with related expression profiles are clustered together (**Figure 12.6**). The dendrogram therefore gives a clear visual indication of the possible functional relationships between genes.

Single-cell studies add greater precision to transcriptomics

Most biological tissues contain a number of cell types, at various stages of differentiation, with the activities occurring in individual cells dependent on the microenvironments within different parts of the tissue. All of this information is lost when RNA is prepared from a tissue sample, because the transcriptome that is cataloged, by microarray analysis or RNA-seq, is a composite with contributions from all of the cells in the sample. A much greater understanding of genome activity would be achieved if transcriptomes could be cataloged from different parts of a tissue, ideally from individual cells.

The feasibility of **single-cell RNA-seq** (**scRNA-seq**) as a means of cataloging a transcriptome from an individual mammalian cell was first demonstrated with mouse blastomeres that had been microdissected from four-cell embryos. The mRNA fraction was converted into cDNA, amplified by PCR and short sequence reads obtained by the SOLiD method. Transcripts were detected for almost 12,000 of the 20,000 protein-coding genes in the mouse genome and, importantly, alternative splicing variants could be distinguished from the sequences of reads that spanned the junctions between adjacent exons.

Can we go beyond the scRNA-seq of single microdissected cells and carry out transcriptome analysis of multiple cells obtained from a tissue sample? Physical and enzymatic methods can be used to disrupt most types of tissue in order to release the cells, so the challenge lies mainly with devising a high throughput procedure that enables a large number of cells to be typed in a single experiment. The most popular approach makes use of a microfluidic device to place each cell in an oil-enclosed droplet that also contains the reagents for RT-PCR along with barcoded primers immobilized on metal beads (**Figure 12.7**). The cells are then lysed within their beads and the mRNAs are converted into cDNAs that include the barcode sequences. The beads are collected, the cDNAs released, pooled, and sequenced, and the barcodes are then used to assign the reads to the individual cells. Because of the huge amount of short-read data that can be obtained in a single sequencing experiment, transcriptomes can be cataloged for between 500 and 10,000 cells in a single experiment.

Sometimes it is possible to assign transcriptomes to cell types by examining the expression profiles of those genes that are most variable within the dataset (**Figure 12.8**). However, there is no spatial information: it is not possible to refer back to images of the original tissue sample and identify which cell gave rise to which transcriptome. In some cases, it might be possible to use the identities of the genes whose transcripts are present to deduce which cell type a particular group of transcriptomes belongs to. Alternatively, individual transcriptomes can be mapped within a tissue by choosing one or more transcripts that are characteristic of a particular transcriptome, and then using fluorescent *in situ* hybridization (FISH) with a sample of the original tissue to identify those cells that contain the marker transcripts. Previously we have encountered FISH as a means

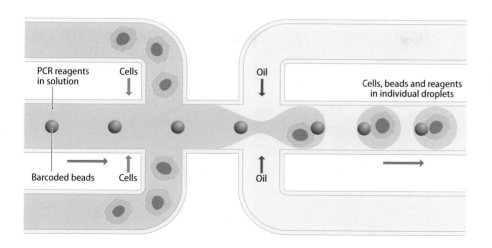

Figure 12.7 Using microfluidics to isolate single cells for scRNA-seq. The microfluidic device places each cell in an oil-enclosed droplet along with the reagents for RT-PCR along with barcoded primers immobilized on metal beads.

of detecting genes and other specific sequences in DNA molecules (Sections 3.5 and 10.1). FISH can also be directed at RNA and so can identify cells that contain particular transcripts (**Figure 12.9**). The presence of RNA in a nucleus does not interfere with the use of FISH to map DNA sequences, because the prehybridization steps include a heat treatment to denature the double-stranded DNA, which degrades most of the RNA. Conversely, if the heat step is omitted, the DNA remains in its double-stranded form and so cannot bind the FISH probes, which can now be directed specifically at the RNA content. **RNA-FISH** is not a high throughput method, as either a different fluorescent label has to be used for each transcript that is being monitored, or if a single or small number of labels is used, then multiple hybridizations have to be performed on the same tissue. The method cannot therefore be used to obtain extensive information on transcriptomes, but it can be used to identify marker transcripts in order to correlate the information from scRNA-SEQ with the structure of the tissue being studied.

There are, of course, many species that are naturally single-celled, both eukaryotes and prokaryotes, and it might be imagined that with these organisms scRNA-seq would be the method of choice for transcriptome studies. Unfortunately, RNA content becomes an issue. A typical mammalian cell contains 10–30 pg of RNA, which is sufficient to catalog a transcriptome by RNA-seq with a detection limit of 5–10 transcript copies, even if only the mRNA fraction is used. However, in a yeast cell, there is probably no more than 1.5 pg of RNA, and for a typical bacterium the content is 100 times less, in the range of 5–20 fg. Additionally, the copy numbers of individual transcripts tend to be lower in bacteria, and a detection limit down to one or two transcripts per cell is necessary in order to generate a comprehensive transcriptome catalog. Most transcriptome

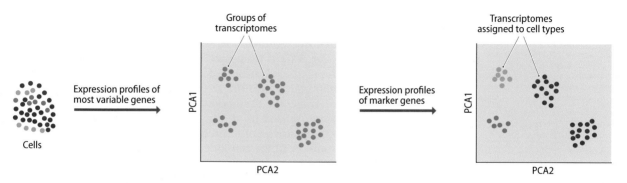

Figure 12.8 Identification of cell types from scRNA-seq data. The initial scRNA-seq dataset is filtered to remove reads for genes that have similar levels of expression in all the transcriptomes. The expression profiles for the more variable genes are then analyzed by a clustering method such as principal components analysis (PCA). In this example, PCA identifies five clusters. It may then be possible to use the expression profiles for marker genes (e.g., ones known to be highly expressed in a particular cell type) to assign the transcriptome clusters to the cell types that were present in the original tissue.

Figure 12.9 RNA-FISH. This experiment shows the importance of the Sir4 histone deacetylase in silencing parts of the yeast genome (Section 10.2). The gene from the bacterial Cre protein has been inserted into the *HML* locus of the *Saccharomyces cerevisiae* genome by homologous recombination (Section 6.2). Expression of the *cre* gene is then monitored by RNA-FISH with probes specific to this sequence. (A) In wild-type cells, the *HML* locus is silent and no *cre* RNA is synthesized. (B) In cells lacking the gene for the Sir4 histone deacetylase, the HML locus is no longer silent and *cre* RNA can be detected. (From Dodson and Rine. *eLife* 2015;4:e05007. DOI: 10.7554/eLife.05007.)

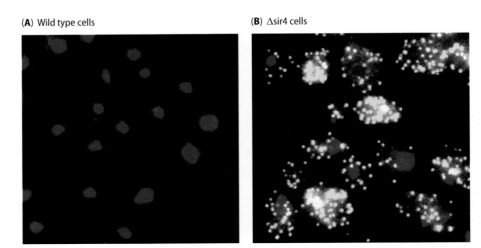

(**A**) Wild type cells

(**B**) Δsir4 cells

studies with microorganisms are therefore still done with bulk RNA from multiple cells, although there has been progress in using scRNA-seq to study heterogeneity within *Saccharomyces cerevisiae* populations, revealing, for example, variations in the response of individual cells to salt stress. With bacteria, an additional problem is that the mRNAs are not polyadenylated and so cannot be specifically captured from the total RNA content of a cell. Sequence databases from bacterial transcriptomes are therefore dominated by reads derived from noncoding RNA, predominantly from the rRNAs and tRNAs, which can make up 95% of the total. Despite these difficulties, transcriptome data have been obtained for individual *Salmonella enterica* and *Pseudomonas aeruginosa* bacteria, with transcripts of 100–200 genes detected, raising the possibility that in the future single-cell transcriptomics will provide a way of studying heterogeneity in bacterial communities, for example with regards to antibiotic resistance and metabolic variations that arise in response to nutrient limitation.

Spatial transcriptomics enables transcripts to be mapped directly in tissues and cells

We have seen how it is possible to use a combination of scRNA-seq and RNA-FISH to assign transcriptome data to individual cell types within a tissue. An additional group of techniques, collectively referred to as **spatial transcriptomics**, take the approach a step further by obtaining transcriptome catalogs directly from cells present in undisrupted tissue samples.

The key to these methods is the conversion of the RNA transcripts in fixed tissue into cDNA in a manner that results in the immobilization of the cDNAs and retention of spatial information. One approach is to prepare a glass slide that is coated with barcoded oligonucleotides, the barcodes identifying the position of each oligonucleotide in the array (**Figure 12.10**). The oligonucleotides also contain a series of T nucleotides, and so will hybridize to the poly(A) tails present on most mRNAs. A fixed tissue section is then laid over the glass slide and imaged so that the spatial distribution of the barcodes can be correlated with positional informational within the tissue. The section is treated with collagenase and pepsin to make the tissue permeable, enabling the mRNAs within the cells to become immobilized on to the oligonucleotide array. Complementary DNA synthesis is then carried out and the cDNAs are amplified by PCR to give products that contain both the barcode and the mRNA sequences. The cDNAs are collected and sequenced, the individual sequences revealing which mRNA is attached to which barcode, and hence providing spatial information on which mRNAs were present at which positions in the original cells. This procedure is called ***in situ* tissue profiling**. The resolution of the method is dependent on the size of each spot containing the barcoded oligonucleotides. In an early version, these had a diameter of 100 μm, which is comparable to the largest human cells

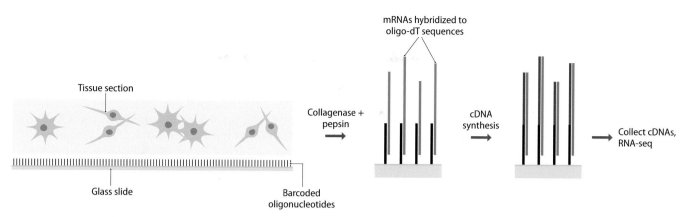

Figure 12.10 *In situ* tissue profiling. A fixed tissue section is laid over a glass slide that is coated with barcoded oligonucleotides. After imaging, the section is treated with collagenase and pepsin to make the tissue permeable, so the mRNAs become immobilized on the oligonucleotide array. Complementary DNA synthesis is then carried out and the cDNAs are collected and sequenced.

but in practical terms results in each spot capturing mRNA from tens of adjacent cells. In a more recent iteration, the spot diameter has been reduced to 10 μm, making single-cell resolution possible.

A different approach is taken by **fluorescent *in situ* RNA sequencing (FISSEQ)**. In this method, the RNAs are converted to cDNA within the fixed tissue, using random hexamer primers, which means that all RNAs will be copied, not just mRNAs, as is the case with *in situ* tissue profiling. The cDNA is then amplified with a modified nucleotide included in the reaction mix, the modification providing a chemical group that can be used to crosslink the PCR products to one another and also to protein molecules, ensuring that they remained immobilized in the tissue (**Figure 12.11**). SOLiD sequencing with labeled oligonucleotides whose incorporation can be followed by fluorescent imaging is then used to reveal the identities of the transcripts at different positions in the tissue. When used with mouse fibroblasts, the mRNA reads made up over 40% of the sequence library and transcripts of over 4000 genes could be detected. With this method, resolution depends on the size of the 'nanoballs' that result from crosslinking the PCR products from a single cDNA. In the original method, the nanoballs were 200–400 nm in diameter, enabling transcripts to be assigned to approximate subcellular locations. A second issue is the extent to which the imaging system can distinguish fluorescent signals from the individual nanoballs, which are densely packed on the surface of the tissue section. One possible

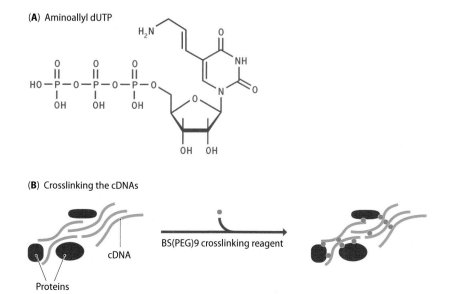

Figure 12.11 Crosslinking cDNAs during FISSEQ. (A) Complementary DNA synthesis is carried out with aminoallyl-dUTP added to the reaction mix. This modified nucleotide has an allylamine group attached to carbon 5. (B) Treatment with PEGylated bis(sulfosuccinimidyl) suberate [BS(PEG)9] forms crosslinks between the allyamine groups, and will also crosslink the cDNAs to –NH₂ groups on adjacent proteins.

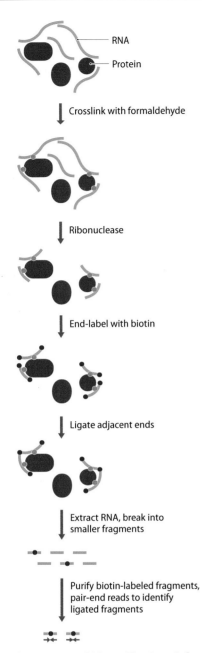

Figure 12.12 RIC-seq. The tissue is fixed with formaldehyde which forms crosslinks between RNA molecules and adjacent proteins. Treatment with ribonuclease leaves short fragments still attached to proteins: non-attached fragments are washed away at this step. The RNA ends are labeled with biotin and an RNA ligase added so that cut ends that are close to one another are ligated together. The RNAs are then extracted from the tissue, broken into smaller fragments, and those fragments containing a biotin label captured and sequenced.

way to reduce this molecular crowding is to pretreat the tissue with acrylic acid, which polymerizes into a gel that expands the tissue by up to four times, so that transcripts that are densely packed in the natural state become more spread out and hence easier to resolve by FISSEQ.

A final innovation in spatial transcriptomics enables pairs and groups of RNAs that interact with one another to be identified. **RNA *in situ* conformation sequencing (RIC-seq)** is conceptually similar to the HiC version of chromosome conformation capture, which is used to identify segments of DNA that are close to one another in the nucleus (Sections 4.2 and 10.1). Fixation of the tissue with formaldehyde forms crosslinks between adjacent RNA molecules and proteins. The resulting networks are then treated with a ribonuclease which makes random cuts in the RNAs and leaves short fragments still attached to proteins (**Figure 12.12**). The RNA ends are labeled with biotin and an RNA ligase is added, so that cut ends that are close to one another are ligated together, with this step favoring the ligation of pairs of fragments that are attached to a single protein. The RNAs are then extracted from the tissue, broken into smaller fragments, and those fragments containing a biotin label captured and sequenced. The resulting reads will include chimeric sequences that are derived from two parts of a single RNA, or parts of two different RNAs, which were adjacent in the cell and hence became crosslinked to the same protein and subsequently ligated together. As well as validating known interactions, RIC-seq has identified a number of transcripts in cultured human cells that form multiple interactions with other RNAs. These 'interaction hubs' (**Figure 12.13**) include the lncRNAs *MALAT1*, which appears to regulate transcription of genes involved in cancer metastasis and cell migration, and *NEAT1*, which forms the core component of the nuclear structures called **paraspeckles**, which are thought to have a role in the stress response by sequestering particular mRNAs, presumably to prevent these from being translated.

12.3 SYNTHESIS OF THE COMPONENTS OF THE TRANSCRIPTOME

The composition of a transcriptome is determined by the balance between the synthesis and degradation of the individual RNAs that it contains. For any RNA, a steady state is reached when its rate of synthesis (i.e., the number of copies made per unit time) equals its rate of degradation (**Figure 12.14**). To increase the amount of an RNA in a transcriptome, the rate of synthesis must increase, or the rate of degradation decrease. To reduce the amount of an RNA, the rate of synthesis must decrease, or the rate of degradation must increase. Understanding how RNAs are synthesized and turned over is therefore central to understanding how the composition of a transcriptome responds to external stimuli such as the presence of hormones or an alteration in the environment, and also provides us with an appreciation of how gene expression patterns change during differentiation, development and disease. In the next few pages, we will begin to address these issues by examining how RNAs, especially the mRNAs that code for proteins, are synthesized.

RNA polymerases are molecular machines for making RNA

In Section 1.2, we learned that the enzymes responsible for transcription of DNA into RNA are called DNA-dependent RNA polymerases. These enzymes synthesize RNA in the 5´→3´ direction, base-pairing between the RNA and the DNA template ensuring that the sequence of nucleotides in the transcript is complementary to the sequence of the DNA molecule that is being transcribed (see Figure 1.13).

Transcription of eukaryotic nuclear genes requires three different RNA polymerases: **RNA polymerase I**, **RNA polymerase II**, and **RNA polymerase III**. Each works on a different set of genes, with no interchangeability (**Table 12.1**). Most research attention has been directed at RNA polymerase II, as this is the

(A) *NEAT1*   **(B)** *MALAT1*

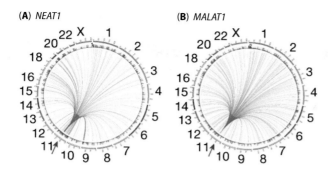

Figure 12.13 Interaction hubs identified by RIC-seq. These circos plots show the interactions between two lncRNAs and other RNAs in human HeLa cells. The outer circle comprises each of the human chromosomes and the arrows indicate the positions of the genes for (A) *NEAT1* and (B) *MALAT1*, both on chromosome 11. The blue lines within the plot lead to the genes coding for the RNAs that interact with *NEAT1* and *MALAT1*. The inner red circle shows the relative amount of contacts detected for each target RNA. (From Cai Z., Cao J., Ji L. et al. [2020] *Nature* 432-437. With permission from Springer Nature.)

enzyme that transcribes genes that code for proteins. It also synthesizes the Sm subgroup of snRNAs, as well as some snoRNAs, siRNAs, miRNAs, piRNAs and most lncRNAs. RNA polymerase III transcribes a variety of sncRNA genes, including those for tRNAs. RNA polymerase I transcribes the multicopy repeat units containing the 28S, 5.8S, and 18S rRNA genes. Plants have two additional RNA polymerases, IV and V, which are related to RNA polymerase II and transcribe particular classes of siRNA genes.

Each of the eukaryotic RNA polymerases is a multisubunit protein with a molecular mass in excess of 500 kDa. Structurally, they are quite similar to one another (**Figure 12.15**). RNA polymerase II, with 12 subunits in most eukaryotes, represents the basic design. RNA polymerases I and III have two and five additional subunits, respectively, which are homologs of proteins that aid the RNA polymerase II transcription process but are not considered permanent subunits of this enzyme. Archaea possess a single RNA polymerase that is structurally very similar to RNA polymerase II. But this is not typical of the prokaryotes in general because the bacterial RNA polymerase is very different, consisting of just six subunits, with its composition described as $\alpha_2\beta\beta'\omega\sigma$ (two α subunits, one each of β and the related β', one of ω, and one of σ). The α, β, β', and ω subunits are structural homologs of subunits present in the eukaryotic RNA polymerases, but the σ subunit has its own special properties, both in terms of its structure and, as we will see later, its function.

The chloroplast genome encodes an RNA polymerase that is very similar to the bacterial enzyme, reflecting the bacterial origins of these organelles (Section 8.3). In addition, the chloroplasts of monocotyledonous and dicotyledonous plants

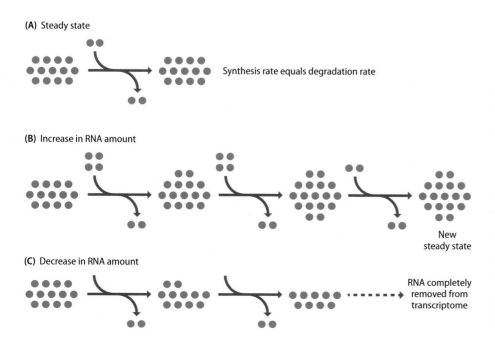

(A) Steady state

Synthesis rate equals degradation rate

(B) Increase in RNA amount

New steady state

(C) Decrease in RNA amount

RNA completely removed from transcriptome

Figure 12.14 The effect of synthesis rate and degradation rate on the amount of an RNA in a transcriptome. (A) If the synthesis rate and degradation rate are equal, then the RNA is in a steady state, and the number of molecules remains constant. (B) An increase in synthesis rate (and/or a decrease in degradation rate) results in the number of RNA molecules increasing. This increase will continue until the synthesis rate decreases, so a second steady state is achieved. (C) A decrease in synthesis rate (and/or an increase in degradation rate) results in the number of RNA molecules decreasing. A new steady state might be reached or, as shown in this example, the RNA might be completely removed from the transcriptome.

TABLE 12.1 FUNCTIONS OF THE THREE EUKARYOTIC NUCLEAR RNA POLYMERASES	
Polymerase	**Types of RNA**
RNA polymerase I	28S, 5.8S, and 18S rRNAs
RNA polymerase II	mRNA, Sm family of snRNAs, some snoRNAs, siRNA*, miRNA, piRNA, lncRNA
RNA polymerase III	tRNAs, 5S rRNA, Lsm family of snRNAs, some snoRNAs, 7SL RNA, 7SK RNA

*In plants, some siRNAs are transcribed by RNA polymerase IV or V.

import from the cytoplasm one or two, respectively, single-subunit RNA polymerases that are coded by nuclear genes. Many genes can be transcribed by both the chloroplast and nuclear-encoded RNA polymerases, though the chloroplast version appears to be used more frequently in photosynthetic tissue. A few genes can only be transcribed by the nuclear-encoded polymerases; interestingly, these include the *rpoB* gene, which specifies the β subunit of the 'chloroplast' RNA polymerase. Mitochondria also have their own RNA polymerases, in this case a single-subunit enzyme coded by a nuclear gene. There is no endogenous 'mitochondrial' RNA polymerase.

The bacterial RNA polymerase can synthesize RNA at a rate of several hundred nucleotides per minute. The average *Escherichia coli* gene, which is just a few thousand nucleotides in length, can therefore be transcribed in a few minutes. The eukaryotic RNA polymerase II has a more rapid synthesis rate than the bacterial polymerase, up to 2000 nucleotides per minute, but can take hours to synthesize a single RNA because many eukaryotic genes are much longer than bacterial ones. For example, the 2400 kb transcript of the human dystrophin gene takes over 16 hours to synthesize. Most RNA polymerases have an accuracy of one error per 10^4–10^5 nucleotide additions, this level of inaccuracy being tolerated because RNAs are usually multicopy and transcripts that contain errors form only a small proportion of the overall pool. The error rate is minimized by **backtracking**, which is stimulated by the bulge in a DNA–RNA duplex that occurs at a position that is not base-paired (**Figure 12.16**). The polymerase undergoes a slight structural rearrangement which enables it to slide back along the template and cut the RNA upstream of the error. The polymerase then reverts to its synthesis structure and moves forward again to retranscribe the previously faulty segment.

Transcription start-points are indicated by promoter sequences

It is essential that transcription initiates at the correct positions on DNA molecules, immediately upstream of the individual genes that must be copied into RNA. These positions are marked by target sequences, loosely called **promoters**,

RNA polymerase I   RNA polymerase II   RNA polymerase III

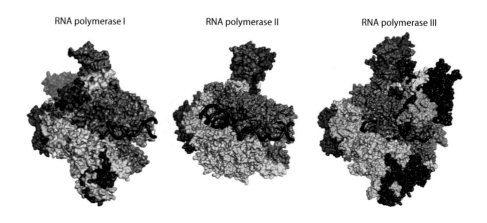

Figure 12.15 Similarities between the structures of the three eukaryotic RNA polymerases. Equivalent subunits are given the same colors in each structure. The two strands of the DNA double helix are shown in blue and brown. (From Khatter H., Vorlander M.K. & Muller C.W. [2017] *Curr. Opin. Chem. Biol.* 88–94. Published under CC BY 4.0.)

that are recognized either by the RNA polymerase itself or by a DNA-binding protein which, once attached to the DNA, forms a platform onto which the RNA polymerase binds (**Figure 12.17**).

The bacterial RNA polymerase recognizes a bipartite promoter sequence, whose two components are located approximately 10 bp and 35 bp upstream of the position where transcription should begin. In *Escherichia coli*, the promoter consensus sequences are:

–35 box	5´-TTGACA-3´
–10 box	5´-TATAAT-3´

The spacing between the two boxes is important because it places the two motifs on the same face of the double helix, facilitating their interaction with the σ subunit, which is the DNA-binding component of the bacterial RNA polymerase. The consensus sequences given above refer to the 'standard' version of the bacterial polymerase, which contains the σ^{70} subunit, so-called because its molecular mass is approximately 70 kDa. *E. coli*, and other bacteria, can also make a variety of other σ subunits, each one specific for a different promoter sequence. An example is the σ^{32} subunit, which is synthesized when the bacterium is exposed to a heat shock. This subunit recognizes a sequence which is found upstream of genes coding for proteins that help the bacterium withstand high temperatures (**Figure 12.18**); these proteins include chaperones that protect other proteins from heat degradation, and enzymes that repair heat-induced DNA damage. Other σ subunits are used during nutrient starvation and nitrogen limitation. Through the use of these alternative σ subunits, *E. coli* is able to remodel its transcriptome in response to changing environmental and nutritional demands. Alternative σ subunits are also used by other bacteria: for example, *Klebsiella pneumoniae* uses a σ^{54} subunit to switch on genes involved in nitrogen fixation, and *Bacillus* species use a whole range of different σ subunits to switch on and off groups of genes during the changeover from normal growth to formation of spores (Section 14.3).

In eukaryotes, the term 'promoter' is used to describe all the sequences that are important in initiation of transcription of a gene. For some genes, these sequences can be numerous and diverse in their functions, including not only the **core promoter**, sometimes called the **basal promoter**, which is the site at which the RNA polymerase attaches, but also one or more **upstream promoter elements**, which, as their name implies, lie upstream of the core promoter. Transcription initiation can usually occur in the absence of the upstream elements, but only in an inefficient way. This indicates that the proteins that bind to the upstream elements include at least some that are activators of transcription, and which therefore 'promote' gene expression. Inclusion of these sequences in the 'promoter' is therefore justified.

Each of the three eukaryotic RNA polymerases recognizes a different type of promoter sequence; indeed, it is the difference between the promoters that defines which genes are transcribed by which polymerase. The details for vertebrates are as follows (**Figure 12.19**):

- RNA polymerase I promoters consist of a core promoter spanning the transcription start point, between nucleotides –45 and +20, and an **upstream control element** (UCE) about 100 bp further upstream.

- RNA polymerase II promoters are variable and can stretch for several kilobases upstream of the transcription start site. The core promoter consists of two main segments: the –25 or **TATA box** (consensus 5´-TATAWAAR-3´, where W is A or T, and R is A or G) and the **initiator (Inr) sequence** (mammalian consensus 5´-YYANWYY-3´, where Y is C or T, and N is any nucleotide) located around nucleotide +1. Some genes transcribed by RNA polymerase II have only one of these two components of the core promoter, and some, surprisingly, have neither. The latter are called 'null' genes. They are still transcribed, although the start position for

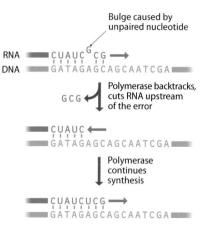

Figure 12.16 Backtracking enables an RNA polymerase to correct a transcription error.

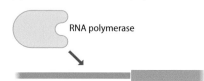

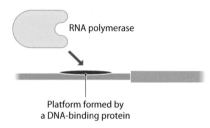

Figure 12.17 Two ways in which RNA polymerases bind to their promoters. (A) Direct recognition of the promoter by the RNA polymerase, as occurs in bacteria. (B) Recognition of the promoter by a DNA-binding protein, which forms a platform onto which the RNA polymerase binds.

(A) An *E. coli* heat-shock gene

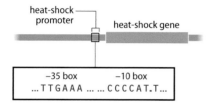

(B) Recognition by the σ³² subunit

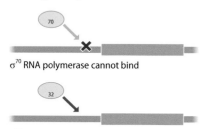

σ⁷⁰ RNA polymerase cannot bind

σ³² RNA polymerase binds to the
heat-shock promoter

Figure 12.18 Regulation of gene expression by the σ³² subunit in *E. coli*. (A) The consensus sequence of the promoter upstream of genes involved in the heat shock response of *E. coli*. The sequence of the –35 box is similar to the equivalent sequence for the σ⁷⁰ subunit, but the –10 sequence is substantially different. (B) The heat shock promoter is not recognized by the normal *E. coli* RNA polymerase containing the σ⁷⁰ subunit, but is recognized by the σ³² subunit.

transcription is more variable than for a gene with a TATA box and/or Inr sequence. Some genes also have additional sequences that can be looked on as part of the core promoter. Examples are:

- The downstream promoter element (DPE; located at positions +28 to +32), which has a variable sequence but has been identified through its ability to bind TFIID, a protein complex that plays a central role in the initiation of transcription.

- A 7 bp GC-rich motif called the B response element (BRE), located immediately upstream or downstream of the TATA box. This motif is recognized by TFIIB, another component of the initiation complex.

- The proximal sequence element (PSE), which is located between positions –45 and –60 upstream of those snRNA genes that are transcribed by RNA polymerase II.

As well as the components of the core promoter, genes transcribed by RNA polymerase II have various upstream promoter elements, usually located within 2 kb of the transcription start site, the functions of which are described later in this chapter.

- RNA polymerase III promoters fall into at least three categories. Two of these categories are unusual in that the important sequences are located within the genes whose transcription they promote. Usually, these sequences span 50–100 bp and comprise two conserved boxes separated by a variable region. The third category of RNA polymerase III promoter is similar to those for RNA polymerase II, having a TATA box and a range of additional promoter elements (sometimes including the PSE mentioned above) located upstream of the target gene. Interestingly, this arrangement is seen with the Lsm class of snRNA genes, which are transcribed by RNA polymerase III, which means that these genes have similar promoter sequences to the Sm group that are transcribed by RNA polymerase II.

An additional level of complexity is seen with some eukaryotic genes that have **alternative promoters** that give rise to different versions of the transcript specified by the gene. An example is provided by the human dystrophin gene, which has been extensively studied because defects in this gene result in the genetic disease called Duchenne muscular dystrophy. The dystrophin gene is one of the largest known in the human genome, stretching over 2.4 Mb and containing 78 introns. It has at least seven alternative promoters, which are used in different tissues and which direct synthesis of mRNAs of different lengths (**Figure 12.20**). Alternative promoters are also used to generate related versions of some proteins at different stages in development, and to enable a single cell to synthesize similar proteins with slightly different biochemical properties. The last point indicates that although usually referred to as 'alternative' promoters these are, more correctly, 'multiple' promoters as more than one may be active at a single

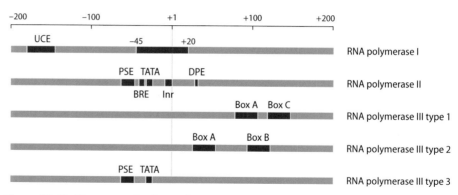

Figure 12.19 Structures of eukaryotic promoters.

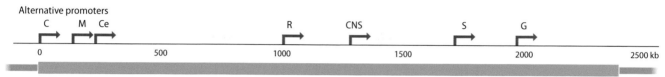

Figure 12.20 Alternative promoters. The positions of seven alternative promoters for the human dystrophin gene are shown.

time. Indeed, this may be the normal situation for many genes. For example, a genome-wide survey has revealed that 10,500 promoters are active in human fibroblast cells, but that these promoters are driving expression of fewer than 8000 genes, indicating that a substantial number of genes in these cells are being expressed from two or more promoters simultaneously.

Synthesis of bacterial RNA is regulated by repressor and activator proteins

The consensus sequence for a bacterial promoter is quite variable, with a range of different motifs being permissible at both the –35 and –10 boxes (Table 12.2). These variations, together with less well-defined sequence features around the transcription start site and in the first 50 or so nucleotides of the transcription unit, affect the efficiency of the promoter. Efficiency is defined as the number of productive initiations that are promoted per second, a productive initiation being one that results in the RNA polymerase clearing the promoter and beginning synthesis of a full-length transcript. Different promoters vary 1000-fold in their efficiencies, with the most efficient promoters (called **strong promoters**) directing 1000 times as many productive initiations as the weakest promoters. We refer to these as differences in the **basal rate of transcription initiation**.

Promoter structure determines the basal rate of transcription initiation but does not provide any general means by which transcription of a gene can respond to changes in the environment or to the biochemical requirements of the cell. Instead, these transient changes are brought about by regulatory proteins which have a negative (repressing) or positive (activating) effect on the basal transcription rate for a particular gene.

The foundation of our understanding of regulatory control over transcription initiation in bacteria was laid in the early 1960s by François Jacob, Jacques Monod, and other geneticists who studied the lactose operon – the group of three genes whose protein products convert lactose to glucose and galactose (see Figure 8.10A). These genetic experiments identified a locus, called the **operator**, adjacent to the promoter, which acts as the binding site for the **lactose repressor** protein (**Figure 12.21**). The original model envisaged that the lactose repressor attached to the operator and prevented the RNA polymerase from binding to the promoter, simply by denying it access to the relevant segment of DNA. Whether

TABLE 12.2 SEQUENCES OF *E. COLI* PROMOTERS

Genes	Protein product	–35 box	–10 box
Consensus	–	5´–TTGACA–3´	5´–TATAAT–3´
argF	ornithine transcarbamoylase	5´–TTGTGA–3´	5´–AATAAT–3´
can	carbonic anhydrase	5´–TTTAAA–3´	5´–TATATT–3´
dnaB	DnaB helicase	5´–TCGTCA–3´	5´–TAAAGT–3´
gcd	glucose dehydrogenase	5´–ATGACG–3´	5´–TATAAT–3´
gltA	citrate synthase	5´–TTGACA–3´	5´–TACAAA–3´
ligB	DNA ligase	5´–GTCACA–3´	5´–TAAAAG–3´

(A) The lactose operator

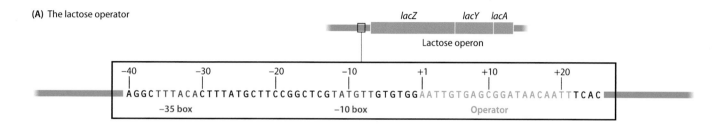

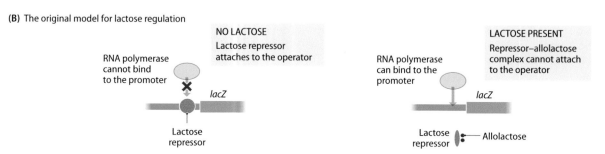

(B) The original model for lactose regulation

Figure 12.21   Regulation of the *E. coli* lactose operon. (A) The operator sequence lies immediately downstream of the promoter for the lactose operon. Note that this sequence has inverted symmetry: when read in the 5→3′ direction, the sequence is the same in both strands. This enables two subunits of the tetrameric repressor protein to make contact with the same operator sequence. (B) In the original model for lactose regulation, the lactose repressor is looked on as a simple blocking device that binds to the operator and prevents the RNA polymerase from gaining access to the promoter. The three genes in the operon are therefore switched off. This is the situation in the absence of lactose, although transcription is not completely blocked because the repressor occasionally detaches, allowing a few transcripts to be made. The bacterium therefore always possesses a few copies of each of the three enzymes coded by the operon, probably less than five of each. This means that when the bacterium encounters a source of lactose, it is able to transport a few molecules into the cell and split these into glucose and galactose. An intermediate in this reaction is allolactose, an isomer of lactose, which induces expression of the lactose operon by binding to the repressor, causing a change in the conformation of the latter, so it is no longer able to attach to the operator. This allows the RNA polymerase to bind to the promoter and transcribe the three genes. When fully induced, approximately 5000 copies of each protein product are present in the cell. When the lactose supply is used up, and allolactose is no longer present, the repressor reattaches to the operator and the operon is switched off. The transcripts of the operon, which have a half-life of fewer than three minutes, decay and the enzymes are no longer made.

the repressor binds depends on the presence in the cell of allolactose, an isomer of lactose, which is an **inducer** of the operon. When allolactose is present it binds to the lactose repressor, causing a slight structural change which prevents the helix–turn–helix motifs of the repressor from recognizing the operator as a DNA-binding site. The allolactose–repressor complex therefore cannot bind to the operator, enabling the RNA polymerase to gain access to the promoter. When the supply of lactose is used up and there is no allolactose left to bind to the repressor, the repressor reattaches to the operator and prevents transcription. The operon is therefore transcribed only when the enzymes coded by the operon are needed.

Most of the original scheme for regulation of the lactose operon has been confirmed by DNA sequencing of the control region and by structural studies of the repressor bound to its operator. The one complication has been the discovery that the repressor has three potential binding sites, centered on nucleotide positions –82, +11, and +412. The operator defined by genetic studies is the sequence located at +11 (see **Figure 12.21A**), and this is the only one of the three sites whose occupancy by the repressor would be expected to prevent access of the RNA polymerase to the promoter. But the other two sites also play some role in repression as their removal, individually or together, significantly impairs the ability of the repressor to switch off transcription. The repressor is a tetramer of four identical subunits that work in pairs to attach to a single operator, so the repressor has the capacity to bind to two of the three operator sites at once. It is likely that the binding of one pair of subunits to the +11 site is enhanced or stabilized by attachment of the other pair of subunits to the –82 or +412 site. It is also possible that the repressor can bind to a pair of operator sequences in such a way that it does not block attachment of the polymerase to the promoter, but prevents the polymerase from leaving the promoter region.

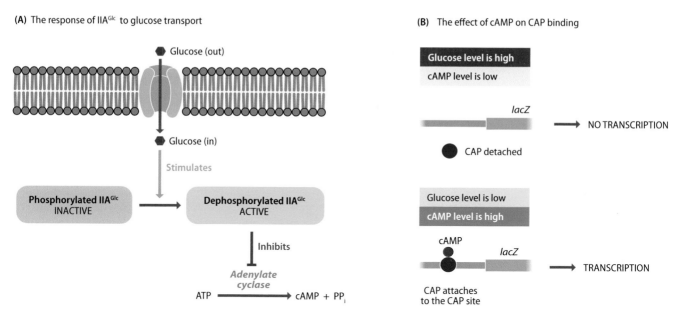

(A) The response of IIA^Glc to glucose transport

(B) The effect of cAMP on CAP binding

Figure 12.22 The role of the catabolite activator protein. (A) The transport of glucose into the bacterium results in dephosphorylation of IIA^Glc. Dephosphorylated IIA^Glc inhibits adenylate cyclase, reducing the conversion of ATP into cyclic AMP. (B) The catabolite activator protein (CAP) can attach to its DNA-binding sites only in the presence of cAMP. If glucose is present, the cAMP level is low, so CAP does not bind to the DNA and does not activate the RNA polymerase. Once the glucose has been used up, the cAMP level rises, allowing CAP to bind to the DNA and activate transcription of the target genes.

As well as down-regulation by repressors, transcription of many bacterial operons and individual genes can be up-regulated by activating proteins. An example is provided by the **catabolite activator protein** (also called the CRP activator). This protein binds to a recognition sequence at various sites in the bacterial genome and activates transcription initiation at downstream promoters, by interacting with the α subunit of the RNA polymerase. Inherent in this activation is the creation of a sharp, 90° bend in the double helix in the region of the binding site when the catabolite activator protein is attached. Productive initiation of transcription at these promoters is dependent on the presence of bound catabolite activator protein: if the protein is absent, then the genes controlled by the promoter are not transcribed.

Attachment of the catabolite activator protein to its binding site only occurs when the bacterium has an inadequate supply of glucose. The availability of glucose is monitored by a protein called IIA^Glc, which is a component of a multiprotein complex that transports sugars into the bacterium. When glucose is being transported into the cell, IIA^Glc becomes dephosphorylated (i.e., additional phosphate groups previously added to the protein by posttranslational modification are removed). The dephosphorylated version of IIA^Glc inhibits adenylate cyclase, the enzyme that converts ATP into cyclic AMP (**Figure 12.22**). This means that if glucose levels are high, the cAMP content of the cell is low. The catabolite activator protein can bind to its target sites only in the presence of cAMP, so when glucose is present the protein remains detached and the operons it controls are switched off. One of these target sites is adjacent to the promoter of the lactose operon. If the bacterium has supplies of both glucose and lactose, the binding site for the catabolite activator protein is unoccupied, and in the absence of activation the lactose operon is not transcribed, even though the lactose repressor is not bound. If the bacterium uses up all the glucose, then the cAMP level rises, and the catabolite activator protein binds to its target sites, including the site upstream of the lactose operon. Transcription of the lactose genes is therefore activated, enabling the bacterium to start using lactose as its primary energy source. The catabolite activator protein therefore remodels the *E. coli* transcriptome so the bacterium makes the most efficient use of the available sugars. This ability of *E. coli* and other bacteria to metabolize one sugar in preference to another, and then switch to that second sugar when the preferred one is used up, was first discovered in 1941 by Jacques Monod, who used the French word **diauxie** to describe it (**Figure 12.23**).

Synthesis of bacterial RNA is also regulated by control over transcription termination

Although initiation of transcription is looked on as the main control point in the process leading to synthesis of a bacterial RNA, it is not the only step that can be influenced in order to adjust the composition of a transcriptome.

Current thinking views transcription as a discontinuous process, with the polymerase pausing regularly and making a 'choice' between continuing elongation by adding more nucleotides to the transcript, or terminating by dissociating from the template. Which choice is selected depends on which alternative is more favorable in thermodynamic terms. This model emphasizes that, in order for termination to occur, the polymerase has to reach a position on the template where dissociation is more favorable than continued RNA synthesis. Consistent with this model, many of the positions in a bacterial genome where transcription terminates are marked by an inverted palindrome, a sequence that can be folded into a stem-loop structure after transcription into RNA (**Figure 12.24**). Formation of this stem-loop is thought to weaken the overall interaction between the polymerase and the template, favoring dissociation over continued transcription. At **intrinsic terminator sequences**, attachment of the polymerase is further weakened by a run of As in the template, which lead to the transcript being held to the DNA by a series of A–U base pairs, which have only two hydrogen bonds each, compared with three for each G–C pair (**Figure 12.25A**). Alternatively, at **Rho-dependent terminators**, a helicase enzyme breaks the base pairs between the DNA and transcript when the polymerase pauses at the stem-loop structure (**Figure 12.25B**).

The ability of stem-loop structures to terminate bacterial transcription is exploited by a regulatory process called **attenuation**. The tryptophan operon of *E. coli* (see Figure 8.10B) illustrates how it works. In this operon, two stem-loops can form in the region between the start of the transcript and the beginning of *trpE*, the first gene in the operon (**Figure 12.26A**). The smaller of these loops acts as a Rho-dependent terminator, but the larger one does not. The larger loop overlaps with the termination hairpin, so only one of the two hairpins can form at any one time. Which loop forms depends on the relative positioning between the RNA polymerase and a ribosome that attaches to the 5′ end of the transcript as soon as it is synthesized, in order to translate the coding parts of the RNA into protein. If the ribosome stalls so that it does not keep up with the polymerase, then the larger stem-loop forms and transcription continues. However, if the ribosome keeps pace with the RNA polymerase, then it disrupts the larger

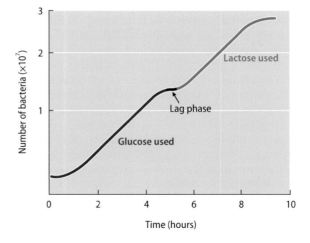

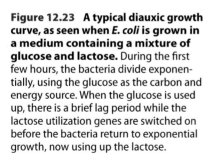

Figure 12.23   A typical diauxic growth curve, as seen when *E. coli* is grown in a medium containing a mixture of glucose and lactose. During the first few hours, the bacteria divide exponentially, using the glucose as the carbon and energy source. When the glucose is used up, there is a brief lag period while the lactose utilization genes are switched on before the bacteria return to exponential growth, now using up the lactose.

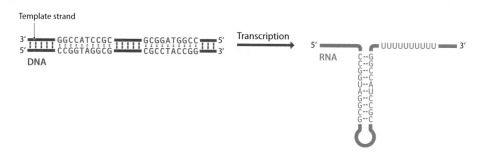

Figure 12.24 A bacterial termination structure. The presence of an inverted palindrome in the DNA sequence results in formation of a stem-loop in the transcript.

stem-loop, which means the terminator structure is able to form, and transcription stops. Ribosome stalling can occur because upstream of the termination signal is a short open reading frame coding for a 14-amino-acid peptide that includes two tryptophans. If the amount of free tryptophan is limiting, then the ribosome stalls as it attempts to synthesize this peptide, enabling the polymerase to continue making its transcript (**Figure 12.26B**). Because the transcript contains copies of the genes coding for the biosynthesis of tryptophan, its continued elongation addresses the requirement that the cell has for this amino acid. When the amount of tryptophan in the cell reaches a satisfactory level, the attenuation system prevents further transcription of the tryptophan operon, because now the ribosome does not stall while making the short peptide, and instead keeps pace with the polymerase, allowing the termination signal to form (**Figure 12.26C**).

The *E. coli* tryptophan operon is controlled not only by attenuation but also by a repressor protein. Exactly how attenuation and repression work together to regulate expression of the operon is not known, but it is thought that repression provides the basic on-off switch and attenuation modulates the precise level of gene expression that occurs. Attenuation is used to regulate transcription of several operons involved in amino-acid biosynthesis in *E. coli* and other bacteria, and is also used to remodel transcriptomes in response to other types of nutrient limitation and to environmental changes such as temperature fluctuations. A common theme among these different attenuation systems is the presence of alternative stem-loop structures in the transcript, but the mechanisms used to influence formation of the terminator stem-loop are more variable. Some attenuators use ribosome stalling in the same way as the *E. coli* tryptophan operon, but others, called **riboswitches**, are regulated directly by the attachment of small molecules, such as amino acids, which disrupt the larger of the two stem-loops, so termination occurs, and a third class are regulated by attachment of an RNA-binding protein.

Synthesis of eukaryotic RNA is regulated primarily by activator proteins

The key lesson that we have learned from our examination of transcriptional control in bacteria is that transcription initiation can be influenced by DNA-binding proteins that recognize specific sequences located near the attachment site for the RNA polymerase. This is also the basis for transcriptional control in

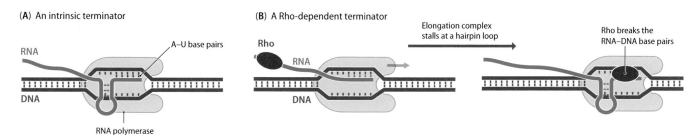

Figure 12.25 Alternative modes for termination of transcription in bacteria. (A) At an intrinsic terminator, attachment of the polymerase to the DNA is weakened by a run of As in the template. (B) Termination at a Rho-dependent terminator. Rho is a helicase that follows the RNA polymerase along the transcript. When the polymerase stalls at a stem-loop structure, Rho catches up and breaks the RNA–DNA base pairs, releasing the transcript.

Figure 12.26 Attenuation at the tryptophan operon. (A) Either of two stem-loops can form in the region between the start of the transcript and the beginning of *trpE*, the first gene in the operon. The smaller of these loops acts as a Rho-dependent terminator. Immediately upstream of the loops is an open reading frame coding for a 14-amino-acid peptide that includes two tryptophans. (B) When tryptophan is limiting, the ribosome stalls at the open reading frame, allowing the larger, non-terminating stem-loop to form. Transcription therefore continues. (C) When there is an ample supply of tryptophan, the ribosome does not stall and disrupts formation of the larger stem-loop. This enables the termination structure to form, so transcription terminates.

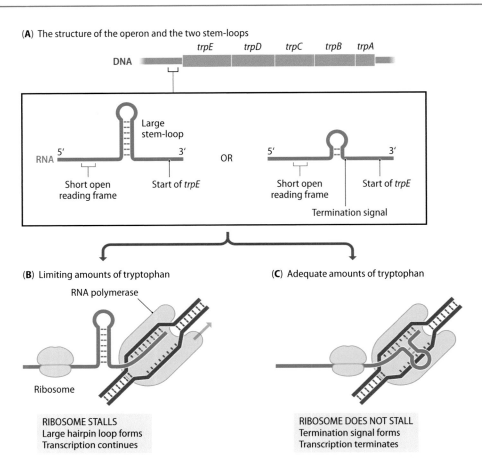

(A) The structure of the operon and the two stem-loops

(B) Limiting amounts of tryptophan

RIBOSOME STALLS
Large hairpin loop forms
Transcription continues

(C) Adequate amounts of tryptophan

RIBOSOME DOES NOT STALL
Termination signal forms
Transcription terminates

eukaryotes, but with one difference. The bacterial RNA polymerase has a strong affinity for its promoter and the basal rate of transcription initiation is relatively high for all but the weakest promoters. With most eukaryotic genes, the reverse is true. The RNA polymerase II and III initiation complexes do not assemble efficiently, and the basal rate of transcription initiation is therefore very low, regardless of how 'strong' the promoter is. In order to achieve effective initiation, formation of the complex must be activated by additional proteins. This means that, compared with bacteria, eukaryotes use different strategies to control transcription initiation, with activators playing a much more prominent role than repressor proteins.

Unlike the bacterial RNA polymerase, eukaryotic polymerases do not bind directly to their core promoter sequences. For genes transcribed by RNA polymerase II, the initial contact is made by the **general transcription factor (GTF)** TFIID, which is a complex made up of the **TATA-binding protein (TBP)** and at least 12 **TBP-associated factors (TAFs)**. TBP is a sequence-specific protein with an unusual DNA-binding domain that makes contact with the minor groove in the region of the TATA box. It induces a conformational change at its target site from B- to A-DNA that facilitates its binding (Section 11.3). The TAFs assist in attachment of TBP to the TATA box and, in conjunction with other proteins called **TAF- and initiator-dependent cofactors (TICs)**, possibly also participate in recognition of the Inr sequence, especially at those promoters that lack a TATA box.

X-ray crystallography studies of TBP show that it has a saddle-like shape that wraps partially around the double helix, forming a platform onto which the RNA polymerase is subsequently positioned (**Figure 12.27**). Recruitment of RNA polymerase is aided by three more transcription factors, TFIIA, TFIIB, and TFIIF, with the addition of TFIIE and TFIIH completing what is called the **preinitiation complex**. Completion of this complex is followed by addition of phosphate groups to the **C-terminal domain (CTD)** of the largest subunit of the

RNA polymerase. In mammals, this domain consists of 52 repeats of the seven amino-acid sequence Tyr–Ser–Pro–Thr–Ser–Pro–Ser. Two of the three serines in each repeat unit can be modified by addition of a phosphate group, catalyzed by a kinase activity provided by two of the subunits of TFIIH. Phosphorylation causes a substantial change in the ionic properties of the polymerase, enabling the polymerase to leave the initiation complex and begin synthesizing RNA.

The above account describes how the transcription of a eukaryotic protein-coding gene is initiated, but does not explain how the rate of transcription is set at the appropriate level between zero and maximal. This is the role of DNA-binding proteins called **transcription factors**, which are separate and distinct from the general transcription factors, such as TFIID, that are contained within the preinitiation complex. An example of a regulatory transcription factor is the CREB protein, which activates its target genes by binding to CRE sequences in response to elevated cyclic AMP levels (Section 11.1). Other transcription factors transduce signals from sources as diverse as steroid hormones and heat shock. The binding sites for transcription factors are located either close to or more distant from the gene whose expression they control (**Figure 12.28**):

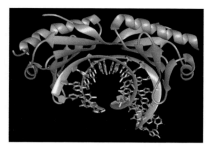

Figure 12.27 TBP attachment to the TATA box forms a platform onto which the initiation complex can be assembled. The dimer of TBP proteins is shown in brown and the DNA is in silver and green. (Courtesy of Song Tan, Penn State University.)

- **Proximal binding sites** are usually located within 2 kb of the transcription start site of the target gene. They are often described as upstream promoter elements and hence looked on as components of the eukaryotic promoter. Attachment of a transcription factor to one of these sites only influences transcription of the gene within whose promoter the element is located.

- The more distant binding sites are located within **enhancers**, which can be positioned anywhere within the topologically associating domain that contains their target gene. A single enhancer can influence transcription of multiple genes with its domain, but is prevented by insulator sequences from affecting genes in neighboring domains (Section 10.1).

How does the attachment of a transcription factor to its binding site activate or repress transcription initiation? The answer is that a bound transcription factor makes physical contact with a multisubunit protein called the **mediator**, which in turn makes contacts with different components of the preinitiation complex. The mediator is thought to influence several events involved in assembly of the complex, including positioning of the polymerase on TBP and activation of the polymerase by CTD phosphorylation. The mediator therefore, as its name implies, mediates the signal from the transcription factor to the preinitiation complex.

The traditional view has been that construction of the preinitiation complex is a stepwise process, occurring on the platform provided by TBP bound to the TATA site, involving the sequential addition of TFIIA, TFIIB, RNA polymerase II, TFIIF, TFIIE, and, finally, TFIIH. According to this model, the preinitiation complex is then activated by the mediator in response to attachment of one or more transcription factors to upstream promoter elements and/or enhancer sequences (**Figure 12.29A**). It now seems more likely that at least some of the transcription factors form an assembly with RNA polymerase prior to attachment of this larger complex to the platform provided by TBP. One possibility is that this assembly is initiated by the presence of a transcription factor at an enhancer site, and that most of the preinitiation complex, including both RNA polymerase II and the mediator, is actually built up at the enhancer before being transferred to the TBP platform (**Figure 12.29**).

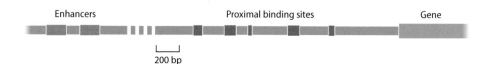

Enhancers   Proximal binding sites   Gene

200 bp

Figure 12.28 Binding sites for transcription factors. Proximal binding sites are located within 2 kb of the transcription start site. Enhancers can be at more distant sites.

Figure 12.29 Assembly of the preinitiation complex at an enhancer followed by transfer to the TBP platform.

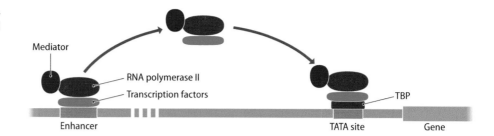

Because of the low level of basal initiation occurring at an RNA polymerase II promoter, most transcription factors for this polymerase are activators. Only a few proteins that repress transcription initiation are known, with these proteins binding to upstream promoter elements or to more distant sites in **silencers**. Some repressors influence genome expression in a general way through histone deacetylation (Section 10.2) or DNA methylation (Section 10.3), but others have more specific effects on individual promoters. A few proteins activate or repress transcription depending on the circumstances. Pit-1, which is the first of the three proteins after which the POU domain is named (Section 11.2), activates some genes and represses others, depending on the sequence of the DNA-binding site. The presence in this site of two additional nucleotides induces a change in the conformation of Pit-1, enabling it to interact with a second protein called N-CoR and repress transcription of the target gene (**Figure 12.30**).

12.4 THE INFLUENCE OF RNA SPLICING ON THE COMPOSITION OF A TRANSCRIPTOME

Following their synthesis, many RNAs are processed by cutting and, in some cases, joining reactions that convert the initial transcripts into the functional molecules. For many types of RNA, these processing events are part of their routine synthesis pathway, and are not looked on as post-transcriptional control points that could influence the composition of the transcriptome. We believe that this is true for the processing of pre-rRNAs and pre-tRNAs in both eukaryotes and prokaryotes (see Figure 1.16), for the splicing pathways that result in removal of introns from the precursors of some eukaryotic tRNAs, rRNAs, and organellar transcripts, and for the splicing pathways for the very small number of prokaryotic mRNAs that contain introns.

Figure 12.30 Conformation of the POU domains of the Pit-1 activator bound to its target sites upstream of the prolactin and growth hormone genes. Pit-1 is a dimer, and each monomer has two POU domains. The two domains of one monomer are shown in red and the two domains of the other monomer are shown in blue. The barrels are α-helices, with α3 being the recognition helix of each domain. Note the difference between the conformations of the domains when attached to the two binding sites. The more open structure adopted at the growth hormone site enables the Pit-1 dimer to interact with N-CoR and other proteins to repress transcription of the growth hormone gene. Pit-1 therefore activates the prolactin gene but represses the growth hormone gene. (Adapted from Scully K.M., Jacobson E.M., Jepsen K. et al. [2000] *Science* 290:1127–1131. With permission from American Association for the Advancement of Science.)

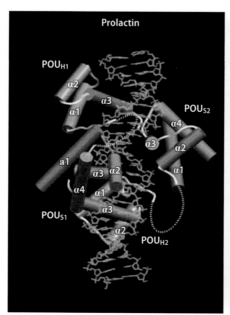
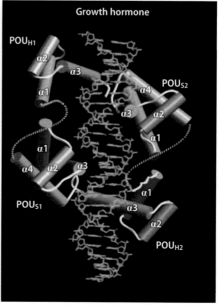

One aspect of RNA processing that is of definite importance in influencing transcriptome composition is the splicing of eukaryotic pre-mRNA. This is because alternative splicing pathways can generate a range of mRNAs from a single pre-mRNA. We must therefore study how the splicing of pre-mRNAs is regulated in order to understand how these events contribute to the composition of a eukaryotic transcriptome.

The splicing pathway for eukaryotic pre-mRNA introns

With the vast majority of pre-mRNA introns, the first two nucleotides of the intron sequence are 5′–GU–3′ and the last two are 5′–AG–3′. They are therefore called **GU–AG introns**. A small minority do not fall into this class and, based on their boundary sequences, were originally called **AU–AC introns**, although now that more examples have been characterized it has become clear that these sequences are variable, and not all introns of this type have the AU and/or AC motifs. The splicing pathway is similar for both types of intron, with differences only in the details, and we will therefore focus our attention on the GU–AG group.

The conserved GU and AG motifs were recognized soon after introns were discovered and it was immediately assumed that they must be important in the splicing process. As intron sequences started to accumulate in the databases, it was realized that these motifs are merely parts of the longer consensus sequences, spanning the 5′ and 3′ splice sites, that we met in Section 5.1 when we considered the methods used to identify intron boundaries when a eukaryotic genome sequence is annotated. Other conserved sequences are present in the introns of some but not all eukaryotes. Introns in higher eukaryotes usually have a **polypyrimidine tract**, which is a pyrimidine-rich region located just upstream of the 3′ end of the intron sequence (**Figure 12.31**). This tract is less frequently seen in yeast introns, but these have an invariant 5′–UACUAAC–3′ sequence, located between 18 and 140 nucleotides upstream of the 3′ splice site, which is not present in higher eukaryotes.

The conserved sequence motifs indicate important regions of GU–AG introns, regions that we would anticipate either acting as recognition sequences for RNA-binding proteins involved in splicing, or playing some other central role in the process. Early attempts to understand splicing were hindered by technical problems (in particular, difficulties in developing a cell-free splicing system with which the process could be probed in detail), but during the 1990s there was an explosion of information. This work showed that the splicing pathway can be divided into two steps (**Figure 12.32**):

- Cleavage of the 5′ splice site, also called the **donor site**, occurs by a transesterification reaction promoted by the hydroxyl group attached to the 2′ carbon of an adenosine nucleotide located within the intron sequence, at a position called the branch site. In yeast, this adenosine is the last one in the conserved UACUAAC sequence. The result of the hydroxyl attack is cleavage of the phosphodiester bond at the donor site, accompanied by formation of a new 5′–2′ phosphodiester bond linking the first nucleotide of the intron (the G of the GU motif) with the internal adenosine. This means that the intron loops back on itself to create a **lariat** structure.

- Cleavage of the 3′ splice site (the **acceptor site**) and joining of the exons result from a second transesterification reaction, this one promoted by the 3′–OH group attached to the end of the upstream exon. This group attacks the phosphodiester bond at the acceptor site, cleaving it and releasing the

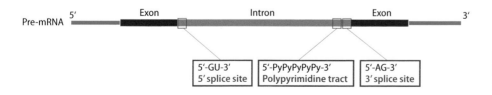

Figure 12.31 Conserved sequences in vertebrate introns. Abbreviation: Py, pyrimidine nucleotide (U or C).

Figure 12.32 Splicing in outline.
Cleavage of the 5′ splice site (the donor site) is promoted by the hydroxyl (–OH) attached to the 2′–carbon of an adenosine nucleotide within the intron sequence. This results in the lariat structure and is followed by the 3′–OH group of the upstream exon inducing cleavage of the 3′ splice site (the acceptor site). This enables the two exons to be ligated, with the released intron being debranched and degraded.

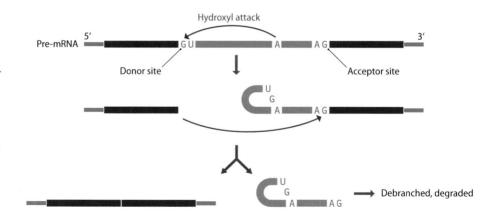

intron as the lariat structure, which is subsequently converted back to a linear RNA and degraded. At the same time, the 3′ end of the upstream exon joins to the newly formed 5′ end of the downstream exon, completing the splicing process.

In a chemical sense, intron splicing is not a great challenge for the cell. The difficulty lies with the topological problem caused by the substantial distance that might lie between splice sites, possibly a few tens of kilobases, representing 100 nm or more if the mRNA is in the form of a linear chain. A means is therefore needed to bring the splice sites into proximity. This is the role of small nuclear ribonucleoproteins (snRNPs) (Section 12.1) which, together with other accessory proteins, attach to specific positions in the pre-mRNA and form a series of complexes, the most important of which is the **spliceosome**, the structure within which the actual splicing reactions occur.

Splicing initiates with formation of **complex E** (**Figure 12.33**). This complex comprises U1-snRNP, which binds to the donor site, partly by RNA–RNA base-pairing, and the protein factors SF1, U2AF1, and U2AF2, which make protein–RNA contacts with the branch site, the polypyrimidine tract, and possibly the acceptor site. The next step is conversion of complex E to **complex A**, also called the **prespliceosome complex**, which occurs when U2-snRNP attaches to the branch site. The U1- and U2-snRNPs have an affinity for each other, and this draws the donor site toward the branch point. **Complex B** (the **precatalytic spliceosome**) is then formed when the U4-, U5-, and U6-snRNPs attach to the intron. Their arrival results in additional interactions that bring the acceptor site close to the donor site and the branch point. The U1- and U4-snRNPs then leave the complex, giving rise to the spliceosome. All three key positions in the intron are now in proximity and the cutting and joining reactions can take place, catalyzed by the U2- and U6-snRNPs. The initial product of the splicing reaction is a **post-spliceosome complex**, which dissociates into the spliced mRNA and the intron lariat, the latter still attached to the U2-, U5-, and U6-snRNPs.

The splicing process must have a high degree of precision

Precision is essential in splicing. If a cut is made just one nucleotide away from the actual exon–intron boundary position, then the open reading frame of the transcript will be disrupted and the mRNA that is made will be nonfunctional. At a different level, precision is needed in the selection of splice sites. All donor sites have a similar sequence, as do all acceptor sites, so if a pre-mRNA contains two or more introns then there is the possibility that the wrong splice sites could be joined, resulting in **exon skipping** – the loss of an exon from the mature mRNA (**Figure 12.34A**). Equally unfortunate would be the selection of a **cryptic splice site**, a site within an intron or exon that has sequence similarity with the consensus motifs of real splice sites (**Figure 12.34B**). Cryptic sites are present in most pre-mRNAs and must be ignored by the splicing apparatus.

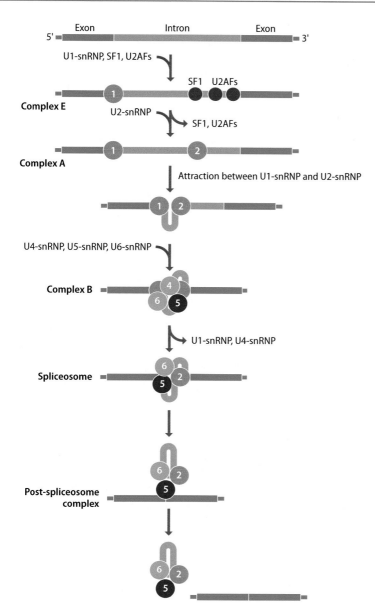

Figure 12.33 The roles of snRNPs and associated proteins during splicing. There are several unanswered questions about the series of events occurring during splicing and it is unlikely that the scheme shown here is entirely accurate. The key point is that associations between the snRNPs are thought to bring the three critical parts of the intron – the two splice sites and the branch point – into close proximity. The protein factors SF1, U2AF1, and U2AF2 are only shown in complex E, as these are thought to detach during formation of complex A.

Key players in the control of splice-site selection are short sequences called **exonic splicing enhancers (ESEs)**, **exonic splicing silencers (ESSs)**, **intronic splicing enhancers (ISEs)**, and **intronic splicing silencers (ISEs)**. These sequences act as binding sites for proteins that, when attached, activate or repress the selection of the adjacent splice site. Most of the proteins that bind to splicing enhancers are **SR proteins**, so-called because they contain a domain that is rich in serine (abbreviation S) and arginine (R). SR proteins attach to the C-terminal domain of the largest subunit of RNA polymerase II soon after the CTD becomes phosphorylated during the initiation stage of transcription. They ride with the polymerase as it synthesizes the transcript, and are deposited at their splicing enhancer sequences as soon as these are transcribed. Electron microscopy studies have shown that transcription and splicing occur together, and the discovery of splicing factors that have an affinity for RNA polymerase II provides a biochemical basis for this observation. When attached to a splicing enhancer, SR proteins appear to participate in several steps in the splicing pathway, including the establishment of a connection between bound U1–snRNP and the bound U2AF proteins in complex E. This is perhaps the clue to their role in splice-site selection, formation of Complex E being the critical stage in the splicing process, as this is the event that identifies which sites will be linked.

Figure 12.34 Two aberrant forms of splicing. (A) In exon skipping, the aberrant splicing results in an exon being lost from the mRNA. (B) When a cryptic splice site is selected, part of an exon might be lost from the mRNA, as shown here, or if the cryptic site lies within an intron, then a segment of that intron will be retained in the mRNA.

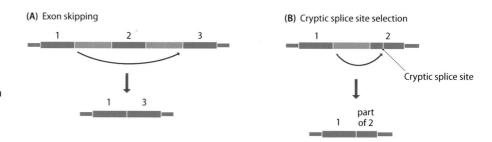

Less is known about the mode of action of exonic and intronic splicing silencers. These sequences act as binding sites for **heterogeneous nuclear ribonucleoproteins (hnRNPs)**. However, as their name implies, these are a broad group of RNA-protein complexes, which play several roles in the nucleus, most of which involve binding to RNAs. The particular types called hnRNP1 and hnRNPL appear to bind at specific positions in introns and exons and are associated with splicing regulation. Their mode of action might simply be to block access to one or more of the splice sites, branch site and/or polypyrimidine site, interfering with the assembly of Complex E and therefore preventing splicing.

Enhancer and silencer elements specify alternative splicing pathways

Alternative splicing is now looked on as a major component of the processes that determine the composition of a transcriptome. As we discussed in Section 7.2, 75% of all human protein-coding genes, representing 95% of those with two or more introns, undergo alternative splicing, giving rise to an average of four different spliced mRNAs, or **isoforms**, per gene. These alternative splicing events can be placed in four categories:

- **Exon skipping**, which results in one or more exons being left out of the final mRNA (see Figure 12.34A).

- **Alternative site selection**, in which the usual donor or acceptor site is ignored and a second site used in its place. This is equivalent to cryptic splice-site selection (see Figure 12.34B).

- **Alternative exons**, where the mRNA contains either of a pair of exons, but not both at the same time.

- **Intron retention**, in which an intron that is usually spliced out of the pre-mRNA is retained in the final mRNA. The retained intron might be in the upstream or downstream untranslated region of the mRNA, but some examples are known where the intron contains an open reading frame and so contributes to the amino-acid sequence of the protein coded by the mRNA.

An example will illustrate the importance and complexity of alternative splicing. Sex determination, a fundamental aspect of the biology of any organism, is specified by an alternative splicing cascade in *Drosophila*. The first gene in this cascade is *sxl*, whose transcript contains an exon which, when spliced to the one preceding it, results in an inactive version of protein SXL. In females, the splicing pathway is such that this exon is skipped so that functional SXL is made (**Figure 12.35**). SXL promotes selection of an alternative splice site in a second transcript, *tra*, by directing U2AF2 away from its normal acceptor site to a second site further downstream. The resulting female-specific TRA protein is again involved in alternative splicing, this time by interacting with SR proteins to form a multifactor complex that attaches to an enhancer element within an exon of a third pre-mRNA, *dsx*, promoting selection of a female-specific splice site in this transcript. The female and male versions of the DSX proteins are the primary determinants of *Drosophila* sex.

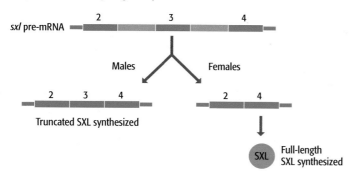

(A) Sex-specific alternative splicing of *sxl* pre-mRNA

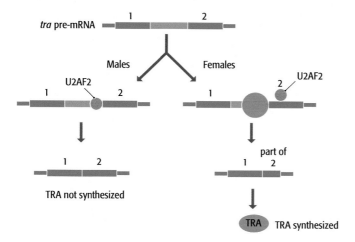

(B) SXL induces cryptic splice site selection in the *tra* pre-mRNA

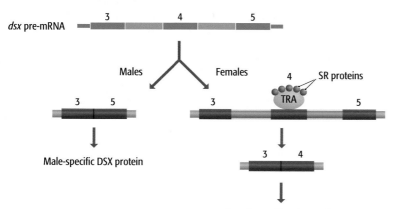

(C) TRA induces alternative splicing of the *dsx* pre-mRNA

Figure 12.35 Regulation of splicing during expression of genes involved in sex determination in *Drosophila*. (A) The cascade begins with sex-specific alternative splicing of the *sxl* pre-mRNA. In males, all exons are present in the mRNA, but this means that a truncated protein is produced because exon 3 contains a termination codon. In females, exon 3 is skipped, leading to a full-length, functional SXL protein. (B) In females, SXL blocks the acceptor site in the first intron of the *tra* pre-mRNA. U2AF2 is unable to locate this site and instead directs splicing to a cryptic site in exon 2. This results in an mRNA that codes for a functional TRA protein. In males, there is no SXL, so the acceptor site is not blocked and a dysfunctional mRNA is produced. (C) In females, TRA stabilizes the attachment of SR proteins to an exonic splicing enhancer located within exon 4 of the *dsx* pre-mRNA, resulting in this exon being attached to exon 3, giving the mRNA that codes for the female-specific DSX protein. In males, absence of TRA means that exon 4 is skipped. The two versions of DSX are the primary determinants of the female and male physiologies. The female *dsx* mRNA ends with exon 4 because the intron between exons 4 and 5 has no donor site, meaning that exon 5 cannot be ligated to the end of exon 4. Instead, a polyadenylation site at the end of exon 4 is recognized in females.

Some pre-mRNAs undergo such extensive alternative splicing that hundreds or even thousands of different transcripts are generated from a single gene. An extreme example is the *Drosophila* gene called *Dscam*, which has 95 variable exons that are spliced into 38,016 distinct mRNAs, with up to 50 different mRNAs per cell. The DSCAM protein is a cell-surface receptor that contributes to the ability of nerve cells to adhere to one another. The diversity of *Dscam* mRNAs is such that each nerve cell could synthesize a different repertoire of DSCAM proteins, possibly as a means of establishing individual cell identities. Multiple splicing pathways of such huge extent are less common in higher eukaryotes, but the human *Megf11* gene, which has a similar role to *Dscam* being involved in cell–cell recognition during development of the retina, gives rise to at least 234 alternatively spliced mRNAs, and it is suspected that the overall diversity is even greater than this. The complexity and specificity of alternative splicing suggests that intron and/or exon sequences contain a **splicing code** of some description, which dictates the various interactions that can occur between enhancers, silencers and their binding proteins and hence specifies the splicing pathways for a transcript and the conditions under which each pathway is followed. At present, our knowledge of the code is rudimentary, and as such there are still important gaps in our understanding of how alternative splicing is controlled.

Backsplicing gives rise to circular RNAs

Another surprising discovery of the last decade has been the realization that eukaryotic cells contain substantial amounts of circular RNA molecules. It has been known for some time that virusoids and viroids (Section 9.1) are circular RNAs, and that a small number of eukaryotic viruses have circular RNA genomes, such as hepatitis delta virus, which is thought to be derived from a viroid. It has also been recognized since the 1990s that the human *Sry* gene, which is the main determinant of the male sex, gives rise to a circular transcript. Other circular RNAs that were occasionally observed in mammalian cells by electron microscopy were assumed to be rare aberrant products of RNA processing, such as intron lariats that had not become debranched.

Our views regarding circular RNA changed dramatically with the advent of RNA-seq and in particular when long-read sequencing was applied to eukaryotic RNA extracts. Circular transcripts were found to make up substantial parts of some transcriptomes, with 15,000 different circular RNAs in human cells. Most of these circular RNAs contain one or more exons, occasionally with the presence also of an intron. They are believed to arise by a process called **backsplicing**, which results in the donor site of a downstream exon becoming attached to an upstream acceptor site (**Figure 12.36A**). This process can occur as an adjunct to the normal splicing pathway, possibly when one or other component of the spliceosome is depleted and the normal splicing process becomes slowed down. In this regard, circular RNAs are often derived from genes with a high transcription rate, whose processing could conceivably place demands on the cell's repertoire of snRNPs and associated splicing proteins. Backsplicing might also be favored by the presence of repeated sequences such as Alu elements (Section 9.2) in the transcript, which could base-pair to one another, or by binding sites for proteins that can dimerize and bring the donor and acceptor sites into proximity (**Figure 12.36B**). If the circularized region contains one or more introns, then these are usually removed, probably by the standard spliceosome pathway. About 50 of the human circular RNAs are more abundant than the linear transcripts of the gene from which they are derived, but the vast majority have much lower copy numbers and most have less than 1% the abundance of their linear partners.

Circular RNAs are classed as long noncoding transcripts. Although they contain exons they lack the sequences that act as ribosome binding sites (Section 13.3) and so, with a few exceptions, cannot be translated into protein. The same questions regarding functionality that we previously considered with lncRNAs therefore apply also to circular RNAs. Circular RNAs are long-lived in the cell, but this might simply reflect the fact that most of the cellular processes for RNA degradation involve exonuclease activity (Section 12.6) and so cannot break down circular molecules. Most circular RNAs are moved to the cytoplasm, though the ones that contain unspliced introns usually remain in the nucleus, where they can act as decoy RNAs in the same way as lncRNAs. Some circular RNAs appear to be more abundant in particular tissues and in certain disease states, but as yet no convincing biological function has been assigned to any backspliced circular RNA. However, this is the position we were in just a few

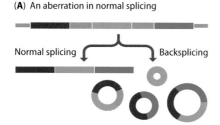

(A) An aberration in normal splicing

Normal splicing   Backsplicing

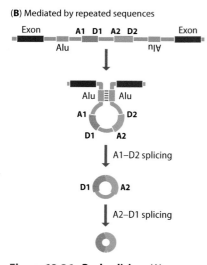

(B) Mediated by repeated sequences

Exon   A1 D1 A2 D2   Exon
Alu   Alu

Alu   Alu
A1   D2
D1   A2

A1–D2 splicing

D1   A2

A2–D1 splicing

Figure 12.36 Backsplicing. (A) Backsplicing can occur as an adjunct to the normal splicing pathway, and can give a variety of circular products. (B) The presence of repeat sequences in the transcript can promote backsplicing. In this example, base-pairing between two Alu sequences brings together acceptor site A1 and donor site D2, which normally would not form a splice pair. Aberrant splicing between A1 and D2 creates a circular RNA that contains an intron, which may then be removed by normal splicing between acceptor site A2 and donor site D1.

years ago with regards to linear lncRNAs, and we are now aware of functions for many of these. The possibility that circular RNAs represent yet another group of important RNAs therefore has to be considered.

12.5 THE INFLUENCE OF CHEMICAL MODIFICATION ON THE COMPOSITION OF A TRANSCRIPTOME

As well as the cutting and joining events that constitute splicing, many RNAs are also processed by chemical modification. Over 170 different types of modification can occur, giving rise to a range of nucleotide variants whose novel chemical properties can influence the structure and/or function of the RNA (see Figure 1.18).

Chemical modifications were first identified in tRNAs and rRNAs (Section 1.2), but these modifications are looked on as part of the standard synthesis pathway for noncoding RNA and hence do not affect the functional capacity of a transcriptome. The same is true for capping and polyadenylation of eukaryotic mRNAs (see Figure 1.17): almost all mRNAs are capped or polyadenylated, and although the cap and poly(A) tail might influence the translation of an mRNA they are looked on a standard features of these transcripts, rather than playing a regulatory role in the composition of the transcriptome or proteome.

Other chemical modifications are more important as they have a specific impact on the translation of individual mRNAs. These modifications fall into two categories (**Figure 12.37**):

- **RNA editing**, which refers to a chemical modification that affects the base-pairing properties of a nucleotide and so changes the sequence of a mRNA (e.g., the conversion of a C into a U). RNA editing therefore has the capacity to alter the amino-acid sequence of the protein coded by an mRNA.

- Modifications that add a novel chemical group to a nucleotide base or sugar but do not affect base-pairing (e.g., conversion of adenosine into N^6-methyladenosine, which still base-pairs with U). Although these modifications do not alter the amino-acid sequence of the encoded protein, they might influence the way in which the mRNA is translated.

The entire collection of nucleotide modifications displayed by the RNAs in a transcriptome is referred to as the **epitranscriptome**. Strictly speaking, the epitranscriptome includes modifications to all RNAs, both coding and noncoding, but it is the mRNA component of the epitranscriptome that is of most interest as at least some of the mRNA modifications affect the translation of the mRNA into protein. The mRNA epitranscriptome is therefore equivalent to the epigenome, the latter comprising the DNA methylations and nucleosome modifications that affect transcription of the genome into RNA (Sections 10.2 and 10.3).

RNA editing alters the coding properties of some transcripts

RNA editing involves a chemical modification that alters the base-pairing properties of a nucleotide and hence changes the sequence of the mRNA. A notable example of RNA editing occurs with the human mRNA for apolipoprotein B. There are two versions of this protein, called apolipoprotein B48, which is synthesized by intestinal cells, and apolipoprotein B100, which is about twice the size of B48 and is made in the liver. Both proteins are involved in the transport of lipids around the body, but their exact roles are different, with the B48 protein forming part of the transport structure called a chylomicron, and B100 combining with other proteins to form a complex called a very low-density lipoprotein. Both types of apolipoprotein B are specified by the same gene, but in intestinal cells a cytosine at position 6666 in the 14,000 nucleotide mRNA is edited by deamination. This change is sufficient to convert the mRNA for apolipoprotein B100 into an mRNA specifying apolipoprotein B48 (**Figure 12.38**). The deamination

(A) RNA editing

(B) Modifications that do not affect base-pairing

Me   Me   Ac

Figure 12.37 Chemical modifications to RNA. (A) RNA editing, which results in a change in the nucleotide sequence of the mRNA. (B) Modifications that do not affect base-pairing and so do not change the nucleotide sequence.

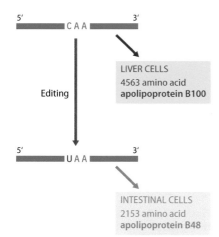

Figure 12.38 Editing of the human apolipoprotein B mRNA. Deamination of a C at position 6666 of the mRNA converts this nucleotide into U, creating a termination codon (a signal for termination of translation). A shortened form of apolipoprotein B is therefore synthesized in intestinal cells.

is carried out by an RNA-binding enzyme which, in conjunction with a set of auxiliary protein factors, binds to a sequence immediately downstream of the modification position within the mRNA.

Although not common, RNA editing occurs in a number of different organisms and includes a variety of different nucleotide changes (Table 12.3). Some editing events have a significant impact on the organism: in humans, editing is partly responsible for the generation of antibody diversity and has also been implicated in the development of some cancers. One particularly interesting type of editing is the deamination of adenosine to inosine, which is carried out by the adenosine deaminases acting on RNA (ADAR) enzymes. Some of the target mRNAs for these enzymes are selectively edited at a limited number of positions. These positions are specified by segments of the pre-mRNA that take up a left-handed double-stranded structure, formed by base-pairing between the modification site and sequences from adjacent introns (recall that ADARs also bind to left-handed Z-DNA – Section 11.3). This type of editing occurs, for example, during processing of the mRNAs for mammalian glutamate receptors, proteins that play important roles in the transmission of signals between nerve cells. There is evidence that ADAR editing is closely linked with RNA synthesis, as some nucleotides within introns are edited (indicating that editing occurs before intron splicing) and editing efficiency is reduced if changes are artificially made to the CTD of RNA polymerase II.

Selective editing contrasts with the second type of modification carried out by ADARs, in which the target molecules become extensively deaminated, with over 50% of the adenosines in the RNA becoming converted to inosines. **Hyperediting** was initially observed with double-stranded viral RNAs, which adopt base-paired structures that act as substrates for ADAR. Hyperediting may have an antiviral effect, but in some cases the editing changes the nature of a viral disease rather than preventing it: with measles, hyperediting might contribute to immune evasion, resulting in a persistent measles infection, as opposed to the more usual transient version of the disease. Large-scale RNA-seq studies have revealed that many eukaryotic transcripts are also hyperedited, but these are predominantly noncoding double-stranded RNAs, for example, ones containing two copies of a repeat element that base-pair to one another. In this context, hyperediting might be part of the RNA degradation process.

The above examples of RNA editing are relatively straightforward events which, with the exception of hyperediting, lead to nucleotide changes at a single or limited number of positions in selected mRNAs. More complex types of RNA editing are also known:

- **Pan-editing** involves the extensive insertion of nucleotides into abbreviated RNAs in order to produce functional molecules. It is particularly common in the mitochondria of trypanosomes. Many of the RNAs transcribed in trypanosome mitochondria are specified by **cryptogenes** – sequences lacking some of the nucleotides present in the mature RNAs. The pre-RNAs transcribed from these cryptogenes are processed by

TABLE 12.3 EXAMPLES OF RNA EDITING IN MAMMALS

Tissue	Target RNA	Change	Comments
Intestine	Apolipoprotein B mRNA	C→U	Converts a glutamine codon to a stop codon
Muscle	α-galactosidase mRNA	U→A	Converts a phenylalanine codon into a tyrosine codon
Testis, tumors	Wilms tumor-1 mRNA	U→C	Converts a leucine codon into a proline codon
Tumors	Neurofibromatosis type-1 mRNA	C→U	Converts an arginine codon into a stop codon
B lymphocytes	Immunoglobulin mRNA	Various	Contributes to the generation of antibody diversity
Brain	Glutamate receptor mRNA	A→inosine	Multiple positions leading to various codon changes

multiple insertions of U nucleotides, at positions defined by short **guide RNAs**. These are short RNAs that can base-pair to the pre-RNA and which contain As at the positions where Us must be inserted (**Figure 12.39**).

- Less extensive **cotranscriptional editing** occurs with some viral RNAs. For example, the paramyxovirus P gene gives rise to three different proteins because of the insertion of Gs at specific positions in the mRNA. These insertions are not specified by guide RNAs: instead, they are added by the RNA polymerase as the mRNA is being synthesized.

- **Polyadenylation editing** is seen with many animal mitochondrial mRNAs. Five of the mRNAs transcribed from the human mitochondrial genome end with just a U or UA, rather than with one of the termination codons (UAA and UAG in the human mitochondrial genetic code). Polyadenylation converts the terminal U or UA into UAAAA…A, and so creates a termination codon. This is just one of several features that appear to have evolved in order to make vertebrate mitochondrial genomes as small as possible.

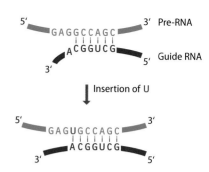

Figure 12.39 The role of the guide RNA in pan-editing. The guide RNA contains an A at the position where a U must be inserted in the pre-RNA.

Chemical modifications that do not affect the sequence of an mRNA

RNA editing can be detected simply by comparing the sequence of an mRNA with that of the gene from which it is transcribed. If the RNA-seq reads have nucleotide differences compared with the DNA sequence, then editing has occurred. Detecting chemical modifications that do not alter the base-pairing properties of a nucleotide is much more difficult because the RNA-seq reads will map on to the DNA sequences without any mismatches. The difficulty in detecting this type of modification explains why, until recently, the widespread existence of modified nucleotides in eukaryotic mRNAs was unsuspected.

Modified nucleotides can be distinguished and identified by liquid chromatography followed by mass spectrometry (LC-MS). The RNA must be broken down into its constituent nucleotides prior to LC-MS, so the technique identifies the modified nucleotides that are present but does not provide any information on the particular mRNAs that contain those nucleotides nor on the positions of the modifications in the transcripts. The use of LC-MS with preparations from different sources has identified seven modified nucleotides that are common in eukaryotic mRNA. The most abundant of these is N^6-methyladenosine (m^{6}A) (**Figure 12.40**).

The difficulties in obtaining information on the location of modified nucleotides in individual mRNAs has been solved, at least in part, by the development of RNA-seq methods coupled with pre-treatments that enable modified positions to be identified in the reads that are obtained. Several different approaches have been used (**Figure 12.41**):

- For m^{6}A, the mRNAs are broken into fragments, and those containing the modification are collected by immunoprecipitation with a m^{6}A-specific antibody. Immunoprecipitation is applicable to any modified nucleotide for which there is a specific antibody, but is less successful with low abundance modifications. This is because some unmodified fragments will also be present in the immunoprecipitate, and these false positives can swamp the genuinely modified fragments if the modification is relatively rare in the transcriptome.

- **Bisulfite sequencing** can detect 5-methylcytidine (m^{5}C) in both RNA and DNA: pre-treatment with bisulfite converts unmethylated cytosines to uracils so the sequence reads reveal the positions of the methylated versions.

- Positions where the ribose component of a nucleotide is methylated cause the reverse transcriptase to pause as it copies the RNA into cDNA at the beginning of the RNA-seq procedure. If reverse transcription is

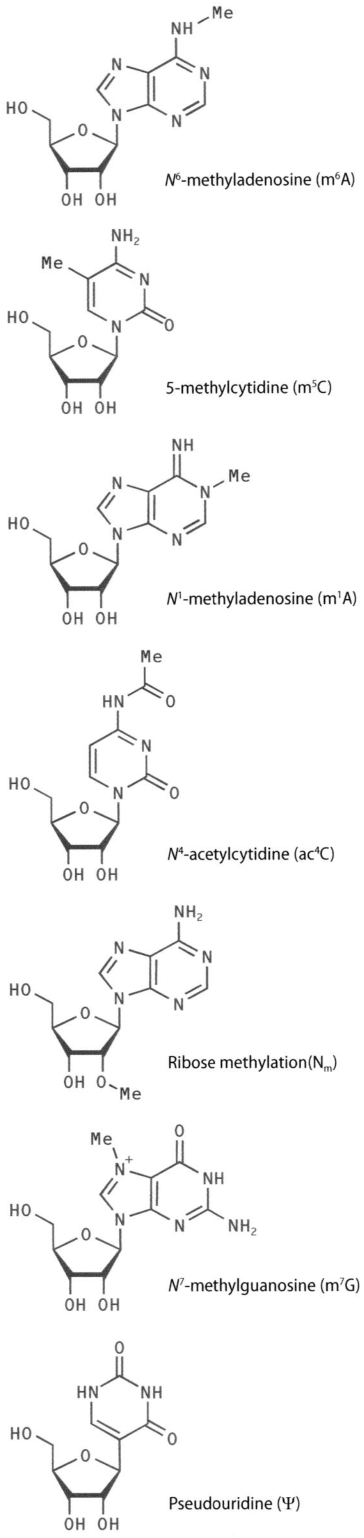

Figure 12.40 The seven modified nucleotides that are common in eukaryotic mRNA.

carried out with low amounts of dNTPs, then the pause results in chain termination. The methylribose positions will therefore appear at high frequency at the ends of sequence reads. This method has been called **MeTH-seq**.

More generally, there is optimism that modified RNA nucleotides could be detected directly by nanopore sequencing, because the modification, or a more bulky derivative obtained by specific chemical treatment, will affect the way in which the nucleotide occludes the nanopore and hence result in a recognizable disruption in the pattern of ion flow, from which the modification could be detected and identified (see Figure 4.15). This approach has been used to map methylation sites in DNA (Section 4.3) and is becoming more feasible as RNA-seq methods are developed for the nanopore platform.

Our knowledge of the enzymes that modify mRNAs is developing rapidly. The most important m⁶A 'writer' – the enzyme that coverts A into m⁶A – is a heterodimeric protein comprising a METTL3 subunit, which possesses the methyltransferase activity, and a METTL14 subunit which aids in attachment of the dimer to the transcript and possibly stimulates METTL3 activity. The METTL dimer works in conjunction with several other proteins, whose roles are not fully understood, but which collectively direct the complex to the methylation sites in the target mRNAs. These sites have the consensus sequence 5′–DRACH–3′, where D is A, G or U; R is A or G; and H is A, C or U. Such sites are quite common in most mRNAs, but only a few are methylated. It is assumed that the accessory proteins direct METTL to particular sites, but how this is achieved is not yet known. The end result is that between zero and ten m⁶A modifications are made to every mammalian mRNA, the majority of these close to the stop codon or in the 3′-untranslated region (3′-UTR) of the transcript. The m⁶A modification can subsequently be removed by either of two 'eraser' enzymes. Between writing and erasing, a variety of 'reader' proteins are able to attach to the modification, either directly or via a separate RNA-binding protein, presumably to influence the processing and/or translation of the mRNA in a specific manner. The various functions that have been ascribed to m⁶A in different transcripts include control over translation rate, mRNA stability, and splice-site selection. The last of these roles occurs in the mRNA for the MAT2A gene, which codes for an enzyme that converts methionine into S-adenosylmethionine, which is a substrate for many important biochemical reactions. The presence or absence of m⁶A in a stem-loop structure located in the 3′-UTR of the MAT2A mRNA appears to regulate splicing of an intron in response to the cellular levels of methionine.

So far, specific writers for other types of mRNA modification have not been identified, the only enzymes implicated in this regard also being responsible for modifying tRNA or rRNA. It is possible that these enzymes target mRNA segments that have sequences or base-paired structures similar to the modification sites in the noncoding transcripts. Some of these mRNA modifications could therefore be 'noise' accompanying the modification of tRNA and rRNA, and have no function in the coding transcripts. Various possible roles have been ascribed to these modifications, but in contrast to m⁶A, a detailed mechanistic description of the link between modification and function has yet to be made.

12.6 DEGRADATION OF THE COMPONENTS OF THE TRANSCRIPTOME

Now that we understand how the components of a transcriptome are synthesized and processed, we must turn our attention to how they are degraded. As we have already discussed, the processes responsible for the synthesis and degradation of individual transcripts act in unison to determine the composition of a transcriptome and to enable that composition to change in response to the prevailing conditions (see Figure 12.14).

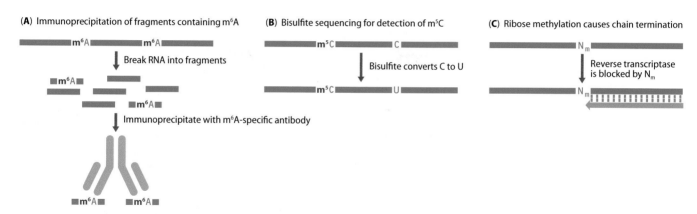

Figure 12.41 Three methods for detection of modified nucleotides in mRNA. (A) Fragments containing N^6-methyladenosine (m6A) can be purified by immunoprecipitation with a specific antibody prior to RNA-seq. (B) Bisulfite converts unmethylated cytosines to uracils which are read as Ts by RNA-seq. Methylated versions such as 5-methylcytidine (m5C) are unaffected so still read as Cs. (C) Ribose methylation (N_m) blocks the progress of reverse transcriptase when dNTPs are limiting. Methylribose positions will therefore appear at the ends of RNA-seq reads.

Several processes are known for nonspecific RNA turnover

The rate of degradation of an RNA can be estimated by determining its **half-life** in a transcriptome. This is the period of time required for the amount of an individual type of RNA to fall to half its initial value, assuming that there is no new synthesis of the molecule. Half-lives can be measured by **pulse labeling**. The cells being studied are briefly provided with a labeled substrate for RNA synthesis, such as radioactive 4-thiouracil, in which the oxygen atom attached to carbon 4 of the normal uracil molecule is replaced with a ^{35}S atom. RNAs that are synthesized during the period of pulse labeling will contain the labeled nucleotide, but those made before or after will not. The degradation rates of the labeled molecules are then followed by measuring the amounts of label present in RNA extracts prepared at intervals after the period of pulse labeling. Experiments of this type have shown that there are considerable variations between and within organisms. Bacterial mRNAs are generally turned over very rapidly, their half-lives rarely being longer than a few minutes, a reflection of the rapid changes in protein synthesis patterns that can occur in an actively growing bacterium with a generation time of 20 minutes or so. Eukaryotic mRNAs are longer lived, with half-lives of, on average, 10–20 minutes for yeast and several hours for mammals. There have been fewer studies of noncoding RNA half-lives, but those that have been reported suggest that, in eukaryotes, tRNAs and rRNAs are both turned over more slowly than mRNAs, with tRNA half-lives between nine hours and several days, and rRNA half-lives up to eight days. The figures for long noncoding RNAs are also variable, from less than two hours to over 16 hours in mouse neuroblastoma cells.

The half-lives measured by pulse labeling indicate that most types of RNA are subject to continual turnover. Much of this turnover is probably nonspecific, acting on all RNAs of a particular type, and not discriminating between the transcripts of individual genes. In bacteria, nonspecific mRNA degradation is carried out by the **degradosome**, a multiprotein structure whose components in *Escherichia coli* include:

- Polynucleotide phosphorylase (PNPase), which removes nucleotides sequentially from the 3′ end of an mRNA.

- RNA helicase B, which opens up stem-loop structures such as terminator sequences and hence aids the progression of PNPase along the mRNA.

- An endonuclease called RNase E, which makes internal cuts in RNA molecules.

The eukaryotic equivalent of the degradosome is the **exosome**, which comprises a ring of six proteins, each of which has ribonuclease activity, with three

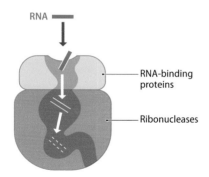

Figure 12.42   RNA degradation in a eukaryotic exosome. The RNA is initially captured by the RNA-binding proteins at the top of the exosome, and then passed into the channel within the ring of ribonucleases. Within the channel, the RNA is degraded by a combination of exonuclease and endonuclease activities.

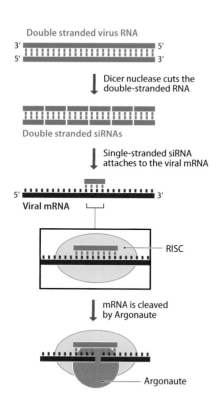

Figure 12.43   The RNA silencing pathway. The double-stranded virus RNA is cut by Dicer, giving double-stranded siRNAs. Single-stranded versions of the siRNAs then base-pair to viral mRNAs, inducing assembly of an RNA-induced silencing complex, which contains an Argonaute endonuclease that cleaves and hence silences the mRNA.

RNA-binding proteins attached to the top of the ring. Other ribonucleases associate with the exosome in a transient manner. It is thought that RNAs to be degraded are initially captured by the binding proteins and then threaded through the channel in the middle of the ring, where they are exposed to the ribonuclease activities of the ring proteins (**Figure 12.42**).

Exosomes are present in both the cytoplasm and the nucleus of a eukaryotic cell. The main role of the nuclear exosomes appears to be the rapid turnover of aberrant RNAs that have not been transcribed or processed correctly and which are therefore not released into the cytoplasm. Aberrant mRNAs are detected by a **surveillance mechanism**, which identifies ones that lack a termination codon, which might occur if the DNA has been copied incorrectly, or have a termination codon at an unexpected position, indicating that the exons have been joined together incorrectly during RNA splicing. Surveillance involves a complex of proteins which scans mRNAs for these errors and directs aberrant transcripts to the exosome or some other degradation pathway. Other surveillance systems look for errors in the chemical modification of tRNAs. Some modification errors can be tolerated, but others appear to be more critical and tRNAs containing them are rapidly degraded, either by the exosome or by a second turnover pathway specific to tRNAs.

RNA silencing was first identified as a means of destroying invading viral RNA

One of the major areas of development in our understanding of genomes over the last 20 years has been the growing realization that most types of organism have mechanisms that enable individual transcripts, in particular specific mRNAs, to be degraded, possibly resulting in their rapid and complete removal from a transcriptome. For many years it has been known that eukaryotes possess RNA degradation mechanisms that protect their cells from attack by foreign RNAs, such as the genomes of viruses. Originally called **RNA silencing**, this process is already familiar to us under its alternative name of RNA interference, as its underlying mechanism has been utilized by genome researchers as a means of inactivating selected genes in order to study their function (Section 6.2).

The target for RNA silencing must be double-stranded, which excludes cellular mRNAs but encompasses viral genomes, many of which are either double-stranded RNA in their native state or replicate via a double-stranded RNA intermediate (Section 9.1). The double-stranded RNA is recognized by binding proteins that form an attachment site for a ribonuclease called **Dicer**, which cuts the molecule into short interfering RNAs (siRNAs) of 20–25 nucleotides in length (**Figure 12.43**). This inactivates the virus genome, but what if the virus genes have already been transcribed? If this has occurred, then the harmful effects of the virus will already have been initiated, and RNA silencing would appear to have failed in its attempt to protect the cell from damage. One of the more remarkable discoveries of recent years has revealed a second stage of the interference process that is directed specifically at the viral mRNAs. The siRNAs produced by cleavage of the viral genome are separated into individual strands, one strand of each siRNA subsequently base-pairing to any viral mRNAs that are present in the cell. The double-stranded regions that are formed are target sites for assembly of the **RNA-induced silencing complex** (**RISC**), which includes an endonuclease of the Argonaute family, which cleaves and hence silences the mRNA.

The work which resulted in the initial description of the molecular process underlying RNA interference was carried out in the late 1990s with *Caenorhabditis elegans*. Since then, RNA interference has been shown to occur in most eukaryotes, the exceptions including *Saccharomyces cerevisiae*, and interference has been linked to various events that involve RNA degradation but which were previously thought to be unrelated. For example, the movement of some types of transposable element involves a double-stranded RNA intermediate which can be degraded by a process now known to be RNA interference. This is one way in which eukaryotes prevent the wholesale proliferation of transposons within

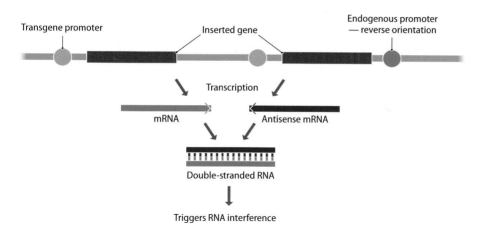

Figure 12.44 RNA interference explains why transgenes are sometimes inactive. For clarity, the mRNA and antisense RNA are shown to be transcribed from different copies of the inserted transgene. They could also come from a single transgene that is transcribed from both its own promoter and an endogenous promoter.

their genomes. Genetic engineers had also been puzzled by the ability of some organisms, especially plants, to silence new genes that had been inserted into their genomes by cloning techniques. We now know that this type of silencing can occur if the transgene is inserted, by chance, upstream of a promoter that directs synthesis of an antisense RNA copy of all or part of the gene. This RNA then base-pairs with the sense mRNA produced from the transgene's own promoter to form a double-stranded RNA that triggers the RNA interference pathway (**Figure 12.44**). Other phenomena in diverse organisms, variously known as quelling, cosuppression, and post-transcriptional gene silencing, are all now known to be different guises of RNA interference.

MicroRNAs regulate genome expression by causing specific target mRNAs to be degraded

A link between silencing of viral RNAs and specific degradation of endogenous mRNAs was made when it was discovered that many eukaryotes have more than one type of Dicer protein. The fruit fly, *Drosophila melanogaster*, for example, has two Dicer enzymes. It turns out that the second type of Dicer in *Drosophila* works not with viral RNAs, but with endogenous molecules called **foldback RNAs**, which are coded by the fruit-fly DNA and synthesized by RNA polymerase II. The name 'foldback' is given to these RNAs because they can form intrastrand base pairs giving rise to a stem-loop structure (**Figure 12.45**). The stem can be cut by Dicer, releasing short double-stranded molecules, each about 21 bp, called microRNAs (miRNAs). One strand of each double-stranded molecule is degraded, giving the functional miRNAs. A few miRNAs are obtained by a slightly different method, not by transcription of a gene for a precursor miRNA, but from an intron cut out of the mRNA of a protein-coding gene. Part of the intron RNA folds up to form the stem-loop structure, which is then processed by Dicer as described above.

Each miRNA is complementary to part of a cellular mRNA and hence base-pairs with this target, stimulating assembly of a RISC. Often the miRNA annealing site is present in the 3′ untranslated region of the target mRNA, sometimes in multiple copies (**Figure 12.46**). Cleavage by Argonaute therefore does not disrupt the coding region of the mRNA, but will lead to detachment of the poly(A) tail. Loss of the poly(A) tail might interfere with the initiation of translation, which is inefficient if this part of the mRNA is absent. Alternatively, removal of the poly(A) tail might target the mRNA for degradation by the exosome or one of the other nonspecific pathways for mRNA turnover. Whatever the precise mechanism, cleavage by Argonaute leads to the mRNA being silenced.

The first miRNA silencing system to be characterized involved the *C. elegans* genes called *lin-4* and *let-7*, both of which code for foldback RNAs that generate miRNAs after cleavage by Dicer. A mutation in either of these two genes causes defects in the worm's development pathway, indicating that this type of RNA

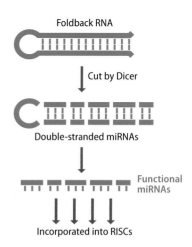

Figure 12.45 Synthesis of miRNAs from a foldback RNA precursor. The stem-loop of the foldback RNA is cut by Dicer, releasing short double-stranded miRNAs, one strand of which is degraded, giving the functional miRNAs.

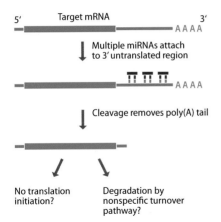

Figure 12.46 Binding sites for miRNAs are often in the 3′ untranslated region of the target mRNA.

degradation is not simply a means of getting rid of aberrant or potentially harmful mRNAs, but instead plays a fundamental role in regulating composition of the transcriptome. Further support for this notion was provided by other studies of *C. elegans* miRNAs, which revealed that these molecules are involved in biological events as diverse as cell death, specification of neuron cell types, and control of fat storage. Genome analysis shows that humans have the capacity to synthesize about 2300 miRNAs, but together these can target mRNAs from over 6000 genes, possibly because mRNAs from different genes share the same miRNA binding sequence, or possibly because a precise match between miRNA and mRNA is not needed in order for the mRNA to be captured by a RISC. Some miRNA genes are located close to the protein-coding genes whose mRNAs are targeted by the miRNA. In these cases, it is possible that the same regulatory proteins control both mRNA and miRNA synthesis. This would allow synthesis of the miRNA to be directly coordinated with repression of the protein-coding gene. The mRNA would therefore be degraded immediately after its synthesis is switched off. But in many other cases, the miRNA and protein genes are not co-located, and the way in which mRNA synthesis and degradation are coordinated is not clear.

SUMMARY

- The transcriptome is the collection of RNA molecules present in a cell.

- The mRNA fraction makes up a relatively small part of the transcriptome of most cells, but comprises many different transcripts.

- Transcriptomes contain a variety of short noncoding RNAs with diverse functions.

- Transcriptomes also contain long noncoding RNAs whose functions, if any, are not well understood.

- The content of a transcriptome can be cataloged by microarray studies and by RNA sequencing. Spatial information on transcriptome variations within a tissue can be obtained by single-cell studies and RNA-seq carried out *in situ*.

- The components of a transcriptome are synthesized by RNA polymerases. In eukaryotes, there are three different RNA polymerases, each responsible for transcribing a different set of nuclear genes.

- The positions at which transcription must begin are marked by promoter sequences. Some genes have alternative promoters that give rise to different versions of the transcript specified by the gene.

- Initiation of transcription in bacteria is regulated by the combined action of activator and repressor proteins.

- Initiation of transcription in eukaryotes is regulated mainly by activator proteins called transcription factors, which interact with the initiation complex via the mediator protein.

- Removal of introns from eukaryotic pre-mRNAs has an important impact on transcriptome composition as many transcripts have alternative splicing pathways, giving rise to different mRNAs and hence different protein products.

- During alternative splicing, splice-site selection is controlled by short enhancer and silencer sequences, which are the binding sites for regulatory proteins.

- Circular RNAs are generated by backsplicing, which occurs during the processing of some pre-mRNAs.

- Chemical modifications can change the coding properties of an mRNA or influence the way in which the mRNA is translated.

- RNA can be turned over by nonspecific processes and by RNA silencing, the latter enabling specific mRNAs to be targeted for degradation.

SHORT ANSWER QUESTIONS

1. Describe the various types of small noncoding RNAs that are found in eukaryotic transcriptomes.

2. What evidence is there to suggest that at least some long noncoding RNAs are functional?

3. Outline how a transcriptome is characterized by (A) microarray analysis and (B) RNA sequencing.

4. Describe the methods that are used to study transcriptomes in single cells and in fixed tissue samples.

5. Compare the structures of the three eukaryotic RNA polymerases with that of the bacterial enzyme. What are the key features of the RNA polymerases that transcribe organelle genes?

6. Distinguish between the structures of the promoters for (A) the bacterial RNA polymerase, (B) RNA polymerase I, (C) RNA polymerase II, and (D) RNA polymerase III.

7. How is the synthesis of bacterial mRNAs regulated?

8. Describe the role of the mediator protein during initiation of eukaryotic transcription.

9. Outline the splicing pathway for a eukaryotic pre-mRNA.

10. What is alternative splicing and how is it regulated?

11. Describe how circular RNAs are generated during pre-mRNA splicing.

12. Describe the different ways in which mRNAs are processed by chemical modification. Which of these processes affect the coding properties of mRNAs?

13. Outline the processes for nonspecific degradation of RNA in bacteria and eukaryotes.

14. Explain how specific RNAs are targeted for degradation by siRNAs and miRNAs.

IN-DEPTH PROBLEMS

1. Construct a hypothesis to explain why eukaryotes have three RNA polymerases. Can your hypothesis be tested?

2. A model for control of transcription of the lactose operon in *Escherichia coli* was first proposed by François Jacob and Jacques Monod in 1961 (Jacob, F. and Monod, J. [1961]. Genetic regulatory mechanisms in the synthesis of proteins. *J. Mol. Biol.* 3:318–356). Explain the extent to which their work, which was based almost entirely on genetic analysis, provided an accurate description of the molecular events that are now known to occur.

3. To what extent is *E. coli* a good model for the regulation of transcription initiation in eukaryotes? Justify your opinion by providing specific examples of how extrapolations from *E. coli* have been helpful and/or unhelpful in the development of our understanding of equivalent events in eukaryotes.

4. Introns are common in higher organisms but virtually absent in bacteria. Discuss the possible reasons for this.

5. How close are we to understanding how the mRNA of a single gene is targeted for degradation in bacteria and eukaryotes?

FURTHER READING

Components of transcriptomes

Clark, M.B., Choudhary, A., Smith, M.A., et al. (2013) The dark matter rises: The expanding world of regulatory RNAs. *Essays Biochem.* 54:1–16.

Clark, M.B., Johnston, R.L., Inostroza-Ponta, M., et al. (2012) Genome-wide analysis of long noncoding RNA stability. *Genome Res.* 22:885–898.

Kung, J.T.Y., Colognori, D. and Lee, J.T. (2013) Long noncoding RNAs: Past, present, and future. *Genetics* 193:651–669.

Qian, Y., Shi, L. and Luo, Z. (2020) Long non-coding RNAs in cancer: Implications for diagnosis, prognosis, and therapy. *Front. Med.* 7:612393.

Statello, L., Guo, C.-J., Chen, L.L., et al. (2021) Gene regulation by long non-coding RNAs and its biological functions. *Nat. Rev. Mol. Cell Biol.* 22:96–118.

Wagatsuma, A., Sadamoto, H., Kitahashi, T., et al. (2005) Determination of the exact copy numbers of particular mRNAs in a single cell by quantitative real-time RT-PCR. *J. Exp. Biol.* 208:2389–2398.

Wan, L., Zhang, L., Fan, K., et al. (2016) Knockdown of long non-coding RNA PCAT6 inhibits proliferation and invasion in lung cancer cells. *Oncol. Res.* 24:161–170.

Wang, H.-L. V. and Chekanova, J.A. (2017) Long noncoding RNAs in plants. *Adv. Exp. Med. Biol.* 1008:133–154.

Methods for studying transcriptomes

Cai, Z., Cao, C., Ji, L., et al. (2020) RIC-seq for global in situ profiling of RNA-RNA spatial interactions. *Nature* 582:432–437.

Cao, C., Cai, Z., Ye, R., et al. (2021) Global in situ profiling of RNA-RNA spatial interactions with RIC-seq. *Nat. Protoc.* 16:2916–2946.

Hwang, B., Lee, J.H. and Bang, D. (2018) Single-cell RNA sequencing technologies and bioinformatics pipelines. *Exp. Mol. Med.* 50:96.

Imdahl, F., Vafadarnejad, E., Homberger, C., et al. (2020) Single-cell RNA-sequencing reports growth-condition-specific global transcriptomes of individual bacteria. *Nat. Microbiol.* 5:1202–1206.

Larsson, L., Frisén, J. and Lundeberg, J. (2021) Spatially resolved transcriptomics adds new dimension to genomics. *Nat. Methods* 18:15–17.

Lee, J.H., Daugharthy, E.R., Scheiman, J., et al. (2014) Highly multiplexed subcellular RNA sequencing in situ. *Science* 343:1360–1363.

Leung, Y.F. and Cavalieri, D. (2003) Fundamentals of cDNA microarray data analysis. *Trends Genet.* 19:649–659.

Shen, X., Zhao, Y., Wang, Z. and Shi, Q. (2022) Recent advances in high-throughput single-cell transcriptomics and spatial transcriptomics. *Lab Chip.* 22:4774–4791.

Ståhl, P.L., Salmén, F., Vickovic, S., et al. (2016) Visualization and analysis of gene expression in tissue sections by spatial transcriptomics. *Science* 353:78–82.

Tang, F., Barbacioru, C., Wang, Y., et al. (2009) mRNA-seq whole-transcriptome analysis of a single cell. *Nat. Methods* 6:377–382.

Wilhelm, B.T. and Landry, J.-R. (2009) RNA-Seq – Quantitative measurement of expression through massively parallel RNA-sequencing. *Methods* 48:249–257.

Young, A.P., Jackson, D.J. and Wyeth, R.C. (2020) A technical review and guide to RNA fluorescence in situ hybridization. *PeerJ* 8:e8806.

RNA polymerases, promoters, and RNA synthesis

Aibara, S., Schilbach, S. and Cramer, P. (2021) Structures of mammalian RNA polymerase II pre-initiation complexes. *Nature* 594:124–128.

Allen, B.L. and Taatjes, D.J. (2015) The mediator complex: A central integrator of transcription. *Nat. Rev. Mol. Cell Biol.* 16:155–166.

Börner, T., Aleynikova, A.Y., Zubo, Y.O., et al. (2015) Chloroplast RNA polymerases: Role in chloroplast biogenesis. *Biochim. Biophys. Acta* 1847:761–769.

Kadonaga, J.T. (2012) Perspectives on the RNA polymerase II core promoter. *Wiley Interdiscip. Rev. Dev. Biol.* 1:40–51.

Kandiah, E., Trowitzsch, S., Gupta, K., et al. (2014) More pieces to the puzzle: Recent structural insights into class II transcription initiation. *Curr. Opin. Struct. Biol.* 24:91–97.

Khatter, H., Vorländer, M.K. and Müller, C.W. (2017) RNA polymerase I and III: Similar yet unique. *Curr. Opin. Struct. Biol.* 47:88–94.

Kim, T.H., Barrera, L.O., Zheng, M., et al. (2005) A high-resolution map of active promoters in the human genome. *Nature* 436:876–880. *Reveals the extent to which alternative promoters are used in the human genome.*

Mathew, R. and Chatterji, D. (2006) The evolving story of the omega subunit of bacterial RNA polymerase. *Trends Microbiol.* 14:450–455.

Patel, A.B., Greber, B.J. and Nogales, E. (2020) Recent insights into the structure of TFIID, its assembly, and its binding to core promoter. *Curr. Opin. Struct. Biol.* 61:17–24.

Saecker, R.M., Record, M.T. and deHaseth, P.L. (2011) Mechanism of bacterial transcription initiation: RNA polymerase-promoter binding, isomerization to initiation-competent open complexes, and initiation of RNA synthesis. *J. Mol. Biol.* 412:754–771.

Turowski, T.W. and Boguta, M. (2021) Specific features of RNA polymerases I and III: Structure and assembly. *Front. Mol. Biosci.* 8:680090.

Control of RNA synthesis

Baek, I., Friedman, L.J., Geles, J. et al. (2021) Single-molecule studies reveal branched pathways for activator-dependent assembly of RNA polymerase II pre-initiation complexes. *Mol. Cell* 81:3576–3588. *Describes the possible assembly of the pre-initiation complex at an enhancer site.*

Browning, D.F. and Busby, S.J.W. (2016) Local and global regulation of transcription initiation in bacteria. *Nat. Rev. Microbiol.* 14:638–650.

Kadonaga, J.T. (2004) Regulation of RNA polymerase II transcription by sequence-specific DNA binding factors. *Cell* 116:247–257.

Koo, B.M., Rhodius, V.A., Campbell, E.A., et al. (2009) Dissection of recognition determinants of Escherichia coli s³² suggest a composite –10 region with an 'extended –10' motif and a core –10 element. *Mol. Microbiol.* 72:815–829. *The E. coli heat shock promoter.*

Naville, M. and Gautheret, D. (2009) Transcription attenuation in bacteria: Themes and variations. *Brief. Funct. Genomic. Proteomic.* 8:482–492.

Oehler, S., Eismann, E.R., Krämer, H., et al. (1990) The three operators of the lac operon cooperate in repression. *EMBO J.* 9:973–979.

Swigon, D., Coleman, B.D. and Olson, W.K. (2006) Modeling the Lac repressor-operator assembly: The influence of DNA looping on Lac repressor conformation. *Proc. Natl Acad. Sci. USA* 103:9879–9884.

Intron splicing and alternative splicing pathways

Black, D.L. (2003) Mechanisms of alternative pre-messenger RNA splicing. *Annu. Rev. Biochem.* 72:291–336.

Blencowe, B.J. (2000) Exonic splicing enhancers: Mechanism of action, diversity and role in human genetic diseases. *Trends Biochem. Sci.* 25:106–110.

Graveley, B.R. (2005) Mutually exclusive splicing of insect *Dscam* pre-mRNA directed by competing intronic RNA secondary structures. *Cell* 123:65–73.

Kristensen, L.S., Anderson, M.S., Stagsted, L.V.W., et al. (2019) The biogenesis, biology and characterization of circular RNAs. *Nat. Rev. Genet.* 20:675–691.

Ray, T.A., Cochran, K., Kozlowski, C., et al. (2020) Comprehensive identification of mRNA isoforms reveals the diversity of neural cell-surface molecules with roles in retinal cell development. *Nat. Commun.* 11:3328. *Describes the role of alternative splicing in retinal cells, including the multiple spice pathways of Megf11.*

Ruegg, M.A. (2005) Structural and functional diversity generated by alternative mRNA splicing. *Trends Biochem. Sci.* 30:515–521.

Stetefeld, J. and Lynch, K.W. and Maniatis, T. (2016) Assembly of specific SR protein complexes on distinct regulatory elements of the *Drosophila doublesex* splicing enhancer. *Genes Dev.* 10:2089–2101.

Tarn, W.-Y. and Steitz, J.A. (1997) Pre-mRNA splicing: The discovery of a new spliceosome doubles the challenge. *Trends Biochem. Sci.* 22:132–137. *AU-AC introns.*

Valcárcel, J. and Green, M.R. (1996) The SR protein family: Pleiotropic functions in pre-mRNA splicing. *Trends Biochem. Sci.* 21:296–301.

Wilkinson, M.E., Charenton, C. and Nagai, K. (2020) RNA splicing by the spliceosome. *Annu. Rev. Biochem.* 89:359–388.

Chemical modification of RNA

Anreiter, I., Mir, Q., Simpson, J.T., et al. (2021) New twists in detecting mRNA modification dynamics. *Trends Biotechnol.* 39:72–89.

Christofi, T. and Zaravinos, A. (2019) RNA editing in the forefront of epitranscriptomics and human health. *J. Transl. Med.* 17:319.

Douglas, J., Drummond, A.J., and Kingston, R.L. (2021) Evolutionary history of cotranscriptional editing in the paramyxoviral phosphoprotein gene. *Virus Evol.* 7:veab028.

Filippini, A., Bonini, D., La Via, L., et al. (2017) The good and the bad of glutamate receptor RNA editing. *Mol. Neurobiol.* 54:6795–6805.

Kung, C.-P., Maggi, L.B. and Weber, J.D. (2018) The role of RNA editing in cancer development and metabolic disorders. *Front. Endocrinol.* 9:762.

Wang, Y., Zhang, X, Liu, H., et al. (2021) Chemical methods and advanced sequencing technologies for deciphering mRNA modifications. *Chem. Soc. Rev.* 50:13481–13497.

Weiner D. and Schwartz, S. (2021) The epitranscriptome beyond m⁶A. *Nat. Rev. Genet.* 22:119–131.

Zaccara, S., Ries, R.J. and Jaffrey, S.R. (2019) Reading, writing and erasing mRNA methylation. *Nat. Rev. Mol. Cell Biol.* 20:608–624.

Degradation of RNA

Alles, J., Fehlmann, T., Fischer, U., et al. (2019) An estimate of the total number of true human miRNAs. *Nucl. Acids Res.* 47:3353–3364.

Carpousis, A.J., Vanzo, N.F. and Raynal, L.C. (1999) mRNA degradation: A tale of poly(A) and multiprotein machines. *Trends Genet.* 15:24–28.

Cho. K.H. (2017) The structure and function of the gram-positive bacterial RNA degradosome. *Front. Microbiol.* 8:154.

Defoiche, J., Zhang, Y., Lagneaux, L., et al. (2009) Measurement of ribosomal RNA turnover *in vivo* by use of deuterium-labeled glucose. *Clin. Chem.* 55:1824–1833.

Mello, C.C. and Conte, D. (2004) Revealing the world of RNA interference. *Nature* 431:338–342.

Pratt, A.J. and MacRae, I.J. (2009) The RNA-induced silencing complex: A versatile gene-silencing machine. *J. Biol. Chem.* 284:17897–17901.

Sontheimer, E.J. and Carthew, R.W. (2005) Silence from within: Endogenous siRNAs and miRNAs. *Cell* 122:9–12.

Vanacova, S. and Stefl, R. (2007) The exosome and RNA quality control in the nucleus. *EMBO Rep.* 8:651–657.

Wilson, R.C. and Doudna, J.A. (2013) Molecular mechanisms of RNA interference. *Annu. Rev. Biophys.* 42:217–239.

Zinder, J.C. and Lima, C.D. (2017) Targeting RNA for processing or destruction by the eukaryotic RNA exosome and its cofactors. *Genes Dev.* 31:88–100.

PROTEOMES

The proteome is the collection of protein molecules present in a cell. The proteome is therefore the final link between the genome and the biochemical capability of the cell, and characterization of the proteomes of different cells is one of the keys to understanding how the genome operates and how dysfunctional genome activity can lead to disease. Transcriptome studies can only partly address these issues. Examination of the transcriptome gives an accurate indication of which genes are active in a particular cell, but gives a less accurate indication of the proteins that are present. This is because the factors that influence protein content include not only the amount of an mRNA that is available but also the rate at which each mRNA is translated into protein and the rate at which the protein is degraded. Additionally, the protein that is the initial product of translation may not be active, as some proteins must undergo physical and/or chemical modification before becoming functional. Determining the amount of the *active* form of a protein is therefore critical to understanding how the genome specifies the biochemistry of a cell or tissue.

The issues that we must examine regarding the proteome are very similar to the issues that interested us in the previous chapter when we studied transcriptomes. First, we will explore the various methods that are used to catalog the components of a proteome and to identify proteins that interact with one another within a cell. Then we will study the events involved in the synthesis, degradation and processing of the components of a proteome. Finally, we will examine more closely the link between the proteome and the biochemistry of the cell.

13.1 STUDYING THE COMPOSITION OF A PROTEOME

The methodology used to study proteomes is called **proteomics**. Strictly speaking, proteomics is a collection of diverse techniques that are related only in their ability to provide information on a proteome, that information encompassing not only the identities of the constituent proteins that are present but also factors such as the functions of individual proteins and their localization within the cell. We will focus on the two aspects of proteomics that are most directly relevant to an understanding of genome activity: cataloging the composition of a proteome (this section) and identifying pairs and groups of proteins that interact with one another (Section 13.2).

The particular technique that is used to study the composition of a proteome is called **protein profiling** or **expression proteomics**. Protein profiling is usually carried out in two stages:

- In the first stage, the individual proteins in a proteome are separated from one another.

- In the second stage, the proteins are identified, usually by **mass spectrometry**.

DOI: 10.1201/9781003133162-13

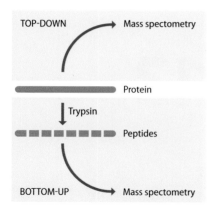

Figure 13.1 Top-down and bottom-up proteomics. In top-down proteomics, the intact protein is examined by mass spectrometry. In bottom-up proteomics, peptides derived from the protein are examined. In this example, peptides are obtained by treating the protein with trypsin, which cuts immediately after arginine or lysine amino acids.

This basic format encompasses two different approaches called **top-down** and **bottom-up proteomics**, the difference being that in top-down proteomics, individual proteins are directly examined by mass spectroscopy, whereas in bottom-up proteomics the proteins are broken into peptides by treatment with a sequence-specific protease, such as trypsin, prior to mass spectrometry (**Figure 13.1**).

The separation stage of a protein profiling project

In order to characterize a proteome, it is first necessary to prepare pure samples of its constituent proteins. How difficult this is depends on the complexity of the proteome, with the separation of the 10,000–20,000 proteins in some mammalian proteomes requiring more sophisticated methods than are needed for the less complex proteomes of bacteria or mammalian cell fractions (e.g., the mitochondria), which might contain fewer than 1000 proteins. The choice of separation technique is therefore dictated in part by the complexity of the proteome that is being studied.

Polyacrylamide gel electrophoresis (PAGE) is the standard method for separating the proteins in a complex mixture. Depending on the composition of the gel and the conditions under which the electrophoresis is carried out, different chemical and physical properties of proteins can be used as the basis for their separation. One technique makes use of the detergent called sodium dodecyl sulfate (SDS), which denatures proteins and confers a negative charge that is roughly equivalent to the length of the unfolded polypeptide. Under these conditions, the proteins separate according to their molecular masses, the smallest proteins migrating more quickly toward the positive electrode. Alternatively, proteins can be separated by **isoelectric focusing** in a gel that contains chemicals that establish a pH gradient when the electrical charge is applied. In this type of gel, a protein migrates to its **isoelectric point**, the position in the gradient where its net charge is zero.

When a complex proteome is being studied, the two versions of PAGE are often combined in **two-dimensional gel electrophoresis**. In the first dimension, the proteins are separated by isoelectric focusing. The gel is then soaked in sodium dodecyl sulfate, rotated by 90°, and a second electrophoresis, separating the proteins according to their sizes, is carried out at right angles to the first (**Figure 13.2**). This approach can separate several thousand proteins in a single gel, the proteins revealed as a complex pattern of spots when the gel is stained (**Figure 13.3**). Individual spots can therefore be cut out of the gel and the proteins they contain purified. Cutting out 20,000 spots would clearly be a laborious process, and in practice two-dimensional PAGE is not used if the aim is to catalog every protein in a proteome, and instead is restricted to identification of interesting proteins, such as those that have different abundances in two or more related proteomes – the healthy and diseased versions of a tissue, for example.

An alternative approach to protein separation by PAGE is provided by **column chromatography**. This method involves passing the protein mixture through a column packed with a solid matrix. The proteins in the mixture move through the matrix at different rates, and so become separated into bands. The solution emerging from the column can then be collected as a series of fractions, with each individual protein present in a different fraction (**Figure 13.4**). The identity

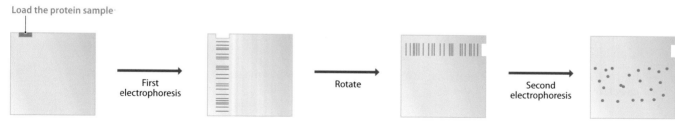

Figure 13.2 Two-dimensional gel electrophoresis.

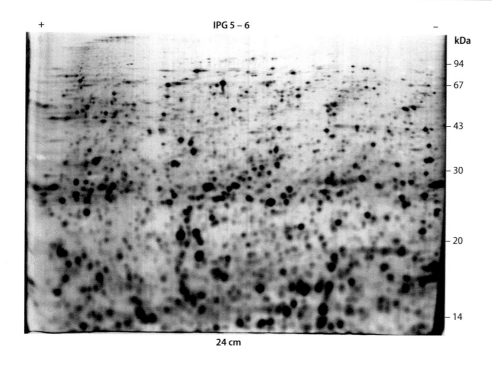

+   IPG 5 – 6   –

kDa

— 94

— 67

— 43

— 30

— 20

— 14

24 cm

Figure 13.3 The result of two-dimensional gel electrophoresis. Mouse liver proteins have been separated by isoelectric focusing on the pH 5–6 range in the first dimension and according to molecular mass in the second dimension. The protein spots have been visualized by staining with silver solution. (From Görg A., Obermaier C., Boguth G. et al. [2000] *Electrophoresis* 21:1037–1053. With permission from John Wiley & Sons, Inc.)

of the **solid phase** (the matrix) and the composition of the **mobile phase** (the liquid used to move the proteins through the column) specify which of the variable physicochemical properties of proteins are used to achieve separation. In protein profiling, the two most commonly used types of column chromatography are:

- **Reverse phase liquid chromatography** (**RPLC**), in which the solid phase is a matrix of silica particles whose surfaces are covered with nonpolar chemical groups such as hydrocarbons. The mobile phase is a mixture of water and an organic solvent such as methanol or acetonitrile. Most proteins have hydrophobic areas on their surfaces, which bind to the nonpolar matrix, but the stability of this attachment decreases as the organic content of the liquid phase increases. Gradually changing the ratio of the aqueous and organic components of the mobile phase therefore results in the elution of proteins according to their degree of surface hydrophobicity.

- **Ion exchange chromatography** separates proteins according to their net electric charges. The matrix consists of polystyrene beads that carry either positive or negative charges. If the beads are positively charged, then proteins with a net negative charge will bind to them, and vice versa. The proteins are eluted with a salt gradient, set up by gradually increasing the salt concentration of the buffer being passed through the column. The charged salt ions compete with the proteins for the binding sites on the resin, so proteins with low charges are eluted at low-salt concentration, and ones with higher charges at higher salt concentrations. The salt gradient therefore separates proteins according to their net charges. Alternatively, a pH gradient can be used. The net charge of a protein depends on the pH, and, as described above, is zero at the pH corresponding to that protein's isoelectric point. Gradually changing the pH of the mobile phase will result in the elution of proteins with different isoelectric points, again achieving their separation.

Compared with two-dimensional PAGE, column chromatography is less laborious to carry out, and has the advantage that individual proteins can be collected as they elute from the column, avoiding the post-separation purification step needed to obtain a protein from its spot on a PAGE gel. Column chromatography

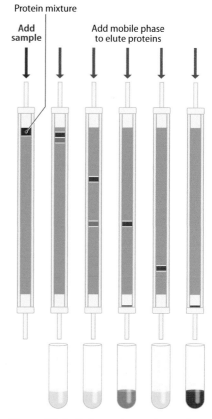

Protein mixture

Add sample

Add mobile phase to elute proteins

Figure 13.4 Column chromatography, The proteins separate into bands as they move through the column. In practice, tens or hundreds of proteins can be separated in this way.

Figure 13.5   High-performance liquid chromatography (HPLC). The diagram shows a typical HPLC apparatus. The protein mixture is injected and pumped through the column along with the mobile phase solution. Proteins are detected as they elute from the column, usually by measuring UV absorbance at 210–220 nm. The fractions that are collected can be of equal volume, or the data from the detector can be used to control the fractionation so that each protein peak is collected as a single sample of minimum volume.

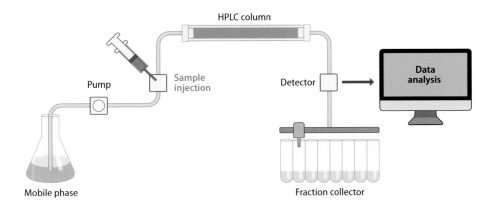

is usually carried out in a capillary tube with an internal diameter of less than 1 mm, with the liquid phase being pumped at high pressure. This procedure, called **high-performance liquid chromatography** (**HPLC**), has a high resolving power and enables proteins with very similar chromatographic properties to be separated (**Figure 13.5**).

To increase the resolving power, different types of chromatography column can be linked together, each consecutive fraction from one column being fed into a second column, in which a further round of separation using a different procedure is carried out (**Figure 13.6A**). In this way, quite complex mixtures of proteins can be fully separated. Alternatively, proteins can initially be separated by one-dimensional PAGE, using either the SDS version or isoelectric focusing, and the resulting gel is cut into segments (**Figure 13.6B**). The set of proteins present in each segment are then be entered sequentially into the column chromatography system. If the top-down approach is being used, then rather than collecting protein fractions as they emerge from the column – the 'offline' mode – the column can be directly attached to the mass spectrometer. Each protein is therefore analyzed by the spectrometer as it elutes from the column (**Figure 13.6C**). This online mode cannot be used with the standard version of bottom-up proteomics, because each protein must be treated with a protease to cut it into fragments prior to injection into the mass spectrometer. The online mode, is however, possible with the modification of the bottom-up approach called **shotgun proteomics**. With this method, the proteins are treated with the protease before the column chromatography step. These proteins could be the entire proteome, if it is not overly complex, or the mixture obtained from a segment of a one-dimensional PAGE gel. In either case, the column now separates

Figure 13.6   Three configurations for the separation phase of protein profiling. (A) Two chromatography columns are linked in series. (B) Fractions from a PAGE gel are entered into the chromatography column. (C) Online mode with the chromatography column directly linked to the mass spectrometer. Combinations of these three formats are also possible.

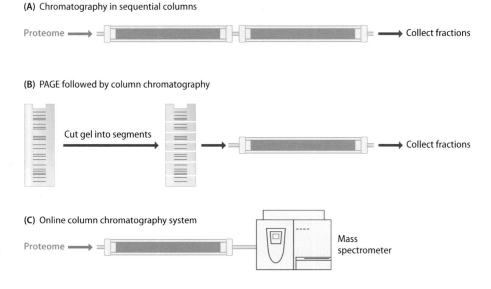

peptides rather than intact proteins, and the online mode is used to inject the eluted molecules directly into the mass spectrometer.

The identification stage of a protein profiling project

Separation of the components of a proteome is followed by identification of the individual proteins, either directly in top-down proteomics, or from the peptides resulting from proteolytic cleavage if a bottom-up method is being used. Identification of proteins and peptides used to be a difficult proposition, but advances in mass spectrometry, driven in parts by the requirements of proteomics, have largely solved this problem.

Mass spectrometry is a means of identifying a compound from the **mass-to-charge ratio** (designated m/z) of the ions that are produced when molecules of the compound are exposed to a high-energy field. The first type of mass spectrometry to be used widely in protein profiling was **matrix-assisted laser desorption ionization time-of-flight** (**MALDI-TOF**). This technique forms the basis to **peptide mass fingerprinting**, the bottom-up procedure that identifies the individual peptides in the mixture obtained by protease cleavage of a protein purified by two-dimensional PAGE or column chromatography. The first step in MALDI-TOF is ionization of the peptides. This is achieved by absorbing the mixture into an organic crystalline matrix, usually made of a phenylpropanoid compound called sinapinic acid, which is excited with a UV laser. The excitation initially ionizes the matrix, with protons then donated to or removed from the peptide molecules, to give the **molecular ions** $[M+H]^+$ and $[M–H]^-$, respectively, where 'M' is the peptide. Ionization also results in vaporization of the peptides, which are then accelerated along the tube of the mass spectrometer by an electric field. The flight path can be direct from the ionization source to a detector, but often the ions are initially directed at a **reflectron**, which reflects the ion beam toward the detector (**Figure 13.7**). As well as enabling a longer flight path to be built into a machine of a defined size, the reflectron acts as a focusing device, ensuring that all ions with the same m/z value travel through the mass spectrometer at the same speed. This is critical because this 'time-of flight' spectrometer uses the time that an ion takes to reach the detector in order to calculate the mass-to-charge ratio for that ion. As the charge is always +1 or –1, the time-of-flight can easily be converted into a molecular mass, which in turn allows the amino acid composition of the peptide to be

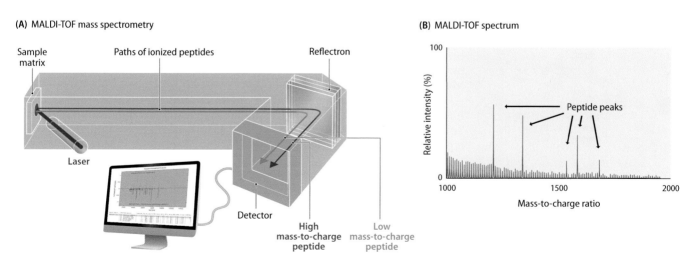

(A) MALDI-TOF mass spectrometry

(B) MALDI-TOF spectrum

Figure 13.7 The use of MALDI-TOF in protein profiling. (A). In the mass spectrometer, the peptides are ionized by a pulse of energy from a laser and then accelerated down the column to the reflectron and onto the detector. The time-of-flight of each peptide depends on its mass-to-charge ratio. (B) The data are visualized as a spectrum indicating the m/z values of the peptides. The computer converts the m/z values into molecular masses and compares these masses with the predicted masses of all the peptides that would be obtained by protease treatment of all of the proteins encoded by the genome of the organism under study. The protein that gave rise to the detected peptides can therefore be identified.

deduced. If a number of peptides from a single protein spot in a two-dimensional gel are analyzed, then the resulting compositional information can be related to the genome sequence in order to identify the gene that specifies that protein. The amino acid compositions of the peptides derived from a single protein can also be used to check that the gene sequence is correct, and in particular to ensure that exon–intron boundaries have been correctly located. This not only helps to ensure that the genome annotation is accurate, but also allows alternative splicing pathways to be identified in cases where two or more proteins are derived from the same gene.

MALDI-TOF is usually able to identify a protein that has been purified from a gel or by column chromatography. It is less suitable for shotgun proteomics, because the larger number of peptides that are produced when a mixture of proteins is treated with a protease increases the possibility that two peptides will have similar m/z values and hence be indistinguishable when examined by MALDI-TOF. Two innovations have been introduced in recent years to improve the resolution of peptide mass spectrometry and hence provide support for the shotgun methods. The first of these is the use of **electrospray ionization**, which can be performed online between HPLC and mass spectrometry. A high voltage is applied to the solution emerging from the HPLC, generating an aerosol of charged droplets which evaporate, transferring their charges to the peptides dissolved within them. The advantage of this ionization method is that, as well as the $[M+H]^+$ and $[M–H]^-$ molecular ions, each with a single ionized group, multiply ionized molecules – such as $[M+nH]^{n+}$ – are also obtained. Generation of multiple ions with different m/z values from individual peptides increases the amount of information that can be used to infer the composition of that peptide.

The second innovation is to break peptides down into smaller fragments within the mass spectrometer. Fragmentation can be induced during the ionization step, by using a 'hard' ionization method, one which injects greater quantities of energy into the molecules being ionized, causing bonds within those molecules to break. However, in peptide mass spectrometry, fragmentation is usually delayed until a later stage, by inducing collisions between the peptide molecular ions and inert atoms such as helium. These collisions cause peptide bonds to break, resulting in a variety of **fragment ions** whose m/z values reveal the composition of the original peptide. If sufficient fragment ions are obtained, then it might be possible to use the data to work out the sequence of the peptide. Knowing the sequence of the peptide enables a much more precise identification of the parent protein than is possible simply from compositional information. Collision-induced fragmentation is also utilized in top-down proteomics, because analysis of molecular ions derived from intact proteins is usually insufficient to distinguish all of the different proteins in a proteome. Fragment ions must therefore be obtained before a protein can be identified unambiguously.

Innovations involving ionization methods and the use of fragmentation have been accompanied by a diversification in the types of mass spectrometry employed in proteomics research. As well as time-of-flight configurations, other types of **mass analyzer** used with peptides and proteins include (**Figure 13.8**):

- The **quadrupole** mass analyzer, which has four magnetic rods placed parallel to one another, surrounding a central channel through which the ions must pass. Oscillating electrical fields are applied to the rods, deflecting the ions in a complex way so that their trajectories 'wiggle' as they pass through the quadrupole. Gradually changing the field strengths enables ions with different m/z values to pass through the quadrupole without colliding with the rods.

- The **Fourier transform ion cyclotron resonance (FT-ICR)** mass analyzer includes an ion trap that captures individual ions and further excites them within a cyclotron, so they accelerate along an outward spiral, the vector of this spiral revealing the m/z ratio.

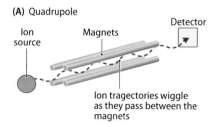

(A) Quadrupole

Ion source

Magnets

Detector

Ion tragectories wiggle as they pass between the magnets

(B) Fourier transform ion cyclotron resonance

Ions follow a spiral trajectory within the cyclotron

Figure 13.8 Two types of mass analyzer. (A) A quadrupole mass analyzer. (B) A Fourier transform ion cyclotron resonance mass analyzer.

Mass analyzers can also be linked in series, further increasing the amount of information that can be obtained regarding a single peptide or protein. This is called **tandem mass spectrometry**. A typical configuration involves analysis of the molecular ions in the first mass analyzer, followed by fragmentation and analysis of the fragment ions in the second mass analyzer.

Comparing the compositions of two proteomes

Often the aim of a protein profiling project is not to identify every protein in a proteome, but to understand the differences between the protein compositions of two different proteomes. If the differences are relatively large, then they will be apparent simply by looking at the stained gels after two-dimensional electrophoresis. However, important changes in the biochemical properties of a proteome can result from relatively minor changes in the amounts of individual proteins, and methods for detecting small-scale changes are therefore essential.

One possibility is to label the constituents of two proteomes with different fluorescent markers, and then run them together in a single two-dimensional gel. This is the same strategy as is used for comparing pairs of transcriptomes (see Figure 12.5). Visualization of the two-dimensional gel at different wavelengths enables the intensities of equivalent spots to be judged more accurately than is possible when two separate gels are obtained.

A more accurate alternative in bottom-up proteomics is to label peptides with an **isotope-coded affinity tag** (**ICAT**). These are chemical groups that can be attached to a peptide. In one system, the tags are short hydrocarbon chains that contain either the common ^{12}C isotope of carbon, or the less common ^{13}C isotope (**Figure 13.9**). The proteins in the proteomes are separated in the normal manner, and equivalent proteins from each proteome are recovered and treated with protease. One set of peptides is then labeled with ^{12}C tags and the other with the ^{13}C tags. Because the ^{12}C and ^{13}C tags have different masses, the m/z ratio of the molecular ion obtained from a peptide labeled with a ^{12}C tag will be different from that of an identical peptide labeled with a ^{13}C tag. The peptides from the two proteomes are therefore run through the mass spectrometer together. A pair of identical peptides (one from each proteome) will occupy slightly different positions on the resulting mass spectrum, because of their distinctive m/z ratios (**Figure 13.10**). Comparison of the peak heights allows the relative abundances of each peptide to be estimated.

The main disadvantage of ICAT labeling is that peptides with ^{12}C and ^{13}C tags can have slightly different chromatographic properties, especially during RPLC, so they might emerge from the column at slightly different times. The different masses are also a hindrance in tandem mass spectrometry, as the ^{12}C- and ^{13}C-labeled peptides will pass through the first mass analyzer at different rates, so their fragment ions will be collected by the second mass analyzer at different times. These problems are avoided by **isobaric labeling**. In this method, each tag consists of three parts: a reactive region, that forms the attachment with the peptide; and a balance region and a reporter region, both of which contain some ^{13}C atoms (**Figure 13.11**). Each tag has the same overall amount of ^{13}C, so all tags have the same mass – in other words, they are isobaric. Pairs of tagged peptides from two proteomes therefore give molecular ions that have the same m/z values

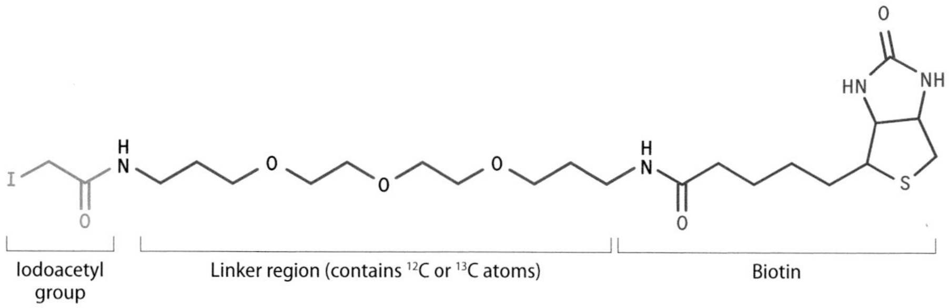

Figure 13.9 A typical isotope-coded affinity tag for proteome studies. The iodoacetyl group reacts with cysteine and hence forms the attachment to the peptide. The linker region contains either ^{12}C or ^{13}C atoms, and so provides the isotope coding.

Iodoacetyl group

Linker region (contains ^{12}C or ^{13}C atoms)

Biotin

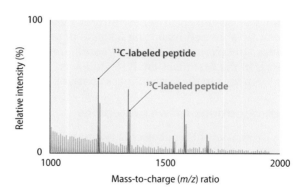

Figure 13.10 Analyzing two proteomes by ICAT. In the mass spectrum, peaks resulting from peptides containing ^{12}C atoms are shown in red, and those from peptides containing ^{13}C are shown in blue. The protein under study is approximately 1.5-fold more abundant in the proteome that has been labeled with ^{12}C-ICATs.

and so behave in the same way in the first mass analyzer. However, in each tag the distribution of ^{13}C between the balance and reporter regions is different. The tags are designed so that the reporter region is cleaved off when the peptide is fragmented, giving reporter ions of different masses. The relative amounts of two reporter ions, as detected in the second mass analyzer, gives the relative amounts of the peptides in the two proteomes.

A final strategy that can sometimes be used with microorganisms and eukaryotic cell lines is **metabolic labeling**. If a microbial or cell culture is supplied with nutrients that contain ^{13}C rather than ^{12}C atoms, then all the proteins in the proteome will become labeled with the heavy isotope. If one of the proteomes that is being studied is obtained in this way, then there is no need to add tags to individual peptides, as all the proteins will be pre-labeled. This approach therefore enables a rapid, high throughput means of comparing the relative amounts of all the proteins in a proteome, albeit with the possible drawbacks described above regarding differential mobility during column chromatography and the first stage of tandem mass spectrometry.

Analytical protein arrays offer an alternative approach to protein profiling

Gel and/or column separation followed by mass spectrometry offers an effective but laborious and expensive means of profiling the contents of a proteome. These approaches are necessary during the initial characterization of a proteome, but in many research projects the objective is not to catalog the entire content of a proteome, but to understand changes that occur within a proteome, for example, in response to extracellular stimuli and during the transition from a healthy tissue to a diseased one. For these applications, a more rapid method of assessing the relative abundances of different proteins is desirable.

Protein arrays provide the main alternative to the top-down and bottom-up approaches to protein profiling. A protein array is similar to a DNA microarray (Section 12.2), the difference being that the immobilized molecules are proteins rather than oligonucleotides. There are several types of protein array, including a version used to detect protein–protein interactions, which we will study in Section 13.2. The particular type of protein array used in protein profiling is called an **analytical protein array** or an **antibody array**, the second name indicating that the array carries a series of antibodies, each one specific for a different protein in the proteome for which the microarray is designed. When a sample of the proteome is applied to the array, individual proteins bind to their antibodies and become captured on the array; the amount of binding at each position is dependent on the abundance of that protein in the proteome. The captured proteins are usually detected with a second, polyclonal antibody that binds to all the proteins in the proteome. This antibody is fluorescently labeled and so gives signals for those positions on the array where a protein has been captured (**Figure 13.12**). As with a DNA microarray, the intensities of the resulting fluorescent signals can be used to assay the amounts of each protein in the proteome.

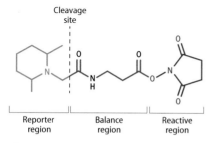

Cleavage site

Reporter region | Balance region | Reactive region

Figure 13.11 A typical isobaric tag. The reactive group on this tag enables the amines in lysine R groups and at the N terminus of a protein to be labeled. Although all tags are labeled with the same amount of ^{13}C, the distribution between the reporter and balance regions is different, so cleavage of the tag at the position shown releases a reporter ion of specific mass.

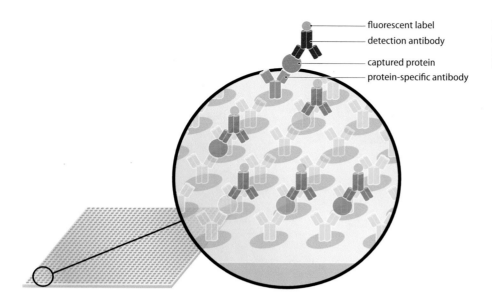

Figure 13.12 Protein detection on an antibody array. Captured proteins are detected with a polyclonal antibody that is fluorescently labeled.

fluorescent label
detection antibody
captured protein
protein-specific antibody

The main difficulty in designing an analytical protein array is ensuring that each antibody is specific for its target protein and does not crossreact with any other proteins. Cross reaction will occur if the epitope recognized by an antibody is a common feature shared by two or more distinct proteins. However, once a noncrossreacting array has been designed, then multiple copies can be fabricated, and its actual usage is relatively straightforward. Although hundreds of thousands of antibodies can be accommodated on a single chip, most antibody arrays are designed for the assay of particular components of a proteome, and hence carry less than 1000 antibodies. Typical applications would be screening for the presence or absence of cytokines in different human tissues, and measuring their relative abundances, for which commercial arrays targeting 640 proteins are available.

13.2 IDENTIFYING PROTEINS THAT INTERACT WITH ONE ANOTHER

Important data pertaining to the function of a proteome can be obtained by identifying pairs and groups of proteins that interact with one another. At a detailed level, this information is often valuable when attempts are made to assign a function to a newly discovered gene or protein (Chapter 6) because an interaction with a second, well-characterized protein can often indicate the role of an unknown protein. For example, an interaction with a protein that is located on the cell surface might indicate that the unknown protein is involved in cell-cell signaling (Section 14.1).

Identifying pairs of interacting proteins

There are several methods for studying protein–protein interactions, the two most useful being **phage display** and the **yeast two-hybrid system**. In phage display, a special type of cloning vector is used, one based on λ bacteriophage or one of the filamentous bacteriophages such as f1 or M13. The vector is designed so that a new gene that is cloned into it is expressed in such a way that its protein product becomes fused with one of the phage coat proteins (**Figure 13.13A**). The phage protein therefore carries the foreign protein into the phage coat, where it is 'displayed' in a form that enables it to interact with other proteins that the phage encounters. There are several ways in which phage display can be used to study protein interactions. In one method, the test protein is displayed, and interactions are sought with a series of purified proteins or protein fragments of known function. This approach is limited

Figure 13.13 Phage display. (A) The cloning vector used for phage display is a bacteriophage genome with a unique restriction site located at the start of a gene for a coat protein. The technique was originally carried out with the gene III coat protein of the filamentous phage called f1, but has now been extended to other phages, including λ. To create a display phage, the DNA sequence coding for the test protein is ligated into the restriction site so that a fused reading frame is produced – one in which the series of codons continues unbroken from the test gene into the coat protein gene. After transformation of *Escherichia coli*, this recombinant molecule directs synthesis of a hybrid protein made up of the test protein fused to the coat protein. Phage particles produced by these transformed bacteria therefore display the test protein in their coats. (B) Using a phage display library. The test protein is immobilized within a well of a microtiter tray and the phage display library is added. After washing, the phages that are retained in the well are those displaying a protein that interacts with the test protein.

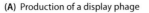
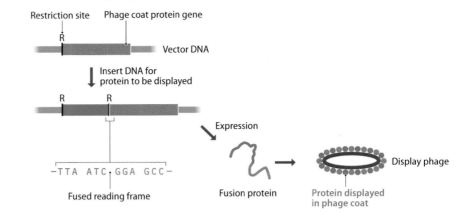

(A) Production of a display phage

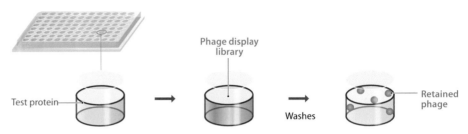

(B) Using a phage display library

because it takes time to carry out each test, so is feasible only if some prior information has been obtained about likely interactions. A more powerful strategy is to prepare a **phage display library**, a collection of clones displaying a range of proteins, and identify which members of the library interact with the test protein (**Figure 13.13B**).

The yeast two-hybrid system detects protein interactions in a more complex way. In Section 12.3, we learned that transcription factors are responsible for controlling the expression of genes in eukaryotes. To carry out this function, a transcription factor must bind to a DNA sequence upstream of a gene and stimulate the mediator protein that regulates the initiation of transcription. These two abilities – DNA binding and mediator activation – are specified by different parts of the transcription factor. Some transcription factors can be cleaved into two segments, one segment containing the DNA-binding domain and one containing the activation domain. In the cell, the two segments interact to form the functional transcription factor.

The two-hybrid system makes use of a *Saccharomyces cerevisiae* strain that lacks a transcription factor for a reporter gene. This gene is therefore switched off. An artificial gene that codes for the DNA-binding domain of the transcription factor is ligated to the gene for the protein whose interactions we wish to study. This protein can come from any organism, not just yeast: in the example shown in **Figure 13.14A**, it is a human protein. After introduction into yeast, this construct specifies synthesis of a fusion protein made up of the DNA-binding domain of the transcription factor attached to the human protein. The recombinant yeast strain is still unable to express the reporter gene because the modified transcription factor only binds to DNA; it cannot influence the mediator protein. Activation only occurs after the yeast strain has been cotransformed with a second construct, one comprising the coding sequence for the activation domain fused to a DNA fragment that specifies a protein able to interact with the human protein that is being tested (**Figure 13.14B**). As with phage display, if there is some prior knowledge about possible interactions,

(A) The two-hybrid system

(B) Screening for protein interactions using the two-hybrid system

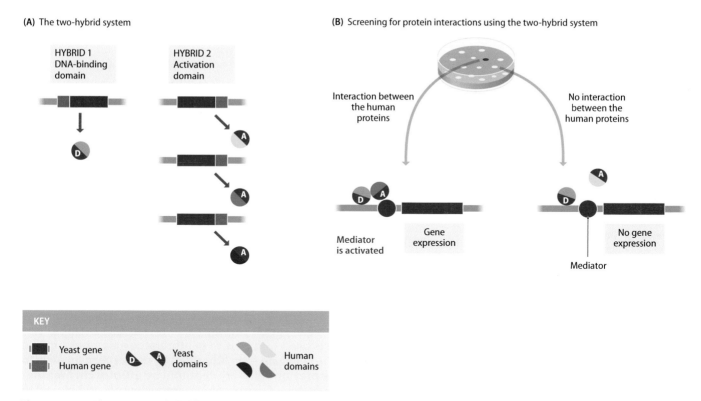

Figure 13.14 **The yeast two-hybrid system.** (A) On the left, a gene for a human protein has been ligated to the gene for the DNA- binding domain of a yeast transcription factor. After transformation of yeast, this construct specifies a fusion protein, part human protein and part yeast transcription factor. On the right, various human DNA fragments have been ligated to the gene for the activation domain of the transcription factor: these constructs specify a variety of fusion proteins. (B) The two sets of constructs are mixed and cotransformed into yeast. A colony in which the reporter gene is expressed contains fusion proteins whose human segments interact, thereby bringing the DNA-binding and activation domains into proximity and stimulating the mediator protein. Abbreviations: A, Activation domain; D, DNA-binding domain.

then individual DNA fragments can be tested one by one in the two-hybrid system. Usually, however, the gene for the activation domain is ligated with a mixture of DNA fragments so that many different constructs are made. After transformation, cells are plated out and those that express the reporter gene are identified. These are cells that have taken up a copy of the gene for the activation domain fused to a DNA fragment that encodes a protein able to interact with the test protein.

It is also possible to use a protein array to study protein–protein interactions. Unlike the arrays used in protein profiling, the immobilized proteins are not antibodies but instead are the actual proteins whose possible interactions we wish to assay. A fluorescently labeled version of the test protein is applied to the array, with the positions of the resulting signals indicating proteins with which the test protein interacts (**Figure 13.15**). Although this approach enables interactions between the test protein and a broad range of other proteins to be checked in a single experiment, it is not usually the first choice for this type of work, with phage display and the two-hybrid system remaining the standard methods for studying protein–protein interactions. Protein arrays are more popular for testing interactions with DNA fragments – for example, to identify proteins that bind to a particular DNA sequence – and with small molecules such as some types of drug.

Identifying the components of multiprotein complexes

Phage display and the yeast two-hybrid system are effective methods for identifying pairs of proteins that interact with one another, but identifying such links reveals only the basic level of protein–protein interactions. Many cellular

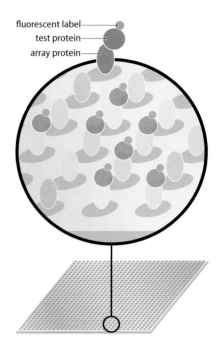

Figure 13.15 **Using a protein array to test protein–protein interactions.** The array carries a series of different proteins. Detection of the fluorescent signal indicates which proteins bind to the test protein.

Multiprotein
complex

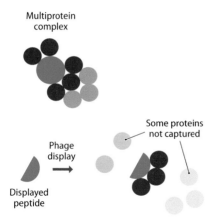

Figure 13.16 Phage display may fail to detect all members of a multiprotein complex. The complex consists of a central protein (blue) that makes direct contact with five additional proteins (red) and indirect contact with three more proteins (green). A peptide from the central protein is used in a phage display experiment. This peptide lacks attachment sites for two of the red proteins, which are therefore undetected, and cannot capture any of the green proteins.

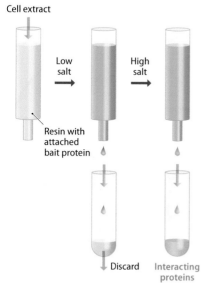

Figure 13.17 Affinity chromatography. The bait protein is attached to the resin, and the cell extract is applied in a low-salt buffer, so the other members of the multiprotein complex bind to the bait. The proteins are then eluted in a high-salt buffer.

activities are carried out by multiprotein complexes, such as the mediator protein (Section 12.3) or the spliceosome (Section 12.4). Complexes such as these typically comprise a set of core proteins, which are present at all times, along with a variety of ancillary proteins that associate with the complex under particular circumstances. Identifying the core and ancillary proteins is a critical step toward understanding how these complexes carry out their functions. In principle, a phage display library can be used to identify the members of a multiprotein complex, as in this procedure all proteins that interact with the test protein are identified in a single experiment (see Figure 13.13B). The problem is that large proteins are displayed inefficiently as they disrupt the phage replication cycle, and to circumvent this problem it is generally necessary to display a short peptide, representing part of a cellular protein, rather than the entire protein. The displayed peptide may therefore be unable to interact with all members of the complex within which the intact protein is located, because the peptide lacks some of the protein–protein attachment sites present in the intact form (**Figure 13.16**). A second problem arises if the complex contains proteins that do not interact directly with the test protein, as these will be undetected by phage display. A phage display library can therefore identify groups of proteins that are present in a complex, but does not necessarily provide the total protein complement of the complex.

A more efficient way of identifying the members of a protein complex is by **affinity purification mass spectrometry** (**AP-MS**). In this method, one member of the complex is used as a **bait** to capture other members, called **prey**; the captured prey proteins are then identified by mass spectrometry. In one format of AP-MS, the prey proteins are captured by **affinity chromatography**. The bait protein is attached to a chromatography matrix and placed in a column (**Figure 13.17**). The cell extract is passed through the column in a low-salt buffer, which allows formation of the hydrogen bonds that hold proteins together in a complex. The proteins that interact with the bait are therefore retained in the column, while all the others are washed away. The prey proteins are then eluted with a high-salt buffer. Alternatively, rather than using chromatography, the bait can be immobilized on agarose or magnetic beads, mixed with the cell extract, and complexes recovered by centrifugation or with a magnet.

Affinity chromatography requires pure samples of the bait protein, which might be difficult to obtain in the amounts needed for a large screening program. This problem can be circumvented by modifying the gene for the bait protein so that the latter now has a short extension, called an **epitope tag**, that provides a convenient means of immobilizing the bait. An example is the **his-tag**, which is simply a peptide comprising six histidines (**Figure 13.18**). Histidines form coordination complexes with divalent transition metal ions, so a bait protein with a his-tag will attach to a chromatography matrix or other support that has been coated with Ni^{2+} or Co^{2+} ions. It is no longer necessary to purify the bait, as the bait proteins in a cell extract will attach to the matrix via their his-tags and then capture the prey proteins. This is called **immobilized metal affinity chromatography** (**IMAC**).

AP-MS methods use an intact protein as the bait, so all of the prey proteins that attach to the bait can be captured, avoiding the problems that arise when a peptide is used with a phage display library. However, the complex is built up on the bait protein during the capture procedure, which means that proteins that form only indirect and hence unstable contacts with the bait might be lost (see Figure 13.16). An alternative approach called **co-immunoprecipitation** (**co-IP**) provides a better opportunity of capturing all of the proteins that interact directly or indirectly with the bait, because it isolates complexes that are preformed in the cell extract. An antibody specific to the bait protein is added to the cell extract. The baits plus attached prey are then captured on beads coated either with **protein A**, a bacterial protein that binds to immunoglobulins, or with a **secondary antibody**, which binds specifically to the **primary antibody** that recognizes the bait (**Figure 13.19**). The beads can then be collected by centrifugation or with a magnetic device, and the proteins removed and identified by mass spectrometry.

Identifying proteins with functional interactions

Proteins do not need to form physical associations with one another in order to have a functional interaction. For example, in bacteria such as *Escherichia coli*, the enzymes lactose permease and β-galactosidase have a functional interaction in that they are both involved in utilization of lactose as a carbon source. But there is no physical interaction between these two proteins: the permease is located in the cell membrane and transports lactose into the cell, and β-galactosidase, which splits lactose into glucose and galactose, is present in the cell cytoplasm (see Figure 8.10A). Many enzymes that work together in the same biochemical pathway never form physical interactions with one another, and if studies were to be based solely on detection of physical associations between proteins, then many functional interactions would be overlooked.

Several methods can be used to identify proteins that have functional interactions. Most of these do not involve direct study of the proteins themselves and hence, strictly speaking, do not come under the general heading of 'proteomics.' Nonetheless, it is convenient to consider them here because the information they yield is often considered alongside the results of proteomics studies. These methods include the following:

- Comparative genomics can be used in various ways to identify groups of proteins that have functional relationships. One approach is based on the observation that pairs of proteins that are separate molecules in some organisms are fused into a single polypeptide in others. An example is provided by the yeast gene *HIS2*, which codes for an enzyme involved in histidine biosynthesis. In *E. coli*, two genes are homologous to *HIS2*. One of these, itself called *his2*, has sequence similarity with the 5′ region of the yeast gene, and the second, *his10*, is similar to the 3′ region (**Figure 13.20**). The implication is that the proteins coded by *his2* and *his10* interact within the *E. coli* proteome to provide part of the histidine biosynthesis activity. Analysis of the sequence databases reveals many examples of this type, where two proteins in one organism have become fused into a single protein in another organism. A similar approach is based on examination of bacterial operons. An operon consists of two or more genes that are transcribed together and which usually have a functional relationship (Section 8.2). For example, the genes for the lactose permease and β-galactosidase of *E. coli* are present in the same operon, along with the gene for a third protein involved in lactose utilization (see Figure 8.10A). The identities of genes in bacterial operons can therefore be used to infer functional interactions between the proteins coded by homologous genes in a eukaryotic genome.

- Transcriptome studies can identify functional interactions between proteins, as the mRNAs for functionally related proteins often display similar expression profiles under different conditions.

- Gene inactivation studies (Section 6.2) can be informative. If a change in phenotype is observed only when two or more genes are inactivated together, then it can be inferred that those genes function together in generation of the phenotype.

Protein interaction maps display the interactions within a proteome

The information from phage display, two-hybrid analyses, co-immunoprecipitation, AP-MS and other methods for identifying pairs and groups of proteins that associate with one another enables **protein interaction maps** to be constructed.

(A) The his-tag

his-his-his-his-his-his

(B) Immobilized metal affinity chromatography

Cell extract + his-tag bait protein

Tris-salt buffer

500 mM imidazole

Matrix coated with Ni²⁺ ions

Discard   His-tagged bait + captured prey proteins

Figure 13.18 The use of a his-tagged bait protein. (A) The his-tag is shown here attached to the N terminus of the bait protein. C-terminal his-tags can also be used. (B) Immobilized metal affinity chromatography. The cell extract and bait protein are applied to the column and washed with a Tris-salt buffer. The his-tagged bait protein is immobilized by the Ni²⁺ ions on the affinity matrix, and any prey proteins that interact with the bait are captured. For a Ni²⁺ matrix, the bait and prey are eluted with 500 mM imidazole.

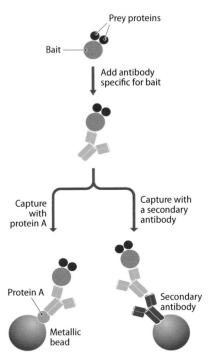

Figure 13.19 Using an antibody to capture bait and prey proteins. An antibody specific to the bait protein is added to the cell extract. The baits plus attached prey are captured on beads coated with protein A, a bacterial protein that binds to immunoglobulins, or with a secondary antibody, which binds specifically to the primary antibody that recognizes the bait.

Figure 13.20 Using homology analysis to deduce protein–protein interactions. The 5′-region of the yeast *HIS2* gene is homologous to *E. coli his2*, and the 3′-region is homologous to *E. coli his10*.

In this type of map, each protein is depicted by a dot, or **node**, with pairs of interacting proteins linked by lines, or **edges**, the resulting network displaying all of the interactions that occur between the components of a proteome. The first of these maps was constructed in 2001 for relatively simple proteomes, almost entirely from two-hybrid experiments. These included maps for the bacterium *Helicobacter pylori*, comprising over 1200 interactions involving almost half of the proteins in the proteome, and for 2240 interactions between 1870 proteins from the *Saccharomyces cerevisiae* proteome (**Figure 13.21**). More recently, the application of additional techniques has led to more detailed versions of the *S. cerevisiae* map, as well as maps for humans and other eukaryotes. Each protein interaction map forms part of the broader **interactome** for the species, the interactome comprising all of the molecular interactions that occur, including those involving small molecules which regulate protein activity and between DNA-binding proteins and the genes whose expression they control.

What interesting features have emerged from these protein interaction maps? The most intriguing discovery is that most maps are built up around a small number of proteins that have many interactions, and which form **hubs** in the network, along with a much larger number of proteins with few individual connections. This modular structure, which was clearly illustrated in a later version of the *S. cerevisiae* map (**Figure 13.22A**), is thought to minimize the impact on the proteome of the disruptive effects of mutations that might inactivate individual proteins. Only if a mutation affects one of the proteins at a highly interconnected node will the network as a whole be damaged. This hypothesis is consistent with the discovery, from gene inactivation studies, that a substantial number of yeast proteins are apparently redundant (Section 6.2), meaning that if the protein activity is destroyed, the proteome as a whole continues to function normally, with no discernible impact on the phenotype of the cell. From these early studies of interaction maps, it was suggested that the hubs could be divided into two groups. The first group are those hub proteins that interact with all their partners simultaneously. These have been called 'party' hubs, and their removal has little effect on the overall structure of the network (**Figure 13.22B**). In contrast, removal of the second group, called the 'date' hubs, which interact with different partners at different times, breaks the network into a series of small subnetworks (**Figure 13.22C**). The implication is that the party hubs work within individual biological processes, and do not contribute greatly to the overall organization of the proteome. The date hubs, on the other hand, are the key players that provide an organization to the proteome by linking biological processes to one another. However, the division into party and date hubs has been challenged by more recent research, which has emphasized that the structure of a map is influenced by the limitations of the methods used in its construction. Most protein interaction maps are incomplete, simply because not all of the interactions occurring in the proteome under study have been identified. It is possible that the omission of weaker and less easy-to-identify interactions increases the apparent modularity of a map, and that the supposed distinctions between different types of hub are artifacts caused by this incompleteness.

Protein interaction maps provide an interesting visual display of the functional structure of a proteome, but are they of any practical value? One important application has been in understanding the interactions between severe acute respiratory syndrome coronavirus 2 (SARS-CoV-2), the causative agent of COVID-19, and the proteomes of infected cells. The SARS-CoV-2 genome contains 14 genes that code for 29 proteins, several of these initially translated as polyproteins (Section 9.1). Open reading frames for 26 of these proteins were modified so that the genes could be expressed in cultured human cells, with the proteins carrying N- and C-terminal **Strep tags**. The Strep-tag is the amino acid sequence Trp–Ser–His–Pro–Gln–Phe–Glu–Lys, which binds to a modified form of streptavidin. Complexes formed in the cultured cells between the tagged viral proteins and components of the human proteome were therefore captured by affinity purification with magnetic beads coated with the modified streptavidin, and the members of each complex identified by mass spectrometry. The resulting maps revealed 332 interactions between SARS-CoV-2 and human proteins involved in functions such as RNA and protein processing, intra- and intercellular signaling, and mitochondrial activity (**Figure 13.23**). As well as providing insights into the biochemical activities that are affected by SARS-CoV-2 infection, the maps identify proteins that might be potential targets for drug treatments aimed at alleviating COVID-19 in infected individuals. Sixty-two of the 332 human proteins that interact with SARS-CoV-2 are already the targets of 69 approved drugs or other compounds currently undergoing clinical trials. These 69 compounds were tested for their ability to inhibit SARS-CoV-2 infection of cell lines, revealing two groups of compounds with promising effects. These compounds include an inhibitor of protein synthesis called zotatifin, which is now being tested as a potential antiviral treatment for COVID-19, and hydroxychloroquine, which is an anti-malarial drug that already has a checkered history as a possible treatment for the disease. Hydroxychloroquine inhibits a protease called cathepsin L, which cleaves the SARS-CoV-2 spike protein, cleavage being essential for entry of the virus into a cell. However, hydroxychloroquine does not prevent SARS-CoV-2 infection because there are other proteases that are also able to cleave the spike protein, and its use in the treatment of COVID-19 patients has been discouraged by the US National Institutes of Health and other bodies.

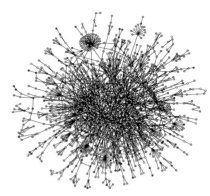

Figure 13.21 The initial version of the *Saccharomyces cerevisiae* protein interaction map. This map was published in 2001. Each dot represents a protein, with connecting lines indicating interactions between pairs of proteins. Red dots are essential proteins: an inactivating mutation in the gene for one of these proteins is lethal. Mutations in the genes for proteins indicated by green dots are nonlethal, and mutations in genes for proteins shown in orange lead to slow growth. The effects of mutations in genes for proteins shown as yellow dots were not known when the map was constructed. (A, From Jeong H, Mason SP, Barabási AL & Oltvai ZN [2001] *Nature* 411: 41–42. With permission from Springer Nature. B, From Stelzl U, Worm U, Lalowski M et al. [2005] 122: 957–968. With permission from CellPress.)

13.3 SYNTHESIS AND DEGRADATION OF THE COMPONENTS OF THE PROTEOME

The composition of a proteome is determined by the balance between the synthesis and degradation of the individual proteins that it contains. Except for the words 'proteome' and 'proteins,' this is the same sentence that we used to describe the composition of the transcriptome (Section 12.3). The principles are exactly the same, and the dynamics of synthesis and degradation illustrated in Figure 12.14 for RNA could equally well refer to proteins. To understand how a proteome is maintained and how a proteome changes in response to external stimuli and during differentiation, development and disease we must therefore study the same topics as in Chapter 12 – synthesis, degradation and processing – but this time with reference to protein rather than RNA.

Ribosomes are molecular machines for making proteins

Proteins are synthesized by the large RNA-protein complexes called **ribosomes**. An *Escherichia coli* cell contains approximately 20,000 ribosomes, distributed throughout its cytoplasm. The average human cell contains more than a million, some free in the cytoplasm and some attached to the outer surface of the endoplasmic reticulum, the membranous network of tubes and vesicles that permeates the cell.

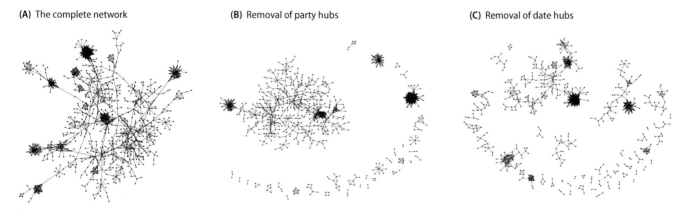

(A) The complete network   **(B)** Removal of party hubs   **(C)** Removal of date hubs

Figure 13.22 Hubs in the *Saccharomyces cerevisiae* protein interaction map. The hubs are clearly visible in this version of the yeast protein interaction map, which was published in 2004. (A) The complete map. (B) After removal of the party hubs the network remains almost intact. (C) However, after the date hubs are removed, the network splits into detached subnetworks. (From Han J-DJ, Bertin N, Hao T et al. [2004] *Nature* 430: 88–93. With permission from Springer Nature.)

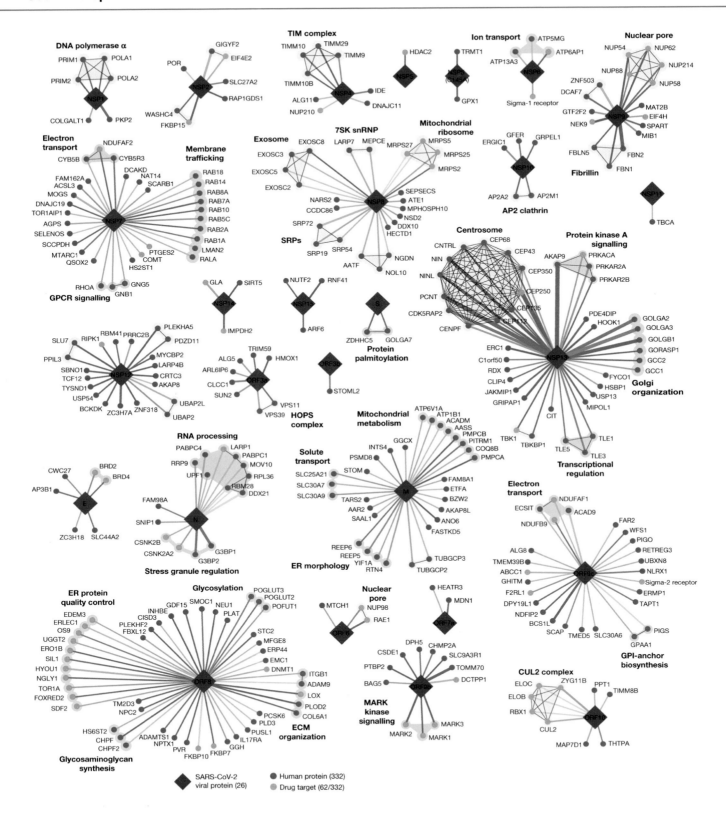

Figure 13.23 Protein interaction maps for 26 SARS-CoV-2 proteins. A total of 332 high-confidence interactions are shown between 26 SARS-CoV-2 proteins (red diamonds) and human proteins (circles). Human proteins are highlighted orange if they are potential drug targets, yellow if members of a protein complex, and blue if involved in the same biological process. Additional interactions between human proteins are also shown. Abbreviations: ECM, extracellular matrix; ER, endoplasmic reticulum; GPI, glycophosphatidylinositol; snRNP, small nuclear ribonucleoprotein; SRP, signal recognition particle. The TIM complex is located in the inner mitochondrial membrane, the HOPS complex is involved in fusion between membrane-bound vesicles, and CUL2 is a ubiquitin ligase complex. (From Gordon DE, Jang GM, Bouhaddou M et al [2020] *Nature* 583:459–468. With permission from Springer Nature.)

Originally, ribosomes were looked on as passive partners in protein synthesis, merely the structures on which mRNAs were translated into polypeptides. This view has changed over the years and ribosomes are now considered to play two active roles:

- Ribosomes *coordinate* protein synthesis by placing the mRNA, tRNAs, and associated protein factors in their correct positions relative to one another.

- Components of ribosomes, including the rRNAs, act as ribozymes and *catalyze* at least some of the chemical reactions occurring during protein synthesis, including the central reaction that results in synthesis of the peptide bond that links two amino acids together.

When the involvement of ribosomes in protein synthesis became clear in the 1950s, biologists quickly realized that a detailed knowledge of ribosome structure would be necessary in order to understand how mRNAs are translated into polypeptides. Originally called 'microsomes,' ribosomes were first observed in the early decades of the twentieth century as tiny particles almost beyond the resolving power of light microscopy. In the 1940s and 1950s, the first electron micrographs showed that bacterial ribosomes are oval-shaped, with dimensions of 29 × 21 nm, rather smaller than eukaryotic ribosomes, the latter varying a little in size depending on species but averaging about 32 × 22 nm. Compositional studies then revealed that a ribosome comprises two subunits, referred to as 'large' and 'small,' each subunit made up of one or more rRNAs and a collection of **ribosomal proteins (Figure 13.24)**. We now know that ribosomes dissociate into their subunits when not actively participating in protein synthesis, the subunits remaining in the cytoplasm until being used for a new round of translation.

Once the basic composition of eukaryotic and bacterial ribosomes had been worked out, attention became focused on the way in which the various rRNAs and proteins fit together. Important information was provided by the first rRNA sequences, comparisons between these identifying conserved regions that can base pair to form complex two-dimensional structures (**Figure 13.25**). This suggested that the rRNAs provide a scaffolding within the ribosome, to which the proteins are attached, an interpretation that underemphasizes the active role that rRNAs play in protein synthesis but which nonetheless was a useful foundation on which to base subsequent research.

Much of that subsequent research has concentrated on the bacterial ribosome, which is smaller than the eukaryotic version and available in large amounts from extracts of cells grown to high density in liquid cultures.

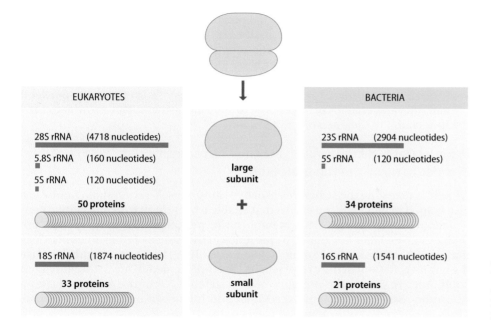

Figure 13.24 The composition of eukaryotic and bacterial ribosomes. The details given are for human ribosomes and those in *Escherichia coli*. There are some variations in the number of ribosomal proteins in different species.

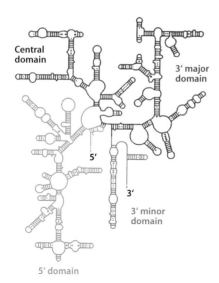

Figure 13.25 The base-paired structure of the *Escherichia coli* 16S rRNA. In this representation, standard base pairs (G–C, A–U) are shown as bars; nonstandard base pairs (e.g., G–U) are shown as dots.

A number of technical approaches have been used to study the bacterial ribosome:

- **Nuclease protection experiments** (Section 7.1) enabled contacts between rRNAs and proteins to be identified.

- **Protein–protein cross-linking** identified pairs or groups of proteins that are located close to one another in the ribosome.

- **Electron microscopy** gradually became more sophisticated, enabling the overall structure of the ribosome to be resolved in greater detail. For example, innovations such as **immunoelectron microscopy**, in which ribosomes are labeled with antibodies specific to individual ribosomal proteins, have been used to locate the positions of these proteins on the surface of the ribosome.

- **Site-directed hydroxyl radical probing**, which makes use of the ability of Fe(II) ions to generate hydroxyl radicals that cleave RNA phosphodiester bonds located within 1 nm of the site of radical production, has been used to determine the exact positioning of ribosomal proteins in the *E. coli* ribosome. For example, to determine the position of S5, different amino acids within this protein were labeled with Fe(II) and hydroxyl radicals induced in reconstituted ribosomes. The positions at which the 16S rRNA was cleaved were then used to infer the topology of the rRNA in the vicinity of the S5 protein (**Figure 13.26**).

In recent years these techniques have been supplemented by X-ray crystallography (Section 11.1), which has been responsible for the most exciting insights into ribosome structure. Analyzing the massive amounts of X-ray diffraction data that are produced by crystals of an object as large as a ribosome is a huge task, particularly at the level needed to obtain a structure that is detailed enough to be informative about the way in which the ribosome works. This challenge has been met, and detailed structures are now known for entire ribosomes, including ones attached to mRNA and tRNAs (**Figure 13.27A**). These studies have shown that, in bacteria, attachment of the two ribosome subunits to one another results in the formation of two sites at which an **aminoacyl-tRNA** (a tRNA with an attached amino acid) can bind. These are called the **P** or **peptidyl site** and the **A** or **aminoacyl site**. The P site is occupied by the aminoacyl-tRNA whose amino acid has just been attached to the end of the growing polypeptide, and the A site is entered by the aminoacyl-tRNA carrying the next amino acid that will be used. There is also a third site, the **E** or **exit site**, through which the tRNA departs after its amino acid has been attached to the polypeptide (**Figure 13.27B**). The structures revealed by X-ray diffraction analysis show that these sites are located in a cavity between the large and small subunits of the ribosome, the mRNA threading through a channel formed mainly by the small subunit. After each amino acid addition, the ribosome adopts a less compact structure, with the two subunits rotating slightly in opposite directions. This opens up the space between the subunits and enables the ribosome to slide along the mRNA in order to read the next codon in the open reading frame.

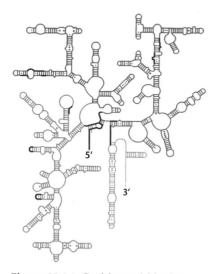

Figure 13.26 Positions within the *Escherichia coli* 16S rRNA that form contacts with ribosomal protein S5. The distribution of the contact positions (shown in red) for this single ribosomal protein emphasizes the extent to which the base-paired secondary structure of the rRNA is further folded within the three-dimensional structure of the ribosome.

During stress, bacteria inactivate their ribosomes in order to downsize the proteome

As well as revealing the structures of active ribosomes, X-ray crystallography has also helped to elucidate the events that enable a bacterium to bring about a general reduction in the size of its proteome during periods of stress, such as nutrient limitation. The latter is signaled by presence in the A sites of ribosomes of tRNAs that do not have attached amino acids; these tRNAs lack amino acids because the amino acid pool in the cytoplasm has become depleted due to the starvation conditions. In *E. coli,* the presence of deacylated tRNA in the A site activates a ribosome-associated protein called RelA and initiates the **stringent** response.

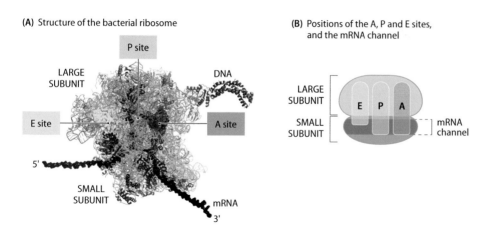

(A) Structure of the bacterial ribosome

P site

LARGE SUBUNIT

DNA

E site

A site

5'

SMALL SUBUNIT

mRNA

3'

(B) Positions of the A, P and E sites, and the mRNA channel

LARGE SUBUNIT

SMALL SUBUNIT

E P A

mRNA channel

Figure 13.27 Detailed structure of a bacterial ribosome. (A) Structure of a ribosome in the process of translating an mRNA. The tRNAs positioned in the A, P and E sites are indicated in red, green and yellow, respectively. (B) A diagram showing the relative positions of the A, P and E sites and the channel through which the mRNA is translocated. (From Schmeing TM & Ramakrishnan V [2009] *Nature* 461: 1234–1242. With permission from Springer Nature.)

RelA converts guanosine 5′-triphosphate (GTP) to guanosine 5′-triphosphate 3′-diphosphate (pppGpp) by transferring a diphosphate from adenosine 5′-triphosphate (ATP) to a GTP molecule. A second enzyme then converts pppGpp to ppGpp (**Figure 13.28**), which is an **alarmone**, a stress response molecule that modifies a broad range of cellular activities in order to help the cell deal with the stress. In starving bacteria, one of these responses is a general decrease in transcription, but an increase in transcription of genes involved in amino acid biosynthesis. These changes are brought about by ppGpp binding to the β and β′ subunits of the bacterial RNA polymerase and altering the affinity of the polymerase for different types of promoter.

By switching on amino acid biosynthesis, the bacterium is able to carry out essential maintenance of its proteome while its rides out the stress conditions and waits for the external nutrient supply to increase. Because its overall metabolic activity has declined, the bacterium also decreases the size of its proteome by globally reducing the rate of protein synthesis. This is achieved, at least in part, by ppGpp binding to the translation **initiation factor** IF-2. This factor is required during the first stage of protein synthesis, when the **initiation complex** is assembled at the **ribosome binding site**. The latter is the position on the mRNA, usually 3–10 nucleotides upstream of the initiation codon, where the bacterial ribosome attaches in order to begin the process of translating the mRNA. The initiation complex comprises the two subunits of the ribosome, the aminoacylated tRNA that recognizes the initiation codon, and three protein initiation factors. The specific roles of IF-2 include hydrolysis of GTP to provide the energy needed to complete assembly of the initiation complex. However, IF-2 has a higher affinity for ppGpp compared with GTP, so it binds the former when it is present. In the absence of GTP hydrolysis, translation stalls during this initiation phase, resulting in a global decrease in protein synthesis and hence a reduction In the size of the proteome.

The downshift in protein synthesis that occurs during starvation means that the bacterium now has an excess of ribosomes. The surplus are not broken down but instead stored in inactive form. The A and P sites become blocked by a protein called the **ribosome modulation factor**, and pairs of ribosomes interact with a second protein, the **hibernation promotion factor**, forming ribosome dimers. The dimers are inactive but can be reconverted to functional ribosomes so protein synthesis can be up-regulated as soon as the environmental conditions improve.

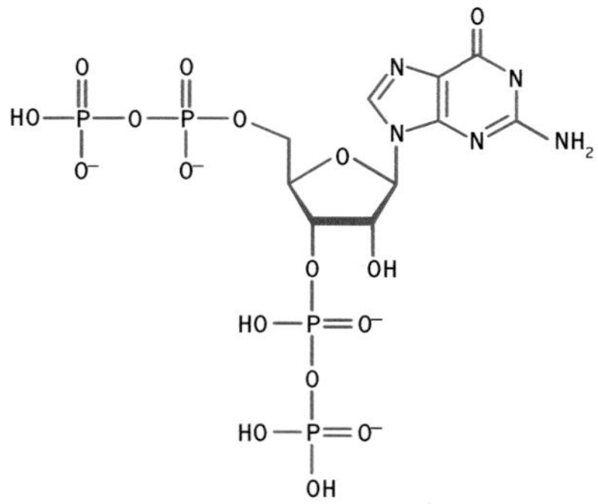

Initiation factors mediate large-scale remodeling of eukaryotic proteomes

Initiation factors are also the mediators of large-scale changes in proteome composition in eukaryotes. Although ribosomal architecture is similar in bacteria and eukaryotes, there are distinctions in the way in which protein synthesis is carried out in the two types of organism, in particular during the initiation phase. Rather

Figure 13.28 The structure of the alarmone guanosine 5′-diphosphate 3′-diphosphate (ppGpp).

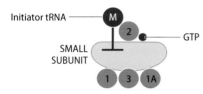

Figure 13.29 The preinitiation complex for eukaryotic protein synthesis. The initiation factors are labeled 1, 1A, 2, and 3. The initiator tRNA is shown attached to a methionine (M) amino acid. The diagram is schematic: the overall configuration of the complex is not known.

than being assembled close to the initiation codon, the eukaryotic initiation complex is constructed at the 5′ end of the transcript and then **scans** along the mRNA until it locates the initiation codon. The first component of the initiation structure, called the **preinitiation complex**, comprises the small subunit of the ribosome, a 'ternary complex' made up of the initiation factor eIF-2 bound to the initiator tRNA and a molecule of GTP, and three additional initiation factors, eIF-1, eIF-1A, and eIF-3 (**Figure 13.29**). After assembly, the preinitiation complex associates with the cap structure at the 5′ end of the mRNA (see Figure 1.17A), and additional initiation factors are recruited prior to commencement of the scanning process.

The initiation process can be repressed by phosphorylation of eIF-2, which prevents this initiation factor from binding the molecule of GTP that it needs before it can transport the initiator tRNA to the small subunit of the ribosome. Phosphorylation of eIF-2 occurs during stresses such as heat shock, and results in a decrease in the overall level of protein synthesis. This is not, however, a means of reducing the size of the proteome, as occurs during the bacterial stringent response. Instead, eIF-2 phosphorylation results in proteome remodeling, with an increase in the amounts of various proteins involved in protection of the cell against stress. This occurs because of a second route to initiation of translation in eukaryotes, which avoids the scanning process and hence circumvents the inactivation of eIF-2. This alternative pathway involves assembly of the initiation complex at an **internal ribosome entry site** (**IRES**), which is similar in function to the ribosome binding site of bacteria, although the positions of IRESs relative to the initiation codon are more variable than the bacterial versions. The ability of eukaryotic translation to initiate in this way was first recognized with the picornaviruses, a group of viruses with RNA genomes, including the human poliovirus and rhinovirus, the latter responsible for the common cold. Transcripts from these viruses are not capped but instead have an IRES, meaning that picornaviruses can block protein synthesis in the host cell without affecting translation of their own transcripts. Researchers were initially puzzled to discover that no virus proteins are required for recognition of an IRES by a host ribosome. In other words, the normal eukaryotic cell possesses proteins and/or other factors that enable it to initiate translation by the IRES method. A search was therefore made for IRESs in eukaryotic mRNAs, revealing the presence of these sequences in the transcripts of over 600 human genes, enabling these mRNAs to undergo preferential translation at times of stress when eIF-2 and the scanning process are inactivated.

Although eIF-2 is the best-studied example of the global regulation of eukaryotic protein synthesis, recent research has shown that there are several other mechanisms for repressing cap-dependent initiation and switching translation to IRES-mRNAs. One of these occurs during apoptosis when several initiation factors and other proteins involved in translation are phosphorylated or cleaved into segments. An example is eIF-4G, which is cleaved by the enzyme caspase 3. eIF-4G is a part of the **cap-binding complex**, which aids attachment of the preinitiation complex to the 5′ end of the mRNA. Cleavage of eIF-4G prevents the cap-binding complex from attaching to the mRNA, repressing cap-dependent translation and promoting translation of IRES-mRNAs coding for proteins involved in apoptosis. These proteins are unlikely to be identical to those required during stress conditions, indicating that repression of cap-dependent initiation does not simply result in an increase in translation of IRES-mRNAs. Instead, there must be mechanisms for modulating the IRES process so that specific groups of transcripts are translated according to the needs of the cell. How the translation of different sets of IRES-mRNAs is regulated is currently not understood, but the discovery of **IRES-transacting factors** (**ITAFs**), which appear to control the affinity of the ribosome for different IRESs, points a way toward a possible explanation.

The translation of individual mRNAs can also be regulated

As well as the large-scale changes to proteome size or composition that can be brought about by global regulation of translation initiation, the translation rates of individual mRNAs can also be regulated. In bacteria, several of these

transcript-specific regulation processes involve attachment of an RNA-binding protein to the target mRNA, the attached protein interfering with the ribosome and hence inhibiting formation of the translation initiation complex. The most frequently cited example involves the operons for the ribosomal protein genes of *E. coli* (**Figure 13.30A**). The leader region of the mRNA transcribed from each operon contains a sequence that acts as a binding site for one of the proteins coded by the operon. When this protein is synthesized it can either attach to its position on the ribosomal RNA, or bind to the leader region of the mRNA. The rRNA attachment is favored and occurs if there are free rRNAs in the cell. Once all the free rRNAs have been assembled into ribosomes, the ribosomal protein binds to its mRNA, blocking access to the ribosome binding site and preventing translation initiation. As a result, there is no further synthesis of the ribosomal proteins coded by that particular mRNA. Similar events involving other mRNAs ensure that synthesis of each ribosomal protein is coordinated with the amount of free rRNA in the cell.

A second example of transcript-specific regulation by RNA-binding proteins, one occurring in mammals, involves the mRNA for ferritin, an iron-storage protein (**Figure 13.30B**). In the absence of iron, ferritin synthesis is inhibited by proteins that bind to sequences called **iron-response elements** located in the leader region of the ferritin mRNA. The bound proteins block the ribosome as it attempts to scan along the mRNA in search of the initiation codon. When iron is present, the binding proteins detach and the mRNA is translated. Interestingly, the mRNA for a related protein – the transferrin receptor involved in the uptake of iron – also has iron-response elements, but in this case detachment of the binding proteins in the presence of iron results not in translation of the mRNA but in its degradation. This is logical because when iron is present in the cell, there is less requirement for transferrin receptor activity because there is less need to import iron from outside.

Initiation of translation of some bacterial mRNAs can also be regulated by short RNAs, which attach to recognition sequences within the mRNAs, but this does not always result in translation being prevented, as some short RNAs can also activate translation of one or more of their target mRNAs. An example is the *E. coli* RNA called OxyS, which is 109 nucleotides in length and regulates translation of 40 or so mRNAs. Synthesis of OxyS is activated by hydrogen peroxide and other reactive oxygen compounds, which can cause oxidative damage to the cell. Once synthesized, OxyS switches off translation of mRNAs, whose products would be deleterious under these circumstances, and switches on translation of other mRNAs, whose products help protect the bacterium from oxidative damage. Prevention of translation occurs in a similar way to the examples we have discussed above involving RNA-binding proteins, attachment of the regulatory RNA simply blocking access to the ribosome binding site. Activation of translation involves a more subtle mechanism. In these cases, the target mRNA is able to form a stem-loop structure, with the ribosome binding site present in the stem region and hence inaccessible to the components of the initiation complex. Attachment of the regulatory RNA disrupts the stem-loop, exposing the ribosome binding site so that translation initiation can now occur.

Stem-loops and other secondary structures can also form in the 5′-untranslated regions of many eukaryotic mRNAs. During scanning of the initiation complex along the mRNA, these secondary structures are disrupted by the initiation factor eIF-4A, which is a helicase and so is able to break the intramolecular base pairs. However, the presence of secondary structure delays progress of the initiation complex and hence reduces the **translational efficiency** – the rate at which proteins are synthesized from the mRNA. Because of differences in the translational efficiencies of different mRNAs, the proteome is not simply a reflection of the transcriptome, with the relative abundances of each protein matching the relative abundances of their mRNAs. Instead, those mRNAs with high translational efficiency gives rise to greater amounts of protein per transcript. Current research is exploring the possibility that transcript-specific regulation of eukaryotic protein synthesis can be exerted by regulatory proteins or RNAs that

(A) Autoregulation of ribosomal protein synthesis

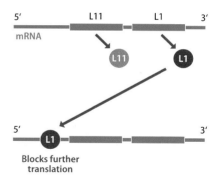

(B) Regulation by iron-response elements

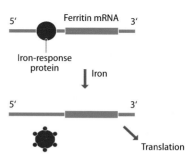

Figure 13.30 Transcript-specific regulation of translation initiation. (A) Regulation of ribosomal protein synthesis in bacteria. The L11 operon of *Escherichia coli* is transcribed into an mRNA carrying copies of the genes for the L11 and L1 ribosomal proteins. When the L1 binding sites on the available rRNA molecules have been filled, L1 binds to the 5′ untranslated region of the mRNA, blocking further initiation of translation. (B) Regulation of ferritin protein synthesis in mammals. The iron-response protein binds to the 5′ untranslated region of the ferritin mRNA when iron is absent, preventing the synthesis of ferritin.

influence the formation of secondary structures in mRNAs and hence increase or decrease their translational efficiency.

Degradation of the components of the proteome

The ability of a proteome to change over time requires the removal of proteins whose functions are no longer required. This removal must be highly selective so that only the correct proteins are degraded, and must also be rapid in order to account for the abrupt changes that occur under certain conditions, for example, during key transitions in the cell cycle.

For many years, protein degradation was an unfashionable subject, and it was not until the 1990s that real progress was made in understanding how specific proteolysis events are linked with processes such as the cell cycle and differentiation. Even now, our knowledge centers largely on descriptions of general protein breakdown pathways and less on the regulation of the pathways and the mechanisms used to target specific proteins. There appear to be a number of different types of breakdown pathway whose interconnectivities have not yet been traced. This is particularly true in bacteria, which seem to have a range of proteases that work together in controlled degradation of proteins. In eukaryotes, most breakdown occurs via a single system, involving **ubiquitin** and the **proteasome**.

A link between ubiquitin and protein degradation was first established in 1975 when it was shown that this abundant 76-amino-acid protein is involved in energy-dependent proteolysis reactions in rabbit cells. Subsequent research identified enzymes that attach ubiquitin molecules, singly or in chains, to lysine amino acids in proteins that are targeted for breakdown (**Figure 13.31**). There are also ubiquitin-like proteins, such as SUMO, which act in the same way as ubiquitin. Ubiquitination is a three-step process in which a ubiquitin molecule is initially attached to an activator protein, then transferred to a conjugating enzyme, and finally transferred to the target protein by a ubiquitin ligase enzyme. How does this process recognize the correct proteins, those that must be degraded? The answer appears to lie with the specificity of the conjugating and ligase enzymes. Most species possess multiple versions of the conjugating enzymes and many types of ubiquitin ligase. In humans, for example, there are over 50 conjugating enzymes and several hundred ligases. It is thought that different pairs of conjugating enzyme and ligase have specificity for different proteins. Activation of different enzyme pairs, in response to intra- or extracellular signals, is probably the key to the specific degradation of particular proteins and groups of proteins.

The second component of the ubiquitin-dependent degradation pathway is the proteasome, the structure within which ubiquitinated proteins are broken down. In eukaryotes, the proteasome is a large, multisubunit structure comprising a hollow cylinder with a 'cap' at either end (**Figure 13.32**). The cylinder has four rings, each made up of seven proteins. The proteins present in the two inner rings are proteases, whose active sites are located on the inner ring surface. Archaea also have proteasomes of about the same size, but these are less complex, being composed of multiple copies of just two proteins. Ubiquitinated proteins might interact directly with the cap structure of the proteasome, or the interaction might be via a **ubiquitin-receptor protein**. Before entry into the proteasome, the protein to be degraded must be at least partially unfolded, and its ubiquitin labels must be removed. These steps require energy, obtained from the hydrolysis of ATP, catalyzed by proteins present in the cap. The protein can then enter the proteasome, within which it is cleaved into short peptides of 4–10 amino acids in length. These are released back into the cytoplasm, where they are broken down into individual amino acids, which can be reutilized in protein synthesis.

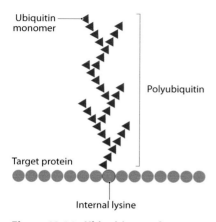

Figure 13.31 Ubiquitin attachment targets proteins for degradation. To act as a label for degradation, chains of linked ubiquitin molecules are linked to lysine amino acids in the target protein.

13.4 THE INFLUENCE OF PROTEIN PROCESSING ON THE COMPOSITION OF THE PROTEOME

Translation is not the end of the genome expression pathway. The polypeptide that emerges from the ribosome is inactive, and before taking on its functional role in the cell must undergo post-translational processing. The most fundamental of these processing events is **protein folding**, the vast majority of polypeptides being inactive until they have adopted their correct tertiary structure. Some polypeptides are also cleaved in various ways to give the functional proteins, and many undergo chemical modification. The proteome therefore contains not only functional proteins, but also a large variety of pre-processed forms, some of which can be retained in their inactive state for substantial periods before processing is completed and the functional protein is produced. We must therefore study protein processing in order to gain a complete understanding of the components of a proteome.

The amino acid sequence contains instructions for protein folding

One of the central principles of molecular biology is that all of the information that a polypeptide needs in order to fold into its correct three-dimensional structure is contained within its amino acid sequence. This link between sequence and structure was first established in the 1950s by experiments with ribonuclease that had been purified from cow pancreas and resuspended in buffer. Ribonuclease is a small protein of just 124 amino acids, containing four disulfide bonds, with a tertiary structure that is made up predominantly of β-sheet, with very little α-helix. Addition of urea, a compound that disrupts hydrogen bonding, resulted in a decrease in the activity of the enzyme (measured by testing its ability to cut RNA) and an increase in the viscosity of the solution (**Figure 13.33**), indicating that the protein was being denatured by unfolding into an unstructured polypeptide chain. The critical observation was that when the urea was removed by dialysis, the viscosity decreased and the enzyme activity reappeared. The conclusion is that the protein refolds spontaneously when the denaturant (in this case, urea) is removed. In these initial experiments, the four disulfide bonds remained intact because they were not disrupted by urea, but the same result occurred when the urea treatment was combined with addition of a reducing agent to break the disulfide bonds: the activity was still regained on renaturation. This shows that the disulfide bonds are not critical to the protein's ability to refold; they merely stabilize the tertiary structure once it has been adopted.

It was quickly realized that the folding process cannot be random, as it would take too long for a protein to explore all the possible conformations that it can take up before finally finding the correct one. It was estimated that a polypeptide

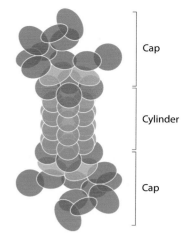

Figure 13.32　The eukaryotic proteasome. The protein components of the two caps are shown in orange, red and pink and those forming the cylinder in blue.

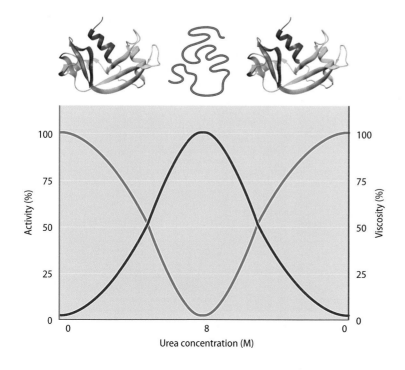

Figure 13.33　Denaturation and spontaneous renaturation of ribonuclease. As the urea concentration increases to 8 M, the protein becomes denatured by unfolding: its activity decreases and the viscosity of the solution increases. When the urea is removed by dialysis, the protein readopts its folded conformation. The activity of the protein increases back to the original level, and the viscosity of the solution decreases.

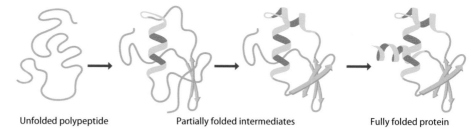

Unfolded polypeptide   Partially folded intermediates   Fully folded protein

Figure 13.34 A protein folding pathway.

of 100 amino acids would take about 10^{87} seconds to check all conformations, which is much longer than the age of the universe. The implication is that a protein follows a **folding pathway**, each step in the process directing the protein toward the following step, so the protein is led toward its correct tertiary structure (**Figure 13.34**). Researchers currently favor the **molten globule** model, in which the initial step is the rapid collapse of the polypeptide into a compact structure, with slightly larger dimensions than the final protein, driven by the desire of the hydrophobic amino acid side chains to avoid water. Collapse into this molten globule might automatically fold some of the polypeptide into its α-helices and β-sheets. Because the globule is 'molten,' it can change conformation rapidly, identifying additional folds so that the correct tertiary structure gradually emerges. For larger proteins, this step might involve construction of correctly folded subdomains which are then brought together to make the final tertiary structure. The whole process can take just a few seconds. In thermodynamic terms, the polypeptide is looked on as passing down a funnel, gradually taking up less random conformations (**Figure 13.35**). The funnel narrows because at each stage in the folding pathway there are fewer available options for the next steps. There are, however, side funnels into which the protein can be diverted, leading to an incorrect structure. If an incorrect structure is sufficiently unstable, then partial or complete unfolding may occur, allowing the protein to return to the main funnel and pursue a productive route toward its correct conformation.

Experiments conducted *in vitro* have been useful in establishing the basic principles of protein folding, but they may not be a good model for the folding of proteins in the cell. In particular, a cellular protein might begin to fold before

Figure 13.35 A thermodynamic representation of protein folding. The folding pathway is looked on as a funnel. The top of the funnel is wide because the unfolded polypeptide can initially adopt any one of many intermediate structures. The funnel gradually narrows as the protein becomes more folded, and its options for future folding are reduced. Gradually those options decrease as the protein becomes more completely folded, and eventually the fully folded protein emerges from the spout at the bottom. Diversion into a side funnel results in an incorrectly folded structure. If the protein enters one of these side funnels, then it has to partially unfold in order to return to the main funnel.

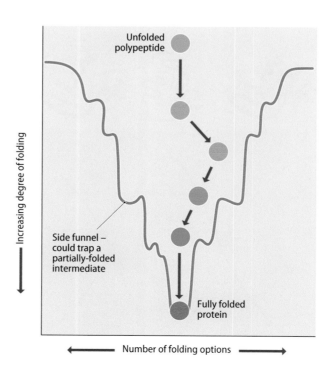

Unfolded polypeptide

Side funnel – could trap a partially-folded intermediate

Fully folded protein

Increasing degree of folding

← Number of folding options →

it has been fully synthesized, a scenario that is very difficult to replicate in a test tube experiment. These considerations have prompted extensive research into protein folding in living cells, with particular focus on the **molecular chaperones**, which are proteins that help other proteins to fold. The molecular chaperones can be divided into various groups, the most important of which are:

- The **Hsp70 chaperones**, which include the *E. coli* Hsp70 protein coded by the *dnaK* gene and sometimes called DnaK protein.

- The **chaperonins**, the main version of which is the GroEL/GroES complex present in bacteria and eukaryotic organelles, and TRiC found in eukaryotic cytoplasm and in archaea.

Molecular chaperones do not specify the tertiary structure of a protein, they merely help the protein find that correct structure. The two types of chaperone do this in different ways. The Hsp70 family bind to hydrophobic regions of unfolded proteins, including proteins that are still being translated (**Figure 13.36**). They hold the protein in an open conformation and aid folding, presumably by modulating the association between those parts of the polypeptide which form interactions in the folded protein. Exactly how this is achieved is not understood, but it involves repeated binding and detachment of the Hsp70 protein, each cycle of which requires energy provided by the hydrolysis of ATP. As well as protein folding, the Hsp70 chaperones are also involved in other processes that require shielding of hydrophobic regions in proteins, such as transport through membranes, association of proteins into multisubunit complexes, and disaggregation of proteins that have been damaged by heat stress.

The chaperonins work in a quite different way. GroEL and GroES form a multisubunit structure that looks like a hollowed-out bullet with a central cavity (**Figure 13.37**). A single, unfolded protein enters the cavity and emerges folded. The mechanism for this is not known, but it is postulated that the GroEL/GroES complex acts as a cage that prevents the unfolded protein from aggregating with other proteins, and that the inside surface of the cavity changes from hydrophobic to hydrophilic in such a way as to promote the burial of hydrophobic amino acids within the protein. This is not the only hypothesis: other researchers hold that the cavity unfolds proteins that have folded incorrectly, passing these unfolded proteins back to the cytoplasm so they can have a second attempt at adopting their correct tertiary structure. A second unanswered question concerns the overall importance of chaperonins in protein folding, with reports that only 5% of bacterial proteins are folded exclusively by GroEL/GroES, with another 5% folded by a combination of Hsp70 proteins and GroEL/GroES, and the remaining 90% folded by Hsp70 proteins without any involvement of the chaperonin. The picture is similar in eukaryotes, with only 10% of the proteome, mainly components of the cytoskeleton and proteins involved in the cell cycle, being folded by TRIC.

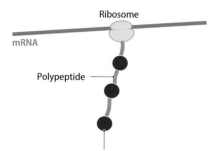

Figure 13.36 The Hsp70 chaperone system. Hsp70 chaperones bind to hydrophobic regions in unfolded polypeptides, including those that are still being translated, and hold the protein in an open conformation to aid its folding.

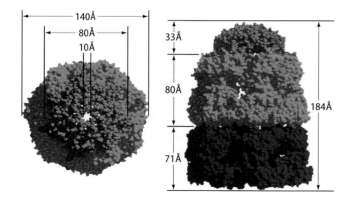

Figure 13.37 The GroEL/GroES chaperonin. On the left is a view from the top, and on the right a view from the side. 1 Å is equal to 0.1 nm. The GroES part of the structure is made up of seven identical protein subunits and is shown in gold. The GroEL components consist of 14 identical proteins arranged into two rings (shown in red and green), each containing seven subunits. The main entrance into the central cavity is through the bottom of the structure shown on the right. (From Xu Z, Horwich AL & Sigler PB [1997] *Nature* 388: 741–750. With permission from Springer Nature.)

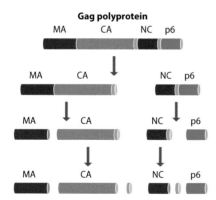

Figure 13.38   Processing of the HIV-1 Gag polyprotein. The protein products are the matrix protein (MA), capsid protein (CA), nucleocapsid protein (NC), and the p6 protein involved virus release. The two spacer peptides are shown in green.

Some proteins undergo proteolytic cleavage

Some proteins are initially synthesized as precursors that are then cut by proteases into smaller, active polypeptides. This type of processing is common in eukaryotes but less frequent in bacteria. The polyproteins coded by some viral genomes are examples (Section 9.1). Several types of virus that infect eukaryotic cells use polyproteins as a way of reducing the sizes of their genomes, a single polyprotein gene with one promoter and one termination sequence taking up less space inside the virus capsid than a series of individual genes. The proteomes of cells infected with one of these viruses might therefore contain a complex mixture of precursor proteins and mature, processed products. The human immunodeficiency virus HIV-1 is an example. During its replication cycle, HIV-1 synthesizes the Gag polyprotein, which is cleaved into four proteins and two short spacer peptides (**Figure 13.38**). Three of these four proteins form structural components of the HIV capsid, and the fourth, called p6, is involved in the process by which virus particles are released from the cell. HIV-1 also synthesizes an extended version of Gag, called Gag-Pol, the additional segment being processed to give two enzymes involved in replication of the HIV genome, as well as the protease that makes the cuts in the Gag and Gag-Pol polyproteins. A few molecules of this protease are stored in each of the new virus particles that are produced, and hence are available to cut up the polyproteins synthesized during the next round of virus replication.

Some polyproteins are coded by eukaryotic genes, the initial translation product being cut into segments, giving rise to a range of active proteins, often with different functions. For example, the polyprotein called proopiomelanocortin, made in the vertebrate pituitary gland, contains at least 11 different peptide hormones. Proopiomelanocortin is initially made as a precursor of 267 amino acids. The first segment to be removed, the 26 amino acids from the N terminus, is a **signal peptide**, a highly hydrophobic stretch of amino acids that aids transfer of the newly-synthesized protein into the endoplasmic reticulum. This is the first stage of the transport pathway that eventually results in secretion of the proopiomelanocortin peptides from the cell. Proopiomelanocortin has a variety of internal cleavage sites recognized by proteolytic enzymes, but not all of these sites are cleaved in every tissue, the combinations and hence the products formed depending on the identities of the proteolytic enzymes that are present (**Figure 13.39**). For example, in the corticotropic cells of the anterior pituitary gland, adrenocorticotropic hormone and the lipotropins are produced. In the melanotropic cells of the intermediate lobe of the pituitary, a different set of cleavage sites are used, generating the melanotropins. Altogether, 11 different peptides can be obtained by alternative patterns of proteolytic cleavage of proopiomelanocortin. The contribution that proopiomelanocortin makes to the proteome is therefore tissue-specific.

Cleavage is also used as a means of activating a secreted polypeptide whose biochemical activities might be deleterious to the cell producing the protein. These proteins are synthesized in an inactive form and then activated after secretion, so the cell is not harmed. The cellular proteome therefore contains the intact but inactive version of the polypeptide, and the secreted proteome contains the cleaved, active version. An example is provided by melittin, the most abundant protein in bee venom and the one responsible for causing cell lysis after injection of the bee sting into the person or animal being stung. Melittin lyses cells in bees as well as animals and so must initially be secreted as an inactive precursor. This precursor, promelittin, has 22 additional amino acids at its N terminus. The presequence is removed by an extracellular protease that cuts at 11 positions, releasing the active venom protein. The protease does not cleave within the active sequence because its mode of action is to release dipeptides with the sequence X–Y, where X is alanine, aspartic acid, or glutamic acid, and Y is alanine or proline; these motifs do not occur in the active sequence (**Figure 13.40**).

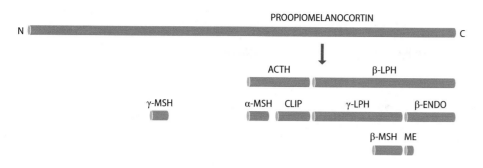

Figure 13.39 Processing of the proopiomelanocortin polyprotein. Abbreviations: ACTH, adrenocorticotropic hormone; CLIP, corticotropin-like intermediate lobe protein; ENDO, endorphin; LPH, lipotropin; ME, met-enkephalin; MSH, melanotropin. Two additional peptides are not shown: one of these is an intermediate in the processing events leading to γ-MSH, and the function of the other is unknown. Note that although met-enkephalin can theoretically be obtained by processing of proopiomelanocortin, as shown here, most met-enkephalin made by humans is probably obtained from a different peptide hormone precursor called proenkephalin.

Important changes in protein activity can be brought about by chemical modification

Genomes have the capacity to code for 22 different amino acids: the 20 specified by the standard genetic code, selenocysteine, and (at least in archaea) pyrrolysine, the latter two being inserted into polypeptides by context-dependent reassignment of 5′–UGA–3′ and 5′–UAG–3′, respectively (Section 1.3). This repertoire is increased dramatically by post-translational chemical modification of proteins, which results in a vast array of different amino acid types. The simpler types of modification occur in all organisms; the more complex ones, especially glycosylation, are rare in bacteria.

The simplest types of chemical modification involve addition of a small chemical group (e.g., an acetyl, methyl, or phosphate group; Table 13.1) to an amino acid side chain, or to the amino or carboxyl groups of the terminal amino acids in a polypeptide. Over 150 different modified amino acids have been documented in different proteins, with each modification carried out in a highly specific manner, the same amino acids being modified in the same way in every copy of the protein. A more complex type of modification is **glycosylation**, the attachment of large carbohydrate side chains, called **glycans**, to polypeptides. There are two general types of glycosylation (**Figure 13.41**):

- **O-linked glycosylation** is the attachment of a glycan to the hydroxyl group of a serine or threonine amino acid.

- **N-linked glycosylation** involves attachment of a glycan to the amino group on the side chain of asparagine.

Glycosylation can result in attachment to the protein of structures comprising branched networks of 12 or more sugar units of various types. These glycans help to target proteins to particular sites in cells and increase the stability of proteins circulating in the bloodstream. Another type of large-scale modification involves attachment of long-chain lipids, often to serine or cysteine amino acids. This process is called **acylation** and occurs with many proteins that become associated with membranes. A less common modification is **biotinylation**, in which a molecule of biotin is attached to a lysine. Biotinylation occurs with a small number of enzymes that catalyze the carboxylation of organic acids, such as acetate and propionate.

Identifying the chemical modifications possessed by individual proteins is an essential part of protein profiling. This is because chemical modification often plays a central role in determining the precise biochemical activity of the target protein. Phosphorylation is particularly important in this regard, especially in signal transduction pathways, where it is used to activate proteins that

Cut sites

Figure 13.40 Processing of promelittin, the bee-sting venom. Arrows indicate the cut sites. Melittin is initially synthesized as a 70-amino-acid precursor called prepromelittin. Removal of a 21-amino-acid signal peptide when this precursor is secreted gives the promelittin sequence shown here.

TABLE 13.1 EXAMPLES OF POST-TRANSLATIONAL CHEMICAL MODIFICATIONS		
Modification	**Amino acids that are modified**	**Examples of proteins**
Addition of small chemical groups		
Acetylation	Lysine	Histones
Methylation	Lysine	Histones
Phosphorylation	Serine, threonine, tyrosine	Some proteins involved in signal transduction
Hydroxylation	Proline, lysine	Collagen
Carbamoylation	Lysine	Ribulose bisphosphate carboxylase
N-formylation	N-terminal glycine	Melittin
Addition of sugar side chains		
O-linked glycosylation	Serine, threonine	Many membrane proteins and secreted proteins
N-linked glycosylation	Asparagine	Many membrane proteins and secreted proteins
Addition of lipid side chains		
Acylation	Serine, threonine, cysteine	Many membrane proteins
N-myristoylation	N-terminal glycine	Some protein kinases involved in signal transduction
Addition of biotin		
Biotinylation	Lysine	Various carboxylase enzymes

carry signals from cell-surface receptors to transcription factors and other regulatory proteins (Section 14.1). During protein profiling, a chemical modification can be detected because it will result in a characteristic change in the m/z value of the peptide or protein that carries the new chemical group. Sometimes, however, the modified form of the protein is present in relatively low amounts, in which case an enrichment procedure might be needed before the modified protein can be detected. Phosphorylated proteins fall into this category, as signal transduction can be effective even if only a few copies of the relevant proteins have been activated. Phosphorylated proteins can be partially purified from a proteome by a version of immobilized metal affinity chromatography (Section 13.2), making use of the interaction that occurs between a phosphate group and metal ions such as Fe^{3+}, Ca^{3+}, and Zr^{3+}. Purification is not complete, because nonphosphorylated proteins can also be retained in the affinity column, but the recovered sample is sufficiently enriched in the phosphorylated versions to enable their detection. Affinity chromatography can also be used to enrich proteomes for other types of modified protein. Glycosylated proteins, for example, can be captured by **lectins**, plant or animal proteins with specific sugar-binding properties; an example is concanavalin A from jack bean (*Canavalis ensiformis*), which binds to terminal glucose and mannose units in O-linked glycans.

13.5 BEYOND THE PROTEOME

Protein synthesis is traditionally looked on as the final stage of genome expression, but this view obscures the true role of the proteome as part of the final link that connects the genome with the biochemistry of the cell. Exploring the nature of this link is proving to be one of the most exciting and productive areas of biological research, and has led to new concepts surrounding the **metabolome** and **systems biology**.

The metabolome is the complete set of metabolites present in a cell

Often in biology the most important steps forward do not result from some groundbreaking experiment but instead arise because biologists devise a new way of thinking about a problem. The introduction of the concept of the metabolome is an example. The metabolome is defined as the complete collection of metabolites present in a cell or tissue under a particular set of conditions. In other words, a metabolome is a biochemical blueprint, and its study, which is called **metabolomics** or **biochemical profiling**, gives a precise description of the biochemistry underlying different physiological states, including disease states, which can be adopted by a cell or tissue. By converting the biochemistry of a cell into an itemized set of metabolites, metabolomics provides a dataset that can be directly linked to the equivalent, itemized information which emerges from proteomics and other studies of genome expression.

A metabolome can be characterized by chemical techniques such as infrared spectroscopy, mass spectrometry, and nuclear magnetic resonance spectroscopy which, individually and in combination, can identify and quantify the various small molecules that make up the metabolites in a cell. When these data are combined with knowledge about the reaction rates for the various steps in well-characterized biochemical pathways such as glycolysis and the tricarboxylic acid cycle, it is possible to model the **metabolic flux**, the rate of flow of metabolites through the network of pathways that make up the cellular biochemistry. Changes in the metabolome can then be defined in terms of perturbations in the flux of metabolites through one or more parts of the network, providing a very sophisticated description of the biochemical basis for changes in the physiological state. For example, metabolic flux analysis illustrates how *Escherichia coli* adapts its metabolome to glucose starvation by reducing the flow of metabolites through the energy-generating tricarboxylic acid cycle and increasing the flow through the glyoxylate pathway (**Figure 13.42**). This enables the cells to convert acetate into succinate, which in turn can be converted into a variety of other carbohydrates, including glucose, and also reduces the amount of carbon that the bacterium loses as CO_2. Studies of metabolic flux lead to the possibility of **metabolic engineering**, in which changes are made to the genome by mutation or recombinant DNA techniques in order to influence the cellular biochemistry in a predetermined way, for example, to increase the synthesis of an antibiotic by a microorganism.

At present, metabolomics is most advanced with organisms such as bacteria and yeast, whose biochemistries are relatively simple. Considerable research is currently being directed toward the human metabolome, with the objective of describing the metabolic profiles of healthy tissues, of disease states, and of tissues in patients undergoing drug treatments. It is hoped that when these studies reach maturity, it will be possible to use the metabolic information to design drugs that reverse or mitigate the particular flux abnormalities that occur in the disease state. Biochemical profiling could also indicate any unwanted side effects of drug treatment, enabling modifications to be made to the chemical structure of the drug, or to its mode of use, so that these side effects are minimized.

Systems biology provides an integrated description of cellular activity

The emphasis that is now placed on concepts such as metabolic flux illustrates the importance of understanding not simply the molecules – RNAs, proteins, and metabolites – whose synthesis the genome directs, but also the biological systems that result from the coordinated activity of those molecules. This new emphasis on systems is a direct result of the leap that has been made in recent years from genes to genomes. One of the underlying principles of the pregenome era of molecular biology was the 'one gene, one enzyme' hypothesis first put forward by George Beadle and Edward Tatum in the 1940s. By 'one gene, one enzyme,' Beadle and Tatum were emphasizing that a single gene codes for

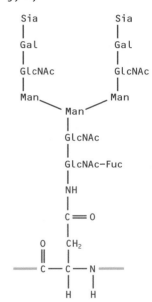

(A) O-linked glycosylation

(B) N-linked glycosylation

Figure 13.41 Glycosylation. (A) O-linked glycosylation. The structure shown is found in a number of glycoproteins. It is drawn here attached to a serine amino acid, but it can also be linked to a threonine. (B) N-linked glycosylation usually results in larger sugar structures than are seen with O-linked glycosylation. The drawing shows a typical example of a complex glycan attached to an asparagine amino acid. Abbreviations: Fuc, fucose; Gal, galactose; GalNAc, *N*-acetylgalactosamine; GlcNAc, *N*-acetylglucosamine; Man, mannose; Sia, sialic acid.

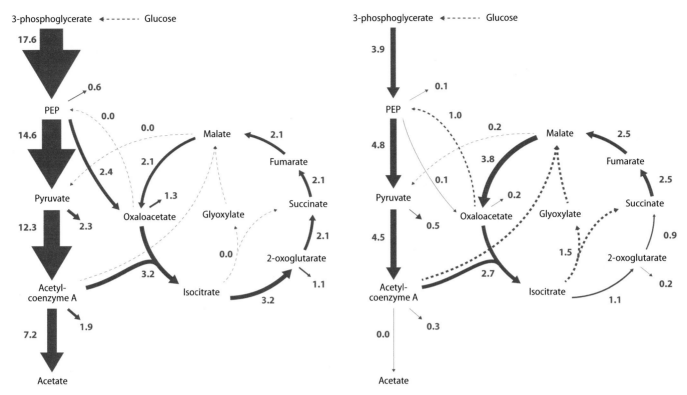

Figure 13.42 Examples of metabolic flux diagrams. These two diagrams illustrate the flow of metabolites through the tricarboxylic acid cycle and glyoxylate pathway in *Escherichia coli* cells grown in (A) complete medium, and (B) under glucose-limited conditions. The numbers indicate molar flux (moles of substrate converted to product expressed as mmol g^{-1} h^{-1}), and the sizes of the arrows are proportional to the flux at each step. (From Fischer E & Sauer U [2003] *Journal of Biological Chemistry* 278: P46446-46451. Published under CC BY 4.0.)

a single protein which, if an enzyme, directs a single biochemical reaction. The *trpC* gene of *Escherichia coli*, for example, codes for the enzyme indole-3-glycerol phosphate synthase, which converts 1-(2-carboxyphenylamino)-1´-deoxyribulose-5´-phosphate into indole-3-glycerol phosphate. However, this enzyme does not work in isolation: its activity forms part of the biochemical pathway which results in synthesis of tryptophan, the other enzymes in this pathway being specified by the genes *trpA*, *trpB*, *trpD*, and *trpE*, which together with *trpC* form the tryptophan operon of *E. coli* (see Figure 8.10B). The tryptophan biosynthesis pathway is therefore a simple biological system and the tryptophan operon is the set of genes that specify that pathway. But simply transcribing and translating the genes in the operon will not lead to the synthesis of tryptophan. Successful operation of the system requires that the biosynthetic enzymes be present at the appropriate place in the cell, in the appropriate relative amounts, at the appropriate time. The system is therefore dependent on factors such as the rate of synthesis of the proteins coded by the genes, the correct folding of these proteins into the functional enzymes, the rate of degradation of the enzyme molecules, their localization in the cell, and the presence of the necessary amounts of the metabolites that act as substrates and cofactors for tryptophan synthesis. This simple biological system is starting to assume quite considerable complexity. And yet, with this system we are considering only 5 of the 4609 genes in the *E. coli* genome.

How do we move beyond the traditional study of isolated components of a system and achieve a meaningful understanding of the system as a whole? A complete description of a system requires an understanding of the integration between the following components:

- The metabolic pathways that make up the system under study, including the enzymes that catalyze the individual steps in the pathway, their reaction rates and the flux of metabolites through the pathway.

- The genes coding for the protein components of the system, as well as the regulatory networks (e.g., transcription factors, histone modifications) that control expression of those genes.

- The protein–protein interactions occurring within the system.

- The signal transduction pathways that regulate the operation of the system in response to the internal and external cellular environments.

- The spatial organization of the system within the cell, tissue and/or organism: are all the components of the system located together or are some components in a different cellular compartment or even a different cell or tissue?

- In many applied settings, information on how the system is perturbed by the presence of a drug or inhibitor.

Systems biology is therefore a very interdisciplinary science as it requires the combined efforts of biochemists, chemists, geneticists, and cell biologists to generate the various types of data that are needed to describe the system, as well as computational biologists, mathematical modelers and bioinformaticians to combine the data into an integrated representation of the system.

An underlying principle of systems biology is that information about a process should be presented in a consistent manner so that all researchers, of whatever background, can understand what is known about the system without the need to sift through the raw data. This has led to establishment of the **systems biology graphical notation** (**SBGN**), which comprises a set of rules and symbols that are used to depict a system in a graphical format. Three different aspects of a biological system can be depicted by SBGN:

- The **process description** shows the biochemical reactions that occur over time.

- The **entity relationship** depicts the interactions that occur between the different components of the system.

- The **activity flow** shows how information flows within the system.

To illustrate the use of SBGN we will consider the simple biological system involving the human transcription factor ELK1, which regulates expression of the *Fos* gene. The c-FOS protein coded by *Fos* is itself a transcription factor involved in the control of cell proliferation. The activity of ELK1 is regulated by post-translational modification: ELK1 is suppressed by attachment of one or more SUMO groups to lysine amino acids by the SUMO-conjugating enzyme UBC9; ELK1 is activated by phosphorylation of a pair of serines by the ERK protein kinase. The process description for ELK1 (**Figure 13.43A**) depicts the system

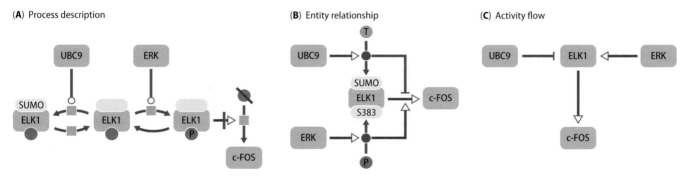

Figure 13.43 An example of the use of the systems biology graphical notation (SBGN). (A) Process description, (B) entity relationship, and (C) activity flow for the biological system involving the human transcription factor ELK1. (Le Novère N [2015] *Nat. Rev. Genet.* 16: 146–158. With permission from Springer Nature.)

Figure 13.44 Process description of the *Drosophila melanogaster* cell cycle. (Touré V, Le Novère N, Waltemath D & Wolkenhauer O [2018] *PLoS Comput Biol* 14: e1005740. https://doi.org/10.1371/journal.pcbi.1005740. Published under CC BY 4.0.)

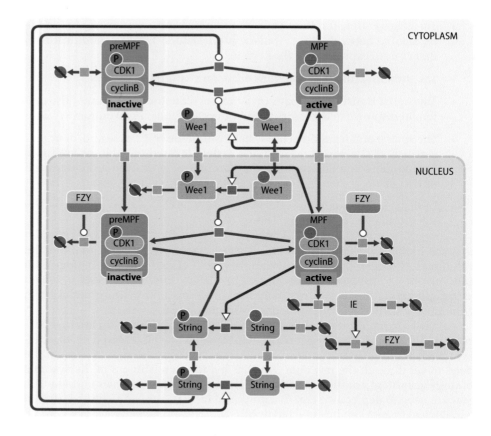

as a flow chart showing the role of the transcription factor in synthesis of c-FOS and the control of the transcription factor by UBC9 and ERK. The entity relationship (**Figure 13.43B**) redraws the flow chart to show the precise interactions between ELK1, UBC9, ERK, and c-FOS, and in particular emphasizes the post-translational modifications of ELK1 that are catalyzed by UBC9 and ERK. Finally, the activity flow (**Figure 13.43C**) shows the positive (activating) roles of ERK and ELK1, and negative (inhibiting) role of UBC9. Because of the standardized notation used to draw these three graphs, any researcher familiar with the SBGN language will be able to understand the different aspects of the ELK1 biological system, even if they have no direct knowledge of the transcription factor or the other components of the system. The same is true for more complex biological systems (**Figure 13.44**).

SUMMARY

- The proteome is the collection of protein molecules present in a cell.

- The composition of a proteome can be examined by top-down and bottom-up processes. In top-down proteomics, individual proteins are directly examined by mass spectrometry, whereas in bottom-up proteomics the proteins are broken into peptides by treatment with a sequence-specific protease, such as trypsin, prior to mass spectrometry.

- The initial stage in both proteomic approaches is separation of the proteins in the proteome. This can be achieved by two-dimensional poly-acrylamide gel electrophoresis or by column chromatography.

- Two proteomes can be compared by labeling the components with different fluorescent markers or with isotope-coded affinity tags.

- Protein arrays can also be used to study proteomes.

- Pairs of proteins that interact with one another can be identified by phage display and the yeast two-hybrid system.

- The members of multiprotein complexes can be identified by affinity purification or co-immunoprecipitation followed by mass spectrometry.

- Protein interaction maps display the detailed interactions between the components of a proteome.

- Proteins are synthesized by ribosomes, which coordinate the events occurring during protein synthesis and also catalyze at least some of the chemical reactions. Bacteria are able to inactivate their ribosomes during periods of stress.

- In eukaryotes, initiation factors mediate the large-scale remodeling of proteome structure. For example, phosphorylation of eIF-2 occurs during stresses such as heat shock and results in a decrease in the overall level of protein synthesis. The translation of individual mRNAs can also be regulated.

- Proteins must be turned over so that proteomes can respond to changing conditions. The degradation pathway in eukaryotes involves tagging the targeted protein with ubiquitin followed by breakdown in a proteasome.

- Most polypeptides are inactive until they have been folded into their three-dimensional structures. The amino acid sequence contains the information for folding, but folding of large proteins is a complex process that can result in diversion of the partially folded protein down an incorrect side pathway. Molecular chaperones help other proteins to fold, reducing the number of folding errors that occur.

- Polyproteins are cut at internal sites to give multiple protein products, and some proteins are cleaved as a means of converting an inactive protein into the active form.

- Chemical modification by glycosylation, acylation, or addition of small chemical groups can bring about changes in protein activity.

- The metabolome is defined as the complete collection of metabolites present in a cell or tissue under a particular set of conditions.

- Systems biology attempts to provide an integrated description of cellular activity.

SHORT ANSWER QUESTIONS

1. Distinguish between the methods used in top-down and bottom-up proteomics.

2. Describe how the components of a proteome are separated prior to the mass spectrometry stage of protein profiling.

3. Outline the different mass spectrometric methods that are used in protein profiling.

4. How can differences between the compositions of two proteomes be identified?

5. What methods are available for identification of proteins that interact with one another as pairs or in multiprotein complexes?

6. Compare the compositions of the bacterial and eukaryotic ribosomes.

7. Explain how bacteria downsize their proteomes during periods of stress.

8. Describe how large and small changes in eukaryotic proteomes can be brought about by initiation factor phosphorylation and by transcript-specific regulation.

9. How are proteins degraded?

10. Outline the key features of the protein folding pathway and describe the role of molecular chaperones in protein folding.

11. Summarize the roles in protein processing of (A) proteolytic cleavage and (B) chemical modification.

12. Describe the objectives of metabolomics and systems biology.

IN-DEPTH PROBLEMS

1. Under what circumstances might a pair of proteins have a functional relationship but no physical interaction? Are there possible scenarios where the reverse might be true – a pair of proteins display a physical interaction but have no functional relationship?

2. Discuss the roles of the hubs in a protein interaction map.

3. There appears to be no biological reason why a DNA polynucleotide could not be directly translated into protein, without the intermediate role played by mRNA. What advantages do eukaryotic cells gain from the existence of mRNA?

4. To what extent have studies of ribosome structure been of value in understanding the detailed process by which proteins are synthesized?

5. Describe the biological system depicted in Figure 13.44.

FURTHER READING

Methods for studying the composition of a proteome

Catherman, A.D., Skinner, O.S. and Kelleher, N.L. (2014) Top down proteomics: Facts and perspectives. *Biochem. Biophys. Res. Commun.* 445:683–693.

Dupree, E.J., Jayathirtha, M., Yorkey, H., et al. (2020) A critical review of bottom-up proteomics: The good, the bad, and the future of this field. *Proteomes* 8:14.

Görg, A., Weiss, W. and Dunn, M.J. (2004) Current two-dimensional electrophoresis technology for proteomics. *Proteomics* 4:3665–3685.

Li, J., Van Vranken, J.G., Vaites, L.P., et al. (2020) TMTpro reagents: A set of isobaric labeling mass tags enables simultaneous proteome-wide measurements across 16 samples. *Nat. Methods* 17:399–404.

Mann, M., Hendrickson, R.C. and Pandey, A. (2001) Analysis of proteins and proteomes by mass spectrometry. *Annu. Rev. Biochem.* 70:437–473.

Noor, Z., Ahn, S.B., Bajer, M.S., et al. (2020) Mass spectrometry-based protein identification in proteomics – A review. *Brief. Bioinformatics* 22:1620–1638.

Phizicky, E., Bastiaens, P.I.H., Zhu, H., et al. (2003) Protein analysis on a proteomic scale. *Nature* 422:208–215. *Reviews all aspects of proteomics.*

Sutandy, F.X.R., Qian, J., Chen, C.-S., et al. (2013) Overview of protein microarrays. *Curr. Protoc. Protein Sci.* 27:Unit 27.1.

Walton, H.F. (1976) Ion exchange and liquid column chromatography. *Anal. Chem.* 48:52R–66R.

Zhang, Y., Fonslow, B.R., Shan, B., et al. (2013) Protein analysis by shotgun/bottom-up proteomics. *Chem. Rev.* 113:2343–2394.

Zhu, H., Bilgin, M. and Snyder, M. (2003) Proteomics. *Annu. Rev. Biochem.* 72:783–812.

Identifying protein interactions

Agarwal, S., Deane, C.M., Porter, M.A., et al. (2010) Revisiting date and party hubs: Novel approaches to role assignment in protein interaction networks. *PLoS Comput. Biol.* 6:e1000817.

Dunham, W.H., Mullin, M. and Gingras, A.-C. (2012) Affinity-purification coupled to mass spectrometry: Basic principles and strategies. *Proteomics* 12:1576–1590.

Gordon, D.E., Jang, G.M., Bouhaddou, M., et al. (2020) A SARS-CoV-2 protein interaction map reveals targets for drug repurposing. *Nature* 583:459–468.

Han, J.-D.J., Bertin, N., Hao, T., et al. (2004) Evidence for dynamically organized modularity in the yeast protein–protein interaction network. *Nature* 430:88–93. *Defines party and date hubs.*

Jeong, H., Mason, S.P., Barabási, A.-L., et al. (2001) Lethality and centrality in protein networks. *Nature* 411:41–42. *The first version of the yeast protein interaction map.*

Pande, J., Szewczyk, M.M. and Grover, A.K. (2010) Phage display: Concept, innovations, applications and future. *Biotechnol. Adv.* 28:849–858.

Parrish, J.R., Gulyas, K.D. and Finley, R.L. (2006) Yeast two-hybrid contributions to interactome mapping. *Curr. Opin. Biotechnol.* 17:387–393.

Snider, J., Kotlyar, M., Saraon, P., et al. (2015) Fundamentals of protein interaction network mapping. *Mol. Syst. Biol.* 11:848.

Vandereyken, K., Van Leene, J., De Coninck, B., et al. (2018) Hub protein controversy: Taking a closer look at plant stress response hubs. *Front. Plant Sci.* 9:694. *Discusses the roles of hubs in the interaction map of proteins involved in the plant stress response.*

Protein synthesis
Kapp, L.D. and Lorsch, J.R. (2004) The molecular mechanics of eukaryotic translation. *Annu. Rev. Biochem.* 73:657–704.

Klinge, S. and Woolford, J.L. (2019) Ribosome assembly coming into focus. *Nat. Rev. Mol. Cell. Biol.* 20:116–131.

Rodnina, M.V. (2016) The ribosome in action: Tuning of translational efficiency and protein folding. *Prot. Sci.* 25:1390–1406.

Schmeing, T.M. and Ramakrishnan, V. (2009) What recent ribosome structures have revealed about the mechanism of translation. *Nature* 461:1234–1242.

Steitz, T.A. and Moore, P.B. (2003) RNA, the first macromolecular catalyst: The ribosome is a ribozyme. *Trends Biochem. Sci.* 28:411–418.

Wilson, D.N. and Cate, J.H.D. (2012) The structure and function of the eukaryotic ribosome. *Cold Spring Harb. Perspect. Biol.* 4:a011536.

Factors influencing the composition of a proteome
Diez, S., Ryu, J., Caban, K., et al. (2020) The alarmones (p)ppGpp directly regulate translation initiation during entry into quiescence. *Proc. Natl. Acad. Sci. USA* 117:15565–15572.

Hershey, J.W.B., Sonenberg, N. and Mathews, M.B. (2012) Principles of translational control: An overview. *Cold Spring Harb. Perspect. Biol.* 4:a011528.

Hinnebusch, A.E., Ivanov, I.P. and Sonenberg, N. (2016) Translational control by 5'- untranslated regions of eukaryotic mRNAs. *Science* 352:1413–1416.

Irving, S.E., Choudhury, N.R. and Corrigan, R.M. (2021) The stringent response and physiological roles of (pp)pGpp in bacteria. *Nat. Rev. Microbiol.* 19:256–271.

Komar, A.A. and Hatzoglou, M. (2011) Cellular IRES-mediated translation: The war of ITAFs in pathophysiological states. *Cell Cycle* 10:229–240.

Loveland, A.B., Bah, E., Madireddy, R., et al. (2016) Ribosome•RelA structures reveal the mechanism of stringent response activation. *eLIFE* 5:e17029.

Polikanov, Y.S., Blaha, G.M. and Steitz, T.A. (2012) How hibernation factors, RFF, HPF, and YfiA turn off protein synthesis. *Science* 336:915–918.

Protein degradation
Kudryavtseva, S.S., Pichkur, E.B., Yaroshevich, I.A., et al. (2021) Novel cryo-EM structure of an ADP-bound Gro-EL-GroES complex. *Sci. Rep.* 11:18241.

Varshavsky, A. (1997) The ubiquitin system. *Trends Biochem. Sci.* 22:383–387.

Voges, D., Zwickl, P. and Baumeister, W. (1999) The 26S proteasome: A molecular machine designed for controlled proteolysis. *Annu. Rev. Biochem.* 68:1015–1068.

Protein folding and other protein processing events
Anfinsen, C.B. (1973) Principles that govern the folding of protein chains. *Science* 181:223–230. *The first experiments on protein folding.*

Chapman-Smith, A. and Cronan, J.E. (1999) The enzymatic biotinylation of proteins: A post-translational modification of exceptional specificity. *Trends Biochem. Sci.* 24:359–363.

Daggett, V. and Fersht, A.R. (2003) Is there a unifying mechanism for protein folding? *Trends Biochem. Sci.* 28:18–25.

Drickamer, K. and Taylor, M.E. (1998) Evolving views of protein glycosylation. *Trends Biochem. Sci.* 23:321–324.

Rosenzweig, R., Nillegoda, N.B, Mayer, M.P., et al. (2019) The Hsp70 chaperone network. *Nat. Rev. Mol. Cell Biol.* 20:665–680.

Smith, A.I. and Funder, J.W. (1988) Proopiomelanocortin processing in the pituitary, central nervous system, and peripheral tissues. *Endocr. Rev.* 9:159–179.

Yébenes, H., Mesa, P., Muñoz, I.G., et al. (2011) Chaperonins: Two rings for folding. *Trends Biochem. Sci.* 36:424–432.

Metabolomics and systems biology
Beger R.D. (2013) A review of applications of metabolomics in cancer. *Metabolites* 3:552–574.

Covert, M.W., Schilling, C.H., Famili, I., et al. (2001) Metabolic modeling of microbial strains *in silico*. *Trends Biochem. Sci.* 26:179–186. *Explains the concept of metabolic flux.*

Fischer, E. and Sauer, U. (2003) A novel metabolic cycle catalyzes glucose oxidation and anaplerosis in hungry *Escherichia coli*. *J. Biol. Chem.* 278:46446–46451.

Kirschner, M.W. (2005) The meaning of systems biology. *Cell* 121:503–504.

Le Novère, N. (2015) Quantitative and logic modeling of molecular and gene networks. *Nat. Rev. Genet.* 16:146–158. *Describes the graphical notation used in systems biology.*

Online resource
Systems Biology Graphical Notation. https://sbgn.github.io/

GENOME EXPRESSION IN THE CONTEXT OF CELL AND ORGANISM

During the previous two chapters we have studied how a genome directs the synthesis of a transcriptome and proteome and thereby specifies the biochemical activities occurring in the cell in which it resides, enabling the cell to generate energy and to grow and divide. To ensure that these cellular activities are in tune with the available nutrient supply and the prevailing physical and chemical conditions, the genome must be responsive to the extracellular environment. Cells in multicellular organisms must also respond to hormones and growth factors, which may signal changes in the environment to which the organism must adapt, but which also, importantly, coordinate the biochemical activities of different cells so that a multicellular organism functions as a unified entity. The response of the genome to these extracellular signals is usually transient, the expression pattern reverting to its original form when the signals cease, or taking on a new pattern if one set of signals is replaced by a second set.

Other changes in genome activity are permanent, or at least semi-permanent, and result in the cell's biochemical signature becoming altered in a way that is not readily reversible. These changes lead to **differentiation**, the adoption by the cell of a specialized physiological role. Differentiation pathways are known in many unicellular organisms, but we more frequently associate differentiation with multicellular organisms, in which a variety of specialized cell types (almost 3000 in humans) are organized into tissues and organs. The assembly and modification of these complex multicellular structures over time constitutes the **development** pathway for the organism, the successful completion of which requires temporal control over genome activity, so that the developmental events occur in the correct order and at the appropriate times.

In this chapter, we will explore how the genome acts within the context of cell and organism by responding to extracellular signals and by specifying the biochemical changes that underlie differentiation and development. The basis for all of these processes is the regulation of genome expression, which has been a theme throughout Part 3 of *Genomes*. In Chapter 10, we studied how genome expression is influenced by the organization of chromosomal DNA in the nucleus, by the chemical modification of histones and by DNA methylation, and in Chapters 12 and 13 we examined various mechanisms by which the composition of the transcriptome and proteome can be changed by modulating the synthesis, processing and/or degradation of individual RNAs and proteins. Our objective now is not simply to reiterate this material, but instead, as the title of the chapter indicates, to understand how these regulatory systems operate within the cell and organism.

14.1 THE RESPONSE OF THE GENOME TO EXTERNAL SIGNALS

For unicellular organisms, the most important external signals relate to nutrient availability, these cells living in variable environments in which the identities and relative amounts of nutrients change over time. The genomes of unicellular organisms therefore include genes for uptake and utilization of a range of nutrients, and changes in nutrient availability are shadowed by changes in genome activity, so that at any one time only those genes needed to utilize the available

DOI: 10.1201/9781003133162-14

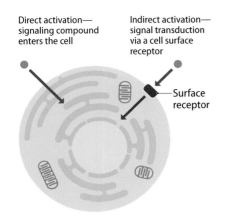

Figure 14.1 Two ways in which an extracellular signalling compound can influence events within a cell.

nutrients are expressed. The response of the lactose operon of *Escherichia coli* to the available glucose and lactose supply (Section 12.3) provides a typical example of the way in which expression of a bacterial genome can be influenced by the nutrient status of the external environment. Most cells in multicellular organisms live in less variable environments, but the maintenance of these environments requires coordination between the activities of different cells. For these cells, the major external stimuli are therefore hormones, growth factors, and related compounds that convey signals within the organism and stimulate coordinated changes in genome activity.

To exert an effect on genome activity, the nutrient, hormone, growth factor or other extracellular compound that represents the external signal must influence events within the cell. There are two ways in which it can do this (**Figure 14.1**):

- Directly, by acting as a signaling compound that is transported across the cell membrane and into the cell.

- Indirectly, by binding to a cell surface receptor which transmits a signal into the cell.

Signal transmission, by direct or indirect means, is one of the major research areas in cell biology, with attention focused in particular on its relevance to the abnormal biochemical activities that underlie cancer. Many examples of signal transmission have been discovered, some found in a variety of organisms and others restricted to just a few species. In the next few pages, we will survey the most important of these signaling pathways.

Signal transmission by import of the extracellular signaling compound

First, we will consider those signaling compounds that are able to pass through the cell membrane and into the cell. We might imagine that the simplest type of imported signaling compound would be a protein that acted in the same way as one of the various regulatory proteins that we have met in previous chapters, for example, by activating or repressing assembly of the transcription initiation complex (Section 12.3), or by interacting with a splicing enhancer or silencer (Section 12.4). This might appear to be an attractively straightforward way of regulating genome activity, but it is not a common mechanism. The reason for this is not clear but probably relates, at least partly, to the difficulty in designing a protein that combines the hydrophobic properties needed for effective transport across a membrane with the hydrophilic properties needed for migration through the aqueous cytoplasm to the protein's site of action in the nucleus or elsewhere in the cell.

The more normal mode of action is therefore for an imported signaling compound to influence the activity of regulatory proteins that are already present in the cell. We encountered one example of this type of regulation in Section 12.3 when we studied the lactose operon of *Escherichia coli*. This operon responds to extracellular levels of lactose, the latter acting as a signaling molecule that enters the cell where, after conversion to its isomer allolactose, it influences the DNA-binding properties of the lactose repressor and hence determines whether or not the lactose operon is transcribed (see Figure 12.21). Many other bacterial operons coding for genes involved in sugar utilization are controlled in a similar way. Interaction between a signaling compound and a transcription factor is also a common means of regulating genome activity in eukaryotes. A good example is provided by the control system that maintains the intracellular metal ion content at an appropriate level. Cells need metal ions such as copper and zinc as cofactors in biochemical reactions, but these metals are toxic if they accumulate in the cell above a certain level. Their uptake therefore has to be carefully controlled so that the cell contains sufficient metal ions when the environment is lacking in metal compounds, but does not overaccumulate metal ions when the environmental concentrations are high. The strategies used are illustrated by the copper control system of *Saccharomyces cerevisiae*. This yeast has two copper-dependent

Figure 14.2 Copper-regulated gene expression in *Saccharomyces cerevisiae*. Yeast requires low amounts of copper because a few of its enzymes (e.g., cytochrome c oxidase and tyrosinase) are copper-containing metalloproteins, but too much copper is toxic for the cell. When copper levels are low, the Mac1p transcription factor is activated by copper binding and switches on expression of genes for copper uptake. When the copper levels are too high, a second factor, Ace1p, is activated, switching on expression of a different set of genes, these coding for proteins involved in copper detoxification.

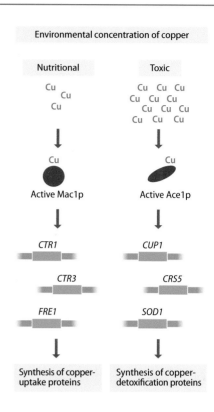

transcription factors, Mac1p and Ace1p. Both of these proteins bind copper ions, the binding inducing a conformational change that enables the factor to stimulate expression of its target genes (**Figure 14.2**). For Mac1p these target genes code for copper-uptake proteins, whereas for Ace1p they are genes coding for proteins such as superoxide dismutase that are involved in copper detoxification. The balance between the activities of Mac1p and Ace1p, controlled by the metal ions acting as signaling molecules, ensures that the copper content of the cell remains within acceptable levels.

Transcription activators are also the targets for **steroid hormones**, which are signaling compounds that coordinate a range of physiological activities in the cells of higher eukaryotes. They include the sex hormones (estrogens for female sex development, androgens for male sex development), and the glucocorticoid and mineralocorticoid hormones. Steroids are hydrophobic and so easily penetrate the cell membrane. Once inside the cell, each hormone binds to a specific **steroid receptor** protein, which is usually located in the cytoplasm. After binding, the activated receptor migrates into the nucleus, where it attaches to a **hormone response element** upstream of a target gene. Once bound, the receptor acts as a transcription activator. Response elements for each receptor are located upstream of 50–100 genes, often within enhancers, so a single steroid hormone can induce a large-scale change in the biochemical properties of the cell. All steroid receptors are structurally similar, not just with regard to their DNA-binding domains but also in other parts of their protein structures (**Figure 14.3**). Recognition of these similarities has led to the identification of a number of putative or orphan steroid receptors whose hormonal partners and cellular functions are not yet known. The structural similarities have also shown that a second set of receptor proteins, the **nuclear receptor superfamily**, belongs to the same general class as steroid receptors, although the hormones that they work with are not steroids. As their name suggests, these receptors are located in the nucleus rather than the cytoplasm. They include the receptors for vitamin D_3, whose roles include control of bone development, and thyroxine, which stimulates the tadpole-to-frog metamorphosis.

Most steroid and nuclear receptors are homodimers, each subunit possessing one of the treble clef zinc fingers that are characteristic of this group of proteins (see Figure 11.13). Each of these zinc fingers recognizes and binds to a 6 bp sequence in its hormone response element. For most receptors, the 6 bp sequence is 5´–AGAACA–3´ or 5´–AGGTCA–3´ arranged as a direct or inverted repeat separated by a 0–5 bp spacer (**Figure 14.4**). The spacer serves simply to ensure that the distance between the recognition sequences is appropriate for the orientation of the two zinc fingers in the dimeric receptor protein. This means that different receptor proteins can possess the same pair of zinc fingers but recognize different response elements, specificity being maintained by the orientation of the fingers and the spacing between the recognition sequences.

Receptor proteins transmit signals across cell membranes

Many extracellular signaling compounds are unable to enter the cell because they are too hydrophilic to penetrate the lipid membrane and the cell lacks a specific

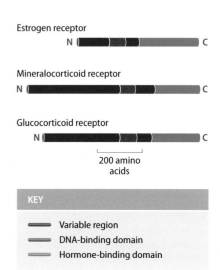

Figure 14.3 All steroid hormone receptor proteins have similar structures. Three receptor proteins are compared. The DNA-binding domain is very similar in all steroid receptors, displaying 50–90% amino acid sequence identity. The hormone-binding domain is less well conserved, with 20–60% sequence identity.

Response element

AGAACA	NNN	TGTTCT	Glucocorticoid
TCTTGT	NNN	ACAAGA	receptor
AGGTCA	NNN	TGACCT	Estrogen
TCCAGT	NNN	ACTGGA	receptor
AGGTCA		TGACCT	Thyroxine
TCCAGT		ACTGGA	receptor
AGGTCA	NNNNN	AGACCA	Retinoic
TCCAGT	NNNNN	TCTGGT	acid receptor

Figure 14.4 The sequences of typical steroid and nuclear receptor response elements. Some retinoic acid receptors are unusual as they are heterodimers and the two subunits recognize different repeat sequences, as shown here.

transport mechanism for their uptake. In order to influence genome activity, these signaling compounds must bind to cell surface receptors that transfer the signal across the cell membrane and into the cell (see Figure 14.1). A cell surface receptor is a protein that spans the membrane, with a site for binding the signaling compound on the outer surface. Binding of the signaling compound results in a conformational change in the receptor, inducing a biochemical event within the cell, that event forming the first step in the intracellular stage of the **signal transduction** pathway.

There are several types of cell surface receptor, but most of those mediating changes in genome expression are **kinase** or **kinase-associated receptors**. As the names indicate, these receptors induce the intracellular stage of the signal transduction pathway by adding a phosphate group to a cytoplasmic protein. The most important examples are the **tyrosine kinase receptors**, which add phosphates to tyrosine amino acids in their target proteins. Most tyrosine kinase receptors are dimers of identical subunits, each subunit comprising an extracellular binding domain and intracellular kinase activity separated by a hydrophobic transmembrane region of some 25–35 amino acids (**Figure 14.5**). Tyrosine kinase receptors recognize a variety of extracellular signaling compounds, including cytokines and other growth factors, as well as hormones such as insulin. In the absence of the signaling compound, the two subunits of a receptor are disassociated. Attachment of the signal results in the subunits coming together to form the dimer, which activates the internal kinase activity, initiating the intracellular signal transduction pathway.

Tyrosine is not the only amino acid that can be phosphorylated in order to activate an intracellular protein. The **serine-threonine kinase receptors** act in a similar way to tyrosine kinase receptors but phosphorylate serines and/or threonines in their intracellular target proteins. The signaling compounds recognized by this type of receptor include transforming growth factor β (TGFβ) and members of the bone morphogenetic protein family of cytokines.

Both the tyrosine and serine-threonine groups of receptors possess kinase activity and so directly phosphorylate their internal target proteins. The **tyrosine kinase-associated receptors** are slightly different, as they do not themselves have kinase activity. Instead, they act indirectly by influencing the activity of intracellular tyrosine kinases. We will study an example of this type of receptor in the next section.

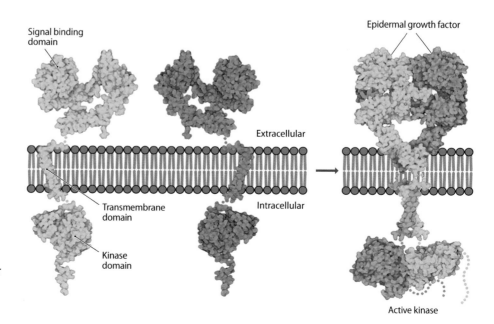

Figure 14.5 A tyrosine kinase receptor. This is the epidermal growth factor (EGF) receptor, which is a typical example of a tyrosine kinase receptor. Attachment of an EGF molecule to each of the receptor subunits induces dimerization, which activates the kinase specified by the intracellular domains of the two monomers.

Some signal transduction pathways have few steps between receptor and genome

With some signal transduction systems, stimulation of the cell surface receptor by attachment of the extracellular signaling compound has a direct effect on the activity of a transcription factor. This is the simplest system by which an extracellular signal can be transduced into a genomic response.

The direct system is used by the cell surface receptors for many cytokines, such as interleukins and interferons, which are extracellular polypeptides that control cell growth and division, and that influence genome expression via the **JAK/STAT pathway**. This pathway is found in all vertebrates and related pathways are present in many invertebrates, including *Drosophila melanogaster* and *Caenorhabditis elegans*. In vertebrates, the receptors are members of the tyrosine kinase-associated family, each receptor protein being associated with an internal **Janus kinase** (**JAK**). Cytokine binding induces dimerization of the receptor, moving its pair of JAKs close enough together to phosphorylate one another (**Figure 14.6**). Phosphorylation activates the JAKs, which now phosphorylate transcription factors called STATs (signal transducers and activators of transcription). Phosphorylation causes pairs of STATs to form dimers which move to the nucleus, where they activate expression of a variety of genes.

Seven STATs have so far been identified in mammals, called STAT1, STAT2, STAT3, STAT4, STAT5A, STAT5B, and STAT6. Each of these except STAT2 can form homodimers, and heterodimers can form between STAT1 and STAT2, STAT1, and STAT3, and STAT5A and STAT5B. The composition of the dimers that are formed depends on the identity of the extracellular compound, and also its concentration. With just nine possible dimers, the JAK/STAT system might appear to have limited flexibility, but the expectation that a particular dimer will always activate the same set of genes turns out to be incorrect. Interleukin-6 and interleukin-10, for example, have opposite effects on genome expression in human myeloid cells, interleukin-6 inducing an inflammatory response and interleukin-10 suppressing inflammation, even though both exert their response by attaching to receptors that construct STAT3 dimers. How the same dimer can mediate different genome responses depending on the nature of the original signal is not yet understood.

The DNA-binding domain of a STAT protein is made up of three loops emerging from a barrel-shaped β-sheet. This structure, called the **immunoglobulin fold**, is found in many proteins but is not usually associated with DNA binding. Other proteins that use the immunoglobulin fold for this purpose include the transcription factors NK-κB and Rel. The similarities between these proteins refer only to the tertiary structures of the DNA-binding domains because STATs, NK-κB, and Rel, as a whole, have very little amino acid sequence identity. The consensus sequence of the DNA-binding sites for most STAT dimers was originally defined as 5′–TTN$_{5-6}$AA–3′, largely from studies in which purified STATs were tested for their ability to bind to oligonucleotides of known sequence. Variations in the surrounding nucleotides and in the internal series of Ns are thought to specify the actual binding sites for different types of dimers. This particular type of binding site is called the **IFN-γ stimulated gene response** (**GAS**) **element**. Some heterodimers attach to additional proteins, such as p48, and then recognize a different binding site, called the **IFN-stimulated response element** (**ISRE**), whose consensus sequence is 5′–AGTTTNNNTTTCC–3′. Chromatin immunoprecipitation sequencing (ChIP-seq; see Section 11.1) has revealed thousands of binding sites for each STAT dimer in the human genome, at positions adjacent to target genes and also in enhancers. However, many of these binding sites lack a GAS or ISRE sequence, suggesting that the association of STATs with the genome is more complicated than originally thought. This added complexity is underlined by studies that show that unphosphorylated STATs also influence genome expression under some conditions, and that the phosphorylated versions can form polymeric oligomers made up of more than two individual STATs. These observations may go some way toward explaining how the same STAT dimer can

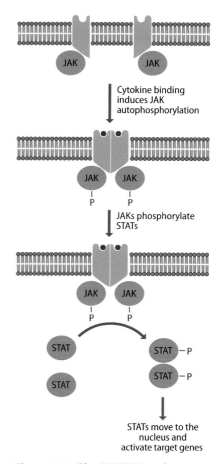

Cytokine binding induces JAK autophosphorylation

JAKs phosphorylate STATs

STATs move to the nucleus and activate target genes

Figure 14.6 The JAK/STAT pathway.

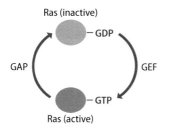

Figure 14.7 Activation and inactivation of Ras. Ras is a G-protein and is inactive when bound to a molecule of GDP. A guanine nucleotide exchange factor can replace the GDP with GTP, activating Ras. A GTPase activating protein has the opposite effect, stimulating Ras to convert the GTP and GDP and hence becoming inactive.

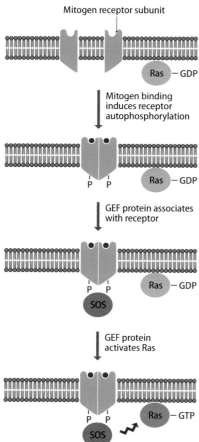

Figure 14.8 Activation of Ras at the start of the MAP kinase pathway. Binding of the mitogen results in receptor dimerization and autophosphorylation. A guanine nucleotide exchange factor, such as the SOS protein associates with the receptor and activates Ras by replacing its GDP molecule with a GTP.

have different effects depending on the identity of the signaling compound, as described above for interleukin-6 and interleukin-10.

Some signal transduction pathways have many steps between receptor and genome

The relative simplicity of the JAK/STAT pathway contrasts with the more prevalent forms of signal transduction, in which the receptor represents just the first in a series of steps that eventually lead to the activation of one or more transcription factors. A number of these **cascade** pathways have been delineated in different organisms, the most important being the **MAP** (mitogen-activated protein) **kinase** or **MAPK/ERK pathway**.

The MAP kinase pathway responds to many extracellular signals, including mitogens, which are compounds that have similar effects to cytokines but more specifically stimulate cell division. The initial steps in the pathway involve the Ras proteins, three of which are known in mammalian cells (H-, K-, and N-Ras). Ras is a **G-protein**, a type of small protein that binds either a molecule of GDP or GTP. When GDP is bound, Ras is inactive, but if the GDP is replaced with GTP, Ras becomes activated. Which nucleotide is bound to Ras depends on the balance between the activities of **GEFs** (**guanine nucleotide exchange factors**), which activate Ras by replacing the GDP with GTP, and **GAPs** (**GTPase activating proteins**) which inactivate Ras by stimulating Ras to convert its bound GTP to GDP (**Figure 14.7**).

How exactly does the presence of the extracellular signal result in Ras activation or inactivation? Ras is attached to the inner surface of the cell membrane in the vicinity of the mitogen receptor. Binding of the signaling compound results in dimerization of the receptor and mutual phosphorylation of the internal parts of its two subunits (**Figure 14.8**). GEFs such as the SOS protein bind to the phosphorylated receptor and activate Ras by replacing its GDP with GTP. The phosphorylated receptor might also recruit GAPs such as RasGAP, which have the opposite effect on Ras, deactivating it by stimulating the conversion of GTP to GDP. The balance between GEF and GAP activity therefore determines the nucleotide status of Ras and modulates further transmission of the signal along the transduction pathway.

Ras, when bound to GTP, is able to activate Raf, the next protein in the MAP kinase pathway. Raf is itself a kinase, but is usually inactive because the protein is folded in such a way that the N-terminal region of its polypeptide blocks access to the catalytic site. The Ras-GTP complex is able to bind to Raf, causing a conformational change in the latter that unblocks the active site. Raf then initiates a cascade of phosphorylation reactions (**Figure 14.9**). It phosphorylates Mek, activating this protein so that it, in turn, phosphorylates a MAP kinase. The activated MAP kinase now moves into the nucleus where it activates, again by phosphorylation, a series of transcription factors. The MAP kinase also phosphorylates another protein kinase, this one called Rsk, which phosphorylates and activates a second set of factors. Additional flexibility is provided by the possibility of replacing one or more of the proteins in the MAP kinase pathway with related proteins, ones with slightly different specificities, and so switching on another suite of transcription activators.

The MAP kinase pathway is used by vertebrate cells, and equivalent pathways, using intermediates similar to those identified in mammals, are known in other organisms. Each of the steps in these cascade pathways involves a physical interaction between two proteins, often resulting in the downstream member of the pair becoming phosphorylated. Phosphorylation activates the downstream protein, enabling it to form a connection with the next protein in the cascade. These interactions involve special protein–protein binding domains, such as the ones called SH2 and SH3, which bind to receptor domains on their partner proteins. The receptor domains contain one or more tyrosines, which must be phosphorylated in order for docking to take place. Hence the upstream protein contains the receptor domain, whose phosphorylation status determines whether the protein can bind its downstream partner and thereby propagate the signal (**Figure 14.10**).

Some signal transduction pathways operate via second messengers

Some signal transduction cascades do not involve the direct transfer of the external signal to the genome but instead utilize an indirect means of influencing transcription. These pathways make use of **second messengers**, which are less specific internal signaling compounds that transduce the signal from a cell surface receptor in several directions so that a variety of cellular activities, not just transcription, respond to the one signal.

We are already familiar with one class of second messenger, the cyclic nucleotides such as cAMP, which also play important signaling roles in bacteria (Section 12.3). Some cell surface receptors in eukaryotic cells have guanylate cyclase activity, and so convert GTP to cGMP, but most receptors in this family work indirectly by influencing the activity of cytoplasmic cyclases and decyclases. These cyclases and decyclases determine the cellular levels of cAMP and cGMP, which in turn control the activities of various target enzymes. The latter include protein kinase A, which is stimulated by cAMP. One of the functions of protein kinase A is to phosphorylate, and hence activate the transcription factor called CREB. As well as activating its specific target genes, CREB stimulates the activity of a variety of other genes by interacting with a second protein, p300, to form a complex, called p300/CBP, which is able to modify histone proteins and so affect chromatin structure (Section 10.2).

As well as being activated indirectly by cAMP, p300/CBP responds to another second messenger, calcium. The calcium ion concentration within the lumen of the endoplasmic reticulum is higher than that in the rest of the cell, so proteins that open calcium channels in the endoplasmic reticulum allow calcium ions to flow into the cytoplasm. This can be induced by extracellular signals that activate tyrosine kinase receptors which in turn activate phospholipases that cleave phosphatidylinositol-4,5-bisphosphate (PtdIns(4,5)P_2), a lipid component of the inner cell membrane, into inositol-1,4,5-trisphosphate (Ins(1,4,5)P_3) and 1,2-diacylglycerol (DAG). Ins(1,4,5)P_3 opens the calcium channels in the endoplasmic reticulum (**Figure 14.11**). Ins(1,4,5)P_3 and DAG are themselves second messengers that can initiate other signal transduction cascades. Both the calcium- and lipid-induced cascades target transcription factors, but only indirectly: the primary targets are other proteins. Calcium, for example, binds to and activates the protein called calmodulin, which regulates a variety of enzymes, including protein kinases, ATPases, phosphatases, and nucleotide cyclases.

14.2 CHANGES IN GENOME ACTIVITY RESULTING IN CELLULAR DIFFERENTIATION

Transient changes in genome activity are, by definition, readily reversible, the genome expression pattern reverting to its original state when the external stimulus is removed. In contrast, the permanent and semi-permanent changes in genome activity that underlie cellular differentiation must persist for long periods, and ideally should be maintained even when the stimulus that originally induced them has disappeared. We therefore anticipate that the regulatory mechanisms bringing about these longer-term changes will involve systems additional to the modulation of transcription factors. This expectation is correct; the mechanisms that result in differentiation include changes in chromatin structure and physical rearrangement of the genome.

Some differentiation processes involve changes to chromatin structure

We studied some of the effects that chromatin structure can have on genome expression in Section 10.2. These effects range from the modulation of transcription initiation at an individual promoter by nucleosome repositioning, through

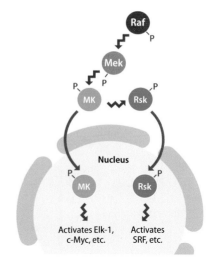

Figure 14.9 The second part of the MAP kinase pathway. 'M' is the MAP kinase. Elk-1, c-Myc, and SRF (serum response factor) are examples of transcription factors activated at the end of the pathway.

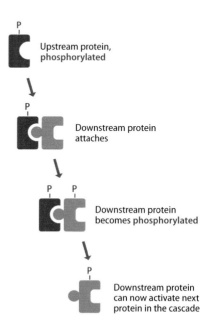

Figure 14.10 A scheme for the interaction of proteins in a signaling cascade. The upstream protein is phosphorylated and hence able to bind its downstream partner. Binding leads to phosphorylation of the receptor domain in the downstream protein, propagating the signal.

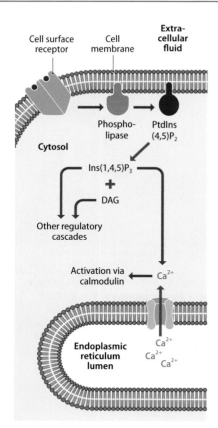

Figure 14.11 Induction of the calcium second messenger system. Abbreviations: DAG, 1,2-diacylglycerol; Ins(1,4,5)P$_3$, inositol-1,4,5- trisphosphate; PtdIns(4,5)P$_2$, phosphatidylinositol-4,5-bisphosphate.

to the silencing of large segments of DNA locked up in higher-order chromatin structures. The latter is an important means of bringing about long-term changes in genome activity and is implicated in a number of differentiation events. One example involves the **Polycomb group** (**PcG**) proteins, which were first discovered in *Drosophila melanogaster* but are now known to have homologs in other organisms, including mammals and plants. In *Drosophila*, Polycomb proteins recognize DNA sequences called **Polycomb response elements**, their attachment inducing localized formation of heterochromatin, the condensed form of chromatin that prevents transcription of the genes that it contains (**Figure 14.12**). Each response element is approximately 10 kb in length and contains multiple copies of the DNA-binding sites for PcG proteins such as Pleiohomeotic (PHO) and Pleiohomeotic-like (PHOL), which form a PhoRC complex that probably acts as the primary DNA recognition component of the Polycomb system (**Figure 14.13**). Once PhoRC is bound to the DNA, it recruits a second group of PcG proteins, called Polycomb Repressive Complex 2 (PRC2). One of the members of this complex, EZH2, is histone methyltransferase (Section 10.2) that trimethylates lysine 27 of histone H3, and possibly also methylates lysine 9. These are repressive histone modifications (see Table 10.1) and hence induce heterochromatin formation. The methylations are then recognized by Polycomb Repressive Complex 1 (PRC1), which contains a second histone modifying enzyme, this one adding a ubiquitin group to lysine 119 of Histone H2A, leading to further compaction of the heterochromatin, which propagates along the DNA for tens of kilobases in either direction. The EZH2 protein also attracts DNA methyltransferases into the bound PcG complex, so the genome region becomes further silenced by DNA methylation (Section 10.3).

In *Drosophila*, the regions that become silenced contain **homeotic selector genes**, which, as we will see in Section 14.3, specify the development of the individual body parts of the fly. As only one body part must be specified at a particular position in the fruit fly, it is important that a cell expresses only the correct homeotic gene. This is ensured by the action of the PcG proteins, which permanently silence the homeotic genes that must be switched off. PcG proteins do not, however, determine which genes will be silenced: expression of these genes has already been repressed before the proteins bind to their response elements. The role of the PcG proteins is therefore to *maintain* rather than *initiate* gene silencing. An important point is that the heterochromatin that is formed is heritable: after division, the two new cells retain the heterochromatin established in the parent cell. This type of regulation of genome activity is therefore permanent not only in a single cell, but also in a cell lineage.

The **trithorax group** (**trxG**) proteins act in a similar manner to the PcG proteins, but have the opposite effect: they maintain an open chromatin state in the regions of active genes, the targets including the same homeotic genes as those that are silenced, in different body parts, by PcG proteins. How the opposing actions of the PcG and trxG proteins are controlled is not yet known. The notion that there is a straightforward toggle switch, PcG switching genes off and trxG switching them on again, is probably overly simplistic. This would imply that PcG proteins would be associated with the target regions of the genome when those regions are repressed, and trxG proteins would be present when the regions are active. In fact, both groups of protein appear to be present all the

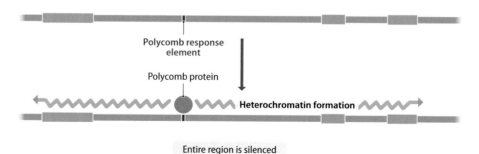

Figure 14.12 Polycomb group proteins maintain silencing in regions of the *Drosophila* genome by initiating heterochromatin formation.

time. One hypothesis is that the heterochromatin induced by the PcG proteins is the default state, and the role of trxG is to modify PcG activity in the region of those genes whose expression is needed in a particular cell.

Yeast mating types are determined by gene conversion events

A permanent change in the pattern of genome expression can also be brought about by changing the physical structure of the genome by DNA rearrangement or by **gene conversion**. A DNA rearrangement results in parts of the genome that were previously separate becoming linked to one another, and gene conversion involves deletion of one part of the genome and replacement with a copy of a segment from a different part of the genome (**Figure 14.14**). These events constitute an effective, if drastic, way of bringing about the permanent change in genome expression needed to maintain a differentiated state.

Yeast **mating types** provide an example of the use of gene conversion to bring about cellular differentiation. Mating types in yeasts and other eukaryotic microorganisms are the equivalent to the male and female sexes in higher eukaryotes and enable eukaryotic microorganisms to engage in sexual reproduction. However, these organisms reproduce mainly by vegetative cell division, which means that it is possible that a population, being derived from just one or a few ancestral cells, will be largely or completely composed of a single mating type and so will not be able to reproduce sexually. In *Saccharomyces cerevisiae* and some other species this problem is avoided because cells are able to change sex by the process called **mating-type switching**.

The two *S. cerevisiae* mating types are called a and α. Both mating types secrete a short polypeptide pheromone (12 amino acids in length for a and 13 for α) that binds to receptors on the surfaces of cells of the opposite mating type. Binding of the pheromone initiates a MAP kinase signal transduction pathway that alters the genome expression profile within the cell, leading to subtle morphological and physiological changes that convert the cell into a gamete able to participate in sexual reproduction. Mixing two haploid strains of opposite mating type therefore stimulates formation of gametes which fuse to produce a diploid zygote. Meiosis occurs within the zygote, giving rise to a tetrad of four haploid **ascospores**, contained in a structure called an **ascus**. The ascus bursts open, releasing the ascospores, which then divide by mitosis to produce new haploid vegetative cells (**Figure 14.15**).

The mating type is specified by the *MAT* gene, located on chromosome III. This gene has two alleles, *MATa* and *MATα*, a haploid yeast cell displaying the mating type corresponding to whichever allele it possesses. Elsewhere on chromosome III are two additional MAT-like genes, called *HMRa* and *HMLα*. These have the same sequences as *MATa* and *MATα*, respectively, but neither gene is expressed because upstream of each one is a silencer that represses transcription initiation. These two genes are called 'silent mating-type cassettes.' Their silencing involves the Sir proteins, several of which have histone deacetylase activity (Section 10.2), indicating that silencing involves changes in the chromatin structure in the region of *HMRa* and *HMLα*.

Mating-type switching is initiated by the HO endonuclease, which makes a double-strand cut at a 24 bp sequence located within the *MAT* gene (**Figure 14.16**). This enables the gene conversion event to take place. We will examine the details of gene conversion in Section 16.1; all that concerns us at the moment is that one of the free 3′ ends produced by the endonuclease can be extended by DNA synthesis, using one of the two silent cassettes as the template.

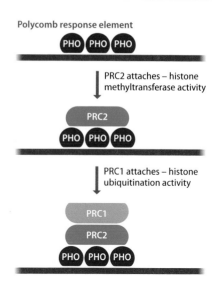

Figure 14.13 Assembly of Polycomb group proteins at a Polycomb response element in *Drosophila*. The element is recognized by PHO and PHOL proteins, which form the platform for attachment of Polycomb Repressive Complexes 1 and 2 (PRC1 and PRC2). Humans and other vertebrates have equivalent complexes to PRC1 and PRC2 but appear to lack clearly defined response elements, and how the Polycomb group proteins recognize their binding sites in these organisms is not understood.

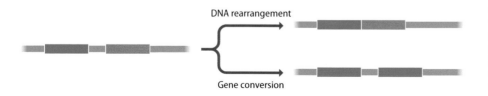

Figure 14.14 DNA rearrangement and gene conversion. In this simple example, the DNA rearrangement results in the relative positions of the orange and blue genes changing so that they become linked to one another, and the gene conversion event replaces the orange gene with a second copy of the blue gene.

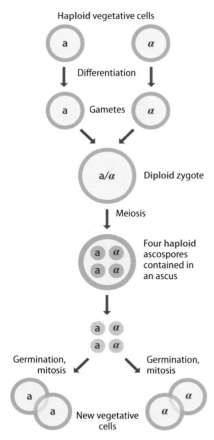

Figure 14.15 The life cycle of the yeast *Saccharomyces cerevisiae*.

The newly synthesized DNA subsequently replaces the DNA currently at the *MAT* locus. The silent cassette chosen as the template is usually the one that is different to the allele originally at *MAT*, so replacement with the newly synthesized strand converts the *MAT* gene from *MATa* to *MATα*, or vice versa. This results in mating-type switching.

The *MAT* genes code for regulatory proteins (one in the case of *MATa* and two for *MATα*) that interact with a transcription activator, MCM1. The *MATa* and *MATα* gene products have different effects on MCM1, and so specify different genome expression patterns. These expression patterns are maintained in a semi-permanent fashion until another *MAT* gene conversion occurs.

Genome rearrangements are responsible for immunoglobulin and T-cell receptor diversities

In vertebrates, there are two striking examples of the use of DNA rearrangements to achieve permanent changes in genome activity. These two examples, which are very similar, are responsible for the generation of immunoglobulin and T-cell receptor diversities.

Immunoglobulins and T-cell receptors are related proteins that are synthesized by B and T lymphocytes, respectively. Both types of protein become attached to the outer surfaces of their cells, and immunoglobulins are also released into the bloodstream. The proteins help to protect the body against invasion by bacteria, viruses, and other unwanted substances by binding to these **antigens**, as they are called. During its lifetime, an organism could be exposed to any number of a vast range of antigens, which means that the immune system must be able to synthesize an equally vast range of immunoglobulin and T-cell receptor proteins. In fact, humans can make approximately 10^8 different immunoglobulin and T-cell receptor proteins. But there are only 20,442 protein-coding genes in the human genome, so where do all these immunoglobulins and T-cell receptor proteins come from?

To understand the answer, we will look at the structure of a typical immunoglobulin protein. Each immunoglobulin is a tetramer of four polypeptides linked by disulfide bonds (**Figure 14.17**). There are two long 'heavy' chains and two short 'light' chains. When the sequences of different heavy chains are compared, it becomes clear that the variability between them lies mainly in the N-terminal regions of these polypeptides, the C-terminal parts being very similar, or 'constant,' in all heavy chains. The same is true for the light chains, except that two families, κ and λ, can be distinguished, differing in the sequences of their constant regions.

In vertebrate genomes, there are no complete genes for the immunoglobulin heavy and light polypeptides. Instead, these proteins are specified by gene segments. The heavy-chain segments are located within a 1 Mb region of chromosome 14 and comprise up to 11 constant region (C_H) gene segments, preceded by 123–129 V_H gene segments, 27 D_H gene segments, and 9 J_H gene segments, these

Figure 14.16 Mating-type switching in yeast. In this example, the cell begins as mating type a. The HO endonuclease cuts the *MATa* locus, initiating gene conversion by the *HMLα* locus. The result is that the mating type switches to type α.

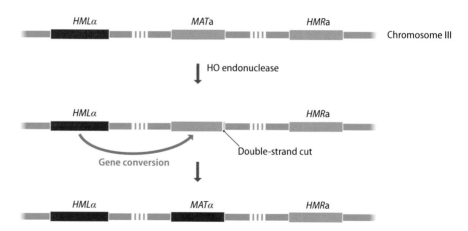

last three types coding for different versions of the V (variable), D (diverse), and J (joining) components of the variable part of the heavy chain (**Figure 14.18**). A similar arrangement is seen with the light-chain loci on chromosomes 2 (κ locus) and 22 (λ locus), the only difference being that the light chains do not have D segments.

During the early stage of B lymphocyte development, the immunoglobulin loci in its genome undergo rearrangements. Within the heavy-chain locus, these rearrangements link one of the V_H gene segments with one of the D_H gene segments, and then link this V–D combination with a J_H gene segment (**Figure 14.19**). These rearrangements occur by an unusual type of recombination, catalyzed by a pair of proteins called RAG1 and RAG2, with the positions at which the breakage and reunion reactions occur in order to link the gene segments marked by a series of 8 bp and 9 bp consensus sequences. The end result is an exon that contains the complete open reading frame specifying the V_H, D_H, and J_H segments of the immunoglobulin protein. This exon becomes linked to a C_H segment exon by splicing during the transcription process, creating a complete heavy-chain mRNA that can be translated into an immunoglobulin protein that is specific for just that one lymphocyte. A similar series of DNA rearrangements results in the lymphocyte's light-chain V–J exon being constructed at either the κ or λ locus, and once again splicing attaches a light-chain C segment exon when the mRNA is synthesized.

Despite its name, the constant region is not identical in every immunoglobulin protein. The small variations that occur result in five different classes of immunoglobulin – IgA, IgD, IgE, IgG, and IgM – each with its own specialized role in the immune system. Initially, each B lymphocyte synthesizes an IgM molecule, the C_H segment of which is specified by the Cμ sequence that lies at the 5′ end of the C_H segment cluster. As shown in Figure 14.19, later in its development the immature cell might also synthesize some IgD proteins, utilizing the second C_H sequence in the cluster (Cδ), the exon for this sequence becoming attached to the V–D–J segment by alternative splicing. Later in their lifetimes, when they have reached maturity, some B lymphocytes undergo **class switching**, which again results in a change in the type of immunoglobulin that the lymphocyte synthesizes. This second class switching requires a further recombination event that deletes the Cμ and Cδ sequence, along with the part of the chromosome between this region and the start of the C_H segment specifying the class of immunoglobulin that the lymphocyte will now synthesize. For example, for the lymphocyte to switch to synthesis of IgG, the most prevalent type of immunoglobulin made by mature lymphocytes, the deletion must place one of the Cγ segments, which specify the IgG heavy chain, at the 5′ end of the cluster (**Figure 14.20**). Class switching is therefore a second example of genome rearrangement occurring during B lymphocyte development. The mechanism is distinct from V–D–J joining and the recombination event does not involve the RAG proteins.

Diversity of T-cell receptors is based on similar rearrangements which link V, D, J, and C gene segments in different combinations to produce cell-specific genes. Each receptor comprises a pair of β molecules, which are similar to the immunoglobulin heavy chain, and two α molecules, which resemble the immunoglobulin κ light chains. The T-cell receptors become embedded in the cell membrane and enable the lymphocyte to recognize and respond to extracellular antigens.

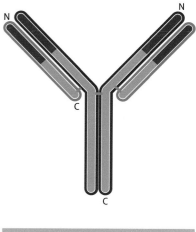

KEY
Variable region   Heavy chain
Constant region   Light chain
Disulfide bond

Figure 14.17 Immunoglobulin structure. Each immunoglobulin protein is made up of two heavy and two light chains, linked by disulfide bonds. Each heavy chain is 446 amino acids in length and consists of a variable region (shown in pink) spanning amino acids 1–108 followed by a constant region. Each light chain is 214 amino acids, again with an N-terminal variable region of 108 amino acids. Additional disulfide bonds form between different parts of individual chains: these and other interactions fold the protein into a more complex three-dimensional structure.

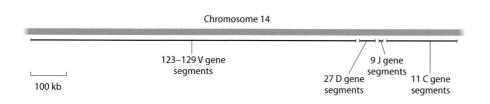

Chromosome 14

123–129 V gene segments

9 J gene segments

27 D gene segments

11 C gene segments

100 kb

Figure 14.18 Organization of the human *IGH* (heavy-chain) locus on chromosome 14. The numbers given refer to the recognizable gene segments at this locus. Some of these segments are likely to be pseudogenes. The active segment numbers are estimated at 38–46 V_H, 23 D_H, 6 J_H, and 5–11 C_H.

Figure 14.19 Synthesis of a specific immunoglobulin protein. DNA rearrangement within the heavy-chain locus links V, D, and J segments, which are then linked to a C segment by splicing of the mRNA. In immature B cells, the V–D–J exon always becomes linked to the $C\mu$ exon (exon 2) to produce an mRNA specifying a class M immunoglobulin. At an early stage in development of the B cell, some immunoglobulin D proteins are also produced by alternative splicing that links the V–D–J exon to the $C\delta$ exon.

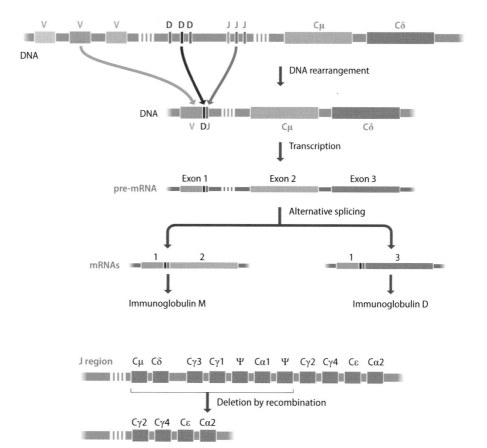

Figure 14.20 Immunoglobulin class switching. In this example, seven C_H segments are deleted, placing the $C\gamma2$ segment adjacent to the J region. This B cell will therefore synthesize an immunoglobulin G molecule, which is secreted by the cell. The two segments labeled ψ are pseudogenes.

14.3 CHANGES IN GENOME ACTIVITY UNDERLYING DEVELOPMENT

The developmental pathway of a multicellular eukaryote begins with a fertilized egg cell and ends with an adult form of the organism. In between lies a complex series of genetic, cellular, and physiological events that must occur in the correct order, in the correct cells, and at the appropriate times if the pathway is to reach a successful culmination. With humans, this developmental pathway results in an adult containing 10^{13} cells differentiated into approximately 3000 specialized types, the activity of each individual cell coordinated with that of every other cell. Developmental processes of such complexity might appear intractable, even to the powerful investigative tools of modern molecular biology, but remarkably good progress toward understanding them has been made in recent years. The research that has underpinned this progress has been designed around three guiding principles:

- It should be possible to describe and comprehend the genetic and biochemical events that underlie differentiation of individual cell types. This in turn means that an understanding of how specialized tissues, and even complex body parts, are constructed should be within reach.

- The signaling processes that coordinate events in different cells should be amenable to study. We saw in Section 14.1 that progress has been made in describing these systems at the molecular level.

- There should be similarities and parallels between developmental processes in different organisms, reflecting common evolutionary origins. This means that information relevant to human development can be obtained from studies of model organisms chosen for the relative simplicity of their developmental pathways.

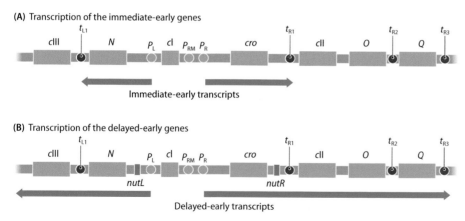

(A) Transcription of the immediate-early genes

Immediate-early transcripts

(B) Transcription of the delayed-early genes

Delayed-early transcripts

Figure 14.21 Transcription of the immediate-early and delayed-early λ genes. (A) Transcription from the promoters P_L and P_R initially results in synthesis of two immediate-early mRNAs, these terminating at positions t_{L1} and t_{R1}. (B) The mRNA transcribed from P_L to t_{L1} codes for the N protein, which attaches to the antitermination sites *nutL* and *nutR*. Now the RNA polymerase continues transcription downstream of t_{L1} and t_{R1}. Transcription from P_R also ignores terminator t_{R2} and continues until t_{R3} is reached.

Developmental biology encompasses areas of genetics, molecular biology, cell biology, physiology, biochemistry, and systems biology. We are concerned only with the role of the genome in development and so will not attempt a wide-ranging overview of developmental research in all its guises. Instead, we will concentrate on four model systems of increasing complexity in order to investigate the types of change in genome activity that occur during development.

Bacteriophage λ: a genetic switch enables a choice to be made between alternative developmental pathways

A bacteriophage that infects *Escherichia coli* might seem an odd place to begin a study of the role of the genome in development. But this is exactly where molecular biologists began the lengthy program of research that today is revealing the underlying genomic basis to development in humans and other vertebrates. We will therefore follow this same progression from the relatively simple to the relatively complex.

In Section 9.1, we learned that lysogenic bacteriophages such as λ can follow two alternative replication pathways after infection of a host cell. As well as the lytic pathway, during which new phages are assembled and released from the cell soon after the initial infection (after 45 minutes for λ), these phages can also pursue a lysogenic cycle characterized by insertion of the phage DNA into the host chromosome. The integrated prophage remains quiescent for many bacterial generations until a chemical or physical stimulus linked to DNA damage induces excision of the λ genome, rapid assembly of phages, and lysis of the host cell (see **Figures 9.3** and **9.4**). How does the phage 'decide' whether to follow the lytic or lysogenic cycle? To answer this question we first need to understand the series of genetic events that underlie lysogeny.

The λ genome contains two promoters, P_L and P_R, which are recognized by the *E. coli* RNA polymerase as soon as λ injects its DNA into the cell. The polymerase therefore transcribes the two 'immediate-early' λ genes, called N and *cro* (**Figure 14.21**). The product of gene N is an **antiterminator protein**, which attaches to the DNA and causes the host RNA polymerase to ignore the termination signals that it encounters immediately downstream of the N and *cro* coding sequences. The polymerase therefore transcribes the 'delayed-early' genes. These genes include *c*II and *c*III, which together activate transcription of *c*I. This is an important gene as it codes for the λ repressor protein, the key master switch that shuts down the lytic cycle and maintains lysogeny. The repressor does this by binding to the operators O_L and O_R, adjacent to P_L and P_R, respectively (**Figure 14.22**). As a result almost the entire λ genome is silenced, because P_L and P_R direct transcription not only of the immediate-early and delayed-early genes, but also of the late genes, which code for the proteins needed for assembly of new phages and host cell lysis. One of the few genes to remain active is *int*, which is transcribed from its own promoter. The integrase protein coded by this gene catalyzes the site-specific recombination, which inserts the λ DNA into the host

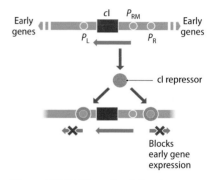

Figure 14.22 The role of the cI repressor. The cI repressor attaches to the operators O_L and O_R and blocks transcription from promoters P_L and P_R.

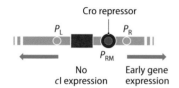

Figure 14.23 The role of the Cro repressor. The Cro repressor prevents transcription of *cI*, so transcription of the early genes is no longer repressed.

genome. Lysogeny is maintained for numerous cell divisions because the *cI* gene is continuously expressed, albeit at a low level, so that the amount of cI repressor present in the cell is always enough to keep P_L and P_R switched off. This continued expression of *cI* occurs because the cI repressor, when bound to O_R, not only blocks transcription from P_R, but also stimulates transcription from its own promoter P_{RM}. The dual role of the cI repressor is therefore the key to lysogeny.

Once *cI* is expressed, the repressor protein prevents entry into the lytic cycle and ensures that lysogeny is set up and maintained. But λ does not always enter the lysogenic cycle – on some occasions an infection proceeds immediately to host lysis. This is because of the activity of the second immediate-early gene, *cro*, which also codes for a repressor, but in this case one that prevents transcription of *cI* (**Figure 14.23**). The decision between lysis and lysogeny is therefore determined by the outcome of a race between *cI* and *cro*. If the cI repressor is synthesized more quickly than the Cro repressor, then genome expression is blocked and lysogeny follows. However, if *cro* wins the race, then the Cro repressor blocks *cI* expression before enough cI repressor has been synthesized to silence the genome. As a result, the phage enters the lytic infection cycle. The decision appears to be random, depending on chance events that lead to either the cI or the Cro repressor accumulating more quickly in the cell, although environmental conditions can have an influence. Growth on a rich medium, for example, shifts the balance toward the lytic cycle, presumably because it is beneficial to produce new phages when the host cells are proliferating. This shift is brought about by activation of proteases that degrade the cII protein, reducing the ability of the cII–cIII combination to switch on transcription of the cI repressor gene.

If the bacteriophage enters the lysogenic cycle, then this state is maintained as long as the cI repressor is bound to the operators O_L and O_R. The prophage will therefore be induced if the level of active cI repressor declines below a certain point. This may happen by chance, leading to spontaneous induction, or may occur in response to physical or chemical stimuli. These stimuli activate a general protective mechanism in *E. coli*, the **SOS response**. Part of the SOS response is expression of an *E. coli* gene, *recA*, whose product inactivates the cI repressor by cleaving it in half. This switches on expression of the early genes, enabling the phage to enter the lytic cycle. Inactivation of the cI repressor also means that transcription of *cI* is no longer stimulated, avoiding the possibility of lysogeny being reestablished through the synthesis of more cI repressor. Inactivation of the cI repressor therefore leads to induction of the prophage.

What do we learn from this model system?

- A simple genetic switch can determine which of two developmental pathways is followed by a cell.

- Genetic switches can involve a combination of activation and repression of different promoters.

- It is possible to reprogram a developmental pathway, and transfer it to an alternative pathway, in response to appropriate stimuli.

Bacillus sporulation: coordination of activities in two distinct cell types

The second system that we will examine is the formation of spores by the bacterium *Bacillus subtilis*. As with λ lysogeny, this is not, strictly speaking, a developmental pathway, merely a type of cellular differentiation, but the process illustrates two of the fundamental issues that have to be addressed when genuine development in multicellular organisms is studied. These issues are how a series of changes in genome activity over time is controlled, and how signaling establishes coordination between events occurring in different cells. The advantages of *Bacillus* as a model system are that it is easy to grow in the laboratory and is amenable to study by genetic and molecular biology techniques such as analysis of mutants and sequencing of its genome.

Bacillus is one of several genera of bacteria that produce endospores in response to unfavorable environmental conditions. These spores are highly resistant to physical and chemical abuse and can survive for decades or even centuries. Resistance is due to the specialized nature of the spore coat, which is impermeable to many chemicals, and to biochemical changes within the spore that retard the breakdown of DNA and other polymers and enable the spore to survive a prolonged period of dormancy.

In the laboratory, sporulation is usually induced by nutrient starvation. This causes the bacteria to abandon their normal vegetative mode of cell division, which involves synthesis of a septum (or cross-wall) in the center of the cell. Instead, the cells construct an unusual septum, one that is thinner than normal, at one end of the cell (**Figure 14.24**). This produces two cellular compartments, the smaller of which is called the prespore and the larger the mother cell. As sporulation proceeds, the prespore becomes entirely engulfed by the mother cell. By now, the two cells are committed to different but coordinated differentiation pathways, the prespore undergoing the biochemical changes that enable it to become dormant, and the mother cell constructing the resistant coat around the spore and eventually dying.

Changes in genome activity during sporulation are controlled largely by the synthesis of special σ subunits that change the promoter specificity of the *Bacillus* RNA polymerase (Section 12.3). We have seen how this simple control

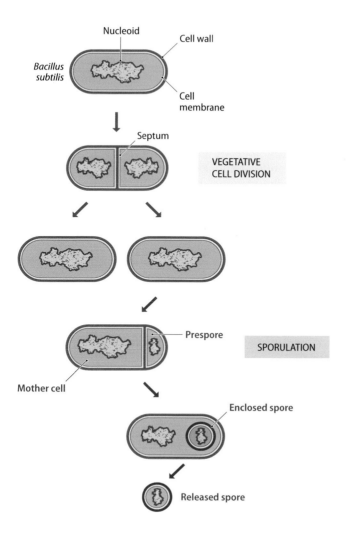

Figure 14.24 Sporulation in *Bacillus subtilis*. The top part of the diagram shows the normal vegetative mode of cell division, involving formation of a septum across the center of the bacterium and resulting in two identical daughter cells. The lower part of the diagram shows sporulation, in which the septum forms near one end of the cell, leading to a mother cell and prespore of different sizes. Eventually, the mother cell completely engulfs the prespore. At the end of the process, the mature, resistant spore is released.

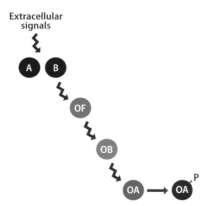

Figure 14.25 The phosphorylation cascade that leads to activation of SpoOA. Abbreviations: A, KinA; B, KinB; OF, SpoOF; OB, SpoOB; OA, SpoOA;

system is used by *E. coli* in response to heat stress (see **Figure 12.18**). It is also the key to the changes in genome activity that occur during sporulation. The standard *B. subtilis* subunits are called σ^A and σ^H. These subunits are synthesized in vegetative cells and enable the RNA polymerase to recognize promoters for all the genes it needs to transcribe in order to maintain normal growth and cell division. In the prespore and mother cell, these subunits are replaced by σ^F and σ^E, respectively, which recognize different promoter sequences and so give rise to large-scale changes in genome expression patterns. The master switch from vegetative growth to spore formation is provided by a protein called SpoOA, which is present in vegetative cells but in an inactive form. This protein is activated by phosphorylation, via a cascade of protein kinases that respond to various extracellular signals that indicate the presence of an environmental stress, such as lack of nutrients. The initial response is provided by two kinases, called KinA and KinB, which phosphorylate themselves and then pass the phosphate via SpoOF and SpoOB to SpoOA (**Figure 14.25**). Activated SpoOA is a transcription factor that changes the expression of various genes transcribed by the vegetative RNA polymerase and hence recognized by the regular σ^A and σ^H subunits. The genes that are switched on include those for σ^F and σ^E, resulting in the switch to prespore and mother-cell differentiation (**Figure 14.26**).

Initially, both σ^F and σ^E are present in both of the two differentiating cells. This is not exactly what is wanted because σ^F is the prespore-specific subunit and so should be active only in this cell, and σ^E is mother-cell specific. A means is therefore needed to activate or inactivate the appropriate subunit in the correct cell. This is thought to be achieved as follows (**Figure 14.27**):

- σ^F is activated by release from a complex with a second protein, SpoIIAB. This is controlled by a third protein, SpoIIAA, which, when unphosphorylated, can also attach to SpoIIAB and prevent the latter from binding to σ^F. If SpoIIAA is unphosphorylated, then σ^F is released and is active; when SpoIIAA is phosphorylated, σ^F remains bound to SpoIIAB and so is inactive. In the mother cell, SpoIIAB phosphorylates SpoIIAA and so keeps σ^F in its bound, inactive state. But in the prespore, SpoIIAB's attempts to phosphorylate SpoIIAA are antagonized by yet another protein, SpoIIE, and so σ^F is released and becomes active. SpoIIE's ability to antagonize SpoIIAB in the prespore but not the mother cell derives from the fact that SpoIIE molecules are bound to the membrane on the surface of the septum. Because the prespore is much smaller than the mother cell, but the septum surface area is similar in both, the concentration of SpoIIE is greater in the prespore, and this enables it to antagonize SpoIIAB.

- σ^E is activated by proteolytic cleavage of a precursor protein. The protease that carries out this cleavage is the SpoIIGA protein, which spans the septum between the prespore and mother cell, with the protease domain on the mother cell side. Activation of SpoIIGA requires the presence of σ^F in the prespore, but the precise nature of this link is not clear. One possibility is that SpoIIGA is activated by binding of SpoIIR to a receptor domain on the prespore side. The gene for SpoIIR is one of those whose promoter is recognized specifically by σ^F, so activation of the protease, and conversion of pre-σ^E to active σ^E, occurs once σ^F-directed transcription is underway in the prespore. If this model is correct, then it would constitute a typical receptor-mediated signal transduction system (Section 14.1).

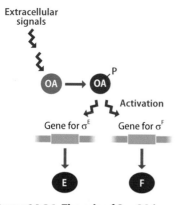

Figure 14.26 The role of SpoOA in *Bacillus* sporulation. SpoOA is phosphorylated in response to extracellular signals derived from environmental stresses, as shown in Figure 14.25. It is a transcription activator with roles that include activation of the genes for the σ^E and σ^F RNA polymerase subunits. Abbreviations: E, σ^E; F, σ^F; OA, SpoOA.

Activation of σ^F and σ^E is just the beginning of the story. In the prespore, about one hour after its activation, σ^F responds to an unknown signal (possibly from the mother cell) which results in a slight change in genome activity in the spore. This includes transcription of a gene for another σ subunit, σ^G, which recognizes promoters upstream of genes whose products are required during the later stages of spore differentiation. One of these proteins is SpoIVB, which activates another septum-bound protease, SpoIVF (**Figure 14.28**). This protease then activates a second mother-cell σ subunit, σ^K, which is coded by

Figure 14.27 Activation of the prespore- and mother-cell-specific σ subunits during *Bacillus* **sporulation.** (A) In the mother cell, σᶠ is inactive because it is bound to SpoIIAB, which phosphorylates SpoIIAA and prevents the latter from releasing σᶠ. Activation of σᶠ in the prespore occurs by its release from its complex with SpoIIAB, which is indirectly influenced by the concentration of membrane-bound SpoIIE. (B) In the mother cell, σᴱ is activated by proteolytic cleavage by SpoIIGA, possibly in response to the presence in the prespore of the σᶠ-dependent protein, SpoIIR. Abbreviations: AA, SpoIIAA; AB, SpoIIAB; E, σᴱ; F, σᶠ; GA, SpoIIGA; R, SpoIIR.

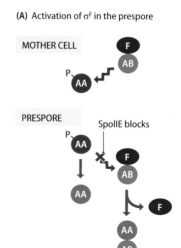

(A) Activation of σᶠ in the prespore

a σᴱ-transcribed gene but retained in the mother cell in an inactive form until the signal for its activation is received from the prespore. σᴷ directs transcription of the genes whose products are needed during the later stages of the mother-cell differentiation pathway.

To summarize, the key features of *Bacillus* sporulation are as follows:

- The master protein, SpoOA, responds to external stimuli via a cascade of phosphorylation events to determine if and when the switch to sporulation should occur.

- A succession of σ subunits in the prespore and mother cell brings about time-dependent changes in genome activity in the two cells.

- Cell–cell signaling ensures that the events occurring in the prespore and mother cell are coordinated.

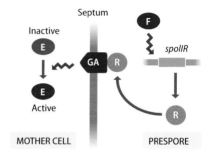

(B) Activation of σᴱ in the mother cell

Caenorhabditis elegans: the genetic basis to positional information and the determination of cell fate

Research with the microscopic nematode worm *C. elegans* (**Figure 14.29**) was initiated by Sydney Brenner in the 1960s with the aim of utilizing it as a simple model for multicellular eukaryotic development. *C. elegans* is easy to grow in the laboratory and has a short generation time, taking just three and a half days for the fertilized egg to develop into a mature adult. The worm is transparent at all stages of its life cycle, so internal examination is possible without killing the animal. This is an important point because it has enabled researchers to follow the entire developmental process of the worm at the cellular level. Every cell division in the pathway from fertilized egg to adult worm has been charted, and every point at which a cell adopts a specialized role has been identified. This pathway is more or less invariant: the pattern of cell division and differentiation is virtually the same in every individual. This appears to be due in large part to cell–cell signaling, which induces each cell to follow its appropriate differentiation pathway. To illustrate this, we will look at development of the *C. elegans* vulva.

Most *C. elegans* worms are hermaphrodites, meaning that they have both male and female sex organs. The vulva is part of the female sex apparatus, being the tube through which sperm enter and fertilized eggs are laid. The adult vulva comprises 22 cells, which are the progeny of three ancestral cells originally located in a row on the undersurface of the developing worm (**Figure 14.30**). Each of these ancestral cells becomes committed to the differentiation pathway that leads to the production of vulva cells. The central cell, called P6.p, adopts the 'primary vulva cell fate' and divides to produce eight new cells. The other two cells – P5.p and P7.p – take on the 'secondary vulva cell fate' and divide into seven cells each. These 22 cells then reorganize their positions to construct the vulva.

A critical aspect of vulva development is that it must occur in the correct position relative to the gonad, the structure containing the egg cells. If the vulva develops in the wrong place, then the gonad will not receive sperm and the egg cells will never be fertilized. The positional information needed by the vulva progenitor cells is provided by a cell within the gonad called the anchor cell. The importance of the anchor cell has been demonstrated by experiments in which it is artificially destroyed in the embryonic worm: in the absence of the anchor cell, a vulva does not develop. The implication is that the anchor cell secretes an

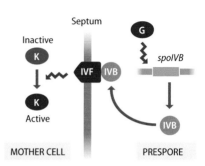

Figure 14.28 Activation of σᴷ during *Bacillus* **sporulation.** Note that the scheme is very similar to the procedure used to activate σᴱ (see Figure 14.27B). Abbreviations: G, σᴳ; K, σᴷ; IVB, SpoIVB; IVF, SpoIVF.

Figure 14.29 The nematode worm *Caenorhabditis elegans*. The micrograph shows an adult hermaphrodite worm, approximately 1 mm in length. The vulva is the small projection located on the underside of the animal, about halfway along. Egg cells can be seen inside the worm's body in the region on either side of the vulva. (From Kendrew J [ed] [1994] *Encyclopedia of Molecular Biology*. With permission from John Wiley and Sons, Inc.)

extracellular signaling compound that induces P5.p, P6.p, and P7.p to differentiate. This signaling compound is the protein called LIN-3, coded by the *lin-3* gene.

Why does P6.p adopt the primary cell fate, whereas P5.p and P7.p take on secondary cell fates? This appears to result from a combination of two different cell–cell signaling systems. The first involves LIN-3, which forms a concentration gradient and therefore has different effects on P6.p, the cell which is closest to the anchor cell, and the more distant P5.p and P7.p, as shown in **Figure 14.31**. The cell surface receptor for LIN-3 is a tyrosine kinase called LET-23, which, when activated by binding LIN-3, initiates a series of intracellular reactions that leads to activation of a MAP kinase-like protein, which in turn switches on a variety of transcription factors. The identity of this MAP kinase, and hence of the transcription factors that are switched on, depends on the number of LET-23 receptors that are activated, which in turn depends on the extracellular LIN-3 concentration, explaining how the primary and secondary fates can be specified by the distance of the recipient cells from the anchor cell.

In the second signaling system, P6.p, having adopted the primary cell fate, synthesizes DSL proteins, some of which become embedded in the cell membrane and some secreted. Both the embedded and secreted versions are able to interact with LIN-12 receptor proteins on P5.p and P7.p, inducing a second intracellular signal transduction pathway that further contributes to adoption of the secondary cell fate. The importance of this pathway is supported by the abnormal features displayed by certain mutants in which more than three cells become committed to vulva development. With these mutants, there is more than one primary cell, but each one is invariably surrounded by two secondary cells, suggesting that in the living worm, adoption of the secondary cell fate is dependent on the presence of an adjacent primary cell.

In summary, the general concepts to emerge from the study of vulva development in *C. elegans* are as follows:

- In a multicellular organism, positional information is important: the correct structure must develop at the appropriate place.

- The commitment of a small number of progenitor cells to particular differentiation pathways can lead to construction of a multicellular structure.

- Cell–cell signaling can utilize a concentration gradient to induce different responses in cells at different positions relative to the signaling cell.

Figure 14.30 Cell divisions resulting in production of the vulva cells of *C. elegans*. Three ancestral cells divide in a programmed manner to produce 22 progeny cells, which reorganize their positions relative to one another to construct the vulva.

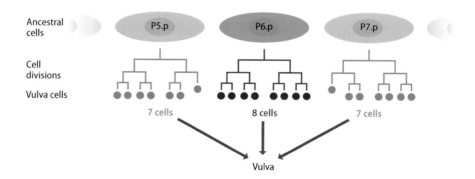

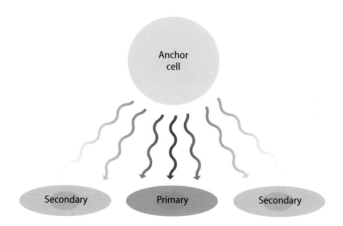

Figure 14.31 The postulated role of the anchor cell in determining cell fate during vulva development in *C. elegans*. It is thought that release of the signaling compound LIN-3 by the anchor cell commits P6.p (shown in pink), the cell closest to the anchor cell, to the primary vulva cell fate. P5.p and P7.p (shown in light orange) are further away from the anchor cell and so are exposed to a lower concentration of LIN-3 and become secondary vulva cells. As described in the text, there is evidence that commitment of the secondary cells to their fates is also influenced by signals from the primary vulva cell.

Fruit flies: conversion of positional information into a segmented body plan

The last organism whose development we will study is *Drosophila melanogaster*. The experimental history of the fruit fly dates back to 1910, when Thomas Hunt Morgan first used this organism as a model system in genetic research. For Morgan, the advantages of *Drosophila* were its small size, enabling large numbers to be studied in a single experiment, its minimal nutritional requirements (the flies like bananas), and the presence in natural populations of occasional variants with easily recognized genetic characteristics such as unusual eye colors. The body plan of the adult fly, as well as that of the larva, is built from a series of segments, each with a different structural role. This is clearest in the thorax, which has three segments, each carrying one pair of legs, and the abdomen, which is made up of eight segments, but is also true for the head, even though in the head the segmented structure is less visible (**Figure 14.32**). The early embryo, on the other hand, is a single **syncytium** comprising a mass of cytoplasm and multiple nuclei (**Figure 14.33**). The major contribution that *Drosophila* has made to our understanding of development has been the insights it has provided into how this undifferentiated embryo acquires positional information that eventually results in the construction of complex body parts at the correct places in the adult organism.

Initially, the positional information that the embryo needs is a definition of which end is the front (anterior) and which is the back (posterior), as well as similar information relating to up (dorsal) and down (ventral). This information is provided by concentration gradients of proteins that become established in the syncytium. The majority of these proteins are not synthesized from genes in the embryo, but are specified by **maternal-effect genes** whose mRNAs are injected into the embryo by the mother: The *bicoid* gene, for example, is transcribed in the maternal nurse cells, which are in contact with the egg cells, and the mRNA is injected into the anterior end of the unfertilized egg. This position is defined by the orientation of the egg cell in the egg chamber. The *bicoid* mRNA remains in the anterior region of the egg cell, attached by its 3′ untranslated region to the cell's cytoskeleton, and the Bicoid protein diffuses through the syncytium, setting up a concentration gradient, highest at the anterior end and lowest at the posterior end (**Figure 14.34**). Additional maternal-effect proteins, such as Hunchback, Nanos, Caudal, and Torso, contribute in a similar way to the anterior–posterior axis, while Dorsal and others set the dorsal–ventral axis. As a result, each point in the syncytium acquires its own unique chemical signature defined by the relative amounts of the various maternal-effect proteins.

This basic positional information is made more precise by expression of the **gap genes**. Three of the anterior–posterior gradient proteins – Bicoid, Hunchback, and Caudal – are transcription factors that target the gap genes in the nuclei that now line the inside of the embryo (see Figure 14.33). The identities of the gap genes expressed in a particular nucleus depend on the relative concentrations

Figure 14.32 The segmentation pattern of the adult *Drosophila melanogaster*. Note that the head is also segmented, but the pattern is not easily discernible from the morphology of the adult fly.

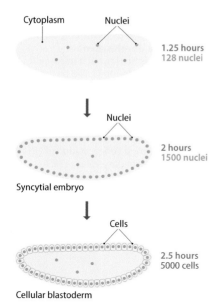

Figure 14.33 Early development of the *Drosophila* embryo. An unusual feature of the early *Drosophila* embryo is that it is not made up of lots of cells, as in most organisms, but instead is a single syncytium. This structure persists until 13 successive rounds of nuclear division have produced some 1500 nuclei. Only then do individual uninucleate cells start to appear around the outside of the syncytium, producing the structure called the **blastoderm**. The embryo is approximately 500 μm in length and 170 μm in diameter.

of the gradient proteins and hence on the position of the nucleus along the anterior–posterior axis. Some gap genes are activated directly by Bicoid, Hunchback, and Caudal, examples being *buttonhead*, *empty spiracles*, and *orthodenticle*, which are activated by Bicoid. Other gap genes are switched on indirectly, as is the case with *huckebein* and *tailless*, which respond to transcription factors that are activated by Torso. This complex interplay results in the positional information in the embryo, now carried by the relative concentrations of the gap gene products, becoming more detailed (**Figure 14.35**).

The next set of genes to be activated, the **pair-rule genes**, establish the basic segmentation pattern. Transcription of these genes responds to the relative concentrations of the gap gene products and occurs in nuclei that have become enclosed in cells. The pair-rule gene products therefore do not diffuse through the syncytium but remain localized within the cells that express them. The result is that the embryo can now be looked upon as comprising a series of stripes, each stripe consisting of a set of cells expressing a particular pair-rule gene. In a further round of gene activation, the **segment polarity genes** become switched on, providing greater definition to the stripes by setting the sizes and precise locations of what will eventually be the segments of the larval fly. Gradually we have converted the imprecise positional information of the maternal-effect gradients into a sharply defined segmentation pattern.

Homeotic selector genes are universal features of higher eukaryotic development

How is the segmentation pattern of the fruit-fly larva converted into the complex body plan of the adult fly? The pair-rule and segment polarity genes establish the segmentation pattern of the embryo but do not determine the actual identities of the individual segments. This is the job of the **homeotic selector genes**, which were first discovered by virtue of the extravagant effects that mutations in these genes have on the appearance of the adult fly. The *antennapedia* mutation, for example, transforms the head segment that usually produces an antenna into one that makes a leg, so the mutant fly has a pair of legs where its antennae should be. The early geneticists were fascinated by these monstrous **homeotic mutants**, and many were collected during the first few decades of the twentieth century.

Genetic mapping of homeotic mutations revealed that in *Drosophila* the selector genes are clustered in two groups on chromosome 3. These clusters are called the Antennapedia complex (ANT-C), which contains genes involved in determination of the head and thorax segments, and the Bithorax complex (BX-C), which contains genes for the abdomen segments (**Figure 14.36**). Some

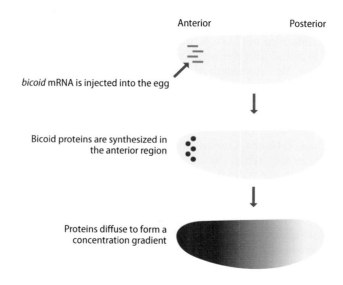

Figure 14.34 Establishment of the Bicoid gradient in a *Drosophila* embryo. The *bicoid* mRNA is injected into the anterior end of the egg. Bicoid proteins, translated from the mRNA, diffuse through the syncytium setting up a concentration gradient.

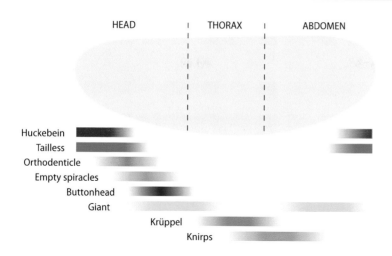

HEAD | THORAX | ABDOMEN

Huckebein
Tailless
Orthodenticle
Empty spiracles
Buttonhead
Giant
Krüppel
Knirps

Figure 14.35 The role of the gap gene products in conferring positional information during embryo development in *D. melanogaster*. The concentration gradient of each gap gene product is denoted by the colored bars. The parts of the embryo that give rise to the head, thorax, and abdomen regions of the adult fly are indicated.

additional non-selector development genes, such as *bicoid*, are also located in ANT-C. One interesting feature of the ANT-C and BX-C clusters, which is still not understood, is that the order of genes corresponds to the order of the segments in the fly, the first gene in ANT-C being *labial palps*, which controls the most anterior segment of the fly, and the last gene in BX-C being *Abdominal B*, which specifies the most posterior segment.

The correct selector gene is expressed in each segment because the activation of each one is responsive to the positional information represented by the distributions of the gap gene and pair-rule gene products. The selector gene products are themselves transcription factors, each containing a homeodomain version of the helix-turn-helix DNA-binding structure (Section 11.2). Each selector gene product switches on the set of genes needed to initiate development of the specified segment. Maintenance of the differentiated state is ensured partly by the repressive effect that each selector gene product has on expression of the other selector genes, and partly by the work of the Polycomb group proteins, which, as we saw in Section 14.2, construct inactive chromatin over the selector genes that are not expressed in a particular cell.

The homeodomain regions of the various *Drosophila* selector genes have strikingly similar nucleotide sequences. This observation led researchers in the 1980s to search for other homeotic genes by using the homeodomain as a probe in hybridization experiments. First, the *Drosophila* genome was searched, resulting in isolation of several previously unknown homeodomain-containing genes. These have turned out not to be selector genes but other genes coding for transcription factors involved in development. Examples include the pair-rule genes *even-skipped* and *fushi tarazu*, and the segment polarity gene *engrailed*. Researchers then probed the genomes of other organisms, and discovered that homeodomains are present in genes in a wide variety of animals, including humans. Examination of these genes showed that some are homeotic selectors organized into clusters similar to ANT-C and BX-C, and that these genes have equivalent functions to the *Drosophila* versions, specifying construction of the body plan. For example, mutations in the HoxC8 gene of mouse results in an animal that has an extra pair of ribs, due to conversion of a lumbar vertebra (normally in the lower back) into a thoracic vertebra (from which the ribs emerge). Other Hox mutations in animals lead to limb deformations, such as absence of the lower arm, or extra digits on the hands or feet.

We now look on the ANT-C and BX-C clusters of selector genes in *Drosophila* as two parts of a single complex, usually referred to as the homeotic gene complex, or HOM-C. In vertebrates, there are four homeotic gene clusters, called HoxA to HoxD. When these four clusters are aligned with one another and with HOM-C (**Figure 14.37**), similarities are seen between the genes at equivalent positions, such that the evolutionary history of the homeotic selector gene clusters can be traced from insects through to humans (Section 18.2). As in *Drosophila*, the order of genes in the vertebrate clusters reflects the order of

Antennapedia complex (ANT-C)

lab   *pb*   *Dfd*   *Scr*   *Antp*

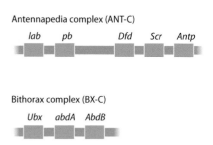

Bithorax complex (BX-C)

Ubx   *abdA*   *AbdB*

Figure 14.36 The Antennapedia and Bithorax gene complexes of *D. melanogaster*. Both complexes are located on the fruit-fly chromosome 3, ANT-C upstream of BX-C. The genes are usually drawn in the order shown, although this means that they are transcribed from right to left. The diagram does not reflect the actual lengths of the genes. The full gene names are as follows: *lab*, labial palps; *pb*, proboscipedia; *Dfd*, Deformed; *Scr*, Sex combs reduced; *Antp*, Antennapedia; *Ubx*, Ultrabithorax; *abdA*, abdominal A; *AbdB*, Abdominal B. In ANT-C, the non-selector genes *zerknüllt* and *bicoid* occur between *pb* and *Dfd*, and *fushi tarazu* lies between *Scr* and *Antp*.

Figure 14.37 Comparison between the *Drosophila* HOM-C gene complex and the four Hox clusters of vertebrates. Genes that code for proteins with related structures and functions are indicated by the colors. The diagram does not reflect the actual lengths of the genes.

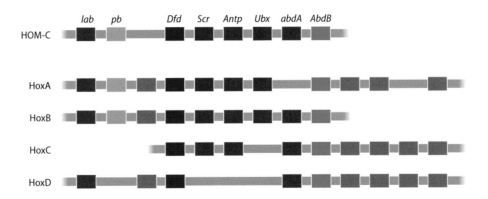

the structures specified by the genes in the adult body plan. This is clearly seen with the mouse HoxB cluster, which controls development of the nervous system (**Figure 14.38**). The remarkable conclusion is that, at this fundamental level, developmental processes in fruit flies and other 'simple' eukaryotes are similar to the processes occurring in humans and other 'complex' organisms. The discovery that studies of fruit flies are directly relevant to human development has opened up vast vistas of research possibilities.

Homeotic genes also underlie plant development

The power of *Drosophila* as a model system for development extends even beyond vertebrates. Developmental processes in plants are, in most respects, very different from those of fruit flies and other animals, but at the genetic level there are certain similarities, sufficient for the knowledge gained about *Drosophila* development to be of value in interpreting similar research carried out with plants. In particular, the recognition that a limited number of homeotic selector genes control the *Drosophila* body plan has led to a model for plant development which postulates that the structure of the flower is determined by a small number of homeotic genes.

All flowers are constructed along similar lines, made up of four concentric whorls, each comprising a different floral organ (**Figure 14.39**). The outer whorl, number 1, contains sepals, which are modified leaves that envelop and protect the bud during its early development. The next whorl, number 2, contains the distinctive petals, and within these are whorls 3 (stamens, the male reproductive organs) and 4 (carpels, the female reproductive organs).

Most of the research on plant development has been carried out with *Antirrhinum* (the snapdragon) and *Arabidopsis thaliana*, a small vetch that has been adopted as a model species, partly because it has a genome of only 135 Mb, one of the smallest known among flowering plants. Although these plants do not appear to contain homeodomain proteins, they do have genes that, when mutated, lead to homeotic changes in the floral architecture, such as replacement of sepals by carpels. Analysis of these mutants has led to the **ABC model**, which states that there are three types of homeotic genes – A, B, and C – that control flower development as follows:

Figure 14.38 Specification of the mouse nervous system by selector genes of the HoxB cluster. The nervous system is shown schematically and the positions specified by the individual HoxB genes (HoxB1 to HoxB9) are indicated by the green bars. The components of the nervous system are: F, forebrain; M, midbrain; r1–r8, rhombomeres 1–8; and the spinal cord. Rhombomeres are segments of the hindbrain seen during development.

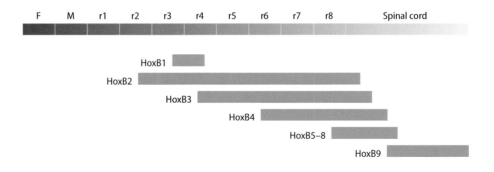

- Whorl 1 is specified by A-type genes: examples in *Arabidopsis* are *apetala1* and *apetala2*.

- Whorl 2 is specified by A genes acting in concert with B genes, examples of the latter including *apetala3* and *pistillata*.

- Whorl 3 is specified by the B genes plus the C gene, *agamous*.

- Whorl 4 is specified by the C gene acting on its own.

As anticipated from the work with *Drosophila*, the A, B, and C homeotic gene products are transcription factors. All except the APETALA2 protein contain the same DNA-binding domain, the **MADS box**, which is also found in other proteins involved in plant development, including SEPALLATA1, 2, and 3, which work with the A, B, and C proteins in defining the detailed structure of the flower. Other components of the flower development system include at least one master gene, called *floricaula* in *Antirrhinum* and *leafy* in *Arabidopsis*, which controls the switch from vegetative to reproductive growth, initiating flower development, and also has a role in establishing the pattern of homeotic gene expression. Plants also have Polycomb group proteins (Section 14.2), with CURLY LEAF and SWINGER of *Arabidopsis* thought to be the key histone methyltransferases responsible for silencing regions of chromatin containing those homeotic genes that are inactive in a particular whorl.

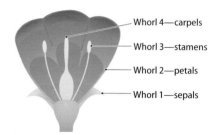

Figure 14.39 Flowers are constructed from four concentric whorls.

Whorl 4—carpels
Whorl 3—stamens
Whorl 2—petals
Whorl 1—sepals

SUMMARY

- Transient alterations in genome expression patterns enable a cell to respond to changes in the external environment, these changes including the presence or absence of signaling compounds that coordinate biochemical activities in different cells.

- More permanent changes in genome expression underlie differentiation and development.

- Steroid hormones enter the cell and influence genome expression via receptor proteins that act as transcription activators.

- Effects of other signaling compounds are mediated by cell surface receptors, many of which dimerize in response to the extracellular signal, initiating a signal transduction pathway that leads to the genome.

- With some signal transduction systems, stimulation of the cell surface receptor by attachment of the extracellular signaling compound has a direct effect on the activity of a transcription factor. An example is the JAK/STAT pathway.

- Other signal transduction pathways, including the MAP kinase pathway, have several steps between receptor and genome. Some of these pathways make use of second messengers, such as cyclic nucleotides and calcium ions, which influence a number of cellular activities, including genome expression.

- Differentiation processes involve semi-permanent changes in genome expression, which can be brought about by changes to chromatin structure, gene conversion events, or genome rearrangements.

- The lysogenic infection cycle of bacteriophage λ demonstrates ways in which simple genetic switches can determine which of two developmental pathways are followed.

- Studies of sporulation in *Bacillus subtilis* have illustrated how time-dependent changes in genome expression can be brought about and how cell–cell signaling can regulate a developmental pathway.

- Mechanisms for the determination of cell fate have been revealed by studies of vulva development in *C. elegans*.

- The most informative pathway for developmental genetics has been embryogenesis in the fruit fly, which has shown how a complex body plan can be specified by controlled patterns of genome expression.

- Work with *Drosophila* has also revealed the existence of homeotic selector genes, which control developmental processes not only in flies but also in vertebrates and in plants.

SHORT ANSWER QUESTIONS

1. Outline the differences between differentiation and development, and describe the basis for these differences.

2. How do steroid hormones influence genome expression?

3. Compare the various types of cell surface receptor proteins that are known.

4. Describe the (A) JAK/STAT and (B) MAP kinase signal transduction pathways.

5. Outline the roles of second messengers in the control of genome expression.

6. Describe how the Polycomb group proteins influence genome expression.

7. How can cells of the vertebrate immune system produce so many different immunoglobulins from a small set of genes?

8. Outline the process by which the choice between the lytic and lysogenic pathways is regulated in bacteriophage λ.

9. During sporulation in *Bacillus*, σ^E and σ^F are present in both the prespore and the mother cell. How is σ^F activated in the prespore?

10. How does the anchor cell of *C. elegans* induce the vulva progenitor cells to differentiate into vulva cells? Why do the vulva progenitor cells follow different pathways upon receiving the signal from the anchor cell?

11. Describe how studies of embryogenesis in *Drosophila* led to discovery of homeotic selector genes in vertebrates.

IN-DEPTH PROBLEMS

1. What methods might be used to identify those parts of a cell surface receptor protein that are exposed on the outer surface of the plasma membrane of an animal cell?

2. Describe how studies of signal transduction have improved our understanding of the abnormal biochemical activities that underlie cancer.

3. Are *Caenorhabditis elegans* and *Drosophila melanogaster* good model organisms for development in higher eukaryotes?

4. What would be the key features of an ideal model organism for development in higher eukaryotes?

5. What can be inferred about genome evolution from the discovery that *Drosophila* has a single homeotic gene complex whereas vertebrates have four? Would you expect any group of organisms to have more than four homeotic gene clusters?

FURTHER READING

Imported extracellular signaling compounds

Mazaira, G.I., Zgajnar, N.R., Lotufo, C.M., et al. (2018) The nuclear receptor field: A historical overview and future challenges. *Nucl. Receptor Res.* 5:101320.

Tsai, M.-J. and O'Malley, B.W. (1994) Molecular mechanisms of action of steroid/thyroid receptor superfamily members. *Annu. Rev. Biochem.* 63:451–486.

Weikum, E.R., Liu, X. and Ortlund, E.A. (2018) The nuclear receptor superfamily: A structural perspective. *Prot. Sci.* 27:1876–1892.

Winge, D.R., Jensen, L.T. and Srinivasan, C. (1998) Metal-ion regulation of gene expression in yeast. *Curr. Opin. Chem. Biol.* 2:216–221.

Cell surface receptor proteins and signal transduction pathways

Cargnello, M. and Roux, P.P. (2011) Activation and function of the MAPKs and their substrates, the MAPK-activated protein kinases. *Microbiol. Mol. Biol. Rev.* 75:50–83.

Karin, M. and Hunter, T. (1995) Transcriptional control by protein phosphorylation: Signal transmission from the cell surface to the nucleus. *Curr. Biol.* 5:747–757.

Lemmon, M.A. and Schlessinger, J. (2010) Cell signaling by receptor tyrosine kinases. *Cell* 141:1117–1134.

Newton, A.C., Bootman, M.D. and Scott, J.D. (2016) Second messengers. *Cold Spring Harb. Perspect. Biol.* 8:a005926.

Robinson, M.J. and Cobb, M.H. (1997) Mitogen-activated protein kinase pathways. *Curr. Opin. Cell Biol.* 9:180–186.

Schlessinger, J. (1993) How receptor tyrosine kinases activate Ras. *Trends Biochem. Sci.* 18:273–275.

Villarino, A.V., Kanno, Y., Ferdinand, J.R., et al. (2015) Mechanisms of Jak/STAT signaling in immunity and disease. *J. Immunol.* 194:21–27.

Wang, Y. and Levy, D.E. (2012) Comparative evolutionary genomics of the STAT family of transcription factors. *JAKSTAT* 1:23–33.

Changes in genome activity during differentiation

Alt, F.W., Blackwell, T.K. and Yancopoulos, G.D. (1987) Development of the primary antibody repertoire. *Science* 238:1079–1087. *Generation of immunoglobulin diversity.*

Blackledge, N.P. and Klose, R.J. (2021) The molecular principles of gene regulation by Polycomb repressive complexes. *Nat. Rev. Mol. Cell Biol.* 22:815–833.

Geisler, S.J. and Paro, R. (2015) Trithorax and Polycomb group-dependent regulation: A tale of opposing activities. *Development* 142:2876–2887.

Haber, J.E. (2012) Mating-type genes and MAT switching in *Saccharomyces cerevisiae*. *Genetics* 191:33–64.

Kim, D.H. and Sung, S. (2014) Polycomb-mediated gene silencing in *Arabidopsis thaliana*. *Mol. Cells* 37:841–850.

Piunti, A. and Shilatifard, A. (2021) The roles of Polycomb repressive complexes in mammalian development and cancer. *Nat. Rev. Mol. Cell Biol.* 22:326–345.

Schuettengruber, B., Bourbon, H.-M., Di Croce, L., et al. (2017) Genome regulation by Polycomb and Trithorax: 70 years and counting. *Cell* 171:34–57.

Simple development pathways

Higgins, D. and Dworkin, J. (2012) Recent progress in *Bacillus subtilis* sporulation. *FEMS Microbiol. Rev.* 36:131–148.

Oppenheim, A.B., Kobiler, O., Stavans, J., et al. (2005) Switches in bacteriophage lambda development. *Annu. Rev. Genet.* 39:409–429.

Riley. E.P., Schwarz, C., Derman, A.I., et al. (2021) Milestones in *Bacillus subtilis* sporulation research. *Microb. Cell* 8:1–16.

Development in *Caenorhabditis elegans*

Aroian, R.V., Koga, M., Mendel, J.E., et al. (1990) The *let-23* gene necessary for *Caenorhabditis elegans* vulval induction encodes a tyrosine kinase of the EGF receptor subfamily. *Nature* 348:693–699.

Katz, W.S., Hill, R.J., Clandinin, T.R., et al. (1995) Different levels of the *C. elegans* growth factor LIN-3 promote distinct vulval precursor fates. *Cell* 82:297–307.

Kornfeld, K. (1997) Vulval development in *Caenorhabditis elegans*. *Trends Genet.* 13:55–61.

Schindler, A.J. and Sherwood, D.R. (2013) Morphogenesis of the *Caenorhabditis elegans* vulva. *Wiley Interdiscip. Rev. Dev. Biol.* 2:75–95.

Sharma-Kishore, R., White, J.G., Southgate, E., et al. (1999) Formation of the vulva in *Caenorhabditis elegans*: A paradigm for organogenesis. *Development* 126:691–699.

Shin, H. and Reiner, D.J. (2018) The signaling network controlling *C. elegans* vulval cell fate patterning. *J. Dev. Biol.* 6:30.

Embryogenesis in fruit flies and homeotic selector genes in vertebrates

Gaunt, S.J. (2015) The significance of Hox gene collinearity. *Int. J. Dev. Biol.* 59:159–170.

Gebelein, B. and Ma, J. (2016) Regulation in the early *Drosophila* embryo. *Rev. Cell Biol. Mol. Med.* 2:140–167.

Ingham, P.W. (1988) The molecular genetics of embryonic pattern formation in *Drosophila*. *Nature* 335:25–34.

Krumlauf, R. (1994) *Hox* genes in vertebrate development. *Cell* 78:191–201.

Maconochie, M., Nonchev, S., Morrison, A., et al. (1996) Paralogous *Hox* genes: Function and regulation. *Annu. Rev. Genet.* 30:529–556. *Describes homeotic selector genes in vertebrates.*

Mahowald, A.P. and Hardy, P.A. (1985) Genetics of *Drosophila* embryogenesis. *Annu. Rev. Genet.* 19:149–177.

Mallo, M., Wellik, D.M. and Deschamps, J. (2010) *Hox* genes and regional patterning of the vertebrate body plan. *Dev. Biol.* 344:7–15.

Zakany, J. and Duboule, D. (2007) The role of *Hox* genes during vertebrate limb development. *Curr. Opin. Genet. Dev.* 17:359–366.

Flower development in plants

Irish, V. (2017) The ABC model of plant development. *Curr. Biol.* 27:R887–R890.

Ma, H. (1998) To be, or not to be, a flower – Control of floral meristem identity. *Trends Genet.* 14:26–32.

Parcy, F., Nilsson, O., Busch, M.A., et al. (1998) A genetic framework for floral patterning. *Nature* 395:561–566.

Robles, P. and Pelaz, S. (2005) Flower and fruit development in *Arabidopsis thaliana*. *Int. J. Dev. Biol.* 49:633–643.

Theissen, G. (2001) Development of floral organ identity: Stories from the MADS house. *Curr. Opin. Plant Biol.* 4:75–85.

PART 4

HOW GENOMES REPLICATE AND EVOLVE

GENOME REPLICATION

The primary function of a genome is to specify the biochemical signature of the cell in which it resides. We have seen that the genome achieves this objective by synthesis and maintenance of a transcriptome and proteome, whose individual RNA and protein components carry out and regulate the cell's biochemical activities. In order to continue performing this function, the genome must replicate every time that the cell divides. This means that the entire DNA content of the cell must be copied at the appropriate period in the cell cycle and the resulting DNA molecules distributed to the daughter cells so that each one receives a complete copy of the genome.

When studying genome replication, it is easy to become absorbed in the molecular details and to lose sight of the broader implications of the process. For example, at the molecular level, we look on the accuracy of DNA replication as vital in ensuring that daughter cells acquire precise copies of the genome, so that those cells can function in the same way as their parent, or can adopt new functions in accordance with the genetic programming contained in the nucleotide sequence of the genome. However, at a higher level, absolute and inviolate identity between parent and daughter genomes would make evolution impossible, because the latter depends on the generation of genome variants that give rise to organisms with modified characteristics and different degrees of fitness to their environment. This variation is responsible not only for the differences between species, but the differences between the members of a single species.

In Part 4 of *Genomes*, we will explore the link between DNA replication and genome evolution. We will begin in this chapter by examining the elaborate process by which the genome is replicated, this process spanning the interface between molecular biology, biochemistry, and cell biology. In Chapters 16 and 17, we will see how variation in genome sequences is introduced by recombination and mutation, and in Chapter 18 we will study the ways in which these processes are thought to have shaped the structures and genetic contents of genomes over evolutionary time.

15.1 THE TOPOLOGY OF GENOME REPLICATION

Genome replication has been studied since Watson and Crick first discovered the double-helix structure of DNA back in 1953. The primary concern in the years from 1953 to 1958 was the **topological problem**. This problem arises from the need to unwind the double helix in order to make copies of its two polynucleotides. The issue assumed center stage in the mid-1950s because it was the main stumbling block to acceptance of the double helix as the correct structure

DOI: 10.1201/9781003133162-15

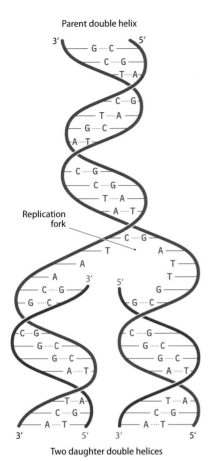

Parent double helix

Replication fork

Two daughter double helices

Figure 15.1 DNA replication, as predicted by Watson and Crick. The polynucleotides of the parent double helix are shown in black. Both act as templates for synthesis of new strands of DNA, shown in red. The sequences of these new strands are determined by base-pairing with the template molecules. The topological problem arises because the two polynucleotides of the parent helix cannot simply be pulled apart: the helix has to be unwound in some way.

for DNA. Before studying the molecular events occurring during genome replication, we must first understand how the cell solves this topological problem.

The double-helix structure complicates the replication process

In their paper in *Nature* announcing the discovery of the double-helix structure of DNA, Watson and Crick made one of the most famous statements in molecular biology:

'It has not escaped our notice that the specific pairing we have postulated immediately suggests a possible copying mechanism for the genetic material.'

The pairing process that they refer to is one in which each strand of the double helix acts as a template for synthesis of a second complementary strand, the end result being that both of the daughter double helices are identical to the parent molecule (**Figure 15.1**). The scheme is almost implicit in the double-helix structure, but it presents problems, as admitted by Watson and Crick in a second paper published in *Nature* just a month after the report of the structure. This paper describes the postulated replication process in more detail, but points out the difficulties that arise from the need to unwind the double helix. The most trivial of these difficulties is the possibility that the daughter molecules get tangled up. More critical is the rotation that would accompany the unwinding: with one turn occurring for every 10 bp of the double helix, complete replication of the DNA molecule in human chromosome 1, which is 249 Mb in length, would require 25 million rotations of the chromosomal DNA. It is difficult to imagine how this could occur within the constrained volume of the nucleus, but the unwinding of a linear chromosomal DNA molecule is not physically impossible. In contrast, a circular double-stranded molecule, for example a bacterial or bacteriophage genome, having no free ends, would not be able to rotate in the required manner and so, apparently, could not be replicated by the Watson–Crick scheme. Finding an answer to this dilemma was a major preoccupation of molecular biology during the 1950s.

The topological problem was considered so serious by some molecular biologists, notably Max Delbrück, that there was initially some resistance to accepting the double helix as the correct structure of DNA. The difficulty relates to the **plectonemic** nature of the double helix, this being the topological arrangement that prevents the two strands of a coil from being separated without unwinding. The problem would therefore be resolved if the double helix was in fact **paranemic**, because this would mean that the two strands could be separated simply by moving each one sideways without unwinding the molecule. It was suggested that the double helix could be converted into a paranemic structure by supercoiling in the direction opposite to the turn of the helix itself, or that within a DNA molecule the right-handed helix proposed by Watson and Crick might be 'balanced' by equal lengths of a left-handed helical structure. The possibility that double-stranded DNA was not a helix at all, but a side-by-side ribbon structure, was also briefly considered, this idea surprisingly being revived in the late 1970s. Each of these proposed solutions to the topological problem were individually rejected for one reason or another, most of them because they required alterations to the double-helix structure, alterations that were not compatible with the X-ray diffraction results and other experimental data pertaining to DNA structure.

The first real progress toward a solution to the topological problem came in 1954 when Delbrück proposed a 'breakage-and-reunion' model for separating the strands of the double helix. In this model, the strands are separated not by unwinding the helix with the accompanying rotation of the molecule, but by breaking one of the strands, passing the second strand through the gap, and rejoining the first strand. This scheme is very close to the correct solution to the topological problem, being one of the ways in which DNA topoisomerases work (see Figure 15.4A), but unfortunately Delbrück overcomplicated the issue by attempting to combine breakage and reunion with the DNA synthesis that occurs during the actual replication process. This led him to propose a model for DNA replication which results in each polynucleotide in the daughter molecule being made up partly of parental DNA and partly of newly synthesized DNA. This **dispersive** mode of replication contrasts with the **semiconservative** system proposed by Watson and Crick (**Figure 15.2**). A third possibility is that replication is fully **conservative**, with one of the daughter double helices being made entirely of newly synthesized DNA and the other comprising the two parental strands. Models for conservative replication are difficult to devise, but one can imagine that this type of replication might be accomplished without unwinding the parent helix.

The Meselson–Stahl experiment proved that replication is semiconservative

Delbrück's breakage-and-reunion model was important because it stimulated experiments designed to test between the three modes of DNA replication illustrated in Figure 15.2. Radioactive isotopes had recently been introduced into molecular biology, so attempts were made to use DNA labeling to distinguish newly synthesized DNA from the parental polynucleotides. Each mode of replication predicts a different distribution of newly synthesized DNA, and hence of radioactive label, in the double helices resulting after two or more rounds of replication. Analysis of the radioactive contents of these molecules should therefore determine which replication scheme operates in living cells. Unfortunately, it proved impossible to obtain a clear-cut result, largely because of the difficulty in measuring the precise amount of radioactivity in the DNA molecules, the analysis being complicated by the rapid decay of the ^{32}P isotope that was used as the label.

The breakthrough was eventually made by Matthew Meselson and Franklin Stahl who, in 1958, carried out the required experiment not with a radioactive label but with ^{15}N, the nonradioactive 'heavy' isotope of nitrogen. Now it was possible to analyze the replicated double helices by density gradient centrifugation (Section 7.3), because a DNA molecule labeled with ^{15}N has a higher buoyant density than an unlabeled molecule. Meselson and Stahl started with a culture of *E. coli* cells that had been grown with ^{15}NH$_4$Cl and whose DNA molecules therefore contained heavy nitrogen. The cells were transferred to a normal medium, and samples were taken after 20 minutes and 40 minutes, corresponding to one and two cell divisions, respectively. DNA was extracted from each sample and the molecules were examined by density gradient centrifugation (**Figure 15.3A**). After one round of DNA replication, the daughter molecules synthesized in the presence of normal nitrogen formed a single band in the density

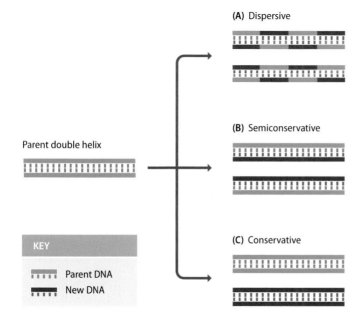

Figure 15.2 Three possible schemes for DNA replication. For the sake of clarity, the DNA molecules are drawn as ladders rather than helices.

Figure 15.3 The Meselson–Stahl experiment. (A) The experiment carried out by Meselson and Stahl involved growing a culture of *Escherichia coli* in a medium containing $^{15}NH_4Cl$ (ammonium chloride labeled with the heavy isotope of nitrogen). Cells were then transferred to normal medium (containing $^{14}NH_4Cl$) and samples taken after 20 minutes (one cell division) and 40 minutes (two cell divisions). DNA was extracted from each sample and the molecules analyzed by density gradient centrifugation. After 20 minutes, all the DNA contained similar amounts of ^{14}N and ^{15}N, but after 40 minutes two bands were seen, one corresponding to hybrid ^{14}N–^{15}N–DNA, and the other to DNA molecules made entirely of ^{14}N. (B) The predicted outcome of the experiment is shown for each of the three possible modes of DNA replication. The banding pattern seen after 20 minutes enables conservative replication to be discounted because this scheme predicts that after one round of replication there will be two different types of double helix, one containing just ^{15}N and the other containing just ^{14}N. The single ^{14}N–^{15}N–DNA band that was actually seen after 20 minutes is compatible with both dispersive and semiconservative replication, but the two bands seen after 40 minutes are consistent only with semiconservative replication. Dispersive replication continues to give hybrid ^{14}N–^{15}N molecules after two rounds of replication, whereas the granddaughter molecules produced at this stage by semiconservative replication include two that are made entirely of ^{14}N–DNA.

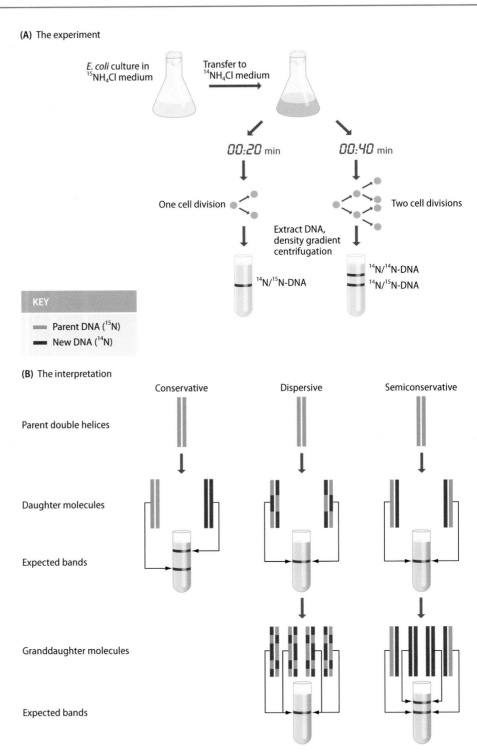

(A) The experiment

E. coli culture in $^{15}NH_4Cl$ medium

Transfer to $^{14}NH_4Cl$ medium

00:20 min   00:40 min

One cell division   Two cell divisions

Extract DNA, density gradient centrifugation

$^{14}N/^{15}N$-DNA

$^{14}N/^{14}N$-DNA
$^{14}N/^{15}N$-DNA

KEY

— Parent DNA (^{15}N)
— New DNA (^{14}N)

(B) The interpretation

Conservative   Dispersive   Semiconservative

Parent double helices

Daughter molecules

Expected bands

Granddaughter molecules

Expected bands

gradient, indicating that each double helix was made up of equal amounts of newly synthesized and parental DNA. This result immediately enabled the conservative mode of replication to be discounted, as this predicts that there would be two bands after one round of replication (**Figure 15.3B**), but did not provide a distinction between Delbrück's dispersive model and the semiconservative process favored by Watson and Crick. The distinction was, however, possible when the DNA molecules resulting from two rounds of replication were examined. Now the density gradient revealed two bands of DNA, the first corresponding to a hybrid composed of equal parts of newly synthesized and old DNA, and the

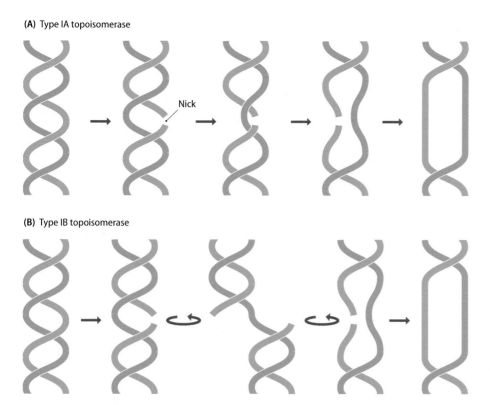

(A) Type IA topoisomerase

Nick

(B) Type IB topoisomerase

Figure 15.4 The mode of action of type I topoisomerases. (A) A type IA topoisomerase makes a cut in one strand and passes the second strand through this gap. (B) A type IB topoisomerase cuts one strand, and then rotates the cut strand around the uncut one.

second corresponding to molecules made up entirely of new DNA. This result agrees with the semiconservative scheme but is incompatible with dispersive replication, the latter predicting that after two rounds of replication all molecules would still be hybrids.

DNA topoisomerases provide a solution to the topological problem

The Meselson–Stahl experiment proved that DNA replication in living cells follows the semiconservative scheme proposed by Watson and Crick, and hence indicated that the cell must have a solution to the topological problem. This solution was not understood by molecular biologists until some 25 years later, when the activities of the enzymes called **DNA topoisomerases** were characterized.

DNA topoisomerases solve the topological problem by counteracting the overwinding that otherwise would be introduced into the molecule by the progression of the replication fork. They do this by carrying out breakage-and-reunion reactions similar but not identical to that envisaged by Delbrück. We recognize two broad classes of DNA topoisomerases, called types I and II, the distinction being that a type I enzyme breaks just one strand of the double helix, whereas type II enzymes break both strands. The type I enzymes are further subdivided into types IA and IB, depending on what they do after making the single-stranded break (**Figure 15.4**):

- A type IA topoisomerase uses the single-strand break as a 'gate' through which it passes the uncut polynucleotide. The two ends of the broken strand are then religated. This mode of action results in the **linking number** (the number of times one strand crosses the other in a circular molecule) being changed by one.

- A type IB topoisomerase acts as a molecular rotor, relaxing the torsional stress in an overwound helix by swiveling the cut strand around the uncut one. The linking number is therefore reduced by multiples of one.

Type II topoisomerases are also divided into A and B subgroups, but both IIA and IIB enzymes act in the same way, the double-strand break creating a gate

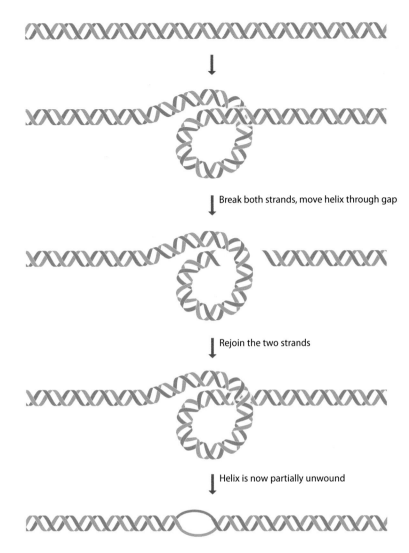

Figure 15.5 The mode of action of a type II topoisomerase.

through which a second segment of the helix is passed prior to religation of the cut ends (**Figure 15.5**). This process changes the linking number by two.

Cutting one or both DNA strands might appear to be a drastic solution to the topological problem, leading to the possibility that the topoisomerase might occasionally fail to rejoin a strand, generating a break that would interfere with the replication process. This possibility is reduced by the mode of action of these enzymes. One end of each cut polynucleotide becomes covalently attached to a tyrosine amino acid at the active site of the enzyme, ensuring that this end of the polynucleotide is held tightly in place while the free end(s) is being manipulated. With type IA and II enzymes, the polynucleotide–tyrosine linkage involves a phosphate group attached to the free 5′ end of the cut polynucleotide, and with the IB enzymes the linkage is via a 3′ phosphate group.

The four groups of topoisomerase – IA, IB, IIA, and IIB – have distinct structures and each probably evolved separately. Type IA topoisomerases were first discovered in *E. coli* and for some time were thought to be specific to prokaryotes, but a type IA enzyme, called Top3, is now known to be present in most eukaryotes. However, the main eukaryotic topoisomerases are type IB enzymes. To date, no definite IB topoisomerase has been identified in prokaryotes, although genome annotations indicate that some archaea possess genes that code for IB-like topoisomerases. There is also a topoisomerase known in just a single archaea species, *Methanopyrus kandleri*, that has the same mode of action as type IB enzymes, but a different structure. This unique topoisomerase is sometimes classified as the sole member of type IC. Type IIA enzymes are known in all species, and IIB enzymes are found in archaea and plants.

Replication is not the only activity that is complicated by the topology of the double helix, and it is becoming increasingly clear that DNA topoisomerases have equally important roles during transcription, recombination, and other processes that can result in over- or underwinding of DNA. In eukaryotes, type II topoisomerases are also responsible for separating DNA molecules that become intertwined during chromosome division. Most topoisomerases are only able to relax overwound DNA, but some prokaryotic enzymes, such as the bacterial DNA gyrase (a IIA topoisomerase) and the archaeal reverse gyrase (a IA enzyme), can carry out the reverse reaction and introduce supercoils into DNA molecules (see Figure 8.2).

Variations on the semiconservative theme

No exceptions to the semiconservative mode of DNA replication are known, but there are several variations on this basic theme. DNA copying via a replication fork, as shown in Figure 15.1, is the predominant system, being used by chromosomal DNA molecules in eukaryotes and by the circular genomes of prokaryotes. Some smaller circular molecules use a slightly different process called **displacement replication**. In these molecules, the point at which replication begins is marked by a **D-loop**, a region of approximately 500 bp where the double helix is disrupted by the presence of an RNA molecule base-paired to one of the DNA strands (**Figure 15.6**). This RNA molecule acts as the starting point for synthesis of one of the daughter polynucleotides. This polynucleotide is synthesized by continuous copying of one strand of the helix, the second strand being displaced and subsequently copied after synthesis of the first daughter genome has been completed. Displacement replication has been studied extensively because it is thought to be the primary mode used to replicate the mitochondrial genomes of humans and other vertebrates.

The advantage of displacement replication compared to the regular form of semiconservative replication is not clear. In contrast, the special type of displacement process called **rolling-circle replication** is an efficient mechanism

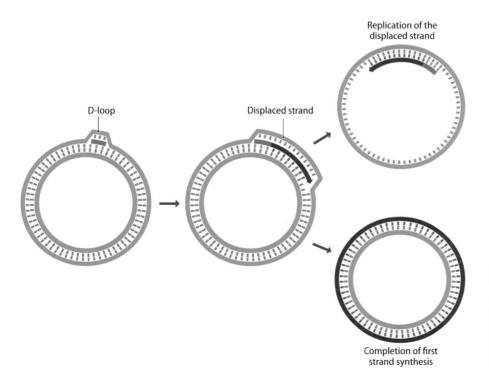

Figure 15.6 Displacement replication. The D-loop contains a short RNA molecule that primes DNA synthesis. After completion of the first strand synthesis a second RNA primer attaches to the displaced strand and initiates replication of this molecule. In this diagram, newly synthesized DNA is shown in red.

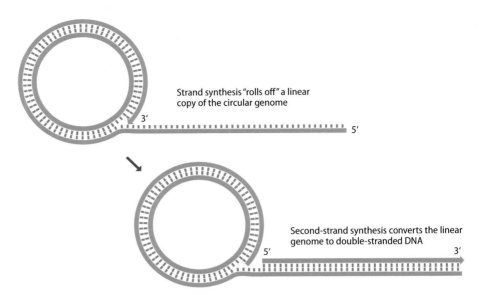

Strand synthesis "rolls off" a linear copy of the circular genome

Second-strand synthesis converts the linear genome to double-stranded DNA

Figure 15.7 Rolling-circle replication.

for the rapid synthesis of multiple copies of a circular genome. Rolling-circle replication, which is used by λ and various other bacteriophages, initiates at a nick which is made in one of the parent polynucleotides. The free 3′ end that results is extended, displacing the 5′ end of the polynucleotide. Continued DNA synthesis 'rolls off' a complete copy of the genome, and further synthesis eventually results in a series of genomes linked head to tail (**Figure 15.7**). These genomes are single-stranded and linear, but can easily be converted to double-stranded circular molecules by complementary strand synthesis, followed by cleavage at the junction points between genomes, and circularization of the resulting segments.

15.2 THE INITIATION PHASE OF GENOME REPLICATION

In order to initiate a round of genome replication, the double helix must be opened up at a particular point, and the replication machinery assembled at the two nascent replication forks that are created. Initiation of replication is not a random process and always begins at the same or similar positions on a DNA molecule, these points being called the **origins of replication**. Once initiated, the replication forks emerge from the origin and progress in opposite directions along the DNA: replication is therefore bidirectional with most genomes (**Figure 15.8**). A circular bacterial genome has a single origin of replication, meaning that several thousand kilobases of DNA are copied by each replication fork. This situation differs from that seen with eukaryotic chromosomes, which have multiple origins and whose replication forks progress for shorter distances. The yeast *Saccharomyces cerevisiae*, for example, has about 400 origins, corresponding to 1 per 15.25 kb of DNA, and humans have 30,000–50,000 origins, or 1 for every 65–110 kb of DNA.

(A) Replication of a circular bacterial chromosome

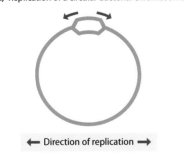

← Direction of replication →

(B) Replication of a linear eukaryotic chromosome

15.25 kb (yeast)
65–110 kb (humans)

Figure 15.8 Bidirectional DNA replication of (A) a circular bacterial chromosome and (B) a linear eukaryotic chromosome.

Initiation at the *E. coli* origin of replication

We know substantially more about the initiation of replication in bacteria than in eukaryotes. The *E. coli* origin of replication is referred to as *oriC*. By transferring segments of DNA from the *oriC* region into plasmids that lack their own origins, it has been estimated that the *E. coli* origin of replication spans approximately 260 bp of DNA. Compared to other bacterial species, the *E. coli* origin is relatively short: the lengths in general range from 250 to 2000 bp. Despite these variations, most bacterial origins have a very similar organization, comprising an AT-rich **DNA unwinding element** (**DUE**) and 5–12 binding sites displaying differing

(A) The structure of *oriC*

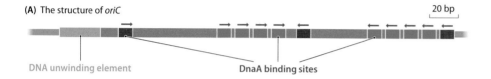

DNA unwinding element   DnaA binding sites

20 bp

(B) Melting of the helix

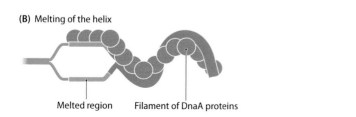

Melted region   Filament of DnaA proteins

Figure 15.9 The *Escherichia coli* origin of replication. (A) The *E. coli* origin of replication is called *oriC* and is approximately 245 bp in length. It contains a DNA unwinding element (DUE) and 11 binding sites for the DnaA protein. These sites have varying binding affinities, shown here as high affinity (red boxes) and low affinity (blue boxes). The high-affinity sites at the extreme left and right of the array have the sequence 5′–TGTGGATAA–3′, and the central high-affinity site has the sequence 5′–TGTGATAA–3′. The low-affinity sites are similar in length but have more variable sequences. (B) Model for the attachment of DnaA proteins to *oriC*, resulting in melting of the helix within the DUE.

degrees of affinity for a protein called DnaA (**Figure 15.9A**). The high-affinity sites, of which there are three in the *E. coli* origin, are permanently occupied by DnaA proteins, the other sites becoming filled immediately before replication commences. Attachment occurs only when the DNA is negatively supercoiled, as is the normal situation for a bacterial chromosome (Section 8.1).

Once attached to their binding sites, the DnaA proteins associated with one another to form a right-handed filament structure with the double-stranded DNA attached to its outer surface (**Figure 15.9B**). This creates torsional stress that opens up, or **melts**, the double helix within the AT-rich DUE. Melting the helix is promoted by members of the HU group of bacterial DNA packaging proteins (Section 8.1).

Melting of the helix initiates a series of events that construct a nascent replication fork at either end of the open region. The first step is the attachment of a **prepriming complex** at each of these two positions. Each prepriming complex initially comprises 12 proteins, 6 copies of DnaB and 6 copies of DnaC, but DnaC has a transitory role and is released from the complex soon after it is formed: its function is probably just to aid the attachment of DnaB. The latter is a helicase, an enzyme which can break base pairs. DnaB begins to increase the size of the melted region within the origin, enabling the enzymes involved in the elongation phase of genome replication to attach. This represents the end of the initiation phase of replication in *E. coli* as the replication forks now start to progress away from the origin and DNA copying begins.

Origins of replication have been clearly defined in yeast

The technique used to delineate the *E. coli oriC* sequence, involving transfer of DNA segments into a nonreplicating plasmid, has also proved valuable in identifying origins of replication in the yeast *Saccharomyces cerevisiae*. Origins identified in this way are called **autonomously replicating sequences**, or **ARSs**. A typical yeast origin is shorter than *E. coli oriC*, usually less than 200 bp in length. Like the *E. coli* origin, the yeast sequence contains discrete regions, called 'subdomains,' each with a different functional role (**Figure 15.10A**). The most important is subdomain A, also called the **autonomous consensus sequence** (**ACS**). This 11 bp sequence is found at over 12,000 positions in the *S. cerevisiae* genome, but only 400 of these act as replication origins under most circumstances. The ACS, along with the adjacent subdomain B1, makes up the **origin recognition sequence**, a stretch of some 40 bp in total that is the binding site for the **origin recognition complex** (**ORC**), a set of six proteins that attach to the origin (**Figure 15.10B**). ORCs are attached to the yeast replication origins at all times and are involved in the regulation of genome replication, acting as mediators between replication origins and the regulatory signals that coordinate the initiation of DNA replication with the cell cycle (Section 15.5).

There is no evidence that the ORC is directly responsible for melting the helix. We must therefore look elsewhere in yeast origins for sequences that provide this

(A) Structure of a yeast origin of replication

B3   B2   B1   A

20bp   Origin recognition sequence

(B) Melting of the helix

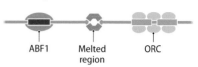

ABF1   Melted region   ORC

Figure 15.10 Structure of a yeast origin of replication. (A) Structure of ARS1, a typical autonomously replicating sequence (ARS) that acts as an origin of replication in *Saccharomyces cerevisiae*. The relative positions of the functional sequences A, B1, B2, and B3 are shown. (B) Melting of the helix occurs within subdomain B2, induced by attachment of the ARS binding factor 1 (ABF1) to subdomain B3. The proteins of the origin replication complex (ORC) are permanently attached to subdomains A and B1.

function. This leads us to the two other conserved sequences in a typical yeast origin, subdomains B2 and B3 (see Figure 15.10A). Our current understanding suggests that melting is induced by attachment of ARS binding factor 1 (ABF1) to subdomain B3 (see Figure 15.10B), and that subdomain B2, which is AT-rich and also acts as the binding site for the MCM helicase, is the position at which the two strands of the helix are first separated. As in *E. coli*, melting of the helix within a yeast replication origin is followed by attachment of the replication enzymes, completing the initiation process and enabling the replication forks to begin their progress along the DNA.

Origins in higher eukaryotes have been less easy to identify

The identification of replication origins in humans and other higher eukaryotes has proved to be less straightforward. **Initiation regions** (parts of the chromosomal DNA where replication initiates) were originally delineated by various biochemical methods, for example, by allowing replication to initiate in the presence of labeled nucleotides, then arresting the process, purifying the newly synthesized DNA, and determining the positions of the labeled strands in the genome. These experiments suggested that there are specific regions in mammalian chromosomes where replication begins, but it quickly became clear that these regions do not contain replication origins equivalent to those in yeast. Further doubts about the specificity of mammalian replication origins were raised by experiments showing that various fragments of mammalian DNA greater than 2 kb in length can confer replicative ability on replication-deficient plasmids introduced into human cells, and that segments of bacterial DNA are only slightly less efficient in this regard. It was also shown that the mammalian ORC, although made of proteins that are homologous to their yeast counterparts, does not have sequence-specific DNA-binding ability and appears to attach to the genome at multiple positions, many of which are not used as replication origins during a particular cell cycle.

The most recent attempts to characterize replication origins in higher eukaryotes have utilized genome-wide screens to identify the positions at which DNA synthesis initiates during genome replication. Various methods have been adopted, the most successful being (**Figure 15.11**):

- **Short nascent strand (SNS) sequencing**, in which DNA is extracted immediately after initiation of a round of genome replication, and the short fragments representing newly synthesized DNA are purified and sequenced. The sequences are then mapped on to the reference genome in order to identify the positions where replication was initiated. A problem with this method is that it can be difficult to distinguish short nascent

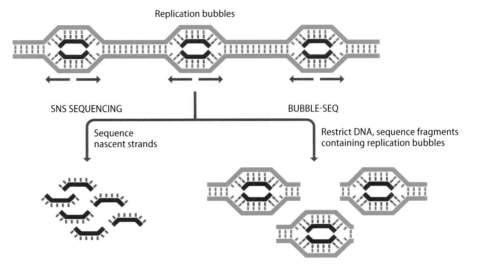

Figure 15.11 Methods for identifying the positions of replication origins. On the left is shown short nascent strand sequencing, in which the short strands representing newly synthesized DNA are sequenced. On the right is bubble-seq, which involves treating the DNA with a restriction endonuclease and sequencing fragments containing replication bubbles. With both methods, the sequence reads are mapped onto the genome annotation in order to identify the positions of replication origins.

fragments from fragments resulting from the random DNA breakage that occurs during DNA extraction.

- **Bubble-seq**, which again involves extracting DNA shortly after genome replication has begun. The difference is that after extraction, the DNA is cut with a restriction endonuclease, and the replication 'bubbles', which form when the two replication forks proceed away from an origin in opposite directions, are purified. Sequencing of these bubbles indicates the replication initiation points.

These approaches have enabled thousands of initiation sites to be identified in various cell types, but there have been concerns about the lack of agreement between the results obtained when different methods are used. In some cases, there is as little as 33% identity between the initiation regions located by SNS sequencing compared with those found by bubble-seq. Results with the same method can also be discordant: there was only a 14% overlap between the initiation regions identified in human HeLa cells when two modifications of SNS sequencing were used. These anomalies might be explained in part by a single cell type using different origins for different rounds of replication, but it is also likely that only a fraction of the active initiation origins are being identified in any single experiment.

Examination of the many initiation regions that have been identified by these high throughput methods has still not revealed any obvious characteristic features among replication origins in higher eukaryotes. It is now accepted that the replication origins of these species are not clearly defined and that the signals that determine where replication initiates are provided largely by the chromatin structure rather than the DNA sequence. In agreement with this model, it is known that ORCs preferentially attach in regions where the DNA displays negative supercoiling, which is induced by removal of nucleosomes. There may also be histone modifications that mark those origins that will be active in a particular round of genome replication.

15.3 EVENTS AT THE REPLICATION FORK

The central players in genome replication are the DNA polymerases that synthesize the daughter strands of DNA. DNA polymerases are a diverse set of enzymes, that can be divided into at least seven groups based on their structural and catalytic properties (Table 15.1). These enzymes include not just the DNA-dependent polymerases that are involved in genome replication and in repair of damaged DNA (Section 17.2), but also reverse transcriptase – an RNA-dependent DNA polymerase – and the template-independent DNA polymerase called terminal deoxynucleotidyl transferase (Section 2.1). To understand the events occurring at the replication fork, we must first understand the properties of the DNA polymerases that carry out genome replication.

DNA polymerases are molecular machines for making (and degrading) DNA

The principal chemical reaction catalyzed by a DNA polymerase is the $5' \rightarrow 3'$ synthesis of a DNA polynucleotide. We learned in Section 2.1 that some DNA polymerases combine this function with at least one exonuclease activity, which means that these enzymes can degrade polynucleotides as well as synthesize them (see Figure 2.7):

- A $3' \rightarrow 5'$ exonuclease activity is possessed by many bacterial and eukaryotic template-dependent DNA polymerases, enabling the enzyme to remove nucleotides from the 3' end of the strand that it has just synthesized. This is looked on as a **proofreading** activity, whose function is to correct the occasional base-pairing error that might occur during strand synthesis.

TABLE 15.1 DNA POLYMERASES

Group	Examples	Organisms	Main roles
A	DNA polymerase I	Bacteria	Completion of lagging-strand synthesis
	DNA polymerase γ	Eukaryotes	Mitochondrial DA replication
	DNA polymerase θ	Eukaryotes	Repair of double-strand breaks
B	DNA polymerases α, δ, ε	Eukaryotes	DNA replication
	DNA polymerase II	Bacteria	Replication of damaged DNA
	DNA polymerase ζ	Eukaryotes	Replication of damaged DNA
C	DNA polymerase III	Bacteria	Main replicating enzyme
D	DNA polymerase D	Archaea	DNA replication
X	DNA polymerases β	Eukaryotes	Base excision repair
	DNA polymerases λ, μ	Eukaryotes	Double-strand break repair
	Terminal deoxynucleotidyl transferase	Eukaryotes	Immunoglobulin and T-cell receptor gene rearrangement
Y	DNA polymerases IV, V	Bacteria	Low-fidelity replication of damaged DNA
	DNA polymerases η, ι, κ	Eukaryotes	Low-fidelity replication of damaged DNA
RT	Reverse transcriptase	Retroelements	Viral replication, retroelement transposition

- A 5′→3′ exonuclease activity is less common but is possessed by some polymerases whose function in replication requires that they must be able to remove at least part of a polynucleotide that is already attached to the template strand that the polymerase is copying.

The search for DNA polymerases began in the mid-1950s, as soon as it was realized that DNA synthesis was the key to replication of genes. It was thought that bacteria would probably have just a single DNA polymerase, and when the enzyme now called **DNA polymerase I** was isolated by Arthur Kornberg in 1957 there was a widespread assumption that this was the main replicating enzyme. The discovery that inactivation of the *E. coli polA* gene, which codes for DNA polymerase I, was not lethal (cells were still able to replicate their genomes) therefore came as something of a surprise, especially when a similar result was obtained with inactivation of *polB*, coding for a second enzyme, **DNA polymerase II**. It was not until 1972 that the main replicating polymerase of *E. coli*, **DNA polymerase III**, was eventually isolated. Both DNA polymerases I and III are involved in genome replication, as we will see later in this section. DNA polymerase II, as well as other bacterial DNA polymerases such as IV and V, are mainly involved in the repair of damaged DNA (Section 17.2).

DNA polymerases I and II are single polypeptides but DNA polymerase III, befitting its role as the main replicating enzyme, is a multisubunit protein. The subunit called α is the one that is responsible for synthesizing the new polynucleotide, with the other subunits playing ancillary roles in the replication process. For example, the ε subunit specifies a 3′→5′ exonuclease activity and the β subunit acts as a 'sliding clamp' holding the polymerase complex tightly to the template strand, but at the same time allowing it to move along that strand as it makes the new polynucleotide.

Eukaryotes have at least 15 DNA polymerases, which in mammals are distinguished by Greek suffixes (α, β, δ, γ, etc.), an unfortunate choice of nomenclature as it tempts confusion with the identically named subunits of *E. coli* DNA polymerase III. The main replicating enzymes are **DNA polymerase δ** and **DNA polymerase ε**, which work in conjunction with an accessory protein called the

proliferating cell nuclear antigen (**PCNA**). PCNA is a homotrimer that encircles the DNA and provides the attachment point for the DNA polymerases and other proteins involved in replication. **DNA polymerase α** also has an important function in genome replication and **DNA polymerase γ**, although coded by a nuclear gene, is responsible for replicating the mitochondrial genome. As with the prokaryotic enzymes, most of the other eukaryotic DNA polymerases are involved in the repair of damaged DNA.

DNA polymerases have limitations that complicate genome replication

DNA polymerases have two limitations that complicate genome replication. The first of these is their ability to synthesize DNA in only the 5′→3′ direction. This means that one strand of the parent double helix, called the **leading strand**, can be copied in a continuous manner, but replication of the **lagging strand** has to be carried out in a discontinuous fashion, resulting in a series of short segments that must be ligated together to produce the intact daughter strand (**Figure 15.12**). The discontinuous nature of lagging-strand replication was confirmed in 1969 when **Okazaki fragments**, as these segments are now called, were first isolated from *E. coli*. In bacteria, Okazaki fragments are 1000–2000 nucleotides in length, but in eukaryotes the equivalent fragments appear to be much shorter, perhaps less than 200 nucleotides in length.

The second limitation of DNA synthesis by DNA polymerases, which is also illustrated in Figure 15.12, is the need for a primer to initiate synthesis of each new polynucleotide. It is not known for certain why DNA polymerases cannot begin synthesis on an entirely single-stranded template, but this may relate to the proofreading activity of these enzymes, which is essential for the accuracy of replication. The logic is as follows. If the nucleotide that has just been added to the 3′ end of the growing polynucleotide is not base-paired to the template, then an error has been made, and to correct this error, the 3′→5′ exonuclease of the polymerase must operate instead of continued 5′→3′ polymerization. In other words, the 5′→3′ polymerase function is active only if the 3′-terminal nucleotide is base-paired to the template. A template that is entirely single-stranded has no base-paired 3′ nucleotide, and so requires a primer, to provide that nucleotide, in order for the polymerase to be activated.

Whatever the reason, priming is a necessity in DNA replication but does not present too much of a problem. Although DNA polymerases cannot deal with an entirely single-stranded template, RNA polymerases have no difficulty in this respect, so the primers for DNA replication are made of RNA. In bacteria, primers are synthesized by **primase**, a special RNA polymerase unrelated to the transcribing enzyme, with each primer being 10–12 nucleotides in length. Once the primer has been completed, strand synthesis is continued by DNA polymerase III (**Figure 15.13A**). In eukaryotes, the situation is slightly more complex because the primase is tightly bound to DNA polymerase α, and cooperates with this enzyme in synthesis of the first few nucleotides of a new polynucleotide. This primase synthesizes an RNA primer of 7–12 nucleotides; the primase is then replaced by DNA polymerase α, which extends the RNA primer by adding about 20 nucleotides of DNA. This DNA stretch often has a few ribonucleotides mixed in, but it is not clear if these are incorporated by DNA polymerase α or by intermittent activity of the primase. After completion of the RNA–DNA primer, DNA synthesis is continued by DNA polymerase ε on the leading strand, and DNA polymerase δ on the lagging strand (**Figure 15.13B**).

Priming needs to occur just once on the leading strand, within the replication origin, because once primed, the leading-strand copy is synthesized continuously until replication is completed. On the lagging strand, priming is a repeated process that must occur every time a new Okazaki fragment is initiated. In *E. coli*, which makes Okazaki fragments of 1000–2000 nucleotides in length, approximately 4000 priming events are needed every time the genome is replicated. In eukaryotes, the Okazaki fragments are much shorter and priming is a highly repetitive event.

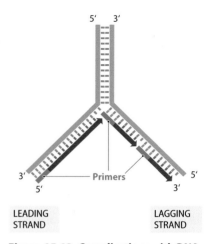

Figure 15.12 Complications with DNA replication. Two complications have to be solved when double-stranded DNA is replicated. First, only the leading strand can be continuously replicated by 5′→3′ DNA synthesis; replication of the lagging strand has to be carried out discontinuously. Second, initiation of DNA synthesis requires a primer.

Figure 15.13 Priming of DNA synthesis in (A) bacteria and (B) eukaryotes. In eukaryotes the primase forms a complex with DNA polymerase α, which is shown synthesizing the RNA primer followed by the first few nucleotides of DNA.

(A) Priming of DNA synthesis in bacteria

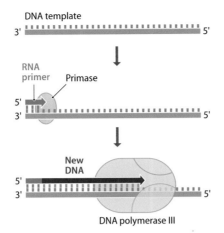

(B) Priming of DNA synthesis in eukaryotes

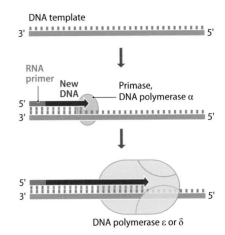

(A) SSBs attach to the unpaired polynucleotides

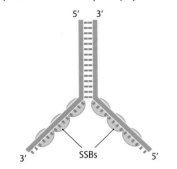

(B) Structure of RPA, a eukaryotic SSB

Figure 15.14 The role of single-strand binding proteins (SSBs) during DNA replication. (A) SSBs attach to the unpaired polynucleotides produced by helicase action and prevent the strands from base-pairing with one another or being degraded by single-strand-specific nucleases. (B) Structure of the eukaryotic SSB called RPA. The protein contains a β-sheet structure that forms a channel in which the DNA (shown in dark orange, viewed from the end) is enclosed. (From Bochkarev, Pfuetzner R.A., Edwards A.M. & Frappier L. [1997] *Nature* 385:176–181. With permission from Springer Nature.)

Okazaki fragments must be joined together to complete lagging-strand replication

Many of the events at the replication fork are similar in both bacteria and eukaryotes. Progress of the replication fork is maintained by helicase activity, with the torsional stress that results from overwinding of the helix ahead of the fork relieved by DNA topoisomerases. Single-stranded DNA is naturally 'sticky' and the two separated polynucleotides produced by helicase action would immediately re-form base pairs after the enzyme has passed, if allowed to. The single strands are also highly susceptible to nuclease attack and are likely to be degraded if not protected in some way. To avoid these unwanted outcomes, **single-strand binding proteins** (**SSBs**) attach to the polynucleotides and prevent them from reassociating or being degraded (**Figure 15.14A**). The *E. coli* SSB is made up of four identical subunits and the major eukaryotic SSB, called **replication protein A** (**RPA**), is a heterotrimer of three different proteins. Both work in a similar way, with the polynucleotide becoming enclosed in a channel formed by a series of SSBs attached side by side on the strand (**Figure 15.14B**). Detachment of the SSBs, which must occur when the replication complex arrives to copy the single strands, is brought about by a second set of proteins called **replication mediator proteins** (**RMPs**). DNA synthesis in *E. coli* is carried out by two linked copies of DNA polymerase III, one for leading-strand replication and one for the lagging strand (**Figure 15.15**), aided by a single **γ complex** (sometimes called the 'clamp loader'). The main role of the γ complex is to interact with the β subunit (the 'sliding clamp') of each polymerase, and hence control the attachment and removal of the enzyme from the template, a function that is required primarily during lagging-strand replication when the enzyme has to attach and detach repeatedly at the start and end of each Okazaki fragment. The combination of these two DNA polymerases, along with the primase needed for repetitive initiation of Okazaki fragments, is called the **replisome**. In eukaryotes, the DNA polymerase ε and δ enzymes that copy the leading and lagging strands, respectively, do not associate into a dimeric complex but instead remain separate. The function performed by the γ complex of the *E. coli* polymerase – controlling attachment and detachment of the DNA polymerase from the lagging strand – is carried out by a multisubunit accessory protein called **replication factor C** (**RFC**).

We are left with just one issue that must be resolved in order to complete the DNA synthesis phase of genome replication. The initial copy of the lagging strand comprises a series of Okazaki fragments. The RNA primers of these fragments must be replaced with DNA, and adjacent fragments joined, in order to generate the final version of the lagging-strand copy. This is one aspect of genome replication where the events occurring in bacteria and eukaryotes are significantly different.

Removal of the RNA primer of an Okazaki fragment can be achieved if the DNA polymerase that is making the next fragment in the series has a 5′→3′ exonuclease activity. This polymerase could continue making its DNA copy of the lagging strand into the region initially occupied by the 5′ end of the adjacent Okazaki fragment, by using the exonuclease activity to remove the RNA nucleotides and then recopying that segment into DNA. Unfortunately, DNA polymerase III, which copies the lagging strand in bacteria, lacks the required 5′→3′ exonuclease activity (Table 15.2). DNA polymerase III therefore detaches from the lagging strand when it reaches the 5′ end of the adjacent Okazaki fragment, and its place is taken by DNA polymerase I, which does have a 5′→3′ exonuclease activity and so removes the primer, and usually the start of the DNA component of the Okazaki fragment as well, extending the 3′ end of the fragment it is synthesizing into the region of the template that is exposed (Figure 15.16A). The two Okazaki fragments now abut, with the terminal regions of both composed entirely of DNA. All that remains is for the missing phosphodiester bond to be put in place by a **DNA ligase**, linking the two fragments and completing replication of this region of the lagging strand.

In eukaryotes, the problem is more extreme as there appears to be no eukaryotic DNA polymerase with the 5′→3′ exonuclease activity needed to degrade the RNA primers of the Okazaki fragments. The solution to this problem is therefore very different to that described for bacteria. The central player is the **flap endonuclease (FEN1)**, which has an unusual endonuclease activity that enables it to cut the phosphodiester bond at the branch point at the base of the 'flap' that is formed when the 5′ end of a polynucleotide is displaced from its template. FEN1 therefore associates with DNA polymerase δ as it approaches the RNA primer of the adjacent Okazaki fragment. The base pairs holding the primer to the lagging strand are broken by a helicase enzyme, enabling the primer to be pushed aside by the polymerase as it extends the adjacent Okazaki fragment into the exposed region, resulting in the flap structure that can be cut off by FEN1 (Figure 15.16B). A DNA ligase can then form the final phosphodiester bond to link the Okazaki fragments. This scheme raises the possibility that both the RNA primer and all of the DNA originally synthesized by the priming enzyme, DNA polymerase α, are removed. This is an attractive option because DNA polymerase α has no 3′→5′ proofreading activity (see Table 15.2) and therefore synthesizes DNA in a relatively error-prone manner. Removal of this region as part of the flap cleaved by FEN1, followed by resynthesis by DNA polymerase δ (which does have a proofreading activity and so makes a highly accurate copy of the template), would prevent any errors made by DNA polymerase α from becoming permanent features of the daughter double helix.

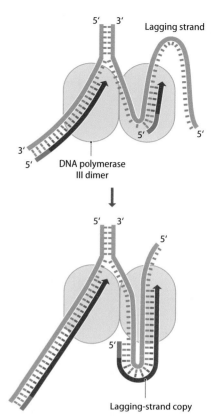

Figure 15.15 A model for parallel synthesis of the leading- and lagging-strand copies by a dimer of DNA polymerase III enzymes. It is thought that the lagging strand loops through its copy of the DNA polymerase III enzyme, in the manner shown, so that both the leading and lagging strands can be copied as the dimer moves along the molecule being replicated. The two components of the DNA polymerase III dimer are not identical because they share a single γ complex, which is made up of subunit γ in association with subunits δ, δ′, χ, and ψ.

TABLE 15.2 EXONUCLEASE ACTIVITIES OF DNA POLYMERASES INVOLVED IN REPLICATION OF BACTERIAL AND EUKARYOTIC GENOMES		
	Exonuclease activities	
Enzyme	**3′→5′**	**5′→3′**
Bacterial DNA polymerases		
DNA polymerase I	Yes	Yes
DNA polymerase III	Yes	No
Eukaryotic DNA polymerases		
DNA polymerase α	No	No
DNA polymerase δ	Yes	No
DNA polymerase ε	Yes	No

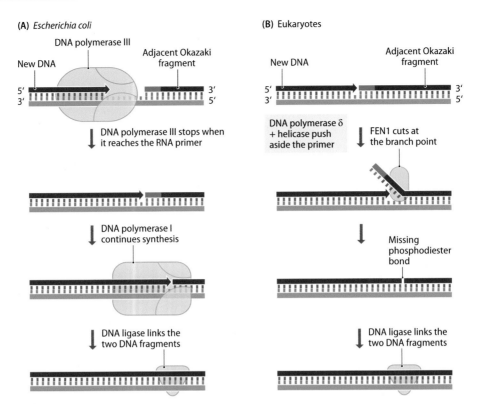

(A) *Escherichia coli*

DNA polymerase III

New DNA

Adjacent Okazaki fragment

↓ DNA polymerase III stops when it reaches the RNA primer

↓ DNA polymerase I continues synthesis

↓ DNA ligase links the two DNA fragments

(B) Eukaryotes

New DNA

Adjacent Okazaki fragment

DNA polymerase δ + helicase push aside the primer

FEN1 cuts at the branch point

↓ Missing phosphodiester bond

↓ DNA ligase links the two DNA fragments

Figure 15.16 The series of events involved in joining up adjacent Okazaki fragments during DNA replication. (A) DNA polymerase III of *E. coli* lacks a 5′→3′ exonuclease activity and so stops making DNA when it reaches the RNA primer of the adjacent Okazaki fragment. At this point, DNA synthesis is continued by DNA polymerase I, which does have a 5′→3′ exonuclease activity, and so can remove the RNA primer and replace it with DNA. DNA polymerase I usually also replaces some of the DNA from the Okazaki fragment before detaching from the template. This leaves a single missing phosphodiester bond, which is synthesized by DNA ligase, completing this step in the replication process. (B) The eukaryotic DNA polymerase δ also lacks a 5′→3′ exonuclease activity, but it continues making DNA into the region occupied by the primer of the adjacent Okazaki fragment, the 5′ end of this fragment being displaced by a helicase. The resulting flap structure is then cut off by FEN1.

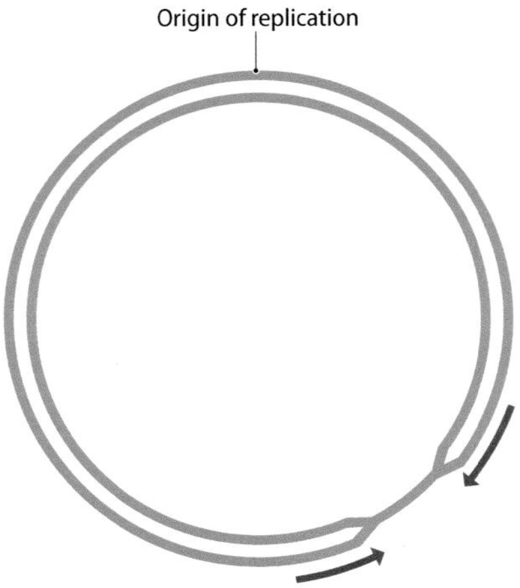

15.4 TERMINATION OF GENOME REPLICATION

Replication forks proceed along linear genomes, or around circular ones, generally unimpeded except when a region that is being transcribed is encountered. DNA synthesis occurs at approximately five times the rate of RNA synthesis, so the replication complex can easily overtake an RNA polymerase, but this probably does not happen: instead, it is thought that the replication fork pauses behind the RNA polymerase, proceeding only when the transcript has been completed.

Eventually, the replication fork reaches the end of the molecule or meets a second replication fork moving in the opposite direction. What happens next is one of the least understood aspects of genome replication.

Replication of the *E. coli* genome terminates within a defined region

Bacterial genomes are replicated bidirectionally from a single point (see Figure 15.8A), which means that the two replication forks should meet at a position diametrically opposite the origin of replication on the genome map. However, if one fork is delayed, possibly because it has to replicate extensive regions where transcription is occurring, then it might be possible for the other fork to overshoot the halfway point and continue replication on the other side of the genome (**Figure 15.17**). It is not immediately apparent why this should be undesirable, the daughter molecules presumably being unaffected, but it is not allowed to happen because of the presence of **terminator sequences**. Ten of these have been identified in the *E. coli* genome (**Figure 15.18A**), each one acting as the recognition site for a sequence-specific DNA-binding protein called the **terminator utilization substance** (**Tus**).

Origin of replication

Figure 15.17 A situation that is not allowed to occur during replication of the circular *E. coli* genome. One of the replication forks has proceeded some distance past the halfway point. This does not happen during DNA replication in *E. coli* because of the action of the Tus proteins.

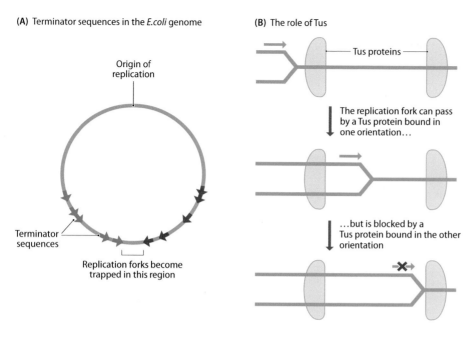

(A) Terminator sequences in the *E.coli* genome

Origin of replication

Terminator sequences

Replication forks become trapped in this region

(B) The role of Tus

Tus proteins

The replication fork can pass by a Tus protein bound in one orientation...

...but is blocked by a Tus protein bound in the other orientation

Figure 15.18 The role of terminator sequences during DNA replication in *E. coli*. (A) The positions of the ten terminator sequences on the *E. coli* genome are shown, with the arrowheads indicating the direction that each terminator sequence can be passed by a replication fork. (B) Bound Tus proteins allow a replication fork to pass when the fork approaches from one direction but not when it approaches from the other direction.

The mode of action of Tus is quite unusual. When bound to a terminator sequence, a Tus protein allows a replication fork to pass if the fork is moving in one direction, but blocks progress if the fork is moving in the opposite direction around the genome. The directionality is set by the orientation of the Tus protein on the double helix (**Figure 15.18B**). Exactly how Tus exerts its effect is not yet understood, but it has been assumed that the Tus protein interacts directly with the DnaB helicase, which is responsible for progression of the replication fork. This assumption has been challenged by research that supports the alternative possibility that the key interactions are between Tus and the replication fork itself. The progress of individual replication forks has been studied in an experimental system in which replisome proteins are absent. Without these proteins, a fork will not move naturally along a DNA double helix. To bring about fork progression in the experimental system, a magnetic bead was attached to the end of one of the polynucleotides, and the end of the second polynucleotide was immobilized by attachment to a solid support. The two polynucleotides were then pulled apart by manipulating the magnetic bead with a **magnetic tweezers** – a device comprising a set of magnets whose positions and field strengths can be varied in such a way that the magnetic bead, and its attached polynucleotide, can be moved about in a controlled manner (**Figure 15.19**). By moving the magnetic bead away from the solid support, the helix becomes opened up, producing a fork that can be moved along the helix simply by pulling the ends of the two polynucleotides further apart. These experiments have shown that if the orientation of the terminator sequence is such that the fork approaches a Tus protein from the permissive direction, then movement of the fork is not impeded, but when the fork approaches from the nonpermissive direction, its progress is prevented by Tus. The implication is that interactions between Tus and the DNA being replicated are at least partly responsible for the ability of Tus to block progression of the replication fork. The results do not exclude a role for a protein–protein interaction between Tus and DnaB, but they do show that such an interaction is unlikely to be the full explanation of the mode of action of Tus. Whatever the mechanism, the orientation of the terminator sequences, and hence of the bound Tus proteins, in the *E. coli* genome is such that both replication forks become trapped within a relatively short region on the opposite side of the genome to the origin (see Figure 15.18A), and it is at this point that the termination events take place.

There appears to be less control over the positions at which replication forks meet in eukaryotes. Fork-blocking proteins that act in a similar manner to Tus are known, but only a small number of terminator sequences, called **replication**

(A) Manipulation of polynucleotides with magnetic tweezers

(B) Effect of Tus proteins on porgression of the replication fork

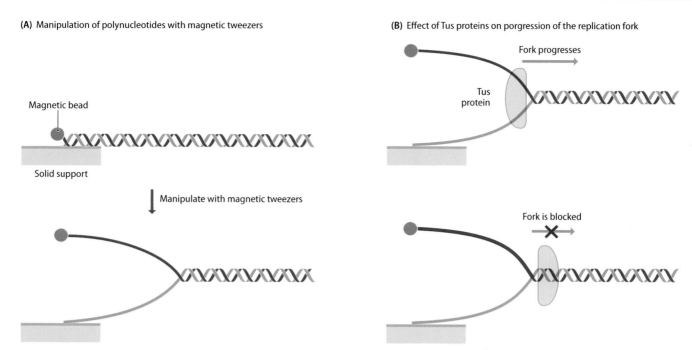

Figure 15.19 Using magnetic tweezers to examine the role of the Tus proteins. (A) The magnetic tweezers enable the magnetic bead to be manipulated so that the two polynucleotides are pulled apart. (B) The resulting replication fork is only able to pass a bound Tus protein when approaching from the permissive direction. As none of the proteins of the replisome are present, the conclusion is that interactions between Tus and the DNA being replicated are at least partly responsible for the ability of Tus to block progression of the replication fork.

fork barriers, are present in a eukaryotic genome. One of these barriers is located adjacent to the ribosomal DNA repeat units and prevents a replication fork from entering this region and colliding head-on with the transcription complex expressing these genes (**Figure 15.20**). The presence of this barrier therefore ensures that the ribosomal DNA region is always replicated by a fork that follows the RNA polymerase. A head-on collision is thought to be undesirable because it can inactivate the helicase component of the replisome, causing the replication fork to stall and possibly becomes destabilized. Head-on collisions are clearly possible in many parts of the genome, including places where there are no replication fork barriers, but presumably the high level of transcription that occurs in the ribosomal RNA regions presents a particular danger to the replication process.

Completion of genome replication

In outline, the termination stage of DNA replication is a straightforward process; a pair of replication forks meet, the components of the replisomes detach, and the ends of the polynucleotides are ligated to create the daughter genomes. There are just two issues that we need to consider.

The first issue is the topological problem that arises when two replication forks approach one another. The torsional stress that builds up ahead of a replication fork as the replisome progresses along a DNA molecule results in the formation of a positive supercoil that is continually unwound by the action of a type I or type II topoisomerase, as described earlier in this chapter (Section 15.1). However, when two replication forks approach to within 150 bp of one another

Figure 15.20 The role of the replication fork barrier in the ribosomal DNA region. The presence of the replication fork barrier ensures that the fork progressing from the right is blocked so that it does not meet the transcription complex head-on. The fork progressing from the left is unimpeded and follows the transcription complex through the ribosomal DNA region.

Transcription complex synthesizing RNA

Replication fork barrier

Replication fork proceeds

Replication fork is blocked

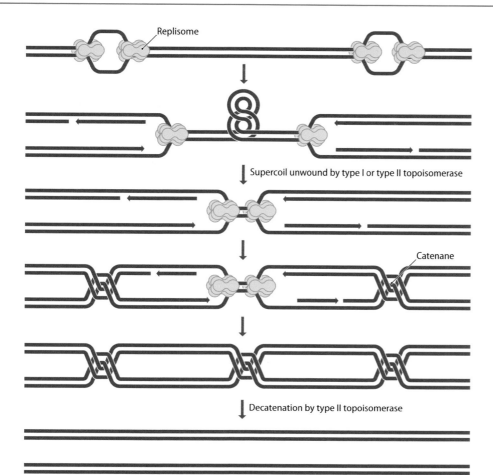

Figure 15.21 Topological issues during termination of replication. The positive supercoil that is generated by the torsional stress that builds up ahead of a replication fork is continually unwound by the action of a type I or type II topoisomerase. When two replication forks approach one another there is insufficient space for a supercoil to form. Instead, the DNA becomes wound into a catenane that is decatenated by a type II topoisomerase.

there is insufficient space for a supercoil to form. Instead, the torsional stress causes the two forks to rotate in opposite directions, resulting in the newly synthesized DNA becoming wound into an interlocked structure called a **catenane** (**Figure 15.21**). Decatenation, which is required to separate the two daughter genomes, cannot be achieved by a type I topoisomerase and requires the action of a type II enzyme, topoisomerase IV, for *E. coli* and Top2 in eukaryotes.

The second question concerns the process by which the replisomes are disassembled. This is not a trivial event because several parts of the replisome are tightly attached to the DNA, and the structure as a whole will not simply fall apart once it has completed the replication process. In eukaryotes, disassembly commences with removal of the helicase from the replisome. This helicase is a multisubunit protein called the MCM2–7 complex. Ubiquitination of the MCM7 subunit by the RNF8 ubiquitin ligase provides a recognition signal for a protein called CDC48 in yeast and p97 in mammals, which acts as a 'segregase,' removing many different types of protein from subcellular structures and entering these into the ubiquitin-directed degradation pathway (Section 13.3). Degradation of the replisome components, rather than their re-use, is to be expected as they will not be needed until the next round of DNA replication, which even in rapidly dividing eukaryotic cells will occur only after a delay of some 24 hours. The MCM2–7 complex makes a number of contacts with other components of the replisome, which are probably detached at the same time, although a few proteins, such as the PCNA, are thought to be left behind and disassembled by different as yet uncharacterized, pathways.

The timing of helicase removal in both *E. coli* and eukaryotes is such that by that this stage there is very little, if any, nonreplicated DNA between the two converging replication forks. Any gap filing that is needed appears to be carried out in the same way as the joining of Okazaki fragments during the elongation phase, by DNA polymerase I in bacteria or DNA polymerase δ plus FEN1 in eukaryotes, with the strands then joined by DNA ligase. Neither DNA polymerase δ nor FEN1 are closely associated with the MCM2–7 helicase, so are likely to remain attached to the DNA after the major components of the replisome have been removed.

Telomerase completes replication of chromosomal DNA molecules, at least in some cells

There is one final problem that we must consider before leaving the replication process. This concerns the steps that have to be taken to prevent the ends of a linear, double-stranded molecule from gradually getting shorter during successive rounds of chromosomal DNA replication. There are two ways in which this shortening might occur:

- The extreme 3′ end of the lagging strand might not be copied because the final Okazaki fragment cannot be primed, the natural position for the priming site being beyond the end of the template (**Figure 15.22A**). The absence of this Okazaki fragment means that the lagging-strand copy is shorter than it should be. If the copy remains this length, then when it acts as a parent polynucleotide in the next round of replication the resulting daughter molecule will be shorter than its grandparent.

- If the primer for the last Okazaki fragment is placed at the extreme 3′ end of the lagging strand, then shortening will still occur, although to a lesser extent, because this terminal RNA primer cannot be converted into DNA by the standard processes for primer removal (**Figure 15.22B**). This is because the methods for primer removal require extension of the 3′ end of an adjacent Okazaki fragment, which cannot occur at the very end of the molecule.

Once this problem had been recognized, attention was directed at the telomeres, the unusual DNA sequences at the ends of eukaryotic chromosomes. We noted in Section 7.1 that telomeric DNA is made up of a type of minisatellite sequence comprised of multiple copies of a short repeat motif, 5′–TTAGGG–3′ in most higher eukaryotes, a few hundred copies of this sequence occurring in tandem repeats at each end of every chromosome. The solution to the end-shortening problem lies in the way in which this telomeric DNA is made. Most of the telomeric DNA is copied in the normal fashion during DNA replication, but this is not the only way in which it can be synthesized. To compensate for the limitations of the replication process, telomeres can be extended by an independent mechanism catalyzed by the enzyme **telomerase**. This is an unusual enzyme in that it consists of both protein and RNA. In the human enzyme, the RNA component is 450 nucleotides in length and contains near its 5′ end the sequence 5′–CUAACCCUAAC–3′, whose central region is the reverse complement of the human telomere repeat sequence 5′-TTAGGG–3′. This complementarity enables telomerase to extend the telomeric DNA at the 3′ end of a polynucleotide by the copying mechanism shown in **Figure 15.23**, in which the telomerase RNA is used as a template for each extension step, the DNA synthesis being carried out by the protein component of the enzyme, which is a reverse transcriptase. The correctness of this model is indicated by comparisons between telomere repeat sequences and the telomerase RNAs of other species (Table 15.3): in all organisms that have been looked at, the telomerase RNA contains a sequence that enables it to make copies of the repeat motif present at the organism's telomeres. An interesting feature is that, in all organisms, the strand synthesized by telomerase has a preponderance of G nucleotides, and is therefore referred to as the G-rich strand.

Telomerase can only synthesize this G-rich strand. It is not clear how the other polynucleotide – the C-rich strand – is extended, but it is presumed that when the G-rich strand is long enough, the primase–DNA polymerase α complex

(A) The final Okazaki fragment cannot be primed

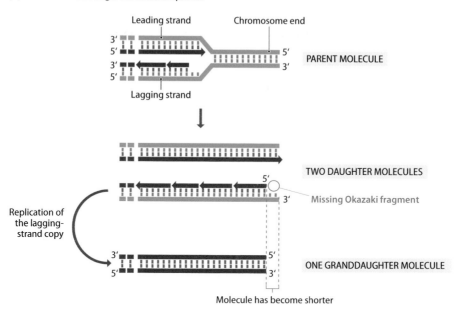

(B) The primer for the last Okazaki fragment is at the extreme 3′ end of the lagging strand

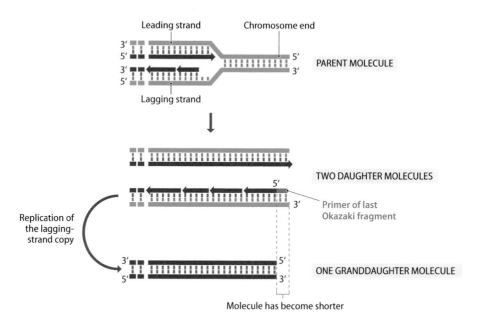

Figure 15.22 Two of the reasons why linear DNA molecules could become shorter after DNA replication. In both examples, the parent molecule is replicated in the normal way. A complete copy is made of its leading strand, but in (A), the lagging-strand copy is incomplete because the last Okazaki fragment is not made. This is because primers for Okazaki fragments are synthesized at positions approximately 200 bp apart on the lagging strand. If one Okazaki fragment begins at a position less than 200 bp from the 3′ end of the lagging strand, then there will not be room for another priming site, and the remaining segment of the lagging strand is not copied. The resulting daughter molecule therefore has a 3′ overhang and, when replicated, gives rise to a granddaughter molecule that is shorter than the original parent. In (B), the final Okazaki fragment can be positioned at the extreme 3′ end of the lagging strand, but its RNA primer cannot be converted into DNA because this would require extension of another Okazaki fragment positioned beyond the end of the lagging strand. It is not clear if a terminal RNA primer can be retained throughout the cell cycle, nor is it clear if a retained RNA primer can be copied into DNA during a subsequent round of DNA replication. If the primer is not retained or is not copied into DNA, then one of the granddaughter molecules will be shorter than the original parent.

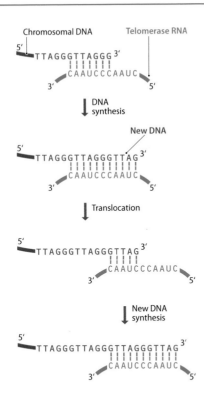

Chromosomal DNA   Telomerase RNA

DNA synthesis

New DNA

Translocation

New DNA synthesis

Figure 15.23 Extension of the end of a human chromosome by telomerase. The 3´ end of a human chromosomal DNA molecule is shown. The sequence comprises repeats of the human telomere motif 5´–TTAGGG–3´. The telomerase RNA base-pairs to the end of the DNA molecule, which is extended a short distance. The telomerase RNA then translocates to a new base-pairing position slightly further along the DNA polynucleotide, and the molecule is extended by a few more nucleotides. The process can be repeated until the chromosome end has been extended by a sufficient amount

attaches at its end and initiates synthesis of complementary DNA in the normal way (**Figure 15.24**). This requires the use of a new RNA primer, so the C-rich strand will still be shorter than the G-rich one, but the important point is that the overall length of the chromosomal DNA has not been reduced. In mammalian cells, after extension, the end of the telomere may form a 't-loop,' in which the free 3´ end loops back, invades the double helix, and forms base pairs with its complementary sequence on the C-rich strand (**Figure 15.25**). This reaction is promoted by the telomere-binding protein, TRF2, and may provide additional stabilization of a chromosome end that does not require further extension.

The activity of telomerase must clearly be controlled very carefully to ensure that the appropriate length extension is made at each chromosome end. Exactly how this is achieved is not yet fully understood. For some years, a 'protein count-ing' model has been favored, in which the proteins that bind to the telomere repeat sequences (Section 7.1) exert a repressive effect on telomerase activity. As the telomere shortens, the number of bound proteins decreases, enabling telom-erase to attach to the end of the chromosome and extend the telomere. As the telomere extends, proteins reattach, so that the telomerase activity is again sup-pressed. In effect, the telomere-binding proteins mediate a negative feedback loop that regulates telomerase activity at a particular chromosome end.

Telomere length is implicated in cell senescence and cancer

Perhaps surprisingly, telomerase is not active in all mammalian cells. The enzyme is functional in the early embryo, but after birth is active only in the reproductive cells and **stem cells**. The latter are progenitor cells that divide con-tinually throughout the lifetime of an organism, producing new cells to maintain organs and tissues in a functioning state. Somatic cells, which lack telomerase activity, undergo chromosome shortening every time they divide. Eventually, after many cell divisions, the chromosome ends could become so truncated that essential genes are lost, but this is unlikely to be a major cause of the defects that can occur in cells lacking telomerase activity. Instead, the critical factor is the need to maintain a protein 'cap' on each chromosome end, to protect these ends from the effects of the DNA repair enzymes that join together the uncapped ends that are produced by accidental breakage of a chromosome (Section 17.2). The proteins that form this protective cap, such as TRF2 in humans, recognize the telomere repeats as their binding sequences, and so have no attachment points after the telomeres have been deleted. If these proteins are absent then the repair enzymes can make inappropriate linkages between the ends of intact, although

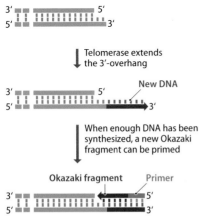

Telomerase extends the 3´-overhang

New DNA

When enough DNA has been synthesized, a new Okazaki fragment can be primed

Okazaki fragment   Primer

Figure 15.24 Completion of the extension process at the end of a chromosome. It is believed that after telomerase has extended the 3´ end by a sufficient amount a new Okazaki frag-ment is primed and synthesized, convert-ing the 3´ extension into a completely double-stranded end.

TABLE 15.3 SEQUENCES OF TELOMERE REPEATS AND TELOMERASE RNAS IN VARIOUS ORGANISMS		
Species	**Telomere repeat sequence**	**Telomerase RNA template sequence**
Human	5´–TTAGGG–3´	5´–CUAACCCUAAC–3´
Oxytricha	5´–TTTTGGGG–3´	5´–CAAAACCCCAAAACC–3´
Tetrahymena	5´–TTGGGG–3´	5´–CAACCCCAA–3´

Oxytricha and *Tetrahymena* are protozoans which are particularly useful for telomere studies because at certain developmental stages their chromosomes break into small fragments, all of which have telomeres: they therefore have many telomeres per cell.

shortened, chromosomes; it is this that is probably the underlying cause of the disruption to the cell cycle that results from telomere shortening.

Telomere shortening will therefore lead to the termination of a cell lineage. For several years, biologists have attempted to link this process with **cell senescence**, a phenomenon originally observed in cell cultures. All normal cell cultures have a limited lifetime: after a certain number of divisions, the cells enter a senescent state in which they remain alive but cannot divide (**Figure 15.26**). With some mammalian cell lines, notably fibroblast cultures (connective tissue cells), senescence can be delayed by engineering the cells so that they synthesize active telomerase. These experiments suggest a clear relationship between telomere shortening and senescence, but the exactness of the link has been questioned, and any extrapolation from cell senescence to aging of the organism is fraught with difficulties.

Not all cell lines display senescence. Cancerous cells are able to divide continuously in culture, their immortality being looked upon as analogous to tumor growth in an intact organism. With several types of cancer, this absence of senescence is associated with activation of telomerase, sometimes to the extent that telomere length is maintained through multiple cell divisions, but often in such a way that the telomeres become longer than normal because the telomerase is overactive. These observations suggest that drugs that inhibit telomerase activity might be useful in cancer treatment. Attempts to inactivate telomerase in cancer cells have focused on both the protein and RNA components of the enzyme. The protein has been used to prepare vaccines that contain antibodies that should bind to and inactivate any telomerase proteins that they encounter. In clinical trials, these vaccines have been able to reduce the number of cancerous cells circulating in the bloodstream of patients, reducing the chances of their cancers spreading to other parts of their body. Whether the vaccines can reduce the growth of existing tumors, or prevent a cancer from becoming established in the first place, is not yet clear. A second approach makes use of a short oligonucleotide that is complementary to part of the RNA component of telomerase. The idea is that this oligonucleotide will inhibit a telomerase enzyme by binding to its RNA molecule (**Figure 15.27**). Vaccines and oligonucleotides targeted at telomerase are among the most promising of the various strategies being used to combat cancer, but a stumbling block at the moment is the ability of some cells to maintain their telomeres, and hence avoid senescence, by an alternative process that does not require telomerase. This involves the transfer of telomere repeats from DNA molecules that have not yet reached a critically-short length to others that are close to the danger point. This process is switched on in some cancer cells, counteracting the effects of telomerase inactivation.

Drosophila has a unique solution to the end-shortening problem

When the amino acid sequences of the protein subunits of telomerase enzymes are compared with those of other reverse transcriptases, the closest similarities are seen with the reverse transcriptases coded by the non-LTR retroelements called retroposons (Section 9.2). This is a fascinating observation when taken in conjunction with the unusual structure of the telomeres of *Drosophila*. These telomeres are not made up of the short repeated sequences seen in most other organisms, but instead consist of tandem arrays of much longer repeats, 6 or 10 kb in length. These repeats are full-length and truncated copies of three *Drosophila* retroposons, called *HeT-A*, *TART*, and *TAHRE*. They are maintained by a process analogous to that carried out by telomerase (**Figure 15.28**), with a template RNA obtained by transcription of the telomeric retroposons being copied by the reverse transcriptase coded by the *TART* and *TAHRE* sequences (*HeT-A* does not have a reverse transcriptase gene).

The unusual structure of the *Drosophila* telomere could simply be a quirk of nature, but the attractive possibility that the telomeres of other organisms are degraded retroposons, as suggested by the similarities between telomerase and retroposon reverse transcriptases, cannot be discounted.

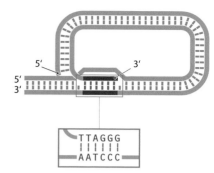

Figure 15.25 The 't-loop.' The t-loop is formed when the free 3´ end of the telomere loops back and invades the double helix.

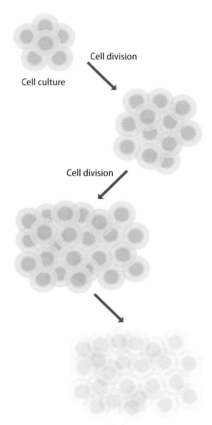

Figure 15.26 Cultured cells become senescent after multiple cell divisions.

Figure 15.27 The use of an oligonucle-otide to inhibit telomerase activity. The oligonucleotide is complementary to the RNA component of telomerase and hence competes with the chromosomal DNA for attachment to the telomerase RNA. Telomere extension is therefore inhibited.

Figure 15.28 A model for maintenance of *Drosophila* telomeres. In this example, a *TART* sequence is located at the extreme end of the telomere. DNA replication has left the typical 3´ overhang, which is extended by reverse transcription of a RNA copy of a *TART* sequence located within the intact region of the telomere. A new Okazaki fragment can then be primed in order to complete the extension process.

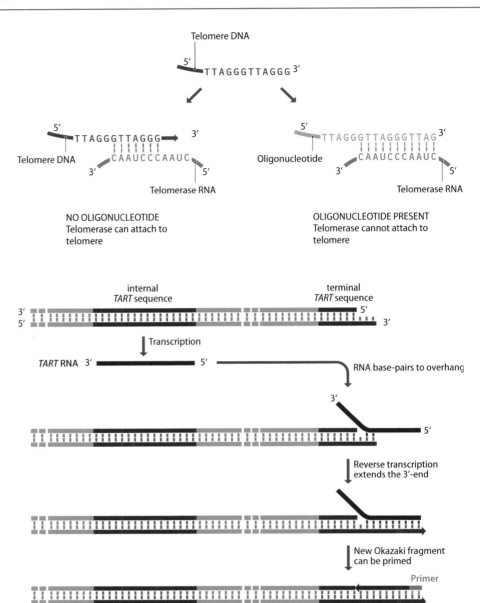

15.5 REGULATION OF EUKARYOTIC GENOME REPLICATION

It is essential that genome replication is regulated so that the process is completed by the time that the cell actually divides. Eukaryotes possess a sophisticated set of controls that coordinate genome replication with the **cell cycle** and which are able to arrest the replication process under certain circumstances, for example, if the DNA is damaged and must be repaired before copying can be completed. We will end this chapter by looking at these regulatory mechanisms.

Genome replication must be synchronized with the cell cycle

The concept of a cell cycle emerged from light microscopy studies carried out by the early cell biologists. Their observations showed that dividing cells pass through repeated cycles of mitosis – the period when nuclear and cell division occurs (see Figure 3.16) – and interphase, a less dramatic period when few dynamic changes can be detected with the light microscope. It was understood that chromosomes divide during interphase, so when DNA was identified as the genetic material, interphase took on a new importance as the period when genome replication takes place. This led to a reinterpretation of the cell cycle as a four-stage process (**Figure 15.29**), comprising:

- **Mitosis**, or **M phase**, the period when the nucleus and cell divide.

- **Gap 1**, or **G1 phase**, an interval when transcription, translation, and other general cellular activities occur.

- **Synthesis**, or **S phase**, when the genome is replicated.

- **Gap 2**, or **G2 phase**, a second interval period.

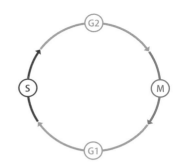

Figure 15.29 The cell cycle. The lengths of the individual phases vary in different cells. Abbreviations: G1 and G2, gap phases; M, mitosis; S, synthesis phase.

It is clearly important that the S phase and M phase are coordinated so that the genome is completely replicated, but replicated only once, before mitosis occurs. To ensure this happens, there are a series of **cell cycle checkpoints** that act as key transition stages, progression past a checkpoint committing the cell to the next phase of the cycle. The most important of these with regards to genome replication is the **G1-S checkpoint**, because only when this point has been passed is the cell able to replicate its DNA. If the cell suffers trauma or the genome has been damaged in some way, for example, by mutation of key genes, then the cell cycle can be arrested at the checkpoint while the damage is repaired. Later in the cell cycle, the **G2-M checkpoint** ensures that the cell is ready to enter mitosis, the central requirement being that the genome has been replicated correctly, every part replicated once and no parts replicated multiple times.

Origin licensing is the prerequisite for passing the G1-S checkpoint

Studies primarily with *Saccharomyces cerevisiae* have led to a model for the regulation of genome replication that defines **origin licensing** as the basis for the preparation of the cell for passage through the G1-S checkpoint. Origin licensing involves construction on a replication origin of a set of proteins, called the **pre-replication complex (pre-RC)**. Each origin is already marked by the six proteins that make up the origin recognition complex, which is assembled onto multiple origins soon after completion of the previous cell division. To become licensed, an origin must initially recruit two additional proteins, Cdc9 and Cdt1, the first of which is an ATPase that releases energy from a molecule of ATP that is present in the ORC (**Figure 15.30**). The energy is used to drive attachment of the MCM2–7 helicase. Cdc9 and Cdt1 depart as the helicase arrives, the attachment of the latter to the ORC completing construction of the pre-RC. These events appear to be similar in most eukaryotes.

Origin licensing enables the cell to pass through the G1-S checkpoint, but the pre-RCs are themselves inactive and unable to initiate genome replication. To become active, each pre-RC must be converted into a **pre-initiation complex (pre-IC)**. In *S. cerevisiae*, this conversion is initiated soon after the beginning of the S phase by a cyclin-dependent kinase (CDK) and by Dbf4-dependent kinase (DDK), which together phosphorylate target proteins, including the MCM2, MCM4 and MCM6 subunits of the inactive helicase, in order to initiate a cascade of reactions that result in addition of Cdc45 and GINS to the pre-RC, converting the latter into a pre-IC (**Figure 15.31**). Cdc45 and GINS, the second of which is itself a complex of four proteins, are the final components of the eukaryotic helicase, which is now active and able to participate in initiation of replication and progression of the replication forks away from the origin.

Identification of the components of the pre-RC and pre-IC takes us some distance toward understanding how genome replication is initiated, but still leaves open the question of how replication is coordinated with other events in the cell cycle. Cell cycle control is a complex process, mediated largely by CDK proteins which phosphorylate and activate enzymes and other proteins that have specific functions during the cell cycle. The activities of these CDKs change throughout the cell cycle, having minimal activity at the beginning of the G1 phase, and increasing rapidly during the S phase (**Figure 15.32**). The CDK activity level is thought to influence the activities of the proteins whose functions are needed

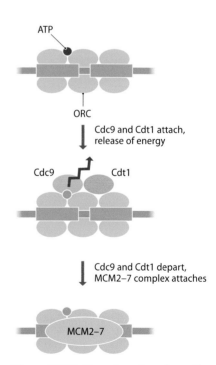

Figure 15.30 Licensing of a eukaryotic replication origin. Cdc9 and Cdt1 attach to the origin recognition complex (ORC), resulting in hydrolysis of the bound ATP molecule and release of energy. Cdc9 and Cdt1 then depart as the MCM2–7 complex attaches. The resulting structure is the prereplication complex (pre-RC).

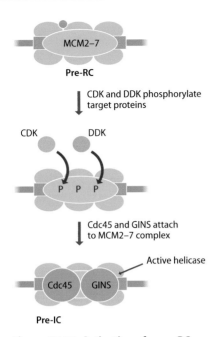

Figure 15.31 Activation of a pre-RC.
Cyclin-dependent kinase (CDK) and Dbf4-dependent kinase (DDK) phosphorylate target proteins, including the MCM2–7 complex. This enables Cdc45 and GINS to attach to the prereplication complex (pre-RC), completing construction of the active helicase and converting the pre-RC into a pre-initiation complex (pre-IC).

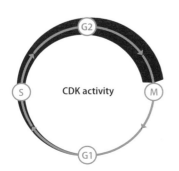

Figure 15.32 Changes in CDK activity during the cell cycle.

at different stages, thereby ensuring that the cell cycle progresses in an orderly manner. Most CDKs are present in the nucleus throughout the cell cycle, so they must themselves be subject to control. This control is exerted partly by proteins called **cyclins**, which vary in abundance at different stages of the cell cycle (**Figure 15.33**), partly by other protein kinases that activate the CDKs, and partly by inhibitory proteins. One of the main inhibitors is geminin, which accumulates during the S, G2 and M phases and prevents the re-replication of genome regions that have already been replicated. It does this by binding to Cdt1 and hence ensuring that the MCM2–7 core helicase cannot be loaded onto the origins that are present on the daughter DNA molecules. Along with many other proteins that are active during the late stages of the cell cycle, geminin is degraded during the anaphase period of mitosis by the **anaphase-promoting complex/cyclosome** (**APC/C**), which is a ubiquitin ligase that targets proteins for degradation by the proteasome (Section 13.3). The APC/C therefore remodels the nuclear proteome so the cell cycle can recommence in the daughter cells resulting from mitosis.

Replication origins do not all fire at the same time

Initiation of replication does not occur at the same time at all replication origins. Instead, during any particular cell cycle, some origins 'fire' early in S phase and some later. Studies with *S. cerevisiae* have suggested that the pattern of origin firing is consistent from cell division to cell division, with euchromatin regions and the centromeres replicated early in the S phase, and heterochromatin and telomeres replicated later on. More recent work has confirmed that this general pattern is correct, but has also indicated that there is little or no correspondence between the detailed patterns of origin firing in different cells, suggesting that the pattern of firing is not inherited in any way.

How can origin firing be followed in living cells? One method involves labeling dividing cells with a brief pulse of a nucleotide analog, such as bromodeoxyuridine (BrdU), which is incorporated into growing polynucleotides in place of thymidine. The DNA that is synthesized during the pulse of labeling will contain BrdU nucleotides as well as Ts, whereas the DNA made before and after the pulse will just contain Ts. The DNA is then extracted and combed on to a glass slide, so the molecules are obtained as linear fibers (Section 3.5). Treatment with an antibody specific for BrdU, this antibody carrying a fluorescent label, will reveal the genome regions that were being replicated at the time of the BrdU pulse. By comparing molecules from cells at different stages in the S phase, the pattern of origin firing can be followed. This technique has been used with *Schizosaccharomyces pombe*, which is a fission yeast, meaning that it divides by the cell splitting into two halves of equal size, rather than via small buds, as is the case with *S. cerevisiae*. The results show that the first origins to fire, at the beginning of the S phase, are distributed randomly throughout the genome. As the S phase continues, more origins fire, these forming clusters at various positions along the individual chromosomes. However, the positions of these clusters are not the same in different cells (**Figure 15.34**), showing that the pattern is not determined by the DNA sequences of individual initiation regions. Instead, it appears likely that the clusters represent regions of euchromatin, which are chosen at random as the starting points for replication of the genome. This hypothesis is supported by experiments with *S. cerevisiae* mutants which are unable to synthesize the RPD3 histone deacetylase. These cells, which are predicted to have a greater than normal degree of histone acetylation and hence a more open chromatin structure, display less control over replication origin firing, with early-firing origins now operating in parts of the genome that are not usually replicated until late in the S phase.

BrdU labeling also enables the migration rates of different replication forks in *S. pombe* to be estimated, from the lengths of the labeled tracts that are synthesized during the pulse period. The mean speed is 2.8 kb per minute, but some forks move much more quickly, up to 11 kb per minute for the most active ones (**Figure 15.35**). These figures are very similar to the migration rates measured for

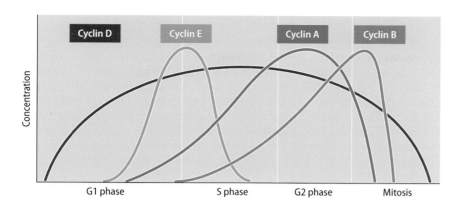

Figure 15.33 Variations in abundance of mammalian cyclins during the cell cycle. The abundance of each of the four cyclins varies during the cell cycle, resulting in stage-specific changes in cyclin-dependent kinase (CDK) activity.

S. cerevisiae and estimated for human cells, but much slower than *E. coli*, which has to copy 116 kb of DNA per minute in order to replicate its entire genome within the 20 minutes that elapse between cell divisions when cultures are grown in rich medium. The fork migration rates are dependent on a number of factors, especially the presence of obstacles such as RNA polymerase complexes that are bound to the DNA in order to transcribe genes, or damaged regions of the genome that must be repaired before they can be replicated. When encountering such an obstacle, a fork might become stalled. After the blockage has been removed, the fork might recommence its journey, or it might remain stalled while the fork approaching from the other direction completes replication of the intervening region of DNA. Occasionally, however, both members of a pair of forks will stall permanently. This situation can be retrieved by activation of additional replication origins within the region between the two stalled forks. This is possible, even in the presence of inhibitors such as geminin, because an excess of origins are always licensed during the G1 phase. This means that the genome contains many activated but unused origins, some of which will remain unused even when the entire genome has been replicated.

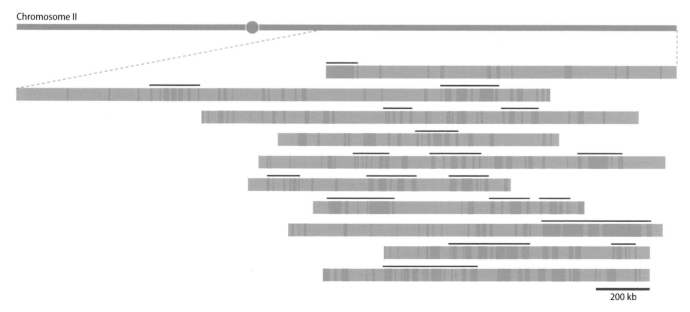

Figure 15.34 Patterns of replication origin firing are not the same in different cells. Ten overlapping molecules from chromosome 2 of different cells of *Schizosaccharomyces pombe* are shown. The molecules were pulse-labeled with bromodeoxyuridine (BrdU), combed into linear strands, and the label detected with a BrdU-specific antibody. The clusters of active origins that are revealed are not the same in each molecule, indicating that the pattern of origin firing is different in different cells. The positions of the main clusters of fired origins are shown as red bars on top of each DNA molecule. (From Kaykov A. & Nurse P. [2015] *Genome Res.* 25:391–401. With permission from Cold Spring Harbor Laboratory Press, published under CC BY 4.0.)

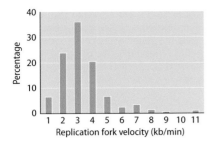

Figure 15.35 Replication fork velocity in *Schizosaccharomyces pombe*. (From Kaykov A. & Nurse P. [2015] *Genome Res.* 25:391–401. With permission from Cold Spring Harbor Laboratory Press, published under CC BY 4.0.)

The cell has various options if the genome is damaged

The cell cycle checkpoints also play an important role in preventing a damaged genome from being replicated. Damage can take many forms, including the dimerization of adjacent nucleotides, which is stimulated by UV irradiation (Section 17.1), or the presence of single- or double-stranded breaks. If damage is detected, then the cell cycle will be arrested at the G1-S checkpoint while the DNA is repaired. There are also additional checkpoints within the S phase that enable genome replication to be halted to repair damage that is detected when one or more replication forks become stalled, and the G2-M checkpoint ensures that post-replicative damage is corrected before mitosis takes place.

Cell cycle arrest occurs because of the activity of signal transduction pathways that are activated by DNA damage, or indicators of damage such as replication fork stalling. Two of these pathways are initiated by the ATM and ATR protein kinases, which either recognize damage directly or are activated by the detection proteins. ATM responds primarily to double-strand breaks in the genome and ATR to various types of damage, including single-strand breaks, which bring about replication fork arrest during the S phase. The ATM pathway targets include the checkpoint kinase Chk2 as well as BRCA1, which when defective gives susceptibility to breast cancer. A second checkpoint kinase, Chk1, is activated by the ATR pathway. The checkpoint kinases act on cell cycle control proteins such as Cdc25, which regulate passage through the G2-M checkpoint. Phosphorylation of Cdc25 results in its degradation, so the cell cycle becomes arrested until the DNA damage is repaired.

If the damage is not excessive, then DNA repair processes are activated (Section 17.2). Alternatively, the cell may be shunted into the pathway of programmed cell death called **apoptosis**, the death of a single somatic cell as a result of DNA damage usually being less dangerous than allowing that cell to replicate its mutated DNA and possibly give rise to a tumor or other cancerous growth. In mammals, a central player in induction of cell cycle arrest and apoptosis is the protein called p53. This is classified as a tumor-suppressor protein, because when this protein is defective, cells with damaged genomes can avoid the S-phase checkpoints and possibly proliferate into a cancer. p53 is among the targets of both the ATM and ATR pathways, and once activated switches on transcription of a number of genes thought to be directly responsible for arrest and apoptosis, and also represses expression of other genes that must be switched off to facilitate these processes.

SUMMARY

- In order to continue carrying out its function, the genome must replicate every time that a cell divides.

- Watson and Crick pointed out, when they first announced their discovery of the structure of DNA, that the specific base-pairing that holds the two strands of the double helix together provides a means for accurate copying of each polynucleotide. They envisaged a semiconservative mode of replication in which each parent strand acts as a template for synthesis of a complementary daughter strand.

- The Meselson–Stahl experiment confirmed that DNA replicates by the semiconservative process, but there were still problems in understanding how the two strands of the helix were separated, especially in circular molecules that have little freedom to rotate.

- The discovery of DNA topoisomerases, which separate the strands of the double helix by repeated breakage and rejoining of one or both polynucleotides, solved the topological problem.

- No exceptions to the semiconservative mode of replication are known, though there are specialized versions such as displacement and rolling-circle replication.

- Initiation of genome replication occurs at discrete origins, which have been well characterized in bacteria and in yeast but are less clearly understood in higher eukaryotes.

- Once replication has been initiated, a pair of replication forks travel in opposite directions along the DNA.

- DNA polymerase can only synthesize DNA in the $5' \rightarrow 3'$ direction, which means that although one strand, called the leading strand, can be replicated in a continuous fashion, the second, lagging strand has to be copied in short segments. These are called Okazaki fragments.

- DNA synthesis must be primed by an RNA polymerase.

- The replicating complex consists of the DNA polymerase enzyme along with ancillary proteins, such as the sliding clamp, which ensures that the connection between the polymerase and the DNA is secure but that the polymerase is still able to move along the DNA.

- Termination of replication occurs at specific regions within a bacterial chromosome but at less well-defined areas in eukaryotic chromosomes.

- Completion of replication involves decatenation of the daughter molecules and disassembly of the replisomes.

- Eukaryotic chromosomes require special processes to maintain their ends, as the replication process results in a gradual shortening of the telomeres. These are elongated by the telomerase enzyme, which has an RNA subunit that acts as the template for synthesis of new telomere repeat units.

- Replication of the genome must be coordinated with the cell cycle. This is achieved by a combination of regulatory proteins, many of which are active only at specific periods of the cell cycle.

- Assembly of the prereplication complex at the origins of replication is a critical step that is regulated to ensure that the genome is replicated just once per cell cycle.

- Once replication is underway, checkpoints during the synthesis phase respond to DNA damage in order to arrest or terminate genome replication.

SHORT ANSWER QUESTIONS

1. Prior to the Meselson–Stahl experiment, it was not known if DNA replication is dispersive, semiconservative, or conservative. Describe the differences in the DNA contents of the daughter molecules resulting from these different modes of replication.

2. Outline the role of DNA topoisomerases in DNA replication.

3. Describe the mechanisms for (A) displacement replication and (B) rolling-circle replication.

4. Where and how does the DnaA protein bind at the origin of replication in *E. coli*?

5. What methods have been used to identify replication origins in eukaryotes?

6. List the key features of the bacterial and eukaryotic DNA polymerases that are involved in DNA replication.

7. How are Okazaki fragments joined together in (A) bacteria and (B) eukaryotes?

8. What is known about the termination of genome replication in *E. coli*? What proteins and sequences are involved in this process?

9. What novel topological problems arise when a pair of replication forks meet?

10. Why do the ends of linear chromosomes get shorter during successive rounds of DNA replication in eukaryotes?

11. How is telomerase activity regulated in eukaryotic cells?

12. Explain what is meant by the term origin licensing and describe the role of licensing in the cell cycle.

13. What general patterns have been observed regarding the timing of replication of different parts of the eukaryotic genome?

IN-DEPTH PROBLEMS

1. Discuss why the semiconservative mode of DNA replication was favored even before the Meselson–Stahl experiment was carried out.

2. Would it be possible to replicate the DNA molecules present in living cells if DNA topoisomerases did not exist?

3. Why is inactivation of the *Escherichia coli polA* gene, coding for DNA polymerase I, not lethal?

4. Construct a hypothesis to explain why all DNA polymerases require a primer in order to initiate synthesis of a new polynucleotide. Can your hypothesis be tested?

5. Our current knowledge of genome replication in eukaryotes is biased toward the events occurring at the replication fork. The next challenge is to convert this DNA-centered description of replication into a model that describes how replication is organized within the nucleus. Devise a research plan to address this issue.

FURTHER READING

The history of research into genome replication

Crick, F.H.C., Wang, J.C. and Bauer, W.R. (1979) Is DNA really a double helix? *J. Mol. Biol.* 129:449–461. *Crick's response to suggestions that DNA has a side-by-side rather than helical conformation.*

Holmes, F.L. (1998) The DNA replication problem, 1953–1958. *Trends Biochem. Sci.* 23:117–120.

Kornberg, A. (1989) *For the Love of Enzymes: The Odyssey of a Biochemist.* Harvard University Press, Boston, MA. *A fascinating autobiography by the discoverer of DNA polymerase.*

Meselson, M. and Stahl, F. (1958) The replication of DNA in *Escherichia coli. Proc. Natl Acad. Sci. USA* 44:671–682. *The Meselson–Stahl experiment.*

Okazaki, T. and Okazaki, R. (1969) Mechanism of DNA chain growth, IV. Direction of synthesis of T4 short DNA chains as revealed by exonucleolytic degradation. *Proc. Natl Acad. Sci. USA* 64:1242–1248. *The discovery of Okazaki fragments.*

Watson, J.D. and Crick, F.H.C. (1953) Genetical implications of the structure of deoxyribonucleic acid. *Nature* 171:964–967. *Describes possible processes for DNA replication, shortly after discovery of the double helix.*

DNA topoisomerases

Berger, J.M., Gamblin, S.J., Harrison, S.C., et al. (1996) Structure and mechanism of DNA topoisomerase II. *Nature* 379:225–232 and 380:179.

Champoux, J.J. (2001) DNA topoisomerases: Structure, function, and mechanism. *Annu. Rev. Biochem.* 70:369–413.

Garnier, F., Couturier, M., Débat, H., et al. (2021) Archaea: A gold mine for topoisomerase diversity. *Front. Microbiol.* 12:661411.

McKie, S.J., Neuman, K.C. and Maxwell, A. (2021) DNA topoisomerases: Advances in understanding of cellular roles and multi-protein complexes via structure-function analysis. *Bioessays* 43:e2000286.

Stewart, L., Redinbo, M.R., Qiu, X., et al. (1998) A model for the mechanism of human topoisomerase I. *Science* 279:1534–1541.

Vos, S.M., Tretter, E.M., Schmidt, B.H., et al. (2011) All tangled up: How cells direct, manage and exploit topoisomerase function. *Nat. Rev. Mol. Cell Biol.* 12:827–841.

Origins of replication

Diffley, J.F.X. and Cocker, J.H. (1992) Protein–DNA interactions at a yeast replication origin. *Nature* 357:169–172.

Ekundayo, B. and Bleichert, F. (2019) Origins of DNA replication. *PLoS Genet.* 15:e1008320.

Hyrien, O. (2015) Peaks cloaked in the mist: The landscape of mammalian replication origins. *J. Cell Biol.* 208:147–160.

Krysan, P.J., Smith, J.G. and Calos, M.P. (1993) Autonomous replication in human cells of multimers of specific human and bacterial DNA sequences. *Mol. Cell. Biol.* 13:2688–2696.

Leonard, A.C. and Méchali, M. (2013) DNA replication origins. *Cold Spring Harb. Perspect. Biol.* 5:a010116.

Li, H. and Stillman, B. (2012) The origin recognition complex: A biochemical and structural view. *Subcell. Biochem.* 62:37–58.

Mott, M.L. and Berger, J.M. (2007) DNA replication initiation: Mechanisms and regulation in bacteria. *Nat. Rev. Microbiol.* 5:343–354.

DNA polymerases and events at the replication fork

Bochkarev, A., Pfuetzner, R.A., Edwards, A.M., et al. (1997) Structure of the single-stranded-DNA-binding domain of replication protein A bound to DNA. *Nature* 385:176–181.

Burgers, P.M.J. (2009) Polymerase dynamics at the eukaryotic DNA replication fork. *J. Biol. Chem.* 284:4041–4045.

Dueva, R. and Iliakis, G. (2020) Replication protein A: A multisubunit protein with roles in DNA replication, repair and beyond. *NAR Cancer* 2:zcaa022.

Finger, L.D., Atack, J.M., Tsutakawa, S., et al. (2012) The wonders of flap endonucleases: Structure, function, mechanism and regulation. *Subcell. Biochem.* 62:301–326.

Hübscher, U., Nasheuer, H.-P. and Syväoja, J.E. (2000) Eukaryotic DNA polymerases, a growing family. *Trends Biochem. Sci.* 25:143–147.

Johnson, A. and O'Donnell, M. (2005) Cellular DNA replicases: Components and dynamics at the replication fork. *Annu. Rev. Biochem.* 74:283–315. *Details of replication in bacteria and eukaryotes.*

Pomerantz, R.T. and O'Donnell, M. (2007) Replisome mechanics: Insights into a twin polymerase machine. *Trends Microbiol.* 15:156–164.

Soultanas, P. and Wigley, D.B. (2001) Unwinding the 'Gordian knot' of helicase action. *Trends Biochem. Sci.* 26:47–54.

Trakselis, M.A. and Bell, S.D. (2004) The loader of the rings. *Nature* 429:708–709. *The sliding clamp and clamp loader.*

Termination of replication and the role of telomerase

Berghuis, B.A., Dulin, D., Xu, Z.-Q., et al. (2015) Strand separation establishes a sustained lock at the Tus–Ter replication fork barrier. *Nat. Chem. Biol.* 11:579–585. *Using molecular tweezers to study the interaction between Tus and the replisome.*

Berguis, B.A, Raducanu, V.-S., Elshenawy, M.M., et al. (2018) What s all this fuss about Tus? Comparison of recent findings from biophysical and biochemical experiments. *Crit. Rev. Biochem. Mol. Biol.* 53:49–63.

Blackburn, E.H. (2000) Telomere states and cell fates. *Nature* 408:53–56.

Casacuberta, E. (2017) Drosophila: Retrotransposons making up telomeres. *Viruses* 9:192.

Dewar, J.M. and Walter, J.C. (2017) Mechanisms of DNA replication termination. *Nat. Rev. Mol. Cell Biol.* 18:507–516.

Fachinetti, D., Bermejo, R., Cocito, A., et al. (2010) Replication termination at eukaryotic chromosomes is mediated by Top2 and occurs at genomic loci containing pausing elements. *Mol. Cell* 39:595–605.

Jafri, M.A., Ansari, S.A., Alqahtani, M.H., et al. (2016) Roles of telomeres and telomerase in cancer, and advances in telomerase-targeted therapies. *Genome Med.* 8:69.

Shay, J.W. and Wright, W.E. (2006) Telomerase therapeutics for cancer: Challenges and new directions. *Nat. Rev. Drug Discov.* 5:577–584. *Methods for inhibiting telomerase in order to treat cancer.*

Shay, J.W. and Wright, W.E. (2019) Telomeres and telomerase: Three decades of progress. *Nat. Rev. Genet.* 20:299–309.

Smogorzewska, A. and de Lange, T. (2004) Regulation of telomerase by telomeric proteins. *Annu. Rev. Biochem.* 73:177–208.

Vaiserman, A. and Krasnienkov, D. (2021) Telomere length as marker of biological age: State-of-the-art, open issues, and future perspectives. *Front. Genet.* 11:630186.

Control of genome replication

Bertoli, C., Skotheim, J.M. and de Bruin, R.A.M. (2013) Control of cell cycle transcription during G1 and S phases. *Nat. Rev. Mol. Cell Biol.* 14:518–528.

Kaykov, A. and Nurse, P. (2015) The spatial and temporal organization of origin firing during the S-phase of fission yeast. *Genome Res.* 25:391–401.

Matthews, H.K., Bertoli, C. and de Bruin, R.A.M. (2022) Cell cycle control in cancer. *Nat. Rev. Mol. Cell Biol.* 23:74–88.

Panagopoulos, A. and Altmeyer, M. (2021) The hammer and the dance of cell cycle control. *Trends Biochem. Sci.* 46:301–314.

Sancar, A., Lindsey-Boltz, L.A., Ünsal-Kaçmaz, K., et al. (2004) Molecular mechanisms of mammalian DNA repair and the DNA damage checkpoints. *Annu. Rev. Biochem.* 73:39–85.

Stillman, B. (1996) Cell cycle control of DNA replication. *Science* 274:1659–1664.

Symeonidou, I.E., Taraviras, S. and Lygerou, Z. (2012) Control over DNA replication in time and space. *FEBS Lett.* 586:2803–2812.

Yekezare, M., Gómez-González, B. and Diffley, J.F.X. (2013) Controlling DNA replication origins in response to DNA damage – Inhibit globally, activate locally. *J. Cell Sci.* 126:1297–1306.

Zhou, B.-B.S. and Elledge, S.J. (2000) The DNA damage response: Putting checkpoints in perspective. *Nature* 408:433–439.

RECOMBINATION AND TRANSPOSITION

Recombination is the term originally used by geneticists to describe the outcome of crossing over between pairs of homologous chromosomes during meiosis. Crossing over results in daughter chromosomes that have different combinations of alleles compared with their parent chromosomes (Section 3.3). In the 1960s, models were proposed for the molecular events that underlie crossing over, and it was realized that a key part of molecular recombination is the breakage and subsequent rejoining of DNA molecules. Biologists now use 'recombination' to refer to a variety of processes that involve the breakage and reunion of polynucleotides. These include:

- **Homologous recombination**, also called **general** (or generalized) **recombination**, which occurs between segments of DNA molecules that share extensive sequence homology. These segments might be present on different chromosomes, or might be two parts of a single chromosome (**Figure 16.1A**). Homologous recombination is responsible for crossing over during meiosis, and was initially studied in this context, but we now believe that its primary cellular role is in DNA repair.

- **Site-specific recombination**, which occurs between DNA molecules that have only short regions of sequence similarity, possibly just a few base pairs (**Figure 16.1B**). Site-specific recombination is responsible for the insertion of phage genomes, such as that of λ, into bacterial chromosomes.

Various other events that we have studied, including mating-type switching in yeast (see Figure 14.16) and construction of immunoglobulin genes (see Figure 14.19), are also the results of recombination.

Transposition is not a type of recombination but a process that often utilizes recombination, the end result being the transfer of a segment of DNA from one position in the genome to another (**Figure 16.2**). Recombination and transposition therefore have similar outcomes, both resulting in the rearrangement of DNA segments within a genome.

Without recombination, genomes would be relatively static structures, undergoing very little change. The gradual accumulation of mutations over a long period of time would result in small-scale alterations in the nucleotide sequence of the genome, but more extensive restructuring would not occur, and the evolutionary potential of the genome would be severely restricted.

16.1 HOMOLOGOUS RECOMBINATION

The study of homologous recombination has presented two significant challenges for molecular biologists, neither of which has yet been fully met. The first challenge has been to describe the series of interactions, involving breakage and reunion of polynucleotides, that occur during recombination. The models for homologous recombination that have resulted from this work are described below. The second challenge relates to the fact that recombination is a cellular process that, like other cellular processes involving DNA (e.g., transcription and replication), is carried out and regulated by enzymes and other proteins.

DOI: 10.1201/9781003133162-16

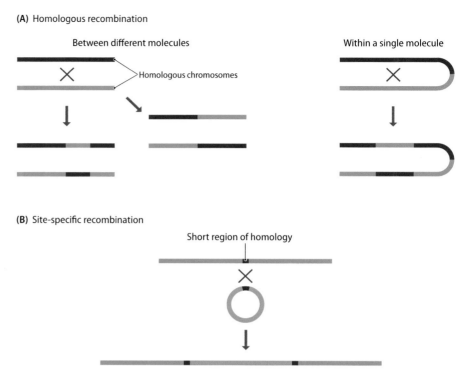

Figure 16.1 Two different types of recombination event.

Biochemical studies have defined a series of related recombination pathways, and have also revealed that homologous recombination underlies several important types of DNA repair, this repair function probably being more important to the cell (especially bacterial cells) than the capacity that homologous recombination provides for crossing over between chromosomes.

The Holliday and Meselson–Radding models for homologous recombination

Many of the breakthroughs in understanding homologous recombination were made by Robin Holliday, Matthew Meselson, and their colleagues in the 1960s and 1970s. This work resulted in a series of models that showed how breakage and reunion of DNA molecules could lead to the exchange of chromosome segments known to occur during crossing over. We will therefore begin our study of homologous recombination by examining these models.

The Holliday and Meselson–Radding models describe recombination between two homologous, double-stranded molecules, ones with identical or nearly identical sequences. The central feature of these models is formation of a **heteroduplex** resulting from the exchange of polynucleotide segments between the two homologous molecules (**Figure 16.3**). The heteroduplex is initially stabilized by base-pairing between each transferred strand and the intact polynucleotides of the recipient molecules, this base-pairing being possible because of the sequence similarity between the two molecules. Subsequently, the gaps

Figure 16.2 Transposition. In replicative transposition, the original transposon remains in place and a new copy appears elsewhere in the genome. In conservative transposition, the transposon moves to a new site.

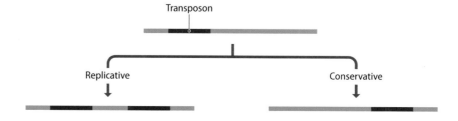

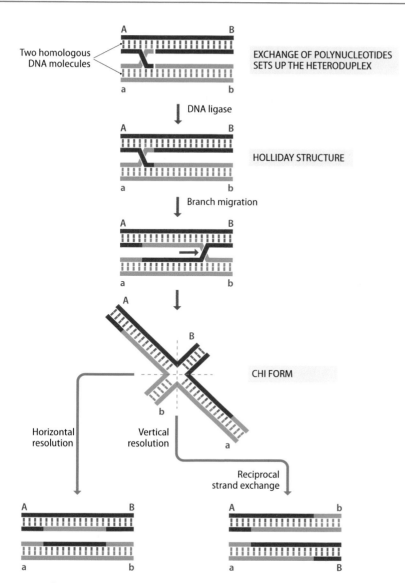

Figure 16.3 The Holliday model for homologous recombination.

are sealed by DNA ligase, giving a **Holliday structure**. This structure is dynamic, with **branch migration** resulting in exchange of longer segments of DNA if the two helices rotate in the same direction.

Separation, or **resolution**, of the Holliday structure back into individual, double-stranded molecules occurs by cleavage across the branch point. This is the key to the entire process because the cut can be made in either of two orientations, as becomes apparent when the three-dimensional configuration, or **chi form**, of the Holliday structure is examined (see Figure 16.3). These two cuts have very different results. If the cut is made left–right across the chi form as drawn in Figure 16.3, then all that happens is that a short segment of polynucleotide, corresponding to the distance migrated by the branch of the Holliday structure, is transferred between the two molecules. On the other hand, an up–down cut results in **reciprocal strand exchange**, with double-stranded DNA being transferred between the two molecules so that the end of one molecule is exchanged for the end of the other molecule. This is the DNA transfer seen in crossing over.

So far, we have ignored one important aspect of this model. This is the way in which the two double-stranded molecules interact at the beginning of the process to produce the heteroduplex. In Holliday's original scheme, the two molecules lined up with one another, and single-strand nicks appeared at equivalent

Figure 16.4 Two schemes for initiation of homologous recombination. (A) Initiation as described by the original model for homologous recombination. (B) The Meselson–Radding modification, which proposes a more plausible series of events for formation of the heteroduplex.

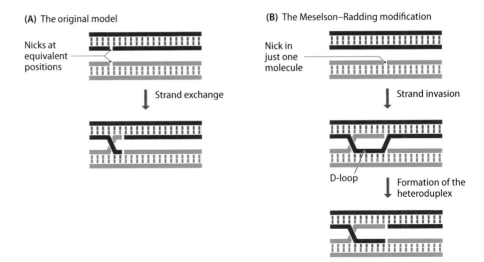

(A) The original model

Nicks at equivalent positions

Strand exchange

(B) The Meselson–Radding modification

Nick in just one molecule

Strand invasion

D-loop

Formation of the heteroduplex

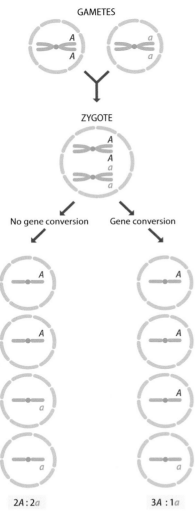

GAMETES

A
A

a
a

ZYGOTE

A
A
a
a

No gene conversion   Gene conversion

A

A

a

a

A

A

A

a

2*A* : 2*a*   3*A* : 1*a*

Figure 16.5 Gene conversion. One gamete contains allele *A* and the other contains allele *a*. These fuse to produce a zygote that gives rise to four haploid spores, all contained in a single ascus. Normally, two of the spores will have allele *A* and two will have allele *a*, as shown on the left, but if gene conversion occurs the ratio will be changed, possibly to 3*A*:1*a* as shown on the right.

positions in each helix. This produced free, single-stranded ends that could be exchanged, resulting in the heteroduplex (**Figure 16.4A**). This feature of the model was criticized because no mechanism could be proposed for ensuring that the nicks occurred at precisely the same position on each molecule. The Meselson–Radding modification proposes a more satisfactory scheme whereby a single-strand nick occurs in just one of the double helices, the free end that is produced 'invading' the unbroken double helix at the homologous position and displacing one of its strands, forming a **D-loop** (**Figure 16.4B**). Subsequent cleavage of the displaced strand at the junction between its single-stranded and base-paired regions produces the heteroduplex.

The double-strand break model for homologous recombination

Although the Holliday model for homologous recombination, either in its original form or as modified by Meselson and Radding, explained how crossing over could occur during meiosis, it had inadequacies which prompted the development of alternative schemes. In particular, it was thought that the Holliday model could not explain **gene conversion**, a phenomenon first described in yeast and fungi but now known to occur with many eukaryotes. In yeast, fusion of a pair of gametes results in a zygote that gives rise to an ascus containing four haploid spores whose genotypes can be individually determined (see Figure 14.15). If the gametes have different alleles at a particular locus, then under normal circumstances two of the spores will display one genotype, and two will display the other genotype, but sometimes this expected 2 : 2 segregation pattern is replaced by an unexpected 3 : 1 ratio (**Figure 16.5**). This is called gene conversion because the ratio can only be explained by one of the alleles 'converting' from one type to the other, presumably by recombination during the meiosis that occurs after the gametes have fused.

The **double-strand break (DSB) model** provides an opportunity for gene conversion to take place during recombination. According to this model, homologous recombination initiates not with a single-strand nick, as in the Meselson–Radding scheme, but with a double-strand cut that breaks one of the partners in the recombination into two pieces (**Figure 16.6**). After the double-strand cut, one strand in each half of the molecule is shortened, so each end now has a 3′ overhang. One of these overhangs invades the homologous DNA molecule in a manner similar to that envisaged by the Meselson–Radding scheme, setting up a Holliday junction that can migrate along the heteroduplex if the invading strand is extended by a DNA polymerase. To complete the heteroduplex, the other broken strand (the one not involved in the Holliday junction) is also extended. Note that both DNA syntheses involve extension of strands from the partner that suffered the double-strand cut, using as templates the equivalent regions of the uncut partner. This is the basis of the gene conversion because it means that the polynucleotide segments removed from the cut partner have been replaced with copies of the DNA from the uncut partner. After ligation, the resulting heteroduplex has a pair of Holliday structures that can be resolved in a number of ways, some resulting in gene conversion and others giving a standard reciprocal strand exchange. An example leading to gene conversion is shown in Figure 16.6.

Although initially proposed as a mechanism for explaining gene conversion in yeast, the DSB model is now looked on as at least a close approximation to the way in which homologous recombination operates in all organisms. Acceptance of this model has come about for two reasons. First, in 1989 it was discovered that during meiosis, chromosomes undergo double-strand breakages at 100–1000 times the rate seen in vegetative cells. The implication that formation of double-strand breaks is an inherent part of meiosis clearly favors the DSB model at the expense of schemes in which recombination is initiated by one or more single-strand nicks. The second factor leading to acceptance of the DSB model was the realization that homologous recombination is involved in DNA repair, and is specifically responsible for repairing double-strand breaks that occur as aberrations in the replication process. The Holliday and Meselson–Radding models do not explain this aspect of homologous recombination, whereas double-strand break repair is implicit in the DSB model. We will return to the role of recombination in DNA repair in Section 17.2.

RecBCD is the most important pathway for homologous recombination in bacteria

Homologous recombination occurs in all organisms, but, as with many aspects of molecular biology, the initial progress in understanding how the process is carried out in the cell was made with *E. coli*. Of course, bacteria do not undergo meiosis, but crossing over by homologous recombination occurs following transfer of DNA from one bacterium to another, and results in integration of the donated DNA into the chromosome of the recipient cell (see Figure 3.25A). The first breakthroughs in understanding the biochemistry of homologous recombination were made when mutation studies identified a number of *E. coli* genes that, when inactivated, give rise to defects in recombination, indicating that their protein products are involved in some way. Two distinct recombination systems have been described, these being the RecBCD and RecFOR pathways, with RecBCD apparently being the most important in the bacterium.

In the RecBCD pathway, recombination is mediated by the **RecBCD complex**, which, as its name implies, is made up of three different proteins. Two of these – RecB and RecD – are helicases. To initiate homologous recombination, one copy of RecBCD attaches to the chromosome at a double-strand break. The DNA is unwound, through the action of RecB, which travels along one strand in the 3′→5′ direction, and RecD, traveling 5′→3′ along the other strand. The RecB protein, as well as being a helicase, also has 3′→5′ exonuclease activity and so progressively degrades the strand it is traveling along – the one with the free 3′ end (**Figure 16.7**).

RecBCD progresses along the DNA molecule at a rate of approximately 1 kb per second until it reaches the first copy of the eight-nucleotide consensus sequence 5′-GCTGGTGG-3′ called the **chi (crossover hotspot initiator) site**, which occurs on average once every 5 kb in the *E. coli* genome. At the chi site, the conformation of RecBCD changes so that the RecD helicase becomes uncoupled, and the progress of the RecBCD complex slows to about half its initial rate. The change in conformation of the complex also reduces or completely abolishes the 3′→5′ exonuclease activity of RecB; this protein now makes a single endonucleolytic cut in the other strand of the DNA molecule, at a position close to the chi site (see Figure 16.7). The result is that the RecBCD complex produces a double-stranded molecule with a 3′ overhang, as envisaged by the DSB model (see the second step in Figure 16.6).

The next step is establishment of the heteroduplex. This stage is mediated by the **RecA** protein, which forms a protein-coated DNA filament that is able

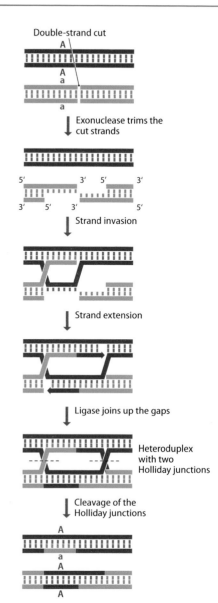

Figure 16.6 The double-strand break model for homologous recombination. This model explains how gene conversion can occur.

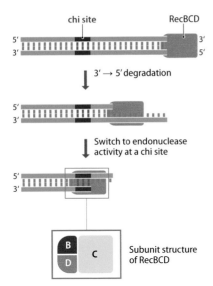

Figure 16.7 The RecBCD pathway for homologous recombination in *E. coli*. Translocation of the RecBCD complex along the DNA is accompanied by 3′→5′ degradation of the upper strand, due to the exonuclease activity of RecB. When a chi site is encountered, the exonuclease activity is suppressed and the RecB endonuclease cleaves the lower strand, giving the 3′ overhang. If RecBCD translocates some distance along the DNA before reaching a chi site, then the 5′ overhang is repeatedly trimmed by the endonuclease, so it is never more than a few tens of bp in length.

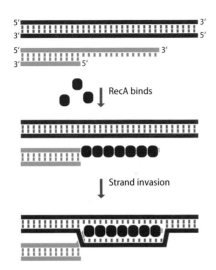

Figure 16.8 The role of the RecA protein in formation of the D-loop during homologous recombination in *E. coli*.

to invade the intact double helix and set up the D-loop (**Figure 16.8**). An intermediate in formation of the D-loop is a **triplex** structure (see the third step in Figure 16.6), a three-stranded DNA helix in which the invading polynucleotide lies within the major groove of the intact helix and forms hydrogen bonds with the base pairs it encounters.

Branch migration is catalyzed by the RuvA and RuvB proteins, both of which attach to the branch point of the heteroduplex formed by invasion of the 3′ overhang into the partner molecule. X-ray crystallography studies suggest that RuvA is a tetramer of four identical proteins, with one or two tetramers binding directly to the branch, forming a core to which two hexameric RuvB rings attach, one to either side (**Figure 16.9**). The resulting structure might act as a molecular motor, rotating the helices in the required manner so that the branch point moves. Branch migration does not appear to be a random process, but instead stops preferentially at the sequence $5' - \frac{A}{T}TT\frac{G}{C} - 3'$, where $\frac{A}{T}$ and $\frac{G}{C}$ denotes that either of two nucleotides can be present at the position indicated. This sequence occurs frequently in the *E. coli* genome, so presumably migration does not always halt at the first instance of the motif that is reached. When branch migration has ended, two RuvC proteins attach to the core of RuvA proteins, possibly displacing one RuvA tetramer if there are two present. RuvC is the **resolvase** that carries out the cleavage that resolves the Holliday structure, making a cut between the second T and the $\frac{G}{C}$ components of the recognition sequence.

Note that the above description provides no precise role for RecC in homologous recombination by the RecBCD pathway. X-ray crystallographic studies have revealed that the RecC protein comprises three structural domains, two of which are similar to the catalytic domains of the SF1 family of helicases, and one of which is similar to a PD-(D/E)xK nuclease domain. This means that RecC has a structural relationship with RecB (which is an SF1 helicase and PD-(D/E)xK nuclease) but, remarkably, key amino acids are absent from the RecC structure, so that RecC has neither helicase nor nuclease activity. The nuclease-like domain retains the ability to make contacts with the DNA molecule, enabling RecC to form a hoop through which one of the DNA strands is fed toward the RecD component of the RecBCD complex. RecC may therefore stabilize the complex and help ensure that RecB and RecD are positioned correctly relative to the DNA. It is also possible that RecC has acquired a scanning function and is responsible for identifying the chi site and hence initiating the conformational change in RecBCD that leads to formation of the heteroduplex.

E. coli has alternative pathways for homologous recombination

Mutants of *E. coli* that lack components of the RecBCD complex are still able to carry out homologous recombination, albeit with lowered efficiency. This is because the bacterium possesses a second homologous recombination pathway, called **RecFOR**. In normal *E. coli* cells most homologous recombination takes place via RecBCD, but if this pathway is inactivated by mutation the RecFOR system is able to take over.

The details of the RecFOR pathway are beginning to emerge, and the general mechanism appears to be similar to that described for RecBCD. The helicase activity for the RecFOR pathway is provided by RecQ, and the 5′ end of the strand is removed by RecJ, leaving a 3′ overhang which becomes coated in RecA proteins through the actions of either RecF or RecO acting in concert with RecR. There is considerable interchangeability between the components of the RecBCD and RecFOR pathways, and it is thought that hybrid systems operate in some mutants that lack one or other components of the standard processes. There are differences, however, as only the RecBCD pathway initiates recombination at the chi sites scattered around the *E. coli* genome, and only RecFOR is able to induce recombination between a pair of plasmids. RecFOR is also the primary pathway responsible for the recombination repair of single-strand gaps resulting from replication of heavily damaged DNA (Section 17.2).

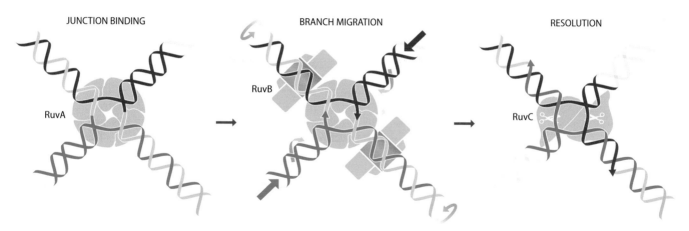

Figure 16.9 The role of the Ruv proteins in homologous recombination in *E. coli*. Branch migration is induced by a structure comprising four or eight copies of RuvA bound to the Holliday junction with an RuvB ring on either side. After branch migration, two RuvC proteins bind to the junction, the orientation of their attachment determining the direction of the cuts that resolve the structure.

As well as RecABC and RecFOR, recent results have indicated that *E. coli* has a third recombination system, called the RecBCD–RecFOR–independent or **RecBFI** pathway. Information on this pathway is still patchy, but it has been shown to be active in transduction, the process that results in integration of a bacteriophage DNA molecule into the *E. coli* chromosome (Section 3.4). Mutants that lack both the RecBCD and RecFOR pathways are still able to participate in transduction, albeit at a fourfold reduced efficiency. The key proteins involved in the RecBFI pathway appear to be RecA, RecJ and RecQ, indicating considerable overlap with the RecFOR system, with various additional proteins also involved, including SbcB15, which protects the 3′ overhangs from degradation and possibly also plays an important role in production of the RecA-DNA filament.

As well as the RecBCD, RecFOR, and RecBFI pathways, whose functions are to set up the heteroduplex structure, *E. coli* also has alternative means for carrying out the branch migration step. Mutants that lack RuvA or RuvB are still able to carry out homologous recombination because the function of RuvAB can also be provided by a helicase called RecG. It is not yet clear whether RuvAB and RecG are simply interchangeable or if they are specific for different recombination scenarios. RuvC mutants are also able to carry out homologous recombination, suggesting that *E. coli* possesses other proteins that are able to resolve Holliday structures, but the identity of these protein(s) is unknown.

Homologous recombination pathways in eukaryotes

The double-strand break model for homologous recombination is thought to hold for all organisms, not just *E. coli* – recall that it was initially devised to explain gene conversion in *Saccharomyces cerevisiae*. The biochemical events underlying the process appear to be similar in all organisms, and a number of yeast proteins have been identified that carry out functions equivalent to those occurring during the RecBCD pathway in *E. coli*. In particular, two proteins called RAD51 and DMC1 are the homologs of RecA of *E. coli*. These proteins have 45% amino acid identity and very similar structures but slightly different biochemical properties. They appear to work together during meiosis, but only RAD51 is present in vegetative cells.

One puzzling aspect of homologous recombination in eukaryotes has been the mechanism by which Holliday structures are resolved, because for many years proteins homologous to *E. coli* RuvC had been sought but not found. In fact, RuvC is not universal in all bacteria, with some species apparently using a totally different type of nuclease to resolve Holliday structures. The first human resolvase to be identified, MUS81, has homologs in both *S. cerevisiae* and *Schizosaccharomyces pombe*, but this protein acts differently to RuvC: although MUS81 can resolve Holliday structures, it does not give rise to cross-overs. Eventually, in 2008, a protein called GEN1, or Yen1 in *S. cerevisiae*, was shown to be the functional equivalent of RuvC in eukaryotes. GEN1 is a member of the Rad2/XPG family of eukaryotic nucleases, whose other members include the FEN1 endonuclease that is involved in lagging-strand replication (Section 15.3), as well as nucleases with roles in mismatch and nucleotide excision repair (Section 17.2). None of these other enzymes are able to cleave Holliday junctions. The resolvase activity of GEN1 might be conferred in part by the presence of a structural motif called a **chromodomain**, which the other Rad2/XPG nucleases do not possess. A chromodomain does not itself have any DNA-binding ability, but the chromodomain within GEN1 is thought to help position the protein onto the Holliday structure, providing GEN1 with its unique ability to make the cuts that give rise to crossovers.

16.2 SITE-SPECIFIC RECOMBINATION

A region of extensive homology is not a prerequisite for recombination: the process can also be initiated between two DNA molecules that have only very short sequences in common. This is called site-specific recombination and it has been extensively studied because of the part that it plays during the infection cycle of bacteriophage λ.

Bacteriophage λ uses site-specific recombination during the lysogenic infection cycle

After injecting its DNA into an *E. coli* cell, bacteriophage λ can follow either of two infection pathways (Section 14.3). One of these, the lytic pathway, results in the rapid synthesis of λ coat proteins, combined with replication of the λ genome, leading to death of the bacterium and release of new phages within about 45 minutes of the initial infection. In contrast, if the phage follows the lysogenic pathway, new phages do not immediately appear. The bacterium divides as normal, possibly for many cell divisions, with the phage in a quiescent form called the prophage. Eventually, possibly as the result of DNA damage or some other stimulus, the phage becomes active again.

During the lysogenic phase, the λ genome becomes integrated into the *E. coli* chromosome. It is therefore replicated whenever the *E. coli* DNA is copied, and is passed on to daughter cells just as though it was a standard part of the bacterium's genome. Integration occurs by site-specific recombination between the attachment or *att* sites, *attP* on the λ genome and *attB* on the *E. coli* chromosome. Each of these attachment sites has at its center an identical 15 bp core sequence referred to as O (**Figure 16.10**), flanked by variable sequences called B and B′ in the bacterial genome and P and P′ in the phage DNA. B and B′ are quite short, just 4 bp each, meaning that *attB* covers just 23 bp of DNA, but P and P′ are much longer, with the entire *attP* sequence spanning over 250 bp. Mutations in the core sequence inevitably lead to inactivation of the *att* site so that it can no longer participate in recombination, but mutations in the flanking sequences have a less severe consequence and only decrease the efficiency of recombination. If *attB*, the attachment site in the *E. coli* genome, is inactivated, then insertion of the λ DNA can occur at secondary sites which share some sequence similarity with the genuine *attB* locus. If a secondary site is being used, then the frequency of lysogeny is greatly reduced, integration possibly occurring at less than 0.01% of the frequency observed with unmutated *E. coli* cells.

Because this is recombination between two circular molecules, the result is that one bigger circle is formed: in other words, the λ DNA becomes integrated into the bacterial genome. The recombination event is catalyzed by a specialized type I topoisomerase (Section 15.1) called **integrase**, a member of a diverse family of **recombinases** present in bacteria, archaea, and eukaryotes. There are at least four binding sites for integrase within *attP*, as well as at least three sites for a second protein, the integration host factor, or IHF. Together these proteins coat the phage attachment site. The integrase then makes a staggered, double-strand cut at equivalent positions in the λ and bacterial *att* sites (see Figure 16.10).

GCTTTTTTATACTAA
CGAAAAAATATGATT

Figure 16.10 The core sequence of the *att* sites present in bacteriophage λ and in the *E. coli* chromosome. The pink line indicates the staggered cut made in each *att* site during integration and excision of the phage genome.

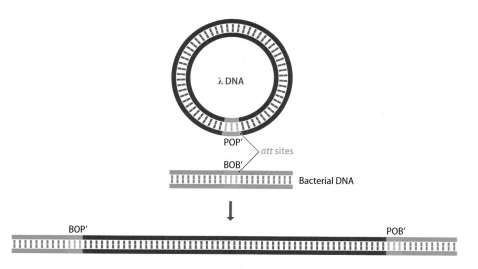

Figure 16.11 Integration of the bacteriophage λ genome into *E. coli* chromosomal DNA. Both λ and *E. coli* DNA have a copy of the *att* site, each one comprising an identical central sequence called O and flanking sequences P and P´ (for the phage *att* site) or B and B´ (bacterial *att* site). Recombination between the O regions integrates the λ genome into the bacterial DNA.

The two short, single-strand overhangs are then exchanged between the DNA molecules, producing a Holliday junction which migrates a few base pairs along the heteroduplex before being cleaved. This cleavage, providing that it is made in the appropriate orientation, resolves the Holliday structure in such a way that the λ DNA becomes inserted into the *E. coli* genome (**Figure 16.11**).

Integration creates hybrid versions of the attachment sites, which are now called *attR* (which has the structure BOP´) and *attL* (whose structure is POB´). A second site-specific recombination between the two *att* sites, now both contained in the same molecule, reverses the original process and releases the λ DNA. This recombination is also catalyzed by the integrase, but in conjunction with a protein called 'excisionase,' coded by the λ *xis* gene, rather than IHF. The functions of Xis and IHF in, respectively, excision and integration are probably quite different, and the two proteins should not be looked on as playing equivalent roles in the two processes. The key point is that the combination of integrase and excisionase is able to draw the *attR* and *attL* sites together in order to initiate the intramolecular recombination that excises the λ genome. After excision, the λ genome returns to the lytic mode of infection and directs synthesis of new phages.

Site-specific recombination is an aid in construction of genetically modified plants

The processes responsible for integration and excision of the λ genome are fairly typical of the strategies used by phages to establish lysogeny, though with some phages the molecular events are less complex than those seen with λ. Integration and excision of the bacteriophage P1 genome, for example, requires just a single enzyme, the Cre recombinase, which recognizes 34 bp target sites, called *loxB* and *loxP*, which are identical to one another and have no flanking sequences equivalent to B, B´, and so forth.

The simplicity of the P1 system has led to its utilization in genetic engineering projects in which site-specific recombination is a requirement. These applications include one that has become important in the generation of genetically modified crops. An area of concern to emerge from the debate over genetically modified plants is the possible harmful effects of the marker genes used with plant cloning vectors. Most plant vectors carry a copy of a gene for kanamycin resistance (see Figure 2.33), enabling transformed plants to be identified during the cloning process. The *kan^R* gene is bacterial in origin and codes for the enzyme neomycin phosphotransferase II. This gene and its enzyme product are present in all cells of an engineered plant. The fear that neomycin phosphotransferase might be toxic to humans has been allayed by tests with animal models, but there are still concerns that the *kan^R* gene contained in a genetically modified foodstuff could be passed to bacteria in the human gut, making these resistant to

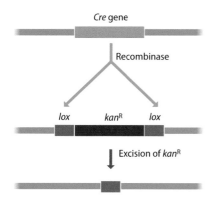

Figure 16.12 The use of the Cre recombinase in plant genetic engineering. Expression of the *Cre* gene results in excision of the *kan*^R gene from the plant DNA.

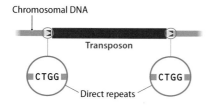

Figure 16.13 Integrated transposable elements are flanked by short direct repeat sequences. This particular transposon is flanked by the tetranucleotide repeat 5´–CTGG–3´. Other transposons have different direct repeat sequences.

kanamycin and related antibiotics, or that the *kan*^R gene could be passed to other organisms in the environment, possibly resulting in damage to the ecosystem.

The fears surrounding the use of *kan*^R and other marker genes have prompted biotechnologists to devise ways of removing these genes from plant DNA after the transformation event has been verified. One of the strategies makes use of Cre recombinase. To use this system, the plant is transformed with two cloning vectors, the first carrying the gene being added to the plant along with its *kanR* selectable marker gene surrounded by the *lox* target sequences, and the second carrying the Cre recombinase gene. After transformation, expression of the *Cre* gene results in excision of the *kan*^R gene from the plant DNA (**Figure 16.12**).

16.3 TRANSPOSITION

Transposition results in the transfer of a segment of DNA from one position in the genome to another. The transposition process results in duplication of the target site, giving rise to a pair of direct repeats that flank the transposable element (**Figure 16.13**).

In Section 9.2, we examined the various types of transposable element known in eukaryotes and prokaryotes and discovered that these could be broadly divided into three categories on the basis of their transposition mechanism:

- DNA transposons that transpose replicatively, the original transposon remaining in place and a new copy appearing elsewhere in the genome (**Figure 16.14A**).

- DNA transposons that transpose conservatively, the original transposon moving to a new site by a cut-and-paste process (**Figure 16.14B**).

- Retroelements, all of which transpose replicatively via an RNA intermediate (see Figure 9.14).

We will now examine the events that are responsible for each of these three types of transposition.

Replicative and conservative transposition of DNA transposons

A number of models for replicative and conservative transposition of DNA transposons have been proposed over the years, but most are modifications of a scheme originally outlined by Shapiro in 1979. According to this model, the replicative transposition of a bacterial element, such as a Tn3-type transposon or a transposable phage (Section 9.2), is initiated by one or more endonucleases that make single-strand cuts on either side of the transposon and at the target site where the new copy of the element will be inserted (**Figure 16.15**). At the target site, the two cuts are separated by a few base pairs, so that the cleaved double-stranded molecule has short 5´ overhangs. Ligation of these 5´ overhangs to the free 3´ ends on either side of the transposon produces a hybrid molecule in which the original two DNAs – the one containing the transposon and the one containing the target site – are linked together by the transposable element flanked by a pair of structures resembling replication forks.

DNA synthesis at the two replication forks copies the transposable element and converts the initial hybrid into a **cointegrate**, in which the two original DNAs are still linked (**Figure 16.16**). Homologous recombination between the

Figure 16.14 Replicative and conservative transposition of DNA transposons. DNA transposons use either the (A) replicative or (B) conservative pathway (some can use both).

two copies of the transposon uncouples the cointegrate, separating the original DNA molecule (with its copy of the transposon still in place) from the target molecule, which now contains a copy of the transposon. Replicative transposition has therefore occurred.

A modification of the process just described changes the mode of transposition from replicative to conservative (**Figure 16.17**). Rather than carrying out DNA synthesis, the hybrid structure is converted back into two separate DNA molecules simply by making additional single-strand nicks on either side of the transposon. This cuts the transposon out of its original molecule, transferring it into the target DNA.

Retroelements transpose replicatively via an RNA intermediate

From the human perspective, the most important retroelements are the retroviruses, which include the human immunodeficiency viruses that cause HIV/AIDS and various other virulent types. Most of what we know about retrotransposition refers specifically to retroviruses, although it is believed that other retroelements, such as retrotransposons of the *Ty1/copia* and *Ty3/gypsy* families, transpose by similar mechanisms.

The first step in retrotransposition is synthesis of an RNA copy of the inserted retroelement (**Figure 16.18**). The long terminal repeat (LTR) at the 5′ end of the element contains a TATA sequence which acts as a promoter for transcription by RNA polymerase II. Some retroelements also have enhancer sequences that are thought to regulate the amount of transcription that occurs. Transcription continues through the entire length of the element, up to a polyadenylation sequence in the 3′ LTR.

The transcript now acts as the template for RNA-dependent DNA synthesis, catalyzed by a reverse transcriptase enzyme coded by part of the *pol* gene of the retroelement (see Figure 9.15). Because this is synthesis of DNA, a primer is required and, as during genome replication, the primer is made of RNA rather than DNA. During genome replication, the primer is synthesized *de novo* by a polymerase enzyme (see Figure 15.13), but retroelements do not code for RNA polymerases and so cannot make primers in this way. Instead, they use one of the cell's tRNA molecules as a primer. Which particular tRNA is used depends on the retroelement: the *Ty1/copia* family of elements always use tRNA^{Met}, but other retroelements use different tRNAs.

The tRNA primer anneals to a site within the 5′ LTR (see Figure 16.18). At first glance, this appears to be a strange location for the priming site because it means that DNA synthesis is directed away from the central region of the retroelement and so results in only a short copy of part of the 5′ LTR. In fact, when the DNA copy has been extended to the end of the LTR, a part of the RNA template is degraded, and the DNA overhang that is produced reanneals to the 3′ LTR of the retroelement, which, being a long terminal repeat, has the same sequence as the 5′ LTR and so can base-pair with the DNA copy. DNA synthesis now continues along the RNA template. Note that the result is a DNA copy of the entire template, including the priming site: the template switching is, in effect, the strategy that the retroelement uses to solve the end-shortening problem, the same problem that chromosomal DNAs address through telomere synthesis (Section 15.4).

Completion of synthesis of the first DNA strand results in a DNA–RNA hybrid. The RNA is partially degraded by an RNase H enzyme, coded by another part of the *pol* gene. The RNA that is not degraded, usually just a single fragment attached to a short polypurine sequence adjacent to the 3′ LTR, primes synthesis of the second DNA strand, again by reverse transcriptase, which is able to act as both an RNA- and DNA-dependent DNA polymerase. As with the first round of

Figure 16.16 Replicative transposition. The two replication forks present in the hybrid molecule move toward one another, as shown by the arrows. The new DNA synthesis results in the cointegrate structure, which can be resolved into two separate molecules, each with a copy of the transposon, by homologous recombination.

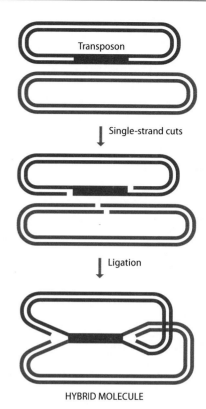

Figure 16.15 Synthesis of the hybrid molecule during the first stage of the Shapiro model for replicative transposition.

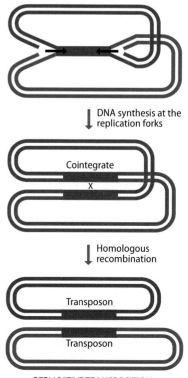

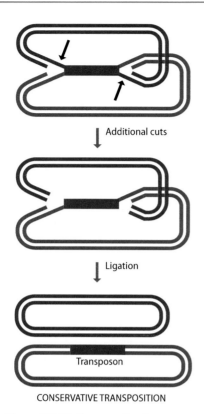

Figure 16.17 Conservative transposition. Additional cuts in the hybrid molecule at the positions indicated by the arrows, followed by ligation of the cut ends, transfers the transposon to the second molecule.

DNA synthesis, second-strand synthesis initially results in a DNA copy of just the LTR, but a second template switch, to the other end of the molecule, enables the DNA copy to be extended until it is full length. This creates a template for further extension of the first DNA strand, so that the resulting double-stranded DNA is a complete copy of the internal region of the retroelement plus the two LTRs.

All that remains is to insert the new copy of the retroelement into the genome. It was originally thought that insertion occurred randomly, but it now appears that although no particular sequence is used as a target site, integration occurs preferentially at certain positions. Insertion involves removal of two nucleotides

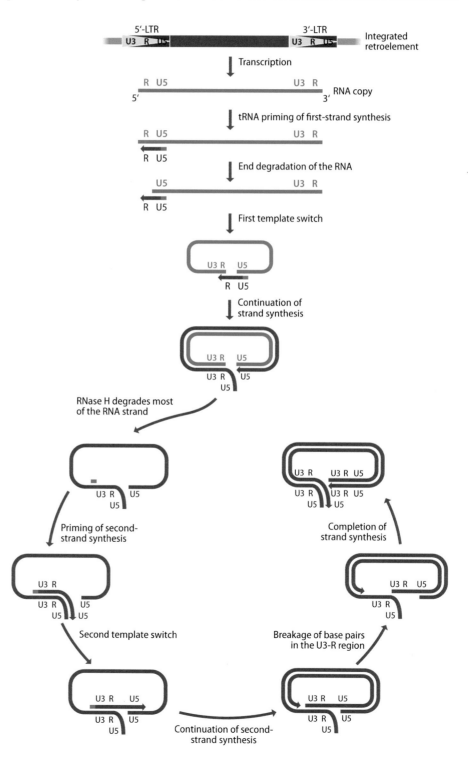

Figure 16.18 RNA and DNA replication during transposition of a retroelement. This diagram shows how an integrated retroelement is copied into a free double-stranded DNA version. The first step is synthesis of an RNA copy, which is then converted to double-stranded DNA by a series of events that involves two template switches, as described in the text.

from the 3´ ends of the double-stranded retroelement by the integrase enzyme (coded by yet another part of *pol*). The integrase also makes a staggered cut in the genomic DNA so that both the retroelement and the integration site now have 5´ overhangs (**Figure 16.19**). These overhangs might not have complementary sequences, but they still appear to interact in some way so that the retroelement becomes inserted into the genomic DNA. The interaction results in loss of the overhangs on the retroelement, and the gaps that are left are filled in, which means that the target site becomes duplicated into a pair of direct repeats, one at either side of the inserted retroelement.

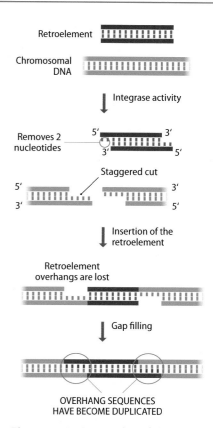

Figure 16.19 Integration of the double-stranded DNA version of a retroelement into the host genome. In this example, integration of the retroelement results in a target site duplication of four nucleotides on either side of the inserted sequence.

SUMMARY

- The term recombination was originally used to describe the outcome of crossing over between pairs of homologous chromosomes during meiosis. The term is now also used to refer to the molecular events that underlie this process.

- Homologous recombination occurs between segments of DNA molecules that share extensive sequence homology.

- The initial models for homologous recombination envisaged the recombination process being initiated by nicks that occur in one or both of the double-stranded molecules, but it is now thought that the start point is a double-strand break in one of these molecules.

- Strand exchange leads to a heteroduplex structure that is resolved by cleavage, possibly leading to exchange of DNA segments or to gene conversion.

- *Escherichia coli* possesses at least three molecular pathways for homologous recombination. These are the RecBCD, RecFOR, and RecBFI pathways.

- The RecBCD pathway involves the unwinding of one partner in the recombination by a pair of helicases, which attach to a double-strand break and progress along the molecule. At a recognition sequence called a chi site, the RecBCD complex initiates the strand exchange, with the RecA protein playing a central role in transfer of the invading strand into the intact double helix.

- Branch migration within the heteroduplex, and resolution of the structure, is catalyzed by the Ruv proteins.

- In eukaryotes, the RAD51 and DMC1 proteins play the equivalent role to RecA. Holliday structures are resolved by GEN1.

- Site-specific recombination does not require regions of extensive homology between the partner molecules. This type of recombination is responsible for insertion of bacteriophage genomes, such as the λ genome, into the host bacterial chromosome.

- Transposition occurs by recombination. With DNA transposons, the process can be either replicative or conservative, both modes occurring by a series of events initially described by Shapiro in 1979.

- Retroelements transpose via an RNA intermediate that is transcribed from the parent copy of the transposon. After copying into double-stranded DNA, the retroelement is reinserted into the host chromosome.

SHORT ANSWER QUESTIONS

1. What is the role of recombination in genome evolution?

2. How can the resolution of a Holliday structure yield two different results?

3. Describe how the double-strand break model explains how gene conversion occurs.

4. Describe the RecBCD pathway for homologous recombination in *E. coli*.

5. What are the distinctive features of the RecFOR pathway for homologous recombination in *E. coli*?

6. Outline our current knowledge of the molecular basis of homologous recombination in eukaryotes.

7. What are the properties of the *attP* and *attB* sites that mediate integration of bacteriophage λ DNA into the *E. coli* genome?

8. Distinguish between the processes that lead to integration and excision of bacteriophage λ DNA into and out of the *E. coli* genome.

9. Describe the model for conservative transposition.

10. How does the model for replicative transposition differ from that for conservative transposition?

11. How is the new copy of a retroelement inserted into a genome?

IN-DEPTH PROBLEMS

1. Determination of the structure of the RecBCD complex was looked on as a key step in understanding the molecular basis of homologous recombination. Explain why knowing the structure of this complex was so important.

2. Some *E. coli* strains that are used for propagating recombinant plasmids contain *recA* mutations. Why might *recA* defects be useful for researchers working with recombinant plasmids?

3. Transposition can have deleterious effects on a genome. These effects go beyond the obvious disruption of gene activity that will occur if a transposable element takes up a new position that lies within the coding region of a gene. Some elements, notably retrotransposons, contain promoter and enhancer sequences that can modify the expression patterns of adjacent genes, and transposition often involves the creation of double-strand breaks. How might cells minimize these deleterious effects by preventing transposition from occurring?

4. Are there circumstances where transposition could be beneficial to a cell?

FURTHER READING

Models for homologous recombination
Holliday, R. (1964) A mechanism for gene conversion in fungi. *Genet. Res.* 5:282–304.

Meselson, M. and Radding, C.M. (1975) A general model for genetic recombination. *Proc. Natl. Acad. Sci. USA* 72:358–361.

Szostak, J.W., Orr-Weaver, T.L., Rothstein, R.J., et al. (1983) The double-strand-break repair model for recombination. *Cell* 33:25–35.

Molecular basis to homologous recombination
Amundsen, S.K. and Smith, G.R. (2003) Interchangeable parts of the *Escherichia coli* recombination machinery. *Cell* 112:741–744. *Hybrid pathways involving parts of the RecBCD and RecFOR systems.*

Bertucat, G., Lavery, R. and Prévost, C. (1999) A molecular model for RecA-promoted strand exchange via parallel triple-stranded helices. *Biophys. J.* 77:1562–1576.

Buljubašić, M., Hlevnjak, A., Repar, J., et al. (2019) RecBCD-RecFOR-independent pathway of homologous recombination in *Escherichia coli*. *DNA Repair* 83:102670.

Henrikus, S.S., Henry, C., Ghodke, H., et al. (2019) RecFOR epistasis group: RecF and RecO have different localizations and functions in *Escherichia coli*. *Nucl. Acids Res.* 47:2946–2965.

Lan, W.-H., Lin, S.-Y., Kao, C.-Y., et al. (2020) Rad51 facilitates filament assembly of meiosis-specific Dmc1 recombinase. *Proc. Natl. Acad. Sci. USA* 117:11257–11264.

Lee, S.-H., Princz, L.N., Klügel, M.F., et al. (2015) Human Holliday junction resolvase GEN1 uses a chromodomain for efficient DNA recognition and cleavage. *eLife* 4:e12256.

Masson, J.-Y. and West, S.C. (2001) The Rad51 and Dmc1 recombinases: A non-identical twin relationship. *Trends Biochem. Sci.* 26:131–136.

Persky, N.S. and Lovett, S.T. (2008) Mechanisms of recombination: Lessons from *E. coli. Crit. Rev. Biochem. Mol. Biol.* 43:347–370.

Punatar, R.S., Martin, M.J., Wyatt, H.D.M., et al. (2017) Resolution of single and double Holliday junction recombination intermediates by GEN1. *Proc. Natl. Acad. Sci. USA* 114:443–450.

Rafferty, J.B., Sedelnikova, S.E., Hargreaves, D., et al. (1996) Crystal structure of DNA recombination protein RuvA and a model for its binding to the Holliday junction. *Science* 274:415–421.

Rigden, D.J. (2005) An inactivated nuclease-like domain in RecC with novel function: Implications for evolution. *BMC Struct. Biol.* 5:9.

Singleton, M.R., Dillingham, M.S., Gaudier, M., et al. (2004) Crystal structure of RecBCD enzyme reveals a machine for processing DNA breaks. *Nature* 432:187–193.

West, S.C. (1997) Processing of recombination intermediates by the RuvABC proteins. *Annu. Rev. Genet.* 31:213–244.

Wyatt, H.D.M. and West, S.C. (2014) Holliday junction resolvases. *Cold Spring Harb. Perspect. Biol.* 6:a023192.

Site-specific recombination

Kwon, H.J., Tirumalai, R., Landy, A., et al. (1997) Flexibility in DNA recombination: Structure of the lambda integrase catalytic core. *Science* 276:126–131.

Van Duyne, G.D. (2015) Cre recombinase. *Microbiol. Spectr.* 3:MDNA3-0014-2014.

Transposition

Bushman, F.D. (2003) Targeting survival: Integration site selection by retroviruses and LTR-retrotransposons. *Cell* 115:135–138.

Derbyshire, K.M. and Grindley, N.D.F. (1986) Replicative and conservative transposition in bacteria. *Cell* 47:325–327.

Shapiro, J.A. (1979) Molecular model for the transposition and replication of bacteriophage Mu and other transposable elements. *Proc. Natl. Acad. Sci. USA* 76:1933–1937.

MUTATIONS AND DNA REPAIR

17.1 THE CAUSES OF MUTATIONS

17.2 REPAIR OF MUTATIONS AND OTHER TYPES OF DNA DAMAGE

Genome sequences change over time as a result of the cumulative effects of small-scale sequence alterations caused by **mutation**. A mutation is a change in the nucleotide sequence of a short region of a genome (**Figure 17.1A**). Many mutations are **point mutations** (also called simple mutations or single-site mutations) that replace one nucleotide with another. Point mutations are divided into two categories: **transitions**, which are purine-to-purine or pyrimidine-to-pyrimidine changes (A→G, G→A, C→T, or T→C), and **transversions**, which are purine-to-pyrimidine or pyrimidine-to-purine changes (A→C, A→T, G→C, G→T, C→A, C→G, T→A, or T→G). Other mutations arise from **insertion** or **deletion** of one or a few nucleotides.

Mutations result either from errors in DNA replication or from the damaging effects of **mutagens**, such as chemicals and radiation, which react with DNA and change the structures of individual nucleotides. All cells possess **DNA repair** enzymes that attempt to reverse the effects of mutations. These enzymes work in two ways. Some are prereplicative and search the DNA for nucleotides with unusual structures, these nucleotides being replaced before replication occurs; others are postreplicative and check newly synthesized DNA for errors, correcting any that they find (**Figure 17.1B**). A possible definition of mutation is therefore *a deficiency in DNA repair*.

Mutations can have dramatic effects on the cell in which they occur; a mutation in a key gene can possibly result in a defective protein, which could lead to death of the cell. Other mutations have a less significant impact on the phenotype of the cell and many have none at all. As we will see in Chapter 18, all events that are not lethal have the potential to contribute to the evolution of the genome, but for this to happen they must be inherited when the organism reproduces. With a single-celled organism, such as a bacterium or yeast, all genome alterations that are not lethal or corrected are inherited by daughter cells, and therefore become permanent features of the lineage that descends from the original cell in which the alteration occurred. In a multicellular organism, only those events that occur in germ cells are relevant to genome evolution. Changes to the genomes of somatic cells are unimportant in an evolutionary sense, but they will have biological relevance if they result in a deleterious phenotype that affects the health of the organism.

In this chapter, we will follow a logical progression, beginning with the causes of mutations and then moving on to the ways in which mutations are repaired.

17.1 THE CAUSES OF MUTATIONS

Mutations arise in two ways: Some mutations are **spontaneous** errors in replication that evade the proofreading function of the DNA polymerases that synthesize new polynucleotides at the replication fork (Section 15.3). These mutations are called **mismatches** because they are positions where the nucleotide that is inserted into the daughter polynucleotide does not match, by base-pairing, the nucleotide at the corresponding position in the template DNA (**Figure 17.2A**). If the mismatch is not corrected in the daughter double helix, then one of the granddaughter molecules produced during the next round of DNA replication will carry a permanent, double-stranded version of the mutation.

DOI: 10.1201/9781003133162-17

(A) A mutation

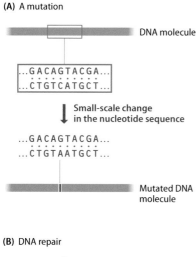

DNA molecule

...G A C A G T A C G A...
...C T G T C A T G C T...

↓ Small-scale change
in the nucleotide sequence

...G A C A G T A C G A...
...C T G T A A T G C T...

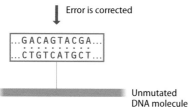

Mutated DNA molecule

(B) DNA repair

↓ Error is corrected

...G A C A G T A C G A...
...C T G T C A T G C T...

Unmutated DNA molecule

Figure 17.1 Mutation and DNA repair.
(A) A mutation is a small-scale change in the nucleotide sequence of a DNA molecule. A point mutation is shown, but there are several other types of mutation, as described in the text. (B) DNA repair corrects mutations that arise as errors in replication and as a result of mutagenic activity.

Other mutations arise because a mutagen has reacted with the parent DNA, causing a structural change that affects the base-pairing capability of the altered nucleotide. Usually, this alteration affects only one strand of the parent double helix, so only one of the daughter molecules carries the mutation, but two of the granddaughter molecules produced during the next round of replication will have it (**Figure 17.2B**).

Errors in replication are a source of point mutations

When considered purely as a chemical reaction, complementary base-pairing is not particularly accurate. Nobody has yet devised a way of carrying out the template-dependent synthesis of DNA without the aid of enzymes, but if the process could be carried out simply as a chemical reaction in a test tube then the resulting polynucleotide would probably have point mutations at 5–10 positions out of every 100. This represents an error rate of 5–10%, which would be completely unacceptable during genome replication. The template-dependent DNA polymerases that carry out DNA replication must therefore increase the accuracy of the process by several orders of magnitude. This improvement is brought about in two ways:

- A DNA polymerase operates a nucleotide selection process that dramatically increases the accuracy of template-dependent DNA synthesis. Exactly how this selection operates is not known, but it involves the nucleotide-binding part of the polymerase switching between an open and closed conformation, the latter placing the selected nucleotide on to the template and checking that it base-pairs correctly. If the pairing is incorrect, then the nucleotide is rejected before it is attached to the 3´ end of the polynucleotide that is being synthesized.

- The accuracy of DNA synthesis is increased still further if the DNA polymerase has proofreading activity (Section 15.3), which means that it possesses a 3´→5´ exonuclease and so is able to remove an incorrect nucleotide that evades the nucleotide selection process and becomes attached to the 3´ end of the new polynucleotide (see Figure 2.7B).

Escherichia coli is able to synthesize DNA with an error rate of only 1 per 10^7 nucleotide additions. Interestingly, these errors are not evenly distributed between the two daughter molecules, with the lagging strand being replicated more accurately than the leading strand. The reason for this is not known but probably relates to the efficiency with which errors in the two strands are repaired, rather than differences between the fidelity of DNA polymerase I, which is involved only in lagging-strand replication (Section 15.3), and that of DNA polymerase III, the main replicating enzyme.

Not all of the errors that occur during DNA synthesis can be blamed on the polymerase enzymes: sometimes an error occurs even though the enzyme adds the 'correct' nucleotide, the one that base-pairs with the template. This is because each nucleotide base can occur as either of two alternative **tautomers**, structural isomers that are in dynamic equilibrium. For example, thymine exists as two tautomers, the *keto* and *enol* forms, with individual molecules occasionally undergoing a shift from one tautomer to the other. The equilibrium is biased very much toward the *keto* form, but every now and then the *enol* version of thymine occurs in the template DNA at the precise time that the replication fork is moving past. This will lead to an error, because *enol*-thymine base-pairs with G rather than A (**Figure 17.3**). The same problem can occur with adenine, the rare *imino*-tautomer of this base preferentially forming a pair with C, and with guanine, *enol*-guanine pairing with thymine. After replication, the rare tautomer will inevitably revert to its more common form, leading to a mismatch in the daughter double helix.

As stated above, the error rate for DNA synthesis in *E. coli* is 1 in 10^7. However, the overall error rate for replication of the *E. coli* genome is only 1 in 10^{10} to 1 in

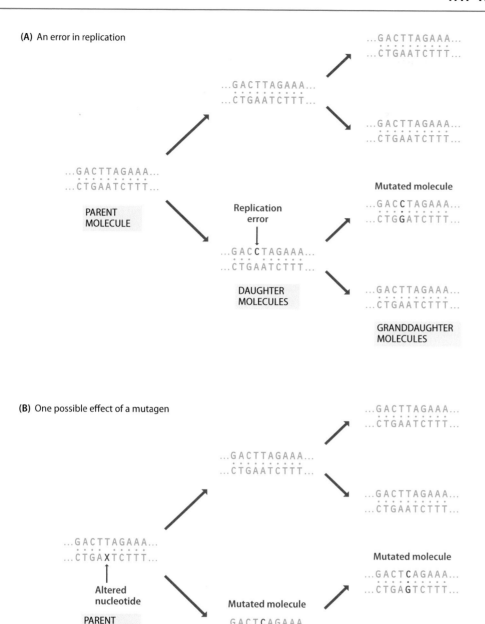

(A) An error in replication

PARENT MOLECULE

Replication error

DAUGHTER MOLECULES

Mutated molecule

GRANDDAUGHTER MOLECULES

(B) One possible effect of a mutagen

Altered nucleotide

PARENT MOLECULE

Mutated molecule

DAUGHTER MOLECULES

Mutated molecule

Mutated molecule

GRANDDAUGHTER MOLECULES

Figure 17.2 Examples of mutations. (A) An error in replication leads to a mismatch in one of the daughter double helices, in this case a T→C change because one of the As in the template DNA was miscopied. When the mismatched molecule is itself replicated, it gives one double helix with the correct sequence and one with a mutated sequence. (B) A mutagen has altered the structure of an A in the lower strand of the parent molecule, giving nucleotide X, which does not base pair with the T in the other strand so, in effect, a mismatch has been created. When the parent molecule is replicated, X base-pairs with C, giving a mutated daughter molecule. When this daughter molecule is replicated, both granddaughters inherit the mutation.

10^{11}, the improvement compared with the polymerase error rate being the result of the mismatch repair system (Section 17.2) that scans newly replicated DNA for positions where the bases are unpaired and hence corrects the few mistakes that the replication enzymes make. This means that, on average, only one uncorrected replication error occurs every 2000 times that the *E. coli* genome is copied.

Replication errors can also lead to insertion and deletion mutations

Not all errors in replication are point mutations. Aberrant replication can also result in small numbers of extra nucleotides being inserted into the polynucleotide being synthesized, or some nucleotides in the template not being copied. An insertion or deletion that occurs within a coding region might result in a

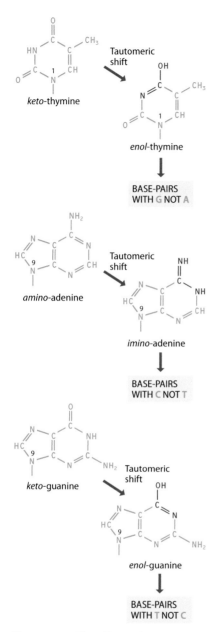

Figure 17.3 The effects of tautomerism on base-pairing. In each of these three examples, the two tautomeric forms of the base have different pairing properties. Cytosine also has *amino* and *imino* tautomers, but both pair with guanine.

frameshift mutation, which changes the reading frame used for translation of the protein specified by the gene (**Figure 17.4**). There is a tendency to use 'frameshift' to describe all insertions and deletions, but this is inaccurate as inserting or deleting three nucleotides, or multiples of three, simply adds or removes codons or parts of adjacent codons without affecting the reading frame. Also, of course, many insertions/deletions occur outside of open reading frames, within introns or in the intergenic regions of a genome.

Insertion and deletion mutations can affect all parts of the genome but are particularly prevalent when the template DNA contains short repeated sequences, such as those found in microsatellites (Section 3.2). This is because repeated sequences can induce **replication slippage**, in which the template strand and its copy shift their relative positions so that part of the template is either copied twice or missed out. The result is that the new polynucleotide has a larger or smaller number, respectively, of the repeat units (**Figure 17.5**). This is the main reason why microsatellite sequences are so variable; replication slippage occasionally generates a new length variant, which can add to the collection of alleles already present in the population.

Replication slippage is probably also responsible, at least in part, for the **trinucleotide repeat expansion diseases** that have been discovered in humans in recent years. Each of these neurodegenerative diseases is caused by a relatively short series of trinucleotide repeats becoming elongated to two or more times its normal length. For example, the human *HTT* gene contains the sequence 5′–CAG–3′ repeated between 6 and 29 times in tandem, coding for a series of glutamines in the protein product. In Huntington's disease this repeat expands to a copy number of 38–180, increasing the length of the polyglutamine tract and resulting in a dysfunctional protein. Several other human diseases are also caused by expansions of polyglutamine codons (Table 17.1). Some diseases associated with mental retardation result from trinucleotide expansions in the 5′ untranslated region of a gene, giving a **fragile site**, a position where the chromosome is likely to break. However, chromosome breakage is not the underlying cause of the disease. Instead, in both fragile X and fragile XE syndromes, the expansion results in aberrant methylation of a CpG island upstream of the affected gene, which is silenced, leading to loss of the protein product. Expansions involving intron and 3′-untranslated regions are also known, most of these expansions affecting transcription of the gene or processing of the mRNA. An exception is the expansion in the 3′ untranslated region of the *DMPK* gene, which does not affect transcription or processing of the *DMPK* mRNA, but does interfere with splicing of other RNAs. The explanation is not known, but it is possible that the expanded part of the *DMPK* mRNA either binds protein splicing factors, preventing the latter from acting on their normal target RNAs, or interferes in some way with the signaling pathways that control attachment of these factors to their targets. Finally, there are a few disease-causing mutations that involve expansions of longer sequences, such as progressive myoclonus epilepsy, caused by a $(CCCCGCCCCGCG)_{2-3}$ to $(CCCCGCCCCGCG)_{30+}$ expansion in the promoter region of the *EPM1* locus.

How triplet expansions are generated is not precisely understood. The size of the expansion is much greater than occurs with normal replication slippage, such as that seen with microsatellite sequences, and once the expansion reaches a certain length it appears to become susceptible to further expansion in subsequent rounds of replication, so that the disease becomes increasingly severe in succeeding generations of an affected family. Expansions also occur between cell divisions, in cells that are not actively replicating their genomes. Initially, it was thought that expansion in non-dividing cells might arise as a by-product of the process used to repair a double-strand break in the genome, but evidence to support this idea is lacking. It now seems likely that expansion can occur during excision repair, which involves removal of a segment of a polynucleotide containing a mutation, followed by resynthesis of the DNA strand to fill in the single-stranded gap that is formed (Section 17.2). If the excised segment includes part or all of a trinucleotide region, then slippage during the DNA synthesis step of the repair process could result in the triplet expansion.

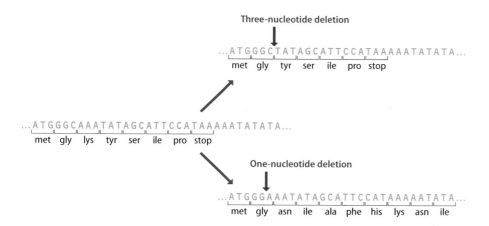

Three-nucleotide deletion

...ATGGGCTATAGCATTCCATAAAAATATATA...
met gly tyr ser ile pro stop

...ATGGGCAAATATAGCATTCCATAAAAATATATA...
met gly lys tyr ser ile pro stop

One-nucleotide deletion

...ATGGGAAATATAGCATTCCATAAAAATATA...
met gly asn ile ala phe his lys asn ile

Figure 17.4 Frameshift mutations. Frameshift mutations resulting from deletions are shown. In the top sequence, three nucleotides comprising a single codon are deleted. This shortens the resulting protein product by one amino acid but does not affect the rest of its sequence. In the lower section, a single nucleotide is deleted. The latter results in a frameshift: all the codons downstream of the deletion are changed, including the termination codon, which is now read through. Note that if a three-nucleotide deletion removes parts of adjacent codons then the result is more complicated than shown here. Consider, for example, deletion of the trinucleotide GCA from the sequence ...ATGGGCAAATAT... coding for met–gly–lys–tyr. The new sequence is ...ATGGAATAT... coding for met–glu–tyr. Two amino acids have been replaced by a single, different one.

Mutations are also caused by chemical and physical mutagens

Many chemicals that occur naturally in the environment have mutagenic properties and these have been supplemented in recent years with other chemical mutagens that result from human industrial activity. Physical agents such as radiation are also mutagenic. Most organisms are exposed to greater or lesser amounts of these various mutagens, their genomes suffering damage as a result.

The definition of the term 'mutagen' is 'a chemical or physical agent that causes mutations.' This definition is important because it distinguishes mutagens from other types of environmental agent that cause damage to cells in ways other than by causing mutations (Table 17.2). There are overlaps between these categories (for example, some mutagens are also carcinogens), but each type of agent has a distinct biological effect. The definition of 'mutagen' also makes a distinction between true mutagens and other agents that damage DNA without causing mutations, for example, by causing breaks in DNA molecules. This type of damage may block replication and cause the cell to die, but it is not a mutation in the strict sense of the term and the causative agents are therefore not mutagens.

Mutagens cause mutations in three different ways:

- Some act as **base analogs** and are mistakenly used as substrates when new DNA is synthesized at the replication fork.

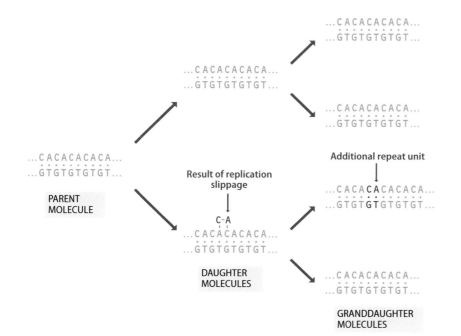

Figure 17.5 Replication slippage. The diagram shows replication of a five-unit CA repeat microsatellite. Slippage has occurred during replication of the parent molecule, inserting an additional repeat unit into the newly synthesized polynucleotide of one of the daughter molecules. When this daughter molecule replicates, it gives a granddaughter molecule whose microsatellite array is one unit longer than that of the original parent.

TABLE 17.1 EXAMPLES OF HUMAN TRINUCLEOTIDE REPEAT EXPANSIONS

Locus	Repeat sequence		Associated disease
	Normal	**Mutated**	
Polyglutamine expansions (all in coding regions of genes)			
AR (exon 1)	$(CAG)_{13-31}$	$(CAG)_{40}$	Spinomuscular bulbar atrophy
ATN1 (exon 5)	$(CAG)_{6-35}$	$(CAG)_{49-88}$	Dentatorubral-pallidoluysian atrophy
ATXN1 (exon 8)	$(CAG)_{6-39}$	$(CAG)_{41-83}$	Spinocerebellar ataxia type 1
ATXN3 (exon 8)	$(CAG)_{12-40}$	$(CAG)_{52-86}$	Machado–Joseph disease
HTT (exon 1)	$(CAG)_{6-29}$	$(CAG)_{38-180}$	Huntington's disease
Fragile site expansions			
AFF2 (5´-UTR)	$(GCC)_{4-39}$	$(GCC)_{over\ 200}$	Fragile XE mental retardation
FMR1 (5´-UTR)	$(CGG)_{6-50}$	$(CGG)_{200-4000}$	Fragile X syndrome
Other expansions			
DMPK (3´-UTR)	$(CTG)_{5-37}$	$(CTG)_{40-50}$	Myotonic dystrophy
FXN (intron 1)	$(GAA)_{5-30}$	$(GAA)_{70-1000}$	Friedreich's ataxia

Abbreviation: UTR, untranslated region

TABLE 17.2 CATEGORIES OF ENVIRONMENTAL AGENT THAT CAUSE DAMAGE TO LIVING CELLS

Agent	Effect on living cells
Carcinogen	Causes cancer – the neoplastic transformation of eukaryotic cells
Clastogen	Causes fragmentation of chromosomes
Mutagen	Causes mutations
Oncogen	Induces tumor formation
Teratogen	Results in developmental abnormalities

- Others react directly with DNA, causing structural changes that lead to miscopying of the template strand when the DNA is replicated. These structural changes are diverse, as we will see when we look at individual mutagens.

- Some mutagens act indirectly on DNA. They do not themselves affect DNA structure, but instead cause the cell to synthesize chemicals such as peroxides that have a direct mutagenic effect.

The range of mutagens is so vast that it is difficult to devise an all-embracing classification. We will therefore restrict our study to the most common types. For chemical mutagens these are as follows:

- **Base analogs** are purine and pyrimidine bases that are similar enough to the standard bases of DNA to be incorporated into nucleotides when these are synthesized by the cell. The resulting unusual nucleotides can then be used as substrates for DNA synthesis during genome replication. For example, **5-bromouracil** (**5-bU**; **Figure 17.6A**) has the same base-pairing

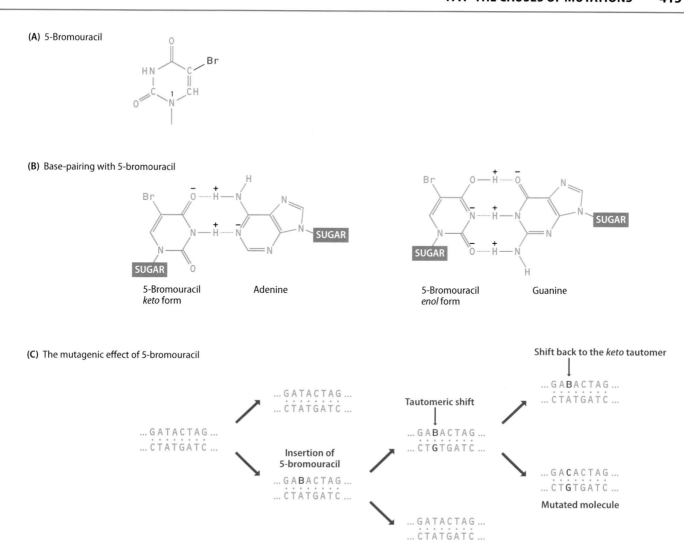

Figure 17.6 5-Bromouracil and its mutagenic effect.

properties as thymine, and nucleotides containing this base can be added to the daughter polynucleotide at positions opposite As in the template. The mutagenic effect arises because the equilibrium between the two tautomers of 5-bU is shifted more toward the rarer *enol* form than is the case with thymine. This means that during the next round of replication there is a relatively high chance of the polymerase encountering *enol*- 5bU, which (like *enol*-thymine) pairs with G rather than A (**Figure 17.6B**). This results in a point mutation (**Figure 17.6C**). **2-Aminopurine** acts in a similar way: it is an analog of adenine with an *amino*-tautomer that pairs with thymine, and an *imino*-tautomer that pairs with cytosine. The *imino* form of 2-aminopurine is more common than *imino*-adenine so 2-aminopurine incorporation increases the frequency of T→C transitions.

- **Deaminating agents** also cause point mutations. A certain amount of base deamination (removal of an amino group) occurs spontaneously in genomic DNA molecules, with the rate being increased by chemicals such as nitrous acid, which is generated in the atmosphere from nitrogen and can accumulate in closed spaces such as poorly ventilated rooms. Nitrous acid deaminates adenine, cytosine, and guanine (thymine has no amino group and so cannot be deaminated). A second deaminating

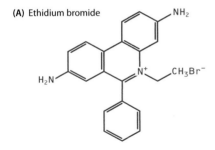

Figure 17.7 **Hypoxanthine is a deaminated version of adenine.** The nucleoside that contains hypoxanthine is called **inosine**.

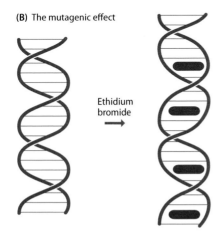

Figure 17.8 **The mutagenic effect of ethidium bromide.** (A) Ethidium bromide is a flat, plate-like molecule that is able to slot in between the base pairs of the double helix. (B) Ethidium bromide molecules are shown intercalated into the helix: the molecules are viewed sideways on. Note that intercalation increases the distance between adjacent base pairs.

agent, sodium bisulfite, acts only on cytosine. Deamination of guanine is not mutagenic because the resulting base, xanthine, blocks replication when it appears in the template polynucleotide. Deamination of adenine gives hypoxanthine (**Figure 17.7**), which pairs with C rather than T, and deamination of cytosine gives uracil, which pairs with A rather than G. Deamination of these two bases therefore results in point mutations when the template strand is copied.

- **Alkylating agents** are a third type of mutagen that can give rise to point mutations. Chemicals such as **ethylmethane sulfonate** (**EMS**), dimethyl-nitrosamine and methyl halides, the latter sometimes used as pesticides, add alkyl groups to nucleotides in DNA molecules. The effect of alkylation depends on the position at which the nucleotide is modified and the type of alkyl group that is added. Some methylations are not mutagenic as they do not alter the base-pairing properties of the nucleotide. An example is the conversion of cytosine to 5-methylcytosine, which occurs extensively in vertebrate and plant genomes and is associated with gene silencing (Section 10.3). Other methylations do alter the base-pairing properties of the modified nucleotide and so lead to point mutations. There are also alkylations that block replication by forming cross-links between the two strands of a DNA molecule or simply by impeding progress of the replisome.

- **Intercalating agents** are usually associated with insertion mutations. The best-known mutagen of this type is **ethidium bromide**, which fluoresces when exposed to ultraviolet (UV) radiation and has been used to reveal the positions of DNA bands after agarose gel electrophoresis. Ethidium bromide and other intercalating agents are flat molecules that can slip between base pairs in the double helix, slightly unwinding the helix and hence increasing the distance between adjacent base pairs (**Figure 17.8**).

Now we move on to the most important types of physical mutagen, which are as follows:

- Ultraviolet radiation of wavelength 260 nm induces dimerization of adjacent pyrimidine bases, especially if these are both thymines (**Figure 17.9A**), resulting in a **cyclobutyl dimer**. Other pyrimidine combinations also form dimers, the order of frequency with which this occurs being 5′–CT–3′ > 5′–TC–3′ > 5′–CC–3′. Purine dimers are much less common. UV-induced dimerization usually results in a deletion mutation when the modified strand is copied. Another type of UV-induced **photoproduct** is the **(6–4) lesion**, in which carbons number 4 and 6 of adjacent pyrimidines become covalently linked (**Figure 17.9B**). (6–4) lesions impede the progress of the replisome when the DNA is being replicated and trigger the ATR damage response pathway (Section 15.5).

- Ionizing radiation has various effects on DNA depending on the type of radiation and its intensity. Point, insertion, and/or deletion mutations might arise, as well as more severe forms of DNA damage that prevent subsequent replication of the genome. Some types of ionizing radiation act directly on DNA, while others act indirectly by stimulating the formation of reactive molecules such as peroxides.

- Heat stimulates the water-induced cleavage of the β-*N*-glycosidic bond that attaches the base to the sugar component of the nucleotide (**Figure 17.10A**). This occurs more frequently with purines than with pyrimidines and results in an **AP** (**apurinic/apyrimidinic**) site, also called a **baseless site**. The sugar–phosphate that is left is unstable and rapidly degrades, leaving a gap if the DNA molecule is double-stranded (**Figure 17.10B**). This reaction is not normally mutagenic because cells have effective systems for repairing gaps (Section 17.2), which is reassuring when one considers that 10,000 AP sites are generated in each human cell per day.

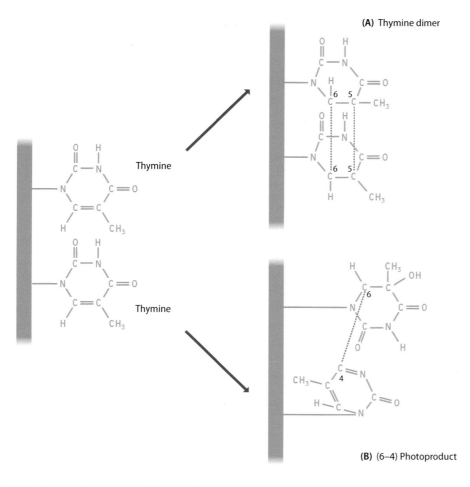

(A) Thymine dimer

(B) (6–4) Photoproduct

Figure 17.9 Photoproducts induced by UV irradiation. A segment of a poly-nucleotide containing two adjacent thymine bases is shown. (A) A thymine dimer contains two UV-induced covalent bonds, one linking the carbons at position 6 and the other linking the carbons at position 5. (B) The (6–4) lesion involves formation of a covalent bond between carbons 4 and 6 of the adjacent nucleotides. In both drawings, the UV-induced covalent bonds are shown as blue dotted lines.

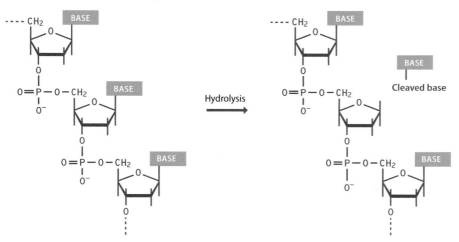

(A) Heat-induced hydrolysis of a β-*N*-glycosidic bond

Figure 17.10 The mutagenic effect of heat. (A) Heat induces hydrolysis of a β-*N*-glycosidic bond, resulting in a baseless site in a polynucleotide. (B) Schematic representation of the effect of heat-induced hydrolysis on a double-stranded DNA molecule. The baseless site is unstable and degrades, leaving a gap in one strand.

(B) The effect of hydrolysis on double-stranded DNA

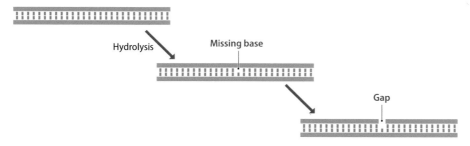

17.2 REPAIR OF MUTATIONS AND OTHER TYPES OF DNA DAMAGE

Having studied the myriad ways in which the genome can become mutated, it is always reassuring to move quickly on to the equally myriad ways in which the genome can be repaired. In view of the thousands of damage events that genomes suffer every day, coupled with the errors that occur when the genome replicates, it is essential that cells possess efficient repair systems. Without these repair systems, a genome would not be able to maintain its essential cellular functions for more than a few hours before key genes became inactivated by DNA damage. Similarly, cell lineages would accumulate replication errors at such a rate that their genomes would become dysfunctional after a few cell divisions.

Most cells possess four different categories of DNA repair system (**Figure 17.11**):

- **Direct repair** systems, as the name suggests, act directly on damaged nucleotides, converting each one back to its original structure.

- **Excision repair** involves excision of a segment of the polynucleotide containing a damaged site, followed by resynthesis of the correct nucleotide sequence by a DNA polymerase.

- **Mismatch repair** corrects errors of replication, again by excising a stretch of single-stranded DNA containing the offending nucleotide and then repairing the resulting gap.

- **Break repair** is used to mend single- and double-strand breaks.

Direct repair systems fill in nicks and correct some types of nucleotide modification

Most of the types of DNA damage that are caused by chemical or physical mutagens can only be repaired by excision of the damaged nucleotide followed by resynthesis of a new stretch of DNA, as shown in **Figure 17.11B**.

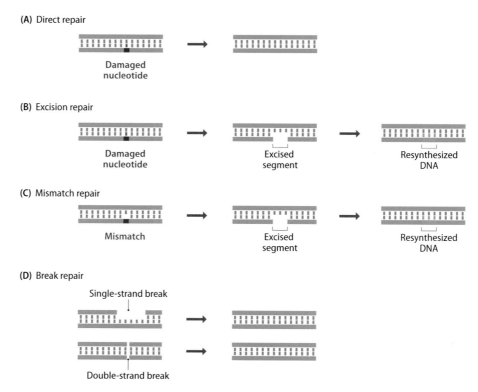

(A) Direct repair

Damaged
nucleotide

(B) Excision repair

Damaged   Excised   Resynthesized
nucleotide   segment   DNA

(C) Mismatch repair

Mismatch   Excised   Resynthesized
                  segment   DNA

(D) Break repair

Single-strand break

Double-strand break

Figure 17.11 Four categories of DNA repair system.

Only a few types of damaged nucleotide can be repaired directly. These include **nicks**, which can be repaired by a DNA ligase if all that has happened is that a phosphodiester bond has been broken, without damage to the 5′–phosphate and 3′–hydroxyl groups of the nucleotides to either side of the nick (**Figure 17.12**). This is often the case with nicks resulting from the effects of ionizing radiation.

Some forms of alkylation damage are also directly reversible. This type of repair is carried out by enzymes that transfer the alkyl group from the nucleotide to their own polypeptide chains. Enzymes capable of doing this are known in many different organisms and include the **Ada enzyme** of *E. coli*, which is involved in an adaptive process that this bacterium is able to activate in response to DNA damage. Ada removes alkyl groups attached to the oxygen atoms at positions 4 and 6 of thymine and guanine, respectively, and can also repair phosphodiester bonds that have become alkylated. The transfer is irreversible, which means that Ada is an example of a **suicide enzyme** – an enzyme that is inactivated once it has carried out its biochemical reaction. However, the alkylated version of Ada now acts as a transcription factor that switches on the *ada* **regulon**, which includes genes for the Ada enzyme and three other repair proteins (**Figure 17.13**). Activation of these genes gives rise to the adaptive response. Eukaryotes also have alkylation repair enzymes, an example being human **MGMT (O^6-methylguanine–DNA methyltransferase)**, which, as its name suggests, removes alkyl groups from position 6 of guanine. The eukaryotic enzymes do not appear to affect transcription, and are simply degraded once they have become alkylated.

The final type of direct repair system involves cyclobutyl dimers, which can be cleaved by a light-dependent direct system called **photoreactivation**. In *E. coli*, the process involves the enzyme called **DNA photolyase** (more correctly named deoxyribodipyrimidine photolyase). When stimulated by light with a wavelength between 300 nm and 500 nm, the enzyme binds to cyclobutyl dimers and converts them back to the original monomeric nucleotides. There are two types of photolyase, one type containing a folate cofactor and the other containing a flavin compound. Both types of cofactor capture light energy which is used to transfer an electron to the cyclobutyl dimer, causing the latter to split into its monomers. Photoreactivation is a widespread but not universal type of repair: it is known in many but not all bacteria and also in quite a few eukaryotes, including some vertebrates, but appears to be absent in humans and other placental mammals. A similar type of photoreactivation involves the **(6–4) photoproduct photolyase** and results in repair of (6–4) lesions. Neither *E. coli* nor humans have this enzyme, but it is possessed by a variety of other organisms.

Base excision repairs many types of damaged nucleotide

Direct repair mechanisms are important, but they form a very minor component of the overall repertoire of repair processes available to most organisms. Excision repair systems are much more prevalent and are able to correct a much broader range of mutations.

Base excision, which is the least complex of this type of repair system, is used to repair many modified nucleotides whose bases have suffered relatively minor damage resulting from, for example, exposure to alkylating agents or ionizing radiation. There are several versions of base excision repair, the main distinction being between 'short patch' repair, in which only the damaged nucleotide is replaced, and 'long patch' repair, which involves removal and resynthesis of up to ten nucleotides, including the damaged one. The pathways are initiated by a **DNA glycosylase**, which cleaves the β-*N*-glycosidic bond between the damaged base and the sugar component of the nucleotide. This reaction has the same effect as the heat-induced creation of an AP site (see Figure 17.10A). Each DNA glycosylase has a limited specificity (**Table 17.3**); the specificities of the glycosylases possessed by a

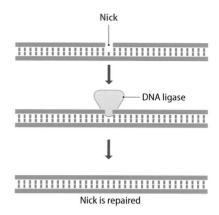

Figure 17.12 Repair of a nick by DNA ligase.

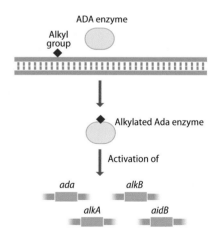

Figure 17.13 The adaptive response stimulated by the Ada enzyme of *E. coli*. Transfer of an alkyl group from a nucleotide to the Ada enzyme converts the latter into an transcription factor that activates the *ada* regulon. A regulon is a group of genes that are controlled by the same transcription factor or group of factors, but are not linked together into a single operon. The *ada* gene codes for the Ada enzyme. The other genes in the regulon are *alkA*, which specifies a DNA glycosylase involved in base excision repair, *alkB*, which codes for a member of a different family of alkylation repair enzymes, and *aidB*, whose function is unknown.

cell determine the range of damaged nucleotides that can be repaired by this pathway. Most organisms are able to deal with deaminated bases such as uracil (deaminated cytosine) and hypoxanthine (deaminated adenine), oxidation products such as 5-hydroxycytosine and thymine glycol, and methylated bases such as 3-methyladenine and 7-methylguanine. Most of the DNA glycosylases involved in base excision repair are thought to diffuse along the minor groove of the DNA double helix in search of damaged nucleotides, but some may be associated with the replication enzymes.

A DNA glycosylase removes a damaged base by 'flipping' the structure to a position outside the helix and then detaching it from the polynucleotide. This creates an AP (baseless) site. In the short patch repair pathway, there are two ways in which this AP site can be converted into a single nucleotide gap:

- An **AP endonuclease**, such as exonuclease III or endonuclease IV of *E. coli* or APE1 of humans, cuts the phosphodiester bond on the 5´ side of the AP site (**Figure 17.14A**). Some AP endonucleases might also remove the ribose from the AP site, this being all that remains of the damaged nucleotide, but others lack this ability and so work in conjunction with a separate **phosphodiesterase** enzyme, which cleaves the phosphodiester bond between the sugar and the next nucleotide. Alternatively, in eukaryotes the ribose can be removed by a **lyase** activity, possessed by DNA polymerase β.

- If the DNA glycosylase is 'bifunctional,' then it can make a cut at the 3´ side of the AP site, probably at the same time that it removes the damaged base. This cut is followed again by excision of the ribose by a phosphodiesterase, by DNA polymerase β, or in *E. coli* by the 5´→3´ exonuclease activity of DNA polymerase I (**Figure 17.14B**).

Whichever method is used, the resulting single nucleotide gap is filled in by a DNA polymerase, using base-pairing with the undamaged base in the other strand of the DNA molecule to ensure that the correct nucleotide is inserted. In *E. coli* the gap is filled in by DNA polymerase I, and in mammals by DNA polymerase β. After gap filling, the final phosphodiester bond is put in place by a DNA ligase.

In the long patch pathway, the AP site created by DNA glycosylase activity is cut on the 5´ side by an AP endonuclease, but this is not followed by further cleavages. Instead, a process similar to lagging-strand replication (Section 15.3) is used to synthesize 2–10 new nucleotides of DNA, the first nucleotide replacing the damaged site (**Figure 17.15**). In bacteria, the synthesis is carried out by DNA polymerase I, which uses its 5´→3´ exonuclease to degrade the segment of polynucleotide that is replaced. In eukaryotes, DNA polymerase δ or ε displaces

TABLE 17.3 EXAMPLES OF MAMMALIAN DNA GLYCOSYLASES

DNA glycosylase	Specific for
MBD4	Uracil
MPG	Ethenoadenine, hypoxanthine, 3-methyladenine, 3-methylguanine, 7-methylguanine
NEIL1	Thymine glycol, formamidopyrimidine, 8-oxoguanine, 5-hydroxyuracil, dihydroxyuracil
NTHL1	5-hydroxycytosine, 5-hydroxyuracil, formamidopyrimidine, thymine glycol
OGG1	Formamidopyrimidine, 8-oxoguanine
SMUG1	Uracil
UNG	Uracil, 5-hydroxyuracil

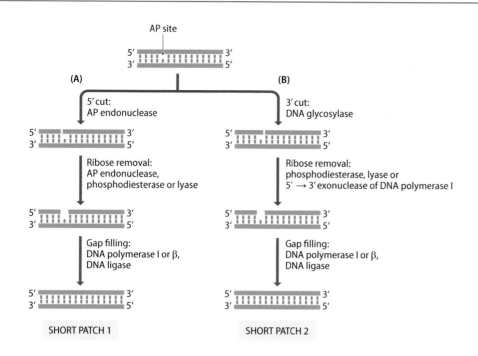

Figure 17.14 The short patch pathways for base excision repair. After removal of the damaged base by DNA glycosylase, the short patch process can follow either of two pathways. (A) An AP endonuclease cuts the phosphodiester bond at the 5´ side of the AP site. The ribose is then removed by the AP endonuclease, by a phosphodiesterase, or by the lyase of DNA polymerase β. (B) The initial cut is made by a bifunctional DNA glycosylase, to the 3´ side of the ribose. The latter is then removed by a phosphodiesterase, the lyase of DNA polymerase β, or the 5´→3´ exonuclease activity of DNA polymerase I.

this segment as the new polynucleotide is made, the resulting flap being cleaved by FEN1 endonuclease. Short patch repair appears to be the main base excision pathway; the choice between the two pathways is influenced at least in part by the identity of the DNA glycosylase that initiates the process.

Nucleotide excision repair is used to correct more extensive types of damage

Nucleotide excision repair has a much broader specificity than the base excision system and is able to deal with more extreme forms of damage, such as intra-strand cross-links and bases that have become modified by attachment of large chemical groups. It is also able to correct cyclobutyl dimers by a **dark repair** process, providing those organisms that do not have the photoreactivation system (such as humans) with a means of repairing this type of damage.

In nucleotide excision repair, a segment of single-stranded DNA containing the damaged nucleotide(s) is excised and replaced with new DNA. The process is therefore similar to base excision repair except that it is not preceded by selective base removal, and a longer stretch of polynucleotide is cut out. The best-studied example of nucleotide excision repair is the short patch process of *E. coli*, so-called because the region of polynucleotide that is excised and subsequently patched is relatively short, usually 12 nucleotides in length. The use of the same terminology for base and nucleotide excision repair is unfortunate, especially as the 'short patch' version of nucleotide excision repair involves replacement of a segment that is similar in length to that removed during 'long patch' base excision repair.

Short patch repair is initiated by a multienzyme complex called the **UvrABC endonuclease**, sometimes also referred to as the 'excinuclease.' In the first stage of the process, a trimer comprising two UvrA proteins and one copy of UvrB attaches to the DNA at the damaged site. How the site is recognized is not known, but the broad specificity of the process indicates that individual types of damage are not directly detected and that the complex must search for a more general attribute of DNA damage, such as distortion of the double helix. UvrA may be the component of the complex most involved in damage location because it dissociates once the site has been found and plays no further part in the repair process. Departure of UvrA allows UvrC to bind (**Figure 17.16**), forming a UvrBC dimer that cuts the polynucleotide on either side of the damaged site. The first cut is

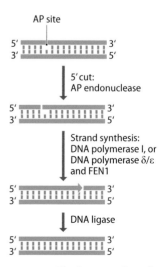

Figure 17.15 The long patch pathway for base excision repair. The long patch pathway begins with a 5´ cut made by an AP endonuclease. The standard process for lagging-strand replication is then used to remove and resynthesize a strand of 2–10 nucleotides, the first of these being the AP site.

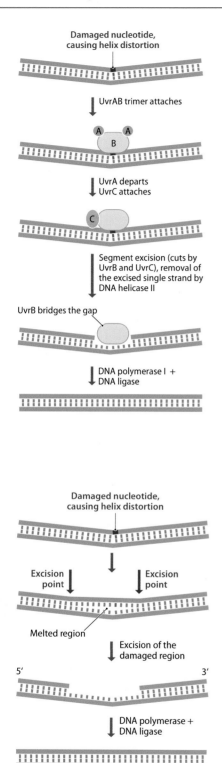

Figure 17.16 Short patch nucleotide excision repair in *E. coli*. The damaged nucleotide is shown distorting the helix because this is thought to be one of the recognition signals for the UvrAB trimer that initiates the short patch process. Abbreviations: A, UvrA; B, UvrB; C, UvrC.

made by UvrB at the fifth phosphodiester bond downstream of the damaged nucleotide, and the second cut is made by UvrC at the eighth phosphodiester bond upstream, resulting in the 12-nucleotide excision, although there is some variability, especially in the position of the UvrB cut site. The excised segment is then removed, usually as an intact oligonucleotide, by DNA helicase II (sometimes called UvrD), which presumably detaches the segment by breaking the base pairs holding it to the second strand. UvrC also detaches at this stage, but UvrB remains in place and bridges the gap produced by the excision. The bound UvrB is thought to prevent the single-stranded region that has been exposed from base-pairing with itself, but the role of UvrB could be to prevent this strand from becoming damaged, or possibly to direct the DNA polymerase to the site that needs to be repaired. As in base excision repair, the gap is filled in by DNA polymerase I and the last phosphodiester bond is synthesized by DNA ligase. *E. coli* also has a long patch nucleotide excision repair system that involves Uvr proteins but differs in that the piece of DNA that is excised can be anything up to 2 kb in length. Long patch repair has been less well studied and the process is not understood in detail, but it is presumed to work on more extensive forms of damage, possibly regions where groups of nucleotides, rather than just individual ones, have become modified.

In eukaryotes, there is a single nucleotide excision repair process that results in replacement of 24–32 nucleotides of DNA. The system is more complex than in *E. coli* and the relevant proteins do not seem to be homologs of the Uvr family. In humans, at least 16 proteins are involved, several of these forming a complex that surveys the DNA for damage. The damaged site is then excised, with the downstream cut typically being made at the same position as in *E. coli* – the fifth phosphodiester bond – but with a more distant upstream cut, 22 nucleotides away, resulting in the longer excision. There is some variability with the cut positions, accounting for the range of fragment lengths that are replaced. Both cuts are made by endonucleases that attack single-stranded DNA, specifically at its junction with a double-stranded region, indicating that before the cuts are made the DNA around the damage site has been melted (**Figure 17.17**). The helicase activity for strand melting is provided by the XPB and XPD proteins, which are associated with TFIIH, which stabilizes the resulting single-stranded region prior to its excision. The gap is then filled by DNA polymerase δ or ε. TFIIH is also one of the components of the RNA polymerase II initiation complex (Section 12.3). At first it was assumed that TFIIH simply had a dual role in the cell, functioning separately in both transcription and repair, but now it is thought that there is a more direct link between the two processes. This view is supported by the discovery of **transcription-coupled repair**, which repairs some forms of damage in the template strands of genes that are being actively transcribed. The only difference between transcription-coupled repair and the 'global' nucleotide excision pathway described above is that the damage that is repaired by the transcription-coupled process is not detected by surveillance proteins, but instead is signaled by the RNA polymerase stalling as it approaches the damaged site.

Mismatch repair corrects replication errors

Each of the repair systems that we have looked at so far – direct, base excision, and nucleotide excision repair – recognize and act upon DNA damage caused by mutagens. This means that they search for abnormal chemical structures such as modified nucleotides, cyclobutyl dimers, and helix distortions. They cannot, however, correct mismatches resulting from errors in replication because the mismatched nucleotide is not abnormal in any way; it is simply an A, C, G, or T that has been inserted at the wrong position. As these nucleotides look exactly like any other nucleotide, the mismatch repair system that corrects replication errors has to detect not the mismatched nucleotide itself but the absence

Figure 17.17 Outline of the events involved during nucleotide excision repair in eukaryotes. The endonucleases that remove the damaged site make cuts at the junction between single-stranded and double-stranded regions of a DNA molecule. The DNA is therefore melted on either side of the damaged nucleotide, as shown in the diagram, by the XPB and XPD helicases.

of base-pairing between the parent and daughter strands. Once it has found a non-base-paired position, the repair system excises part of the daughter polynucleotide and fills in the gap, in a manner similar to base and nucleotide excision repair.

The scheme described above leaves one important question unanswered. The repair must be made in the daughter polynucleotide because it is in this newly synthesized strand that the error has occurred: the parent polynucleotide has the correct sequence. How does the repair process know which strand is which? In *E. coli* the answer is that the daughter strand is, at this stage, undermethylated and can therefore be distinguished from the parent polynucleotide, which has a full complement of methyl groups. *E. coli* DNA is methylated because of the activities of the **DNA adenine methylase (Dam)**, which converts adenines to 6-methyladenines in the sequence 5′–GATC–3′, and the **DNA cytosine methylase (Dcm)**, which converts cytosines to 5-methylcytosines in 5′–CCAGG–3′ and 5′–CCTGG–3′. These methylations are not mutagenic, the modified nucleotides having the same base-pairing properties as the unmodified versions. There is a delay between DNA replication and methylation of the daughter strand, and it is during this window of opportunity that the repair system scans the DNA for mismatches and makes the required corrections in the undermethylated daughter strand (**Figure 17.18**)

E. coli has at least three mismatch repair systems, called 'long patch,' 'short patch,' and 'very short patch,' the names once again indicating the relative lengths of the excised and resynthesized segments of DNA. The long patch system replaces up to a kilobase or more of DNA and requires the MutH, MutL, and MutS proteins, as well as the DNA helicase II that we met during nucleotide excision repair. MutS recognizes the mismatch and MutH distinguishes the two strands by binding to unmethylated 5′–GATC–3′ sequences (**Figure 17.19**). The role of MutL is to coordinate the activities of the other two proteins so that MutH binds to unmethylated 5′–GATC–3′ sequences only in the vicinity of mismatch sites recognized by MutS. After binding, MutH cuts the phosphodiester bond immediately upstream of the G in the methylation sequence, and DNA helicase II detaches the single strand. There does not appear to be an enzyme that cuts the strand downstream of the mismatch: instead, the detached single-stranded region is degraded by an exonuclease that follows the helicase and continues beyond the mismatch site. The gap is then filled in by DNA polymerase III and DNA ligase. Similar events are thought to occur during short patch and very short patch mismatch repair, the difference being the specificities of the proteins that recognize the mismatch. The short patch system, which results in excision of a segment less than ten nucleotides in length, begins when MutY recognizes an A–G or A–C mismatch, and the very short repair system corrects G–T mismatches which are recognized by the Vsr endonuclease.

Eukaryotes have homologs of the *E. coli* MutS and MutL proteins and their mismatch repair processes probably work in a similar way. The one difference is the absence of a homolog of MutH, which suggests that methylation might not be the method used to distinguish between the parent and daughter polynucleotides. Methylation has been implicated in mismatch repair in mammalian cells, but the DNA of some eukaryotes, including fruit flies and yeast, is not extensively methylated. It is therefore thought that in these organisms a different method must be used to identify the daughter strand, the most likely explanation being that the repair enzymes associate with the replisome in such a way that the parent and daughter strands can be distinguished as the latter are synthesized.

Single- and double-strand breaks can be repaired

A single-strand break in a double-stranded DNA molecule, such as is produced by some types of oxidative damage, does not present the cell with a critical problem as the double helix retains its overall intactness. The exposed single strand is coated with PARP1 proteins, which protect the intact strand from breaking and prevent it from participating in unwanted recombination events. The

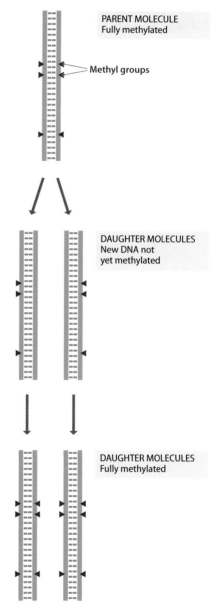

Figure 17.18 Distinction between the parent and daughter polynucleotides in *E. coli*. Methylation of newly synthesized DNA in *E. coli* does not occur immediately after replication, providing a window of opportunity for the mismatch repair proteins to recognize the daughter strands and correct replication errors.

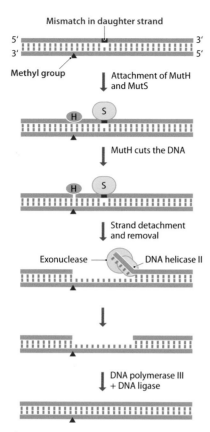

Figure 17.19 Long patch mismatch repair in *E. coli*. This diagram does not include MutL, which is thought to coordinate the activities of MutH and MutS. Abbreviations: H, MutH; S, MutS.

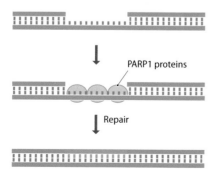

Figure 17.20 Single-strand break repair.

break is then filled in by the enzymes involved in the excision repair pathways (**Figure 17.20**).

A double-strand break is more serious because it converts the original double helix into two separate fragments, which have to be brought back together again in order for the break to be repaired. Approximately ten double-strand breaks occur per day in a mammalian cell, due to the effects of ionizing radiation and some chemical mutagens, and breakages can also occur during DNA replication. The repair process must ensure that the correct ends are joined: if there are two broken chromosomes in the nucleus, then the correct pairs must be brought together so that the original structures are restored. Experimental studies of mouse cells indicate that achieving this outcome is difficult and if two chromosomes are broken then misrepair resulting in hybrid structures occurs relatively frequently. Even if only one chromosome is broken, there is still a possibility that a natural chromosome end could be confused as a break and an incorrect repair made. This type of error is not unknown, despite the presence of shelterin proteins that mark the natural ends of chromosomes (Section 7.1).

Double-strand breaks can be repaired by a system called **nonhomologous end-joining** (**NHEJ**). Progress in understanding NHEJ has been stimulated by studies of mutant human cell lines, which have resulted in the identification of various sets of genes involved in the process. These genes specify a multicomponent protein complex that directs a DNA ligase to the break. A key player in the complex is the **Ku70-Ku80 heterodimer**, which forms a loop structure that encloses the cut end of a DNA molecule (**Figure 17.21A**). The heterodimer binds to the DNA in association with the DNA-PKcs protein kinase, which activates a third protein, XRCC4, which interacts with the mammalian DNA ligase IV, directing this repair protein to the double-strand break (**Figure 17.21B**).

For NHEJ to be successful, it is clearly necessary for the two broken ends to be brought into proximity. Exactly how this is achieved is not understood. Various models have been proposed, suggesting, for example, that the bound Ku70-Ku80 heterodimers have an affinity for one another, and by coming together form a bridge between the broken ends. Similar roles have been ascribed to other components of the NHEJ complex. Whatever the explanation, the process is probably aided by the internal architecture of the nucleus (Section 10.1), which prevents broken ends from moving too far away from one another. A second issue is that a double-strand break rarely leaves two blunt, unmodified ends. Usually, one or both ends have short 5´- or 3´-overhangs, and some of the nucleotides in the break region might be chemically damaged. If caused by ionizing radiation or oxidative damage, then it is also possible that the break will leave a 3´-P group which becomes modified by attachment of a glycolate group ($-OCH_2COOH$). A set of NHEJ subpathways exist, utilizing different proteins, to deal with the various possibilities. Each of these subpathways involves nucleases that trim the ends by a few nucleotides (this is called end **resection**), possibly followed by their re-extension in a template-independent manner so that a short random sequence of nucleotides is added. The repeated action of nuclease and polymerase will, at some point, create overhangs with **microhomology**, where the sequences are not 100% complementary but enable sufficient base-pairing to hold the ends together while they are ligated (**Figure 17.22**). An outcome of this process is that a small deletion or insertion will occur at the point where the two broken ends are rejoined.

As well as the canonical end-joining process described above, eukaryotes also have an **alternative end-joining (a-EJ)** mechanism that does not involve the Ku heterodimer. Elucidating this mechanism has proven difficult, partly because the vast majority of ends are repaired by NHEJ, so a-EJ is a rarely used component of the cell's repair processes, and partly because a-EJ appears to comprise at least three distinct subpathways. Each subpathway involves end resection to create segments displaying microhomology that can be used to anneal the broken ends prior to ligation, but the resected regions are longer than in NHEJ. A complex called MRN, comprising the Mre11, Rad50 and Nbs1 proteins, is involved in resection and either MRN or Rad52 carry out the end-bridging function.

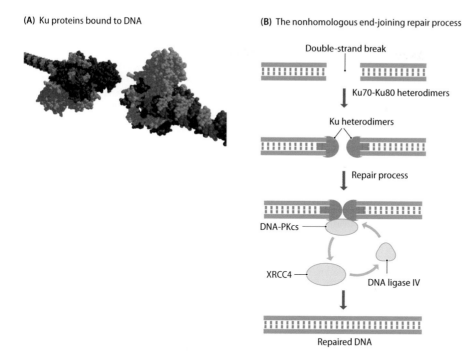

(A) Ku proteins bound to DNA

(B) The nonhomologous end-joining repair process

Double-strand break

Ku70-Ku80 heterodimers

Ku heterodimers

Repair process

DNA-PKcs

XRCC4

DNA ligase IV

Repaired DNA

Figure 17.21 Nonhomologous end-joining (NHEJ) in humans. (A) A space-filling model showing the Ku70-Ku80 heterodimer bound to DNA. The Ku70 subunit is colored red, and Ku80 is yellow. DNA is depicted with one dark gray and one light gray strand. (B) The repair process. (Part A courtesy of Jonathan Goldberg, Howard Hughes Medical Institute.)

NHEJ was originally thought to be restricted to eukaryotes, but genome annotations have uncovered bacterial homologs of the mammalian Ku proteins, and experimental studies have indicated that these act in conjunction with bacterial ligases in a simplified version of the double-strand break repair process.

Some types of damage can be repaired by homologous recombination

A role for homologous recombination in DNA repair became apparent when *E. coli* mutants defective for components of the RecBCD and other recombination pathways were first examined and shown to be unable to correct some types of DNA damage. Homologous recombination is particularly important as a repair process when breaks arise in daughter DNA molecules as a result of aberrations in replication. One such aberration can occur when the replication machinery attempts to copy a segment of the genome that is heavily damaged, in particular regions in which cyclobutyl dimers are prevalent. When a cyclobutyl dimer is encountered, the template strand cannot be copied, and the DNA polymerase simply jumps ahead to the nearest undamaged region, where it restarts the replication process. The result is that one of the daughter polynucleotides has a gap (**Figure 17.23**). One way in which this gap could be repaired is by a recombination event that transfers the equivalent segment of DNA from the parent polynucleotide present in the second daughter double helix. The gap that is now present in the second double helix is refilled by a DNA polymerase, using the undamaged daughter polynucleotide within this helix as the template. In *E. coli*, this type of single-strand gap repair utilizes the RecFOR recombination pathway.

If the damaged site cannot be bypassed, then the daughter polynucleotide, rather than having a gap, will terminate (**Figure 17.24**). There are several ways in which this problem can be overcome. One possibility is that the replication fork stalls and reverses a short distance, so that a duplex is formed between the daughter polynucleotides. The incomplete polynucleotide is then extended by a DNA polymerase, using the undamaged daughter strand as a template. The replication fork then moves forward again, by a process equivalent to the branch migration step of homologous recombination. As a result, the damaged site is bypassed and replication can continue.

A more serious aberration occurs if one of the parent polynucleotides being replicated contains a single-strand nick. Now the replication process leads to a double-strand break in one of the daughter double helices, and the replication

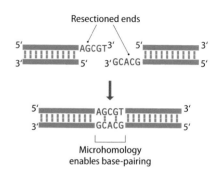

Resectioned ends

5′ AGCGT 3′   5′ 3′
3′ 5′   3′ GCACG 5′

5′ AGCGT 3′
3′ GCACG 5′

Microhomology enables base-pairing

Figure 17.22 Microhomology between resectioned ends enables base-pairing. The sequences of the two single-stranded overhangs are not 100% complementary but enable sufficient base-pairing to hold the ends together while they are ligated.

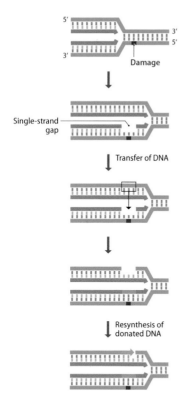

Figure 17.23 Single-strand gap repair by the RecFOR pathway of *E. coli*.

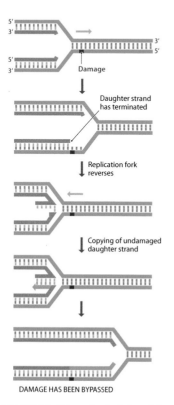

Figure 17.24 A terminated daughter polynucleotide can be rescued by reversal of the replication fork.

fork is lost (**Figure 17.25**). The break can be repaired by a form of homologous recombination between the broken end and the second, undamaged molecule. In the scheme shown in Figure 17.25, the daughter polynucleotide at the double-strand break is extended via a strand exchange reaction which enables it to continue to use the other parent strand as a template. Branch migration followed by resolution of the Holliday structure then restores the replication fork.

If necessary, DNA damage can be bypassed during genome replication

If a region of the genome has suffered extensive damage, then it is conceivable that the repair processes will be overwhelmed. The cell then faces a stark choice between dying or attempting to replicate the damaged region, even though this replication may be error-prone and result in mutated daughter molecules. When faced with this choice, *E. coli* cells invariably take the second option, by inducing one of several emergency procedures for bypassing sites of major damage.

The best studied of these bypass processes occurs as part of the **SOS response**, which enables an *E. coli* cell to replicate its DNA even though the template polynucleotides contain AP sites and/or cyclobutyl dimers and other photoproducts, resulting from exposure to chemical mutagens or UV radiation, that would normally block, or at least delay, the replication complex. Bypass of these sites requires construction of a **mutasome**, comprising DNA polymerase V (also called the UmuD$'_2$C complex, because it is a trimer made up of two UmuD$'$ proteins and one UmuC) and several copies of the RecA protein. The latter is the single-strand binding protein that plays a role in recombination (Section 16.1) and is also important in DNA repair. In this bypass system, RecA coats the damaged polynucleotide strands, enabling DNA polymerase V to displace DNA polymerase III and carry out error-prone DNA synthesis until the damaged region has been passed and DNA polymerase III can take over once again (**Figure 17.26**). DNA polymerase V is therefore an example of a **translesion polymerase**.

As well as acting as a single-strand binding protein that facilitates the mutasome bypass process, RecA also has a second function as an activator of the overall SOS response. The protein is stimulated by chemical signals (not yet identified) that indicate the presence of extensive DNA damage. In response, RecA cleaves a number of target proteins, directly or indirectly, including UmuD, the cleavage converting this protein into its active form UmuD$'$, and initiating the mutasome repair process. RecA also cleaves a repressor protein called LexA, switching on or increasing the expression of a number of genes normally repressed by LexA, these including both the *recA* gene itself (leading to a 50-fold increase in RecA synthesis) and several other genes whose products are involved in DNA repair pathways. RecA also cleaves the cI repressor of λ bacteriophage, so if an integrated λ prophage is present in the genome it can excise and leave the sinking ship (Section 14.3).

The SOS response is primarily looked on as the last best chance that the bacterium has to replicate its DNA and hence survive under adverse conditions. However, the price of survival is an increased mutation rate because the mutasome does not repair damage; it simply allows a damaged region of a polynucleotide to be replicated. When it encounters a damaged position in the template DNA, the polymerase selects a nucleotide more or less at random, although with some preference for placing an A opposite an AP site: in effect, the error rate of the replication process increases. It has been suggested that this increased mutation rate is the purpose of the SOS response, mutation being in some way an advantageous response to extensive DNA damage, but this idea remains controversial.

For some time, the SOS response was thought to be the only damage-bypass process in bacteria, but we now appreciate that at least two other *E. coli* translesion polymerases act in a similar way, although with different types of damage. These are DNA polymerase II, which can bypass AP sites and nucleotides bound to mutagenic chemicals such as *N*-2-deoxyguanosine-acetyl aminofluorene, and DNA polymerase IV (also called DinB), which can replicate through a region of template DNA in which the two parent polynucleotides have become

Figure 17.25 One mechanism for recovery of a collapsed replication fork by homologous recombination. At the single-strand break, the replication fork collapses. The 3′ end of the terminated daughter molecule invades the intact helix and is extended. Branch migration followed by resolution of the Holliday junction results in restoration of the replication fork. The gap that remains in the upper of the two daughter double helices will be filled by the next Okazaki fragment when replication resumes.

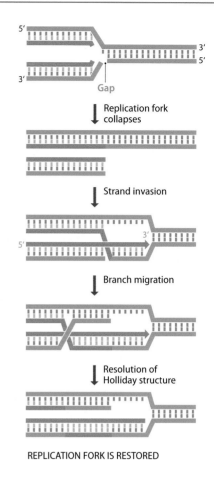

cross-linked. Translesion polymerases have also been discovered in eukaryotic cells. These include DNA polymerases η, which can bypass cyclobutyl dimers, and DNA polymerases ι and ζ, whose functions are less well understood.

Defects in DNA repair underlie human diseases, including cancers

The importance of DNA repair is emphasized by the number and severity of inherited human diseases that have been linked with defects in one or other of the repair processes. One of the best characterized of these is xeroderma pigmentosum, which results from a mutation in any one of several genes for proteins involved in nucleotide excision repair. Nucleotide excision is the only way in which human cells can repair cyclobutyl dimers and other photoproducts, so it is no surprise that the symptoms of xeroderma pigmentosum include hypersensitivity to UV radiation, patients suffering more mutations than normal on exposure to sunlight, and often developing skin cancer. Trichothiodystrophy is also caused by defects in nucleotide excision repair, but this is a more complex disorder that, although not involving cancer, usually includes problems with both the skin and nervous system.

A few diseases have been linked with defects in the transcription-coupled component of nucleotide excision repair. These include breast and ovarian cancers, the *BRCA1* gene that confers susceptibility to these cancers coding for a protein that has been implicated with transcription-coupled repair, and Cockayne syndrome, a complex disease manifested by growth and neurologic disorders. Ataxia telangiectasia, the symptoms of which include sensitivity to ionizing radiation, results from defects in the *ATM* gene, which is involved in the damage-detection process. Other diseases that are associated with a breakdown in DNA repair are Bloom's and Werner's syndromes, which are caused by inactivation of the RecQ DNA helicase, which has a role in NHEJ, the cancer-susceptibility syndrome called HNPCC (hereditary nonpolyposis colorectal cancer) which results from a defect in mismatch repair, and some types of spinocerebellar ataxia, which arise from defects in the pathway used to repair single-strand breaks. Understanding the genetic background of these disorders is important not just for devising therapies for managing the disease. The disorders can also provide novel information on the biochemical basis of DNA repair. An example is provided by Fanconi's anemia, a rare disease first characterized in 1927 with a frequency of only one case per 100,000 individuals. Patients suffering from this type of anemia have an increased susceptibility to chemicals that cause DNA cross-links. Recognition that the disease can arise as a result of mutations in at least 16 different genes has enabled the pathway that repairs this type of DNA damage to be elucidated.

SUMMARY

- A mutation is a change in the nucleotide sequence of a DNA molecule.

- A point mutation affects just a single nucleotide. Mutations can also arise by insertion or deletion of one or more adjacent nucleotides.

- Mutations can result from errors made during DNA replication. DNA polymerases have nucleotide selection and proofreading activities that maintain a high degree of accuracy, but these checking mechanisms can be evaded if an unusual tautomeric version of a nucleotide is present in the template.

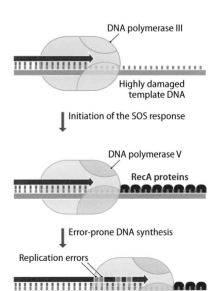

Figure 17.26 The SOS response of *Escherichia coli*.

- A second type of replication error, called slippage, can lead to insertion or deletion mutations.

- There are also many types of chemical or physical agent that can cause mutations.

- Some compounds act as base analogs and cause mutations after being mistaken for genuine nucleotides by the replication machinery.

- Deaminating and alkylating agents directly attack DNA molecules, and intercalating agents such as ethidium bromide slide between base-pairs, causing insertions and deletions when the helix is replicated.

- UV radiation causes adjacent nucleotides to link together into dimers, and ionizing radiation and heat cause various types of damage.

- All cells possess DNA repair processes that enable many mutations to be corrected.

- Direct repair systems are uncommon, but the few that are known correct some types of base damage, including removal of UV-induced nucleotide dimers.

- Excision repair processes involve excision of a segment of a polynucleotide containing a damaged site, followed by resynthesis of the correct nucleotide sequence by a DNA polymerase.

- Mismatch repair corrects errors of replication, again by excising a stretch of single-stranded DNA containing the mutation and repairing the resulting gap.

- Nonhomologous end-joining is used to mend double-strand breaks.

- Homologous recombination plays a major role in the repair of DNA strand breaks.

- There are also processes for bypassing sites of DNA damage during replication, many of these acting as an emergency system for rescuing a genome that has become heavily mutated.

- Defects in DNA repair often result in disease, including several types of cancer.

SHORT ANSWER QUESTIONS

1. Describe how the mode of action of a DNA polymerase maximizes the accuracy with which the enzyme replicates DNA.

2. Outline the way in which tautomers induce errors in replication.

3. Explain what is meant by the term 'replication slippage.'

4. How does the base analog 2-aminopurine produce mutations in DNA?

5. Describe the effect of UV irradiation on DNA structure.

6. How does heat affect the structure of DNA?

7. Describe the direct mutation repair processes that are known in bacterial and eukaryotic cells.

8. Outline the steps that occur during the base excision repair pathway.

9. Describe the key features of the nucleotide excision repair pathways of *E. coli*.

10. How are the parent and daughter strands distinguished during the mismatch repair process of *E. coli*?

11. What is the process by which double-strand breaks in DNA are repaired by the nonhomologous end-joining system?

12. Describe the processes by which homologous recombination is used to repair breaks in DNA molecules.

13. What is the role of the RecA protein in the SOS response of *E. coli*?

IN-DEPTH PROBLEMS

1. Explain why a purine-to-purine or pyrimidine-to-pyrimidine point mutation is called a transition, whereas a purine-to-pyrimidine (or vice versa) change is called a transversion.

2. What would be the anticipated ratio of transitions to transversions in a large number of mutations?

3. A mutation in a protein-coding gene can result in an alteration in the amino acid sequence of the protein product, possibly modifying or inactivating the function of that protein. An individual who inherits a mutated gene from one of their parents might therefore suffer from a genetic disease. However, not all such diseases have an immediate impact. Some are delayed onset and are only expressed later in the individual's life. Others display nonpenetrance in some individuals, never being expressed. Devise mechanisms to explain how mutations can exhibit delayed onset or nonpenetrance.

4. The bacterium *Deinococcus radiodurans* is highly resistant to radiation and to other physical and chemical mutagens. Discuss how these special properties of *D. radiodurans* might be reflected in its genome sequence.

5. Why do defects in DNA repair often lead to cancer?

FURTHER READING

Causes of mutations

Chatterjee, N. and Walker, G.C. (2017) Mechanisms of DNA damage, repair, and mutagenesis. *Environ. Mol. Mutagen.* 58:235–263.

Drake, J.W., Glickman, B.W. and Ripley, L.S. (1983) Updating the theory of mutation. *Am. Sci.* 71:621–630. *A general review of mutation.*

Fijalkowska, I.J., Jonczyk, P., Tkaczyk, M.M., et al. (1998) Unequal fidelity of leading strand and lagging strand DNA replication on the *Escherichia coli* chromosome. *Proc. Natl Acad. Sci. USA* 95:10020–10025.

Hung, K.-F., Sidorova, J.M., Nghiem, P., et al. (2020) The 6–4 photoproduct is the trigger of UV-induce replication blockage and ATR activation. *Proc. Natl. Acad. Sci. USA* 117:12806–12816.

Kunkel, T.A. (2004) DNA replication fidelity. *J. Biol. Chem.* 279:16895–16898. *Covers the processes that ensure that the minimum number of errors are made during DNA replication.*

Trinucleotide repeat expansion diseases

Lee, D.-Y. and McMurray, C.T. (2014) Trinucleotide expansion in disease: Why is there a length threshold? *Curr. Opin. Genet. Dev.* 26:131–140.

McMurray, C.T. (2010) Mechanisms of trinucleotide repeat instability during human development. *Nat. Rev. Genet.* 11:786–799.

Paulson, H. (2019) Repeat expansion diseases. *Handb. Clin. Neurol.* 147:105–123.

Sutherland, G.R., Baker, E. and Richards, R.I. (1998) Fragile sites still breaking. *Trends Genet.* 14:501–506.

Direct repair

Hearst, J.E. (1995) The structure of photolyase: Using photon energy for DNA repair. *Science* 268:1858–1859.

Yi, C. and He, C. (2013) DNA repair by reversal of DNA damage. *Cold Spring Harb. Perspect. Biol.* 5:a012575.

Zhong, D. (2015) Electron transfer mechanisms of DNA repair by photolyase. *Annu. Rev. Phys. Chem.* 66:691–715.

Excision repair

Caldecott, K.W. (2020) Mammalian base excision repair: Dancing in the moonlight. *DNA Repair* 93:102921.

David, S.S., O'Shea, V.L. and Kundu, S. (2007) Base-excision repair of oxidative DNA damage. *Nature* 447:941–950.

Kamileri, I., Karakasilioti, I. and Garinis, G.A. (2012) Nucleotide excision repair: New tricks with old bricks. *Trends Genet.* 28:566–573.

Krokan, H.E. and Bjørås, M. (2013) Base excision repair. *Cold Spring Harb. Perspect. Biol.* 5:a012583.

Krwawicz, J., Arczewska, K.D., Speina, E. et al. (2007) Bacterial DNA repair genes and their eukaryotic homologues: 1. Mutations in genes involved in base excision repair (BER) and DNA-end processors and their implication in mutagenesis and human disease. *Acta Biochim. Pol.* 54:413–434. *Gives details of DNA glycosylases and their substrates.*

Kusajabe, M., Onishi, Y., Tada, H., et al. (2019) Mechanism and regulation of DNA damage recognition in nucleotide excision repair. *Gene Environ.* 41:2.

Roldán-Arjona, T., Ariza, R.R., and Córdoba-Cañero, D. (2019) DNA base excision repair in plants: An unfolding story with familiar and novel characters. *Front. Plant Sci.* 10:1055.

Mismatch repair
Hsieh, P. and Zhang, Y. (2017) The devil is in the detail for DNA mismatch repair. *Proc. Natl. Acad. Sci. USA* 114:3552–3554.

Jiricny, J. (2013) Postreplicative mismatch repair. *Cold Spring Harb. Perspect. Biol.* 5:a012633.

Kolodner, R.D. (1995) Mismatch repair: Mechanisms and relationship to cancer susceptibility. *Trends Biochem. Sci.* 20:397–401.

Kunkel, T.A. and Erie, D.A. (2015) Eukaryotic mismatch repair in relation to DNA replication. *Annu. Rev. Genet.* 49:291–313.

Li, G.-M. (2008) Mechanisms and functions of DNA mismatch repair. *Cell Res.* 18:85–98.

Repair of DNA breaks
Bertrand, C, Thibessard, A., Bruand, C., et al. (2019) Bacterial NHEJ: A never ending story. *Mol. Microbiol.* 111:1139–1151.

Chang, H.H.Y., Pannunzio, N.R., Adachi, N., et al. (2017) Non-homologous DNA end joining and alternative pathways to double-strand break repair. *Nat. Rev. Mol. Cell Biol.* 18:495–506.

Chaplin, A.K., Hardwick, S.W., Stavridi, A.K., et al. (2021) Cryo-EM of NHEJ supercomplexes provide insights into DNA repair. *Mol. Cell* 81:3400–3409.

Chapman, J.R., Taylor, M.R.G. and Boulton, S.J. (2012) Playing the end game: DNA double-strand break repair pathway choice. *Mol. Cell* 47:497–510.

Jasin, M. and Rothstein, R. (2013) Repair of strand breaks by homologous recombination. *Cold Spring Harb. Perspect. Biol.* 5:a012740.

Mehta, A. and Haber, J.E. (2014) Sources of DNA double-strand breaks and models of recombinational DNA repair. *Cold Spring Harb. Perspect. Biol.* 6:a016428.

Sallmyr, A. and Tomkinson, A.E. (2018) Repair of DNA double-strand breaks by mammalian alternative end-joining pathways. *J. Biol. Chem.* 293:10536–10546.

Walker, J.R., Corpina, R.A. and Goldberg, J. (2001) Structure of the Ku heterodimer bound to DNA and its implications for double-strand break repair. *Nature* 412:607–614.

Bypassing DNA damage
Goodman, M.F. and Woodgate, R. (2013) Translesion DNA polymerases. *Cold Spring Harb. Perspect. Biol.* 5:a010363.

Johnson, R.E., Prakash, S. and Prakash, L. (1999) Efficient bypass of a thymine-thymine dimer by yeast DNA polymerase, Polη. *Science* 283:1001–1004.

Sutton, M.D., Smith, B.T., Godoy, V.G., et al. (2000) The SOS response: Recent insights into *umuDC*-dependent mutagenesis and DNA damage tolerance. *Annu. Rev. Genet.* 34:479–497.

Vaisman, A. and Woodgate, R. (2017) Translesion DNA polymerases in eukaryotes: What makes them tick? *Crit. Rev. Biochem. Mol. Biol.* 52:274–303.

Yoon, J.-H., Basu, D., Sellamuthu, K., et al. (2021) A novel role of DNA polymerase λ in translesion synthesis in conjunction with DNA polymerase ζ. *Life Sci. Alliance* 4:e202000900.

Repair and disease
Hanawalt, P.C. (2000) The bases for Cockayne syndrome. *Nature* 405:415–416.

O'Driscoll, M. (2012) Diseases associated with defective responses to DNA damage. *Cold Spring Harb. Perspect. Biol.* 4:a012773.

Sharma, R., Lewis, S. and Wlodarski, M.W. (2020) DNA repair syndromes and cancer: Insights into genetics and phenotype patterns. *Front. Pediatr.* 23:570084.

Walden, H. and Deans, A.J. (2014) The Fanconi anemia DNA repair pathway: Structural and functional insights into a complex disorder. *Annu. Rev. Biophys.* 43:257–278.

Wei, L., Lan, L., Yasui, A., et al. (2011) *BRCA1* contributes to transcription-coupled repair of DNA damage through polyubiquitination and degradation of Cockayne syndrome B protein. *Cancer Sci.* 102:1840–1847.

HOW GENOMES EVOLVE

Recombination and mutation provide the genome with the means to evolve, but we learn very little about the evolutionary histories of genomes simply by studying these events in living cells. Instead, we must combine our understanding of recombination and mutation with comparisons between the genomes of different organisms in order to infer the patterns of genome evolution that have occurred. Clearly, this approach is imprecise and uncertain but, as we will see, it is based on a surprisingly large amount of hard data and we can be reasonably confident that, at least in outline, the picture that emerges is not too far from the truth.

In this chapter we will explore the evolution of genomes from the very origins of biochemical systems through to the present day. We will look at ideas regarding the **RNA world**, prior to the appearance of the first DNA molecules, and then examine how DNA genomes have gradually become more complex. We will examine the differences between the human and chimpanzee genomes in an attempt to identify the evolutionary changes that have occurred during the last 6 million years and which must, somehow, make us what we are. Finally, we will study how the diversity of genome sequences in modern-day populations is being used as a tool in research and in biotechnology.

18.1 GENOMES: THE FIRST 10 BILLION YEARS

Cosmologists believe that the universe began some 14 billion years ago with the gigantic primordial fireball called the Big Bang. Mathematical models suggest that after about 4 billion years galaxies began to fragment from the clouds of gas emitted by the Big Bang, and that within our own galaxy the solar nebula condensed to form the Sun and its planets about 4.6 billion years ago (**Figure 18.1**). The early Earth was covered with water and it was in this vast planetary ocean that the first biochemical systems appeared, cellular life being well established by the time land masses began to appear, some 3.5 billion years ago. We believe this to be the case because tiny microfossils of structures resembling bacteria have been discovered in 3.4-billion-year-old rocks from Australia (**Figure 18.2**). But cellular life was a relatively late stage in biochemical evolution, being preceded by self-replicating polynucleotides that were the progenitors of the first genomes. We must begin our study of genome evolution with these precellular systems.

The first biochemical systems were centered on RNA

The first oceans are thought to have had a similar salt composition to those of today, but the Earth's atmosphere, and hence the dissolved gases in the oceans, was very different. The oxygen content of the atmosphere remained very low until photosynthesis evolved, and to begin with the most abundant gases were probably methane and ammonia. The possibility that organic compounds important in biochemistry could be formed on the early Earth was first demonstrated in 1952, when experiments that recreated the conditions in the ancient atmosphere showed that electrical discharges in a methane–ammonia mixture could result in synthesis of a range of amino acids, including alanine, glycine, valine, and several of the others found in proteins (**Figure 18.3**). Formation of

DOI: 10.1201/9781003133162-18

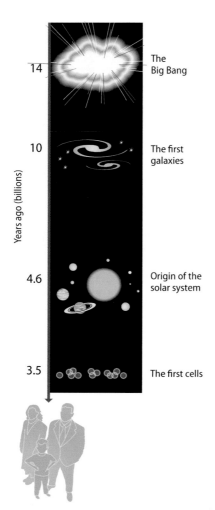

Figure 18.1 The origins of the universe, galaxies, solar system, and cellular life.

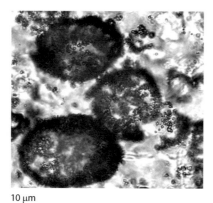

10 μm

Figure 18.2 The earliest known fossils. Microfossils, thought to be the remains of some of the earliest prokaryotic cells, from 3.4 billion-year-old rocks in Western Australia. (From Wacey D., Kilburn M.R., Saunders M. et al. [2011] *Nature Geoscience* 4:698-702. With permission from Springer Nature.)

ribonucleotides has been less easy to understand because of difficulties in envisaging how the first ribose sugars and nucleotide bases could be synthesized, but it has recently been demonstrated that complete ribonucleotides can be produced directly without the prior synthesis of the sugar and base components. For example, pyrimidine nucleotides can be constructed from cyanamide, cyanoacetylene, glycolaldehyde, glyceraldehyde, and inorganic phosphate via an intermediate compound called arabinose amino-oxazoline and anhydronucleoside.

Once ribonucleotides had formed, their polymerization into RNA molecules would have been aided by base-stacking interactions (Section 1.1). These interactions would have been much too weak to hold nucleotides together for any length of time in the early ocean, but might have been sufficient to stabilize structures that formed on solid surfaces such as clay particles or ice. Alternatively, base-stacking could have been promoted by the repeated condensation and drying of droplets of water in clouds. The precise mechanism need not concern us: what is important is that it is possible to envisage purely geochemical processes that could lead to synthesis of polymeric RNA, and that similar scenarios give plausible routes for polypeptide synthesis. It is the next steps that we must worry about. We have to go from a random collection of biomolecules to an ordered assemblage that displays at least some of the biochemical properties that we associate with life. These steps have never been reproduced experimentally, and our ideas are therefore based mainly on speculation tempered by a certain amount of computer simulation. One problem is that the speculations are unconstrained because the global ocean could have contained as many as 10^{10} biomolecules per liter and we can allow a billion years for the necessary events to take place. This means that even the most improbable scenarios cannot be dismissed out of hand.

Progress in understanding the origins of cellular life was initially stalled by the apparent requirement that polynucleotides and polypeptides must work in harness to give rise to a self-reproducing biochemical system. This is because proteins are required to catalyze biochemical reactions but cannot carry out their own self-replication. Polynucleotides, on the other hand, can specify the synthesis of proteins and act as templates for their own self-replication, but it was thought that they could do neither without the aid of other proteins. It appeared that the biochemical system would have to spring fully formed from the random collection of biomolecules because any intermediate stage could not be perpetuated. Many of these problems were swept aside in the 1980s with the discovery of the first catalytic RNA molecules. It was quickly realized that ribozymes can catalyze biochemical reactions such as self-cleavage, as displayed by some modern-day viroid and virusoid genomes (see Figure 9.11), cleavage of other RNAs as carried out by, for example, RNase P during pre-tRNA processing, and synthesis of peptide bonds as performed by the rRNA component of the ribosome (Section 13.3). In the test tube, synthetic RNA molecules have been shown to carry out other biologically relevant reactions such as synthesis of ribonucleotides, synthesis and copying of RNA molecules, and transfer of an RNA-bound amino acid to a second amino acid forming a dipeptide, in a manner analogous to the role of tRNA in protein synthesis. The discovery of these catalytic properties, and of various ways in which the activities of ribozymes might be regulated by riboswitches (Section 12.3), have solved the polynucleotide–polypeptide dilemma by showing that the first biochemical systems could have been centered entirely on RNA.

Various ideas about the RNA world have taken shape in recent years. We now envisage that RNA molecules initially replicated in a slow and haphazard fashion simply by acting as templates for binding of complementary nucleotides, which polymerized spontaneously (**Figure 18.4**). This replication process would have been very inaccurate, so a variety of RNA sequences would have been generated, eventually leading to one or more with nascent ribozyme properties that were able to direct their own, more accurate self-replication. It is possible that a form of natural selection operated so that the most efficient replicating systems began to predominate, as has been shown to occur in experimental systems. A greater accuracy in replication would have enabled RNAs to increase in length without

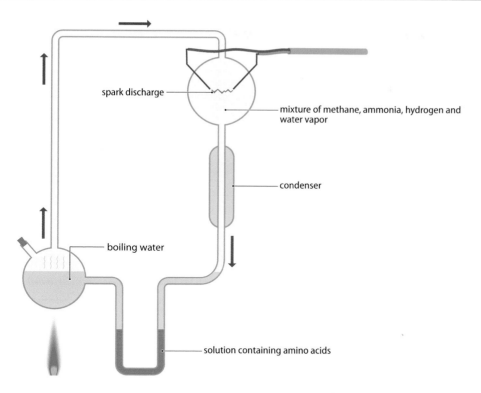

Figure 18.3 An experiment recreating the chemical conditions of the early atmosphere. In this experiment, carried out by Stanley Miller and Harold Urey in 1952, a mixture of methane, ammonia, hydrogen, and water vapor (the last of these from the boiling water) was subjected to electrical discharges simulating lightning. The products were collected by passing the gaseous mixture through the condenser. Analysis of the resulting solution revealed the presence of two amino acids, glycine and alanine. Re-analysis of the solution in 2007, using more sensitive techniques, detected over 20 amino acids.

losing their sequence specificity, providing the potential for more sophisticated catalytic properties, possibly culminating in structures as complex as present-day ribosomal RNAs (see Figure 13.25).

To call the early RNAs 'genomes' is a little fanciful, but the term **protogenome** has attractions as a descriptor for molecules that were self-replicating and able to direct simple biochemical reactions. These reactions might have included energy metabolism based, as today, on the release of free energy by hydrolysis of the phosphate–phosphate bonds in the ribonucleotides ATP and GTP, and the reactions might have become compartmentalized within lipid membranes, forming the first cell-like structures. There are difficulties in envisaging how long-chain unbranched lipids could form by chemical- or ribozyme-catalyzed reactions, but once present in sufficient quantities they would have assembled spontaneously into membranes, possibly encapsulating one or more protogenomes. Within these early cells, the concentration of ribozymes would have been greater than in the unstructured external matrix, possibly giving the cells a selective advantage that enabled them to propagate. Inclusion in the cells of compounds such as citrate might have provided the first steps toward control of the ionic environment, as citrate chelates magnesium ions, which promote the activity of many of today's ribozymes. The early cells could therefore have provided the RNA protogenomes with an enclosed environment in which more controlled biochemical reactions could be carried out.

The first DNA genomes

How did the RNA world develop into the DNA world? The first major change was probably the development of protein enzymes, which supplemented, and eventually replaced, most of the catalytic activities of ribozymes. There are several unanswered questions relating to this stage of biochemical evolution, including the reason why the transition from RNA to protein occurred in the first place. Originally, it was assumed that the 20 amino acids in polypeptides provided proteins with greater chemical variability than the four ribonucleotides in RNA, enabling protein enzymes to catalyze a broader range of biochemical reactions, but this explanation has become less attractive as more and more ribozyme-catalyzed reactions have been demonstrated in the test tube. A more recent

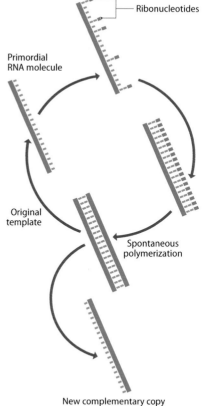

Figure 18.4 Copying of RNA molecules in the early RNA world. Before the evolution of RNA polymerases, ribonucleotides that became associated with an RNA template would have had to polymerize spontaneously. This process would have been inaccurate and many RNA sequences would have been generated.

Figure 18.5 Two scenarios for the evolution of the first coding RNA. A ribozyme could have evolved to have a dual catalytic and coding function (A), or a ribozyme could have synthesized a coding molecule (B). In both examples, the amino acids are shown attaching to the coding molecule via small adaptor RNAs, the presumed progenitors of today's tRNAs.

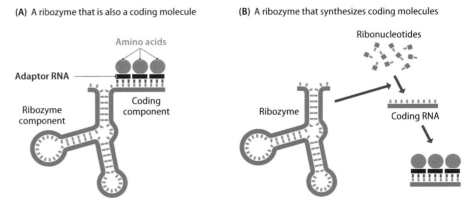

(A) A ribozyme that is also a coding molecule

(B) A ribozyme that synthesizes coding molecules

suggestion is that protein-mediated catalysis is more efficient because of the inherent flexibility of folded polypeptides compared with the greater rigidity of base-paired RNAs. Alternatively, enclosure of RNA protogenomes within membrane vesicles could have prompted the evolution of the first proteins, because RNA molecules are hydrophilic and must be given a hydrophobic coat, for instance, by attachment to peptide molecules, before being able to pass through or become integrated into a membrane.

The transition to protein-mediated catalysis demanded a radical shift in the function of the RNA protogenomes. Rather than being directly responsible for the biochemical reactions occurring in the early cell-like structures, the protogenomes became coding molecules whose main function was to specify the construction of the catalytic proteins. Whether the ribozymes themselves became coding molecules, or coding molecules were synthesized by the ribozymes, is not known, although the most persuasive theories about the origins of protein synthesis and the genetic code suggest that the latter alternative is more likely to be correct (**Figure 18.5**). Whatever the mechanism, the result was the paradoxical situation whereby the RNA protogenomes had abandoned their roles as enzymes, which they were good at, and taken on a coding function for which they were less well suited because of the relative instability of the RNA phosphodiester bond. A transfer of the coding function to the more stable DNA seems almost inevitable and would not have been difficult to achieve, reduction of ribonucleotides giving deoxyribonucleotides, which could then be polymerized into DNA copies of the RNA protogenomes by a reverse-transcriptase-catalyzed reaction (**Figure 18.6**). The replacement of uracil with its methylated derivative thymine probably conferred even more stability on the DNA polynucleotide, and the adoption of double-stranded DNA as the coding molecule was almost certainly prompted by the possibility of repairing DNA damage by copying the partner strand (Section 17.2).

According to this scenario, the first DNA genomes comprised many separate molecules, each specifying a single protein and each therefore equivalent to a single gene. The linking together of these genes into the first chromosomes, which could have begun before the transition to DNA, would have improved the efficiency of gene segregation during cell division, as it is easier to organize the equal distribution of a few large chromosomes than many separate genes. As with most stages in early genome evolution, several different mechanisms by which genes might have become linked have been proposed.

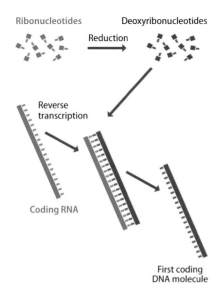

Figure 18.6 Conversion of a coding RNA molecule into the progenitor of the first DNA genome.

How unique is life?

If the experimental simulations and computer models are correct then it is likely that the initial stages in biochemical evolution occurred many times in parallel in the oceans or atmosphere of the early Earth. It is therefore quite possible that 'life' arose on more than one occasion, even though all present-day organisms appear to derive from a single origin. This single origin is indicated by the remarkable similarity between the basic molecular biological and biochemical mechanisms in bacterial, archaeal, and eukaryotic cells. To take just one example, there appears to be no obvious biological or chemical reason why any particular triplet of nucleotides should code for any particular amino acid, but the genetic code, although not universal, is virtually the same in all organisms that have been studied (Section 1.3). If these organisms derived from more than one origin, then we would anticipate two or more very different codes.

If multiple origins are possible, but modern life is derived from just one origin, then at what stage did a single biochemical system begin to predominate? The question cannot be answered precisely, but the most likely scenario is that the predominant system was the first to develop the means to synthesize protein enzymes and therefore probably also the first to adopt a DNA genome. The greater catalytic potential and more accurate replication conferred by protein enzymes and DNA genomes would have given these cells a significant advantage compared to those still containing RNA protogenomes. The DNA–RNA–protein cells would have multiplied more rapidly, enabling them to out-compete the RNA cells for nutrients which, before long, would have included the RNA cells themselves.

What might have been the alternatives to life as we know it among the early biomolecules? One possibility is that base pairs involving nucleotides other than the ones found in today's versions of DNA and RNA were utilized. An interesting example is the pair formed by purine-2,6-dicarboxylate and 3-pyridine, which is coordinated by a Cu^{2+} ion (**Figure 18.7**). Many natural DNA polymerases are able to add 3-pyridine to a growing polynucleotide if the template contains purine-2,6-dicarboxylate, indicating that the polymerizing capability of these enzymes is not limited to the canonical Watson–Crick base pairs. Equally interesting is the possibility that life forms based on informational molecules other than DNA or RNA might have occurred in the past, or might occur on a different planet. In particular, a pyranosyl nucleic acid, which is based on a six-carbon rather than five-carbon sugar (**Figure 18.8**), might be a better choice than RNA for an early protogenome because the base-paired molecules that it forms are more stable. Other possibilities include threose nucleic acid, which uses four-carbon sugars, or a glycerol-derived nucleic acid, which like peptide nucleic acid (see Figure 3.32), has a linear backbone that lacks sugar ring structures. None of these alternative nucleic acids are known in nature, but they can be synthesized in the laboratory and each one forms a stable double-stranded hybrid with RNA. Threose nucleic acids are particularly interesting as they have at least some enzymatic activity, being able to both cleave and ligate RNA molecules. Prior to the discovery of pathways for the abiotic synthesis of ribonucleotides, these nucleic acid analogs were looked on as attractive precursors to the RNA world, in that their spontaneous synthesis might be more plausible than that of RNA. Today, the general consensus is that there is no need to postulate a precursor to RNA as the latter is likely to have evolved directly from the prebiotic soup.

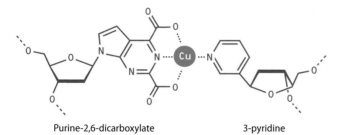

Purine-2,6-dicarboxylate   3-pyridine

Figure 18.7 A base pair formed between purine-2,6-dicarboxylate and 3-pyridine. The two bases are shown attached to 2′-deoxyribose units present in polynucleotide strands. The base pair is coordinated by a Cu^{2+} ion.

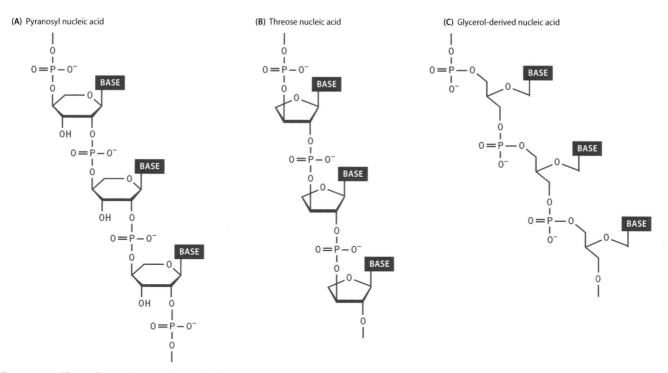

(A) Pyranosyl nucleic acid

(B) Threose nucleic acid

(C) Glycerol-derived nucleic acid

Figure 18.8 Three alternative nucleic acids that could form the basis for life on the early Earth or on other planets.

18.2 THE EVOLUTION OF INCREASINGLY COMPLEX GENOMES

If we follow the planet's history forwards from the earliest bacterial cells, dating to 3.4 billion years ago, we see the first evidence for eukaryotes 2.7 billion years ago, in the form of sterols, thought to be characteristic features of eukaryotes but not prokaryotes, which have been detected in oil shales of this age. The earliest eukaryote visible in the fossil record is currently the organism called *Bangiomorpha*, a red alga from 1.2 billion years ago. *Bangiomorpha* is multicellular and reproduces sexually, so it is clearly a relatively advanced type, indicating that earlier eukaryotes remain undiscovered. Multicellular animals appeared around 640 million years ago, although there are enigmatic fossils suggesting that multicellular protists and sponges lived earlier than this. The Cambrian Revolution, when invertebrate life proliferated into many novel forms, occurred 530 million years ago and ended with the disappearance of many of the novel forms in a mass extinction 490 million years ago. Since then, evolution has continued apace and with increasing diversification: the first terrestrial insects, animals, and plants were established 350 million years ago, the dinosaurs had been and gone by the end of the Cretaceous, 65 million years ago, and the first hominins appeared a mere 4.5 million years ago.

Morphological evolution was accompanied by genome evolution. It is dangerous to equate evolution with 'progress,' but it is undeniable that as we move up the evolutionary tree we see increasingly complex genomes. In this section, we will explore how this complexity evolved.

Genome sequences provide extensive evidence of past gene duplications

One indication of genome complexity is gene number, which varies from less than 1000 in some bacteria to 20,000 in vertebrates such as humans. If we assume that the earliest genomes had fewer genes than the most complex modern ones, then we must ask how these gene numbers increased.

There are two fundamentally different ways in which new genes could be acquired by a genome:

- By duplicating some or all of the existing genes in the genome.

- By acquiring genes from other species.

We have already discussed how prokaryotes acquire new genes by horizontal transfer from other species (Section 8.2), and we will return to this topic later when we consider the less widespread ways in which eukaryotic genomes have been affected by horizontal transfer. First, we will focus on **gene duplication**, which is looked on as having played a central role in genome evolution in all types of organism.

The initial result of gene duplication will be two identical genes. Selective constraints will ensure that one of these genes retains its original nucleotide sequence, or something very similar to it, so that it can continue to provide the protein function that was originally supplied by the single gene copy before the duplication took place. It is possible that the same selective constraints will apply to the second gene, especially if the increase in the rate of synthesis of the gene product, made possible by the duplication, confers a benefit to the organism (**Figure 18.9**). More frequently, however, the second copy will confer no benefit and hence will not be subject to the same selective pressures and so will accumulate mutations at random. Evidence shows that the majority of new genes that arise by duplication acquire deleterious mutations that inactivate them so that they become nonprocessed pseudogenes. Occasionally, however, mutations might not lead to inactivation of the gene but instead result in a new gene function that is useful to the organism (see Figure 18.9). In this case, the gene will be retained and the gene content of the genome will have increased.

The most cursory examination of genome sequences provides ample evidence that many genes have arisen from duplication events. The importance of the first scenario illustrated in Figure 18.9, where the increased amount of gene product resulting from a gene duplication is beneficial and stabilizes the duplication, is supported by the many examples of multigene families made up of genes with identical or near-identical sequences. The prime examples are the rRNA genes, which are multicopy in all but a few bacteria, and which have copy numbers of approximately 400 in the human genome and over 4000 in pea. These multiple copies of identical genes presumably reflect the need for rapid synthesis of rRNAs at certain stages of the cell cycle. Note that the existence of these multigene families indicates not only that gene duplications have occurred in the past but also that there must be a molecular mechanism that ensures that the family members retain their identity over evolutionary time. This is called **concerted evolution**. If one copy of the family acquires an advantageous mutation, then it is possible for that mutation to spread throughout the family until

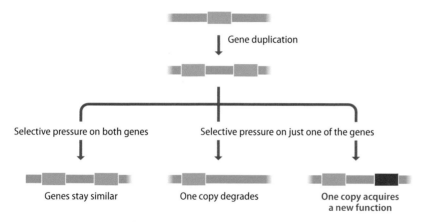

Figure 18.9 Three scenarios for the outcome of a gene duplication.

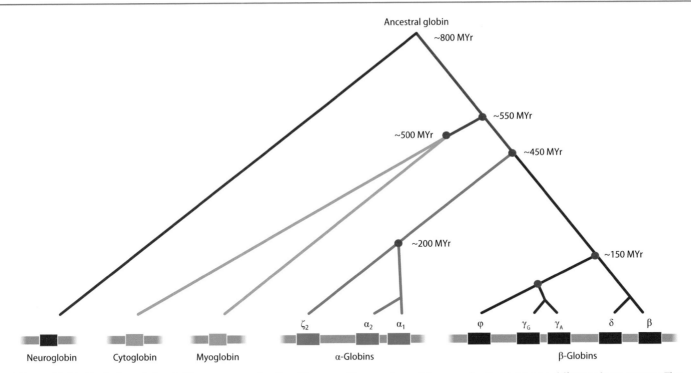

Figure 18.10 Evolution of the globin gene superfamily of humans. The members of the superfamily are now on different chromosomes. The neuroglobin gene is on chromosome 14, the cytoglobin gene is on chromosome 17, and the myoglobin gene is on chromosome 22. The α-globin cluster is on chromosome 16, and the β-globin cluster is on chromosome 11. The relationships between the genes is inferred from the degree of nucleotide similarity between gene pairs, and the estimated rate at which mutations accumulate in human exons is used as a molecular clock to estimate the dates at which gene duplications occurred. Abbreviation: MYr, millions of years ago.

all members possess it. The most likely way in which this can be achieved is by gene conversion, which, as described in Section 16.1, can result in the sequence of one copy of a gene being replaced with all or part of the sequence of a second copy. Multiple gene conversion events could therefore maintain identity among the sequences of the individual members of a multigene family, especially if those members are arranged in tandem arrays.

The third scenario in Figure 18.9 results in the duplicated gene accumulating mutations that give it a new, useful function. Again, multigene families provide many indications that such events have occurred frequently in the past. The α- and β-type globin gene superfamily (Section 7.2) provides an example. This superfamily also includes genes specifying various other proteins which, like the blood globins, have the capacity to bind oxygen molecules. From the degrees of similarity displayed by pairs of genes in the superfamily, it is possible to deduce the pattern of gene duplications that gave rise to the genes we see today, and by applying the **molecular clock** to the data, which tells us how rapidly a pair of sequences will diverge over time, we can estimate how many millions of years ago each duplication took place. These analyses tell us that a duplication some 800 million years ago resulted in a pair of ancestral genes, one of which evolved into the modern gene for the brain protein neuroglobin, and the other of which gave rise to all the other members of the superfamily (**Figure 18.10**). Some 250 million years later, there was a second duplication on the path leading to the blood globins, one of the products of this duplication being a gene that, via another duplication, gave rise to myoglobin, which is active in muscle, and cytoglobin, which is present in many tissues and is thought to protect cells from oxidative stress. The proto-α and proto-β lineages split by a duplication that occurred 450 million years ago, and the duplications within the α- and β-globin gene families took place during the last 200 million years.

We observe similar patterns of evolution when we compare the sequences of other genes. The trypsin and chymotrypsin genes, for example, are related by a common ancestral gene that duplicated approximately 1500 million years ago.

Both now code for proteases involved in protein breakdown in the vertebrate digestive tract, trypsin cutting proteins at arginine and lysine amino acids, and chymotrypsin cutting at phenylalanines, tryptophans, and tyrosines. Genome evolution has therefore produced two complementary protein functions where originally there was just one.

Another striking example of gene evolution by duplication is provided by the homeotic selector genes, the key developmental genes responsible for specification of the body plans of animals. As described in Section 14.3, *Drosophila melanogaster* has a single cluster of homeotic selector genes, called HOM-C, which consists of eight genes, each containing a homeodomain sequence coding for a DNA-binding motif in the protein product (see Figure 14.37). These eight genes, as well as other homeodomain genes in *Drosophila*, are believed to have arisen from a series of gene duplications that began with an ancestral gene that existed about 1000 million years ago. The functions of the modern genes, each specifying the identity of a different segment of the fruit fly, gives us a tantalizing glimpse into how gene duplication and sequence divergence could, in this case, have been the underlying processes responsible for increasing the morphological complexity of the series of organisms in the evolutionary lineage leading to *Drosophila*. If we then move further up the evolutionary tree, we see that most vertebrates have four Hox gene clusters (see Figure 14.37), each a recognizable copy of the *Drosophila* cluster, with sequence similarities between genes in equivalent positions. The implication is that in the vertebrate lineage, there were two duplications, not of individual Hox genes but of the entire cluster (**Figure 18.11**). So far, no intermediate species with two Hox clusters has been found, but there are well-known examples of vertebrates with more than four. Teleosts, a group of ray-finned fish that are probably the most diverse group of vertebrates, with a vast range of different variations of the basic body plan, have seven or eight Hox clusters, presumed to have arisen by duplication of the set of four, followed by the loss of one cluster in the ancestor of those fish that have just seven clusters. A further duplication in the salmonid lineage has resulted in 13 Hox clusters in the Atlantic salmon *Salmo salar* and the rainbow trout *Oncorhynchus mykiss*.

A variety of processes could result in gene duplication

Genome annotations provide extensive evidence that gene duplications have occurred in the past. How did these duplications arise? There are several possibilities, including:

- **Unequal crossing over**, which is a recombination event that is initiated by similar nucleotide sequences that are not at identical places in a pair of homologous chromosomes. As shown in **Figure 18.12A**, the result of unequal crossing over can be duplication of a segment of DNA in one of the recombination products.

- **Unequal sister chromatid exchange**, which occurs by the same mechanism as unequal crossing over, but involves a pair of chromatids from a single chromosome (**Figure 18.12B**).

- **DNA amplification**, which is sometimes used in this context to describe duplication of segments of DNA in bacteria and other haploid organisms, the duplications arising from unequal recombination between the two daughter DNA molecules in a replication bubble (**Figure 18.12C**).

Each of these three processes lead to tandem duplications – ones in which the two duplicated segments lie adjacent to one another in the genome. This is the pattern seen with many multigene families, but it is not the only possibility (Section 7.2). Family members are not always co-located: for example, in the human genome there are five genes for the metabolic enzyme aldolase, each on a different chromosome. These copies might have once been present as a tandem array and subsequently have become dispersed, but it is also possible that

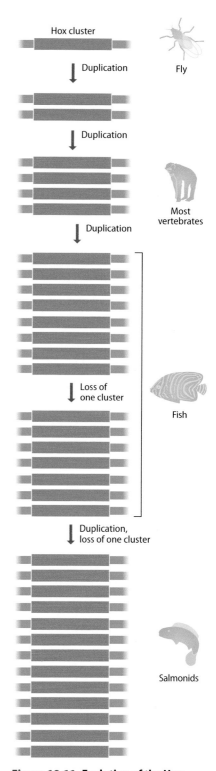

Figure 18.11 Evolution of the Hox gene cluster from flies to salmonid fishes.

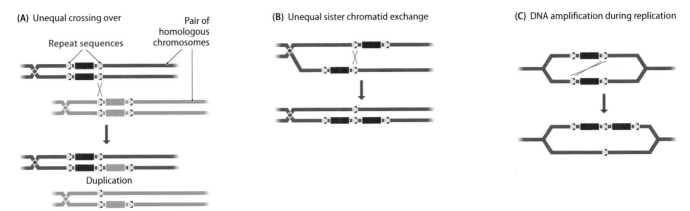

Figure 18.12 Models for gene duplication. (A) Unequal crossing over between homologous chromosomes. (B) Unequal sister chromatid exchange. (C) DNA amplification during replication of a bacterial genome. In each case, recombination occurs between two different copies of a short repeat sequence, leading to duplication of the sequence between the repeats. Unequal crossing over and unequal sister chromatid exchange are essentially the same, except that the first involves chromatids from a pair of homologous chromosomes and the second involves chromatids from a single chromosome. In (C), recombination occurs between two daughter double helices that have just been synthesized by DNA replication.

the distant locations are a consequence of a different type of duplication process, one that occurs by retrotransposition, in a manner similar to that thought to lead to formation of processed pseudogenes (see Figure 7.17). A processed pseudogene arises when the mRNA copy of a gene is converted into cDNA and reinserted into the genome. The resulting structure is a pseudogene because it lacks a promoter sequence, this being absent from the mRNA. But what if the mRNA copy becomes inserted adjacent to the promoter of an existing gene (**Figure 18.13**)? Now it might become active by subverting this promoter for its own use. Gene duplicates that arise in this way are called **retrogenes**.

How do we identify retrogenes when a genome is being annotated? A distinctive feature of a retrogene is that it lacks any introns that were present in the parent copy of the gene, as these are not present in the mRNA. The discovery of a pair of homologous genes, one containing introns and one lacking introns, would therefore be an indication that the latter might be a retrogene. However, when we discussed the possible activity of those nonprocessed pseudogenes that are still transcribed from their original promoters (Section 7.2), we learned that transcription and translation *per se* do not provide sufficient evidence for a function to be assigned to a pseudogene. It must also be established that natural selection is acting in a positive way on that gene; otherwise, it has to be concluded that expression is merely fortuitous and is of no long-term benefit to the organism. The same issue relates to retrogenes. Expression due to insertion of the retrogene adjacent to a novel promoter is an indication that the gene might be functional, but this has to be confirmed by establishing that positive selection is acting upon it. When these issues are taken into account, the best estimates are that there are 600–700 retrogenes in the human genome, a small number of which, 25 or so, are **orphan retrogenes**, whose parent copies have been lost, the retrogene now bearing sole responsibility for providing the protein function.

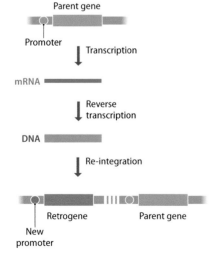

Figure 18.13 Creation of a retrogene. The mRNA from the parent gene is reverse-transcribed into a DNA copy, which is reinserted in the genome adjacent to the promoter of a different gene. This promoter now drives expression of the retrogene.

Whole-genome duplication is also possible

The processes described above give rise to relatively short DNA duplications, perhaps a few tens of kilobases in length. Are larger duplications possible? It seems unlikely that duplication of entire chromosomes has played any major role in genome evolution, because we know that duplication of individual human chromosomes, resulting in a cell that contains three copies of one chromosome and two copies of all the others (the condition called **trisomy**), is either lethal or results in a genetic defect. Similar harmful effects have been observed in artificially generated trisomic mutants of *Drosophila*. Probably, the resulting

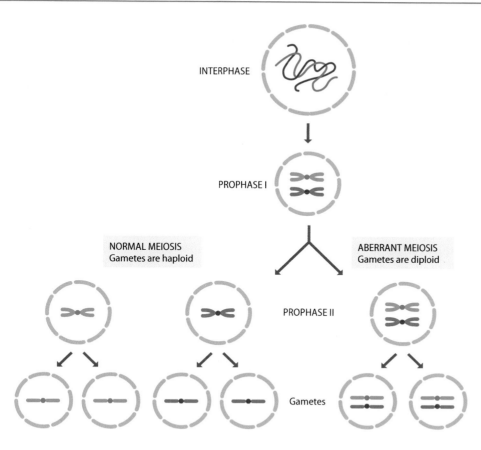

Figure 18.14 The basis of auto-polyploidization. The normal events occurring during meiosis are shown, in abbreviated form, on the left. On the right, an aberration has occurred between prophase I and prophase II and the pairs of homologous chromosomes have not separated into different nuclei. The resulting gametes will be diploid rather than haploid.

increase in copy numbers for some genes but not others leads to an imbalance of the gene products and disruption of the cellular biochemistry.

The harmful effects of trisomy do not mean that duplication of the entire set of chromosomes in a nucleus must be discounted. Genome duplication can occur if an error during meiosis leads to the production of gametes that are diploid rather than haploid (**Figure 18.14**). If two diploid gametes fuse, then the result will be a type of **autopolyploid**, in this case a tetraploid cell whose nucleus contains four copies of each chromosome. Autopolyploidy, as with other types of **polyploidy**, is not uncommon, especially among plants. Autopolyploids are often viable because each chromosome still has a homologous partner and so can form a bivalent during meiosis. This allows an autopolyploid to reproduce successfully, but generally prevents interbreeding with the original organism from which it was derived. This is because a cross between, for example, a tetraploid and diploid would give a triploid offspring which would not itself be able to reproduce because one full set of its chromosomes would lack homologous partners (**Figure 18.15**). Autopolyploidy is therefore a mechanism by which speciation can occur, a pair of species usually being defined as two organisms that are unable to interbreed. The generation of new plant species by autopolyploidy has in fact been observed, notably by Hugo de Vries, one of the rediscoverers of Mendel's experiments. During his work with evening primrose, *Oenothera lamarckiana*, de Vries isolated a tetraploid version of this normally diploid plant,

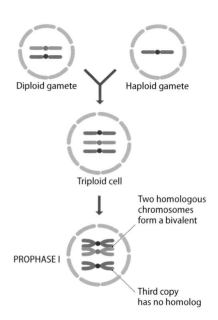

Figure 18.15 Autopolyploids cannot interbreed successfully with their parents. Fusion of a diploid gamete produced by the aberrant meiosis shown in Figure 18.14 with a haploid gamete produced by normal meiosis leads to a triploid nucleus, one that has three copies of each homologous chromosome. During prophase I of the next meiosis, two of these homologous chromosomes will form a bivalent, but the third will have no partner. This has a disruptive effect on the segregation of chromosomes during anaphase and usually prevents meiosis from reaching a successful conclusion. This means that gametes are not produced and the triploid organism is sterile. Note that the bivalent could have formed between any two of the three homologous chromosomes, not just between the pair shown in the diagram.

which he named *Oenothera gigas*. Autopolyploidy among animals is less common, especially in those with two distinct sexes, possibly because of problems that arise if a nucleus possesses more than one pair of sex chromosomes.

Autopolyploidy does not lead directly to an increase in genome complexity because the initial product is an organism that simply has extra copies of every gene, rather than any new genes. It does, however, provide the potential for increased complexity because the extra genes are not essential to the functioning of the cell and so can undergo mutational change without harming the viability of the organism. With this in mind, is there any evidence that whole-genome duplication has been important in the large-scale acquisition of new genes during the evolutionary histories of present-day genomes?

From what we understand about the way in which genomes change over time, we might anticipate that evidence for whole-genome duplication would be quite difficult to obtain. Many of the extra gene copies resulting from genome duplication would decay to the extent that they are no longer visible in the DNA sequence. Those genes that are retained, because their duplicated function is useful to the organism or because they have evolved new functions, should be identifiable, but it would be difficult to distinguish if they have arisen by duplication of the entire genome or by duplication of much smaller segments. For a genome duplication to be signaled, two criteria have to be met:

- Molecular clock analyses should indicate that a substantial fraction of the duplicated genes in a genome arose simultaneously.

- Despite post-duplication rearrangements, sets of duplicated genes should still occur in blocks, and there should be synteny (i.e., similar gene orders) in different copies of each duplicated block.

The first evidence for whole-genome duplication in the evolutionary ancestry of eukaryotes was obtained during the *Saccharomyces cerevisiae* sequencing project. As parts of the sequence were assembled, it became clear that the genome contained syntenic blocks. When the sequence was complete, homology analysis (Section 5.1) was carried out with every yeast gene tested against every other yeast gene. To be considered descendants of a duplication event, two genes had to display at least 25% identity when the predicted amino acid sequences of their protein products were compared. About 800 gene pairs were identified in this way, 376 of which could be placed in 55 duplicate sets, each of these sets containing at least three genes in the same order, possibly with other genes interspersed between them, the sets altogether covering half the genome. These sets could have arisen by duplication of segments rather than the entire genome, but if this was the case then it might be anticipated that some of the genes would have been duplicated more than once. The fact that there were just two copies of each gene, and never three or four, therefore supported the notion that the copies arose by whole-genome duplication. This possibility became more certain when the complete genome sequences of other yeast species were obtained. Comparisons between the genomes of *S. cerevisiae*, *Kluyveromyces waltii*, and *Ashbya gossypii* have been particularly informative. These three species shared a common ancestor that lived over 100 million years ago, previous to the time of the genome duplication event inferred from the homology analysis. If that duplication had indeed occurred in the lineage leading to *S. cerevisiae* then it would be anticipated that this species would have duplicated copies of many genes present as singletons in the *K. waltii* and *A. gossypii* genomes. This turns out to be the case, with this new analysis suggesting that some 10% of the genes in the modern *S. cerevisiae* genome derive from a whole-genome duplication that occurred just under 100 million years ago.

Equivalent work has now been carried out with other genomes, using more sophisticated computational tools to detect syntenic blocks of duplicated genes, and incorporating molecular clock studies to date when duplications occurred. These analyses have uncovered evidence of two whole-genome duplications early in the evolutionary history of vertebrates, some 550 and 450 million years ago (**Figure 18.16**). A subsequent duplication on the lineage leading to teleost fish occurred about 310 million years ago, and more recently the proto-salmonid genome duplicated 80–100 million years ago. Detailed analysis of these two more recent duplications have enabled the events following the duplication to be inferred, suggesting that during the 60 million years following the teleost duplication, 60% of the duplicated genes were lost, reverting these gene pairs back to singletons. Since then, there has been a continued, but much slower, loss of duplicates. Whole-genome duplications have been equally important during plant evolution, comparisons between the genomes of different species revealing multiple events in the lineages of most land plants, dating back to 400 million years ago (**Figure 18.17**).

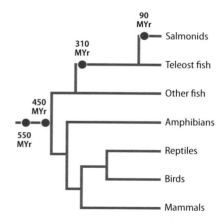

Figure 18.16 Genome duplications in the vertebrate evolutionary lineage. The four genome duplications believed to have occurred in the vertebrate evolutionary lineage are shown. Note that the timings of these duplications explain the different numbers of Hox clusters found in flies, vertebrates, teleosts, and salmonids (see Figure 18.11). Abbreviation: MYr, millions of years ago.

Smaller duplications can also be identified in the human genome and other genomes

Although the most recent whole-genome duplication in the human lineage occurred about 450 million years ago, the human genome has not been quiescent in the intervening period. Indeed, the opposite is true, one surprise arising from detailed analysis of human genome sequences being the discovery that there has been extensive and frequent duplication of short segments of the genome in the relatively recent past. These are called **segmental duplications** or **low-copy repeats**, and are usually defined as duplications between 1–400 kb in length, with greater than 90% sequence identity, that are repeated up to 50 times in the genome, though often occur as just two copies. Although segmental duplications sometimes contain repetitive DNA elements, such as LINEs, SINEs, or HERVs (Section 9.2), these are not themselves classified as segmental duplications because they have high copy numbers.

The impact of segmental duplication on the structure of the human genome is illustrated by **Figure 18.18**, which depicts duplication events giving rise to segments >1 kb with >90% identity. As well as many duplications within individual chromosomes, there are multiple duplications that have placed the second copy on a different chromosome. Duplications occur more frequently in the vicinity of the centromeres and in the subtelomeric regions, but there are still many repeats away from these areas. Altogether, segmental duplications make up 7.0% of the human genome. Most or all of these duplications occurred since the split between the Old World monkeys, which include primates such as humans, and New World monkeys some 35 million years ago.

Segmental duplications are thought to play an important role in genome evolution by giving rise to gene duplicates and by placing duplicated genes under the control of new regulatory sequences. On the other hand, segmental duplications can have a potentially deleterious impact on a genome. Recombination between a pair of intrachromosomal duplicates could result in deletion of the region between the repeated sequences, deletions generated in this way possibly giving rise to genetic disease. An increasing number of disorders are known to have arisen in this way. An example is Prader–Willi syndrome, an incurable disorder associated with developmental and nutritional problems, the latter often leading to obesity. In about 70% of cases, Prader–Willi syndrome is caused by deletion of a segment of the paternal version of chromosome 13, due to recombination between a pair of segmental duplicates. The deleted region contains various protein-coding and noncoding RNA genes, several of which are maternally imprinted. Deletion of the active, paternal versions of these genes therefore leads to a loss of function. Which gene or genes in the deleted region is directly responsible for the disease is not yet known, but attention is currently focused on the *SNORD116* locus, which contains 29 copies of a snoRNA gene. Patients who still have the region between the duplicates, but have smaller deletions that remove the *SNORD116* locus, display many of the symptoms of Prader–Willi syndrome, suggesting the snoRNA genes play a central role in the disease.

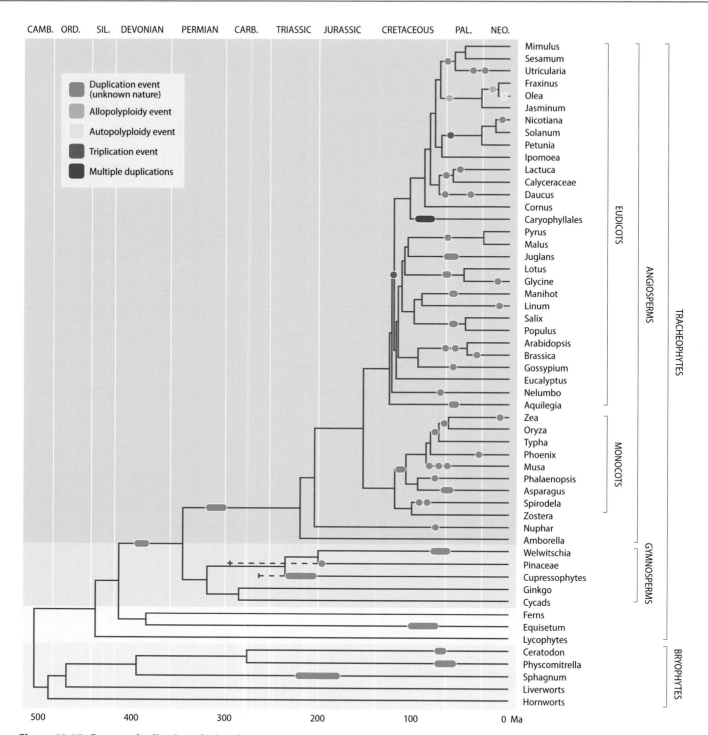

Figure 18.17 Genome duplications during the evolution of land plants. Duplication events of unknown origin are shown in blue, triplications in red, known autopolyploidy events in yellow, and allopolyploidy events in green. The purple bar associated with Caryophyllales represents 26 independent duplication events, some of which are autopolyploidy and some allopolyploidy. Abbreviations: Camb., Cambrian; Carb., Carboniferous; Ord., Ordovician; Neo., Neogene; Pal., Paleozoic; Sil., Silurian. (From Clark J.W. & Donoghue P.C.J. [2018] *Trends in Plant Science* 23:P933-945. With permission from Elsevier.)

Both prokaryotes and eukaryotes acquire genes from other species

Duplication events are not the only way of adding new genes to a genome. Both prokaryotes and eukaryotes can also acquire genes from other species by horizontal gene transfer. In bacteria and archaea, horizontal gene transfer is so frequent that it has had a major impact on the gene catalogs of individual genomes (Section 8.2).

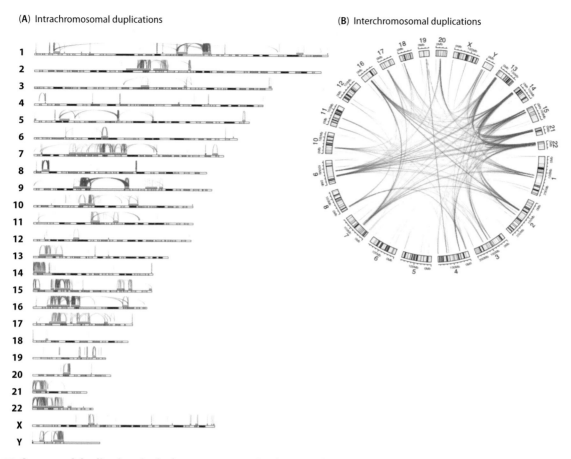

(A) Intrachromosomal duplications

(B) Interchromosomal duplications

Figure 18.18 Segmental duplications in the human genome. Duplications of segments >1 kb in length with >90% identity are shown. (A) Intrachromosomal duplications. The orange bars indicate regions that are flanked by large (>10 kbp), high-identity (>95%) duplications, which have the potential to drive unequal crossing-over events during meiosis. (B) Interchromosomal duplications. (From Vollger M.R., Guitart X., Dishuck P.C. [2022] *Science* 376:eabj6965. With permission from American Association for the Advancement of Science.)

The prevalence of horizontal gene transfer in prokaryotes is due, at least in part, to the ease with which many prokaryotic species can take up DNA from their environment, with some species having cell membrane proteins specifically for this purpose. Eukaryotes do not have equivalent mechanisms for DNA uptake, and it is therefore not surprising that horizontal gene transfer has been much less important during the evolution of eukaryotic genomes. Exactly how important the process has been is not clear; several reports of 'bacterial' genes in eukaryotic genomes have been ascribed to contamination of eukaryotic DNA preparations with bacteria, the latter contributing sequence reads that are mistakenly included in the eukaryotic genome assembly. Conversely, exclusion of 'bacterial' reads from a eukaryotic assembly, simply on the grounds that these reads might be contaminants, risks missing genuine cases of gene transfer from bacteria to eukaryotes.

To become a stable part of the recipient genome, it is necessary for a gene acquired by horizontal transfer to be passed to the progeny. This is straightforward if the recipient species is unicellular and reproduces simply by cell division. It is therefore not surprising that most of the documented cases of horizontal gene transfer into eukaryotes involve microbial forms such as unicellular algae and phytoplankton. Among these organisms, horizontal gene transfer can account for 1% or more of the genes in a genome. Transfers have been implicated in the ability of some species of red algae to withstand high temperatures and are also thought to have conferred cold tolerance on diatoms found in sea ice. If the recipient species is multicellular and has sexual reproduction, then the transferred gene must become inserted in to the DNA of the germ cells in order for it to be passed to the progeny. Despite this complication, there are several authentic examples of horizontal gene transfer involving sexually repro-ducing eukaryotes. One of these concerns a group of plant-parasitic round-worms, including the potato cyst nematode *Globodera pallida*. In order to feed on sucrose provided by the host plant, these animals synthesize enzymes that degrade root cell walls and suppress the plant's defense response. Several com-ponents of this specialized biochemical repertoire appear to have been acquired by the nematodes via horizontal gene transfer from other species that inhabit the root biosphere. The genes for the cell wall degrading enzymes, for example, have closest similarity with homologous genes in the soil bacterium *Ralstonia*, and one gene for sucrose utilization is also probably bacterial in origin. Other plant-parasitic nematodes appear to have obtained cellulase genes from soil fungi.

A more extreme form of horizontal gene transfer is well documented in plant species. We have already seen how autopolyploidization can result in genome duplication in plants (see Figure 18.14). **Allopolyploidy**, which results from interbreeding between two different species, is also common and, like auto-polyploidy, can result in a viable hybrid. Usually, the two species that form the allopolyploid are closely related and have many genes in common, but each par-ent will possess a few novel genes or at least distinctive alleles of shared genes. For example, the bread wheat, *Triticum aestivum*, is a hexaploid that arose by allopolyploidization between cultivated emmer wheat, *T. turgidum*, which is a tetraploid, and a diploid wild grass, *Aegilops squarrosa*. The wild grass nucleus contained novel alleles for the high-molecular-weight glutenin genes, which, when combined with the glutenin alleles already present in emmer wheat, resulted in the superior properties for breadmaking displayed by the hexaploid wheats. Allopolyploidization can therefore be looked upon as a combination of genome duplication and interspecies gene transfer.

Genome evolution also involves rearrangement of existing gene sequences

Duplications do not have to be of entire genes in order to have an impact on the genetic content of a genome. Duplications of gene segments containing one or more exons can alter the coding specificity of existing genes and even create entirely new genes. Rearrangements of this type can result in novel pro-tein functions because most proteins are made up of structural domains, each domain comprising a segment of the polypeptide chain and hence encoded by a contiguous series of nucleotides (**Figure 18.19**). **Domain duplication** occurs when the gene segment coding for a structural domain is duplicated by unequal crossing-over, replication slippage, or one of the other methods that we have considered for duplication of DNA sequences (**Figure 18.20A**). Duplication could result in the structural domain being repeated in the protein, which might itself be advantageous, for example, by making the protein more stable. The duplicated domain might also change over time as its coding sequence becomes mutated, leading to a modified structure that might provide the protein with a new activity. Note that domain duplication causes the gene to become longer. Gene elongation appears to be a general consequence of genome evolution, the genes of higher eukaryotes being longer, on average, than those of lower organ-isms. Alternatively, **domain shuffling** combines segments coding for structural domains from completely different genes to form a new coding sequence that specifies a hybrid or mosaic protein, one that would have a novel combination of structural features and might provide the cell with an entirely new biochemical function (**Figure 18.20B**).

Figure 18.19 Each structural domain is an individual unit in a polypeptide chain and is coded by a contiguous series of nucleotides. In this simplified example, each α-helix and β-sheet in the polypeptide is looked upon as an indi-vidual structural domain. In reality, most structural domains comprise two or more secondary structural units.

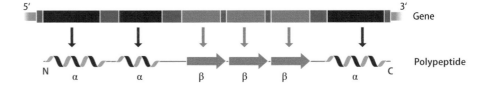

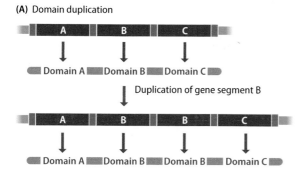

(A) Domain duplication

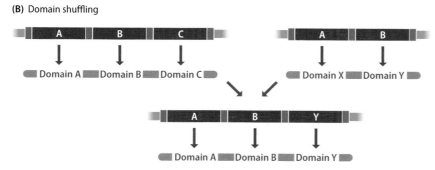

(B) Domain shuffling

Figure 18.20 Creating new genes by (A) domain duplication and (B) domain shuffling.

Implicit in these models of domain duplication and shuffling is the need for the relevant gene segments to be separated so that they can themselves be rearranged and shuffled. This requirement has led to the attractive suggestion that exons might code for structural domains. With some proteins, duplication or shuffling of exons does seem to have resulted in the structures seen today. An example is provided by the α2 Type I collagen gene of vertebrates, which codes for one of the three polypeptide chains of collagen. Each of the three collagen polypeptides has a highly repetitive sequence made up of repeats of the tripeptide glycine–X–Y, where X is usually proline and Y is usually hydroxyproline (**Figure 18.21**). The chicken α2 Type I gene is split into 52 exons, 42 of which cover the part of the gene coding for the glycine–X–Y repeats. Within this region, each exon encodes a set of complete tripeptide repeats. The number of repeats per exon varies but is 5 (in 5 exons), 6 (23 exons), 11 (5 exons), 12 (8 exons), or 18 (1 exon). Clearly, this gene could have evolved by duplication of exons leading to repetition of the structural domains.

Domain shuffling is illustrated by tissue plasminogen activator (TPA), a protein found in the blood of vertebrates and which is involved in the blood clotting response. The TPA gene has four exons, each coding for a different structural domain (**Figure 18.22**). The upstream exon codes for a 'finger' module that enables the TPA protein to bind to fibrin, a fibrous protein found in blood clots and which activates TPA. This exon appears to be derived from a second fibrin-binding protein, fibronectin, and is absent from the gene for a related protein, urokinase, which is not activated by fibrin. The second TPA exon specifies a growth factor domain which has apparently been obtained from the gene for epidermal growth factor and which may enable TPA to stimulate cell proliferation. The last two exons code for 'kringle' structures which TPA uses to bind to fibrin clots; these kringle exons come from the plasminogen gene.

Type I collagen and TPA provide elegant examples of gene evolution but, unfortunately, the clear links that they display between structural domains and exons are exceptional and are rarely seen with other genes. Many other genes appear to have evolved by duplication and shuffling of segments, but in these the structural domains are coded by segments of genes that do not coincide with

Collagen polypeptide

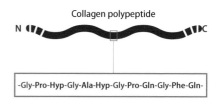

-Gly-Pro-Hyp-Gly-Ala-Hyp-Gly-Pro-Gln-Gly-Phe-Gln-

Figure 18.21 The α2 Type I collagen polypeptide has a repetitive sequence described as Gly–X–Y. Every third amino acid is glycine, X is often proline, and Y is often hydroxyproline (Hyp). Hydroxyproline is synthesized by post-translational modification of proline. The collagen polypeptide has a helical conformation, but one that is more extended than the standard α-helix.

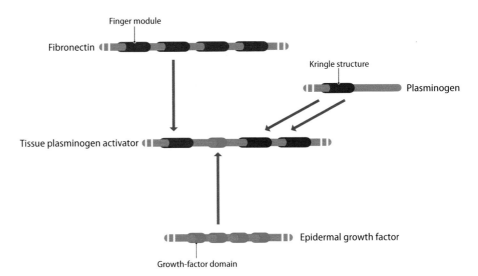

Figure 18.22 The modular structure of the tissue plasminogen activator gene.

individual exons or even groups of exons. Domain duplication and shuffling still occur, but presumably in a less precise manner and with many of the rearranged genes having no useful function. Despite being haphazard, the process clearly works, as indicated by, among other examples, the number of proteins that share the same DNA-binding motifs (Section 11.2). Several of these motifs probably evolved *de novo* on more than one occasion, but it is clear that in many cases the nucleotide sequence coding for the motif has been transferred to a variety of different genes.

One possible mechanism for moving gene segments around a genome is in association with transposable elements. The transposition of a LINE-1 element (Section 9.2) can occasionally result in a short piece of the adjacent DNA being transferred along with the transposon, a process called **3′-transduction**, the transferred segment being located at the 3′-end of the element. LINE-1 elements are sometimes found in introns, so 3′-transduction could conceivably move downstream exons to new sites in a genome. Movement of exons and other gene segments might also be brought about by DNA transposons called ***Mutator*-like transposable elements** (**MULEs**), which are found in many eukaryotes but are especially common in plants. MULEs often contain within their DNA sequence segments of genes captured from the host genome. Transposition of a MULE would therefore move the captured segments to a new location. MULEs can collect segments of different genes as they travel around a genome, assembling new hybrid genes as they go. MULEs therefore provide an attractive way of driving gene evolution, but there are still several unanswered questions about their impact. In particular, it is not yet clear how frequently gene segments are able to escape from MULEs.

There are competing hypotheses for the origins of introns

Ever since introns were discovered in 1977, there has been debate about their origins. Many of the early ideas were influenced by the **exon theory of genes**, which suggests that introns were formed when the first DNA genomes were constructed, soon after the end of the RNA world. These genomes would have contained many short genes, each derived from a single coding RNA molecule and each specifying a very small polypeptide, perhaps just a single structural domain. The polypeptides would probably have had to associate together into larger multidomain proteins in order to produce enzymes with specific and efficient catalytic mechanisms (**Figure 18.23**). To aid the synthesis of a multidomain enzyme, it would have been beneficial for the enzyme's individual polypeptides to become linked into a single protein, such as we see today. This could have been achieved by splicing together the transcripts of the relevant minigenes, a process that would have been aided by rearranging the genome so that groups of minigenes specifying the different parts of individual multidomain proteins were positioned next to each other. In other words, the minigenes became exons and the DNA sequences between them became introns.

According to the exon theory of genes and other **introns early** hypotheses, all genomes originally possessed introns. But we know that bacterial genomes do not have GU–AG introns, so if these hypotheses are correct then we must assume that for some reason introns became lost from the ancestral bacterial genome at an early stage in its evolution. The alternative **introns late** hypothesis avoids this problem by proposing that, to begin with, no genes had introns, these structures invading eukaryotic nuclear genomes and subsequently proliferating into the numbers seen today.

One of the reasons why the debate regarding the origin of GU–AG introns has continued for over 40 years is because evidence in support of the early and late hypotheses has been difficult to obtain and has often been ambiguous. A study of the vertebrate globin genes illustrates the problems. Initially, it was concluded that a globin protein comprises four structural domains, the first corresponding to exon 1 of the globin gene, the second and third to exon 2, and the fourth to exon 3 (**Figure 18.24**). This pattern matches the expectations of the exon theory of genes. The prediction that there should be globin genes with another intron that splits the second and third domains was found to be correct when the leghemoglobin gene of soybean was shown to have an intron at exactly the expected position. Unfortunately, as globin genes were sequenced from other species, more introns were discovered, most of these located at positions that do not correspond to junctions between the protein domains.

There is still no consensus regarding the origins of introns, but the availability today of genome sequences from multiple species does at least enable greater amounts of data to be applied to the problem. Comparisons of intron positions in homologous genes from different species have been particularly informative. These have shown that there is a significant correlation between the positions of introns in fungi, plants, and animals, with over 25% of introns having the same position in genes from at least two of these three groups of organisms. Importantly, intron positions in fungi, plants and animals are also shared with basal eukaryotes – ones whose lineages diverged from other eukaryotes at a very early evolutionary stage, and which root back to the base of the eukaryotic phylogenetic tree. Examples are the excavates, a group of single-celled flagellated organisms, including *Trichomonas vaginalis*, which causes the sexually transmitted disease called trichomoniasis (**Figure 18.25**). The genomes of excavates contain introns, and many of those introns are at the same positions as in other eukaryotes. The implication is that the genome of the **last eukaryotic common ancestor** (**LECA**) contained multiple introns, and that many of those introns have been retained at the same positions in the genes of modern-day organisms.

It has also been discovered that the components of the spliceosome, the structure that removes introns from eukaryotic pre-mRNA (Section 12.4), are very similar in different groups of eukaryotes, suggesting that the splicing process evolved at an early stage in evolution and has not changed greatly since then. Based on these various strands of evidence, many researchers now favor a hypothesis for the origin of introns that is intermediate between the early and late schemes. The absence of GU–AG introns in prokaryotes is looked on as a major difficulty for the exon theory of genes and the very earliest genomes are therefore considered to have lacked introns of this type. Those early genomes might, however, have contained the ancestors of the retroelements called **Group II introns**, which are present in organelle genomes and are known in a few prokaryotes. The splicing pathway for Group II introns is very similar to that of GU–AG introns, but Group II introns do not require a spliceosome, as they are ribozymes and so are able to self-splice. The Group II intron sequence encodes a reverse transcriptase, which enables the excised intron to be copied into DNA and to reinsert at a second position in its host genome, by a process called **retrohoming**. The

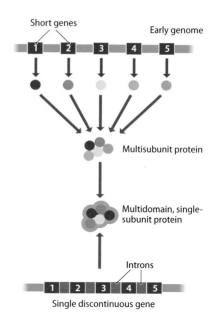

Figure 18.23 The exon theory of genes. The short genes of the first genomes probably coded for single-domain polypeptides that would have had to associate together to form a multisubunit protein to produce an effective enzyme. Later the synthesis of this enzyme could have been made more efficient by linking the short genes together into one discontinuous gene coding for a multidomain single-subunit protein.

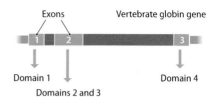

Figure 18.24 A vertebrate globin gene showing the relationship between the three exons and the four domains of the globin protein. The prediction that there should be some globin genes in which an intron separates domains 2 and 3 into separate exons was found to be correct when the soybean leghemoglobin gene was sequenced. Leghemoglobin is a protein found in root nodules. It binds oxygen which otherwise would inhibit the nitrogen-fixing nitrogenase complex.

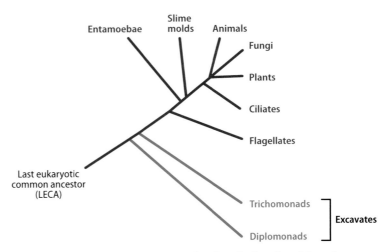

Figure 18.25 The phylogeny of eukaryotes, showing the basal position of excavates.

model that is now emerging is one in which Group II introns were present in the genomes of the endosymbionts that formed the precursors of mitochondria and chloroplasts in the very first eukaryotic cells (Section 8.3). These introns escaped from their organelles and invaded the early eukaryotic nucleus, where they proliferated by retrohoming, the resulting structures evolving into the large numbers of GU–AG introns that were present in LECA. The ability of modern Group II introns to transpose by retrohoming in human cells in experimental systems indicates that the proliferation of Group II introns in early eukaryotic nuclear genomes was possible. Other Group II introns are though to have evolved into non-LTR retroelements such as LINEs. Introns are therefore neither 'early' nor 'late,' their genesis accompanying the origins of the first eukaryotic cells.

The evolution of the epigenome

In this overview of the evolution of genome complexity, we have so far concentrated solely on the DNA sequence features of genomes. We must not forget that in eukaryotic cells, the activity of the genome is inextricably linked with the pattern of histone modifications and DNA methylation (Sections 10.2 and 10.3). These patterns constitute the epigenome and determine which parts of the genome are accessible to the RNA polymerases that transcribe genes in to RNA, and in particular determine which regions of the genome are packaged into heterochromatin and hence silenced in a particular cell or cell lineage.

Evolutionary geneticists are starting to ask how the epigenome evolved or, to phrase the question slightly differently, how the genome evolved its ability to respond to external and internal signals through the use of chromatin modification as a means of silencing those genes that are not needed in a particular tissue or at a particular developmental stage. So far, much of the work in this area has centered on **comparative epigenomics**, which explores the extent to which the equivalent regions of two different genomes display the same pattern of chromatin modification, the resulting data being used to infer the extent to which epigenomic effects are programmed by the DNA sequence. It has already been established that within a single genome, the two copies of a segmental duplication have similar modification patterns. In the human genome, for example, the pattern of methylation within CpG islands tends to be retained after duplication of the segment containing that island, even when the daughter copy has been inserted at a new genomic location distant from the genes whose expression the CpG island normally influences (**Figure 18.26**). This holds true even if the sequence surrounding the island in the daughter duplicate has begun to degrade. Similar patterns of histone modification are also seen in duplicated regions. The implication is that, at least in a single genome, the DNA sequence plays a major role in determining the pattern of chromatin modification.

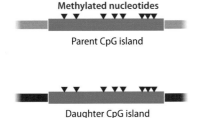

Figure 18.26 Methylation of duplicated CpG islands. The pattern of methylation within the daughter CpG island is the same as in the parent, even though the daughter is at a distant genomic location.

When genomes of different species are compared, the analysis becomes more complicated because of the possibility that equivalent genome regions have non-identical functions in different organisms. In a comparison of the human, chimpanzee, bonobo, gorilla, and orangutan genomes, many examples were found of homologous genes with different methylation patterns, but these included genes with developmental or neurological functions, whose altered methylation patterns might reflect differences in the associated physiological processes in the different primate species. Broader-based studies employing phylogenetic methods have shown that the degree of relatedness between pairs of species as inferred from their methylation patterns is similar to the degree of relatedness as measured by DNA sequence comparisons. A correlation between genome methylation and genome sequence is therefore evident, even if that correlation may be less easy to discern when short regions of the genome are examined.

Understanding the evolution of the epigenome is challenging, but it is a challenge that is worth tackling because it provides one of the keys to linking the evolution of genomes with the evolution of phenotypes. Because of the ease with which DNA sequences can now be obtained we still tend to think of a genome predominantly as a string of As, Cs, Gs, and Ts, displayed in a genome browser with the interesting sequence features highlighted. In order to move on to the next era of genome biology, we must increasingly study the genome as a component of the cell and organism in which it resides. Describing the way in which the epigenome has evolved alongside the genome is one of the first steps toward the establishment of this new and more integrated approach to genome biology.

18.3 GENOMES: THE LAST 6 MILLION YEARS

Bishop Samuel Wilberforce once famously asked Thomas Huxley, one of Charles Darwin's supporters, if his descent from a monkey was on his mother's or father's side. The answer is both, humans and chimpanzees being descended from a common ancestor that lived around 6 million years ago. Since the split, the human lineage has embraced two genera – *Australopithecus* and *Homo* – and a number of species, not all of which were on the direct line of descent to *Homo sapiens* (**Figure 18.27**). The result is us, a novel species in possession of what are, at least to our eyes, important biological attributes that make us very different from all other animals. So, how different are we from the chimpanzees?

The human genome is very similar to that of the chimpanzee

Although a draft of the chimpanzee genome sequence was completed in 2005, just five years after the human sequence was published, it is only during the last few years that it has been possible to make a detailed analysis of the differences between the two genomes. This is because a comparison between two closely related species requires high-quality genome assemblies. The initial drafts, and indeed all of the human and chimpanzee genome assemblies based on short-read sequencing, had two limitations that disguised the differences between the two genomes:

- An assembly based entirely on short reads is likely to give an incomplete and possibly inaccurate description of the repetitive DNA content of a genome, including the numbers and positions of segmental duplications and other large-scale structural variations (see Figure 4.23). These variations can have a significant impact on the copy number and expression profiles of certain genes.

- The human genome was used as a reference for assembly of some parts of the early chimpanzee genome sequences. This leads to a bias, called 'humanization,' whereby structural variations in the chimpanzee genome are missed, and some genes incorrectly annotated, because the short chimpanzee reads contain insufficient information for differences from the human sequence to be recognized.

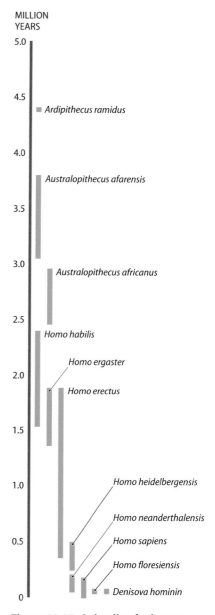

Figure 18.27 A timeline for human evolution. There are many controversies regarding the evolutionary relationships between extinct members of the *Australopithecus* and *Homo* genera. This diagram shows the periods when different species were in existence and does not attempt to distinguish between species that were on the direct line of descent to *Homo sapiens*, and others that lie on branches away from that lineage.

Human

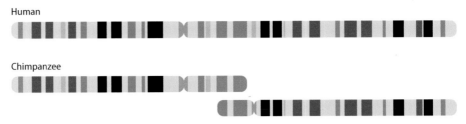

Chimpanzee

Figure 18.28 Human chromosome 2 is a fusion of two chromosomes that are separate in chimpanzees.

Understanding the true nature and extent of the differences between the human and chimpanzee genomes has therefore only been possible since the advent of long-read sequencing (Section 4.1). Long reads have enabled more accurate assemblies to be made of repeat sequences and, importantly, have made it possible to obtain a *de novo* genome sequence for chimpanzees, and other great apes, without recourse to the human genome as a reference.

The comparisons that can now be made between the human and chimpanzee genomes have revealed the extensive similarity that is expected for two species that diverged just 6 million years ago. The degree of nucleotide sequence identity within the coding DNA is 99.1%, with over 30% of the genes in the human genome coding for proteins whose amino acid sequences are identical to the sequences of their counterparts in chimpanzees. Gene order is almost the same in the two genomes, and the chromosomes have very similar appearances. At this level, the most dramatic difference is that human chromosome 2 is two separate chromosomes in chimpanzees (**Figure 18.28**), so chimpanzees, as well as other apes, have 24 pairs of chromosomes, whereas humans have just 23 pairs. However, underlying the near identity of the karyograms is a substantial amount of small-scale structural variation that has evolved in the two genomes since the species diverged, including almost 12,000 insertions of >50 bp in the human genome, as well as another 6000 deletions of >50 bp and over 250 segmental duplications of up to 400 bp in length and >90% identity.

What impact do these structural variations have on the genes present in the human genome? The 6000 deletions result in the loss of 13 initiation codons and 16 termination codons, meaning that the start or end points of these genes have changed since the split between humans and chimpanzees. A further 61 deletions result in loss of an exon, and the regulatory regions of 479 genes are predicted to be affected by insertion or deletion. About half of the segmental duplications contain genes or parts of genes. These include two sets of duplications in chromosome 17 that result in the *TBC1D3* gene, which is present only in a great apes and is a single-copy gene in chimpanzees, becoming multicopy in humans (**Figure 18.29**). The biochemical function of TBC1D3 is not clear though the protein has been implicated in generation of the outer radial glial cells leading to expansion of the frontal cortex of the brain.

Duplications have also been implicated in the generation of other human-specific genome features that might be relevant to brain development. The *SRGAP2* gene, which is involved in cortical development, has undergone two consecutive duplications in the human lineage, giving rise to daughter and granddaughter copies of the parent gene. The granddaughter duplicate codes for a truncated version of the SRGAP2 protein, which can form a heterodimer with the full-length product of the parent gene. Unlike the homodimers formed by two complete SRGAP2 proteins, the heterodimer is nonfunctional. Formation of the heterodimer therefore sequesters full-length SRAP2 copies, modulating the activity of the protein. The result is an increase in synapse density in the human neocortex. The segmental duplication giving rise to the truncated granddaughter gene is thought to have occurred between 2 and 3 million years ago, just before the emergence of the *Homo* genus (see Figure 18.27), at a period when the human brain began to increase in size.

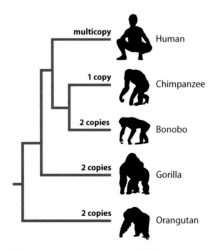

Figure 18.29 Increase in *TBC1D3* copy number in the human lineage. The primate phylogeny shows how the copy number of the *TBC1D3* gene has increased specifically in the human lineage.

Other investigations have shown that a protein domain called DUF1220 has undergone amplification in the human genome. This domain comprises 65 amino acids that are coded by two adjacent exons. The exon pair has a copy number of 212 in the human genome, compared with 34 in chimpanzee and just one in mouse. DUF1220 domains are mainly found in the neuroblastoma breakpoint family of genes, some members of which have been duplicated in the human genome and some of which have undergone expansion because of intragenic amplification of the DUF1220 exons. The functions of the members of the gene family are not known, but they have been associated with brain size and cortical neuron number.

The assumption that the answer to what makes us human must lie in brain development and function has meant that genes involved in neural processes have also been the focus when SNPs in the human and chimpanzee genomes are compared. An example is the gene for the FOXP2 transcription factor. Defects in this protein result in the human disability called dysarthria, characterized by a difficulty in articulating speech, suggesting that the *FOXP2* gene might be involved in the human ability for language. There are two amino acid differences between the FOXP2 proteins in humans and chimpanzees, and there is evidence that there has been positive selection of the gene in the human lineage. The latter is an important observation because genes under positive selection are likely to be ones that have contributed to the recent phenotypic evolution of the species. Further indications of the importance of FOXP2 were obtained from studies of mice whose own *FOXP2* genes had been modified by directed mutations that introduced the amino acid substitutions present in the human protein. These mice had increased neuron growth in the striatum of the basal ganglia, the part of the forebrain associated with speech in humans. Mice use a range of acoustic vocalizations to communicate with one another: most remarkably, the mice containing the humanized *FOXP2* gene generated patterns of ultrasonic vocalizations not recorded from other mice.

As well as changes to the DNA sequence, alterations in gene expression patterns could quite possibly underlie the unique capabilities of the human genome. At least 100 genes, and possibly many more, give rise to human-specific transcripts via novel alternative promoters or splicing pathways. Characterization of these and other changes in gene expression are less advanced than the research described above, but is likely to make an increasing contribution to our understanding of the species-specific features of the human genome during the next few years. Equally important are recent studies of similarities and differences between the organization of DNA in human and chimpanzee nuclei. Chromosome conformation capture (Section 10.1) has revealed that at the detailed level, the contacts made between different segments of DNA are very similar in humans and chimpanzees, with most associations between segments <10 kb apart conserved in the two species. In contrast, there are significant higher-order differences, which means that the boundary positions for topologically associating domains are generally dissimilar. These dissimilarities can be correlated with differences in the expression patterns of genes located in those TADs.

Paleogenomics is helping us understand the recent evolution of the human genome

One of the latest advances in genomics research is the ability to obtain sequences from extinct species by analysis of the 'ancient' DNA fragments that are sometimes preserved in bones and other remains. We have already seen how ancient DNA has been used to obtain a complete sequence of the genome of Neanderthals, the extinct hominins who inhabited many parts of Europe and Asia between 200,000 and 30,000 years ago (Section 4.3). Neanderthals are, arguably, a subspecies of *Homo sapiens*, and it is therefore no surprise that the Neanderthal genome is almost identical to our own genome, with 99.7% nucleotide similarity. What can such a closely related genome tell us about human evolution?

The most remarkable insight into our past that has been revealed by this branch of paleogenomics is that our ancestors interbred with Neanderthals. Evidence of interbreeding was obtained when comparisons were made between the Neanderthal genome and the genomes of present-day humans from Europe and Africa. If there had been no interbreeding, then we would expect the genomes of modern Europeans and Africans to display identical degrees of divergence when compared with the Neanderthal genome. In fact, the divergence between the Neanderthal and modern European genomes is slightly less than that between the genomes of Neanderthals and modern Africans, suggesting that some Neanderthal DNA has found its way into the genomes of modern Europeans. This finding indicates that there was a small amount of interbreeding between Neanderthals and *H. sapiens* during the 15,000 years or so that they were co-resident in Europe.

Ancient DNA sequencing has also identified an Asian version of Neanderthals, called the 'Denisovans.' The Denisovan genome sequence gives additional evidence for interbreeding with *H. sapiens*, in this case specifically with the ancestors of modern inhabitants of Oceania. The most recent estimates are that about 2.0% of the DNA of modern humans from outside of Africa is of Neanderthal origin, and 1.0% of the genomes of modern inhabitants of Oceania is derived from Denisovans (**Figure 18.30**). There is also evidence of interbreeding between Neanderthals and Denisovans, and between Denisovans and an unidentified extinct lineages of humans.

Has the influx of Neanderthal and Denisovan DNA had any significant effect on the human genome? Of course, the three genomes have the same genes, but it is possible that Neanderthals and/or Denisovans possessed alleles that were not present in the ancestral human population, and these alleles might have been acquired by humans via interbreeding. One intriguing possibility is that the ability of people from Tibet to withstand low oxygen conditions, such as those found in the high altitudes of the Tibetan plateau, were inherited from Denisovans. This hypothesis is prompted by the discovery that an allele of the *EPAS1* gene that is common in Tibetans is absent in human genomes from all other parts of the world. The EPAS1 protein is a transcription factor that controls the expression of

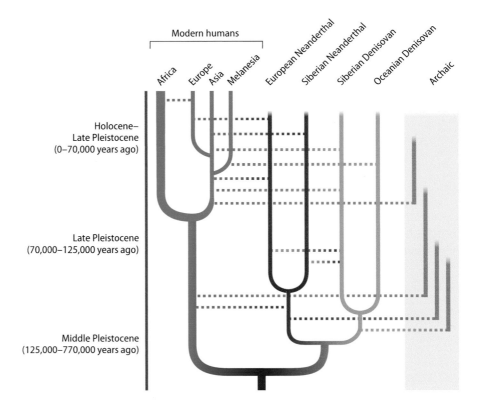

Figure 18.30 Interbreeding between modern and extinct lineages of humans. Anatomically modern human lineages are shown in blue, Neanderthals in red and Denisovans in green. The archaic lineages represent unidentified extinct humans. The horizontal lines indicate gene flow between populations.

genes for proteins that help the body tolerate anoxic conditions. The allele possessed by Tibetans, although absent in non-Tibetans, is present in the Denisovan genome, raising the possibility that its presence in Tibetans is the outcome of interbreeding between Denisovans and *H. sapiens* in the distant past. Other human phenotypes that may have been affected by introgression of Neanderthal or Denisovan alleles include skin pigmentation, the immune response, and lipid metabolism.

18.4 GENOMES TODAY: DIVERSITY IN POPULATIONS

We conclude our survey of genome evolution by exploring the snapshot of evolutionary time represented by the present day. Genomes continue to evolve through recombination and mutation, creating new variants of the DNA sequence and new arrangements of different sequence units. One outcome of this continuing process is that the genomes of most species display intraspecific variability. The human genome, for example, is not a single DNA sequence. Every individual, with the possible exception of identical twins, possesses their own unique genome sequence, the uniqueness determined by the identities of the nucleotides present at each of the $3–4 \times 10^8$ SNP positions in the genome, as well as the repeat lengths of each of the millions of SSLPs. A small minority of these personal variations fall in genes and so specify the phenotype of the individual, generating the characteristic features that distinguish one person from another. The vast majority of the polymorphisms are silent, but all of the variations, whether in coding or intergenic regions, provide an evolutionary record of the genome, comparisons between the variations enabling the relationships between individuals and groups of individuals to be inferred. In this section, we will explore how the diversity of genome sequences is used in research and in biotechnology. We will examine three case studies, on the origin of HIV/AIDS, the prehistoric migrations of *Homo sapiens*, and the breeding of new varieties of crop plant.

The origins of HIV/AIDS

The global epidemic of human immunodeficiency virus infection and acquired immune deficiency syndrome (HIV/AIDS) has touched everyone's lives. AIDS is the final stage of infection by human immunodeficiency virus 1 (HIV-1), a retrovirus (Section 9.1) that attacks cells involved in the immune response. The suppression of the immune system that occurs in patients who progress to this final stage increases the risk of opportunistic infections and tumor growth that can lead to death.

The demonstration in the early 1980s that HIV-1 is responsible for AIDS was quickly followed by speculation about the origin of the disease. Speculation centered around the discovery that similar immunodeficiency viruses are present in primates such as the chimpanzee, sooty mangabey, mandrill, and various monkeys. These simian immunodeficiency viruses (SIVs) are not pathogenic in their normal hosts, but it was thought that if one had become transferred to humans then within this new species the virus might have acquired new properties, such as the ability to cause disease and to spread rapidly through the population.

Retrovirus genomes accumulate mutations relatively quickly because reverse transcriptase, the enzyme that copies the RNA genome contained in the virus particle into the DNA version that integrates into the host genome (Section 9.1), lacks an efficient proofreading activity and so tends to make errors when it carries out RNA-dependent DNA synthesis. This means that retrovirus genomes that diverged quite recently display sufficient nucleotide dissimilarity for an evolutionary analysis to be carried out.

Various methods can be used to deduce the evolutionary relationships between genomes. If the genome sequences are relatively short, as is the case with retroviruses, and they have not been rearranged by recombination, then

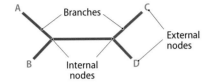

Figure 18.31 A simple phylogenetic tree.

a **phylogenetic tree** is usually constructed. The tree comprises a set of **external nodes**, each representing one of the genome sequences, linked by branches to **internal nodes** representing ancestral genome sequences (**Figure 18.31**). The lengths of the branches indicate the degrees of difference between the sequences represented by the nodes. The tree is generated by aligning the various sequences that are being examined (**Figure 18.32**) and then converting this **multiple alignment** into numerical data that can be analyzed mathematically in order to produce the tree. The simplest method involves conversion of the sequence data into a **distance matrix**, which is a table showing the evolutionary distances between all pairs of sequences; these distances are calculated from the number of nucleotide differences between each pair. The matrix is used by the tree-building software to establish the lengths of the branches connecting pairs of sequences in the tree that is drawn.

What do we discover when phylogenetic analysis is applied to HIV and SIV genome sequences? The tree that is obtained has a number of interesting features (**Figure 18.33**). First, it shows that different samples of HIV-1 have slightly different sequences, the samples as a whole forming a tight cluster, almost a star-like pattern, that radiates from one end of the tree. This star-like topology implies that the global AIDS epidemic began with a very small number of viruses, perhaps just one, which have spread and diversified since entering the human population. The closest relative to HIV-1 among primates is the SIV$_{cpzPtt}$ strain, which is found in the chimpanzee subspecies *Pan troglodytes*, the implication being that this virus jumped across the species barrier between chimpanzees and humans and initiated the HIV/AIDS epidemic. However, this epidemic did not begin immediately: a relatively long, uninterrupted branch links the center of the HIV-1 radiation with the internal node leading to the SIV$_{cpzPtt}$ sequence, suggesting that after transmission to humans, HIV-1 underwent a latent period when it remained restricted to a small part of the global human population, presumably in Africa, before beginning its rapid spread to other parts of the world. Other primate SIVs are less closely related to HIV-1, but one, the SIV$_{smm}$ from sooty mangabey, clusters in the tree with the second human immunodeficiency virus, HIV-2. It appears that HIV-2 was transferred to the human population independently of HIV-1, and from a different simian host. HIV-2 is also able to cause HIV/AIDS, but has not, as yet, become globally epidemic.

An intriguing addition to the HIV/SIV tree was made in 1998 when the RNA of an HIV-1 isolate from a blood sample taken in 1959 from an African male was sequenced. The RNA was highly fragmented and only a short sequence could be obtained, but this was sufficient for the sample to be placed on the phylogenetic tree (see Figure 18.33). The sequence, called ZR59, attaches to the tree by a short branch that emerges from near the center of the HIV-1 radiation. The positioning indicates that the ZR59 sequence represents one of the earliest versions of HIV-1 and shows that the global spread of HIV-1 was already underway by 1959. A later and more comprehensive analysis of HIV-1 sequences has suggested that the spread began in the period between 1915 and 1941, with a best estimate of 1931. Pinning down the date in this way has enabled epidemiologists to begin an investigation of the historic and social conditions that might have been responsible for the start of the HIV/AIDS epidemic.

Figure 18.32 A simple distance matrix. The matrix shows the evolutionary distance between each pair of sequences in the alignment.

Multiple alignment

1	A G G C C A A G C C A T A G C T G T C C
2	A G G C A A A G A C A T A C C T G A C C
3	A G G C C A A G A C A T A G C T G T C C
4	A G G C A A A G A C A T A C C T G T C C

Distance matrix

	1	2	3	4
1	–	0.20	0.05	0.15
2		–	0.15	0.05
3			–	0.10
4				–

The first migrations of humans out of Africa

Our second case study illustrating the ways in which intraspecific genome diversity is used in research concerns the prehistoric migrations that resulted in the expansion of *Homo sapiens* from our evolutionary homeland in Africa to our current global distribution. Paleoanthropologists refer to our species as **anatomically modern humans** (**AMH**), a term that sidesteps debates about the precise evolutionary relationships between us and our extinct ancestors. Transitional fossils with a mix of modern and archaic features have been found at Jebel Irhoud in Morocco, dating to 315,000 years ago, and Florisbad in South Africa from 259,000 years ago (**Figure 18.34**). The oldest fossils identified as genuine AMH are a collection of skulls, jaws, teeth, and leg bones that are 233,000 years in age and were discovered at the Omo Kibish site in south-western Ethiopia. Other early AMH fossils are present at Herto, also in Ethiopia, from 160,000 years ago, at Laetoli in Tanzania from 120,000 years ago, and at Border Cave, South Africa, about 110,000 years ago. At some point since the emergence of AMH in Africa, our species began the migrations that would take them to Asia, Europe, the Americas, Australasia and Oceania.

Initial attempts to use genomic data to follow the early migrations of humans out of Africa made use of mitochondrial DNA. In mammals such as humans, the molecular clock for mitochondrial DNA is faster than that for DNA in the nucleus, probably because mitochondria lack many of the DNA repair systems that operate in the nucleus, enabling a greater proportion of the mutations that occur to be retained. Each of these retained mutations results in a **substitution**, a permanent change in the nucleotide sequence. This means that within the human population there are a variety of mitochondrial DNA sequences, which on the basis of shared substitutions can be divided into different **haplogroups**. Each haplogroup can then be further subdivided into **haplotypes**, all of which have the characteristic substitutions that define the haplogroup, along with 'private' substitutions that distinguish each haplotype (**Figure 18.35**). The date when a haplogroup originated can be deduced by estimating its **coalescence time**. The reasoning is that the greater the diversity among the haplotypes within the haplogroup, then the larger the number of substitutions that has occurred, and the more ancient the coalescence time.

Mitochondrial DNA analysis has suggested that the first modern human migration out of Africa left from Ethiopia, further south than the area of modern Suez that forms the physical link between Africa and Asia. The basis for this hypothesis is as follows. All the mitochondrial DNA haplogroups that are known in the modern human population can be linked together in a phylogenetic network displaying their sequence relationships. Within this network, all of the haplogroups that are common today in Africa are clustered together, with just two links connecting them with the remainder of the tree. These two links are between haplogroup L3 on the African side and M and N on the non-African side (**Figure 18.36**). Coalescence analyses suggest that the M and N haplogroups originated 60,000–70,000 years ago. The modern Africans whose L3 mitochondrial DNAs have the greatest similarity to M and N live in or close to Ethiopia, so the migration probably originated in the area we now call Ethiopia. The most direct route from Ethiopia to Asia is across the Bab el Mandeb strait, at the entrance of the Red Sea, to the southern coast of Arabia (**Figure 18.37**). We therefore conclude that the initial migration out of Africa was from Ethiopia to southern Arabia, and occurred some 60,000–70,000 years ago.

Mitochondrial DNA studies therefore provide a model for the first human migrations from Africa, but to what extent is this a true interpretation of the past? The mitochondrial genome is just a small part of the human DNA, and its evolutionary pattern is unusual in that it is inherited solely through the maternal line, and does not recombine with paternal DNA during reproduction. Hypotheses based on mitochondrial DNA have been criticized as 'narratives' that may or may not give a true indication of the past evolutionary history of a species or population. So do studies of the complete human genome confirm or contradict

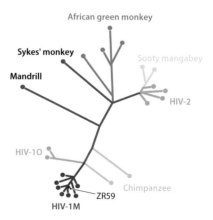

Figure 18.33 The phylogenetic tree reconstructed from HIV and SIV genome sequences. The HIV/AIDS epidemic is due to the HIV-1M type of immunodeficiency virus. The much rarer HIV-1N type (not shown in this tree) is closely related to HIV-1M and was probably also transferred from chimpanzees. HIV-1O, which accounts for 2% of cases from Cameroon, and HIV-1P (not shown), which has been isolated from just a single person, are both related to gorilla SIVs, but it is not known if these were transferred from gorillas to humans, or if both humans and gorillas acquired them from chimpanzees.

Figure 18.34 Locations of sites in Africa where early *Homo sapiens* fossils have been found.

Figure 18.35 Haplogroups and haplotypes. Each haplogroup is characterized by a set of nucleotide substitutions that distinguish it from other haplogroups. The haplotypes additionally have their own private substitutions.

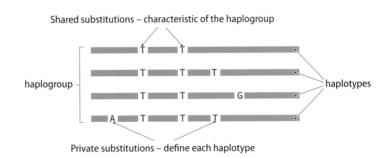

the scheme suggested by mitochondrial DNA? This is a rapidly developing area of research, with genome sequences being collected from more and more individuals from different parts of the world, and increasingly sophisticated methods being developed for the analysis of these sequences. While the datasets and methodologies are still reaching maturity, it is inevitable that different projects will yield conflicting results. However, the general consensus is that the conclusions of the mitochondrial DNA studies are broadly correct. A migration dating to approximately 70,000 years ago has also been identified by studies based on Y chromosome sequences and genome-wide microsatellite and SNP variations. For example, one project examined over 4 million SNPs in genomes from Australia, Africa, Europe, and East Asia. For each SNP, data from primate genomes was used to identify the **ancestral** and **derived** alleles, on the basis that the allele at the equivalent positions in the primate genomes must be the ancestral version. The derived **allele frequency** for each SNP was calculated and the combined data for all the SNPs compared between the different populations and with sequences from the Neanderthal and Denisovan genomes. The conclusion was that non-Africans originate from a single population that left Africa approximately 72,000 years ago, with subsequent divergence between Aboriginal Australians and Eurasians some 59,000 years ago, and a split between Europeans and East Asians 42,000 years ago (**Figure 18.38**).

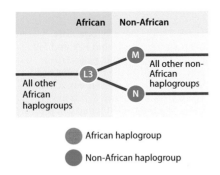

Figure 18.36 Mitochondrial DNA evidence for the first human migration out of Africa. There are just two links between the African and non-African parts of the human mitochondrial DNA haplogroup network. These are between haplogroup L3 on the African side and M and N on the non-African side.

The growing evidence from human genomics therefore suggests that there was a wave of migration out of Africa around 70,000 years ago, although there is still debate about whether this migration was actually across the Bab el Mandeb strait as suggested by the mitochondrial DNA studies, or through the land connection between Africa and Asia in modern Suez, or whether the two routes were used in parallel. However, there is a substantial amount of archaeological evidence suggesting that either this migration has been misdated by the genomic analyses, or there was an earlier migration that has left no record in the genomic data. The evidence includes fossils of modern humans discovered in the Qafzeh and Skhul caves near Nazareth in modern Israel dated to 90,000–100,000 years ago. The caves were reoccupied by Neanderthals at a later date, and modern humans were not seen there again for some time, which has led some researchers to suggest that the Qafzeh and Skhul people either were part of a failed attempt at migration, or merely represent an extension of the African population into Asia during a warm period. This interpretation might be correct for Qafzeh and Skhul, but there are increasing indications of modern human occupation in East Asia at about the same time, in the form of AMH fossils that are older than 70,000 years, with the most ancient dating to over 100,000 years ago. One particularly interesting find is a set of stone tools with features typical of those used by anatomically modern humans in Africa, that was discovered at a site in the Jurreru River valley of southern India. This assemblage is covered in ash from the eruption of the volcano at Toba in northern Sumatra, which can be reliably dated to 74,000 years ago. It seems certain that modern humans must have been widespread in southern and eastern Asia at that time, with the archaeological evidence as a whole implying a substantial migration out of Africa about 100,000 years ago. This conclusion is prompting reappraisal of the methods used to analyze the genomic data, with novel and more sophisticated methods for calculating the divergence time between different populations now being tested.

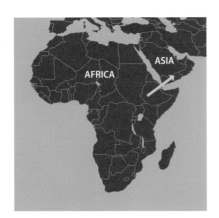

Figure 18.37 The route from Ethiopia to southern Arabia across the Bab el Mandeb strait. Examination of mitochondrial DNA haplotypes in modern Africans and Asians suggests that this route was the trajectory for the first migration of *Homo sapiens* out of Africa.

The diversity of plant genomes is an aid in crop breeding

The genomes of plant species also display intraspecific variability. Geneticists have used this variability to study the 'migrations' of plants, in particular the expansion of species from the **glacial refugia** that they occupied during the Ice Ages. These refugia were relatively warm geographical areas where plants, insects, and animals were able to survive the harsh conditions of the glaciations, prior to spreading back to the adjacent regions when the climate improved. In Europe, there were refugia in Iberia, Italy, and the Balkans, and in North America refugia are believed to have been present in the coastal regions of Alaska and British Columbia.

Population genomics is proving particularly important in research with crop plants. Agriculture began independently in at least three parts of the world – Mesoamerica, southwest Asia and southeast Asia – at about the same time, 10,000 years ago. In the **Fertile Crescent** of southwest Asia the early farmers grew barley, einkorn and emmer wheat, lentils, peas, chickpeas, and bitter vetch. Barley and wheat cultivation then began to expand out of the Fertile Crescent about 9000 years ago and, over the next 3000 years, spread throughout Europe, central and southern Asia, and north and east Africa. Genomic analyses, similar in concept to those used to trace human migrations, have distinguished the trajectories followed by the spread of crops, and identified areas where there was cross-breeding between the crop and local wild plants, introducing new beneficial alleles in to the cultivated population (**Figure 18.39**).

Continued adaptation of early crop plants was essential because the human-driven spread of agriculture exposed crops to a variety of environmental conditions, many of which were quite different from those present in the natural ranges of the wild progenitor species in southwest Asia. These environmental conditions included not only changes to the climate, but exposure to new insect and fungal pests as well as growth on new types of soil. The barley and wheat genomes evolved by adapting to the local environments in which the plants were grown, resulting in what botanists call **landraces**, populations of a crop plants that were locally adapted and which were grown throughout Europe until the 20th century, when they gradually were replaced with the products of modern crop breeding. The landraces were not entirely lost, however, as ardent plant collectors scoured the continent, collecting seeds which have been propagated in germplasm collections from which samples of these landraces can be obtained today.

The adaptations displayed by landraces are now being utilized in breeding programs that aim to improve modern cereal cultivars so that these are better able to withstand the changing climate. A problem in many of these breeding programs is that the genes that specify desirable traits, such as tolerance to drought or resistance to pests, are unidentified. This means that it is not possible to use DNA cloning techniques to transfer the interesting genes from one plant to another. Crop improvement therefore still depends very much on conventional

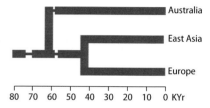

Figure 18.38 Relationships between human populations deduced by SNP analysis. The analysis suggests that all non-African populations originated from a single source population about 72,000 years ago, with subsequent splits between Aboriginal Australians and Eurasians 59,000 years ago and Europeans and East Asians 42,000 years ago. The narrow lines indicate population **bottlenecks**. Abbreviation: KYr, thousands of years ago.

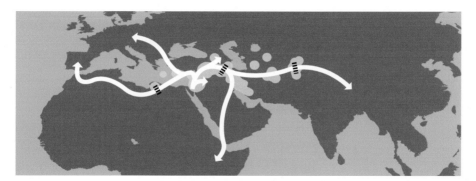

Figure 18.39 The spread of barley cultivation from the Fertile Crescent to Europe, Asia and Africa. Areas where wild barley is found are highlighted in yellow. The cross-hatching indicates places where cross-breeding between cultivated barley and local wild plants occurred as the crop spread into Asia and Africa.

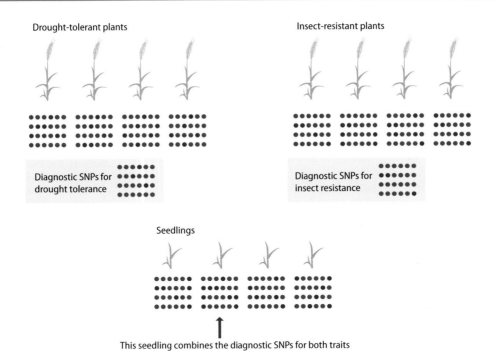

Figure 18.40 A simple association study. SNPs are typed in a set of plants that display drought tolerance. The microarray data identify three SNP alleles that are found in all plants with this trait. An additional three SNPs are then found that associate with insect resistance. Seedlings from a cross between drought-tolerant and insect-resistant parent plants are then examined. One seedling has the six SNPs that together are diagnostic of the two traits. This seedling is retained for further study and the others discarded.

breeding methods, in which two plants with different characteristics are crossed in the hope that at least some of the progeny will inherit the combination of desirable traits from the parents. This approach works, but it is a lengthy process that typically involves thousands of seedlings being generated whose characteristics are often not known until they have been grown to maturity and subjected to a range of physiological and biochemical tests. Genomics is now speeding up the process by making possible the approach called **marker-assisted selection**. Although the genes for a trait might be unknown, SNPs whose positions in the genome are close to the relevant genes can usually be identified by an **association study**, which searches for SNP alleles that are consistently present in plants with a particular trait (**Figure 18.40**). Being closely linked to the genes of interest, these SNPs cosegregate with those genes when different plants are crossed. This means that the SNP alleles can be used as proxies for the genes. A cross is carried out, and when the seedlings are obtained, leaf samples are taken, DNA prepared, and the diagnostic SNPs typed by DNA chip technology or some other appropriate high-throughput method (Section 3.2). The results enable those seedlings that do not contain SNP alleles that are associated with the desirable traits of the two parents to be discarded without additional time wasted on their propagation and phenotypic analysis. The research project can therefore move quickly forward with seedlings that have the diagnostic SNP pattern and which are therefore likely to possess the desired combination of phenotypes.

SUMMARY

- It is thought that the first polynucleotides to evolve, several billion years ago, were made of RNA rather than DNA.

- These RNA molecules probably combined a self-replicating ability with some enzymatic activity, and it is possible that enclosure of these in simple lipid envelopes gave rise to the progenitors of the first cells.

- DNA probably evolved as a more stable version of the initial RNA protogenomes.

- Life forms based on coding molecules other than DNA and RNA are unknown but are thought to be possible.

- Gene duplication is an important event that can result in a genome acquiring new genes. This is illustrated by the globin gene superfamily, which arose from a series of gene duplications whose pattern and timing can be inferred by making comparisons between the sequences of the globin genes in existence today.

- Duplication events have also played an important role during evolution of the homeotic selector genes that specify the body plan of eukaryotes. These duplication events can be linked to whole-genome duplications that have occurred in the past.

- Smaller duplications, 1–400 kb in length, have occurred regularly in the recent evolution of the human genome.

- Horizontal gene transfer results in the acquisition of genes from other species. This has been a regular event in the evolution of prokaryotic genomes and several examples are also known in unicellular eukaryotes.

- Plants can form new polyploid genomes by fusion of gametes from related species.

- Duplications and rearrangements of gene segments have resulted in new combinations of exons. Exons can also be transported around a genome by attachment to transposable elements.

- The origins of introns are unclear, but current evidence suggests that introns proliferated in nuclear genomes shortly after the evolution of the first eukaryotic cell.

- The epigenome has also evolved over time, but we are only just beginning to understand the processes that were involved.

- The genomes of humans and chimpanzees display considerable sequence similarities, with many genes giving rise to identical protein products. It is thought that point mutations in key genes, segmental duplications, and changes in gene expression patterns are responsible for the features of the human genome that give rise to the particular characteristics of our species.

- The ability to sequence ancient DNA from preserved fossils has revealed that *Homo sapiens* interbred with Neanderthals and Denisovans.

- The diversity of genome sequences within a species enables recent evolutionary events to be studied. The origins of HIV/AIDS have been deduced from phylogenetic studies of HIV and SIV genome sequences.

- The diversity of modern human genome sequences enables past migrations of human populations to be mapped.

- Plant breeders make use of marker-assisted selection to identify plants that combine desirable characteristics that have evolved in different populations of a single crop species.

SHORT ANSWER QUESTIONS

1. How might some of the amino acids, nucleotide bases, and sugars have been synthesized before life evolved?

2. Provide a timeline for the evolution of living organisms from the formation of the Earth to the appearance of the first hominins.

3. Describe the evidence indicating that the content of the human genome has been influenced by past gene duplication events.

4. Outline the mechanisms by which gene duplication might occur.

5. Describe the evidence indicating that the content of the human genome has been influenced by past whole-genome duplication events.

6. How have segmental duplications affected the genomes of primates?

7. Give examples of horizontal gene transfer into eukaryotic species.

8. In what ways can new genes arise by rearrangement of existing genes?

9. What is the exon theory of genes? How does this theory compare with modern theories for the origin of introns?

10. What are the key differences between the human and chimpanzee genomes?

11. What has ancient DNA sequencing revealed regarding the relationships between *Homo sapiens* and extinct types of human?

12. Describe how genome sequencing has been used to study the origin of HIV/AIDS.

13. How are genomic data used to infer the trajectories and timings of human migrations out of Africa?

14. Outline the methodology used in marker-assisted selection.

IN-DEPTH PROBLEMS

1. Are the examples of domain duplication and domain shuffling given on pp. 446–448 special cases or are they representative of genome evolution in general?

2. One of the initial publications of the draft human genome sequence (International Human Genome Sequencing Consortium [2001] Initial sequencing and analysis of the human genome. *Nature* 409: 860–921) suggested that between 113 and 223 human genes might have been acquired from bacteria by horizontal gene transfer. Subsequently, it was concluded that this interpretation is incorrect and these genes are not bacterial in origin. What was the evidence that supported horizontal transfer of these genes and why was this evidence subsequently discounted?

3. To what extent do you believe it will be possible to determine the genetic basis of the special attributes of humans from comparisons between the genome sequences of humans and other primates?

4. How reliable are molecular clocks?

5. What is the potential of ancient DNA in studies of human evolution?

FURTHER READING

The RNA world and the origins of genomes

Higgs, P.G. and Lehman, N. (2015) The RNA world: Molecular cooperation at the origins of life. *Nat. Rev. Genet.* 16:7–17.

Joyce, G.F. (2002) The antiquity of RNA-based evolution. *Nature* 418:214–221.

Lohse, P.A. and Szostak, J.W. (1996) Ribozyme-catalysed amino-acid transfer reactions. *Nature* 381:442–444.

Mann, A. (2021) Making headway with the mysteries of life's origins. *Proc. Natl. Acad. Sci. USA* 118:e2105383118.

Miller, S.L. (1953) A production of amino acids under possible primitive earth conditions. *Science* 117:528–529.

Orgel, L.E. (2000) A simpler nucleic acid. *Science* 290:1306–1307. *Pyranosyl nucleic acid.*

Powner, M.W., Gerland, B. and Sutherland, J.D. (2009) Synthesis of activated pyrimidine ribonucleotides in prebiotically plausible conditions. *Nature* 459:239–242.

Robertson, M.P. and Ellington, A.D. (1998) How to make a nucleotide. *Nature* 395:223–225.

Unrau, P.J. and Bartel, D.P. (1998) RNA-catalysed nucleotide synthesis. *Nature* 395:260–263.

Walker, S.I, Packard, N. and Cody, G.D. (2017) Re-conceptualizing the origins of life. *Philos. Trans. R. Soc. A* 375:20160337.

Wang, Y., Wang, Y., Song, D., et al. (2022) An RNA-cleaving threose nucleic acid enzyme capable of single point mutation discrimination. *Nat. Chem.* 14:350–359.

Gene, genome, and segmental duplications

Baertsch, R., Diekhans, M., Kent, W.J., et al. (2008) Retrocopy contributions to the evolution of the human genome. *BMC Genomics* 9:466.

Ciomborowska, J., Rosikiewicz, W., Szklarczyk, D., et al. (2012) 'Orphan' retrogenes in the human genome. *Mol. Biol. Evol.* 30:384–396.

Clark, J.W. and Donoghue, P.C.J. (2018) Whole-genome duplication and plant macroevolution. *Trends Plant Sci.* 23:933–945.

Hardison, R.C. (2012) Evolution of hemoglobin and its genes. *Cold Spring Harb. Perspect. Med.* 2:a011627.

MacKintosh, C. and Ferrier, D.E.K. (2018) Recent advances in understanding the roles of whole genome duplications in evolution. *F1000 Res.* 6:1623.

Pascual-Anaya, J., D'Aniello, S., Kuratani, S., et al. (2013) Evolution of *Hox* gene clusters in deuterostomes. *BMC Dev. Biol.* 13:26.

Sacerdot, C., Louis, A, Bon, C., et al. (2018) Chromosome evolution at the origin of the ancestral vertebrate genome. *Genome Biol.* 19:166.

Vollger, M.R., Guitart, X., Dischuck, P.C., et al. (2022) Segmental duplications and their variation in a complete human genome. *Science* 376:eabj6965.

Wolfe, K.H. (2015) Origin of the yeast whole-genome duplication. *PLoS Biol.* 13:e1002221.

Gene rearrangements

Jiang, N., Bao, Z., Zhang, X., et al. (2004) Pack-MULE transposable elements mediate gene evolution in plants. *Nature* 431:569–573.

Kazazian, H.H. (2000) L1 retrotransposons shape the mammalian genome. *Science* 289:1152–1153.

Horizontal gene transfer and allopolyploidization

Danchin, E.G.J. (2016) Lateral gene transfer in eukaryotes: Tip of the iceberg or of the ice cube? *BMC Biol.* 14:101.

Danchin, E.G.J., Guzeeva, E.A., Mantelin, S., et al. (2016) Horizontal gene transfer from bacteria has enabled the plant-parasitic nematode *Globodera pallida* to feed on host-derived sucrose. *Mol. Biol. Evol.* 33:1571–1579.

Danchin, E.G.J., Rosso, M.-N., Vieira, P., et al. (2010) Multiple lateral gene transfers and duplications have promoted plant parasitism ability in nematodes. *Proc. Natl Acad. Sci. USA* 107:17651–17656.

Feldman, M. and Levy, A.A. (2012) Genome evolution due to allopolyploidization in wheat. *Genetics* 192:763–774.

Van Etten, J. and Bhattacharya, D. (2020) Horizontal gene transfer in eukaryotes: Not if, but how much. *Trends Genet.* 36:915–925.

Origins of introns

de Souza, S.J., Long, M., Schoenbach, L., et al. (1996) Intron positions correlate with module boundaries in ancient proteins. *Proc. Natl Acad. Sci. USA* 93:14632–14636.

Doolittle, W.F. (2014) The trouble with (group II) introns. *Proc. Natl Acad. Sci. USA* 111:6536–6537.

Fedorov, A., Merican, A.F. and Gilbert, W. (2002) Large-scale comparison of intron positions among animal, plant, and fungal genes. *Proc. Natl Acad. Sci. USA* 99:16128–16133.

Gilbert, W. (1987) The exon theory of genes. *Cold Spring Harb. Symp. Quant. Biol.* 52:901–905.

Irimia, M. and Roy, S.W. (2014) Origin of spliceosomal introns and alternative splicing. *Cold Spring Harb. Perspect. Biol.* 6:a016071.

Rogozin, I.B., Carmel, L., Csuros, M., et al. (2012) Origin and evolution of spliceosomal introns. *Biol. Direct* 7:11.

Roy, S.W. (2003) Recent evidence for the exon theory of genes. *Genetica* 118:251–266.

Truong, D.M., Hewitt, F.C., Hanson, J.H., et al. (2015) Retrohoming of a mobile group II intron in human cells suggests how eukaryotes limit group II intron proliferation. *PLoS Genet.* 11:el005422.

Evolution of the epigenome

Hernando-Herraez, I., Prado-Martinez, J., Garg, P., et al. (2013) Dynamics of DNA methylation in recent human and great ape evolution. *PLoS Genet.* 9:el003763.

Lowdon, R.F., Jang, H.S. and Wang, T. (2016) Evolution of epigenetic regulation in vertebrate genomes. *Trends Genet.* 32:269–283.

Prendergast, J.G.D., Chambers, E.V. and Semple, C.A.M. (2014) Sequence-level mechanisms of human epigenome evolution. *Genome Biol. Evol.* 6:1758–1771.

Humans and other primates

Enard, W., Gehre, S., Hammerschmidt, K., et al. (2009) A humanized version of Foxp2 affects cortico-basal ganglia circuits in mice. *Cell* 137:961–971.

Eres, I.E., Luo, K., Hsiao, C.J., et al. (2019) Reorganization of 3D genome structure may contribute to gene regulatory evolution in primates. *PLoS Genet.* 15:e1008278.

Kronenberg, Z,N., Fiddes, I.T., Gordon, D., et al. (2018) High-resolution comparative analysis of great ape genomes. *Science* 360:eaar6343.

Mitchell, C and Silver, D.L. (2018) Enhancing our brains: Genomic mechanisms underlying cortical evolution. *Semin. Cell Dev. Biol.* 76:23–32.

O'Bleness, M.S., Dickens, C.M., Dumas, L.J., et al. (2012) Evolutionary history and genome organization of DUF1220 protein domains. *Genes Genomes Genet.* 2:977–986.

Rogers, J. and Gibbs, R.A. (2014) Comparative primate genomics: Emerging patterns of genome content and dynamics. *Nat. Rev. Genet.* 15:347–359.

Suntsova, M.V. and Buzdin, A.A. (2020) Differences between human and chimpanzee genomes and their implications in gene expression, protein functions and biochemical properties of the two species. *BMC Genom.* 21:535.

Paleogenomics and past human migrations
Ahlquist, K.D., Bañuelos, M.M., Funk, A., et al. (2021) Our tangled family tree: New genomic methods offer insight into the legacy of archaic admixture. *Genome Biol. Evol.* 13:evab115.

Llamas, B., Willerslev, E. and Orlando, L. (2017) Human evolution: A tale from ancient genomes. *Philos. Trans. R. Soc. Lond. B Biol. Sci.* 372:20150484.

Malaspinas, A.S., Westaway, M.C., Muller, C., et al. (2016) A genomic history of aboriginal Australia. *Nature* 538:207–214. *Deductions regarding the first human migrations out of Africa.*

López, S., van Dorp, L. and Hellenthal, G. (2015) Human dispersal out of Africa; a lasting debate. *Evol. Bioinform.* 11(S2):57–68.

Mellars, P. (2006) Going East: New genetic and archaeological perspectives on the modern human colonization of Eurasia. *Science* 313:796–800.

Origins of HIV/AIDS
Sharp. P.M. and Hahn, B.H. (2010) The evolution of HIV-1 and the origin of AIDS. *Philos. Trans. R. Soc. Lond. B Biol. Sci.* 365:2487–2494.

Zhu, T., Korber, B.T., Nahmias, A.J., et al. (1998) An African HIV-1 sequence from 1959 and implications for the origin of the epidemic. *Nature* 391:594–597.

Crop plants and marker-assisted selection
Civáň, P., Drosou, K., Armisen-Gimenez, D., et al. (2021) Episodes of gene flow and selection during the evolutionary history of domesticated barley. *BMC Genom.* 22:227.

Collard, B.C.Y. and Mackill, D.J. (2007) Marker-assisted selection: An approach for precision plant breeding in the twenty-first century. *Philos. Trans. R. Soc. B* 363:557–572.

Das, G., Patra, J.K. and Baek, K.-H. (2017) Insight into MAS: A molecular tool for development of stress resistant and quality of rice through gene stacking. *Front. Plant Sci.* 8:985.

GLOSSARY

–25 Box: A component of the bacterial promoter.

(6–4) lesion: A dimer between two adjacent pyrimidine bases in a polynucleotide, formed by ultraviolet irradiation.

(6–4) photoproduct photolyase: An enzyme involved in photoreactivation repair.

10 nm fiber: An unpacked form of chromatin consisting of nucleosome beads-on-a-string of DNA.

2 μm plasmid: A plasmid found in the yeast *Saccharomyces cerevisiae* and used as the basis for a series of cloning vectors.

2-aminopurine: A base analog that can cause mutations by replacing adenine in a DNA molecule.

2′-deoxyribose: The sugar component of a deoxyribonucleotide.

3′-transduction: Transfer of a segment of genomic DNA from one place to another caused by movement of a LINE element.

3′-OH terminus: The end of a polynucleotide that terminates with a hydroxyl group attached to the 3′-carbon of the sugar.

3′-untranslated region: The untranslated region of an mRNA downstream of the termination codon.

3′→5′ exonuclease: An exonuclease activity that removes nucleotide sequentially from the 3′ end of a polynucleotide.

30 nm fiber: A form of chromatin, seen only in cell extracts, consisting of a possibly helical array of nucleosomes in a fiber approximately 30 nm in diameter.

3C: The original version of chromosome conformation capture, used to determine if particular interactions that are thought to occur between different parts of a DNA molecule actually take place.

454 sequencing: A next-generation method that makes use of the pyrosequencing procedure.

5-bromouracil: A base analog that can cause mutations by replacing thymine in a DNA molecule.

5′-P terminus: The end of a polynucleotide that terminates with a mono-, di- or triphosphate attached to the 5′-carbon of the sugar.

5′-untranslated region: The untranslated region of an mRNA upstream of the initiation codon.

5′→3′ exonuclease: An exonuclease activity that removes nucleotides sequentially from the 5′ of a polynucleotide.

7SK RNA: A component of a protein–RNA complex that is indirectly involved in the control of transcription elongation.

7SL RNA: A component of the eukaryotic signal recognition particle.

A-DNA: A structural configuration of the double helix, present but not common in cellular DNA.

Ab initio **gene prediction**: Identification of putative genes by ORF scanning of a DNA sequence.

ABC model: A model for the genetic control of the identity of the whorls in a flower.

Ac/Ds transposon: A DNA transposon of maize.

Acceptor arm: Part of the structure of a tRNA molecule.

Acceptor site: The splice site at the 3′ end of an intron.

Accessory genome: The component of a prokaryotic genome comprising all those genes not present in the core genome.

Acridine dye: A chemical compound that causes a frameshift mutation by intercalating between adjacent base pairs of the double helix.

Activity flow: A graph that shows how information flows within a biological system.

Acylation: The attachment of a lipid side chain to a polypeptide.

Ada enzyme: An *Escherichia coli* enzyme that is involved in the direct repair of alkylation mutations.

***ada* regulon**: The set of genes that are switched on by the Ada enzyme.

Adaptor: A synthetic, double-stranded oligonucleotide used to attach sticky ends to a blunt-ended molecule.

Adenine: A purine base found in DNA and RNA.

Adenosine deaminase acting on RNA (ADAR): An enzyme that edits various eukaryotic mRNAs by deaminating adenosine to inosine.

Adenylate cyclase: The enzyme that converts ATP to cyclic AMP.

Affinity chromatography: A column chromatography method that makes use of a ligand that binds to the molecule being purified.

Affinity purification mass spectrometry (AP-MS): A method for identifying the members of a protein complex.

Agarose gel electrophoresis: Electrophoresis carried out in an agarose gel and used to separate DNA molecules between 100 bp and 50 kb in length.

Alarmone: One of the stringent response activators, ppGppp and pppGpp.

Alkaline phosphatase: An enzyme that removes phosphate groups from the 5′ ends of DNA molecules.

Alkylating agent: A mutagen that acts by adding alkyl groups to nucleotide bases.

Allele: One of two or more alternative forms of a gene.

Allele frequency: The frequency of an allele in a population.

Allele-specific oligonucleotide (ASO) hybridization: The use of an oligonucleotide probe to determine which of two alternative nucleotide sequences is contained in a DNA molecule.

Allopolyploidy: The result of interbreeding between different species, giving a polyploid nucleus derived from fusion between gametes of those two species.

Alphoid DNA: The tandemly repeated nucleotide sequences located in the centromeric regions of human chromosomes.

Alternative end-joining (a-EJ): A mechanism for repairing double-strand breaks in eukaryotes, distinct from the nonhomologous end-joining process.

Alternative exons: An alternative splicing scenario where the mRNA contains either of a pair of exons, but not both at the same time.

Alternative polyadenylation: The use of two or more different sites for polyadenylation of an mRNA.

Alternative promoter: One of two or more different promoters acting on the same gene.

Alternative site selection: An alternative splicing scenario where the usual donor or acceptor site is ignored and a second site is used in its place.

Alternative splicing: The production of two or more mRNAs from a single pre-mRNA by joining together different combinations of exons.

Alu: A type of SINE found in the genomes of humans and related mammals.

Alu-PCR: A clone fingerprinting technique that uses PCR to detect the relative positions of Alu sequences in cloned DNA fragments.

Amino acid: One of the monomeric units of a protein molecule.

Amino terminus: The end of a polypeptide that has a free amino group.

Aminoacyl or A site: The site in the ribosome occupied by the aminoacyl-tRNA during translation.

Aminoacyl-tRNA: A tRNA attached to an amino acid.

Aminoacyl-tRNA synthetase: An enzyme that catalyzes the aminoacylation of one or more tRNAs.

Aminoacylation: Attachment of an amino acid to the acceptor arm of a tRNA.

Amplification refraction mutation system (ARMS test): A technique for SNP typing, in which PCR is directed by a pair of primers, with one covering the position of the SNP.

Analytical protein array: A type of protein array that is used in protein profiling.

Anaphase-promoting complex/cyclosome (APC/C): A ubiquitin ligase that directs degradation of proteins that are active late in the cell cycle, so that these proteins are not carried over to the daughter cells resulting from mitosis.

Anatomically modern humans (AMH): Fossil members of our own species.

Ancestral allele: The version of an allele that is possessed by the common ancestor of a group of organisms.

Ancient DNA: DNA preserved in ancient biological material.

Annealing: Attachment of an oligonucleotide primer to a DNA or RNA template.

Antibody array: A protein array that carries a series of antibodies.

Anticodon: The triplet of nucleotides, at positions 34–36 in a tRNA molecule, that base-pairs with a codon in an mRNA molecule.

Anticodon arm: Part of the structure of a tRNA molecule.

Antigen: A substance that elicits an immune response.

Antitermination: A bacterial mechanism for regulating the termination of transcription.

Antiterminator protein: A protein that attaches to bacterial DNA and mediates antitermination.

AP (apurinic/apyrimidinic) site: A position in a DNA molecule where the base component of the nucleotide is missing.

AP endonuclease: An enzyme involved in base excision repair.

Apoptosis: Programmed cell death.

Archaea: One of the two main groups of prokaryotes, mostly found in extreme environments.

Artificial gene synthesis: Construction of an artificial gene from a series of overlapping oligonucleotides.

Ascospore: One of the haploid products of meiosis in an ascomycete such as the yeast *Saccharomyces cerevisiae*.

Ascus: The structure which contains the four ascospores produced by a single meiosis in the yeast *Saccharomyces cerevisiae*.

Association study: A method that attempts to identify genetic or DNA markers that are associated with a disease or other phenotype.

Attenuation: A process used by some bacteria to regulate expression of an amino acid biosynthetic operon in accordance with the levels of the amino acid in the cell.

AU–AC intron A type of intron found in eukaryotic nuclear genes:: the first two nucleotides in the intron are 5´–AU–3´ and the last two are 5´–AC–3´.

Autonomous consensus sequence (ACS): An 11 bp subdomain of a yeast origin of replication that is part of the origin recognition complex.

Autonomously replicating sequence (ARS): A DNA sequence, particularly from yeast, that confers replicative ability on a non-replicative plasmid.

Autopolyploid: A polyploid nucleus derived from fusion of two gametes from the same species, neither of which is haploid.

Autoradiography: The detection of radioactively labeled molecules by exposure to an X-ray-sensitive photographic film.

Autosome: A chromosome that is not a sex chromosome.

Auxotroph: A mutant microorganism that can grow only when supplied with a nutrient that is not needed by the wild type.

Avidin: A protein, from egg white, that has a high binding affinity for biotin.

B chromosome: A chromosome possessed by some individuals in a population, but not all.

B-DNA: The commonest structural conformation of the DNA double helix in living cells.

Backsplicing: A process that attaches the donor site of a downstream exon to an upstream acceptor site and gives rise to a circular RNA.

Backtracking: The reversal of an RNA polymerase a short distance along its DNA template strand.

Bacteria: One of the two main groups of prokaryotes.

Bacterial artificial chromosome (BAC): A high-capacity cloning vector based on the F plasmid of *Escherichia coli*.

Bacteriophage: A virus that infects a bacterium.

Bacteriophage P1 vector: A high-capacity cloning vector based on bacteriophage P1.

Bait: (1) A member of a protein complex that is used to capture other members of the complex. (2) One of a set of oligonucleotides used to capture particular DNA fragments during target enrichment.

Barcode: A DNA sequence that is unique to a particular DNA fragment or set of fragments, and which can be created by attaching a short adaptor molecule to the end of each fragment.

Barcode deletion strategy: A method that has been developed for the large-scale screening of deletion mutations in *Saccharomyces cerevisiae*.

Barr body: The highly condensed chromatin structure taken up by an inactivated X chromosome.

Basal promoter: The position within a eukaryotic promoter where the initiation complex is assembled.

Basal rate of transcription initiation: The number of productive initiations of transcription occurring per unit of time at a particular promoter.

Base analog: A compound whose structural similarity to one of the bases in DNA enables it to act as a mutagen.

Base editor: An enzyme capable of changing one nucleotide to another within a DNA molecule.

Base excision repair: A DNA repair process that involves excision and replacement of an abnormal base.

Base pair: The hydrogen-bonded structure formed by two complementary nucleotides. When abbreviated to 'bp,' the shortest unit of length for a double-stranded DNA molecule.

Base-pairing: The attachment of one polynucleotide to another, or one part of a polynucleotide to another part of the same polynucleotide, by base pairs.

Base ratio: The ratio of A to T, or G to C, in a double-stranded DNA molecule. Chargaff showed that the base ratios are always close to 1.0.

Base-stacking: The hydrophobic interactions that occur between adjacent base pairs in a double-stranded DNA molecule.

Baseless site: A position in a DNA molecule where the base component of the nucleotide is missing.

Basic helix–loop–helix: A DNA-binding domain.

Beads-on-a-string: Another name for the 10 nm fiber.

Biobank: A collection of biological material such as blood samples, donated following informed consent by patients and volunteers, often used to study the genetic basis of inherited disease.

Biochemical profiling: The study of metabolomes.

Bioinformatics: The use of computer methods in studies of genomes.

Biolistics: A means of introducing DNA into cells that involves bombardment with high-velocity microprojectiles coated with DNA.

Biological information: The information contained in the genome of an organism and which directs the development and maintenance of that organism.

Biotechnology: The use of living organisms, often, but not always microbes, in industrial processes.

Biotin: A molecule that can be incorporated into dUTP and used as a non-radioactive label for a DNA probe.

Biotinylation: Attachment of a biotin label to a DNA or RNA molecule.

Bisulfite sequencing: A sequencing method that can detect 5-methylcytidine modifications in RNA and DNA.

Bivalent: The structure formed when a pair of homologous chromosomes lines up during meiosis.

BLAST: An algorithm frequently used in homology searching.

Blunt end: An end of a double-stranded DNA molecule where both strands terminate at the same nucleotide position with no single-stranded extension.

Bootstrap analysis: A method for inferring the degree of confidence that can be assigned to a branch point in a phylogenetic tree.

Bootstrap value: The statistical value obtained by bootstrap analysis.

Bottleneck: A temporary reduction in the size of a population.

Bottom-up proteomics: A version of proteomics in which proteins are broken into peptides by treatment with a sequence-specific protease, such as trypsin, prior to mass spectroscopy.

Branch: A component of a phylogenetic tree.

Branch migration: A step in the Holliday model for homologous recombination, involving exchange of polynucleotides between a pair of recombining double-stranded DNA molecules.

Break repair: A process for the repair of single- or double-strand breaks in a DNA molecule.

Bubble-seq: A method used to identify the positions of replication origins in a eukaryotic genome sequence.

Buoyant density: The density possessed by a molecule or particle when suspended in an aqueous salt or sugar solution.

Bystander mutation: Unwanted mutations occurring adjacent to the target nucleotide when base editing is carried out.

C terminus: The end of a polypeptide that has a free carboxyl group.

C-terminal domain (CTD): A component of the largest subunit of RNA polymerase II, important in activation of the polymerase.

C-value paradox: The non-equivalence between genome size and gene number that is seen when comparisons are made between some eukaryotes.

Cajal body: A nuclear structure associated with synthesis of small nuclear and small nucleolar RNAs.

cAMP response element-binding (CREB) protein: A transcription factor that responds to elevated levels of cyclic AMP by binding to the CRE sequence.

Candidate gene: A gene, identified by experimental means, that might be a disease-causing or disease-susceptibility gene.

Cap analysis gene expression (CAGE): A method for the rapid acquisition of RNA-seq data.

Cap binding complex: The complex that makes the initial attachment to the cap structure at the beginning of the scanning phase of eukaryotic translation.

CAP site: A DNA-binding site for the catabolite activator protein.

Cap structure: The chemical modification at the 5′ end of most eukaryotic mRNA molecules.

Capillary electrophoresis: Polyacrylamide gel electrophoresis carried out in a thin capillary tube, providing high resolution.

Capping: Attachment of a cap to the 5′ end of a eukaryotic mRNA.

Capsid: The protein coat that surrounds the DNA or RNA genome of a virus.

Carboxyl terminus: The end of a polypeptide that has a free carboxyl group.

Cas9 endonuclease: A programmable nuclease that is directed to its target site by a 20-nucleotide guide RNA.

Cascade: A pathway comprising a series of proteins or other molecules, which passes a signal from the cell surface to genes and other targets within a cell.

Catabolite activator protein: A regulatory protein that binds to various sites in a bacterial genome and activates transcription initiation at downstream promoters.

Catabolite repression: The means by which extracellular glucose levels dictate whether genes for sugar utilization are switched on or off in bacteria.

Catenane: An interlocked structure comprising one or more intertwined rings.

CCCTC-binding factor (CTCF): A mammalian insulator-binding protein.

cDNA: A double-stranded DNA copy of an mRNA molecule.

cDNA capture or cDNA selection: Repeated hybridization probing of a pool of cDNAs with the objective of obtaining a subpool enriched in certain sequences.

Cell cycle: The series of events occurring in a cell between one division and the next.

Cell cycle checkpoint: A period before entry into the S or M phase of the cell cycle, a key point at which regulation is exerted.

Cell senescence: The period in a cell lineage when the cells are alive but no longer able to divide.

Cell transformation: The alteration in morphological and biochemical properties that occurs when an animal cell is infected by an oncogenic virus.

Cell-free protein synthesizing system: A cell extract containing all the components needed for protein synthesis and able to translate added mRNA molecules.

centiMorgan: The unit used to describe the distance between two genes on a chromosome; 1 cM is the distance that corresponds to a 1% probability of recombination in a single meiosis.

Centromere: The constricted region of a chromosome that is the position at which the pair of chromatids are held together.

Chain-termination method: A DNA sequencing method that involves enzymatic synthesis of polynucleotide chains that terminate at specific nucleotide positions.

Chaperonin: A multisubunit protein that forms a structure that aids the folding of other proteins.

Chemical degradation sequencing: A DNA sequencing method that involves the use of chemicals that cut DNA molecules at specific nucleotide positions.

Chemical modification: Modification of a protein or RNA by addition of novel chemical groups.

Chemical shift: The change in the rotation of a chemical nucleus, used as the basis of NMR.

Chemiluminescent marker: A chemiluminescent chemical group incorporated into or attached to a molecule and whose chemiluminescent emissions are subsequently used to detect and follow that molecule during a biochemical reaction.

Chi (crossover hotspot initiation) site: A repeated nucleotide sequence in the *Escherichia coli* genome that is involved in the initiation of homologous recombination.

Chi form: An intermediate structure seen during recombination between DNA molecules.

Chimera: An organism composed of two or more genetically different cell types.

ChIP-on-chip or ChIP-chip: A microarray-based version of chromatin immunoprecipitation sequencing.

Chloroplast: One of the photosynthetic organelles of a eukaryotic cell.

Chloroplast genome: The genome present in the chloroplasts of a photosynthetic eukaryotic cell.

Chromatid: The arm of a chromosome.

Chromatin: The complex of DNA and proteins found in chromosomes.

Chromatin immunoprecipitation sequencing (ChIP-seq): A method for identifying the positions where individual DNA-binding proteins attach to a genome.

Chromatophore: Photosynthetic organelles that retain a miniature version of a cyanobacterium genome.

Chromatosome: A subcomponent of chromatin made up of a nucleosome core octamer with associated DNA and a linker histone.

Chromid: A bacterial DNA molecule that has the characteristic features of a plasmid but which carries essential genes.

Chromodomain: A protein structural motif that is not a DNA-binding domain but is often found in proteins with DNA-binding ability.

Chromosomal interaction domain (CID): A loop of DNA, attached to a protein core, which forms a structural component of a prokaryotic nucleoid.

Chromosome: One of the DNA–protein structures that contains part of the nuclear genome of a eukaryote. Less accurately, the DNA molecule(s) that contain(s) a prokaryotic genome.

Chromosome conformation capture: A method for identifying regions of chromosomes that are located close to one another in the nucleus.

Chromosome painting: A version of fluorescent *in situ* hybridization in which the hybridization probe is a mixture of DNA molecules, each specific for different regions of a single chromosome.

Chromosome territory: The region of a nucleus occupied by a single chromosome.

Chromosome theory: The theory, first propounded by Sutton in 1903, that genes lie on chromosomes.

Chromosome walking: A technique that can be used to construct a clone contig by identifying overlapping fragments of cloned DNA.

***Cis*-displacement:** Movement of a nucleosome to a new position on a DNA molecule.

Class switching: A process that results in a complete change in the type of immunoglobulin synthesized by a B lymphocyte.

Cleavage and polyadenylation specificity factor (CPSF): A protein that plays an ancillary role during polyadenylation of eukaryotic mRNAs.

Cleavage stimulation factor (CstF): A protein that plays an ancillary role during polyadenylation of eukaryotic mRNAs.

Clone: A group of cells that contain the same recombinant DNA molecule.

Clone contig: A collection of clones whose DNA fragments overlap.

Clone contig approach: A genome sequencing strategy in which the molecules to be sequenced are broken into manageable segments, each a few hundred kb or a few Mb in length, which are sequenced individually.

Clone fingerprinting: Any one of several techniques that compare cloned DNA fragments in order to identify ones that overlap.

Clone library: A collection of clones, possibly representing an entire genome, from which individual clones of interest are obtained.

Cloning vector: A DNA molecule that is able to replicate inside a host cell and therefore can be used to clone other fragments of DNA.

Closed pan-genome: A pan-genome that is no longer increasing in size as new genomes are sequenced.

Closed promoter complex: The structure formed during the initial step in assembly of the transcription initiation complex. The closed promoter complex consists of the RNA polymerase and/or accessory proteins attached to the promoter, before the DNA has been opened up by breakage of base pairs.

Cloverleaf: A two-dimensional representation of the structure of a tRNA molecule.

Clustered regularly interspaced short palindromic repeats (CRISPRs): A type of bacterial repetitive DNA, made up of 20–50 bp sequences found in tandem arrays, with each pair of repeats separated by a spacer of similar length but with a unique sequence.

Co-immunoprecipitation (co-IP): Isolation of all the members of a protein complex with an antibody specific for just one of those proteins.

Coalescence time: An estimate of the time that has elapsed since a haplogroup first came into existence, based on the degree of divergence of the haplotypes in that haplogroup.

Coding RNA: An RNA molecule that codes for a protein; an mRNA.

Codominance: The relationship between a pair of alleles which both contribute to the phenotype of a heterozygote.

Codon: A triplet of nucleotides coding for a single amino acid.

Codon bias: Refers to the fact that not all codons are used equally frequently in the genes of a particular organism.

Codon–anticodon recognition: The interaction between a codon on an mRNA molecule and the corresponding anticodon on a tRNA.

Cohesin: A ring-shaped protein complex that is able to generate loops of DNA.

Cohesive end: An end of a double-stranded DNA molecule where there is a single-stranded extension.

Cointegrate: An intermediate in the pathway resulting in replicative transposition.

Column chromatography: A method for the separation of compounds that makes use of a resin contained in a column.

Comorbidity: The tendency for patients suffering from one disease to display symptoms associated with other diseases.

Comparative epigenomics: Examination of the extent to which the equivalent regions of two different genomes display the same pattern of chromatin modification.

Comparative genomics: A research strategy that uses information obtained from the study of one genome to make inferences about the map positions and functions of genes in a second genome.

Competent: Refers to a culture of bacteria that have been treated, for example, by soaking in calcium chloride, so that their ability to take up DNA molecules is enhanced.

Competing endogenous RNA (ceRNA): An RNA molecule that contains binding sites for miRNAs or regulatory proteins, and competes with other RNAs for attachment of these molecules.

Complementary: Refers to two nucleotides or nucleotide sequences that are able to base-pair with one another.

Complementary DNA (cDNA): A double-stranded DNA copy of an mRNA molecule.

Complex A: The prespliceosome complex.

Complex B: The precatalytic spliceosome.

Complex E: The first protein–RNA complex formed during splicing of a GU–AG intron.

Composite transposon: A DNA transposon comprising a pair of insertion sequences flanking a segment of DNA, usually containing one or more genes.

Concatemer: A DNA molecule made up of linear genomes or other DNA units linked head to tail.

Concerted evolution: The evolutionary process that results in the members of a multigene family retaining the same or similar sequences.

Conjugation: Transfer of DNA between two bacteria that come into physical contact with one another.

Conjugation mapping: A technique for mapping bacterial genes by determining the time it takes for each gene to be transferred during conjugation.

Consensus sequence: A nucleotide sequence that represents an 'average' of a number of related but non-identical sequences.

Conservative replication: A hypothetical mode of DNA replication in which one daughter double helix is made up of the two parental polynucleotides and the other is made up of two newly synthesized polynucleotides.

Conservative transposition: Transposition that does not result in copying of the transposable element.

Constitutive heterochromatin: Chromatin that is permanently in a compact organization.

Context-dependent codon reassignment: Refers to the situation whereby the DNA sequence surrounding a codon changes the meaning of that codon.

Contig: A contiguous set of overlapping DNA sequences or clones.

Contour-clamped homogeneous electric fields (CHEF): An electrophoresis method used to separate large DNA molecules.

Conventional pseudogene: A gene that has become inactive because of the accumulation of mutations.

COOH-terminus: The end of a polypeptide that has a free carboxyl group.

Core genome: The component of a bacterial pan-genome that contains the set of genes possessed by all members of the species.

Core octamer: The central component of a nucleosome, made up of two sub-units each of histones H2A, H2B, H3, and H4, around which DNA is wound.

Core promoter: The position within a eukaryotic promoter where the initiation complex is assembled.

cos **site:** One of the cohesive, single-stranded extensions present at the ends of the DNA molecules of certain strains of λ phage.

Cosmid: A high-capacity cloning vector consisting of the λ *cos* site inserted into a plasmid.

Cotranscriptional editing: Small-scale editing of viral RNAs.

Cotransduction: Transfer of two or more genes from one bacterium to another via a transducing phage.

Cotransformation: Uptake of two or more genes on a single DNA molecule during transformation of a bacterium.

CpG island: A GC-rich DNA region located upstream of approximately 56% of the genes in the human genome.

Crossing over: The exchange of DNA between chromosomes during meiosis.

Cryptic splice site: A site whose sequence resembles an authentic splice site and which might be selected instead of the authentic site during aberrant splicing.

Cryptogene: A gene that lacks some of the nucleotides present in the mature RNA. The mature RNA is generated from the pre-RNA by pan-editing.

Cyanelle: A photosynthetic organelle that resembles an ingested cyanobacterium.

Cyclic AMP (cAMP): A modified version of AMP in which an intramolecular phosphodiester bond links the 5′ and 3′ carbons.

Cyclic AMP response element (CRE): The binding site for the cAMP response element-binding (CREB) protein.

Cyclin: A regulatory protein whose abundance varies during the cell cycle and which regulates biochemical events in a cell-cycle-specific manner.

Cyclobutyl dimer: A dimer between two adjacent pyrimidine bases in a poly-nucleotide, formed by ultraviolet irradiation.

Cys$_2$His$_2$ zinc finger: A type of zinc-finger DNA-binding domain.

Cytochemistry: The use of compound-specific stains, combined with micros-copy, to determine the biochemical content of cellular structures.

Cytokines: Protein involved in cell-cell signaling.

Cytosine: One of the pyrimidine bases found in DNA and RNA.

D arm: Part of the structure of a tRNA molecule.

D-loop: An intermediate structure formed during the Meselson–Radding model for homologous recombination. Also, a region of approximately 500 bp where the double helix is disrupted by the presence of an RNA molecule base-paired to one of the DNA strands, and which acts as the origin for the displacement mode of replication.

Dark repair: A type of nucleotide excision repair process that corrects cyclobutyl dimers.

de Bruijn graph: A computational approach, based on a mathematical concept for identifying overlaps between strings of symbols, used by some types of sequence assembler.

De novo **methylation:** Addition of methyl groups to new positions on a DNA molecule.

De novo **sequencing:** A strategy in which a genome sequence is assembled solely by finding overlaps between individual sequence reads.

Deadenylation-dependent decapping: A process for degradation of eukaryotic mRNAs that is initiated by removal of the poly(A) tail.

Deaminating agent: A mutagen that acts by removing amino groups from nucle-otide bases.

Decoy RNA: An RNA molecule that contains binding sites for regulatory proteins or RNAs, and competes with other RNAs for attachment of these molecules.

Defective retrovirus: A retrovirus whose genome contains a cellular gene, which replaces part or all of a retrovirus gene, so the virus is unable to replicate without the use of proteins from other retroviruses.

Degeneracy: Refers to the fact that the genetic code has more than one codon for most amino acids.

Degradosome: A multienzyme complex responsible for degradation of bacterial mRNAs.

Delayed onset mutation: A mutation whose effect is not apparent until a relatively late stage in the life of the mutant organism.

Deletion cassette: A segment of DNA that is transferred to a yeast chromosome by homologous recombination in order to create a deleted version of a target gene, in order to inactivate that gene and identify its function.

Deletion mutation: A mutation resulting from deletion of one or more nucleotides from a DNA sequence.

Denaturation: Breakdown by chemical or physical means of the noncovalent interactions, such as hydrogen bonding, that maintain the secondary and higher levels of structure of proteins and nucleic acids.

Dendrogram: A tree that is drawn to indicate the relationships between, for example, a group of transcriptomes.

Density gradient centrifugation: A technique in which a cell fraction is centrifuged through a dense solution, in the form of a gradient, so that individual components are separated.

Deoxyribonuclease: An enzyme that cleaves phosphodiester bonds in a DNA molecule.

Deoxyribonuclease (DNase I) hypersensitive site: A short region of eukaryotic DNA that is relatively easily cleaved with deoxyribonuclease I, possibly coinciding with positions where nucleosomes are absent.

Derived allele: An allele that arises in a population by mutation of an existing allele.

Development: A coordinated series of transient and permanent changes that occurs during the life history of a cell or organism.

Diauxie: The phenomenon whereby a bacterium, when provided with a mixture of sugars, uses up one sugar before beginning to metabolize the second sugar.

Dicer: The ribonuclease that plays a central role in RNA interference.

Dideoxynucleotide: A modified nucleotide that lacks the 3′ hydroxyl group and so terminates strand synthesis when incorporated into a polynucleotide.

Differential centrifugation: A technique that separates cell components by centrifuging an extract at different speeds.

Differentiation: The adoption by a cell of a specialized biochemical and/or physiological role.

Dihybrid cross: A sexual cross in which the inheritance of two pairs of alleles is followed.

Dimer: A protein or other structure that comprises two subunits.

Diol: A compound containing two hydroxyl groups.

Diploid: A nucleus that has two copies of each chromosome.

Direct labeling enzyme: An enzyme that transfers a labeled group directly to a DNA molecule.

Direct readout: The recognition of a DNA sequence by a binding protein that makes contacts with the outside of a double helix.

Direct repair : A DNA repair system that acts directly on a damaged nucleotide.

Direct repeat: A nucleotide sequence that is repeated twice or more frequently in a DNA molecule.

Directed acyclic graph (DAG): A device used to give a hierarchical categorization of a molecular function.

Discontinuous gene: A gene that is split into exons and introns.

Dispersive replication: A hypothetical mode of DNA replication in which both polynucleotides of each daughter double helix are made up partly of parental DNA and partly of newly synthesized DNA.

Displacement replication: A mode of replication which involves continuous copying of one strand of the helix, the second strand being displaced and subsequently copied after synthesis of the first daughter strand has been completed.

Distance matrix: A table showing the evolutionary distances between all pairs of nucleotide sequences in a dataset.

Disulfide bond: A covalent bond linking cysteine amino acids on different polypeptides or at different positions on the same polypeptide.

DNA: Deoxyribonucleic acid, one of the two forms of nucleic acid in living cells; the genetic material for all cellular life forms and many viruses.

DNA adenine methylase (Dam): An enzyme involved in methylation of *Escherichia coli* DNA.

DNA bending: A type of conformational change introduced into a DNA molecule by a binding protein.

DNA chip: A high-density array of DNA molecules used for parallel hybridization analyses.

DNA cloning: Insertion of a fragment of DNA into a cloning vector, and subsequent propagation of the recombinant DNA molecule in a host organism.

DNA cytosine methylase (Dcm): An enzyme involved in methylation of *Escherichia coli* DNA.

DNA glycosylase: An enzyme that cleaves the β-*N*-glycosidic bond between a base and the sugar component of a nucleotide as part of the base excision and mismatch repair processes. The name is a misnomer and should be *DNA glycolyase*, but the incorrect usage is now embedded in the literature.

DNA gyrase: A Type II topoisomerase of *Escherichia coli*.

DNA kink: An abrupt alteration in the linearity of the helix that occurs when the stacking between two adjacent bases is disrupted.

DNA labeling: Attachment of a radioactive, fluorescent, or other marker to a DNA molecule.

DNA ligase: An enzyme that synthesizes phosphodiester bonds as part of DNA replication, repair and recombination processes.

DNA marker: A DNA sequence that exists as two or more readily distinguished versions and which can therefore be used to mark a map position on a genetic, physical or integrated genome map.

DNA methylation: Refers to the chemical modification of DNA by attachment of methyl groups.

DNA methyltransferase: An enzyme that attaches methyl groups to a DNA molecule.

DNA photolyase: A bacterial enzyme involved in photoreactivation repair.

DNA polymerase: An enzyme that synthesizes DNA.

DNA polymerase I: The bacterial enzyme that completes synthesis of Okazaki fragments during genome replication.

DNA polymerase II: A bacterial DNA polymerase involved in DNA repair.

DNA polymerase III: The main DNA replicating enzyme of bacteria.

DNA polymerase α: The enzyme that primes DNA replication in eukaryotes.

DNA polymerase γ: The enzyme responsible for replication of the mitochondrial genome.

DNA polymerase δ: The enzyme responsible for replication of the lagging DNA strand in eukaryotes.

DNA polymerase ε: The enzyme responsible for replication of the leading DNA strand in eukaryotes.

DNA profile: The banding pattern revealed after electrophoresis of the products of PCRs directed at a range of microsatellite loci.

DNA repair: The biochemical processes that correct mutations arising from replication errors and the effects of mutagenic agents.

DNA replication: Synthesis of a new copy of the genome.

DNA sequencing: The technique for determining the order of nucleotides in a DNA molecule.

DNA topoisomerase: An enzyme that introduces or removes turns from the double helix by breakage and reunion of one or both polynucleotides.

DNA transposon: A transposon whose transposition mechanism does not involve an RNA intermediate.

DNA tumor virus: A virus with a DNA genome, able to cause cancer after infection of an animal cell.

DNA unwinding element (DUE): The AT-rich component of a bacterial origin of replication; the position at which helix melting occurs.

DNA-binding motif: The part of a DNA-binding protein that makes contact with the double helix.

DNA-binding protein: A protein that attaches to a DNA molecule.

DNA-dependent DNA polymerase: An enzyme that makes a DNA copy of a DNA template.

DNA-dependent RNA polymerase: An enzyme that makes an RNA copy of a DNA template.

Domain: A segment of a polypeptide that folds independently of other segments; also, the segment of a gene coding for such a domain.

Domain duplication: Duplication of a gene segment coding for a structural domain in the protein product.

Domain shuffling: Rearrangement of segments of one or more genes, each segment coding for a structural domain in the gene product, to create a new gene.

Dominant: The allele that is expressed in a heterozygote.

Donor site: The splice site at the 5′ end of an intron.

Double helix: The base-paired double-stranded structure that is the natural form of DNA in the cell.

Double heterozygote: A nucleus that is heterozygous for two genes.

Double homozygote: A nucleus that is homozygous for two genes.

Double restriction: Digestion of DNA with two restriction endonucleases at the same time.

Double-stranded: Comprising two polynucleotides attached to one another by base-pairing.

Double-strand break (DSB) model: A model of the events occurring during homologous recombination.

Double-strand break repair: A DNA repair process that mends double-stranded breaks.

Double-stranded RNA-binding domain (dsRBD): A common type of RNA-binding domain.

Downstream: Toward the 3′ end of a polynucleotide.

Draft sequence: An incomplete chromosome or genome sequence, typically containing some errors, gaps, and ambiguity about the order and/or orientation of some sequence contigs.

Duplicated pseudogene: A nonprocessed pseudogene that is the decayed version of one copy of a duplicated gene.

E site: A position within a bacterial ribosome to which a tRNA moves immediately after deacylation.

Earth BioGenome Project: A project that aims to sequence the genome of every eukaryotic species.

EC number: A four-part number describing the activity of an enzyme in accordance with the nomenclature set by the International Union of Biochemistry and Molecular Biology.

Edge: A line used to link pairs of interacting proteins in a protein interaction map.

Electroendosmosis: The motion of a liquid, such as the buffer in a gel, induced by an electric field.

Electron density map: A plot of the electron density at different positions within a molecule, deduced from an X-ray diffraction pattern.

Electron microscopy: A version of microscopy in which visualization of the sample is provided by a beam of electrons.

Electrophoresis: Separation of molecules on the basis of their net electrical charge.

Electrospray ionization: An ionization method used during mass spectrometry whereby a high voltage is applied to a liquid to create an aerosol.

Electrostatic interactions: Ionic bonds that form between charged chemical groups.

Elution: The unbinding of a molecule from a chromatography column.

Embryonic stem (ES) cell: A totipotent cell from the embryo of a mouse or other organism.

End-labeling: The attachment of a radioactive or other label to one end of a DNA or RNA molecule.

End-modification: The chemical alteration of the end of a DNA or RNA molecule.

End-modification enzyme: An enzyme used in recombinant DNA technology that alters the chemical structure at the end of a DNA molecule.

Endogenous retrovirus (ERV): An active or inactive retroviral genome integrated into a host chromosome.

Endonuclease: An enzyme that breaks phosphodiester bonds within a nucleic acid molecule.

Endosymbiont theory: A theory that states that the mitochondria and chloroplasts of eukaryotic cells are derived from symbiotic prokaryotes.

Enhancer: A regulatory sequence that increases the rate of transcription of a gene or genes located some distance away in either direction.

Ensembl: An online genome browser.

Entity description: A graph that depicts the interactions that occur between the different components of a biological system.

Epigenome: The various chemical modifications that are made to the DNA of a genome and to the nucleosomes attached to the DNA.

Episome: A plasmid that is able to integrate into the host cell's chromosome.

Episome transfer: Transfer between cells of some or all of a bacterial chromosome by integration into a plasmid.

Epitope tag: A short extension that is attached to a protein to provide a convenient means of immobilizing that protein.

Epitranscriptome: The entire collection of nucleotide modifications displayed by the RNAs in a transcriptome.

Error-prone gene editing: A version of genome editing that stimulates nonhomologous end-joining and results in a short insertion or deletion at the target site.

Ethidium bromide: A type of intercalating agent that causes mutations by inserting between adjacent base pairs in a double-stranded DNA molecule.

Ethylmethane sulfonate (EMS): A mutagen that acts by adding alkyl groups to nucleotide bases.

Euchromatin: Regions of a eukaryotic chromosome that are relatively uncondensed, thought to contain active genes.

Eukaryote: An organism whose cells contain membrane-bound nuclei.

Eulerian pathway: A pathway through a graph that visits each edge just once, used by some sequence assemblers to identify the correct sequence in regions of a genome containing repetitive DNA.

Excision repair: A DNA repair process that corrects various types of DNA damage by excising and resynthesizing a region of polynucleotide.

Exit site: A position within a bacterial ribosome to which a tRNA moves immediately after deacylation.

Exome: The sequences of all of the exons in a genome.

Exon: A coding region within a discontinuous gene.

Exon skipping: Aberrant splicing, or an alternative splicing scenario, in which one or more or exons are omitted from the spliced RNA.

Exon theory of genes: An 'introns early' hypothesis that holds that introns were formed when the first DNA genomes were constructed.

Exon trapping: A method, based on cloning, for identifying the positions of exons in a DNA sequence.

Exon–intron boundary: The nucleotide sequence at the junction between an exon and an intron.

Exonic splicing enhancer (ESE): A nucleotide sequence that plays a positive regulatory role during splicing of GU–AG introns.

Exonic splicing silencer (ESS): A nucleotide sequence that plays a negative regulatory role during splicing of GU–AG introns.

Exonuclease: An enzyme that removes nucleotides from the ends of a nucleic acid molecule.

Exosome: A multiprotein complex involved in degradation of mRNA in eukaryotes.

Expressed sequence tag (EST): A cDNA that is sequenced in order to gain rapid access to the genes in a genome.

Expression proteomics: The methodology used to identify the proteins in a proteome.

External node: The end of a branch in a phylogenetic tree, representing one of the organisms or DNA sequences being studied.

Extrachromosomal gene: A gene in a mitochondrial or chloroplast genome.

Extremophile: An organism that is able to live in an environment whose physical and/or chemical conditions are hostile to other organisms.

F plasmid: A fertility plasmid that directs conjugal transfer of DNA between bacteria.

Factor for inversion stimulation (Fis): A nucleoid-associated protein of prokaryotes.

Facultative heterochromatin: A type of chromatin that has a compact organization in some, but not all cells, thought to contain genes that are inactive in some cells or at some periods of the cell cycle.

FEN1: The 'flap endonuclease' involved in replication of the lagging strand in eukaryotes.

Fertile Crescent: The region of southwest Asia where barley and wheat are thought to have been domesticated.

Fiber-FISH: A specialized form of FISH that enables high marker resolution.

Field inversion gel electrophoresis (FIGE): An electrophoresis method used to separate large DNA molecules.

Filamentous: One of the capsid structures of a bacteriophage or virus.

Finished sequence: A chromosome or genome sequence that is almost complete, but typically still has some unsequenced gaps between contigs and an average of up to one error per 10^4 nucleotides.

Flap endonuclease (FEN1): An enzyme involved in replication of the lagging strand in eukaryotes.

Flow cell: A reaction chamber in which DNA sequencing, or some other biochemical reaction, is carried out.

Flow cytometry: A method for the separation of chromosomes.

Fluorescence recovery after photobleaching (FRAP): A technique used to study the mobility of nuclear proteins.

Fluorescent *in situ* hybridization (FISH): A technique for locating markers on chromosomes by observing the hybridization positions of fluorescent labels.

Fluorescent *in situ* RNA sequencing (FISSEQ): A spatial transcriptomics method that involves direct sequencing of RNAs in fixed tissue.

Fluorescent marker: A fluorescent chemical group incorporated into or attached to a molecule and whose fluorescent emissions are subsequently used to detect and follow that molecule during a biochemical reaction.

Flush end: An end of a double-stranded DNA molecule where both strands terminate at the same nucleotide position with no single-stranded extension.

fMet: *N*-formylmethionine, the modified amino acid carried by the tRNA that is used during the initiation of translation in bacteria.

Foldback RNA: The precursor RNA molecules which are cleaved to produce microRNAs.

Folding pathway: The series of events, involving partially folded intermediates, that results in an unfolded protein attaining its correct three-dimensional structure.

Footprinting: A range of techniques used for locating bound proteins on DNA molecules.

Förster resonance energy transfer (FRET): Energy transfer between two molecules, resulting in dye-quenching when a reporter probe is used.

Forward genetics: The conventional approach to genetics, where the researcher starts with a phenotype and attempts to discover the gene or genes responsible for that phenotype.

Forward sequence: One of the two directions in which a double-stranded DNA molecule can be sequenced.

Fosmid: A high-capacity vector carrying the F plasmid origin of replication and a λ *cos* site.

Fourier transform ion cyclotron resonance (FT-ICR) mass analyzer: A mass analyzer that incorporates an ion trap that captures individual ions and further excites them within a cyclotron, so they accelerate along an outward spiral, the vector of this spiral revealing the *m/z* ratio.

Fragile site: A position in a chromosome that is prone to breakage, possibly because it contains an expanded trinucleotide repeat sequence.

Fragment ion: An ion resulting from fragmentation of a molecule during the ionization phase of mass spectrometry.

Frameshift mutation: A mutation resulting from insertion or deletion of a group of nucleotides that is not a multiple of three and which therefore changes the frame in which translation occurs.

Functional RNA: RNA that has a functional role in the cell, i.e., RNAs other than mRNA.

Fusion protein: A protein that consists of a fusion of two polypeptides, or parts of polypeptides, normally coded by separate genes.

G-protein: a small protein that binds either a molecule of GDP or GTP, the replacement of GDP with GTP activating the protein.

G1 phase: The first gap period of the cell cycle.

G1-S checkpoint: A cell cycle checkpoint that a cell must pass before it is able to replicate its DNA.

G2 phase: The second gap period of the cell cycle.

G2-M checkpoint: A cell cycle checkpoint that can only be passed when a cell is ready to enter mitosis.

Gamete: A reproductive cell, usually haploid, that fuses with a second gamete to produce a new cell during sexual reproduction.

Gap genes: Developmental genes that play a role in establishing positional information within the *Drosophila* embryo.

Gap period: One of two intermediate periods within the cell cycle.

GATA zinc finger: A type of zinc-finger DNA-binding domain.

GC content: The percentage of nucleotides in a genome that are G or C.

Gel electrophoresis: Electrophoresis performed in a gel so that molecules of similar electrical charge can be separated on the basis of size.

Gel retardation analysis: A technique that identifies protein-binding sites on DNA molecules by virtue of the effect that a bound protein has on the mobility of the DNA fragments during gel electrophoresis.

Gel stretching: A technique for preparing restricted DNA molecules for optical mapping.

GenBank: An online repository of DNA sequences.

Gene: A DNA segment containing biological information and hence coding for an RNA and/or polypeptide molecule.

Gene cloning: Insertion of a fragment of DNA, containing a gene, into a cloning vector, and subsequent propagation of the recombinant DNA molecule in a host organism.

Gene conversion: A process that results in the four haploid products of meiosis displaying an unusual segregation pattern.

Gene desert: A region of genome in which here are few if any genes.

Gene duplication: The duplication of a gene to give two daughter copies, which initially will have identical nucleotide sequences, but whose sequences might subsequently change due to mutation.

Gene expression: The series of events by which the biological information carried by a gene is released and made available to the cell.

Gene flow: The transfer of a gene from one organism to another.

Gene fragment: A gene relic consisting of a short, isolated region from within a gene.

Gene ontology (GO): A scheme for describing gene function.

Gene space: A version of the barley genome, comprising the sequences of the vast majority of barley genes anchored on to a detailed genome map.

Gene superfamily: A group of two or more evolutionarily related multigene families.

Gene therapy: A clinical procedure in which a gene or other DNA sequence is used to treat a disease.

General recombination: Recombination between two homologous double-stranded DNA molecules.

General transcription factor (GTF): A protein or protein complex that is a transient or permanent component of the initiation complex formed during eukaryotic transcription.

Genes-within-genes: Refers to a gene whose intron contains a second gene.

Genetic code: The rules that determine which triplet of nucleotides codes for which amino acid during protein synthesis.

Genetic cross: A breeding experiment designed to establish the pattern of inheritance of one or more markers.

Genetic linkage: The physical association between two genes that are on the same chromosome.

Genetic mapping: The use of genetic techniques to construct a genome map.

Genetic marker: A gene that exists as two or more readily distinguished alleles and whose inheritance can therefore be followed during a genetic cross, enabling the map position of the gene to be determined.

Genetic modification: The use of experimental techniques to produce DNA molecules containing new genes or new combinations of genes in order to alter the phenotype of an organism in a defined way.

Genetic redundancy: The situation that occurs when two genes in the same genome perform the same function.

Genetics: The branch of biology devoted to the study of genes.

Genome: The entire genetic complement of a living organism.

Genome annotation: The process by which the genes, control sequences and other interesting features are identified in a genome sequence.

Genome browser: A software package or online system for display of an annotated genome sequence.

Genome editing: A method that enables directed changes to be made in a target gene or at some other position in a genome.

Genome expression: The series of events by which the biological information carried by a genome is released and made available to the cell.

Genome map: A chart showing the positions of genetic and/or physical markers in a genome.

Genome resequencing: Sequencing of multiple versions of a genome that has already been sequenced, in order to study the sequence variations that occur within a species or within populations of that species.

Genome streamlining: Reduction in genome size which may confer a reproductive advantage compared to the non-streamlined competitors in an ecosystem.

Genome-wide association study (GWAS): A method that attempts to identify all of the markers, from all over the genome, that are associated with a disease or other phenotype.

Genome-wide repeat: A sequence that recurs at many dispersed positions within a genome.

Genomic imprinting: Inactivation by methylation of a gene on one of a pair of homologous chromosomes.

Genotype: A description of the genetic composition of an organism.

Gigabase pair: 1,000,000 kb; 1,000,000,000 bp.

Glacial refugia: A geographical region occupied by various species during the Last Glacial Maximum, enabling those species to survive the glaciations and eventually recolonize adjacent regions following the warning of the planet.

Glycan: The oligosaccharide at a single glycosylated position in a glycoprotein.

Glycosylation: The attachment of sugar units to a polypeptide.

Greedy algorithm: A computational approach, based on making the most logical choice at each step in an iterative process, used by some types of sequence assembler.

Green fluorescent protein: A protein that is used to label other proteins and whose gene is used as a reporter gene.

Group I intron: A type of intron found mainly in organelle genes.

Group II intron: A type of intron found in organelle genes.

Group III intron: A type of intron found in organelle genes.

GTPase activating protein (GAP): A protein which inactivates a G-protein by stimulating it to convert its GTP molecule into a GDP molecule.

GU–AG intron: The commonest type of intron in eukaryotic nuclear genes. The first two nucleotides of the intron are 5′–GU–3′ and the last two are 5′–AG–3′.

Guanine: One of the purine nucleotides found in DNA and RNA.

Guanine methyltransferase: The enzyme that attaches a methyl group to the 5′ end of a eukaryotic mRNA during the capping reaction.

Guanine nucleotide exchange factor (GEF): A protein which activates a G-protein by replacing its GDP molecule with a GTP molecule.

Guanylyl transferase: The enzyme that attaches a GTP to the 5′ end of a eukaryotic mRNA at the start of the capping reaction.

Guide RNA: An RNA that attaches to a DNA or RNA target site and acts as a guide for the enzyme involved in an editing process.

Hairpin: A stem-loop structure made up of a base-paired stem and non-base-paired loop, which can form in a single-stranded polynucleotide that contains an inverted-repeat.

Half-life: The time needed for half the atoms or molecules in a sample to decay or be degraded.

Hammerhead: An RNA structure with ribozyme activity that is found in some viruses.

Haplogroup: One of the major sequence classes of mitochondrial DNA present in the human population.

Haploid: A nucleus that has a single copy of each chromosome.

Haploinsufficiency: The situation where inactivation of a gene on one of a pair of homologous chromosomes results in a change in the phenotype of the mutant organism.

Haplotagging: The use of linked-read sequencing to distinguish the DNA sequences of the two members of a homologous pair of chromosomes.

Haplotype: An individual mitochondrial DNA sequence.

Head-and-tail: One of the capsid structures of a bacteriophage.

Helicase: An enzyme that breaks base pairs in a double-stranded DNA molecule.

Helix–turn–helix motif: A common structural motif for attachment of a protein to a DNA molecule.

Heterochromatin: Chromatin that is relatively condensed and is thought to contain DNA that is not being transcribed.

Heteroduplex: A DNA–DNA or DNA–RNA hybrid.

Heteroduplex analysis: Transcript mapping by analysis of DNA–RNA hybrids with a single-strand-specific nuclease such as S1.

Heterogeneous nuclear ribonucleoproteins (hnRNPs): A broad group of RNA-protein complexes, which play several roles in the nucleus, most of which involve binding to RNAs.

Heterogeneous nuclear RNA (hnRNA): The nuclear RNA fraction that comprises unprocessed transcripts synthesized by RNA polymerase II.

Heterozygosity: The probability that a person chosen at random from the population will be heterozygous for a particular marker.

Heterozygous: A diploid nucleus that contains two different alleles for a particular gene.

Hexaploid: An auto- or allopolyploid with three diploid genome copies.

Hi-C sequencing: A version of chromosome conformation capture that is used to identify sequence contigs that are located close to one another in a genome sequence.

Hibernation promotion factor: A protein involved in the inactivation of surplus ribosomes in *Escherichia coli*.

Hierarchical clustering: A method for analyzing transcriptomes based on comparisons between the expression levels of pairs of genes.

Hierarchical shotgun sequencing: A DNA sequencing strategy which involves a pre-sequencing phase during which the genome is broken into large fragments, which are cloned and each sequenced individually by the shotgun method.

High mobility group (HMG) box: A DNA-binding domain.

High-performance liquid chromatography (HPLC): A column chromatography method with many applications in biochemistry.

His-tag: An epitope tag comprising six histidines, which enables a tagged protein to be purified with a chromatography matrix or other support that has been coated with Ni^{2+} or Co^{2+} ions.

Histone: One of the basic proteins found in nucleosomes.

Histone acetylation: Modification of chromatin structure by attachment of acetyl groups to core histones.

Histone acetyltransferase (HAT): An enzyme that attaches acetyl groups to core histones.

Histone code: The hypothesis that the pattern of chemical modification on histone proteins influences various cellular activities.

Histone deacetylase (HDAC): An enzyme that removes acetyl groups from core histones.

Histone-like nucleoid structuring protein (H-NS): A nucleoid protein that binds specifically to AT-rich regions, which are thought to be present at the boundaries of the supercoiled loops of a bacterial chromosome.

Holliday structure: An intermediate structure formed during recombination between two DNA molecules.

Holocentric chromosome: A chromosome that does not have a single centromere but instead has multiple kinetochores spread along its length.

Homeodomain: A DNA-binding motif found in many proteins involved in developmental regulation of gene expression.

Homeotic mutation: A mutation that results in the transformation of one body part into another.

Homeotic selector gene: A gene that establishes the identity of a body part, such as a segment of the *Drosophila* embryo.

Homologous chromosomes: Two or more identical chromosomes present in a single nucleus.

Homologous genes: Genes that share a common evolutionary ancestor.

Homologous recombination: Recombination between two homologous double-stranded DNA molecules, i.e., ones which share extensive nucleotide sequence similarity.

Homologous sequences: DNA sequences that share a common evolutionary ancestor.

Homology searching: A technique in which genes with sequences similar to that of an unknown gene are sought, the objective being to gain an insight into the function of the unknown gene.

Homology-directed repair: The version of gene editing that enables individual nucleotides in a target gene to be changed.

Homopolymer tailing: The attachment of a sequence of identical nucleotides (e.g., AAAAA) to the end of a nucleic acid molecule, usually referring to the synthesis of single-stranded homopolymer extensions on the ends of a double-stranded DNA molecule.

Homozygous: A diploid nucleus that contains two identical alleles for a particular gene.

Hoogsteen base pairs: Base pairs that involve the same combinations (A–T and G–C) as Watson–Crick base pairs, but are held together by hydrogen bonds between different groups.

Horizontal gene transfer: Transfer of a gene from one species to another.

Hormone response element: A nucleotide sequence upstream of a gene that mediates the regulatory effect of a steroid hormone.

Housekeeping protein: A protein that is continually expressed in all or at least most cells of a multicellular organism.

Hsp70 chaperone: A family of proteins that bind to hydrophobic regions in other proteins in order to aid their folding.

HU family: A family of nucleoid proteins that have some amino acid sequence similarity with the eukaryotic histone H2B.

Hub: A protein that has many interactions within a protein interaction map.

Human Genome Project: The publicly funded project responsible for one of the draft human genome sequences and which continues to study the functions of human genes.

Hybrid dysgenesis: The event that occurs when females from laboratory strains of *Drosophila melanogaster* are crossed with males from wild populations, the offspring resulting from such crosses being sterile and having chromosomal abnormalities and other genetic malfunctions.

Hybridization: The attachment to one another, by base-pairing, of two complementary polynucleotides.

Hybridization probe: A labeled nucleic acid molecule used as a probe to identify complementary or homologous molecules to which it base-pairs.

Hybridize: Base-pairing between nucleic acid molecules that may or may not have entirely complementary sequences.

Hydatidiform mole: A small growth that results when an egg cell that lacks a nucleus is fertilized by a sperm and implants in the uterus wall. The cells in a hydatidiform mole contain only the paternal chromosomes.

Hydrogen bond: A weak electrostatic attraction between an electronegative atom such as oxygen or nitrogen and a hydrogen atom attached to a second electronegative atom.

Hydrophobic effects: Chemical interactions that result in hydrophobic groups becoming buried inside a protein.

Hyperediting: The extensive conversion of adenosines to inosines in an RNA.

Icosahedral: One of the capsid structures of a bacteriophage or virus.

IFN-stimulated response element (ISRE): One type of DNA-binding site for STAT dimers.

IFN-γ stimulated gene response (GAS) element: One type of DNA-binding site for STAT dimers.

Illumina sequencing: A short-read sequencing method utilizing reversible terminator sequencing of immobilized fragments.

Immobilized metal affinity chromatography (IMAC): The method by which his-tagged proteins are purified with a chromatography matrix or other support that has been coated with Ni^{2+} or Co^{2+} ions.

Immunocytochemistry: A technique that uses antibody probing to locate the position of a protein in a tissue.

Immunoelectron microscopy: An electron microscopy technique that uses antibody labeling to identify the positions of specific proteins on the surface of a structure such as a ribosome.

Immunofluorescence microscopy: An electron microscopic method that utilizes a fluorescently labeled antibody to visualize the location of particular proteins within a cell.

Immunoglobulin fold: A DNA-binding domain made up of three loops emerging from a barrel-shaped β-sheet.

Immunoscreening: The use of an antibody probe to detect a polypeptide synthesized by a cloned gene.

Imprint control element: A DNA sequence found within a few kb of clusters of imprinted genes, which mediate the methylation of the imprinted regions.

***In situ* sequencing**: A sequencing method that enables the nuclear location of DNA sequences to be identified.

***In situ* tissue profiling**: A spatial transcriptomics method that involves transfer of mRNAs from a tissue to a glass slide, where barcodes are attached in order to retain positional information prior to RNA-seq.

***In vitro* mutagenesis**: Techniques used to produce a specified mutation at a predetermined position in a DNA molecule.

***In vitro* packaging**: Synthesis of infective λ phages from a preparation of λ proteins and a concatemer of λ DNA molecules.

Incomplete dominance: Refers to a pair of alleles, neither of which displays dominance, the phenotype of a heterozygote being intermediate between the phenotypes of the two homozygotes.

Indel: A position in an alignment between two DNA sequences where an insertion or deletion has occurred.

Inducer: A molecule that induces expression of a gene or operon by binding to a repressor protein and preventing the repressor from attaching to the operator.

Inducible operon: An operon that is switched on by an inducer molecule.

Induction: (1) Of a gene: the switching on of the expression of a gene or group of genes in response to a chemical or other stimulus. (2) Of λ phage: the excision of the integrated form of λ and accompanying switch to the lytic mode of infection, in response to a chemical or other stimulus.

Informational problem: The problem tackled by early molecular biologists concerning the nature of the genetic code.

Inherited disease: A disease caused by a defect in a gene.

Initiation codon: The codon, usually but not exclusively 5′–AUG–3′, found at the start of the coding region of a gene.

Initiation complex: The complex of proteins that initiates transcription. Also, the complex that initiates translation.

Initiation factor: A protein that plays an ancillary role during initiation of translation.

Initiation of transcription: The assembly upstream of a gene of the complex of proteins that will subsequently copy the gene into RNA.

Initiation region: A region of eukaryotic chromosomal DNA within which replication initiates at positions that are not clearly defined.

Initiator (Inr) sequence: A component of the RNA polymerase II core promoter.

Initiator tRNA: The tRNA, aminoacylated with methionine in eukaryotes or N-formylmethionine in bacteria, that recognizes the initiation codon during protein synthesis.

Inosine: A modified version of adenosine, sometimes found at the wobble position of an anticodon.

Insertion mutation: A mutation that arises by insertion of one or more nucleotides into a DNA sequence.

Insertion sequence: A short transposable element found in bacteria.

Insertion vector: A λ vector constructed by deleting a segment of nonessential DNA.

Insertional inactivation: A cloning strategy whereby insertion of a new piece of DNA into a vector inactivates a gene carried by the vector.

Insulator: A segment of DNA that acts as the boundary point between two chromosomal domains.

Insulator-binding protein: A protein that attaches to an insulator site in a eukaryotic genome.

Integrase: A type I topoisomerase that catalyzes insertion of the λ genome into *Escherichia coli* DNA.

Integration host factor (IHF): A nucleoid-associated protein of prokaryotes.

Integron: A set of genes and other DNA sequences that enable plasmids to capture genes from bacteriophages and other plasmids.

Interactome: The whole set of molecular interactions occurring in a cell.

Intercalating agent: A compound that can enter the space between adjacent base pairs of a double-stranded DNA molecule, often causing mutations.

Interferon: A type of cytokine.

Intergenic DNA: The regions of a genome that do not contain genes.

Internal node: A branch point within a phylogenetic tree, representing an organism or DNA sequence that is ancestral to those being studied.

Internal ribosome entry site (IRES): A nucleotide sequence that enables the ribosome to assemble at an internal position in some eukaryotic mRNAs.

Interphase: The period between cell divisions.

Interspersed repeat: A sequence that recurs at many dispersed positions within a genome.

Interspersed repeat element PCR (IRE-PCR): A clone fingerprinting technique that uses PCR to detect the relative positions of genome-wide repeats in cloned DNA fragments.

Intramolecular base-pairing: Base-pairing that occurs between two parts of the same DNA or RNA polynucleotide.

Intrinsic terminator: A position in bacterial DNA where termination of transcription occurs without the involvement of Rho.

Intron: A noncoding region within a discontinuous gene.

Intron retention: An alternative splicing scenario in which an intron that is usually spliced out of the pre-mRNA is retained in the final mRNA.

Intronic splicing enhancer (ISE): A nucleotide sequence that plays a positive regulatory role during splicing of GU–AG introns.

Intronic splicing silencer (ISE): A nucleotide sequence that plays a negative regulatory role during splicing of GU–AG introns.

Introns early: The hypothesis that introns evolved relatively early and are gradually being lost from eukaryotic genomes.

Introns late: The hypothesis that introns evolved relatively late and are gradually accumulating in eukaryotic genomes.

Inverted repeat: Two identical nucleotide sequences repeated in opposite orientations in a DNA molecule.

Ion exchange chromatography: A method for separating molecules according to how tightly they bind to electrically charged particles present in a chromatographic matrix.

Ion torrent sequencing: A short-read sequencing method that reads a sequence by detection of the hydrogen ions that are released every time a nucleotide is incorporated into the growing strand.

Ion-sensitive field effect transistor (ISFET): The component of an ion torrent sequencer that detects the hydrogen ions that are released during strand synthesis.

IRES-transacting factors (ITAFs): Proteins that regulate the usage of internal ribosome entry sites.

Iron-response element: A type of response module.

Isoaccepting tRNAs: Two or more tRNAs that are charged with the same amino acid.

Isobaric labeling: The use of labeled tags that have equal mass but during mass spectrometry give rise to reporter fragment ions that are differentially labeled.

Isoelectric focusing: Separation of proteins in a gel that contains chemicals which establish a pH gradient when the electrical charge is applied.

Isoelectric point: The position in a pH gradient where the net charge of a protein is zero.

Isoforms: The products of the alternative splicing pathways of a single gene.

Isopycnic centrifugation: A centrifugation method used to separate molecules or structures on the basis of their buoyant densities.

Isotope: One of two or more atoms that have the same atomic number but different atomic weights.

Isotope-coded affinity tag (ICAT): Markers, containing normal hydrogen and deuterium atoms, used to label individual proteomes.

JAK/STAT pathway: A relatively noncomplex type of signal transduction pathway found in many vertebrates.

Janus kinase (JAK): A type of kinase that plays an intermediary role in some types of signal transduction pathways involving STATs.

Junk DNA: One interpretation of the intergenic DNA content of a genome.

K homology (KH) domain: An RNA-binding domain.

k-mers: Sequence reads of length k.

Karyogram: The entire chromosome complement of a cell, with each chromosome described in terms of its appearance at metaphase.

Kilobase pair (kb): 1000 base pairs.

Kinase receptor: A type of cell surface receptor that has kinase activity.

Kinase-associated receptors: A type of cell surface receptor that works in conjunction with a protein that has kinase activity.

Kinetochore: The part of the centromere to which spindle microtubules attach.

Klenow polymerase: A DNA polymerase enzyme, obtained by chemical modification of *Escherichia coli* DNA polymerase I, used primarily in chain-termination DNA sequencing.

Knockout mouse: A mouse that has been engineered so that it carries an inactivated gene.

Kornberg polymerase: The DNA polymerase I enzyme of *Escherichia coli*.

Ku70-Ku80 heterodimer: A key component of the protein complex responsible for nonhomologous end-joining.

Lac selection: A means of identifying recombinant bacteria containing vectors that carry the *lacZ'* gene. The bacteria are plated on a medium that contains an analog of lactose that gives a blue color in the presence of β-galactosidase activity.

Lactose operon: The cluster of three genes that code for enzymes involved in utilization of lactose by *Escherichia coli*.

Lactose repressor: The regulatory protein that controls transcription of the lactose operon in response to the presence or absence of lactose in the environment.

Lagging strand: The strand of the double helix which is copied in a discontinuous fashion during genome replication.

Lambda (λ): A bacteriophage that infects *E. coli*, derivatives of which are used as cloning vectors.

Landrace: Populations of a crop plants that are locally adapted and which were grown by farmers until replaced by the products of modern breeding programs during the 20th century.

Lariat: Refers to the lariat-shaped intron RNA that results from splicing a GU–AG intron.

Last eukaryotic common ancestor (LECA): The archaic organism from which all modern DNA eukaryotes are descended.

Latent period: The period between injection of a phage genome into a bacterial cell and the time when cell lysis occurs.

Lateral gene transfer: Transfer of a gene from one species to another.

Leader segment: The untranslated region of an mRNA upstream of the initiation codon.

Leading strand: The strand of the double helix which is copied in a continuous fashion during genome replication.

Lectin: A plant or animal protein with specific sugar-binding properties.

Leucine zipper: A dimerization domain commonly found in DNA-binding proteins.

Ligase: An enzyme that synthesizes phosphodiester bonds as part of DNA replication, repair and recombination processes.

LINE (long interspersed nuclear element): A type of genome-wide repeat, often with transposable activity.

LINE-1: One type of human LINE.

Linkage: The physical association between two genes that are on the same chromosome.

Linkage analysis: The procedure used to assign map positions to genes by genetic crosses.

Linkage disequilibrium: The situation where a particular combination of alleles at linked loci occur more frequently or less frequently than expected in a population.

Linkage group: A group of genes that display linkage. With eukaryotes a single linkage group usually corresponds to a single chromosome.

Linked-read sequencing: A method used to identify sequence contigs that are located close to one another in a genome sequence.

Linker: A synthetic, double-stranded oligonucleotide used to attach sticky ends to a blunt-ended molecule.

Linker DNA The DNA that links nucleosomes:: the 'string' in the 'beads-on-a-string' model for chromatin structure.

Linker histone: A histone, such as H1, that is located outside of the nucleosome core octamer.

Linking number: The number of times one strand crosses the other in a circular molecule.

Lod score: A statistical measure of linkage as revealed by pedigree analysis.

Long intergenic noncoding RNAs (lincRNAs): A long noncoding RNA that is located entirely within an intergenic region.

Long noncoding RNA (lncRNA): Noncoding RNAs longer than 200 nucleotides in length.

Long terminal repeat (LTR): A repeated DNA sequence found at the ends of some retroelements.

Long-read sequencing: A sequencing method that generates reads tens or thousands of kb in length.

Low-copy repeat: A duplication between 1–400 kb in length, with greater than 90% sequence identity, that is repeated up to 50 times in a genome.

LSm fold: An RNA-binding domain comprising a five-stranded β–sheet with an N-terminal α-helix.

LTR element: A type of genome-wide repeat typified by the presence of long terminal repeats (LTRs).

Lyase: An enzyme that breaks chemical bonds by processes other than oxidation and hydrolysis.

Lysis: The disruption of a bacterial cell by lysozyme, such as occurs at the end of the infection cycle of a lytic bacteriophage.

Lysogenic infection cycle: The type of bacteriophage infection that involves integration of the phage genome into the host DNA molecule.

Lysozyme: A protein used to destabilize the bacterial cell wall prior to DNA purification.

Lytic infection cycle: The type of bacteriophage infection that involves lysis of the host cell immediately after the initial infection, with no integration of the phage DNA molecule into the host genome.

M phase: The stage of the cell cycle when mitosis or meiosis occurs.

M13 bacteriophage: A bacteriophage that infects *E. coli*, derivatives of which are used as cloning vectors.

Macrochromosome: One of the larger gene-deficient chromosomes seen in the nuclei of chickens and various other species.

Macrodomain: One of the four distinct regions of *Escherichia coli* nucleoid DNA.

MADS-box: A DNA-binding domain found in several transcription factors involved in plant development.

Magnetic tweezers: A set of magnets whose positions and field strengths can be varied to move magnetic particles, such as magnetic beads, in a controlled manner to study the mechanical properties of biomolecules.

Maintenance methylation: Addition of methyl groups to positions on newly synthesized DNA strands that correspond with the positions of methylation on the parent strand.

Major groove: The larger of the two grooves that spiral around the surface of the B-form of DNA.

MAP (mitogen-activated protein) kinase or **MAPK/ERK pathway**: An important signal transduction pathway, found in many organisms.

MAP kinase: One of the components of the MAP kinase signal transduction pathway.

Map unit: A unit used to describe the distance between two genes on a chromosome, now superseded by centiMorgan.

Mapping reagent: A collection of DNA fragments spanning a chromosome or the entire genome and used in STS mapping.

Marker-assisted selection: A DNA screening method that enables individuals with a particular characteristic to be identified from their possession of DNA markers associated with that characteristic.

Mass analyzer: The component of a mass spectrometer that measures the mass-to-charge ratios of the ions that are being studied.

Mass spectrometry: An analytical technique in which ions are separated according to their mass-to-charge ratios.

Mass-to-charge ratio: The basis for separation of ions by mass spectrometry.

Massively parallel: A high throughput sequencing strategy in which many individual sequences are generated in parallel.

Massively parallel array: An array of DNA fragments immobilized in a format suitable for DNA sequencing

Maternal-effect gene: A *Drosophila* gene that is expressed in the parent and whose mRNA is subsequently injected into the egg, after which it influences development of the embryo.

Mating type: The equivalent of male and female for a eukaryotic microorganism.

Mating-type switching: The ability of yeast cells to change from a to α mating type, or vice versa, by gene conversion.

Matrix-assisted laser desorption ionization time-of-flight (MALDI-TOF): A type of mass spectrometry used in proteomics.

Mediator: A protein complex that forms a contact between various activators and the C-terminal domain of the largest subunit of RNA polymerase II.

Megabase pair (Mb): 1000 kb; 1,000,000 bp.

Meiosis: The series of events, involving two nuclear divisions, by which diploid nuclei are converted to haploid gametes.

Melting: Denaturation of a double-stranded DNA molecule.

Melting temperature (T_m): The temperature at which the two strands of a double-stranded nucleic acid molecule or base-paired hybrid detach as a result of complete breakage of hydrogen bonding.

Meselson–Stahl experiment: The experiment which showed that cellular DNA replication occurs through the semiconservative process.

Messenger RNA (mRNA): The transcript of a protein-coding gene.

Metabolic engineering: The process by which changes are made to the genome by mutation or recombinant DNA techniques in order to influence the cellular biochemistry in a predetermined way.

Metabolic flux: The rate of flow of metabolites through the network of pathways that make up the cellular biochemistry.

Metabolic labeling: A labeling method that involves growing cells in the presence of labeled nutrients.

Metabolome: The complete collection of metabolites present in a cell under a particular set of conditions.

Metabolomics: The study of metabolomes.

Metagenomics: Studies of the mixture of genomes present in a particular habitat.

Metaphase chromosome: A chromosome at the metaphase stage of cell division, when the chromatin takes on its most condensed structure and features such as the banding pattern can be visualized.

MeTH-seq: A method used to detect methylribose positions in a RNA, based on the tendency of reverse transcriptase to pause when it reaches a methylribose.

Methyl-CpG-binding protein (MeCP): A protein that binds to methylated CpG islands and may influence acetylation of nearby histones.

MGMT (O^6-methylguanine-DNA methyltransferase): An enzyme involved in the direct repair of alkylation mutations.

Microarray: A low-density array of DNA molecules used for parallel hybridization analysis.

Microbiome: The microorganisms that live on or within the human body.

Microchromosome: One of the shorter gene-rich chromosomes seen in the nuclei of chickens and various other species.

Microhomology: The relationship between the single-stranded overhangs at the ends of a pair of double-stranded molecules, where the sequences are not 100% complementary but enable sufficient base-pairing to hold the ends together while they are ligated.

MicroRNA: A class of short RNAs involved in regulation of gene expression in eukaryotes, and which act by a pathway similar to RNA interference.

Microsatellite: A type of simple sequence length polymorphism comprising tandem copies of, usually, di-, tri- or tetranucleotide repeat units. Also called a short tandem repeat (STR).

Miniature inverted-repeat transposable element (MITE): A general term for the truncated relic of a DNA transposon.

Minigene: The name given to the pair of exons carried by a cloning vector used in the exon-trapping procedure.

MinION: A device for carrying out nanopore sequencing.

Minisatellite: A type of simple sequence length polymorphism comprising tandem copies of repeats that are a few tens of nucleotides in length. Also called a variable number of tandem repeats (VNTR).

Minor groove: The smaller of the two grooves that spiral around the surface of the B-form of DNA.

Miscoding lesion: A sequence error caused by chemical alteration of a nucleotide, resulting from the partial degradation of an ancient DNA molecule.

Mismatch: A position in a double-stranded DNA molecule where base-pairing does not occur because the nucleotides are not complementary; in particular, a non-base-paired position resulting from an error in replication.

Mismatch repair: A DNA repair process that corrects mismatched nucleotide pairs by replacing the incorrect nucleotide in the daughter polynucleotide.

Mitochondrial genome: The genome present in the mitochondria of a eukaryotic cell.

Mitochondrion: One of the energy-generating organelles of eukaryotic cells.

Mitosis: The series of events that results in nuclear division.

Mobile phase: The movable phase in a chromatography system, usually a liquid in which the compounds have been dissolved or a gas in which they have been vaporized.

Model organism: An organism which is relatively easy to study and hence can be used to obtain information that is relevant to the biology of a second organism that is more difficult to study.

Modification assay : A range of techniques used for locating bound proteins on DNA molecules.

Modification interference : A technique used to identify nucleotides involved in interactions with a DNA-binding protein.

Modification protection: A technique used to identify nucleotides involved in interactions with a DNA-binding protein.

Molecular beacon: A method, based on dye-quenching, used to type SNPs.

Molecular biologist: A person who studies the molecular life sciences.

Molecular chaperone: A protein that helps other proteins to fold.

Molecular clock: A device based on the inferred mutation rate that enables times to be assigned to the branch points in a gene tree.

Molecular combing: A technique for preparing restricted DNA molecules for optical mapping.

Molecular evolution: The gradual changes that occur in genomes over time as a result of the accumulation of mutations and structural rearrangements resulting from recombination and transposition.

Molecular ion An ion formed during peptide mass fingerprinting:: $[M+H]^+$ and $[M–H]^-$, where 'M' is the peptide.

Molecular life sciences: The area of research comprising molecular biology, biochemistry and cell biology, as well as some aspects of genetics and physiology.

Molten globule: An intermediate in protein folding, formed by the rapid collapse of a polypeptide into a compact structure, with slightly larger dimensions than the final protein.

Monogenic: A characteristic that is specified by a single gene.

Monohybrid cross: A sexual cross in which the inheritance of one pair of alleles is followed.

Multicopy: A gene, cloning vector or other genetic element that is present in multiple copies in a single cell.

Multicysteine zinc finger: A type of zinc-finger DNA-binding domain.

Multigene family: A group of genes, clustered or dispersed, with related nucleotide sequences.

Multiple alignment: An alignment of three or more nucleotide sequences.

Multiple alleles: The different alternative forms of a gene that has more than two alleles.

Mutagen: A chemical or physical agent that can cause a mutation in a DNA molecule.

Mutagenesis: Treatment of a group of cells or organisms with a mutagen as a means of inducing mutations.

Mutant: A cell or organism that possesses a mutation.

Mutasome: A protein complex that is constructed during the SOS response of *Escherichia coli*.

Mutation: An alteration in the nucleotide sequence of a DNA molecule.

Mutator-**like transposable element (MULE)**: A type of DNA transposon that is able to capture exons and other gene segments.

N terminus: The end of a polypeptide that has a free amino group.

N-linked glycosylation: The attachment of sugar units to an asparagine in a polypeptide.

N50 size: A measure the degree of completeness of a genome sequence.

Nanopore sequencing: A sequencing method that reads the order of nucleotides from the perturbations in electric current that occur as a DNA molecule passes through a small pore in a membrane.

Next-generation sequencing: A collection of DNA sequencing methods, each involving a massively parallel strategy.

NG50 size: A measure the degree of completeness of a genome sequence.

NH$_2$-terminus: The end of a polypeptide that has a free amino group.

Nick: A position in a double-stranded DNA molecule where one of the polynucleotides is broken as a result of the absence of a phosphodiester bond.

Nick translation: The repair of a nick with DNA polymerase I, usually to introduce labeled nucleotides into a DNA molecule.

Nicking endonuclease: A type of restriction endonuclease that recognizes a specific nucleotide sequence, but rather than making a double-stranded break in the DNA cuts just one strand, resulting in a sequence-specific nick.

Nitrogenous base: One of the purines or pyrimidines that form part of the molecular structure of a nucleotide.

Node: The depiction of a protein in a protein interaction map.

Nonhomologous end-joining (NHEJ): Another name for the double-strand break repair process.

Nonchromatin region: The space separating the chromosome territories within a nucleus.

Noncoding RNA: An RNA molecule that does not code for a protein.

Nonpenetrance: The situation whereby the effect of a mutation is never observed during the lifetime of a mutant organism.

Nonpolar: A hydrophobic (water-hating) chemical group.

Nonprocessed pseudogene: A gene that has become inactive because of the accumulation of mutations.

Northern blotting: The transfer of RNA from an electrophoresis gel to a membrane prior to northern hybridization.

Northern hybridization: A technique used for detection of a specific RNA molecule against a background of many other RNA molecules.

Nuclear genome: The DNA molecules present in the nucleus of a eukaryotic cell.

Nuclear lamina: A network of filaments on the internal side of the nuclear membrane.

Nuclear magnetic resonance (NMR) spectroscopy: A technique for determining the three-dimensional structure of large molecules.

Nuclear matrix: A proteinaceous scaffold-like network that is thought to permeate the nucleus.

Nuclear receptor superfamily: A family of receptor proteins that bind hormones as an intermediate step in modulation of genome activity by these hormones.

Nuclease: An enzyme that degrades a nucleic acid molecule.

Nuclease protection experiment: A technique that uses nuclease digestion to determine the positions of proteins on DNA or RNA molecules.

Nucleic acid: The term first used to describe the acidic chemical compound isolated from the nuclei of eukaryotic cells. Now used specifically to describe a polymeric molecule comprising nucleotide monomers, such as DNA and RNA.

Nucleic acid hybridization: Formation of a double-stranded hybrid by base-pairing between complementary polynucleotides.

Nucleoid: The DNA-containing region of a prokaryotic cell.

Nucleoid-associated proteins: The protein component of a bacterial nucleoid.

Nucleolus: The region of the eukaryotic nucleus in which rRNA transcription occurs.

Nucleoside: A purine or pyrimidine base attached to a five-carbon sugar.

Nucleosome: The complex of histones and DNA that is the basic structural unit in chromatin.

Nucleosome remodeling: A change in the conformation of a nucleosome, associated with a change in access to the DNA to which the nucleosome is attached.

Nucleotide: A purine or pyrimidine base attached to a five-carbon sugar, to which a mono-, di-, or triphosphate is also attached. The monomeric unit of DNA and RNA.

Nucleotide excision repair: A repair process that corrects various types of DNA damage by excising and resynthesizing a region of a polynucleotide.

Nucleus: The membrane-bound structure of a eukaryotic cell in which the chromosomes are contained.

O-linked glycosylation: The attachment of sugar units to a serine or threonine in a polypeptide.

Off-target editing: Editing of sites other than the target site during a genome editing experiment.

Okazaki fragment: One of the short segments of RNA-primed DNA synthesized during replication of the lagging strand of the double helix.

Oligonucleotide: A short synthetic single-stranded DNA molecule.

Oligonucleotide hybridization analysis: The use of an oligonucleotide as a hybridization probe.

Oligonucleotide ligation assay (OLA): A technique for SNP typing, which depends on ligation of two oligonucleotides which anneal adjacent to one another, one covering the position of the SNP.

Oligonucleotide-directed mutagenesis: An *in vitro* mutagenesis technique in which a synthetic oligonucleotide is used to introduce a predetermined nucleotide alteration into the gene to be mutated.

Oncogene: A gene which, when defective, can give rise to cancer.

One-step growth curve: A single infection cycle for a lytic bacteriophage.

Oocyte: An unfertilized female egg cell.

Open pan-genome: A pan-genome whose gene number continues to increase as more genomes are sequenced.

Open promoter complex: A structure formed during assembly of the transcription initiation complex consisting of the RNA polymerase and/or accessory proteins attached to the promoter, after the DNA has been opened up by breakage of base pairs.

Open reading frame (ORF): A series of codons starting with an initiation codon and ending with a termination codon. The part of a protein-coding gene that is translated into protein.

Operator: The nucleotide sequence to which a repressor protein binds to prevent transcription of a gene or operon.

Operon: A set of adjacent genes in a bacterial genome, transcribed from a single promoter and subject to the same regulatory regime.

Optical mapping: A technique for the direct visual examination of restricted DNA molecules.

ORF scanning: Examination of a DNA sequence for open reading frames in order to locate the genes.

Origin licensing: The construction of prereplication complexes on replication origins.

Origin of replication: A site on a DNA molecule where replication initiates.

Origin recognition complex (ORC): A set of proteins that binds to the origin recognition sequence.

Origin recognition sequence: A component of a eukaryotic origin of replication.

Orphan retrogene: A retrogene whose parent copy has been lost.

Orthogonal field alternation gel electrophoresis (OFAGE): An electrophoresis system in which the field alternates between pairs of electrodes set at an angle of 45°, used to separate large DNA molecules.

Orthologous: Refers to homologous genes located in the genomes of different organisms.

Overlap graph: The output of a sequence assembler, comprising overlapping sequence reads.

Overlapping genes: Two genes whose coding regions overlap.

P element: A DNA transposon of *Drosophila*.

P1-derived artificial chromosome (PAC): A high-capacity vector that combines features of bacteriophage P1 vectors and bacterial artificial chromosomes.

p300/CBP: A protein complex that is able to modify histone proteins and so affect chromatin structure and nucleosome positioning.

PacBio sequencing: A version of single-molecule real-time sequencing.

Pair-rule genes: Developmental genes that establish the basic segmentation pattern of the *Drosophila* embryo.

Paired-end reads: Mini-sequences from the two ends of a single cloned fragment.

Paleogenomics: The study of the genomes of extinct species.

Pan-editing: The extensive insertion of nucleotides into abbreviated RNAs in order to produce functional molecules.

Pan-genome concept: The concept that views a bacterial genome as a combination of a core and accessory genome.

Paralogous: Refers to two or more homologous genes located in the same genome.

Paranemic: Refers to a helix whose strands can be separated without unwinding.

Pararetrovirus: A viral retroelement whose encapsidated genome is made of DNA.

Paraspeckle: Nuclear structures thought to have a role in the stress response by sequestering particular mRNAs, presumably to prevent these from being translated.

Parental genotype: The genotype possessed by one or both of the parents in a genetic cross.

Partial linkage: The type of linkage usually displayed by a pair of genetic and/or physical markers on the same chromosome, the markers not always being inherited together because of the possibility of recombination between them.

Partial restriction: Digestion of DNA with a restriction endonuclease under limiting conditions so that not all restriction sites are cut.

Pedigree: A chart showing the genetic relationships between the members of a human family.

Pedigree analysis: The use of pedigree charts to analyze the inheritance of a genetic or DNA marker in a human family.

Pentose: A sugar comprising five carbon atoms.

Peptide bond: The chemical link between adjacent amino acids in a polypeptide.

Peptide mass fingerprinting: Identification of a protein by examination of the mass spectrometric properties of peptides generated by treatment with a sequence-specific protease.

Peptide nucleic acid (PNA): A polynucleotide analog in which the sugar–phosphate backbone is replaced by amide bonds.

Peptidyl or P site: The site in the ribosome occupied by the tRNA attached to the growing polypeptide during translation.

Peptidyl transferase: The enzyme activity that synthesizes peptide bonds during translation.

Personalized medicine: The use of individual genome sequences to make accurate diagnoses of a person's risk of developing a disease, and the use of that person's genetic characteristics to plan effective therapies and treatment regimes.

Pfam: A protein structure database.

Phage: A virus that infects a bacterium.

Phage display: A technique for identifying proteins that interact with one another.

Phage display library: A collection of clones carrying different DNA fragments, used in phage display.

Phenotype: The observable characteristics displayed by a cell or organism.

Phenotype ontology: The characteristics of an organism that can be distinguished by visual inspection or by biochemical tests.

Philadelphia chromosome: An abnormal chromosome resulting from a translocation between human chromosomes 9 and 22, a common cause of chronic myeloid leukemia.

Phosphate group: One of the components of a nucleotide.

Phosphodiester bond: The chemical link between adjacent nucleotides in a polynucleotide.

Phosphodiesterase: A type of enzyme that can break phosphodiester bonds.

Phosphorylase: An enzyme that adds a phosphate group to another molecule.

Photobleaching: A component of the FRAP technique for studying protein mobility in the nucleus.

Photolithography: A technique that uses pulses of light to construct an oligonucleotide from light-activated nucleotide substrates.

Photolyase: An *Escherichia coli* enzyme involved in photoreactivation repair.

Photoproduct: A modified nucleotide resulting from treatment of DNA with ultraviolet radiation.

Photoreactivation: A DNA repair process in which cyclobutyl dimers and (6–4) photoproducts are corrected by a light-activated enzyme.

Phylogenetic tree: A tree depicting the evolutionary relationships between a set of DNA sequences, species or other taxa.

Physical mapping: The use of molecular biology techniques to construct a genome map.

Pilus: A structure involved in bringing a pair of bacteria together during conjugation; possibly the tube through which DNA is transferred.

Piwi protein: A type of protein that forms a complex with piRNA to form structures that regulate gene expression during various developmental processes.

Piwi-interacting RNA (piRNA): A type of snRNA, 25–30 nucleotides in length, which associates with piwi proteins.

Plant GDB: An online plant genome browser.

Plaque: A zone of clearing on a lawn of bacteria caused by lysis of the cells by infecting bacteriophages.

Plasmid: A usually circular piece of DNA often found in bacteria and some other types of cell.

Plectonemic: Refers to a helix whose strands can only be separated by unwinding.

Point centromere: A type of centromere, found in *Saccharomyces cerevisiae*, which does not contain repetitive DNA and instead is defined by a single-copy sequence.

Point mutation: A mutation that results from a single nucleotide change in a DNA molecule.

Polar: A hydrophilic (water-loving) chemical group.

Polishing: A method in which short reads are aligned to a long-read assembly to ensure that all sequence errors have been accounted for.

Poly(A) polymerase: The enzyme that attaches a poly(A) tail to the 3′ end of a eukaryotic mRNA.

Poly(A) tail: A series of A nucleotides attached to the 3′ end of a eukaryotic mRNA.

Polyacrylamide gel electrophoresis: Electrophoresis carried out in a polyacrylamide gel and used to separate DNA molecules between 10 and 1500 bp in length.

Polyadenylate-binding protein: A protein that aids poly(A) polymerase during polyadenylation of eukaryotic mRNAs, and which plays a role in maintenance of the tail after synthesis.

Polyadenylation: The addition of a series of As to the 3′ end of a eukaryotic mRNA.

Polyadenylation editing: Conversion of a terminal U or UA into UAAAA…A, creating a termination codon, as occurs with many animal mitochondrial RNAs.

Polycomb group (PcG) : A group of proteins that induce localized formation of heterochromatin.

Polycomb response element: The DNA recognition sequences for Polycomb group proteins.

Polymer: A compound made up of a long chain of identical or similar units.

Polymerase chain reaction (PCR): A technique that results in exponential amplification of a selected region of a DNA molecule.

Polynucleotide: A single-stranded DNA or RNA molecule.

Polynucleotide kinase: An enzyme that adds phosphate groups to the 5′ ends of DNA molecules.

Polypeptide: A polymer of amino acids.

Polyploidy: Having two or more diploid genome copies.

Polyprotein: A translation product consisting of a series of linked proteins which are processed by proteolytic cleavage to release the mature proteins.

Polypyrimidine tract: A pyrimidine-rich region near the 3′ end of a GU–AG intron.

Positional effect: Refers to the different levels of expression that result after insertion of a gene at different positions in a eukaryotic genome.

Post-spliceosome complex: The immediate product of the splicing reaction for a GU–AG intron, which dissociates into the spliced mRNA and the intron lariat.

POU domain: A DNA-binding motif found in a variety of proteins.

Precatalytic spliceosome: An intermediate in the splicing pathway for a GU–AG intron, the immediate precursor of the spliceosome.

Pre-mRNA: The primary transcript of a protein-coding gene.

Pre-RNA: The initial product of transcription of a gene or group of genes, subsequently processed to give the mature transcript(s).

Pre-rRNA: The primary transcript of a gene or group of genes specifying rRNA molecules.

Pre-tRNA: The primary transcript of a gene or group of genes specifying tRNA molecules.

Preinitiation complex: (1) The structure comprising the small subunit of the ribosome and the initiator tRNA plus ancillary factors that forms the initial association with the mRNA during protein synthesis. (2) The structure that forms at the core promoter of a gene transcribed by RNA polymerase II. (3) The activated version of a prereplication complex.

Prepriming complex: A complex of proteins formed during initiation of replication in bacteria.

Prereplication complex (pre-RC): A protein complex that is constructed at a eukaryotic origin of replication and enables initiation of replication to occur.

Prespliceosome complex: An intermediate in the splicing pathway for a GU–AG intron.

Prey: The proteins captured by the bait protein when the members of a protein complex are being purified.

Pribnow box: A component of the bacterial promoter.

Primary antibody: An antibody that recognizes a specific target protein.

Primary structure: The sequence of amino acids in a polypeptide.

Primary transcript: The initial product of transcription of a gene or group of genes, subsequently processed to give the mature transcript(s).

Primase: The RNA polymerase enzyme that synthesizes RNA primers during bacterial DNA replication.

Prime editing: A version of base editing that has a low frequency of off-target events.

Prime editing guide RNA (pegRNA): The guide RNA used in a prime editing experiment, which is able to base-pair to both of the DNA strands at the target site.

Primer: A short oligonucleotide that is attached to a single-stranded DNA molecule in order to provide a starting point for strand synthesis.

Primosome: A protein complex involved in genome replication.

Prion: An unusual infectious agent that consists purely of protein.

Process description: A graph that shows the biochemical reactions that occur over time in a biological system.

Processed pseudogene: A pseudogene that results from integration into the genome of a reverse-transcribed copy of an mRNA.

Processivity: Refers to the amount of DNA synthesis that is carried out by a DNA polymerase before dissociation from the template.

Programmable nuclease: A nuclease that can be directed to a specific site in a genome.

Prokaryote: An organism whose cells lack a distinct nucleus.

Proliferating cell nuclear antigen (PCNA): An accessory protein involved in genome replication in eukaryotes.

Promiscuous DNA: DNA that has been transferred from one organelle genome to another.

Promoter: The nucleotide sequence, upstream of a gene, to which RNA polymerase binds in order to initiate transcription.

Promoter clearance: The completion of successful initiation of transcription that occurs when the RNA polymerase moves away from the promoter sequence.

Promoter escape: The stage in transcription during which the polymerase moves away from the promoter region and becomes committed to making a transcript.

Proofreading: The 3′→5′ exonuclease activity possessed by some DNA polymerases, which enables the enzyme to replace a misincorporated nucleotide.

Prophage: The integrated form of the genome of a lysogenic bacteriophage.

PROSITE: An online protein structure database.

Protease: An enzyme that degrades protein.

Proteasome: A multisubunit protein structure that is involved in the degradation of other proteins.

Protein: The polymeric compound made of amino acid monomers.

Protein A: A protein from the bacterium *Staphylococcus aureus* that binds specifically to immunoglobulin G (i.e., antibody) molecules.

Protein array: A microarray comprising immobilized proteins.

Protein electrophoresis: Separation of proteins in an electrophoresis gel.

Protein engineering: Various techniques for making directed alterations in protein molecules, often to improve the properties of enzymes used in industrial processes.

Protein folding: The adoption of a folded structure by a polypeptide.

Protein interaction map: A map showing the interactions between all or some of the proteins in a proteome.

Protein profiling: The methodology used to identify the proteins in a proteome.

Protein–protein crosslinking: A technique that links together adjacent proteins in order to identify proteins that are positioned close to one another in a structure such as a ribosome.

Proteome: The collection of proteins synthesized by a living cell.

Proteomics: A variety of techniques used to study proteomes.

Protogenome: An RNA genome that existed during the RNA world.

Protomer: One of the protein subunits in a bacteriophage or virus capsid.

Protoplast: A cell from which the cell wall has been completely removed.

Protospacer adjacent motif (PAM): A component of the target site for a Cas9 endonuclease.

Proximal binding sites: Components of a eukaryotic promoter, usually located within 2 kb of the transcription start site of the target gene, that act as binding sites for transcription factors.

Pseudogene: An inactivated and hence nonfunctional copy of a gene.

PSI-BLAST: A modified and more powerful version of the BLAST algorithm.

Pulse labeling: A brief period of labeling carried out at a defined period during the progress of an experiment.

Punctuation codon: A codon that specifies either the start or the end of a gene.

Purine: One of the two types of nitrogenous base found in nucleotides.

Pyrimidine: One of the two types of nitrogenous base found in nucleotides.

Pyrosequencing: A DNA sequencing method in which addition of a nucleotide to the end of a growing polynucleotide is detected directly by conversion of the released pyrophosphate into a flash of chemiluminescence.

Qrr RNA: A type of RNA involved in quorum sensing in *Vibrio* species.

Quadrupole mass analyzer: A mass spectrometer in which the mass analyzer has four magnetic rods placed parallel to one another, surrounding a central channel through which the ions must pass.

Quantitative PCR: A PCR method that enables the number of DNA molecules in a sample to be estimated.

Quantitative trait locus (QTL): A region of a genome, each possibly containing several genes, which controls a variable trait.

Quaternary structure: The structure resulting from the association of two or more polypeptides.

RACE (rapid amplification of cDNA ends): A PCR-based technique for mapping the end of an RNA molecule.

Radiation hybrid: A collection of rodent cell lines that contain different fragments of a second genome, constructed by a technique involving irradiation and used as a mapping reagent, for example, in studies of the human genome.

Radioactive marker: A radioactive atom incorporated into a molecule and whose radioactive emissions are subsequently used to detect and follow that molecule during a biochemical reaction.

Radiolabeling: The technique for attaching a radioactive atom to a molecule.

Random genomic sequences: Sequence-tagged sites obtained by sequencing random pieces of cloned genomic DNA.

Read: A single sequence from the output of a sequencing experiment.

Reading frame: A series of triplet codons in a DNA sequence.

Real-time PCR: A modification of the standard PCR technique in which synthesis of the product is measured as the PCR proceeds through its series of cycles.

RecA: An *Escherichia coli* protein involved in homologous recombination.

RecBCD complex: An enzyme complex involved in homologous recombination in *Escherichia coli*.

RecBFI pathway: A pathway for homologous recombination in *Escherichia coli*.

Recessive: The allele that is not expressed in a heterozygote.

RecFOR pathway: A pathway for homologous recombination in *Escherichia coli*.

Reciprocal strand exchange: The exchange of DNA between two double-stranded molecules, occurring as a result of recombination, such that the end of one molecule is exchanged for the end of the other molecule.

Recognition helix: An α-helix in a DNA-binding protein, one that is responsible for recognition of the target nucleotide sequence.

Recombinant: A progeny member that possesses neither of the combinations of alleles displayed by the parents.

Recombinant DNA molecule: A DNA molecule created in the test tube by ligating pieces of DNA that are not normally joined together.

Recombinant DNA technology: The techniques involved in the construction, study and use of recombinant DNA molecules.

Recombinant genotype: A genotype not possessed by either of the parents in a genetic cross.

Recombinant plasmid: A plasmid that contains an inserted piece of DNA.

Recombinase: A diverse family of enzymes that catalyze site-specific recombination events.

Recombination: A large-scale rearrangement of a DNA molecule.

Recombination frequency: The proportion of recombinant progeny arising from a genetic cross.

Recombination hotspot: A region of a chromosome where crossovers occur at a higher frequency than the average for the chromosome as a whole.

Recombination repair: A DNA repair process that mends strand breaks.

Reductive evolution: A process displayed by some prokaryotic genomes that involves extensive gene loss and a decrease in the lengths of intergenic regions so that the genome becomes as compact as possible.

Reference sequence: An existing genome sequence that is used to aid assembly of the reads obtained by sequencing of a related genome.

Reflectron: An ion mirror used in some types of mass spectrometer; also used to denote a mass spectrometer that contains an ion mirror.

Regional centromere: A typical eukaryotic centromere, associated with a region of repetitive DNA.

Renaturation: The return of a denatured molecule to its natural state.

Repetitive DNA: A DNA sequence that is repeated two or more times in a DNA molecule or genome.

Repetitive DNA fingerprinting: A clone fingerprinting technique that involves determining the positions of genome-wide repeats in cloned DNA fragments.

Repetitive DNA PCR: A clone fingerprinting technique that uses PCR to detect the relative positions of genome-wide repeats in cloned DNA fragments.

Repetitive extragenic palindromic (REP) sequences: A type of bacterial repetitive DNA, made up of motifs most of which are 20–35 bp in length and occur singly or in arrays.

Replacement vector: A λ vector designed so that insertion of new DNA is by replacement of part of the nonessential region of the λ DNA molecule.

Replication factor C (RFC): A multisubunit accessory protein involved in eukaryotic genome replication.

Replication fork: The region of a double-stranded DNA molecule that is being opened up to enable DNA replication to occur.

Replication fork barrier: A sequence in a eukaryotic genome that impedes the progress of a replication fork and might be involved in termination of replication at some sites.

Replication mediator protein (RMP): A protein responsible for detachment of single-strand binding proteins during genome replication.

Replication origin: A site on a DNA molecule where replication initiates.

Replication protein A (RPA): The main single-strand binding protein involved in replication of eukaryotic DNA.

Replication slippage: An error in replication that leads to an increase or decrease in the number of repeat units in a tandem repeat, such as a microsatellite.

Replicative transposition: Transposition that results in copying of the transposable element.

Replisome: A complex of proteins involved in genome replication.

Reporter gene: A gene whose phenotype can be assayed and which can therefore be used to determine the function of a regulatory DNA sequence.

Reporter probe: A short oligonucleotide that gives a fluorescent signal when it hybridizes with a target DNA.

Repressible operon: An operon that is controlled by a product of the pathway catalyzed by the gene products.

Resection: The process by which a few nucleotides are removed from one of the polynucleotides at the end of a double-stranded DNA molecule, creating a short single-stranded overhang.

Resin: A chromatography matrix.

Resolution: Separation of a pair of recombining double-stranded DNA molecules.

Resolvase: A protein capable of resolving a Holliday structure.

Resonance frequency: The energy difference between the α and β spin states of a nucleus.

Restriction endonuclease: An enzyme that cuts DNA molecules at a limited number of specific nucleotide sequences.

Restriction fragment length polymorphism (RFLP): A restriction fragment whose length is variable because of the presence of a polymorphic restriction site at one or both ends.

Restriction mapping: Determination of the positions of restriction sites in a DNA molecule by analyzing the sizes of restriction fragments.

Restriction pattern: The set of fragments obtained after digestion of a DNA molecule with a restriction endonuclease, referring to the pattern of bands obtained after separation of the fragments by gel electrophoresis.

Retroelement: A genetic element that transposes via an RNA intermediate.

Retrogene: A gene duplicate that arises by insertion of a pseudogene adjacent to the promoter of an existing gene.

Retrohoming: A process during which an excised intron, comprising single-stranded RNA, inserts directly into an organelle genome prior to being copied into double-stranded DNA.

Retron: The commonest type of bacterial retroelement.

Retroposon: A retroelement that does not have LTRs.

Retrotransposition : Transposition via an RNA intermediate.

Retrotransposon: A genome-wide repeat with a sequence similar to an integrated retroviral genome and possibly with retrotransposition activity.

Retrovirus: A virus with an RNA genome that integrates into the genome of its host cell.

Reverse genetics: The strategy by which the function of a gene is identified by mutating that gene and identifying the phenotypic change that results.

Reverse phase liquid chromatography (RPLC): A column chromatography method that separates proteins according to their degree of surface hydrophobicity.

Reverse sequence: One of the two directions in which a double-stranded DNA molecule can be sequenced.

Reverse transcriptase: A polymerase that synthesizes DNA on an RNA template.

Reverse transcriptase PCR (RT-PCR): PCR in which the first step is carried out by reverse transcriptase, so RNA can be used as the starting material.

Reversible terminator sequencing: A DNA sequencing method in which the sequence is read by detection of the fluorescent label attached to each nucleotide that is added to a growing polynucleotide.

Rho: A protein involved in termination of transcription of some bacterial genes.

Rho-dependent terminator: A position in bacterial DNA where termination of transcription occurs with the involvement of Rho.

Ribbon–helix–helix motif: A type of DNA-binding domain.

Ribonuclease: An enzyme that degrades RNA.

Ribose: The sugar component of a ribonucleotide.

Ribosomal protein: One of the protein components of a ribosome.

Ribosomal RNA (rRNA): The RNA molecules that are components of ribosomes.

Ribosome: One of the protein–RNA assemblies on which translation occurs.

Ribosome binding site: The nucleotide sequence that acts as the attachment site for the small subunit of the ribosome during initiation of translation in bacteria.

Ribosome modulation factor: A protein involved in the inactivation of surplus ribosomes in *Escherichia coli.*

Riboswitch: A segment of an mRNA that can bind a small molecule, attachment of this molecule affecting the translation or processing of the mRNA.

Ribozyme: An RNA molecule that has catalytic activity.

RNA: Ribonucleic acid, one of the two forms of nucleic acid in living cells; the genetic material for some viruses.

RNA editing: A process by which nucleotides not coded by a gene are introduced at specific positions in an RNA molecule after transcription.

RNA *in situ* conformation sequencing (RIC-seq): A spatial transcriptomics method that involves crosslinking of RNAs in fixed tissue so that RNAs adjacent to one another can be identified by RNA-seq.

RNA-induced silencing complex (RISC): A complex of proteins which cleaves and hence silences an mRNA as part of the RNA interference pathway.

RNA interference (RNAi): An RNA degradation process in eukaryotes.

RNA polymerase: An enzyme that synthesizes RNA on a DNA or RNA template.

RNA polymerase I: The eukaryotic RNA polymerase that transcribes ribosomal RNA.

RNA polymerase II: The eukaryotic RNA polymerase that transcribes protein-coding and snRNA genes.

RNA polymerase III: The eukaryotic RNA polymerase that transcribes tRNA and other short genes.

RNA recognition domain: An RNA-binding domain.

RNA silencing: An RNA degradation process in eukaryotes.

RNA transcript: An RNA copy of a gene.

RNA world: The early period of evolution when all biological reactions were centered on RNA.

RNA-dependent DNA polymerase: An enzyme that makes a DNA copy of an RNA template; a reverse transcriptase.

RNA-dependent RNA polymerase: An enzyme that makes an RNA copy of an RNA template.

RNA-FISH: Fluorescent *in situ* hybridization directed specifically at the RNA content of a cell or tissue.

RNA-seq: Short- or long-read sequencing of RNA.

Rolling circle replication: A replication process that involves continual synthesis of a polynucleotide which is 'rolled off' of a circular template molecule.

S phase: The stage of the cell cycle when DNA synthesis occurs.

S value: The unit of measurement for a sedimentation coefficient.

S1 nuclease: An enzyme that degrades single-stranded DNA or RNA molecules, including single-stranded regions in predominantly double-stranded molecules.

Satellite DNA: Repetitive DNA that forms a satellite band in a density gradient.

Satellite RNA: An RNA molecule some 320–400 nucleotides in length, which does not encode its own capsid proteins, instead moving from cell to cell within the capsid of a helper virus.

Scaffold: A series of sequence contigs separated by sequence gaps.

Scaffold/matrix attachment regions (S/MARs): Nucleotide sequences, within chromosomes, that bind to the proteins of the nuclear matrix.

Scanning: A system used during initiation of eukaryotic translation, in which the preinitiation complex attaches to the 5′-terminal cap structure of the mRNA and then scans along the molecule until it reaches an initiation codon.

Second messenger: An intermediate in a certain type of signal transduction pathway.

Secondary antibody: An antibody that binds specifically to a primary antibody.

Secondary structure: The conformations, such as α-helix and β-sheet, taken up by a polypeptide.

Sedimentation analysis: The centrifugal technique used to measure the sedimentation coefficient of a molecule or structure.

Sedimentation coefficient: The value used to express the velocity at which a molecule or structure sediments when centrifuged in a dense solution.

Segment polarity genes: Developmental genes that provide greater definition of the segmentation pattern of the *Drosophila* embryo established by the action of the pair-rule genes.

Segmental duplication: A duplication between 1–400 kb in length, with greater than 90% sequence identity, that is repeated up to 50 times in the genome.

Segmented genome: A virus genome that is split into two or more DNA or RNA molecules.

Selectable marker: A gene carried by a vector and conferring a recognizable characteristic on a cell containing the vector or a recombinant DNA molecule derived from the vector.

Selfish DNA: DNA that appears to have no function and apparently contributes nothing to the cell in which it is found.

Semiconservative replication: The mode of DNA replication in which each daughter double helix is made up of one polynucleotide from the parent and one newly synthesized polynucleotide.

Sequence assembler: A software package that converts sequence reads into contigs.

Sequence assembly: Assembly of the many short reads obtained by sequencing into a contiguous DNA sequence.

Sequence contig: A contiguous DNA sequence obtained as an intermediate in a genome sequencing project.

Sequence coverage or **sequence depth**: The average number of reads that cover each nucleotide position in a DNA sequence.

Sequence read: A single sequence from the output of a sequencing experiment.

Sequence-tagged site (STS): A DNA sequence that is unique in the genome.

Sequence-tagged site (STS) content mapping: A clone fingerprinting technique.

Sequencing by oligonucleotide ligation and detection (SOLiD): A short-read sequencing method in which the sequence is deduced by hybridization of a series of oligonucleotides whose sequences are complementary to that of the template.

Sequencing library: A set of DNA fragments that have been immobilized on a solid support in such a way that multiple sequencing reactions can be carried out side by side in a massively parallel array format.

Serial analysis of gene expression (SAGE): A method for studying the composition of a transcriptome.

Serine-threonine kinase receptors: A type of cell surface receptor that has serine-threonine kinase activity.

Sex cell: A reproductive cell; a cell that divides by meiosis.

Sex chromosome: A chromosome that is involved in sex determination.

Shelterin: A structure, comprising telomere-binding proteins, that protects the telomeres from degradation by nuclease enzymes, and mediates the enzymatic activity that maintains the length of each telomere during DNA replication.

Shine–Dalgarno sequence: Another name for the prokaryotic ribosome binding site.

Short interfering RNA (siRNA): An intermediate in the RNA interference pathway.

Short nascent strand (SNS) sequencing: A method used to identify the positions of replication origins in a eukaryotic genome sequence.

Short noncoding RNA (sncRNA): Noncoding RNAs less than 200 nucleotides in length.

Short tandem repeat (STR): A type of simple sequence length polymorphism comprising tandem copies of, usually, di-, tri- or tetranucleotide repeat units. Also called a microsatellite.

Short-read sequencing: A sequencing method that generates millions of sequence reads <600 bp in length.

Shotgun method: A genome sequencing strategy in which the molecules to be sequenced are randomly broken into fragments which are then sequenced individually.

Shotgun proteomics: A version of proteomics in which a mixture of proteins is fragmented.

Shuttle vector: A vector that can replicate in the cells of more than one organism (e.g., in *Escherichia coli* and in yeast).

Signal peptide: A short sequence at the N terminus of some proteins that directs the protein across a membrane.

Signal transduction: Control of cellular activity, including genome expression, via a cell surface receptor that responds to an external signal.

Silencer: A regulatory sequence that reduces the rate of transcription of a gene or genes located some distance away in either direction.

Simple sequence length polymorphism (SSLP): An array of repeat sequences that display length variations.

SINE (short interspersed nuclear element): A type of genome-wide repeat, typified by the Alu sequences found in the human genome.

Single nucleotide polymorphism (SNP): A point mutation that is carried by some individuals of a population.

Single-stranded: A DNA or RNA molecule that comprises just a single polynucleotide.

Single-cell RNA-seq (scRNA-seq): The use of RNA-seq to study the transcriptome of a single cell.

Single-copy DNA: A DNA sequence that is not repeated elsewhere in the genome.

Single-molecule real-time (SMRT) sequencing: A DNA sequencing method which uses an advanced optical system to observe the addition of individual nucleotides to a growing polynucleotide.

Single-strand binding protein (SSB): One of the proteins that attach to single-stranded DNA in the region of the replication fork, preventing base pairs forming between the two parent strands before they have been copied.

Site-directed hydroxyl radical probing: A technique for locating the position of a protein in a protein–RNA complex, such as a ribosome, by making use of the ability of Fe(II) ions to generate hydroxyl radicals which cleave nearby RNA phosphodiester bonds.

Site-directed mutagenesis: Techniques used to produce a specified mutation at a predetermined position in a DNA molecule.

Site-specific recombination: Recombination between two double-stranded DNA molecules that have only short regions of nucleotide sequence similarity.

Small Cajal body-specific RNA (scaRNA): A snoRNA associated with a Cajal body.

Small nuclear ribonucleoprotein (snRNP): Structures involved in splicing GU–AG and AU–AC introns and in other RNA processing events, comprising one or two snRNA molecules complexed with proteins.

Small nuclear RNA (snRNA): A type of short eukaryotic RNA molecule involved in splicing GU–AG and AU–AC introns and in other RNA processing events.

Small nucleolar RNA (snoRNA) : A type of short eukaryotic RNA molecule involved in chemical modification of rRNA.

Small ubiquitin-like modifier (SUMO): A protein related to ubiquitin that is added to some proteins as a post-translational modification.

SOLiD: A short-read sequencing method in which the sequence is deduced by hybridization of a series of oligonucleotides whose sequences are complementary to that of the template.

Solid phase: The immobile phase in a chromatography system.

Solution hybridization: Hybridization between nucleic acid molecules carried out in a solution.

Somatic cell: A non-reproductive cell; a cell that divides by mitosis.

Sonication: A procedure that uses ultrasound to cause random breaks in DNA molecules.

Sorting sequence: An amino acid sequence that directs a protein to an organelle such as the nucleus or mitochondria, or might specify that the protein is secreted from the cell.

SOS response: A series of biochemical changes that occur in *Escherichia coli* in response to damage to the genome and other stimuli.

Southern hybridization: A technique used for detection of a specific restriction fragment against a background of many other restriction fragments.

Spatial transcriptomics: Techniques used to study transcriptomes directly in cells present in undisrupted tissue samples.

Speckle: A nuclear structure associated with mRNA splicing.

Spliceosome: The protein–RNA complex involved in splicing GU–AG or AU–AC introns.

Splicing: The removal of introns from the primary transcript of a discontinuous gene.

Splicing code: A hypothetical code that would explain the impact on a splicing pathway of the various interactions that can occur between enhancers, silencers and their binding proteins.

Splicing pathway: The series of events that converts a discontinuous pre-mRNA into a functional mRNA.

Spm element: A DNA transposon of maize.

Sponge RNA: An RNA molecule that contains binding sites for regulatory proteins or RNAs, and competes with other RNAs for attachment of these molecules.

Spontaneous mutation: A mutation that arises from an error in replication.

Squiggle: The initial output of a nanopore sequencing experiment, showing the electrical changes that occur as the DNA molecule passes through the nanopore.

SR protein: A protein that plays a role in splice-site selection during splicing of GU–AG introns.

STAT (signal transducer and activator of transcription): A type of protein that responds to binding of an extracellular signaling compound to a cell surface receptor by activating a transcription factor.

Stem cell: A progenitor cell that divides continually throughout the lifetime of an organism.

Stem-loop structure: A structure made up of a base-paired stem and non-base-paired loop, which can form in a single-stranded polynucleotide that contains an inverted-repeat.

Steroid hormone: A type of extracellular signaling compound.

Steroid receptor: A protein that binds a steroid hormone after the latter has entered the cell, as an intermediate step in modulation of genome activity.

Sticky end: An end of a double-stranded DNA molecule where there is a single-stranded extension.

Strep tag: The amino acid sequence Trp–Ser–His–Pro–Gln–Phe–Glu–Lys, which binds to a modified form of streptavidin.

Streptavidin: A protein, from the bacterium *Streptomyces avidinii*, which has a high binding affinity for biotin.

Stringent response: A biochemical and genetic response initiated in *Escherichia coli* when the bacterium encounters poor growth conditions, such a low levels of essential amino acids.

Strong promoter: A promoter that directs a relatively large number of productive initiations per unit of time.

Structural variant: Variations that result in the genomes of two individuals of the same species differing in the numbers and locations of repeat units, the presence or absence of certain sequences of 50–2000 bp in length, and the positioning of some sequences.

STS mapping: A physical mapping procedure that locates the positions of sequence-tagged sites (STSs) in a genome.

Stuffer fragment: A DNA fragment contained within a λ vector that is replaced by the DNA to be cloned.

Substitution : A point mutation that escapes the repair processes and results in a permanent change in a DNA sequence.

Sugar pucker: Alternative conformations of a sugar ring structure.

Suicide enzyme: An enzyme that is inactivated once it has carried out its biochemical reaction.

SUMO: A protein related to ubiquitin.

Supercoiling: A conformational state in which a double helix is overwound or underwound so that superhelical coiling occurs.

Surveillance mechanism: A process that identifies mRNAs that lack a termination codon, or have a termination codon at an unexpected position, and which therefore should be degraded.

Syncytium: A cell-like structure comprising a mass of cytoplasm and many nuclei.

Synteny: Refers to a pair of genomes in which at least some of the genes are located at similar map positions.

Synthesis, or S phase: The phase of the cell cycle when DNA replication takes place.

Systems biology: An approach to biology that attempts to link metabolic pathways and subcellular processes with genome expression.

Systems biology graphical notation (SBGN): A set of rules and symbols that are used to depict a biological system in a graphical format.

T-DNA: The portion of the Ti plasmid that is transferred to the plant DNA.

T4 polynucleotide kinase: An enzyme that adds phosphate groups to the 5´ ends of DNA molecules.

TAF and initiator-dependent cofactor (TIC): A type of protein involved in initiation of transcription by RNA polymerase II.

Tandem mass spectrometry: A type of mass spectrometry that uses two or more mass analyzers linked in series.

Tandem repeat: Direct repeats that are adjacent to each other.

Tandemly repeated DNA: DNA sequence motifs that are repeated head to tail.

Target enrichment: A method for enriching a DNA sequencing library for fragments derived from particular genes of interest.

TATA box: A component of the RNA polymerase II core promoter.

TATA-binding protein (TBP): A component of the general transcription factor TFIID, the part that recognizes the TATA box of the RNA polymerase II promoter.

Tautomeric shift: The spontaneous change of a molecule from one structural isomer to another.

Tautomers: Structural isomers that are in dynamic equilibrium.

TBP domain: The binding domain of the TATA-binding protein, which forms a saddle-like structure that encloses part of the double helix.

TBP-associated factor (TAF): One of several components of the general transcription factor TFIID, playing ancillary roles in recognition of the TATA box.

Telomerase: The enzyme that maintains the ends of eukaryotic chromosomes by synthesizing telomeric repeat sequences.

Telomere: The end of a eukaryotic chromosome.

Telomere-binding protein (TBP): A protein that binds to and regulates the length of a telomere.

Temperate bacteriophage: A bacteriophage that is able to follow a lysogenic mode of infection.

Template: The polynucleotide that is copied during a strand-synthesis reaction catalyzed by a DNA or RNA polymerase.

Template-dependent DNA polymerase: An enzyme that synthesizes DNA in accordance with the sequence of a template.

Template-dependent DNA synthesis: Synthesis of a DNA molecule on a DNA or RNA template.

Template-dependent RNA polymerase: An enzyme that synthesizes RNA in accordance with the sequence of a template.

Template-dependent RNA synthesis: Synthesis of an RNA molecule on a DNA or RNA template.

Template-independent DNA polymerase: An enzyme that synthesizes DNA without the use of a template.

Template-independent RNA polymerase: An enzyme that synthesizes RNA without the use of a template.

Terminal deoxynucleotidyl transferase: An enzyme that adds one or more nucleotides to the 3´ end of a DNA molecule.

Termination codon: One of the three codons that mark the position where translation of an mRNA should stop.

Terminator sequence: One of several sequences on a bacterial genome involved in termination of genome replication.

Territory: The region of a nucleus occupied by a single chromosome.

Tertiary structure: The structure resulting from folding the secondary structural units of a polypeptide.

Test cross: A genetic cross between a double heterozygote and a double homozygote.

Thermal cycle sequencing: A DNA sequencing method that uses PCR to generate chain-terminated polynucleotides.

Thermostable: Able to withstand high temperatures.

Three-point cross: A genetic cross in which the inheritance of three markers is followed.

Thymine: One of the pyrimidine bases found in DNA.

Ti plasmid: The large plasmid found in those *Agrobacterium tumefaciens* cells able to direct crown gall formation on certain species of plants.

Tiling array: A collection of oligonucleotide probes, each targeting a different position along a chromosome or a part of a chromosome.

T_m: Melting temperature.

Tn3-type transposon: A type of DNA transposon that does not have flanking insertion sequences.

Top-down proteomics: A version of proteomics in which individual proteins are directly examined by mass spectroscopy.

Topological problem: Refers to the need to unwind the double helix in order for DNA replication to occur, and the difficulties that the resulting rotation of the DNA molecule would cause.

Topologically associating domain (TAD): A contiguous segment of chromatin folded into coils and loops.

Topology: The branching pattern of a phylogenetic tree.

Totipotent: Refers to a cell that is not committed to a single developmental pathway and can hence give rise to all types of differentiated cell.

Trailer segment: The untranslated region of an mRNA downstream of the termination codon.

***Trans*-displacement:** Transfer of a nucleosome from one DNA molecule to another.

Transcript: An RNA copy of a gene.

Transcript-specific regulation: Regulatory mechanisms that control protein synthesis by acting on a single transcript or a small group of transcripts coding for related proteins.

Transcription: The synthesis of an RNA copy of a gene.

Transcription factor: A protein that activates or represses the initiation of transcription.

Transcription initiation: The assembly, upstream of a gene, of the complex of proteins that will subsequently copy the gene into RNA.

Transcription-coupled repair: A nucleotide excision repair process that results in repair of the template strands of genes.

Transcriptional noise: Transcription directed by sequences that resemble promoters and which results in synthesis of RNAs that the cell does not need.

Transcriptome: The entire mRNA content of a cell.

Transcriptomics: The methods used to catalog a transcriptome.

Transduction: Transfer of bacterial genes from one cell to another by packaging in a phage particle.

Transduction mapping: The use of transduction to map the relative positions of genes in a bacterial genome.

Transfection: The introduction of purified phage DNA molecules into a bacterial cell.

Transfer RNA (tRNA): A small RNA molecule that acts as an adaptor during translation and is responsible for decoding the genetic code.

Transfer-messenger RNA (tmRNA): A bacterial RNA involved in protein degradation.

Transformant: A cell that has become transformed by the uptake of naked DNA.

Transformation: The acquisition by a cell of new genes by the uptake of naked DNA.

Transformation mapping: The use of transformation to map the relative positions of genes in a bacterial genome.

Transforming principle: The compound, now known to be DNA, responsible for transformation of an avirulent *Streptococcus pneumoniae* bacterium into a virulent form.

Transgenic mouse: A mouse that carries a cloned gene.

Transition: A point mutation that replaces a purine with another purine, or a pyrimidine with another pyrimidine.

Translation: The synthesis of a polypeptide, the amino acid sequence of which is determined by the nucleotide sequence of an mRNA in accordance with the rules of the genetic code.

Translational efficiency: The rate at which proteins are synthesized from an mRNA.

Translesion polymerase: A DNA polymerase that can carry out error-prone replication of a damaged region of DNA.

Translocation: (A) The attachment of a segment of one chromosome to another chromosome. (B) The movement of a ribosome along an mRNA molecule during translation.

Transposable element: A genetic element that can move from one position to another in a DNA molecule.

Transposable phage: A bacteriophage that transposes as part of its infection cycle.

Transposase: An enzyme that catalyzes transposition of a transposable genetic element.

Transposition: The movement of a genetic element from one site to another in a DNA molecule.

Transposon: A genetic element that can move from one position to another in a DNA molecule.

Transposon tagging: A gene isolation technique that involves inactivation of a gene by movement of a transposon into its coding sequence, followed by the use of a transposon-specific hybridization probe to isolate a copy of the tagged gene from a clone library.

Transversion: A point mutation that involves a purine being replaced by a pyrimidine, or vice versa.

Treble clef finger: A type of zinc finger.

Trinucleotide repeat expansion disease: A disease that results from the expansion of an array of trinucleotide repeats in or near to a gene.

Triplex: A DNA structure comprising three polynucleotides.

Trisomy: The presence of three copies of a homologous chromosome in a nucleus that is otherwise diploid.

Trithorax group (trxG): A group of proteins that maintain an open chromatin state in the regions of active genes.

Truncated gene: A gene relic that lacks a segment from one end of the original, complete gene.

Tudor domain: A five-stranded β-sheet structure encoded by an approximately 60–amino acid sequence that binds to methylated arginine and/or lysine amino acids contained in other proteins.

Tus (terminator utilization substance) protein: The protein that binds to a bacterial terminator sequence and mediates termination of genome replication.

Two-dimensional gel electrophoresis: A method for separation of proteins used especially in studies of the proteome.

Type 0 cap: The basic cap structure, consisting of 7-methylguanosine attached to the 5′ end of an mRNA.

Type 1 cap: A cap structure comprising the basic 5′-terminal cap plus an additional methylation of the ribose of the second nucleotide.

Type 2 cap: A cap structure comprising the basic 5′-terminal cap plus methylation of the riboses of the second and third nucleotides.

Tyrosine kinase receptor: A type of cell surface receptor that has tyrosine kinase activity.

Tyrosine kinase-associated receptor: A type of cell surface receptor that works in conjunction with a protein that has tyrosine kinase activity.

TψC arm: Part of the structure of a tRNA molecule.

U-RNA: A uracil-rich nuclear RNA molecule including the snRNAs and snoRNAs.

Ubiquitin: A 76-amino-acid protein which, when attached to a second protein, acts as a tag directing that protein for degradation.

Ubiquitin-receptor protein: A protein that directs ubiquitinated proteins into the proteasome.

Ubiquitination: The attachment of ubiquitin to a protein.

UCSC Genome Browser: An online genome browser.

Unequal crossing over: A recombination event that results in duplication of a segment of DNA.

Unequal sister chromatid exchange: A recombination event that results in duplication of a segment of DNA.

Unit factor: Mendel's term for a gene.

Unit transposon: A Tn3-type transposon.

Unitary pseudogene: A nonprocessed pseudogene that arises by decay of a gene that is not a member of gene family, and whose function is lost as a result of the inactivation of the gene.

Universal primer: A sequencing primer that is complementary to the part of the vector DNA immediately adjacent to the point into which new DNA is ligated.

Untranslated region (UTR): The parts of an mRNA, upstream and downstream of the ORF, that are not translated into protein.

Upstream: Toward the 5´ end of a polynucleotide.

Upstream control element: A component of an RNA polymerase I promoter.

Upstream promoter element: Components of a eukaryotic promoter that lie upstream of the position where the initiation complex is assembled.

Upstream regulatory sequence: A regulatory sequence, usually a binding site for a transcription factor, found upstream of a gene.

Uracil: One of the pyrimidine bases found in RNA.

UvrABC endonuclease: A multienzyme complex involved in the short patch repair process of *Escherichia coli*.

V loop: Part of the structure of a tRNA molecule.

van der Waals forces: A particular type of attractive or repulsive noncovalent bond.

Variable number of tandem repeats (VNTR): A type of simple sequence length polymorphism comprising tandem copies of repeats that are a few tens of nucleotides in length. Also called a minisatellite.

Vault RNA: A type of snRNA found in protein–RNA complexes called vaults, which are found in most eukaryotic cells but whose functions are not known.

Vegetative cell: A non-reproductive cell; a cell that divides by mitosis.

Vertebrate Genomes project: A project that aims to produce at least one high-quality genome sequence for every one of the 71,656 vertebrate species.

Viral retroelement: A virus whose genome replication process involves reverse transcription.

Viroid: An RNA molecule 240–375 nucleotides in length which contains no genes and never becomes encapsidated, spreading from cell to cell as naked DNA.

Virulent bacteriophage: A bacteriophage that follows the lytic mode of infection.

Virus: An infective particle, composed of protein and nucleic acid, that must parasitize a host cell in order to replicate.

Virusoid: An RNA molecule of 320–400 nucleotides in length, which does not encode its own capsid proteins, instead moving from cell to cell within the capsid of a helper virus.

Weak promoter: A promoter that directs relatively few productive initiations per unit of time.

Wild type: A gene, cell or organism that displays the typical phenotype and/or genotype for the species and is therefore adopted as a standard.

Winged helix–turn–helix: A type of DNA-binding domain.

X inactivation: Inactivation by methylation of most of the genes on one copy of the X chromosome in a female nucleus.

X-ray crystallography: A technique for determining the three-dimensional structure of a large molecule.

X-ray diffraction: The diffraction of X-rays that occurs during passage through a crystal.

X-ray diffraction pattern: The pattern obtained after diffraction of X-rays through a crystal.

Yeast two-hybrid system: A technique for identifying proteins that interact with one other.

Z-binding domain: A binding domain that enables a protein to attach to Z-DNA and to double-stranded RNAs that have adopted a left-handed helical conformation.

Z-DNA: A conformation of DNA in which the two polynucleotides are wound into a left-handed helix.

Zero-mode waveguide: A nanostructure that enables individual molecules to be observed.

Zinc finger: A common structural motif for attachment of a protein to a DNA molecule.

Zoo blotting: A technique that attempts to determine if a DNA fragment contains a gene by hybridizing that fragment to DNA preparations from related species, on the basis that genes have similar sequences in related species and so give positive hybridization signals.

Zygote: The cell resulting from fusion of gametes during meiosis.

Zα domain: A binding domain that enables a protein to attach to Z-DNA and to double-stranded RNAs that have adopted a left-handed helical conformation.

α-helix: One of the commonest secondary structural conformations taken up by segments of polypeptides.

β-*N*-glycosidic bond: The linkage between the base and sugar of a nucleotide.

β-sheet: One of the commonest secondary structural conformations taken up by segments of polypeptides.

β-turn: A sequence of four amino acids, the second usually glycine, which causes a polypeptide to change direction.

γ-complex: A component of DNA polymerase III comprising subunit γ in association with δ, δ', χ, and ψ.

π–π interactions: The hydrophobic interactions that occur between adjacent base pairs in a double-stranded DNA molecule.

INDEX

M